动物传染病防控技术的研究与应用

Research and Application of Animal Infectious Diseases Prevention and Control Technology

谢芝勋　主　编

广西科学技术出版社

·南宁·

图书在版编目（CIP）数据

动物传染病防控技术的研究与应用：英文 / 谢芝勋
主编. -- 南宁：广西科学技术出版社，2024. 11.
ISBN 978-7-5551-2261-6

Ⅰ. S855

中国国家版本馆 CIP 数据核字第 2024VU4275 号

Dongwu Chuanranbing Fangkong Jishu De Yanjiu Yu Yingyong

动物传染病防控技术的研究与应用

谢芝勋　主编

策　　划：黎志海	封面设计：梁　良
责任编辑：覃　艳	责任校对：吴书丽
责任印制：陆　弟	

出 版 人：岑　刚
出版发行：广西科学技术出版社　　　　　地　　址：广西南宁市东葛路 66 号
邮政编码：530023　　　　　　　　　　　网　　址：http://www.gxkjs.com

经　　销：全国各地新华书店
印　　刷：广西民族印刷包装集团有限公司

开　　本：889mm×1194mm　　1/16
印　　张：71.75　　彩插 20　　　　　　字　　数：1062 千字
版　　次：2024 年 11 月第 1 版　　　　　印　　次：2024 年 11 月第 1 次印刷
书　　号：ISBN 978-7-5551-2261-6
定　　价：598.00 元（上、下册）

《动物传染病防控技术的研究与应用》
编辑委员会

主 编

谢芝勋

编 委

谢丽基	罗思思	陈忠伟	谢志勤	李　孟
刘加波	庞耀珊	范　晴	邓显文	黄娇玲
张艳芳	曾婷婷	王　盛	张民秀	韦　悠
万丽军	任红玉	李小凤	李　丹	阮志华
华　俊	谢宇舟	吴媛琼	王　粲	牙侯勋

致　谢

本书出版得到"国家高层次人才特殊支持计划"领军人才项目（W02060083）、"新世纪百千万人才工程"项目（945200603）和"广西八桂学者"项目（2019A50）经费的支持，特此致谢！

Thanks

The publication of this book has been supported by the leading talents of the "National Special Support Program for High-level Talents" (W02060083), the "New Century National Hundred, Thousand and Ten Thousand Talents Project" (945200603) and the "Guangxi BaGui Scholars Program" (2019A50). I would like to express my deep gratitude.

主编简介

谢芝勋，二级研究员，国际知名学者和禽病学专家。享受国务院政府特殊津贴专家、"国家高层次人才特殊支持计划"领军人才、"新世纪百千万人才工程"国家级人选、广西第五批八桂学者、广西第一批自治区特聘专家、广西优秀专家、广西先进工作者、全国有突出贡献的回国留学人员和全国农业农村系统先进个人。曾任广西壮族自治区兽医研究所副所长，广西兽医生物技术重点实验室主任，农业农村部中国（广西）—东盟跨境动物疫病防控重点实验室主任，国家国际科技合作基地、博士后科研工作站和国家引进国外智力成果示范推广基地负责人。

谢芝勋研究员 1983 年从广西农学院畜牧兽医系（现为广西大学动物科学技术学院）毕业，曾在美国农业部东南家禽研究所和美国康涅狄格大学做访问学者，积累了丰富的国际前沿学术经验。他长期致力于动物传染病综合防控技术的研究，主持包括国家重点研发计划，国家自然科学基金，科技部、农业农村部、广西科技攻关重大专项和广西自然科学基金重点项目等在内的 40 多项研究。凭借出类拔萃的科研能力和不懈的努力，谢芝勋研究员在动物传染病防控技术的多个前沿领域，包括病原生态学、致病机理与免疫机制、快速检测诊断技术和新型高效疫苗等方面，均取得一系列重要突破和具有国际影响力及国际先进水平的科研创新成果。由他主持完成的项目获国家级和省级科学技术进步奖共 15 项，其中包括国家科学技术进步奖三等奖 1 项，广西壮族自治区党委、政府重奖 1 项，广西科学技术进步奖一等奖 2 项、二等奖 8 项、三等奖 3 项。发表论文 660 多篇，其中 SCI 收录 120 余篇。出版专著 6 部（主编 2 部英文版专著），获授权发明专利 119 件（美国授权 3 件），取得国家新兽药证书 1 个，制定广西地方技术标准 23 项。

谢芝勋研究员受聘为广西大学兼职教授，为博士后和硕士研究生导师，共培养出 10 位博士后和 39 位硕士研究生。同时，担任国家重点研发计划和广西科技重大专项的首席专家，国家重点研发计划和国家自然科学基金项目的评审专家，国家级和多个省级科技奖的评审专家，中国工程院咨询专家，广西自然科学基金会委员。此外，他还是世界禽病协会会员、国际支原体组织会员、中国畜牧兽医学会动物传染病学分会常务理事、中国畜牧兽医学会禽病学分会理事、中国畜牧兽医学会畜牧兽医生物技术学分会理事、中国畜牧业协会理事以及广西畜牧兽医学会养禽与禽病学分会副理事长等，在学术交流与合作方面发挥着重要的引领作用。

谢芝勋研究员是国际著名期刊 *Frontiers in Microbiology*（*Virology*）的副主编，中文核心期刊《病毒学报》《中国兽医科学》和《基因组学与应用生物学》的编委会成员，*Emerging Infectious Diseases*、*Veterinary Microbiology*、*Frontiers in Immunology*、*Cells*、*Analytica Chimica Acta*、*Scientific Reports*、*Virology*、*Frontiers in Microbiology*、*TALANTA*、*Infection, Genetics and Evolution*、*PLOS ONE*、*BMC Microbiology*、*Transboundary and Emerging Diseases*、*Virology Journal*、*Biomed Research International*、*Journal of Agricultural and Food Chemistry*、*Journal of Virological Methods*、*Applied Biochemistry and Biotechnology* 和 *Molecular and Cellular Probes* 等 20 多种 SCI 期刊的审稿专家。他以严谨的科学态度和无私的奉献精神，持续引领着学术前沿，激励着后辈学者不断探索与前行。

Brief Introduction of Professor Xie Zhixun

Professor Xie Zhixun, a second-level professor, is an internationally renowned scholar and expert in avian diseases. He was the former vice-director of the Guangxi Veterinary Research Institute and the director of Guangxi Key Laboratory of Veterinary Biotechnology and the Key Laboratory of Cross-border Animal Disease Prevention and Control of China (Guangxi)-ASEAN. At the same time, he is also the person in charge of the international scientific and technological cooperation base of the Ministry of Science and Technology, the postdoctoral research workstation of the Ministry of Human Resources and Social Security, and the demonstration base for the introduction of foreign intellectual achievements of the State Administration of Foreign Experts Affairs. Professor Xie has received many honors. He is an expert enjoying the special government allowances of the State Council, was selected as a leading talent in the "National Special Support Program for High-level Talents", and is also a national candidate of the "New Century National Hundred, Thousand and Ten Thousand Talents Project". In addition, he is also the fifth batch of Guangxi Bagui Scholar, the first batch of specially invited experts in Guangxi, an outstanding expert in Guangxi, an advanced worker in Guangxi, a nationally outstanding returned overseas trainee who has made outstanding contributions, and a national advanced individual in agriculture.

In 1983, Xie Zhixun graduated from the Department of Animal Husbandry and Veterinary Medicine at Guangxi Agricultural College (It has now become the College of Animal Science and Technology, Guangxi University). He worked as a visiting scholar at the USDA Southeast Poultry Research Laboratory and the University of Connecticut, accumulating rich international cutting-edge academic experience. Professor Xie has long been dedicated to in-depth research on comprehensive prevention and control technologies for animal infectious diseases. He has presided over as many as more than 40 research projects, including the National Key Research and Development Program, the National Natural Science Foundation of China, the Ministry of Science and Technology, the Ministry of Agriculture and Rural Affairs of the People's Republic of China, major special projects for scientific and technological breakthroughs in Guangxi, and key projects of the Natural Science Foundation of Guangxi. With his outstanding scientific research ability and persistent efforts, he has made a series of significant breakthroughs and achieved innovative results with international influence and at the international advanced level in multiple cutting-edge fields of animal infectious disease prevention and control technologies, including pathogen ecology, pathogenic mechanism and immune mechanism, rapid detection and diagnosis technology, and new and efficient vaccines. The projects he presided over have won a total of 15 national and provincial-level science and technology progress awards, specifically including 1 third prize of the National Science and Technology Progress Award, 1 major science and technology award of Guangxi, 2 first prizes of the Guangxi Science and Technology Progress Award, 8 second prizes of the Guangxi Science and Technology Progress Award, and 3 third prizes of the Guangxi Science and Technology Progress Award. He has published over 660 papers at home and abroad, among which more than 120 were included in SCI. He has published 6 monographs (edited 2 monographs in English), obtained 119 authorized invention patents (including 3 authorized invention patents in the United States), successfully obtained 1 national new veterinary drug certificate, formulated 23 local technical standards in Guangxi.

In addition, Professor Xie has also been hired as an adjunct professor at Guangxi University, a supervisor for postdoctoral students and master's students, and trained 10 postdoctoral fellows and 39 master's degree students. He is the chief expert of the National Key Research and Development Program of China and the major scientific and technological projects in Guangxi, a reviewer of the National Key Research and Development Program and the National Natural Science Foundation of China projects, a reviewer of national and several provincial-level science and technology awards, China Academy of Engineering consulting expert, and a member of the Guangxi Natural Science Foundation Committee. At the same time, he has been hired as a member of The World Veterinary Poultry Disease Association and The International Organization for Mycoplasmology, a standing director of the Animal Infectious Disease Branch of the Chinese Association of Animal Science and Veterinary Medicine, a director of the Poultry Disease Branch of the Chinese Association of Animal Science and Veterinary Medicine, a director of the Animal Husbandry and Veterinary Biotechnology Branch of the Chinese Association of Animal Science and Veterinary Medicine, a director of the China Animal Husbandry Association, and the vice-chairman of the Poultry and Poultry Disease Branch of the Guangxi Association of Animal Husbandry and Veterinary Medicine, playing an important leading role in academic exchanges and cooperation.

Moreover, Professor Xie is also the associate editor of the internationally renowned journal *Frontiers in Microbiology* (Virology), and an editorial board member of the Chinese core journals *Chinese Journal of Virology*, *Chinese Veterinary Science* and *Genomics and Applied Biology*. More than twenty SCI journals selected him as their peer reviewer, including *Emerging Infectious Diseases*, *Veterinary Microbiology*, *Frontiers in Immunology*, *Cells*, *Analytica Chimica Acta*, *Scientific Reports*, *Virology*, *Frontiers in Microbiology*, *TALANTA*, *Infection, Genetics and Evolution*, *PLOS ONE*, *BMC Microbiology*, *Transboundary and Emerging Diseases*, *Virology Journal*, *Biomed Research International*, *Journal of Agricultural and Food Chemistry*, *Journal of Virological Methods*, *Applied Biochemistry and Biotechnology* and *Molecular and Cellular Probes*. As a reviewer for numerous high-level journals, he continues to lead the academic frontier with his rigorous scientific attitude and selfless dedication, inspiring younger scholars to constantly explore and move forward.

内容简介

    本书重点介绍国内外重要动物传染病防控技术的研究与应用。全书分为七章，第一章为综述，其余各章分别介绍动物传染病的流行病学、分子病原学、致病机理与免疫机制、快速诊断技术、高效疫苗与抗体、畜禽遗传资源的研发等。本书由长期在动物传染病防控技术科研第一线从事研究的中青年学者精心编写而成，贯彻理论联系实际的原则，力求原理清晰、简明规范。书中内容指导性和操作性强，不仅介绍动物传染病防控的新技术和新理论，还对新技术的临床实践应用进行深入阐述。

    本书以百余篇论文的形式对相关动物传染病防控的新技术或新方法及其应用进行详细的介绍，内容丰富新颖，图文并茂，既有理论性，又具实践性和可操作性。不仅可供科研工作者、中青年教师和临床及实验室检测诊断技术人员等参考，而且对在读研究生等的论文写作、学习与研究具有重要的参考价值。

Introduction

    This book focuses on the research and application of important animal infectious disease prevention and control technology. The book consists of seven chapters, with the first chapter providing reviews and the remaining chapters respectively introducing studies on molecular epidemiology, molecular etiology, pathogenesis and immune mechanism, rapid diagnostic technology, efficient vaccines and antibodies, and livestock and poultry genetic resources. All the authors contributed to this book are middle-aged scholars, working for many years at the frontline of research for prevention and control of animal infectious diseases. With the principle of "integration of theory with practice", the main advantages of this book are precise principles, concise, standardization, strong guidance and operatability. This book not only introduces the latest cutting-edge technologies and new theory on prevention and control of animal infectious diseases, but also elaborates on clinical practice of these new techniques.

    This book describes in detail by more than 100 papers the latest techniques and assays currently applied to prevention and control of animal infectious diseases. In addition, it has ample information, illustrations and pictures, involving theories and practice. Not only can it serve as a reference for researchers, middle-aged and young teachers, and technical personnel engaged in clinical and laboratory testing and diagnosis, but it also holds significant reference value for graduate students in their writing, learning and research.

自　序

　　悠悠四十余载科研漫途，岁月如诗，韵味难书；时光似梦，情思难住。幸得各级领导与专家的热忱鼓励、鼎力相助，历经三年的辛勤笔耕，《动物传染病防控技术的研究与应用》（*Research and Application of Animal Infectious Diseases Prevention and Control Technology*）终告编写完成，交出版社付梓。身为这部著作的主编，此刻我欣喜之情溢于言表，只因将四十余载科研生涯的深厚积淀与珍贵积累汇集成册，为奋战于动物传染病防控一线的科研人员、高校师生以及兽医防疫工作者提供一部实用参考书的夙愿即将成真。

　　"科学无国界，但科学家有祖国"。忆往昔，1988年1月至1989年7月以及1996年3月至1997年4月，我两度负笈赴美留学进修，亲睹中美科研基础条件与水平的悬殊差距，内心深受触动，震撼难平。我遂立志学成归国，投身改革开放的时代洪流，愿以所学知识为科技强国奉献心力，报效祖国，建设家园。

　　"路漫漫其修远兮，吾将上下而求索"。科研之路，荆棘丛生，艰辛备尝，一路走来，并非通途坦道。我曾深陷迷茫的雾霭，曾困于困惑的阴霾，亦曾饱受失落的风雨，然而从未有过丝毫退缩之意。每一次实验的挫败，皆成砥砺成长的基石；每一组数据的偏差，皆为触发思考的新起点。恰似于黑暗中艰辛摸索，却陡然迎来柳暗花明的转机。每一次的峰回路转，皆让我愈加坚信，只要胸怀信念，勇往直前，必能邂逅那渴盼已久的璀璨曙光。

　　"千磨万击还坚劲，任尔东西南北风"。正是对科学的赤诚热爱与坚定执着，使我在时光滔滔的长河中破浪勇进，于困境中坚守，在挫折中奋然崛起。四十余载科研生涯，我的初心坚若磐石，始终专注于禽流感病毒、血清4型禽腺病毒和禽呼肠孤病毒等病原生态学奥秘，深入探究禽呼肠孤病毒和血清4型禽腺病毒致病与免疫机制，精心钻研快速诊断技术（如高通量GeXP鉴别诊断技术、多重PCR诊断技术、多重LAMP诊断技术、数字PCR诊断技术、纳米电化学传感器诊断技术、ELISA鉴别诊断技术等），倾尽全力研发禽多杀性巴氏杆菌B_{26}-T_{1200}弱毒活疫苗和血清4型禽腺病毒（GX2019-014株）弱毒疫苗等，斩获一系列具有国际影响力及国际先进水平的科研创新成果，引领相关领域的创新发展与技术进步。

　　"长风破浪会有时，直挂云帆济沧海"。如今，我将四十余载的研究成果汇集于《动物传染病防控技术的研究与应用》一书中，愿此书宛如科研道路上的熠熠明灯，为后来者照亮前行的方向，守护动物的健康福祉，推动动物传染病防控领域的科学技术不断发展。

在编写与校审过程中，每位作者皆秉持严谨之态度，齐心协力，在此谨表衷心谢忱！

因时间仓促，书中或存疏漏之处乃至错讹之嫌，诚望各位领导、专家及读者不吝斧正，予以批评赐教。

谢芝勋

2024 年 8 月

Contents

Chapter Three   Studies on Molecular Etiology

Chapter Four　Studies on Pathogenesis and Immune Mechanism ···············409

Appendix ···545

Chapter One
Reviews

Biological features of fowl adenovirus serotype-4

Rashid Farooq, Xie Zhixun, Wei You, Xie Zhiqin, Xie Liji, Li Meng, and Luo Sisi

Abstract

Fowl adenovirus serotype 4 (FAdV-4) is highly pathogenic to broilers aged 3 to 5 weeks and has caused considerable economic loss in the poultry industry worldwide. FAdV-4 is the causative agent of hydropericardium-hepatitis syndrome (HHS) or hydropericardium syndrome (HPS). The virus targets mainly the liver, and HPS symptoms are observed in infected chickens. This disease was first reported in Pakistan but has now spread worldwide, and over time, various deletions in the FAdV genome and mutations in its major structural proteins have been detected. This review provides detailed information about FAdV-4 genome organization, physiological features, epidemiology, coinfection with other viruses, and host immune suppression. Moreover, we investigated the role and functions of important structural proteins in FAdV-4 pathogenesis. Finally, the potential regulatory effects of FAdV-4 infection on ncRNAs are also discussed.

Keywords

FAdV-4, epidemiology, coinfection, structural proteins, mutations

Introduction

Fowl adenoviruses (FAdVs) are the causative agents of many clinical diseases in poultry and have caused substantial economic loss to the poultry industry worldwide[1]. FAdVs are classified under the family Adenoviridae and genus *Aviadenovirus* and are further divided into five species, from A to E, based on their restriction enzyme digestion profile[2]. The International Committee on Taxonomy of Viruses, based on cross-neutralization tests results, has divided FAdVs into 12 serotypes[3-5]. The typical diseases in chickens caused by adenoviruses include hepatitis-hydropericardium syndrome (HHS), which is caused by FAdV-4 (species FAdV-C)[6, 7]; inclusion body hepatitis (IBH), which is caused by FAdV-8 and -11 (species FAdV-E and FAdV-D, respectively)[8, 9]; and adenoviral gizzard erosion (GE), which is caused by FAdV-1 infections (belonging to FAdV-A)[10, 11]. It has been reported that FAdVs are distributed worldwide, and different serotypes have been found in different geographical locations and subsequently spread to various regions[12, 13].

FAdV-4 was first reported in the Angara region of Pakistan in 1987. This disease subsequently spread to other regions of the world and caused enormous economic losses in the poultry sector[14]. The virus is nonenveloped, and its genome is double-stranded DNA of approximately 45 kb. HHS or hydropericardium syndrome (HPS) mainly affects chickens aged 3- to 6-weeks[4]. Ruffled feathers, lethargy, prostration and asitia are common clinical symptoms. The accumulation of clear, straw-colored fluid in the pericardial sac is a typical sign of a necrotic lesion. The liver also shows multifocal areas of necrosis and hepatitis[15]. Moreover, HPS has also been observed in other birds, such as pigeons, quails, ducks and mandarin ducks[19].

In China, FAdV-4 outbreaks were reported earlier[7], but since June 2015, dramatic increases in the incidence of FAdV-4 outbreaks have been reported in several provinces, including Anhui, Henan, Hubei, Jiangsu, Jiangxi, Shandong, and Zhejiang[20], with higher mortality rates in 3- to 5-week-old broiler chickens than in previous HPS outbreaks caused by FAdVs. Infection with these hypervirulent isolates of FAdV-4 resulted in 80%~100% mortality[21]. The Chinese FAdV-4 isolates share about 99.9%~100% genetic identities among themselves, and isolates from other countries (KR5, ON1, MXSHP95, and B1-7) also share nucleotide identities ranging from 97.7%~98.9% among themselves. However, lower identities were observed between isolates from the other countries (KR5, ON1, MXSHP95, and B1-7) and Chinese isolates. In the current review, we discuss the different aspects of FAdV-4 infection, e.g., epidemiology, physiological features, coinfections with other viruses, comparisons of nonpathogenic and highly pathogenic FAdV-4 strains, major proteins and their functions, and effects on host immune responses. Moreover, we have discussed the regulatory effect of FAdV-4 infection on non coding RNAs (ncRNAs).

Epidemiology of FAdV-4

Studies have confirmed that broiler chickens are the most affected targets of FAdV-4 infection. When chickens are infected with FAdV-4, they most likely will develop HPS[22]. The first outbreak of FAdV-4 was reported in Pakistan in Angara Goth, near Karachi, in 1987, in which the mortality rate was high in broiler chickens aged 3~5 weeks[22, 23]. In India, HHS was known as "lychee" disease, and subsequently, the disease spread to other geographical regions, including India[24], Iraq[25], Japan[26], Central and South America[27-29], Russia[30], Slovakia[31], Bangladesh[32], the Republic of Korea[33], China[20, 34] Poland[35], Iran[36], Brazil[37], Croatia[38], Australia[39], Greece[40], the United Arab Emirates[41], Saudi Arabia[42], Morocco[43], and South Africa[44], with a mortality rate of 20%~80%. FAdV-4 infections are mostly endemic during humid and hot weather[45, 46]; however, during other weather conditions, sporadic outbreaks can also occur. In China, surveillance of FAdV-4 during 2006 and 2014 revealed sporadic or clustered distributions of the virus[34], but since July 2015, epidemics have been observed in several provinces of China, such as Anhui, Xinjiang, Shandong, Liaoning, Hebei, Jilin, Heilongjiang, Jiangxi, Henan, Jiangsu, and Hubei, where the virus tends to spread rapidly[28, 34]. The mortality in these regions was approximately 40%, reaching 90% in serious cases and causing substantial economic loss. Phylogenetic analysis of the hexon gene revealed that a FAdV-4 outbreak was associated with two Chinese isolates (PK-01 and PK-06) that resided on a comparatively independent branch of the tree. This analysis revealed that Chinese strains might have originated from Indian strains[47, 48].

The FAdV-4 Chinese isolates that have a truncated ORF19 gene are considered responsible for this pandemic[21, 34]. However, further research is needed to determine the mechanism of ORF19 deletion in highly pathogenic strains. Furthermore, compared with those of a 2013 Chinese isolate, the ORF29 genes of these Chinese isolates contained 33-nt or 66-nt deletions[21], suggesting that these FAdV-4 strains have adapted to their hosts. Similarly, FAdV-4 detected in ducks, geese and ostriches caused the same symptoms as HPS in chickens[34, 49-50]. In ducks aged 25~40 days, the mortality rate is 15%~30%[50]. In goslings, morbidity and mortality begin at 8~9 days and continue until 24~25 days of age[29]. In ostriches, the mortality rate is 15%~30%, and death occurs at a young age[28]. These epidemiological data are important because they are helpful in determining the relationships of this virus in different host species and thus can help in the control of HPS.

Physiological features of FAdV-4

Various reports have shown that FAdV-4 cells are approximately 80 to 90 nanometers (nm) in diameter and filterable through a 0.1 μm pore size membrane. Transmission electron microscopy (TEM) of the virus from liver extract indicated that the virus was isometric and spherical[51]. The virus can be inactivated by heating at 60 ℃ for 1 h, 80 ℃ for 10 min, or 100 ℃ for 5 min or by treatment with 5% chloroform or 10% ether. In addition, it can tolerate pH values ranging from 3 to 10[14]. Under natural conditions, wild birds are carriers of FAdV-4[52], which can also be transmitted to healthy birds through subcutaneous or intramuscular injection of virus-infected liver homogenate from infected birds[53]. The highly pathogenic FAdV-4 serotype of broilers is naturally transmitted both vertically and horizontally[54]. In the former case, FAdV-4 infection of broiler chickens prior to egg laying was most likely congenital because the virus survives in the eggs[55]. In the latter case, the virus from the infected chicken is transferred to healthy flocks through the fecal-oral route because of the high shedding titer in feces[51]; however, airborne and aerosol transmission of FAdV-4 is negligible. The severity of the pathogenicity of FAdV-4 infection is reportedly related to the mode of infection[56]. Infection with HN/151025 in 7-, 21-, and 35-d-old chickens led to 90%, 30%, and 28.6% mortality, respectively, when administered intranasally or intramuscularly. These results indicate that the intramuscular route is the most sensitive for detecting FAdV-4 infection in chickens[47, 56-57]. HHS is characterized by depression, decreased feed intake, neck feather ruffling, and lethargy[57].

The common gross lesions of HPS in chickens include the accumulation of clear or yellowish jelly-like fluid in the pericardial sac and hydropericardium and a yellow-brown-brown color and swollen, congested, enlarged, and friable with foci of necrosis and hemorrhages in the liver. In some cases, hemorrhages occur in the kidneys, spleens and lungs[58]. Additionally, an enlarged bursa of Fabricius has also been observed[59]. FAdV-4 target the chicken hepatic cells and therefore this virus is highly pathogenic to chickens[60]. The incubation period of FAdV-4 disease has been reported to be 24 to 48 hours after infection trough natural routes[61].

The genome of nonpathogenic FAdV-4 consists of 23.3% A, 27.7% C, 26.9% G and 22.1% T, with a G+C content of 54.6%[62]. The FAdV genome encodes 10 structural proteins, namely, hexon, penton base, fiber-1, fiber-2, terminal protein (TP); and proteins Ⅵ, Ⅶ, Ⅷ, Ⅲ, and μ[20], and 11 nonstructural proteins. The major structural proteins include hexon, which forms the facets of the icosahedral structure; the penton base, which covers the vertices of the capsid; and two antenna-like fiber proteins (fiber-1 and fiber-2) anchored in each penton base protein[63]. Similarly, the four core structural proteins include Tp, μ, pV, and pⅦ, which form a complex with the genomic DNA inside the viral particles[64]. Similarly, three minor capsid proteins, protein Ⅲa (pⅢa), protein Ⅵ (pⅥ), and protein Ⅷ (pⅧ), are embedded in the inner surface of the capsid structure and play major roles in virion structural integrity[65]. The nonstructural proteins include DNA-binding Protein (DBP; also known as E2A), E1A, E1B, ADP (also known as E3), E4, EP, pol, pIVaII, 33 K, 100 K and 52/55 K[62-64, 66].

Compared with non-Chinese isolates, FAdV-4 isolates from China have been reported to have significant deletions and mutations in their genome[21, 57]. All the pathogenic strains isolated in China have 10 bp and 3 bp insertions in tandem repeat region B (TR-B) and a 55 kDa protein-coding sequence, respectively, while a 6 bp deletion was detected in the 33 kDa protein-coding fragment. Moreover, a 1 966 bp deletion has been observed in isolates from China (Figure 1-1-1), resulting in the removal of open reading frame (ORF) 19, ORF27, and ORF48. All the isolates from China also have longer GA repeats between the pX and pⅥ genes compared to the isolates from other countries (Figure 1-1-1). In addition, amino acid mutations were also observed in structural genes of Chinese isolates compared to those in isolates from other countries.

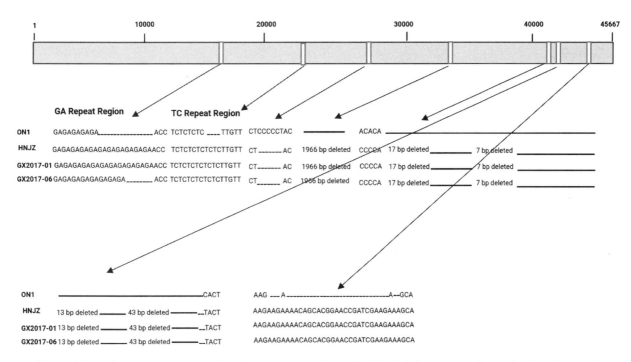

Figure 1-1-1   Schematic representation of the genome of three FAdV-4 isolates representing major insertions and deletions at various sites when compared to ON1 strain of FAdV-4

Coinfection of FAdV-4 with other serotypes and avian viruses

In the poultry industry, coinfections of FAdV with other pathogens have been reported regularly. Several reports have indicated that coinfections of FAdV-4 with different serotypes of FAdVs as well as with other immunosuppression pathogens produce various synergistic effects, as exemplified by coinfections of FAdV-4 with the infectious bursal disease virus (IBDV), which induces immunosuppression and enhances the pathogenicity of FAdV-4[67]. Similarly, coinfection of chickens with FAdV-4 and chicken infectious anemia virus (CIAV) and *Avibacterium paragallinarum* resulted in significantly increased mortality compared to with infection caused by FAdV-4 alone[68]. However, coinfection of FAdV-4 with serotype FAdV-8a in specific-pathogen-free (SPF) chickens resulted in decreased proliferation and replication of FAdV-4 while enhancing the proliferation and replication of FAdV-8a in chickens[69]. A recent study compared coinfection with FAdV-4 and duck circovirus (DuCV) and monoinfection[70]. DuCV is a prevalent infectious virus in the duck industry in China. The incidence of FAdV-4 infection in ducks has increased in recent years. A systematic study of Cherry valley ducks revealed more serious clinical signs and symptoms in those coinfected with FAdV-4 and DuCV than in those with a monoinfection. These pronounced signs included pericardial effusion, immunosuppression, and hepatitis. Moreover, the different organs showed greater viral loads, immune indices and biochemical indices in these ducks than in those with a monoinfection. According to further experimental findings, the coinfected ducks exhibited significantly greater viral replication and more tissue damage than did the monoinfected ducks. The data obtained from coinfection studies indicate that strict monitoring and preventive measures need to be taken in the field to control the severity of coinfections caused by FAdV-4 and other infectious agents.

Comparison of nonpathogenic and highly pathogenic FAdV-4 isolates

When comparing nonpathogenic strains of FAdV-4 with pathogenic strains, a large difference has been observed. A typical example of a nonpathogenic strain of FAdV-4 is ON1, which was isolated in 2004 from a broiler breeder flock in Canada that showed no clinical signs of HPS[72]. The genome size of ON1 is 45 667 bp, containing 23.3%, 27.7%, 26.9% and 22.1% A, C, G, and T, respectively, with a G+C content of 54.6%. This strain contains 46 protein-coding ORFs. To evaluate the pathogenic potential of ON1, a trial was conducted on chickens, and the virus was administered either orally or intramuscularly. The usual signs of HHS, such as huddling of chickens in corners with ruffled feathers, depression, or a specific posture of chickens with their chest and beak resting on the ground, were not observed. Furthermore, the dissected chickens also did not exhibit hydropericardium or any gross lesions. The liver seemed normal, and no inclusion bodies were detected in hepatocytes.

Severe FAdV-4 infection cases with IBH and HPS have been observed in China since 2013[65], and since then, highly pathogenic FAdV-4 strains have been found throughout China. There was a greater number of changes in the highly pathogenic isolates from China during different years than in the ON1 strain from Canada. The major changes observed were as follows: a 6 bp deletion at positions 28 784 to 28 789 and a 1 966 bp deletion at the 3' end of the genome at positions 35 389 to 37 354. The 1 966 bp deletion has been reported in all FAdV-4 isolates from China thus far (Figure 1-1-1)[71-73]; a 43 bp deletion at positions 41 766 to 41 809 was also observed (Figure 1-1-1). Furthermore, other deletions at 41 675 bp to 41 691 bp, 41 704 bp to 41 710 bp, and 41 748 bp to 41 760 bp have also been observed. In addition to deletions, insertions were also observed in these isolates compared to those in the ON1 strain. A major 27 bp insertion in ORF19 A at position 44 376 was also observed. Since all the highly pathogenic FAdV-4 strains from China had the deletion of 1 966 bp, this deletion was explored for its role in pathogenicity; however, it was determined to be dispensable for the development of highly pathogenic FAdV-4 isolates[18]. The naturally occurring mutations are rare in DNA viruses as their genomes are considered very stable. However, besides FAdV-4, another DNA virus; hepatitis B virus (HBV) was also found to have naturally occurring mutations[74]. The transcription of HBV genome is under the control of four promoters i.e. X, S, preS1, and C. Among these, the core promoter (CP) transcribes the precore (preC) and pregenomic RNA (pgRNA) mRNA, and thus the CP plays significant role in viral replication[75]. Point mutations in CP greatly affect virus expression and replication, and therefore disease development and progression are directly related to mutations in CP[74]. Since, in FAdV-4 also the mutations are natural and therefore must be thoroughly investigated for their roles in viral replication, expression and disease pattern as determined for HBV.

When GX-1, a highly pathogenic strain at 10^4 TCID$_{50}$ and three additional virus doses was inoculated in SPF chickens, 100% mortality was observed at 2 and 4 days post-inoculation (dpi)[61]. After infection, the clinical signs of chickens were huddling in corners, ruffled feathers, depression, and decreased movement. The infected chicken livers were swollen and pale brown and contained necrotic foci. The pericardial sacs contained jelly-like fluid. Swollen and congested kidneys were also observed. In contrast to the nonpathogenic strain of FAdV-4, the pathogenic strain GX1 contains 43 721 bp, 54.87% G+C content, and 38 open reading frames. This strain had 98.3% sequence identity with ON1.

Roles and functions of structural proteins of FAdV-4 in viral pathogenesis

The complete genome sequence analysis of all FAdVs delineated that they share a common genome organization. The major structural proteins and some nonstructural proteins (Nsps) are located in the middle part of the genome.

Hexon protein

Hexon, a monomer of FAdV-4, is the most abundant structural protein, with 937 amino acid residues and a molecular weight of approximately 107 kDa. It is found in an abundance of 40 to 820 copies per virion and plays a vital role not only in genome organization but also in virus-neutralizing activity and serotype specificity[72, 73]. Each hexon protein molecule consists of two conserved pedestal regions, P1 and P2, involved in hexon trimer formation and four loops (L1, L2, L3, and L4), with loops L1, L2 and L4 located on the external side of each hexon monomer; these loops are responsible for antibody binding, while L3 is situated internally and stabilizes the interface between P1 and P2. The loops containing several hypervariable regions (HVRs), in which four HVRs (HVR1-HVR4) are located in loop L1, two in loop L2 and one in loop L4, contain antigenic and immunogenic determinants and thus are used to differentiate 12 serotypes of FAdVs[71]; however, in loop L3, there are no such determinants[76]. The R188I mutation in the right side of HVR1 in L1 changed the conformation of the protein and thus was involved in the pathogenesis of FAdV-4[77]. Hypervariation in the amino acid sequence was reported in the HVRs. Due to its antigenic determinants, hexon has been widely used in molecular epidemiology studies. The L1 loop of the hexon protein is mainly used for serotyping, diagnosis and molecular typing of FAdV strains from different geographical locations worldwide[78-80]. It has been shown that the hexon protein physically binds the chaperone TCP-1- containing subunit (CCT7) and mediates viral replication. This finding highlights the potential of CCT7 to be an effective target protein for controlling FAdV-4 infection, as overexpression of CCT7 results in increased hexon expression, whereas knockdown of CCT7 results in decreased hexon protein expression[81].

The emergence of severe HHS in China due to a novel FAdV-4 strain compelled researchers to determine its virulence factor(s); therefore, in 2018, Zhang et al. investigated this issue[82]. Since the highly virulent isolates of FAdV-4 contain several mutations in the hexon gene, FAdV-4 infectious clones were generated by cloning the complete genome of CH/HNJZ/2015, a highly pathogenic isolate of FAdV-4, and ON1, a nonpathogenic strain, into the p15A-cm vector. A recombinant hexon-containing virus was constructed, and the pathogenesis of the rescued virus was compared with parent viruses, rON1 and rHNJZ in SPF chickens aged 3 weeks. Chickens infected with rescued viruses harboring the hexon gene of HNJZ developed clinical signs similar to those of natural infection, indicating that the hexon protein is related to the pathogenesis of FAdV-4[82]. Afterward, in 2021, Zhang et al. further investigated the specific amino acid residues in the hexon gene responsible for FAdV-4 pathogenesis. They generated a recombinant chimeric virus, the rHN20 strain, from a highly pathogenic strain by replacing the hexon protein of the nonpathogenic FAdV-4. After the inoculation of this chimeric strain into chickens, no mortality or even clinical signs appeared. These findings suggested that the hexon protein regulates the virulence of the novel FAdV-4 strain. Moreover, it was determined that with hexon, at position 188, the amino acid arginine (R) is important for determining the pathogenicity of this virus. The substitution of R with isoleucine (I) was performed to create the R188I mutantstrain, and the rR188I strain and the wild-type strain replicate well *in vitro*; however, the rR188I strain is nonpathogenic according to serum neutralization *in vivo* and provides protection against HHS, suggesting that the R amino acid at position 181 is critical for pathogenesis[82]. After these findings were reported, Wang et al. in 2022 further delineated the function of the amino acid 188 in the hexon protein in both pathogenic and nonpathogenic FAdV-4 isolates[77]. A mutation in the hexon protein of HNJZ (a highly pathogenic FAdV-4 isolate) at amino acid residue 188 results in conformational changes; however, a similar kind of experiment with ON1 (a nonpathogenic FAdV-4 isolate) did not result in any conformational changes. Moreover, these results revealed that the role of amino acid 188 in the hexon protein relative to its virulence varies between different strains. In addition to the amino acid 188 in the hexon protein of FAdV-4, other factors must be present to induce severe HHS in chickens[77].

Fiber proteins

The first step in viral infection is the attachment of virus to host cell surface receptors and can have direct or indirect effects on pathogenesis, and these interactions are mediated by fiber proteins. This protein also plays a role in variations in virulence, virus neutralization, and cellular epitope binding[83]. Among the three capsid proteins of FAdV-4, fiber proteins protrude from one penton base of the viral particle and play an important role in viral infection and pathogenicity[18]. The fiber patterns of the twelve FAdV serotypes showed an extreme degree of diversity. Unlike most serotypes of FAdV, serotypes 1, 4 and 10 have two distinct fiber proteins (fiber-1 and fiber-2) on their surface[84]. Both fiber proteins play important roles in viral pathogenesis[85]. Notably, fiber-1 of a novel hypervirulent FAdV-4 has 431 amino acids, whereas Fiber-2 has 479 amino acids[57], suggesting that fiber-1 might be the short fiber of FAdV-4. Fiber-1 was shown to directly trigger FAdV-4 infection by binding its knob and shaft domains to domain 2 (the coxsackie and adenovirus receptor; CAR) of the host cell[86]. In addition to being the foremost virulence determinant, the FAdV-4 fiber-2 protein is a vaccine target that counters FAdV-4 infection upon vaccination. Numerous studies have shown that recombinant fiber-2 can provide improved defense against a deadly encounter with FAdV-4 compared to that from extra capsid proteins, as indicated by the presence of hexon, penton and the all-important fiber-1. Nonetheless, the fiber-2 protein cannot induce the production of detectable neutralizing antibodies to counter FAdV-4[45]. Although the fiber-1 complete protein or its knob domain trigger the production of neutralizing antibodies against FAdV-4, the efficiency of this protein for inducing the production of efficient antibodies still needs to be determined[85, 87]. The pathogenic potential of the fiber-1 protein was evaluated in a study in which a fiber-1 replacement mutant was constructed using the HNJZ strain. The rescued virus was tested in three-week-old SPF chickens, revealing that fiber-1 from ON1 mediated the induction of clinical signs similar to those of natural hypervirulent FAdV-4 infection, suggesting that the pathogenicity of FAdV-4 was independent of fiber-1[88]. Similarly, recombinant fiber2-knob protein (F2-knob) was found to be immunogenic in 14-day-old SPF chickens[89]. The protective efficacy of F2-knob was evaluated in terms of virus shedding, clinical symptoms, mortality, and histopathology after FAdV-4 challenge. The antibody levels measured by ELISA were greater than those in chickens immunized with an inactivated vaccine against FAdV-4. Further experiments showed that F2-knob immunization significantly reduced viral shedding and provided complete protection against virulent FAdV-4, suggesting that F2-knob could be a vaccine candidate for controlling FAdV-4[89].

Penton protein

Penton, a major structural protein, is responsible for the internalization of viruses during the infection cycle. Penton is composed of two units: a penton base and a projection (fiber)[90]. The function of penton in viral pathogenesis was evaluated by constructing a penton-replacement mutant using the HNJZ strain, and SPF chickens, aged three weeks were infected with the resultant strain. The results showed that the ON1 penton resulted in similar clinical signs as the natural hypervirulent FAdV-4, suggesting that the enhanced FAdV-4 virulence is independent of penton[88].

Structural PX protein

The PX structural protein of FAdV-4 contains approximately 179 amino acid residues and has a molecular weight of 20 kDa; this protein also forms a complex with FAdV-4 genomic DNA. It has been reported that the PX proteins of several virulent FAdV-4 strains, such as HNJZ, SDDZ SXCZ, AHBZ, JSXZ and HuBWH in China are different from that of the nonpathogenic FAdV-4 ON1 strain by only two amino acids, at positions 11 and 129. The virulent strains had alanine substitutions[57] in place of threonine (T) at both positions (A11T, A129T), compared to the nonpathogenic FAdV-4 ON1 strain. It has been found that amino acids at positions 11 and 129 (A11 and A129) are important for PX-induced apoptosis, as A11T and A129T mutations resulted in reduced apoptosis. Similarly, the R/K regions at positions 14 to 16 and 22 to 25 of PX are responsible transport the protein from cytoplasm to the nucleus. Thus, FAdV-4-induced apoptosis mediates viral replication, which enhances FAdV-4 infection[57, 91].

Roles and functions of nonstructural proteins of FAdV-4 in viral pathogenesis

The nonstructural proteins (Nsps) of FAdV-4 play vital roles in its pathogenicity, genome replication and viral assembly. The 100k protein is 95 kDa, and contains approximately 798 amino acid residues. The 100K protein is encoded by the L4 gene during the last stage of infection and is involved in hexon trimerization and thus the efficient generation of progeny viruses[92]. The physical interaction between 100K and the host cellular protein HSC70 was verified through a coimmunoprecipitation (co-IP) assay in LMH cells, and this interaction is important for viral replication[93]. This interaction between 100K of the virus and HSC70 of the host cell should be considered for the development of novel antiviral strategies against FAdV-4.

Maximum studies have focused on well-known proteins, however, studies about the roles and functions of unnamed ORFs is almost absent. Two ORFs i.e. ORF20A and ORF28, located in 3' terminal region and ORF1B, located in the 5' terminal region are unique to FAdV-4 and are absent in *Aviadenovirus*, and even in *Mastadenovirus*. The protein coding potential of ORF20A and ORF28 is extremely low and might act as regulatory genes that control the expression of other proteins[94]. However, ORF1B could be only detected in the infected cells and not in mature virions. Moreover, the expression levels of ORF1B decreased gradually along FAdV-4 infection suggesting that this protein is an early gene transcript[94]. The location of ORF1B is same as the E1A in *Mastadenovirus* members, indicating that this protein might be an early transcript which could counteract host immune responses and thus to regulate virus infection. Further investigations about this protein will dissect its role in the pathogenesis of this virus. The 33K Nsp plays important role in capsid assembly and was found highly immunogenic[95]. The 52/55K protein helps in host immune evasion by degrading PKR protein by a ubiquitin proteasome degradation system[94].

The potential regulatory effect of FAdV-4 infection on ncRNAs

Several studies have indicated that viral infection induces modifications in the transcriptomes of target cells, including coding and noncoding RNAs (ncRNAs). MicroRNAs (miRNAs) are small ncRNAs approximately 20~24 nucleotides long. miRNAs either degrade or repress the translation of target mRNAs[96]. These miRNAs regulate various cellular processes i.e. differentiation[97], apoptosis[98], autophagy[99], development[100], and tumorigenesis[101], in addition to immune responses and virus entry[102]. Host cellular miRNAs indirectly affect viral replication by targeting host cellular proteins involved in viral replication.

To dissect the roles of miRNAs in host, RNA sequencing was performed in LMH cells infected with FAdV-4[103]. During this study it was found that the expression of 552 miRNAs were altered. Twelve (12) miRNAs with altered expression levels were further selected for their roles in viral replication and apoptosis in host cells. Among these miRNAs, gga-miR-30c-5p was found to be involved in the inhibition of invasion and proliferation of glioma by targeting Bcl-2[104]. Therefore, the role of gga-miR-30c-5p was further evaluated in response to FAdV-4 infection in host cells. The expression of gga-miR-30c-5p was significantly decreased during FAdV-4 infection. Furthermore, over expression of gga-miR-30c-5p mimics in LMH cells resulted in significant apoptosis, depicted by flow cytometry experiments whereas the reciprocal experiments delineated that apoptosis was inhibited by blocking the activity of gga-miR-30c-5p by using the inhibitors. These findings suggested that gga-miR-30c-5p acts as pro-apoptotic in LMH cells that is down-regulated during FAdV-4 infection[1].

In the same study[1], gga-miR-30c-5p was found to enhance FAdV-4 replication in LMH cells. The replication of FAdV-4 was assessed by the expression of hexon and PX proteins. The expression of hexon and PX proteins was significantly increased in FAdV-4 infected LMH cells with gga-miR-30c-5p transfection. In reciprocal experiments by inhibiting the activity of gga-miR-30c-5p by using the miRNA inhibitor, the replication of FAdV-4 was reduced[1]. These findings suggested that gga-miR-30c-5p promotes the replication of FAdV-4. Further to dig out the mechanism, Mcl-1 was identified as direct target of gga-miR-30c-5p, as knock-down of this protein increased apoptosis in LMH cells, whereas over expression of Mcl-1 counteracted gga-miR-30c-5p or FAdV-4 induced apoptosis. Therefore, gga-miR-30c-5p promotes FAdV-4 induced apoptosis by targeting Mcl-1 directly and facilitates FAdV-4 replication inside host cells[1].

In another study miRNA-seq analysis was performed in LMH cells in order to understand the entry of FAdV-4 into host cells[59]. During this study the expression of 785 miRNAs was altered, and gga-miR-15c-3p, gga-miR-148a-3p, and gga-miR-148a-5p were identified for the first time to be associated with FAdV-4 infection[59]. Furthermore, the expression of gga-miR-128-2-5p was also altered and luciferase assay showed that OBSL1 was its true target. Overexpression of gga-miR-128-2-5p resulted in the down regulation of OBSL1 and inhibited the entry of FAdV-4 into the cells. These findings suggested that gga-miR-128-2-5p plays a crucial role in the entry of FAdV-4 into host cells by targeting OBSL1[59].

In another study, the role of gga-miR-181a-5p in regulating FAdV-4 replication was elucidated[105]. During FAdV-4 infection, the expression of gga-miR-181a-5p was induced in LMH cells and was found to promote FAdV-4 replication. The stimulator of interferon genes (STING) was found to be the direct target of gga-miR-181a-5p, and knockout of STING increased FAdV-4 replication, whereas its over expression decreased FAdV-4 replication. These findings revealed that gga-miR-181a-5p promotes FAdV-4 replication[105].

Long noncoding RNAs (lncRNAs) are also a class of ncRNAs, are greater than 200 nucleotides in length. Since lncRNAs do not contain an open reading frame, they cannot encode functional proteins[106, 107]. Like miRNAs, lncRNAs affect the transcription and translation of both host and virus genes, the stability of mRNAs and the host antiviral response[108]. It has been reported that viral infection alters the lncRNA transcription profile of infected host cells, which is involved in the establishment and maintenance of persistent infection[109].

It has been reported that FAdV-4 induces apoptosis in LMH cells by modulating the transcription of the BMP4 gene, which is targeted by lncRNA 54 128[110]. The miRNAs and lncRNAs whose transcription is induced during FAdV-4 infection, as well as their target mRNAs, are listed in Table 1-1-1.

Table 1-1-1   MiRNAs, lncRNAs and their targets during FAdV-4 infection

S. No.	ncRNA	Target protein(s)	Protein function	Reference
1	gga-miR-7475-5p	CREBZF/LFNG	Host cell response viral infection	[59], [111]
2	gga-miR-128-2-5p	OBSL1	FAdV-4 entry to the host cell	[59]
3	novel-miR271	EGR1	Virus infection	[59], [112]
4	gga-miR-30c-5p	Mcl-1	Viral replication	[103]
5	gga-miR-205c-3p	FRMD6	Apoptosis	[59], [113]
6	miR-181a-5p	STING	Viral replication	[106]
7	gga-miR-12223-3p	SERTAD2	Carcinogenesis	[59], [114]
8	gga-miR-6654-3p	NFATC1	T-cell regulation	[115]
9	gga-miR-12245-3p	AQP11	Transport function	[116]
10	miR-27	SNAP25/TXN2	Virus entry	[117]
11	ENSGALG00000055015	BMP4	Apoptosis	[108], [118]
12	ENSGALG00000054172	SOCS2	Immune-related cancer	[108], [119]
13	XLOC_026155	FOXO3	Apoptosis	[120]
14	ENSGALG00000050472	FOXO3	Apoptosis	[120]
15	ENSGALG00000048532	SLC40A1	Iron metabolism	[121]
16	XLOC_003734	MKP3	Apoptosis	[108], [122]

The role of FAdV-4 infection in host immune responses

Although limited information is available regarding FAdV infection and the host immune response, it is broadly believed that certain components of innate immunity are crucial for combatting FAdV infection in fowl. One such innate immune component is avian β-defensins (AvBDs), which play a pivotal role in viral infections. It has been reported that most AvBDs (AvBD 5, 7, 8, 9, and 13) were upregulated in specific FAdV-4-infected fowl tissues, and a positive correlation was observed between the FAdV-4 genome levels and upregulation of the aforementioned AvBDs, suggesting the important role of AvBDs in the immune response to FAdV-4 infection[123]. In addition, the host immune system initiates an immune response by inducing the production of antiviral factors after recognizing viral pathogen-associated molecular patterns through pattern recognition receptors (PRRs). Consequently, several immunity-related pathways, including but not limited to the Toll-like receptor (TLR) signaling pathway, MyD88-mediated FAdV-4-induced inflammation and the cytokine-cytokine receptor interaction pathway, are activated after FAdV-4 infection[124, 125]. The cyclic GMP-AMP (cGAMP) synthase (cGAS) is a DNA sensor that triggers the induction of innate immune responses to produce IFN-I and proinflammatory cytokines[126]. The cGAS has been well studied in mammals, however, in the chickens the Cgas (chcGAS) has not been thoroughly studied. The cGAS recognizes pathogens that contain DNA, whereas cells lacking cGAS do not respond efficiently to DNA viruses[127, 128]. Soon after binding the double stranded DNA inside cytosol, the cGAS converts guanosine 5'-triphosphate and adenosine 5'-triphosphate into 2', 3'-cyclic GMP-AMP (2', 3'-cGAMP) that serves to trigger the stimulator of IFN genes (STING)[129, 130]. STING undergoes significant conformational changes when binds to 2', 3'-cGAMP, recruits TANK binding kinase (TBK1), leading to the phosphorylation of IFN-regulatory factor 3 (IRF3) and nuclear factor-κB (NF-κB), and finally expresses IFNs and proinflamatory cytokines[131, 132]. cGAS can also translocate into the nucleus and suppresses homologous-recombination-mediated DNA repair in response to DNA damage[133]. The role of chcGAS was well evaluated by Wang et al. who cloned the complete open reading frame of chcGAS from cDNA derived from ileum[126]. The phylogenetic analysis and multiple sequence alignment delineated that chcGAS was homologous to cGAS of mammals. The chcGAS was found to be highly expressed in ileum and bone marrow. Moreover, chcGAS was highly expressed inside cytoplasm and partially co-localized with endoplasmic reticulum (ER). Over expression of chcGAS resulted in increased transcription of proinflammatory cytokine i.e. interleukin 1β (IL-1β) and IFN-β in a dose dependent manner in CEF and LMH cells. Moreover, transfection of herring sperm DNA (HS-DNA) and poly (dA:dT) but not poly (I:C) further enhanced chcGAS-induced IL-1β and IFN-β expression, indicating that chcGAS responds only to dsDNA in LMH and CEF cells. Furthermore, chcGAS-induced IL-1β and IFN-β activation was found to be dependent on STING signaling pathway. Finally, it was shown that chcGAS was involved in sensing FAdV-4 in LMH cells, and knock down of chcGAS promoted FAdV-4 infection in LMH cells. All this indicate that chcGAS plays an antiviral role during FAdV-4 infection through STING signaling pathway to induce the expression of IL-1β and IFN-β[126]. Cytokines are proteins or glycoproteins that are secreted by immune cells after viral infections. Usually, the pro-inflammatory cytokines and IFN-I are expressed quickly to counteract viral infections. The infection of chickens with FAdV-9 increased the expression levels of IFN-α, IFN-γ, and IL-12 in bursa of Fabricius, spleen, and liver[134]. The infection of chickens with FAdV-8 increased the expression levels of IFNg mRNA and decreased IL-10 levels in the spleen[135]. Similarly, the infection of FAdV-4 has resulted in the increased expression levels of IL-1β, IL-2, IL-6, IL-8, IL-10, IL-18, IFNγ, and TNF-α in the *in vivo* and *in vitro* conditions[22, 125, 136].

In another study, the differences in the innate immune responses both *in vivo* and *in vitro*, between chickens and ducks were delineated by using SD0828 strain of FAdV-4[124]. The mRNA levels of pro-inflammatory cytokines i.e. IL-6 and IL-8 and interferon stimulated genes (ISGs) such as, Mx and OAS were significantly increased in chicken embryo fibroblasts (CEFs) and duck embryo fibroblasts (DEFs). The significantly highly expression of IL-6 and IL-8 also produced severe inflammatory responses in the infected tissues, resulting in deaths at higher rate. Although the mRNA levels of ISGs were also increased significantly, they did not withstand the inflammatory storm in the infected organisms[124]. Another study performed by He et al. delineated that mRNA levels of IFNs were increased from 24 h to 72 h post FAdV-4 infection in LMH cells and livers, however, the FAdV-4 viral load still increased significantly[137]. These findings suggested that FAdV-4 escaped the host innate immune responses. Moreover, in the same study it was revealed that only the protein levels and not the mRNA levels of Protein Kinase R (PKR) were degraded after 48 h of FAdV-4 infection inside the host cells. Furthermore, the 52/55 K protein was ubiquitinylated, leading to the degradation of protein kinase R (PKR) inside host cells through the proteasomal degradation pathway. These results revealed that the 52/55 K protein of FAdV-4 degrades PKR through the ubiquitin-proteasome pathway and helps FAdV-4 evade host immune responses[137].

Cytokines contribute to both adaptive and innate immunity to viral infection and have therefore both local and systemic effects. Higher immune responses sometimes produce negative effects to the host, for example, malaise, fever, and inflammatory damage are observed when increased levels of cytokines are produced[138]. High inflammatory damage has been observed in FAdV-4 infected chickens, and a few studies have elaborated the cytokine secretion. High expression of cytokinesis observed in the organs targeted by the virus, however, function of each cytokine is still to be determined. Interleukins (ILs), Interferons (IFs), transforming growth factors (TGFs), TNGs, chemokines and colony stimulating factors (CSFs) all are inflammatory cytokines. In case of FAdV-4, the well-studied cytokines are: Tumor necrosis factors (TNFs), IFNs, and ILs. Usually cytokines affect the production of their own as well other cytokines production to make a complex cytokine network. The effect of cytokines mostly depends on timing, duration and the target organ, however, is still difficult to make a final conclusion whether a cytokine protects or contributes to hosts damage caused by FAdV-4 infection. When the MDA5, NLRP3, and TLRs recognize the FAdV-4, cytokines expression is triggered by translocation of NF-κB from cytoplasm to nucleus. The TNFs and IL-1β trigger the activation of NLRP3 inflamasomes, which in turn enhances the production of proinflammatory cytokines. FAdV-4 is known to trigger both pro- and anti-inflammatory cytokines in the main targeted organs. The production of pro-inflammatory cytokines has been found to cause organ failure and even death of the host[139, 140].

Pericardial effusion is also observed in chickens infected by FAdV-4, which could lead to a conclusion that excessive inflammatory response may result in organ exhaustion and finally to death. It has been observed experimentally that chicken infected with pathogenic FAdV-4 strain, CH/HNJZ/2015 showed that expression levels of mRNA encoding pro-inflammatory cytokines i.e. IL-8 and IL-18 and antiviral cytokine, IFN-β were significantly higher compared to the chickens infected with non-pathogenic FAdV-4 strain ON1 in various organs[22]. The role of cytokines in host immune responses is important in determining their action as friend or enemy. Previously IL-1β, IL-2, IL-4, IL-6, and IL-18 have showed potential treatment or vaccine adjuvants for avian influenza viruses (AIV), Newcastle disease, and infectious bursal disease virus (IBDV), therefore, the possible therapeutic application of these cytokines against FAdV-4 infection warrants future investigations[138].

Furthermore, infection with AD234, a virulent FAdV-4 strain, decreased the number of CD3-, CD4-, and CD8-positive T cells in the spleen, as well as the number of CD4- and CD8-positive T lymphocytes in the thymus. Lymphocyte numbers were also decreased in the bursa of Fabricius. These observations suggested that FAdV-4 escapes the adaptive immune response of host chickens[123]. Infection of SPF chickens with FAdV-4 resulted in atrophy of the bursa of Fabricius, a decrease in lymphocyte count and a decrease in the number of B and T cells in lymphoid organs[141]. The PX protein is critical for apoptosis, and lymphocyte apoptosis causes immunosuppression. However, the detailed mechanism by which FAdV-4 induces apoptosis leading to immunosuppression still needs to be investigated[123]. Species immunity and genetic background might also be factors involved in immune evasion by FAdV-4.

Conclusions

FAdV-4 has spread globally, is highly pathogenic and has caused substantial economic loss in the associated industry. FAdV-4 proteins, especially structural proteins, play important roles in pathogenesis; however, mutations in these proteins have been identified through continuous surveillance, and these mutations may cause the development of even more hypervirulent strains of FAdV-4. Therefore, continuous screening for new mutations in viral proteins needs to be performed to help in the design of effective vaccines against FAdV-4. In recent years, ncRNAs have been used for therapeutic purposes to treat different diseases; therefore, the therapeutic potential of ncRNAs that are regulated after FAdV-4 infection must be further explored. Regarding the immune responses, identification of specific genes of FAdV-4 regulating cytokines production will help in the robust understanding FAdV-4 infection.

References

[1] SCHACHNER A, MATOS M, GRAFL B, et al. Fowl adenovirus-induced diseases and strategies for their control—a review on the current global situation.Avian Pathol, 2018, 47(2):111-126.

[2] ZSÁK L, KISARY J. Grouping of fowl adenoviruses based upon the restriction patterns of DNA generated by BamHI and HindⅢ. Intervirology, 1984, 22(2):110-114.

[3] HESS M. Detection and differentiation of avian adenoviruses: a review. Avian Pathol, 2000, 29(3):195-206.

[4] BALAMURUGAN V, KATARIA J M. The hydropericardium syndrome in poultry—a current scenario. Vet Res Commun. 2004, 28(2):127-148.

[5] KIM M S, LIM T H, LEE D H, et al. An inactivated oil-emulsion fowl Adenovirus serotype 4 vaccine provides broad cross-protection against various serotypes of fowl Adenovirus. Vaccine, 2014, 32(28): 3564-3568.

[6] METTIFOGO E, NUNEZ L, SANTANDER PARRA S H, et al. Fowl adenovirus Group I as a causal agent of inclusion body hepatitis/hydropericardium syndrome outbreak in Brazilian broiler flocks. Pesqui Vet Bras, 2014, 34(8): 733-737.

[7] LI H, WANG J, QIU L, et al. Fowl adenovirus species C serotype 4 is attributed to the emergence of hepatitis-hydropericardium syndrome in chickens in China. Infect Genet Evol, 2016, 45: 230-241.

[8] NAKAMURA K, MASE M, YAMAMOTO Y, et al. Inclusion body hepatitis caused by fowl adenovirus in broiler chickens in Japan, 2009-2010. Avian Dis, 2011, 55(4): 719-723.

[9] SCHACHNER A, MAREK A, GRAFL B, et al. Detailed molecular analyses of the hexon loop-1 and fibers of fowl aviadenoviruses reveal new insights into the antigenic relationship and confirm that specific genotypes are involved in field outbreaks of inclusion body hepatitis. Veterinary Microbiol, 2016, 186: 13-20.

[10] DOMANSKA-BLICHARZ K, TOMCZYK G, SMIETANKA K, et al. Molecular characterization of fowl adenoviruses isolated from chickens with gizzard erosions. Poultry Sci, 2011, 90(5): 983-989.

[11] THANASUT K, FUJINO K, TAHARAGUCHI M, et al. Genome sequence of fowl aviadenovirus a strain JM1/1, which caused gizzard erosions in Japan. Genome Announcements, 2017, 5(41): e00749-17.

[12] ALEMNESH W, HAIR-BEJO M, AINI I, et al. Pathogenicity of fowl adenovirus in specific pathogen free chicken embryos. Journal of Comparative Pathology, 2012, 146(2-3): 223-229.

[13] MASE M, HIRAMATSU K, NISHIJIMA N, et al. Fowl adenoviruses Type 8b isolated from chickens with inclusion body hepatitis in Japan. Avian Dis, 2020, 64(3): 330-334.

[14] AFZAL M, MUNEER R, STEIN G .Studies on the aetiology of hydropericardium syndrome (Angara disease) in broilers. Veterinary Record, 1991, 128(25): 591-593.

[15] ABE T, NAKAMURA K,TOJO H. Histology, immunohistochemistry and ultrastructure of hydropericardium syndrome in adult broiler breeders and broiler chicks. Avian Dis, 1998.42(3):606-612.

[16] HESS M, PRUSAS C, VEREECKEN M, et al. Isolation of fowl adenoviruses serotype 4 from pigeons with hepatic necrosis. Berl Much Tierarztl Wochenschr, 1998, 111(4): 140-142.

[17] ROY P, PURUSHOTHAMAN V, VAIRAMUTHU S, et al. Hydropericardium syndrome in Japanese quail (Coturnix coturnix japonica). Veterinary Record, 2004, 155(9): 273-274.

[18] PAN Q, LIU L, WANG Y, et al. The first whole genome sequence and pathogenicity characterization of a fowl adenovirus 4 isolated from ducks associated with inclusion body hepatitis and hydropericardium syndrome. Avian Pathol, 2017, 46(5): 571-578.

[19] SHEN Z, XIANG B, LI S, et al. Genetic characterization of fowl adenovirus serotype 4 isolates in southern China reveals potential cross-species transmission. Infection Genet. Evol, 2019: 75, 103928.

[20] LI P H, ZHENG P P, ZHANG T F, et al. Fowl adenovirus serotype 4: Epidemiology, pathogenesis, diagnostic detection, and vaccine strategies. Poult Sci, 2017, 96(8): 2630-2640.

[21] YE J, LIANG G, ZHANG J, et al. Outbreaks of serotype 4 fowl adenovirus with novel genotype, China. Emerg Microbes Infect, 2016, 5(5): e50.

[22] GRGIĆ H, POLJAK Z, SHARIF S, et al. Pathogenicity and cytokine gene expression pattern of a serotype 4 fowl adenovirus isolate. PLOS ONE, 2013, 8(10): e77601.

[23] KHAWAJA D A, SATTAR A, AKBAR M, et al. A report on cross protection of IBD vaccine strain D-78 against adenovirus isolated from hydropericardium syndrome in broiler chicks. Pakistan J Veterinary Res, 1998, 1: 51-52.

[24] GOWDA R S, SATYANARAYANA M. Hydropericardium syndrome in poultry. Indian J Veterinary Pathol, 1994, 18: 159-161.

[25] ABDUL-AZIZ T A, AL-ATTAR M A. New syndrome in Iraqi chicks. Vet Rec, 1991, 129(12): 272.

[26] MASE M, NAKAMURA K, MINAMI F. Fowl adenoviruses isolated from chickens with inclusion body hepatitis in Japan, 2009-2010. J Vet Med Sci, 2012, 4(8):1087-1089.

[27] COWEN B S, LU H, WEINSTOCK D, et al. Pathogenicity studies of fowl adenoviruses isolated in several regions of the world. Int. Symposium Adenovirus Infection Poultry, 1996, 79-88.

[28] SHANE S M. Hydropericardium-Hepatitis Syndrome the current world situation. Zootecnica Int1996, 19: 20-27.

[29] TORO H, PRUSAS C, RAUE R, et al. Characterization of fowl adenoviruses from outbreaks of inclusion body hepatitis/ hydropericardium syndrome in Chile. Avian Dis, 1999, 43(2): 262-270.

[30] BORISOV V V, BORISOV A V, GUSEV A A. Hydropericardium syndrome in chickens in Russia//Proceedings of 10th International Congress of the WVPA. Budapest Hungary: Poult Assoc, 1997: 258.

[31] JANTOSOVIC J, KONARD J, SALY J, et al. Hydropericardium syndrome in chicks. Veterinastvi, 1991, 41: 261-263.

[32] BISWAS P K, SIL B K, FARUQUE R, et al. Adenovirus induced hydropericardium-hepatitis syndrome in broiler parent chickens in Chittagong, Bangladesh. Pakistan Journal of Biological Sciences, 2002, 5: 994-996.

[33] CHOI K S, KYE S J, KIM J Y, et al. Epidemiological investigation of outbreaks of fowl adenovirus infection in commercial chickens in Korea[*]. Poult Sci, 2012, 91(10): 2502-2506.

[34] LI L, LUO L, LUO Q, et al. Genome sequence of a fowl adenovirus serotype 4 strain lethal to chickens, isolated from China. Genome Announ, 2016, 4(2): e00140-16.

[35] NICZYPORUK J S. Phylogenetic and geographic analysis of fowl adenovirus field strains isolated from poultry in Poland. Arch Virol, 2016, 161(1):33-42.

[36] MIRZAZADEH A, ASASI K, MOSLEH N, et al. A primary occurrence of inclusion body hepatitis in absence of predisposing agents in commercial broilers in Iran: a case report. Iran J Vet Res, 2020, 21(4): 314-318.

[37] DE LA TORRE D, NUÑEZ L F N, SANTANDER PARRA S H, et al. Molecular characterization of fowl adenovirus group I in commercial broiler chickens in Brazil. Virusdisease, 2018, 29(1): 83-88.

[38] ZADRAVEC M, SLAVEC B, KRAPEŽ U, et al. Inclusion body hepatitis (IBH) outbreaks in broiler chickens in Slovenia// IX Simpozij Peradarski dani 2011. Sibenik, Croatia: S Međunarodnim Sudjelovanjem, 2011: 36-39.

[39] STEER P A, O'ROURKE D, GHORASHI S A, et al. Application of high-resolution melting curve analysis for typing of fowl adenoviruses in field cases of inclusion body hepatitis. Aust Vet J, 2011, 89(5):184-192.

[40] FRANZO G, PRENTZA Z, PAPAROUNIS T, et al. Molecular epidemiology of Fowl adenoviruses (FAdV) in Greece. Poultry Science, 2020, 99(11): 5983-5990.

[41] ISHAG H Z A, TERAB A M A, EL TIGANI-ASIL E T A, et al. Pathology and molecular epidemiology of fowl adenovirus serotype 4 outbreaks in broiler chicken in Abu Dhabi Emirate, UAE. Veterinary Sci, 2022, 9(4): 154.

[42] HEMIDA M, ALHAMMADI M. Prevalence and molecular characteristics of fowl adenovirus serotype 4 in eastern Saudi Arabia. Turkish Journal of Veterinary and Animal Sciences, 2017, 41(4): 506-513.

[43] ABGHOUR S, ZRO K, MOUAHID M, et al. Isolation and characterization of fowl aviadenovirus serotype 11 from chickens with inclusion body hepatitis in Morocco. PLOS ONE, 2019, 14(12): e0227004.

[44] JOUBERT H W, AITCHISON H, MAARTENS L H, et al. Molecular differentiation and pathogenicity of Aviadenoviruses isolated during an outbreak of inclusion body hepatitis in South Africa. J S Afr Vet Assoc, 2014, 85(1): 1058.

---

* "Korea" refers to the Republic of Korea. The same applies to other occurrences in the book.

[45] SCHACHNER A, MAREK A, JASKULSKA B, et al. Recombinant FAdV-4 fiber-2 protein protects chickens against hepatitis-hydropericardium syndrome (HHS). Vaccine, 2014, 32(9):1086-1092.

[46] SHAH M S, ASHRAF A, KHAN M I, et al. Molecular cloning, expression and characterization of 100K gene of fowl adenovirus-4 for prevention and control of hydropericardium syndrome. Biologicals, 2016, 44(1): 19-23.

[47] LI C, LI H, WANG D, et al. Characterization of fowl adenoviruses isolated between 2007 and 2014 in China. Vet Microbiol, 2016, 197: 62-67.

[48] ZHANG T, JIN Q, DING P, et al. Molecular epidemiology of hydropericardium syndrome outbreak-associated serotype 4 fowl adenovirus isolates in central China. Virol J, 2016, 13(1):188.

[49] IVANICS E, PALYA V, MARKOS B, et al. Hepatitis and hydropericardium syndrome associated with adenovirus infection in goslings. Acta Vet Hung, 2010, 58(1): 47-58.

[50] CHEN H, DOU Y, ZHENG X, et al. Hydropericardium hepatitis syndrome emerged in Cherry valley ducks in China. Transbound Emerg Dis, 2017, 64(4):1262-1267.

[51] CHANDRA R, SHUKLA S K, KUMAR M. The hydropericardium syndrome and inclusion body hepatitis in domestic fowl. Tropical Animal Health & Production, 2000, 32(2):99-111.

[52] MANZOOR S, HUSSAIN Z, RAHMAN S U, et al. Identification of antibodies against hydropericardium syndrome in wild birds. Br Poult Sci, 2013, 54(3): 325-328.

[53] NAEEM K, RAHIM A, MAJEED I U. Post infection dissemination pattern of avian adenovirus involved in hydropericardium syndrome. Pak Vet J, 2001, 21(3): 152-156.

[54] ASTHANA M, CHANDRA R, KUMAR R. Hydropericardium syndrome: current state and future developments. Arch Virol. 2013, 158(5): 921-931.

[55] TAHIR R, MUHAMMAD K, RABBANI M, et al. Transovarian transmission of hydropericardium syndrome virus in experimentally infected poultry birds. Hosts and Viruses, 2017, 4(6): 88-91.

[56] MAZAHERI A, PRUSAS C, VOSS M, et al. Some strains of serotype 4 fowl adenoviruses cause inclusion body hepatitis and hydropericardium syndrome in chickens. Avian Pathol, 1998, 27(3): 269-276.

[57] LIU Y, WAN W, GAO D, et al. Genetic characterization of novel fowl aviadenovirus 4 isolates from outbreaks of hepatitis-hydropericardium syndrome in broiler chickens in China. Emerging Microbes & Infections, 2016, 5(11): e117.

[58] SUN J, ZHANG Y, GAO S, et al. Pathogenicity of fowl adenovirus serotype 4 (FAdV-4) in chickens. Infection Genet Evol 2019, 75: 104017.

[59] WU N, YANG B, WEN B, et al. Interactions among expressed microRNAs and mRNAs in the early stages of fowl adenovirus aerotype 4-infected leghorn male hepatocellular cells. Front Microbiol, 2020, 11: 831.

[60] CHEN L, YIN L, ZHOU Q, et al. Epidemiological investigation of fowl adenovirus infections in poultry in China during 2015-2018. BMC Veterinary Research, 2019, 15(1): 1-7.

[61] REN G, WANG H, HUANG M, et al. Transcriptome analysis of fowl adenovirus serotype 4 infection in chickens. Virus Genes, 2019, 55(5):619-629.

[62] GRIFFIN B D, NAGY E. Coding potential and transcript analysis of fowl adenovirus 4: insight into upstream ORFs as common sequence features in adenoviral transcripts.The Journal of General Virology, 2011, 92(Pt 6):1260.

[63] HESS M, CUZANGE A, RUIGROK R W, et al. The avian adenovirus penton: two fibers and one base. J Mol Biol, 1995, 252(4):379-385.

[64] NEMEROW G R, PACHE L, REDDY V, et al. Insights into adenovirus host cell interactions from structural studies. Virology, 2009, 384(2):380-388.

[65] ZHAO J, ZHONG Q, ZHAO Y, et al. Pathogenicity and complete genome characterization of fowl adenoviruses isolated from chickens associated with inclusion body hepatitis and hydropericardium syndrome in China. PLOS ONE, 2015, 10(7): e0133073.

[66] XIE Z, LUO S, FAN Q, et al. Detection of antibodies specific to the non-structural proteins of fowl adenoviruses in infected

chickens but not in vaccinated chickens. Avian Pathol, 2013, 42(5): 491-496.

[67] XU A H, SUN L, TU K H, et al. Experimental co-infection of variant infectious bursal disease virus and fowl adenovirus serotype 4 increases mortality and reduces immune response in chickens. Veterinary Res, 2021, 52(1):61.

[68] MEI C, XIAN H, BLACKALL P J, et al. Concurrent infection of Avibacterium paragallinarum and fowl adenovirus in layer chickens. Poultry Sci, 2020, 99(12):6525-6532.

[69] LIU J, MEI N, WANG Y, et al. Identification of a novel immunological epitope on hexon of fowl adenovirus serotype 4. AMB Express, 2021, 11(1): 153.

[70] SHEN M, GAO P, WANG C, et al. Pathogenicity of duck circovirus and fowl adenovirus serotype 4 co-infection in Cherry valley ducks. Veterinary Microbiology, 2023, 279: 109662.

[71] GANESH K, SURYANARAYANA V, RAGHAVAN R, et al. Nucleotide sequence of L1 and part of P1 of hexon gene of fowl adenovirus associated with hydropericardium hepatitis syndrome differs with the corresponding region of other fowl adenoviruses. Veterinary Microbiology, 2001, 78(1):1-11.

[72] ROBERTS D M, NANDA A, HAVENGA M J, et al. Hexon-chimaeric adenovirus serotype 5 vectors circumvent pre-existing anti-vector immunity. Nature, 2006, 441(7090):239-243.

[73] MATSUSHIMA Y, SHIMIZU H, PHAN T G, et al. Genomic characterization of a novel human adenovirus type 31 recombinant in the hexon gene. Journal of General Virology, 2011, 92(12): 2770-2775.

[74] PENG Y, LIU B, HOU J, et al. Naturally occurring deletions/insertions in HBV core promoter tend to decrease in hepatitis B e antigen- positive chronic hepatitis B patients during antiviral therapy. Antiviral Therapy, 2015, 20(6): 623.

[75] QUARLERI J. Core promoter: a critical region where the hepatitis B virus makes decisions. World J Gastroenterology, 2014, 20(2): 425-435.

[76] NICZYPORUK J S. Deep analysis of Loop L1 HVRs1-4 region of the hexon gene of adenovirus field strains isolated in Poland. PLOS ONE, 2018, 13(11): e0207668.

[77] WANG B, SONG C, YANG P, et al. The role of hexon amino acid 188 varies in fowl adenovirus serotype 4 strains with different virulence. Microbiol Spectr, 2022, 10(3): e0149322.

[78] RAUE R, HESS M. Hexon based PCRs combined with restriction enzyme analysis for rapid detection and differentiation of fowl adenoviruses and egg drop syndrome virus. Journal of Virological Methods, 1998, 73(2):211-217.

[79] MAREK A, GÜNES A, SCHULZ E, et al. Classification of fowl adenoviruses by use of phylogenetic analysis and high-resolution melting-curve analysis of the hexon L1 gene region. Journal of Virological Methods, 2010, 170(1-2): 147-154.

[80] PIZZUTO M S, DE BATTISTI C, MARCIANO S, et al. Pyrosequencing analysis for a rapid classification of fowl adenovirus species. Avian Pathol, 2010, 39(5): 391-398.

[81] GAO J, ZHAO M, DUAN X, et al. Requirement of cellular protein CCT7 for the replication of fowl adenovirus serotype 4 (FAdV-4) in leghorn male hepatocellular cells via interaction with the viral hexon protein. Viruses, 2019, 11(2): 107.

[82] ZHANG Y, LIU A, WANG Y, et al. A single amino acid at residue 188 of the hexon protein is responsible for the pathogenicity of the emerging novel virus fowl adenovirus 4. J Virol, 2021, 95(17): e0060321.

[83] PALLISTER J, WRIGHT P J, SHEPPARD M. A single gene encoding the fiber is responsible for variations in virulence in the fowl adenoviruses. J Virol, 1996, 70(8): 5115-5122.

[84] MAREK A, NOLTE V, SCHACHNER A, et al. Two fiber genes of nearly equal lengths are a common and distinctive feature of Fowl adenovirus C members. Veterinary Microbiol, 2012, 156(3-4): 411-417.

[85] WANG W, LIU Q, LI T, et al. Fiber-1, not fiber-2, directly mediates the infection of the pathogenic serotype 4 fowl adenovirus via its shaft and knob domains. Journal of Virology, 2020, 94(17): e00954-20.

[86] FREIMUTH P, SPRINGER K, BERARD C, et al. Coxsackievirus and adenovirus receptor amino-terminal immunoglobulin V-related domain binds adenovirus type 2 and fiber knob from adenovirus type 12. J Virol, 1999, 73(2): 1392-1398.

[87] PAN Q, WANG J, GAO Y, et al. Development and application of a novel ELISA for detecting antibodies against group I fowl adenoviruses. Appl Microbiol Biotechnol, 2020, 104(2): 853-859.

[88] LIU R, ZHANG Y, GUO H, et al. The increased virulence of hypervirulent fowl adenovirus 4 is independent of fiber-1 and penton. Res Veterinary Sci, 2020, 131: 31-37.

[89] SONG Y, ZHAO Z, LIU L, et al. Knob domain of fiber 2 protein provides full protection against fowl adenovirus serotype 4. Virus Res, 2023, 330: 199113.

[90] RASHID F, XIE Z, ZHANG L, et al. Genetic characterization of fowl aviadenovirus 4 isolates from Guangxi, China, during 2017-2019. Poultry Sci, 2020, 99(9): 4166-4173.

[91] ZHAO M, DUAN X, WANG Y, et al. A novel role for PX, a structural protein of fowl adenovirus serotype 4 (FAdV-4), as an apoptosis-inducer in leghorn male hepatocellular cell. Viruses, 2020, 12(2): 228.

[92] KOYUNCU O O, SPEISEDER T, DOBNER T, et al. Amino acid exchanges in the putative nuclear export signal of adenovirus type 5 L4-100K severely reduce viral progeny due to effects on hexon biogenesis. J Virol, 2013, 87(3): 1893-1898.

[93] GAO S, CHEN H, ZHANG X, et al. Cellular protein HSC70 promotes fowl adenovirus serotype 4 replication in LMH cells via interacting with viral 100K protein. Poultry Sci, 2022, 101(7): 101941.

[94] GAO S, LI R, ZHANG X, et al. Identification of ORF1B as a unique nonstructural protein for fowl adenovirus serotype 4. Microbial Pathogenesis, 2024, 186: 106508.

[95] MORIN N, BOULANGER P. Morphogenesis of human adenovirus type 2: sequence of entry of proteins into previral and viral particles. Virology, 1984, 136(1): 153-167.

[96] BARTEL D P. MicroRNAs: genomics, biogenesis, mechanism, and function. Cell, 2004, 116(2): 281-297.

[97] KORNFELD S F, CUMMINGS S E, FATHI S, et al. MiRNA-145-5p prevents differentiation of oligodendrocyte progenitor cells by regulating expression of myelin gene regulatory factor. J Cell Physiol, 2021, 236(2): 997-1012.

[98] CHEN X, LI A, ZHAN Q, et al. microRNA-637 promotes apoptosis and suppresses proliferation and autophagy in multiple myeloma cell lines via NUPR1. FEBS Open Bio, 2021, 11(2): 519-528.

[99] LI P, HE J, YANG Z, et al. ZNNT1 long noncoding RNA induces autophagy to inhibit tumorigenesis of uveal melanoma by regulating key autophagy gene expression. Autophagy, 2020, 16(7): 1186-1199.

[100] TSAGAKIS I, DOUKA K, BIRDS I, et al. Long non-coding RNAs in development and disease: conservation to mechanisms. J Pathol, 2020, 250(5): 480-495.

[101] DU Z, SUN T, HACISULEYMAN E, et al. Integrative analyses reveal along noncoding RNA-mediated sponge regulatory network in prostate cancer. Nat Commun, 2016, 7: 10982.

[102] XU H, JIANG Y, XU X, et al. Inducible degradation of lncRNA Sros1 promotes IFN-g-mediated activation of innate immune responses by stabilizing Stat1 mRNA. Nat Immunol, 2019, 20(12): 1621-1630.

[103] HAIYILATI A, ZHOU L, LI J, et al. Gga-miR-30c-5p enhances apoptosis in fowl adenovirus serotype 4-infected leghorn male hepatocellular cells and facilitates viral replication through myeloid cell leukemia-1. Viruses, 2022, 14(5):990.

[104] YUAN L Q, ZHANG T, XU L, et al. miR-30c-5p inhibits glioma proliferation and invasion via targeting Bcl2. Transl Cancer Res, 2021, 10(1): 337-348.

[105] YIN D, SHAO Y, YANG K, et al. Fowl adenovirus serotype 4 uses gga-miR-181a-5p expression to facilitate viral replication via targeting of STING. Veterinary Microbiol, 2021, 263: 109276.

[106] PATEL S, HELT G, GANESH M, et al. RNA maps reveal new RNA classes and a possible function for pervasive transcription. Science, 2007, 316(5830): 1484-1488.

[107] CAO J. The functional role of long non-coding RNAs and epigenetics. Biol Proced Online, 2014, 16: 11.

[108] LIU W, DING C. Roles of lncRNAs in viral infections. Front Cell Infect Microbiol, 2017, 7: 205.

[109] MENG X Y, LUO Y, ANWAR M N, et al. Long non-coding RNAs: emerging and versatile regulators in host-virus interactions. Front Immunol, 2017, 8: 1663.

[110] WEN B, WANG X, YANG L, et al. Transcriptome analysis reveals the potential role of long noncoding RNAs in regulating fowl adenovirus serotype 4-induced apoptosis in leghorn male hepatocellular cells. Viruses, 2021, 13(8): 1623.

[111] SMITH A J, LI Q, WIETGREFE S W, et al. Host genes associated with HIV-1 replication in lymphatic tissue. J Immunol, 2010, 185(9): 5417-5424.

[112] WANG X P, WEN B, ZHANG X J, et al. Transcriptome analysis of genes responding to infection of Leghorn male hepatocellular cells with fowl adenovirus serotype 4. Front Vet Sci, 2022, 9: 871038.

[113] ANGUS L, MOLEIRINHO S, HERRON L, et al. Willin/FRMD6 expression activates the Hippo signaling pathway kinases in mammals and antagonizes oncogenic YAP. Oncogene, 2012, 31(2): 238-250.

[114] CHEN Y, JIE X, XING B, et al. REV1 promotes lung tumorigenesis by activating the Rad18/SERTAD2 axis. Cell Death Dis, 2022, 13(2): 110.

[115] GIRI P S, BHARTI A H, BEGUM R, et al. Calcium controlled NFATc1 activation enhances suppressive capacity of regulatory T cells isolated from generalized vitiligo patients. Immunology, 2022, 167(3): 314-327.

[116] GORELICK D A, PRAETORIUS J, TSUNENARI T, et al. Aquaporin-11: a channel protein lacking apparent transport function expressed in brain. BMC Biochem, 2006, 7: 14.

[117] MACHITANI M, SAKURAI F, WAKABAYASHI K, et al. MicroRNA miR-27 inhibits adenovirus infection by suppressing the expression of SNAP25 and TXN2. J Virol, 2017, 91(12): e00159-17.

[118] CAO S, REECE E A, SHEN W B, et al. Restoring BMP4 expression in vascular endothelial progenitors ameliorates maternal diabetes-induced apoptosis and neural tube defects. Cell Death Dis, 2020, 11(10): 859.

[119] SOBAH M L, LIONGUE C, WARD A C. SOCS proteins in immunity, inflammatory diseases, and immune-related cancer. Front Med (Lausanne), 2021, 8: 727987.

[120] LIM H M, LEE J, NAM M J, et al. Acetylshikonin induces apoptosis in human colorectal cancer HCT-15 and LoVo cells via nuclear translocation of FOXO3 and ROS level elevation. Oxid Med Cell Longev, 2021, 2021: 6647107.

[121] WANG Z, HE Y, CUN Y, et al. Transcriptomic analysis identified SLC40A1 as a key iron metabolism-related gene in airway macrophages in childhood allergic asthma. Front Cell Dev Biol, 2023, 11: 1164544.

[122] RÖSSIG L, HAENDELER J, HERMANN C, et al. Nitric oxide down-regulates MKP-3 mRNA levels: involvement in endothelial cell protection from apoptosis. J Biol Chem, 2000, 275(33): 25502-25507.

[123] HAIYILATI A, LI X, ZHENG S J. Fowl adenovirus: pathogenesis and control. Int J Plant Anim Environ Sci, 2021, 11: 566-589.

[124] LI R, LI G, LI N, et al. Fowl adenovirus serotype 4 SD0828 infections causes high mortality rate and cytokine levels in specific pathogen- free chickens compared to ducks. Front Immunol, 2018, 9: 49.

[125] ZHAO W, LI X, LI H, et al. Fowl adenoviruse-4 infection induces strong innate immune responses in chicken. Comp Immunol Microbiol Infect Dis, 2020, 68: 101404.

[126] WANG J, BA G, HAN Y Q, et al. Cyclic GMP-AMP synthase is essential for cytosolic double-stranded DNA and fowl adenovirus serotype 4 triggered innate immune responses in chickens. Int J Biol Macromol, 2020, 146: 497-507.

[127] LAM E, STEIN S, FALCK-PEDERSEN E. Adenovirus detection by the cGAS/ STING/TBK1 DNA sensing cascade. J Virol, 2014, 88(2): 974-981.

[128] MA Z, JACOBS S R, WEST J A, et al. Modulation of the cGAS-STING DNA sensing pathway by gammaherpesviruses. Proc Natl Acad Sci U S A, 2015, 112(31): E4306-4315.

[129] LI X, SHU C, YI G, et al. Cyclic GMP-AMP synthase is activated by double-stranded DNA-induced oligomerization. Immunity, 2013, 39(6): 1019-1031.

[130] SUN L, WU J, DU F, et al. Cyclic GMP-AMP synthase is a cytosolic DNA sensor that activates the type I interferon pathway. Science, 2013, 339(6121): 786-791.

[131] TANAKA Y, CHEN Z J. STING specifies IRF3 phosphorylation by TBK1 in the cytosolic DNA signaling pathway. Sci Signaling, 2012, 5(214): ra20.

[132] ABLASSER A, GOLDECK M, CAVLAR T, et al. cGAS produces a 2, -5, -linked cyclic dinucleotide second messenger that activates STING. Nature, 2013, 498(7454): 380-384.

[133] LIU H, ZHANG H, WU X, et al. Nuclear cGAS suppresses DNA repair and promotes tumorigenesis. Nature, 2018, 563(7729): 131-136.

[134] DENG L, SHARIF S, NAGY E. Oral inoculation of chickens with a candidate fowl adenovirus 9 vector. Clin. Vaccine Immunol, 2013, 20(8):1189-1196.

[135] GRGIĆ H, SHARIF S, HAGHIGHI H R, et al. Cytokine patterns associated with a serotype 8 fowl adenovirus infection, 2013, 26(2): 143-149.

[136] NIU Y, SUN Q, LIU X, et al. Mechanism of fowl adenovirus serotype 4-induced heart damage and formation of pericardial effusion. Poultry Sci, 2019, 98(3): 1134-1145.

[137] HE Q, LU S, LIN Y, et al. Fowl adenovirus serotype 4 52/55k protein triggers PKR degradation by ubiquitin proteasome system to evade effective innate immunity. Veterinary Microbiol, 2023, 278: 109660.

[138] WANG B, GUO H, ZHAO J. The pros and cons of cytokines for fowl adenovirus serotype 4 infection. Arch Virol, 2022, 167(2): 281-292.

[139] CHEN L, DENG H, CUI H, et al. Inflammatory responses and inflammation-associated diseases in organs. Oncotarget, 2017, 9(6): 7204-7218.

[140] NILE S H, NILE A, QIU J, et al. COVID-19: pathogenesis, cytokine storm and therapeutic potential of interferons. Cytokine Growth Factor Rev, 2020, 53:66-70.

[141] EL-SHALL N A, EL-HAMID H S A, ELKADY M F, et al. Corrigendum: epidemiology, pathology, prevention, and control strategies of inclusion body hepatitis and hepatitis-hydropericardium syndrome in poultry: A comprehensive review. Front Vet Sci, 2022, 9: 1075948.

Avian orthoreovirus: a systematic review of their distribution, dissemination patterns and genotypic clustering

Rafique Saba, Rashid Farooq, Wei You, Zeng Tingting, Xie Liji, and Xie Zhixun

Abstract

Avian orthoreoviruses have become a global challenge to the poultry industry, causing significant economic impacts on commercial poultry. Avian reoviruses (ARVs) are resistant to heat, proteolytic enzymes, a wide range of pH values, and disinfectants, so keeping chicken farms free of ARV infections is difficult. This review focuses on the global prevalence of ARVs and associated clinical signs and symptoms. The most common signs and symptoms include tenosynovitis/arthritis, malabsorption syndrome, runting-stunting syndrome, and respiratory diseases. Moreover, this review also focuses on the characterization of ARV into genotypic clusters (I-VI) and their relation to tissue tropism or viral distribution. The prevailing strains of ARV in Africa belong to all genotypic clusters (GCs) except GC VI, whereas all GC are present in Asia and the America. In addition, all ARV strains are associated with or belong to GC I-VI in Europe. Moreover, in Oceania only the genotypic clusters of GC V and VI are prevalent. This review also shows that, regardless of the genotypic cluster, tenosynovitis/arthritis is the predominant clinical manifestation, indicating its universal occurrence across all clusters. Globally, most avian reovirus infections can be prevented by vaccination against four major strains: S1 133, 1 733, 2 408, and 2 177. Nevertheless, these vaccines may not a provide sufficient defense against field isolates. Due to the increase in the number of ARV variants, classical vaccine approaches are being developed depending on the degree of antigenic similarity between the vaccine and field strains, which determines how successful the vaccination will be. Moreover, there is a need to look more closely at the antigenic and pathogenic properties of the reported ARV strains. The information acquired will help aid in the selection of more effective strain in combination with biosecurity and farm management methods to prevent ARV infections.

Keywords

avian reovirus; orthoreovirus; reovirus; ARV; genotypic clustering

---

Introduction

Avian orthoreovirus is a virus that infects birds worldwide and belongs to the genus *Orthoreovirus* and the family Reoviridae. ARVs have been linked to a range of clinical conditions in chickens, including neurological disorder[1, 2], malabsorption syndrome (MAS), stunting syndrome, respiratory/enteric illnesses, pericarditis, myocarditis, hepatitis, immunological depression, and secondary infections by various micro-organisms[3, 4]. The most frequent problems caused by ARV are arthritis and tenosynovitis, which reduces feed conversion and weight gain in birds and makes movement difficult. Flocks infected with ARV had higher mortality rates due to malnutrition and dehydration. However, birds between the ages of 4 and 7 weeks exhibit these clinical signs. ARV poses a major health concern to broiler farmers[3, 5]. Chicken carcasses must be removed at slaughter due to the damaged hock joints, resulting in significant financial losses[6].

ARV horizontal transmission occurs via viral particle shedding in the digestive tract, as most birds become infected via the fecal-oral route, although transmission via the respiratory tract is also possible. Furthermore, reoviruses can enter through the damaged skin of the chicks in the litter, and the virus can spread to the hock joints. Vertical transmission to progeny has also been reported[7]. Infections with low pathogenic strains are normally asymptomatic; however, exposure to virulent strains causes tenosynovitis and MAS in immunocompromised birds[8]. ARV is common in all seasons. Meulemans and Halen[9] discovered that the ARV virus remained active at temperatures as high as 50 ℃, demonstrating the virus's resilience. The incidence of ARV is lower in summer and fall than in winter and spring[5].

ARV is an icosahedral, nonenveloped virus that is approximately 80 nm in diameter and has a double-capsid structure[10], with a size of 23.42 kb[11].The ten double-stranded RNA (dsRNA) genome segments are categorized into three size classes, large, medium, and small, and subcategorized into L1, L2, L3, M1, M2, M3, S1, S2, S3, and S4 (Figure 1-2-1). These genes encode nonstructural proteins and structural proteins that vary in number[12]. Each segment's positive strand is identical to its encoded mRNA and contains a type-1 cap at the 5'-end; however, the negative strand contains a pyrophosphate group[13]. The initial seven nucleotides at the 5'-end (GCUUUUU) and the final five nucleotides at the 3'-end (UCAUC) have been conserved in all ARV-positive strands sequenced thus far. This result shows that these sequences may act as target signals for viral transcript in replication, transcription, and encapsidation[14].

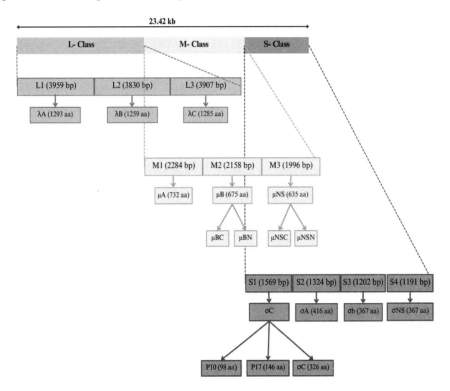

Figure 1-2-1　An overview of avian orthoreovirus genetic structure

L-class segment: The lambda (λ) proteins are those that the L-class genes encode. The ARV structural protein λA, which the L1 gene encodes, is essential for viral replication and assembly. As the λA protein is highly conserved among the many ARV strains, it is a prime target for the development of vaccines and antiviral medications. Although the λB protein is encoded by the L2 gene, its exact sequence is unknown;

this protein is thought to be very important for the function of RNA polymerase, which is necessary for viral replication. λC, which is encoded by the L3 gene, is an ARV structural protein that generates turrets that protrude from the fivefold axis of cores and runs from the inner core to the virion's outer capsid[13].

M-class segment: The micro (μ) class refers to the proteins that the M-class genes encode. The μA is a small subunit of the inner capsid that the M1 gene encodes[15]. μB, the principal translation product of the M2 gene, has an amino terminus that contains a consensus motif for myristoylation[16]. The function of μB in viral replication depends on its myristoylation, and it has been proposed that μB may be involved in the establishment of viral factories[17]. The protein known as μNS is a nonstructural RNA-binding protein that builds up in the viral factories of cells infected with ARV. A more thorough investigation would be necessary to verify the idea that μNS is involved in RNA packaging and replication. The whole μNS sequence contains the five conserved basic residues that are critical for RNA binding[18].

The S-class segment Sigma (σ) refers to the proteins that the S-class genes encode. The minor outer capsid protein σC is a viral cell attachment protein[19]. Antiviral medicines might potentially target σC, which is substantially conserved among ARVs[20]. Several conserved portions of the σC gene, including the 5' end, 3' end, and entire ORF, can be sequenced to classify and genotype avian reoviruses[21].

Despite advances in our knowledge of the biology and variability of these viruses, and efforts of various groups in the United States[22], Europe[23], Canada[4], and China[24] to identify and type ARV variants, the traditional vaccine strains that are used to immunize commercial flocks. Including S1 133, 1 733, and 2 408, have remained unchanged since the 1970s. Due in part to the RNA viruses for mutation and recombination processes, which might result in variations that are only partially or entirely protected by antibodies produced by traditional vaccination strains, these strains have been shown to be ineffective at managing the infection.

The distribution of avian orthoreovirus, clinical manifestation, and genotypic clustering across continents were the main focus of this study. Additionally, a molecular characterization through a phylogenetic analysis of a subset of reoviruses was conducted using partial S1 gene sequences, specifically the sigmaC gene. Sequences were obtained from the National Center for Biotechnology Information (NCBI) and used to trace genotypic clusters worldwide. The evolutionary history was inferred via the maximum likelihood method based on the Tamura-Nei model.

Clinical manifestations based on genotype

The pathogenicity of ARV is highly variable, and different strains of the virus are linked to illnesses such as tenosynovitis, viral arthritis, and malabsorption syndrome (MAS)[25]. Additionally, these viruses can also be isolated from chicks that show no clinical symptoms. There are many different serotypes and phylogenetic classification schemes; the most popular one is based on the Kant classification[21, 26]. Different ARV serotypes and genotypes have been identified under this phylogenetic categorization, and new genotypes such as genotype 7, continue to arise as a natural consequence of mutation, recombination, and reassortment events during virus replication in Canada[27]. Although the connection between reovirus and tenosynovitis has been demonstrated, whether MAS is caused by reovirus remains unclear[10]. Regardless of the genotypic cluster, tenosynovitis/arthritis emerges was the predominant clinical manifestation, highlighting its universal occurrence across all clusters. GC I and IV exhibit a higher prevalence globally, suggesting a widespread impact. GC IV, in particular, presents a diverse range of clinical signs, including tenosynovitis/arthritis, runting-stunting syndrome, malabsorption, and numerous cases that remain unidentified. This diversity underscores the complex nature of GC IV and the potential challenges in identifying specific clinical markers. In contrast, GC

Ⅵ displays a more focused clinical manifestation, with cases primarily characterized by tenosynovitis/arthritis. This distinct clinical profile sets GC Ⅵ apart from others, suggesting a more homogenous impact within this genotypic cluster. Notably, within GC Ⅰ, two isolates also exhibited respiratory signs of infection. This result highlights the variability even within a single genotypic cluster, thus showcasing the importance of considering a broad spectrum of clinical manifestations (Table 1-2-1 and Table1-2-2).

Table 1-2-1　Summary of the avian orthoreovirus prevalence, reference strains, clinical symptoms, and genotypic clusters based on continent

Continent	Country/Region	Reference strains	Clinical symptoms	Genotypic clusters						Reference
				GC Ⅰ	GC Ⅱ	GC Ⅲ	GC Ⅳ	GC Ⅴ	GC Ⅵ	
Africa	Tunisia	TU430 TU105B6 TU5 TU97.2	Viral arthritis, and malabsorption syndrome	Ⅰ						[25]
	Egypt	Chicken/Egypt/Gharbia/1-20/2020 Chicken/Egypt/Gharbia/2-20/2020	Stunted growth and enlarged lemon-shaped proventriculus with reduced gizzard size					Ⅴ		[28]
		D2572/3/2/14EGVAR2 D2248/1/2/13EG D2095/1/4/12EG D2929/2/1/15EG	Tenosynovitis/arthritis and runting-stunting syndrome		Ⅱ	Ⅲ	Ⅳ	Ⅴ		[29]
Asia	Japan	ARV Bro NGN20 7-1 2b ARV Bro GF20 3-7 b ARV Lay NIGT20 30B2b ARV Lay NIGT20 40J1 a ARV Bro NGN20 7-1 b ARV Bro GF20 4-1 a	Viral arthritis/tenosynovitis, malabsorption syndrome, and runting-stunting syndrome		Ⅱ		Ⅳ	Ⅴ		[30]
		JP/Tottori/2016 JP/Nagasaki/2017	N/A	Ⅰ					Ⅵ	[31]
		OS161	Malabsorption	Ⅰ						[32]
	China	ARV LY383 China 2016	Arthritis/tenosynovitis, runting-stunting syndrome, hepatitis, myocarditis, MAS, and central nervous system disease					Ⅴ		[33]
		HeN130728, LN160607-1, GX150816, SD150806, JS170705-1	Foot and joint swelling, hemorrhage, foot scab, paralysis and lameness	Ⅰ	Ⅱ	Ⅲ			Ⅵ	[34]
		750505 T6 916 1017-1	Tenosynovitis, respiratory and MAS	Ⅰ	Ⅱ		Ⅳ			[21]
		T6 601G 918 916	Respiratory disease, viral arthritis, MAS	Ⅰ	Ⅱ	Ⅲ	Ⅳ			[32]
	the Republic of Korea (ROK)	A15-48/Wild bird/ROK 2015 A18-205/Wild bird/ROK 2018	Viral arthritis/tenosynovitis, MAS, runting-stunting syndrome, and respiratory diseases	Ⅰ						[35]
	Iran	ARV2IR018, ARV1IR019	Viral arthritis and MAS		Ⅱ		Ⅳ			[36]

continued

Continent	Country/Region	Reference strains	Clinical symptoms	GC I	GC II	GC III	GC IV	GC V	GC VI	Reference
Europe	France	11-12523	Poor growth rates, lameness, ruptures of gastrocnemicus tendons and nonuniform body weights	I						[23]
		11-17268								
		12-1167								
	Germany	GEL06 97M	Tenosynovitis, and MAS	I	II	III	IV	V		[21]
		GEL13 98M								
		GEL01 96T								
		GEL15 00M								
	the Netherlands	NLI03 92T	Tenosynovitis, and MAS	I	II		IV			[21]
		NLI20 98M								
		NLA13 96T								
	Hungary	HUN131	Lameness, runting-stunting syndrome and uneven growth rate	I	II	III	IV	V		[37]
		HUN392								
		HUN290								
		HUN142								
		HUN385								
	Romania	ROM6	Runting-stunting syndrome		II		IV			
		ROM11								
	Ukraine	UKR1	Mortality							
	Russia	RUS1	Runting-stunting syndrome							
North America	Canada	SK R38	Unilateral or bilateral inflammation of tendons, ruffled feathers, lameness, splayed legs and reluctance to get up and walk		II		IV	V	VI	[4]
		RAM-1								
		05682/12								
		NLI12 96M								
	Pennsylvania	Reo/PA/Broiler/05273a/14	Lesions of pericarditis, swelling, edema, and hemorrhages in the tendons and tendon sheath		II	III	IV	V	VI	[22]
		Reo/PA/Broiler/07634/14								
		Reo/PA/Broiler/30857/11								
		Reo/PA/Broiler/07209a/13								
		Reo/PA/Broiler/03476/12								
	Canada	17-0160-Broiler-AB-2017	Unilateral lameness, subcutaneous hemorrhage, rupture of tendon, and secondary bacterial infection	I	II	III	IV	V	VI	[27]
		14-0041-Broiler-SK-2014								
		16-0711-Broiler-BC-2016								
		12-1009-Broiler-AB-2012								
		15-0157-Broiler-BC-2015								
		17-0025-Broiler-AB-2017								
	California	MK247039	Swollen hock joints, lameness, stunting and lack of uniformity	I	II	III	IV	V	VI	[38]
		MK247050								
		MK246988								
		MK247008								
		MK247040								
		MK247049								

continued

Continent	Country/Region	Reference strains	Clinical symptoms	Genotypic clusters						Reference
				GC I	GC II	GC III	GC IV	GC V	GC VI	
South America	Brazil	Reo/BR_Sc_6996	Synovial membranes hyperplasia, Inflammation, hemorrhages, fibrin deposition and necrosis of muscle fibers		II			V		[6]
		Reo/BR_Sc_7001								
		BR-3118	Tenosynovitis, and MAS	I	II	III		V		[8]
		BR/2290								
		BR-5881								
		BR/3292								
Oceania	Australia	SOM-4	Unclear					V		[21]
		RAM-1	Healthy							
		RAM-1	Healthy					V	VI	[32]
		SOM-4	Viral arthritis							

Table 1-2-2   Clinical manifestation of avian orthoreovirus according to genotype

Clusters	TS	RS	TS/RS	MAS	RES	Other	ND	Healthy	Continent	References
GC I	38	2	0	21	2	12	8	2	Africa, America, Asia, Europe, Middle East	[8], [21], [25], [29], [32]
GC II	9	7	1	2	0	18	6	0	Africa, America, Asia, Europe, Middle East	[8], [21], [29], [32]
GC III	2	1	0	3	0	9	2	0	Africa, America, Asia, Europe, Middle East	[8], [21], [29], [32]
GC IV	17	10	1	16	0	29	12	2	Africa, America, Asia, Europe, Middle East	[21], [29], [32]
GC V	7	3	0	3	0	1	2	2	Africa, America, Asia, Europe, Middle East, Australia	[8], [21], [29], [32]
GC VI	46	0	0	0	0	0	0	0	America, Asia, Australia	[22], [27], [31], [32], [34], [38]

Note: TS, tenosynovitis/arthritis; RS, runting-stunting syndrome; TS/RS, tenosynovitis/arthritis or runting-stunting syndrome; MAS, malabsorption syndrome; RES, respiratory disease; ND, not defined.

Phylogenetic analysis based on sigmaC gene

With the use of partial S1 gene characterization methods, ARV strains have been divided into six genotypic groups[21, 22, 32]. The only available ARV sequences for strains from the USA, Canada, Australia, the Netherlands, Germany, Japan, England, Tunisia, Egypt, Iran, Hungary, Brazil, and China are now available. The gene sequences encoding σC retrieved from the National Center for Biotechnology Information (NCBI), Genbank (Table 1-2-1, Figure 1-2-2).

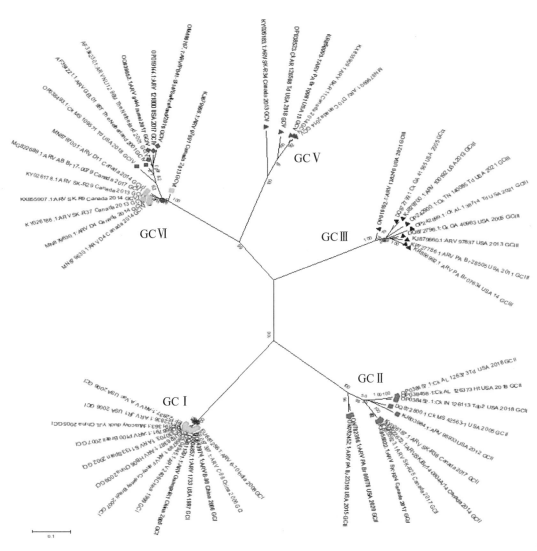

Phylogenetic trees were constructed using MEGA-6 software by the maximum likelihood method with 1 000 bootstrap replicates.

Figure 1-2-2   Phylogenetic tree of avian orthoreovirus strains based on the nucleotide sequence data of the S1 gene (sigmaC)

The vaccine isolates are found to be closely related to each other and can be grouped in GC Ⅰ along with all strains from China, India, and Japan, as well as those isolated from Canada, based on a phylogenetic comparison of the nucleotide sequences of the avian orthoreoviruses worldwide[25, 32]. These strains differ from isolates from the Netherlands and Australia, which are restricted to GC Ⅴ and GC Ⅵ, respectively. The isolates from America and Germany are widely dispersed and are divided into five distinct clusters. This result suggests that, in contrast to the Tunisian isolates, which are grouped into a single cluster, these isolates display genetic diversity and are not closely related to one another[25]. In 2019, a new serotype strain LY383 was isolated and sequence analysis revealed that the vaccination strains and LY383 are not very comparable. Among all the isolates of ARV with chicken origin, LY383 is grouped into GC Ⅴ[33].

In California, ARV strains are associated with six different genotypic clusters (GC Ⅰ to GCⅥ). The prevalence of ARV, GC Ⅰ (51.8%) and GC Ⅵ (24.7%) were the most common clusters, followed by GC Ⅱ (12.9%) and GC Ⅳ (7.1%), with GC Ⅴ (2.4%) and GC Ⅲ (1.2%) having lesser rates. Kant described similar outcomes throughout Europe. Few isolates from clusters Ⅱ and Ⅴ and the majority of isolates linked to malabsorption syndrome were found in clusters Ⅰ and Ⅳ. The molecular characterization and findings of these compounds have also been reported[27]. The genotypic clusters of ARV isolates in California changed between 2015 and 2018. There was a decrease in representation of GC I strains and an increase in that of GC VI. Numberous variables, including the use of autogenous vaccinations, may have contributed to this significant change. The ARV genetic clusters that cause disease may be represented differently in the field due to the use of certain GCs as antigen in autogenous vaccines. A valuable method to determine the missing link between strain diversity and pathogenic characteristics is to sequence the whole genomes of ARV strains on a large scale because this extensive data can be used to identify new genetic variants, mutations, and rearrangements in the genetic code.

Avian orthoreovirus distribution base on continent

Avian reovirus infections in chickens present a significant global issue to the poultry industry. Thus, determining the distribution of avian orthoreovirus by continent is essential in order to develop relevant control strategies.

Africa

In Egypt, ARV was initially discovered in 1984[39], and was subsequently detected serologically in several Egyptian governorates[40]. In addition, in Egypt, both vaccinated and nonvaccinated flocks have a high prevalence of ARV. Moreover, 40 nonvaccinated flock members were tested for ARV resistance using RT-PCR, and the σC gene of the virus was sequenced. ARV strains isolated in Egypt do not correspond to GC I, which is the classification of the vaccination strains, but rather to GC V[28]. Comparably, seven African isolates were sequenced in a different study on ARV genotypic grouping from global strain collections, and none of them aligned with GC I, although all seven isolates are associated with GC I, II, IV, and V[29]. After being molecularly characterized, the ARV isolates from Tunisia were found to belong to GC I, which also included strains identified from China, England, Japan, and Canada[41]. Previous report reveals a seroprevalence of 41.0% in grill flocks from Nigeria.

South America

The prevalence of ARV-related illnesses in South America has increased over the past ten years, and this increase has been attributed to pathogenic strains of various lineages[26]. Reports from Brazil showed that arthritis was the cause of the partial or complete removal of many carcasses at slaughterhouses[6, 42]. Two previously identified strains, GC II and V, contained novel Brazilian ARVs sequences, which were grouped into distinct tree branches. While the GC V sequence of strain BR SC 7 001 was comparable to the ARV sequences from Germany, Israel, and the USA, the four ARVs grouped within GC II exhibited greater similarity to isolates from Canada and the USA. The σC sequence variability of ARVs from clinical cases of tenosynovitis in Brazilian poultry flocks was examined[6]. Nevertheless, in addition to the previously published GC II and V, different stuides conducted in Brazil later in 2023 revealed additional genotypic clusters, namely, I and III. Subgenotypic cluster I (I vaccine, I a, I b), II (II a, II b, II c), and IV (IV a and IV b) might be further separated from this grouping. Four genotypic/subgenotypic clusters comprised Brazilian ARVs: I b (48.2%), II b (22.2%), III (3.7%), and V (25.9%)[8].

North America

Based on nucleotide sequence analysis for molecular characterization of ARVs, the USA isolates S1 133, 1 733, 2 408, and C08 were shown to be closely related and classified into a cluster[21]. Similarly, a phylogenetic study of σC in another study grouped all the isolates from the USA into GC Ⅰ [32]. Six separate clusters of ARV strains with varying genotypes and degrees of amino acid (aa) sequence similarity were found during 2011 and 2014 by molecular analysis of newly emerging ARV variants in Pennsylvania. Standard ARV vaccine strains and 25 field strains are included in GC Ⅰ, whereas 38 field strains from GC Ⅱ, which is very widespread, form three distinct subclusters with varying origins and degrees of aa identity. GC Ⅴ comprises 27 field strains that differ from the vaccine strains and other reference strains in terms of genotype, making them the next most closely related strains. The seven field strains that had a low aa identity with the reference strains were part of the GC Ⅲ and Ⅳ. However, 10 field strains from GC Ⅵ are unique and different from all previously released ARV reference strains[22]. Although ARV infection was common (90.5%) among broiler flocks in Ontario, this was not synonymous with ARV-associated diseases[5].

In contrast to earlier research, a different report revealed genotypic clustering that was divided into four separate: GC Ⅱ, Ⅳ, Ⅴ, and Ⅵ. No single isolate was categorized under GC Ⅲ, nor were any of the isolates grouped with the reference vaccination strains. Circulating ARV strains may have developed and diverged greatly from vaccination strains, as evidenced by the isolates with only 53% aa sequence similarity with the American S1 133 vaccine strain. Previous research has documented the genotypic clustering of ARV isolates originating from distinct geographical areas[4]. In contrast, GC Ⅴ was the most prevalent cluster in different studies using grill chickens from Western Canada, followed by GC Ⅳ. With only five, four, and three isolates each, GC Ⅰ, Ⅱ, and Ⅲ had the fewest number of isolates overall. There was only one GC Ⅵ isolate; however, importantly, four SK sequences that were assigned to that cluster were examined. All five sequences had an identity between 78.2% and 98.6%, and phylogenetic trees classified them as though they were isolated from GC Ⅳ. An autogenous vaccination cannot be effective against a challenge virus that is specific to one cluster because of the differences in the genetic composition of the various clusters. GC Ⅰ and GC Ⅵ are the most common, with frequencies 51.8% and 24.7%, respectively, followed by GC Ⅱ and GC Ⅳ (7%~13%) and GC Ⅴ and GC Ⅲ (1%~2%), which have lower frequencies[27]. Very few isolates belonged to GC Ⅱ and GC Ⅴ, while most isolates belonging to MAS were found in GC Ⅰ and GC Ⅳ[27]. Uncertain viruses belonged to GC Ⅱ, while the majority of tenosynovitis clusters belonged to GC Ⅳ[27], indicating that their most common sequences were from GC Ⅴ, followed by GC Ⅳ and GC Ⅰ [38].

Asia

Tenosynovitis/arthritis syndrome has been more common in the past several years, and many ARV strains have been identified from broilers in China, Republic of Korea, the Middle East, and other nations[29, 33]. Most isolates from Taiwan and Chinese mainland, are grouped in GC Ⅰ. Additionally, isolates from Taiwan were also found in GC Ⅱ and Ⅳ. Few MAS isolates are grouped in GC Ⅱ[21]. There is significant genetic diversity among ARVs in Taiwan. As in another report, GC Ⅰ, Ⅱ, Ⅲ, and Ⅳ are reported in Taiwan and Chinese mainland[32].

Up to 95.83% of broiler breeders in the western provinces of Turkey were positive for ARV antibodies[43], while[44] 98.5% of samples positive for ARV were found in Swiss poultry herds. The Iranian province of Tehran had a 98.3% ARV prevalence, demonstrating the need for ARV immunization in chicken flocks. In India, the

29

total prevalence of ARV was 8.67%[45].

A comparative study using the σC sequence revealed that 14 of the 18 isolates were located in GC Ⅱ, Ⅲ, and Ⅵ, whereas 4/18 isolates were in the same GC Ⅰ as vaccination strains. The strains identified in 2017 were categorized into a new genotype known as GC Ⅵ, whereas the strains isolated between 2013 and 2016 were mostly found in GC Ⅰ to GC Ⅲ[34]. Field strains of ARV belonging to GC V are strongly associated with LY383, an isolate from a Chinese grill flock that received vaccinations. Although these field strains and vaccination strains differ greatly, they have a high level of amino acid similarity with LY383. ARV strains of turkey ancestry created GC Ⅱ, whereas strains of waterfowl origin developed GC Ⅲ and Ⅳ concurrently[33]. The σC protein exhibited a high level of antigenic homogeneity compared to that of isolates from Chinese chicken origin and commercial vaccines (inactivated vaccines). These isolates, however, were not the same as the isolates from Korean chickens. According to the findings, the inferred amino acid substitution patterns in the σC protein of all the isolates were the same as those of ARVs of chicken origin[35]. According to another report in Japan, GC Ⅱ and V were found in broiler breeders, whereas GC Ⅱ and Ⅳ were found in layer breeders[30]. A phylogenetic analysis of the σC protein revealed that two field isolates, ARV1IR018 (MAS isolate) and ARV2IR018 (viral arthritis isolate), were clustered in GC Ⅳ and Ⅱ, respectively[36].

Europe

All five genotypic clusters are present in different countries in Europe, including France, Germany, the Netherlands, Spain, Hungary, Romania, and Ukraine[21, 23, 29, 37, 46]. GC Ⅰ, which includes several closely related strains from various locations, including European, Central and South Asian, and North African nations, is the most prevalent in Europe. A different cluster within GC Ⅱ included European strains originating from the Balkan Peninsula and nearby nations. GC Ⅲ comprises strains from the same countries that have high sequence identity. Chicken reoviruses in GC Ⅳ had the highest genetic diversity and the greatest number of isolates from the worldwide collection. With just a few isolates from Germany and Ukraine sharing high sequence similarities (98%), GC V contained the fewest isolates overall. Additionally, it has been reported that GC V pathogens were only detected in samples taken after 2013. Consequently, GC V is probably less common in Europe than the other clusters[29].

Five distinct genotypic groups were discovered from the categorization of Dutch and German ARVs using σC protein sequencing, with most isolates from Germany falling into GC Ⅰ. Furthermore, the majority of the German and Dutch MAS isolates under study belong to GC Ⅳ. German and Dutch isolates accounted for the majority of GC Ⅳ isolates with uncertain cases, and the majority of Dutch isolates with tenosynovitis were classified as GC Ⅰ. While a small number of MAS isolates from the Netherlands were classified in GC Ⅱ, isolates from Germany were classified as GC V and Ⅲ[21]. In contrast, a recent study examined the evolutionary history and genetic diversity of ARV isolates from French grill chickens. Only GC Ⅰ, which was only tangentially connected to the other clusters and the vaccination strains, contained isolates[23].

Only GC Ⅰ, Ⅲ, and V were found in Hungary. With 43 isolates, GC Ⅱ was the most frequently found cluster in Hungary. Eight strains from Romania and one from Hungary were also discovered by the researchers, and both strains had a significant degree of nucleotide and aa identity similarity with the Hungarian cluster Ⅱ strains. However, samples from the four nations under study—Hungary, Romania, Russia, and Ukraine—all showed the presence of GC Ⅳ. Eleven Hungarian strains that were obtained from the same farm were shown to have strong nucleotide and aa identity similarities with the three Romanian strains. These results demonstrate the notable genetic variation among ARVs in East Central European hens, which may aid in the development of efficient vaccinations and management techniques[37].

Oceania

ARV has been documented within the geographic confines of Australia, although with a lower prevalence in comparison to that in other continents. According to one study, GC Ⅴ has been exclusively identified[21], while another investigation reported the presence of GC Ⅳ and Ⅵ[32]. The absence of any other genotypic clusters inside Australia's boundaries supports the suggestion that ARV transmission in Australia is unlikely to be the result of vaccination and is instead related to migrating birds.

Vaccine challenges

Major genetic changes were detected in 1986, with genetic variants from the "conventional vaccine types" of reovirus. This significant change may have been induced by several factors, including the use of autogenous vaccinations. The many ARV genetic clusters producing disease in the field may be changing as a result of the use of certain GCs as antigens in autogenous vaccinations. Although GC Ⅰ strains of reovirus accounted for most strains in 2016, autogenous vaccines containing isolates of two GC Ⅰ strains and one GC Ⅴ variant were produced for use in breeders that provide chickens to the state of California[38, 47]. The theory underlying the inactivated nonhomologous vaccines is that they provide some protection against the field issue of viral shedding in infected birds, thus allowing strains aside from GC Ⅰ and GC Ⅴ to be selected, hence altering the environmental representation of ARVs. However, this justification falls short of explaining why GC Ⅳ or GC Ⅲ were overlooked. Because GC Ⅵ is more fit than the other genotypes in the current environment, those genotypes were likely not picked because they were not as fit as the others. Reports on surveillance activities typically do not address how GC detection varies from year to year[21, 22, 27]. The high variability in sequences causes the reporting of avian orthoreovirus to fluctuate over time. When choosing autogenous vaccine candidates, considering the prevalence of GCs in addition to their antigens is important.

In addition to determining the temporal GC frequencies, the available data permitted the computation of homologies with a reference strain, in this case, the commercial vaccine strain S1 133. The benefit is being able to track the variability of each cluster's variant and determine whether there are any significant changes over time. GC1 has the most homology, and GC1 is the group that includes the vaccination strains; however, its average homology was 77%. The average homologies of the remaining GCs to S1 133 ranged from 58.5% to 53.1%, which is extremely different from those of viruses found in commercial live and inactivated vaccines (S1 133, 1 733, and 2 408). These findings may explain the ineffectiveness of vaccination in poor defending commercial broilers. Every cluster remained homologous to S1 133 since 2016, according to the homologies reported throughout time[38].

Over the past ten years, pathogenic strains of different lineages have led to an increase in the number of ARV-related infections across North and South America, Europe, Africa, and Asia[4, 26, 29]. To choose the optimal vaccine candidates, thorough and frequent sero- and viro-surveillance is necessary to gain an understanding of the GC homologies.

Conclusions

Avian reovirus is now a moving target, similar to the influenza virus, infectious bronchitis virus, and other RNA-based avian pathogens, due to its genetic nature, especially the recombination, genetic drift, and absence of an RNA proofreading mechanism. The incidence of ARVs in the intestines of wild birds was greater than that of ARVs in their excrement[35]. Despite several vaccinations to birds throughout their lives, the chances of infection and reinfection still exist. Overall, several genotypes circulate among the poultry population, and no significant cross-protection has been reported among different genotypes. At least six distinct genotypes were found when ARVs were genotyped utilizing the σC-encoding gene; however, the relationships between genotypes, pathogenic traits, and serotype classifications are still being determined. ARVs exhibit a broad range of tissue tropisms, and virulent ARVs from free-living birds were genetically linked to ARVs from chickens, suggesting that these species might serve as reservoirs for the spread of ARVs within poultry farms and could become a moving target that needs to be monitored through regular sero-and viro-surveillance.

In addition, GC Ⅰ and Ⅳ have a higher prevalence globally, indicating a widespread impact. The clinical signs of GC Ⅳ are particularly diverse and include tenosynovitis/arthritis, runting-stunting syndrome, malabsorption, and unidentified cases. This diversity underscores the complexity of GC Ⅳ and the challenges in identifying specific clinical markers. In contrast, GC Ⅵ has a more focused clinical manifestation, primarily characterized by tenosynovitis/arthritis. This characteristic sets GC Ⅵ apart from others, suggesting a more homogeneous impact within this genotypic cluster. Moreover, GC Ⅰ isolates also exhibit respiratory signs of infection, thereby highlighting the variability even within a single genotypic cluster. This result emphasizes the importance of considering a broad spectrum of clinical manifestations.

A phylogenetic analysis revealed that the vaccine isolates were closely related to each other and categorized into GC Ⅰ. This grouping includes strains from China, India, Japan, and Canada, in addition to the vaccine isolates. However, these isolates are distinct from those originating from the Netherlands and Australia, which are classified as GC Ⅴ and Ⅵ, respectively. Furthermore, isolates from Germany and the United States are widely distributed and form five distinct clusters based on their genetic similarity. This result suggests a diverse genetic landscape for avian orthoreoviruses in these regions.

This review further indicated, based on continent prevalence, that there is no exact correlation between ARV genotypes and geographic location. Furthermore, point mutation accumulation and reassortment processes play a critical role in the evolution of ARVs. The nonspecific geographic distribution of all six ARV genotyping cluster groups indicated that vaccine formulations containing appropriate antigens from all six genotypes are necessary for the successful prevention of viral-induced arthritis/tenosynovitis. In vaccinated breeders or broiler flocks, novel variant strains lead to vaccine breakthroughs. The antigenic and pathogenic characteristics of a few of the ARV strains that have been identified will be further examined. In an effort to control ARV infections, the information gathered here will facilitate more effective vaccination strain selection with biosecurity and farm management practices.

References

[1] DANDÁR E, BÁLINT Á, KECSKEMÉTI S, et al. Detection and characterization of a divergent avian reovirus strain from a broiler chicken with central nervous system disease. Archives of Virology, 2013, 158(12): 2583-2588.

[2] VAN D E ZANDE S, KUHN E M. Central nervous system signs in chickens caused by a new avian reovirus strain: a pathogenesis study. Veterinary Microbiology, 2007, 120(1/2): 42-49.

[3] JONES R C. Avian reovirus infections. Rev Sci Tech Oie, 2000, 19(2): 614-625.

[4] AYALEW L E, GUPTA A, FRICKE J, et al. Phenotypic, genotypic and antigenic characterization of emerging avian reoviruses isolated from clinical cases of arthritis in broilers in Saskatchewan, Canada. Scientific Reports, 2017, 7(1): 3565.

[5] NHAM E G, PEARL D L, SLAVIC D, et al. Flock-level prevalence, geographical distribution, and seasonal variation of avian

reovirus among broiler flocks in Ontario. Can Vet J, 2017, 58(8): 828-834.

[6] SOUZA S O, DE CARLI S, LUNGE V R, et al. Pathological and molecular findings of avian reoviruses from clinical cases of tenosynovitis in poultry flocks from Brazil. Poult Sci, 2018, 97(10): 3550-3555.

[7] PITCOVSKI J, GOYAL S M. Avian reovirus infections//SWAYNE D V, BOULIANNE M, LOGUE C M, et al. Diseases of poultry, 14th edtion. New Jersey: Wiley-Blackwell, 2019.

[8] CARLI S D, WOLF J M, GRF T, et al. Genotypic characterization and molecular evolution of avian reovirus in poultry flocks from Brazil.Avian Pathol, 2020: 1-26.

[9] MEULEMANS G, HALEN P. Efficacy of some disinfectants against infectious bursal disease virus and avian reovirus. Veterinary Record, 1982, 111(18): 412-413.

[10] HEIDE L V D. The history of avian reovirus. Avian Dis, 2000, 44(3): 638-664.

[11] TENG L, XIE Z, XIE L, et al. Complete genome sequences of an avian orthoreovirus isolated from Guangxi, China. Genome Announc, 2013, 1(4): e00495-13.

[12] SPANDIDOS D A, GRAHAM A F. Physical and chemical characterization of an avian reovirus. Journal of Virology, 1976, 19(3): 968-976.

[13] MARTINEZ-COSTAS J, RUBÉN VARELA, BENAVENTE J. Endogenous enzymatic activities of the avian reovirus S1133: identification of the viral capping enzyme. Virology, 1995, 206(2): 1017-1026.

[14] BENAVENTE J, JOSE MARTÍNEZ-COSTAS. Avian reovirus: structure and biology.Virus Research, 2007, 123(2): 105-119.

[15] PARKER J S L, BROERING T J, KIM J, et al. Reovirus core protein μ2 determines the filamentous morphology of viral inclusion bodies by interacting with and stabilizing microtubules. Journal of Virology, 2002, 76(9): 4483-4496.

[16] VARELA R, MARTÍNEZ-COSTAS J, MALLO M, et al. Intracellular posttranslational modifications of S1133 avian reovirus proteins. Journal of Virology, 1996, 70(5): 2974.

[17] DUNCAN R. The low ph-dependent entry of avian reovirus is accompanied by two specific cleavages of the major outer capsid protein μ2C. Virology, 1996, 219(1): 179-189.

[18] TOURIS-OTERO F, MARTÍNEZ-COSTAS J, VAKHARIA V N, et al. Avian reovirus nonstructural protein μNS forms viroplasm-like inclusions and recruits protein σNS to these structures. Virology, 2004, 319(1): 94-106.

[19] GRANDE A, COSTAS C, BENAVENTE J. Subunit composition and conformational stability of the oligomeric form of the avian reovirus cell-attachment protein σC.Journal of General Virology, 2002, 83(Pt 1): 131-139.

[20] GUARDADO CALVO P, FOX G C, HERMO PARRADO X L, et al. Structure of the carboxy-terminal receptor-binding domain of avian reovirus fibre σC. Journal of Molecular Biology, 2005, 354(1): 137-149.

[21] KANT A, BALK F, BORN L, et al. Classification of dutch and german avian reoviruses by sequencing the σC protein. Veterinary Research, 2003, 34(2): 203-212.

[22] LU H, TANG Y, DUNN P A, et al. Isolation and molecular characterization of newly emerging avian reovirus variants and novel strains in Pennsylvania, USA, 2011-2014. Scientific Reports, 2015, 5: 14727.

[23] TROXLER S, RIGOMIER P, BILIC I, et al. Identification of a new reovirus causing substantial losses in broiler production in France, despite routine vaccination of breeders. Veterinary Record, 2013, 172(21): 556-556.

[24] ZHONG L, GAO L, LIU Y, et al. Genetic and pathogenic characterisation of 11 avian reovirus isolates from northern China suggests continued evolution of virulence. Sci Rep, 2016, 6: 35271.

[25] KORT Y H, BOUROGAA H, GRIBAA L, et al. Molecular characterization of avian reovirus isolates in Tunisia.Virology Journal, 2013, 10(1): 12.

[26] SELLERS H S. Current limitations in control of viral arthritis and tenosynovitis caused by avian reoviruses in commercial poultry. Vet. Microbiol, 2017, 206: 152-156.

[27] PALOMINO-TAPIA V, MITEVSKI D, INGLIS T, et al. Molecular characterization of emerging avian reovirus variants isolated from viral arthritis cases in Western Canada 2012-2017 based on partial sigma (σ)C gene. Virology, 2018, 522: 138-146.

[28] MOSAD S M, ELMAHALLAWY E K, ALGHAMDI A M, et al. Molecular and pathological investigation of avian reovirus

(ARV) in Egypt with the assessment of the genetic variability of field strains compared to vaccine strains. Front Microbiol, 2023, 14: 1156251.

[29] KOVÁCS E, VARGA-KUGLER R, MATÓ T, et al. Identification of the main genetic clusters of avian reoviruses from a global strain collection. Front Vet Sci, 2023, 9: 1094761.

[30] YAMAGUCHI M, MIYAOKA Y, HASAN M A, et al. Isolation and molecular characterization of fowl adenovirus and avian reovirus from breeder chickens in Japan in 2019-2021. J Vet Med Sci, 2022, 84(2): 238-243.

[31] MASE M, GOTOU M, INOUE D, et al. Genetic analysis of avian reovirus isolated from chickens in Japan. Avian Dis, 2021, 65(3): 346-350.

[32] LIU H J, LEE L H, HSU H W, et al. Molecular evolution of avian reovirus: evidence for genetic diversity and reassortment of the S-class genome segments and multiple cocirculating lineages. Virology, 2003, 314(1): 336-349.

[33] CHEN H, YAN M, TANG Y, et al. Pathogenicity and genomic characterization of a novel avian orthoreovius variant isolated from a vaccinated broiler flock in China. Avian Pathol, 2019, 48(4): 334-342.

[34] ZHANG X, LEI X, MA L, et al. Genetic and pathogenic characteristics of newly emerging avian reovirus from infected chickens with clinical arthritis in China. Poult Sci, 2019, 98(11): 5321-5329.

[35] KIM S W, CHOI Y R, PARK J Y, et al. Isolation and genomic characterization of avian reovirus from wild birds in South Korea*. Front Vet Sci. 2022, 9: 794934.

[36] MIRZAZADEH A, ABBASNIA M, ZAHABI H, et al. Genotypic characterization of two novel avian orthoreoviruses isolated in Iran from broilers with viral arthritis and malabsorption syndrome. Iran J Vet Res. 2022, 23(1): 74-79.

[37] GÁL B, VARGA-KUGLER R, IHÁSZ K, et al. Marked genotype diversity among reoviruses isolated from chicken in selected East-Central European countries. Animals (Basel), 2023, 13(13): 2137.

[38] EGAÑA-LABRIN S, HAUCK R, FIGUEROA A, et al. Genotypic characterization of emerging avian reovirus genetic variants in California. Sci Rep, 2019, 9(1): 9351.

[39] TANTAWI H H, AMINA N, YOUSSEF Y I, et al. Infectious tenosynovitis in broilers and broiler breeders in Egypt. Vet Res Commun. 1984, 8(3): 229-235.

[40] AL-EBSHAHY E, MOHAMED S, ABAS O. First report of seroprevalence and genetic characterization of avian orthoreovirus in Egypt. Trop Anim Health Prod, 2020, 52(3): 1049-1054.

[41] OWOADE A A, DUCATEZ M F, MULLER C P. Seroprevalence of avian influenza virus, infectious bronchitis virus, reovirus, avian pneumovirus, infectious laryngotracheitis virus, and avian leukosis virus in Nigerian poultry. Avian Dis, 2006, 50(2): 222-227.

[42] RECK C, MENIN Á, CANEVER M F, et al. Molecular detection of Mycoplasma synoviae and avian reovirus infection in arthritis and tenosynovitis lesions of broiler and breeder chickens in Santa Catarina State, Brazil. J S Afr Vet Assoc, 2019, 90(0): e1-e5.

[43] EROL N, ŞENGÜL S S.Seroprevalence of avian reovirus infections in chickens in western provinces of Turkey. Kafkas Universitesi Veteriner Fakultesi Dergisi, 2012, 18(4): 653-656.

[44] WUNDERWALD C, HOOP R K. Serological monitoring of 40 Swiss fancy breed poultry flocks.Avian Pathol, 2002, 31(2): 157-162.

[45] BAKSI S, RAO N, CHAUHAN P, et al. Sero-prevalence of avian reovirus in broiler breeders in different parts of India. PSM Vet Res, 2018, 3(2): 22-25.

[46] LOSTALÉ-SEIJO I, MARTÍNEZ-COSTAS J, BENAVENTE J. Interferon induction by avian reovirus.Virology, 2016, 487: 104-111.

[47] SELLERS H S. Avian reoviruses from clinical cases of tenosynovitis: an overview of diagnostic approaches and 10-year review of isolations and genetic characterization. Avian Dis, 2022, 66(4): 420-426.

---

* "South Korea"refers to the Republic of Korea (ROK). The same applies to other occurrences in the book.

Roles and functions of IAV proteins in host immune evasion

Rashid Farooq, Xie Zhixun, Li Meng, Xie Zhiqin, Luo Sisi, and Xie Liji

Abstract

Inluenza A viruses (IAVs) evade the immune system of the host by several regulatory mechanisms. Their genomes consist of eight single-stranded segments, including nonstructural proteins (NS), basic polymerase 1 (PB1), basic polymerase 2 (PB2), hemagglutinin (HA), acidic polymerase (PA), matrix (M), neuraminidase (NA), and nucleoprotein (NP). Some of these proteins are known to suppress host immune responses. In this review, we discuss the roles, functions and underlying strategies adopted by IAV proteins to escape the host immune system by targeting different proteins in the interferon (IFN) signaling pathway, such as tripartite motif containing 25 (TRIM25), inhibitor of nuclear factor κB kinase (IKK), mitochondrial antiviral signaling protein (MAVS), Janus kinase 1 (JAK1), type I interferon receptor (IFNAR1), interferon regulatory factor 3 (IRF3), IRF7, and nuclear factor-κB (NF-κB). To date, the IAV proteins NS1, NS2, PB1, PB1-F2, PB2, HA, and PA have been well studied in terms of their roles in evading the host immune system. However, the detailed mechanisms of NS3, PB1-N40, PA-N155, PA-N182, PA-X, M42, NA, and NP have not been well studied with respect to their roles in immune evasion. Moreover, we also highlight the future perspectives of research on IAV proteins.

Keywords

IAVs, IAV proteins, immune evasion, host immune system, IFNs

Introduction

The RNA genome of influenza viruses is segmented, negative sense, and belongs to the family Orthomyxoviridae. Influenza viruses are divided into four genera, A, B, C, and D[1]. Influenza viruses A and B cause all seasonal epidemics, whereas influenza viruses C and D cause mild disease in specific hosts[2, 3]. Their genomes consist of eight single-stranded segments, including nonstructural proteins (NS), basic polymerase 1 (PB1), basic polymerase 2 (PB2), hemagglutinin (HA), acidic polymerase (PA), matrix (M), neuraminidase (NA), and nucleoprotein (NP) [4]. The NS proteins encode at least 14 proteins and have a wide range of functions, including antagonism of host immune responses[5]. PB1, PB2, and PA are polymerase components and are responsible for IAV genome replication and transcription[6-8]. The HA protein recognizes host cells and mediates IAV entry into host cells[9, 10], M provides structural support to virions, NA is asialidase and mediates virion release from cells, and NP is an RNP component and is responsible for RNA encapsidation.

Influenza A virus (IAV) is a serious risk to both animals and humans because it causes annual seasonal epidemics and has resulted in severe pandemic outbreaks[11, 12]. A novel IAV strain can be generated after the coinfection of a host with two or more different viruses followed by a reassortment process, after which, the strain expresses new combinations of HA and NA subtypes; therefore, IAVs are subtyped by the antigenic characteristics of their HA and NA glycoproteins[13]. To date, all human pandemics have occurred due to genetic reassortment between human, swine, and avian influenza viruses (AIVs) [14]. The fast and frequent mutations and recombination process of influenza viruses have resulted in highly pathogenic strains of IAVs, e.g., H5N1, H5N6, H5N8, and H7N9[15]. To date, eighteen different HA subtypes (H1-H18) and eleven NA subtypes (N1-N11) have been reported, and the most recently identified H17N10 and H18N11 subtypes have been discovered in New World bats and have not been found in humans or any avian species[16, 17]. The most lethal example of a new influenza virus subtype was recorded in 1918 when the H1N1 (H1 subtype, N1 subtype) virus (Spanish flu) caused the death of approximately 100 million people[18]. This subtype was replaced by H2N2 (H2 subtype, N2 subtype) virus (Asian flu) in 1957, which had novel H2, the N2, and PB1 genes derived from a Eurasian avian virus source[19]. In 1968, the H3N2 (H3 subtype, N2 subtype retained) virus (Hong Kong flu) emerged, and in 1977, a small but notorious H1N1 virus of the 1950s reemerged in Russia (Russian flu). A novel pandemic H1N1 virus (swine flu) emerged in 2009 in Mexico, which spread around the globe and replaced the previous H1N1 viruses that circulated for approximately thirty years until 2009[14, 18, 19].

Interferons (IFNs) are the first line of host defense against invading viruses. The innate immune response is activated by the recognition of viral pathogen-associated molecular patterns (PAMPs) by retinoic acid-inducible gene-I (RIG-I) and melanoma differentiation-associated gene 5 (MDA5) [12] (Figure 1-3-1). After activation, RIG-I and MDA5 interact with mitochondrial antiviral signaling protein (MAVS) and eventually activate MAVS. After MAVS activation, downstream signaling proteins are recruited to the mitochondria, leading to the activation of an inhibitor of NF-κB kinase (IKKε) and TANK-binding kinase (TBK1) and finally resulting in the phosphorylation of the transcription factors interferon regulatory factor 3 (IRF3) and IRF7. Phosphorylated IRF3 and IRF7, as well as activated nuclear factor kappa-light-chain-enhancer of activated B cells (NF-κB), trigger the expression of interferons (IFNs) and cytokines[20]and thus create an antiviral state to counteract the pathogenesis of IAVs[12, 21-24].

TLR3, TLR7, TLR8, and TLR9 recognize PAMPs derived from several viruses, including IAVs[25-27]. TLR3 recognizes dsRNA in endosomes, while TLR7 recognizes the ssRNA of influenza viruses and is activated after interaction[20, 28, 29]. Activated TLR7 leads to the activation of IFN-I or proinflammatory cytokines[20], while TLR3 activation leads to IFN-β activation[30].

The expressed IFN-I interacts with IFN-α/β receptors (IFNAR), and IFN-Ⅲ interacts with IFNL receptors (IFNLR) in an autocrine or paracrine manner[31]. Secretion of IFN-I and -Ⅲ leads to the activation of Janus kinase 1 (JAK1) and tyrosine kinase 2 (Tyk2), followed by the phosphorylation of signal transducer and activator of transcription 1 and 2 (STAT1 and STAT2) [32, 33]. Phosphorylated STAT1 and STAT2 form heterodimers and interact with IRF9 to form the IFN-stimulated growth factor 3 (ISGF3) complex. This complex is translocated into the nucleus, where at the ISG promoter, it interacts with IFN-stimulated response elements (ISREs) to transcribe its downstream genes, which counteract viral infection (Figure 1-3-1).

However, IAVs have developed mechanisms to counteract IFNs and cultivate successful infection. The different proteins of IAVs act in different manners by targeting different steps of the IFN signaling pathway. To date, several IAV proteins have been identified that evade the host immune system through various strategies. The following sections elaborate on how the different IAV proteins adopt different strategies to target different proteins of the IFN signaling pathway and establish successful infection.

NS1 inhibits IFN-I activation by multiple mechanisms

NS1, a nonstructural protein of IAV, is considered the most potent inhibitor of the host innate immune system during viral replication[1, 34]. The smallest RNA segment of IAV encodes the NS1 protein in conjunction with NEP, which is an NS mRNA splice variant[35]. NS1 promotes IAV replication by using its functional

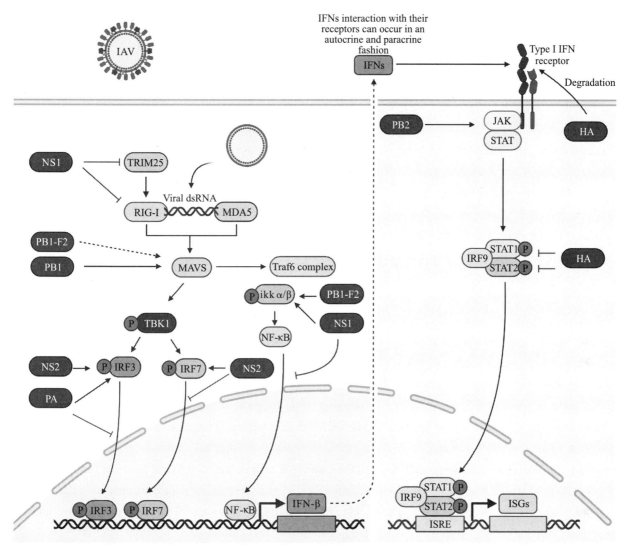

The viral genome is recognized by RIG-I and MDA5. IRF3, IRF7, and NF-κB trigger the production of IFNs. IFNs are secreted to induce the JAK/STAT pathway, which express ISGs. Different IAV proteins antagonize different proteins in the host innate immune signaling pathway. NS1 binds with IKK, RIG-I, and TRIM25 to escape the host immune system, NS2 interacts with IRF7 to suppress its nuclear translocation, the PB1 protein interacts with and degrades MAVS, and PB1-F2 probably interacts with MAVS to inhibit IFN production (dotted lines indicate that the interaction between these two proteins has not yet been confirmed). PB1-F2 also binds with IKKβ to suppress NF-κB signaling. The HA protein degrades IFNAR1, and PA binds with and inhibits the nuclear translocation of IRF3. The dotted arrow indicates that the interaction between the proteins has not yet been confirmed.

Figure 1-3-1 IFN signaling pathway activation and evasion by IAV proteins

domains that interact with both viral and cellular proteins. The RNA-binding domain (RBD) of NS1 at the amino terminal end interacts with dsRNA molecules to block the induction of RLR sensors such as OASL and PKR[36]. This mechanism of host immune response inhibition by NS1 was considered the main strategy; however, later, several other interactions involved in these mechanisms were later found, and these interactions are discussed below[37].

Tripartite motif containing 25 (TRIM25) is a ubiquitin ligase that mediates lysine 63-linked ubiquitination of the N-terminal CARD domain of RIG-I to trigger IFN-I production[38]. NS1 of IAV interacts with TRIM25 and thus blocks TRIM25-mediated RIG-I signaling. A domain in NS1 that contains residues E96/97 is responsible for interacting with the TRIM25 coil-coil domain to inhibit RIG-I CARD ubiquitination. IAV with mutations in the NS1 domain at residues E96/97, i.e., E96A/E97A, was unable to block IFN responses mediated by TRIM25, and virulence in mice was lost[38]. In 2012, Rajsbaum et al. further studied the mechanisms of host tropism and viral adaptation, and for this purpose, they studied the ability of NS1 from avian (HK156), human (CaI04), mouse (PR8) and swine adapted (SwTx98) influenza viruses to interact with TRIM25 encoded by avian and mammalian species[39]. Coimmunoprecipitation experiments indicated that TRIM25 in humans interacts with all NS1 proteins under study; however, TRIM25 in chickens interacts preferentially with the avian virus NS1. In addition, none of the NS1 proteins interact with TRIM25 in mice. Furthermore, the researchers determined the effect of NS1 and TRIM25 on RIG-I ubiquitination in mouse cells and found that NS1 inhibited the ubiquitination of full-length mouse RIG-I in a mouse TRIM25-independent manner. They further wanted to know if Riplet interacts with NS1, as Riplet is already known to ubiquitinate RIG-I. NS1 binds with Riplet from mice in murine cells and blocks the function of Riplet to induce IFN-β. Only the human NS1 protein was able to bind to human Riplet to suppress RIG-I ubiquitination; however, the NS1 of avian or swine viruses was unable to bind Riplet from humans. Therefore, influenza NS1 targets Riplet and TRIM25 in a species-specific manner to suppress RIG-I ubiquitination and IFN production[39]. A previous study[40] demonstrated that the NS1 protein binds with RIG-I and therefore suppresses the RIG-I-mediated activation of IFN-β. Moreover, they found that the NS1 of IAV binds with RIG-I and inhibits IRF3 activation and its nuclear translocation[40]. This study was further elaborated upon by Jureka et al.[41], who delineated through biophysical and structural evidence that the binding between RIG-I and the IAV NS1 protein is direct and that the RNA binding domain (RBD) of NS1 and the second caspase activation and recruitment domain of RIG-I are responsible for this interaction[41].

NS1 also inhibits IFN activation by interacting with the protein kinase PKR[42, 43]. Additionally, NS1 blocks host antiviral activity by interacting with the 30 kDa subunit of the cleavage and polyadenylation specificity factor (CPSF30) to process cellular mRNA[44, 45]. Furthermore, NS1 inhibits the NK-κB pathway by targeting the inhibitor of κB kinase (IKK), and therefore, the expression of antiviral genes is blocked (Figure 1-3-3, Table 1-3-1) [46]. The phosphorylation of IKKα/IKKβ is affected in both compartments of cells, i.e., the cytoplasm and nucleus when interacting with NS1. Furthermore, NF-κB translocates into the nucleus, and the expression of its downstream target genes is also inhibited. The phosphorylation of histone H3 Ser 10, which is mediated by IKK, is also impaired by NS1[46].

Table 1-3-1　IAV proteins that interfere with IFN induction and signaling

Protein	Mechanism	Experimental approach	Cellular model	References
IFN production inhibition				
NS1	Interacts with TRIM25 and Riplet to block RIG-I activation, interacts with RIG-I to inhibit nuclear translocation of IRF3, interacts with IKK to inhibit NF-κB nuclear translocation and suppress the expression of NF-κB target genes	Coimmunoprecipitation, Western blotting, immunofluorescence assay, mouse experiments, biophysical approaches	HEK293T cells, A549 cells, Hepa 1.6 (Mouse), Vero, L929 cells	[38-41, 46]
NS2	Interacts with and blocks nuclear translocation of IRF7 to inhibit IFN production	Coimmunoprecipitation, Western blotting, immunofluorescence assay	HEK293 cells, A549 cells	[12]
PB1	Interacts with and degrades MAVS	Coimmunoprecipitation, Western blotting	HEK293 cells, A549 cells, HeLa cells	[47]
PB1-F2	Inhibits IFN production at MAVS level, PB1-F2 interacts with IKKβ and inhibits NF-κB signaling	Luciferase assay, immunofluorescence assay, Western blotting, coimmunoprecipitation, yeast two-hybrid assays	293T cells, African green monkey kidney Verocells	[48, 49]
PA	Interacts with and blocks nuclear translocation of IRF3 to inhibit IFN production	Coimmunoprecipitation, Western blotting, immunofluorescence assay	HEK293T cells, A549 cells	[50]

continued

Protein	Mechanism	Experimental approach	Cellular model	References
		IFN signaling inhibition		
PB2	Interacts with JAK1 and inhibits JAK1/STAT signaling	Luciferase assay, immunofluorescence assay, coimmunoprecipitation, mouse experiments	HEK293T cells, DF-1 cells, A549 cells	[51]
HA	Degrades IFNAR1 and IFNGR1, decreases STAT1/STAT2 phosphorylation	Western blotting, immunoprecipitation, ubiquitination assays, luciferase assay	HEK293 cells, A549 cells, African green monkey kidney Vero cells	[52, 53]

Another mechanism that involves the obstruction of the host innate immune response is NS1 induction of the degradation of sphingosine 1-phosphate lyase (SPL) [54]. SPL was previously shown to enhance IKKε-mediated IFN-I responses[55]. IAV infections or NS1 expression caused ubiquitination and downregulation of SPL expression; however, IAV deficient in NS1 failed to downregulate SPL expression[54].

The Hippo pathway regulates homeostasis and organ development[56, 57]. The important proteins of this pathway in mammals are the kinase cascade involving mammalian STE20-like protein kinase 1 (MST1) and mammalian STE20-like protein kinase 2 (MST2); the large tumor suppressor 1 (LATS1) and large tumor suppressor 2 (LATS2); the adaptor proteins Salvador homolog 1 (SAV1) for MST1/2 and MOB kinase activators (MOB1A/MOB1B) for LATS1/2; downstream effectors Yes-associated protein (YAP) and its analog protein, transcriptional coactivator with PDZ-binding motif (TAZ); and TEA domain transcription factors (TEADs) [58-62]. Once this pathway is activated, a series of phosphorylation events start via MST and LATS kinases, and finally, YAP/TAZ are phosphorylated. Phosphorylated YAP/TAZ is polyubiquitinated and degraded in the proteasome[63, 64]. Reports have shown that the Hippo pathway also regulates the innate immune pathway[65, 66].

Another study demonstrated the relationship between the Hippo pathway and IAV infection in A549 cells[67]. The YAP/TAZ effectors of the Hippo pathway were dephosphorylated, their expression was upregulated, and they were translocated into the nucleus after IAV infection. Furthermore, in this study, the researchers investigated whether the NS1 protein was responsible for the transcriptional activities of YAP/TAZ via the physical interaction between NS1 and YAP/TAZ. This study further dissected how YAP/TAZ suppress IAV-induced innate immune responses. Knockdown of YAP/TAZ induced the activation of proinflammatory and antiviral factors such as CXCL8, IFNB1, and IRF7, which are triggered by virus infection. Similarly, YAP overexpression decreased the activation of IFNB promoter activity, and the TAD domain at the C-terminus of YAP was indispensable for innate immune signaling. Furthermore, overexpression of YAP suppressed TLR3 expression, and deletion of the C-terminal TAD domain of YAP abolished TLR3 suppression. YAP binds to TEAD binding sites on the TLR3 promoter region. Furthermore, elimination of acetylated histone H3 occupancy in the TLR3 promoter led to its transcriptional silencing, and treatment with the histone deacetylase (HDAC) inhibitor trichostatin A reversed TLR3 expression inhibition mediated by YAP/TAZ. These results delineated the novel mechanism of innate immune system regulation by IAV, where YAP/TAZ inhibits TL3-mediated innate immunity[67].

NS1 of IAV is thus a multifunctional virulence factor, as it regulates the host machinery to facilitate viral replication. NS1 is under constant evolutionary pressure due to its high mutation rate and its infection of and replication in different hosts.

NS2 interacts with IRF7 and suppresses its nuclear translocation to inhibit IFN-I production

H1N1-NS2 significantly inhibits the production of IFN-β, IFN-stimulated gene 56 (ISG56), and IFN-induced protein with tetratricopeptide repeats 2 (IFIT2) [12]. NS2 colocalized with IRF3 and IRF7 and even physically interacted with these two proteins. The same study also found that NS2 significantly suppressed the dimerization of IRF7 and its nuclear translocation (Figure 1-3-1, Table 1-3-1). Furthermore, the indispensable domain in NS2 was determined based on its interaction with IRF7. Amino acid residues 1~53 in the N-terminal domain were found to be responsible for interacting with IRF7, while amino acid residues 54~121 in the C-terminal domain did not interact with IRF7; hence, amino acid residues 1~53 in the N-terminal domain of NS2 are responsible for inhibiting IFN-I production[12].

PB1 interacts with and degrades MAVS to suppress the innate immune response

IAV polymerase plays a role in regulating innate immune responses in the host. PB1 suppresses the Sendai virus (SeV) -or poly (I : C) -triggered activation of the IFN-β promoter; however, it does not inhibit STAT1 activation induced by IFN-β. Overexpression of PB1 inhibited Sev-or poly (I : C) -induced transcription of the following genes: IFNB1, IFN-stimulated gene 15 (ISG15), IFN-induced protein with tetratricopeptide repeats 1 (IFIT1), regulated upon activation normal T-cell expressed and secreted factor (RANTES), and oligoadenylate synthetase-like protein (OASL). This study further showed that PB1 inhibited SeV-induced phosphorylation of IκBα, IRF3, and TBK1[47]. PB1 inhibition of virus-induced antiviral responses was evaluated, and overexpression of PB1 resulted in inhibition of the activation of the IFN-β promoter induced by the expression of MAVS, MDA5-N, and RIG-IN[47]. Further investigations confirmed the colocalization of MAVS and PB1 inside the cytoplasm, and coimmunoprecipitation experiments showed that MAVS interacts with PB1 via its transmembrane domain. The interaction of these proteins leads to MAVS degradation in an autophagy-dependent manner, indicating that MAVS is a direct PB1 target. Moreover, this study also showed that PB1 promotes E3 ligase ring-finger protein 5 (RNF5) to catalyze K27-linked polyubiquitination of MAVS at Lys362 and Lys461. Additionally, PB1 senses and transfers ubiquitinated MAVS to autophagosomes for degradation. PB1 promotes H7N9 infection (used in this study as an IAV model) by inhibiting RIG-I/MAVS-mediated host innate immune responses (Figure 1-3-1, Table 1-3-1) by regulating the degradation cascade[47]. In summary, upon IAV infection, RIG-I senses viral RNA, and hence, RLR-mediated signaling is activated. PB1 regulates the K27-linked ubiquitination of MAVS that is mediated by the E3 ligase RNF5. Furthermore, PB1 interacts with NBR1 to recognize ubiquitinated MAVS, and the PB1-RNF5-MAVS-NBR1 complex fuses with lysosomes for MAVS degradation; therefore, the MAVS-mediated innate signaling pathway is disrupted, leading to IFN-I suppression.

PB1-F2 inhibits NF-κB signaling by binding IKKβ

The PB1-F2 protein contains 90 amino acid residues that are encoded by the + 1 alternate open reading frame (ORF) in the PB1 gene of some IAVs[48, 68]. The influenza viruses that caused the 1918, 1957 and 1968 pandemics express the PB1-F2 protein[48]. In animal models, this protein delayed the clearance of virus by killing immune cells by apoptosis, thus contributing to the pathogenesis of the virus[69]. Replacement of Aspargin (N) by serine (S) (N66S) in this protein at amino acid 66 of the 1918 H5N1 IAV pandemic strain increased the virulence of this virus[70]. Surprisingly, infection of mice with viruses expressing PB1-F2 with the N66S substitution led to the inhibition of ISGs at early stages of infection[71]. In vitro assays verified the in vivo experimental data. Overexpression of the A/Puerto Rico/8/1934 (PR8) PB1-F2 protein in 293T cells inhibited RIG-I-regulated activation of the IFN-β reporter and the secretion of IFNs. Overexpression of the PB1-F2 N66S protein in 293T cells resulted in increased IFN inhibition compared to wild-type PB1-F2. The same results were obtained in the context of viral infection with the PB1-F2 N66S virus. Immunofluorescence assays demonstrated that both PB1-F2 66 N and 66S colocalized with MAVS, and therefore, PB1-F2 (both 66-N and 66S) was thought to inhibit IFN induction at the MAVS level, probably by direct binding[48].

Coimmunoprecipitation and yeast two-hybrid assays showed that PB1-F2 binds with IKKβ and thus suppresses downstream NF-κB signaling (Figure 1-3-1, Table 1-3-1) [49]. An electrophoretic mobility shift assay demonstrated that in cells overexpressing this protein, NF-κB binding to DNA was significantly impaired and thus NF-κB signaling was suppressed; however, IKKβ kinase activity or NF-κB translocation to the nucleus was not affected. Moreover, the complete protein was indispensable for inhibiting NF-κB signaling, as neither the C-terminus nor the 57 amino acids of the N-terminus decreased NF-κB signaling[49].

As the IAV NS1 protein is a major inhibitor of IFN activation[72], this protein uses several strategies to create an anti-IAV state in virus-infected cells[73]. Among the different strategies, one strategy is masking viral RNA from recognition by RIG-I[74, 75]. The amino acid residues R38 and K41 at the N-terminus of NS1 are crucial for mediating the interaction with dsRNA species[74]. Mutations of these residues abolished the binding between TRIM25 and NS1, a binding that is indispensable for suppressing the activation of RIG-I mediated by E3 ligase[38]. Therefore, Varga et al. examined IFN activation by viruses that express TRIM25/dsRNA binding mutants of NS1 and PB1-F2 66 N or 66S[48]. These findings indicated that even in the presence of NS1, which is deficient in dsRNA and TRIM25 binding, 66S, compared to 66 N, reduced the activation of IFN. Similarly, they also found that N66S, but not wild-type PB1-F2, increased the IFN inhibition of NS1[48].

PB2 interacts with and inhibits JAK1/ STAT signaling to inhibit host innate immune signaling

The PB2 protein is important for IAV replication. The E627K substitution is responsible for replication in mammalian cells and is a dominant adaptation marker in human-adapted IAVs[51]. PB2 is a negative regulator of the IFN-stimulated antiviral response and interacts with JAK1, downregulates its expression and degrades it by using proteasome machinery, thus inhibiting IFN signaling. Furthermore, the possibility of JAK1 polyubiquitination was explored by Yang et al.[51], and they found that overexpression of PB2 significantly promoted K48-linked (but not K11-, K27-, K63-linked) ubiquitination of JAK1. The amino acids Lys859 and Lys860 (K859 and K860) of JAK1 were determined to be critical for PB2 action, as substitution of these two amino acids with arginine (R) (K859R; K860R) resulted in the loss of JAK1 degradation mediated by PB2. These results indicated that the degradation of JAK1, mediated by ubiquitination at the K859/K860 residues, is important for the replication of IAVs. A past study demonstrated that in mice, 283 M/526R of PB2 from the CZ is a virulence marker for A/duck/eastern China/JY/2014 (JY) [76]. Notably, the H5 subtype of highly pathogenic AIV (HPAIV) with I283M/K526R mutations in PB2 enhanced the ability to degrade mammalian JAK1 and thus replicate with higher efficiency in mammalian cells but not in avian cells. This study provided a mechanistic explanation of the host immune evasion strategy employed by IAV that involves the degradation of JAK1 by PB2 (Figure 1-3-1, Table 1-3-1) [51].

HA degrades IFNAR1 and IFNGR1 to facilitate IAV replication

Overexpression of IFNAR1 exerts anti-IAV effects by blocking virus replication. The HA protein activates IFNAR1 ubiquitination, thereby decreasing IFNAR1 levels and hence facilitating IAV replication inside cells. During maturation, HA is cleaved into the HA1 and HA2 subunits, and only the HA1 subunit significantly decreases IFNAR1 levels. Moreover, HA overexpression does not affect the mRNA levels of IFNAR1, and only the protein levels of IFNAR1 were decreased. The IFNAR1 levels on the cell surface affect cellular sensitivity to IFN-α/β[77-80], and HA decreases the surface IFNAR1 levels; therefore, the status of STAT1 and STAT2 was determined in the presence or absence of HA[52]. HA overexpression significantly inhibited IFN-induced STAT1 and STAT2 phosphorylation (Figure 1-3-1, Table 1-3-1). IAV HA degrades IFNAR1, which helps viruses escape the host innate immune system. After this study, Xia et al.[53] thoroughly investigated the mechanism of IFNAR1 degradation by IAV HA[53]. They found that, in addition to IFNAR1, HA could also degrade type Ⅱ IFN (IFN-γ) receptor 1 (IFNGR1) through casein kinase 1α (CK1α). Knockdown of CK1α by small interfering RNA (siRNA) repressed the degradation of both IFNAR1 and IFNGR1 induced by IAV infection. These studies suggested that the HA protein of IAV activates the degradation of IFN receptors via CK1α, thereby facilitating viral replication[52, 53].

The molecular mechanism of IFN receptor degradation by the IAV HA protein was delineated further by Xia et al.[81], who performed mass spectrometry (MS) to identify the host protein that binds to viral HA. MS was used to identify a host protein, poly (ADP-ribose) polymerase 1 (PARP1), that interacts with the HA of IAV. PARP1 belongs to the PARP family, which regulates the differentiation and proliferation of cells[82-84]. The MS result was confirmed by a coimmunoprecipitation experiment. The colocalization experiment showed that endogenous PARP1 localization was altered after IAV infection or HA overexpression. IFNAR1 expression is critical for antiviral responses mediated by IFN-I, and PARP1 promoted the replication of IAV by controlling IFNR1 levels. Knockdown of PARP1 rescued IFNAR1 levels upon HA overexpression or IAV infection, indicating the importance of PARP1 for IAV-or HA-induced reduction in IFNAR1. This mechanistic study revealed that PARP1 facilitates IAV replication by regulating HA-induced degradation of IFNR1 and thus impacts the IFN signaling pathway[81].

PA interacts with and inhibits IRF3 nuclear translocation to suppress IFN-β activation

The polymerase acid (PA) subunit protein enhances influenza virus pathogenicity, transmission, its capability to infect a broader range of hosts, and polymerase activity and participates in restricting IFN-β production[85-88]. The PA protein suppresses the production of IFN-β by interacting with IRF3, subsequently suppressing its phosphorylation, dimerization and nuclear translocation (Figure 1-3-1, Table 1-3-1). The PA protein has two functional domains, i.e., a PAN domain of 30 kDa at the N-terminal end, which contains amino acid residues 1~257, and a PAC domain of 53 kDa at the C-terminal end, which contains amino acid residues 258~716[89]. PAN conveys endonuclease activity, while PAC is a PB1 binding domain[90]. The PAN domain of PA was sufficient to reduce the expression levels of IFN-β and ISG-56, similar to the complete PA protein[50], while the PAC domain failed to reduce IFN-β or ISG-56 levels[50]. The coimmunoprecipitation experiments showed that the PAN domain interacts with IRF3, and thus, this fragment was responsible for suppressing the IFN-β signaling pathway. Since the PAN domain conveys endonuclease activity, a previous study demonstrated that when aspartic acid (D) was substituted with alanine (A) at amino acid 108 (D108A) in the PAN domain, the endonuclease activity of PA was abolished[91]. Therefore, this mutation was explored in terms of its ability to suppress IFN-β activity[50]. The D108A mutation failed to inhibit IFN-β activation, and this mutation abolished the interaction between PA and IRF3.

Conclusions and future perspectives

A complex and dynamic interplay exists between IAVs and host innate immune responses. IAVs use various proteins that target different proteins in the host to evade the host innate immune system. During the last two decades, virus-host interactions have been thoroughly studied, and these studies provide further explanations for the mechanisms that IAVs use against the host to establish successful infection.

Several NSPs have been discovered, including PA-X, PB1-N40, PA-N155, PA-N182, and M42 NS3, whose roles in host immune evasion need to be determined[92-96]. Additionally, some gaps also exist in the existing research, e.g., NS2 interacts with both IRF3 and IRF7 and inhibits the nuclear translocation of IRF7[12]; however, the mechanism of IRF3 nuclear translocation still needs to be determined. Similarly, the PB1-F2 protein colocalizes with MAVS to inhibit IFN induction[48]; however, the physical interaction between PB1-F2 and MAVS still needs to be evaluated.

The conserved nature of the host innate immune system and high diversity of IAVs demands further thorough investigation to provide a deeper understanding and to enable the development of effective anti-IAV treatments that could be used against diverse IAV strains. In the future, more extensive efforts are needed to apply the current research to increase host innate immunity and control disease. As IAVs are under severe evolutionary pressure due to mutations and recombination processes, their protein characterization and functions warrant repeated thorough investigations, especially with respect to the evasion of innate host processes. Bridging the different gaps will provide directions for designing better vaccines and novel antiviral agents.

References

[1] MUÑOZ-MORENO R, MARTÍNEZ-ROMERO C, GARCÍA-SASTRE A. Induction and evasion of type-I interferon responses during influenza A virus infection. Cold Spring Harb Perspect Med, 2021, 11(10): a038414.

[2] WAGAMAN P C, SPENCE H A, O'CALLAGHAN R J. Detection of influenza C virus by using an in situ esterase assay. J Clin Microbiol, 1989, 27: 832-836.

[3] FERGUSON L, OLIVIER A K, GENOVA S, et al. Pathogenesis of influenza D virus in cattle. J Virol, 2016, 90: 5636-5642.

[4] EISFELD A J, NEUMANN G, KAWAOKA Y. At the centre: influenza A virus ribonucleoproteins. Nat Rev Microbiol, 2015, 13: 28-41.

[5] HALE B G, ALBRECHT R A, GARCÍA-SASTRE A. Innate immune evasion strategies of influenza viruses. Future Microbiol, 2010, 5: 23-41.

[6] LAMB R A, CHOPPIN W. The gene structure and replication of influenza virus. Annu Rev Biochem, 1983, 52: 467-506.

[7] KLUMPP K, RUIGROK R W, BAUDIN F. Roles of the influenza virus polymerase and nucleoprotein in forming a functional RNP structure. EMBO J, 1997, 16:1248-1257.

[8] LUO W, ZHANG J, LIANG L, et al. Phospholipidscramblase 1 interacts with influenza impairing its nuclear import and thereby suppressing virus replication. PLOS Pathog, 2018, 14: e1006851.

[9] SHINYA K, EBINA M, YAMADA S, et al. Influenza virus receptors in the human airway. Nature, 2006, 440: 435-436.

[10] VAN RIEL D, DEN BAKKER M A, LEIJTEN L M E, et al. Seasonal and pandemic human influenza viruses attach better to human upper respiratory tract epithelium than avian influenza viruses. Am J Pathol, 2010, 176: 1614-1618.

[11] HORIMOTO T, KAWAOKA Y. Influenza: lessons from past pandemics, warnings from current incidents. Nat Rev

Microbiol, 2005, 3: 591-600.

[12] ZHANG B, LIU M, HUANG J, et al. H1N1 influenza A virus protein NS2 inhibits innate immune response by targeting IRF7. Viruses, 2022, 14: 2411.

[13] MEDINA R A, GARCÍA-SASTRE A. Influenza A viruses: new research developments.Nat Rev Microbiol, 2011, 9: 590-603.

[14] LIU W J, BI Y, WANG D, et al. On the centenary of the spanish flu: being prepared for the next pandemic. Virol Sin, 2018, 33: 463-466.

[15] YIN X, DENG G, ZENG X, et al. Genetic and biological properties of H7N9 avian influenza viruses detected after application of the H7N9 poultry vaccine in China. PLOS Pathog, 2021, 17: e1009561.

[16] YANG W, SCHOUNTZ T, MA W. Bat influenza viruses: current status and perspective. Viruses, 2021, 13: 547.

[17] FEREIDOUNI S, STARICK E, KARAMENDIN K, et al. Genetic characterization of a new candidate hemagglutinin subtype of influenza A viruses. Emerg Microbes Infect, 2023, 12: e2225645.

[18] PALESE P, WANG T T. Why do influenza virus subtypes die out? A hypothesis. MBio, 2011, 2(5): e00150-11.

[19] LIU W J, WU Y, BI Y, et al. Emerging hxNy influenza A viruses. Cold Spring Harb Perspect Med, 2022, 12: a038406.

[20] LUND J M, ALEXOPOULOU L, SATO A, et al. Recognition of single-stranded RNA viruses by Toll-like receptor 7. Proc Natl Acad Sci, 2004, 101: 5598-5603.

[21] SETH R B, SUN L, EA C K, et al. Identification and characterization of MAVS, a mitochondrial antiviral signaling protein that activates NF-kappaB and IRF 3. Cell, 2005, 122: 669-682.

[22] XU L G, WANG Y Y, HAN K J, et al. VISA is an adapter protein required for virus-triggered IFN-beta signaling. Mol Cell, 2005, 19: 727-740.

[23] NEGISHI H, TANIGUCHI T, YANAI H. The interferon (IFN) class of cytokines and the IFN regulatory factor (IRF) transcription factor family. Cold Spring Harb Perspect Biol, 2018, 10(11): a028423.

[24] CHEN Y, LEI X, JIANG Z, et al. Cellular nucleic acid-binding protein is essential for type I interferon-mediated immunity to RNA virus infection. Proc Natl Acad Sci USA, 2021, 118(26): e2100383118.

[25] KAWAI T, AKIRA S. The role of pattern-recognition receptors in innate immunity: update on Toll-like receptors. Nat Immunol, 2010, 11: 373-384.

[26] TAKEUCHI O, AKIRA S. Pattern recognition receptors and inflammation. Cell, 2010, 140: 805-820.

[27] KUMAR H, KAWAI T, AKIRA S. Pathogen recognition by the innate immune system. Int Rev Immunol, 2011, 30: 16-34.

[28] SCHULZ O, DIEBOLD S S, CHEN M, et al. Toll-like receptor 3 promotes cross-priming to virus-infected cells. Nature, 2005, 433: 887-892.

[29] GOUBAU D, SCHLEE M, DEDDOUCHE S, et al. Antiviral immunity via RIG-I-mediated recognition of RNA bearing 5'-diphosphates. Nature, 2014, 514: 372-375.

[30] LE GOFIC R, POTHLICHET J, VITOUR D, et al. Cutting edge: influenza A virus activates TLR3-dependent inflammatory and RIG-I-dependent antiviral responses in human lung epithelial cells. J Immunol, 2007, 178: 3368-3372.

[31] WANG J, OBERLEY-DEEGAN R, WANG S, et al. Differentiated human alveolar type II cells secrete antiviral IL-29 (IFN- λ 1) in response to influenza A infection. J Immunol, 2009, 182: 1296-1304.

[32] DARNELL J E, LAN M, STARK G R. Jak-STAT pathways and transcriptional activation in response to IFNs and other extracellular signaling proteins. Science, 1994, 264: 1415-1421.

[33] SCHINDLER C, LEVY D E, DECKER T. JAK-STAT signaling: from interferons to cytokines. J Biol Chem, 2007, 282: 20059-20063.

[34] KOCHS G, GARCÍA-SASTRE A, MARTÍNEZ-SOBRIDO L. Multiple anti-interferon actions of the influenza A virus NS1 protein. J Virol, 2007, 81:7011-7021.

[35] BAEZ M, TAUSSIG R, ZAZRA J J, et al. Complete nucleotide sequence of the influenza A/PR/8/34 virus NS gene and comparison with the NS genes of the A/Udorn/72 and A/FPV/Rostock/34 strains. Nucleic Acids Res, 1980, 8: 5845-5858.

[36] LIU J, LYNCH P A, CHIEN C Y, et al. Crystal structure of the unique RNA-binding domain of the influenza virus NS1 protein. Nat Struct Biol, 1997, 4: 896-899.

[37] KLEMM C, BOERGELING Y, LUDWIG S, et al. Immunomodulatory nonstructural proteins of influenza A viruses. Trends Microbiol, 2018, 26: 624-636.

[38] GACK M U, ALBRECHT R A, URANO T, et al. Influenza A virus NS1 targets the ubiquitin ligase TRIM25 to evade recognition by the host viral RNA sensor RIG-I. Cell Host Microbe, 2009, 5: 439-449.

[39] RAJSBAUM R, ALBRECHT R A, WANG M K, et al. Species-specific inhibition of RIG-I ubiquitination and IFN induction by the influenza A virus NS1 protein. PLOS Pathog, 2012, 8: e1003059.

[40] MIBAYASHI M, MARTÍNEZ-SOBRIDO L, LOO Y M, et al. Inhibition of retinoic acid-inducible gene I-mediated induction of beta interferon by the NS1 protein of influenza A virus. J Virol, 2007, 81: 514-524.

[41] JUREKA A S, KLEINPETER A B, CORNILESCU G, et al. Structural basis for a novel interaction between the NS1 protein derived from the 1918 influenza virus and RIG-I. Structure, 2015, 23: 2001-2010.

[42] LI S, MIN J Y, KRUG R M, et al. Binding of the influenza A virus NS1 protein to PKR mediates the inhibition of its activation by either PACT or double-stranded RNA. Virology, 2006, 349:13-21.

[43] MIN J-Y, LI S, SEN G C, et al. A site on the influenza A virus NS1 protein mediates both inhibition of PKR activation and temporal regulation of viral RNA synthesis. Virology, 2007, 363: 236-243.

[44] NEMEROFF M E, BARABINO S M, LI Y, et al. Influenza virus NS1 protein interacts with the cellular 30 kDa subunit of CPSF and inhibits 3'end formation of cellular pre-mRNAs. Mol Cell, 1998, 1: 991-1000.

[45] NOAH D L, TWU K Y, KRUG R M. Cellular antiviral responses against influenza A virus are countered at the post-transcriptional level by the viral NS1A protein via its binding to acellular protein required for the 3' end processing of cellular pre-mRNAS. Virology, 2003, 307: 386-395.

[46] GAO S, SONG L, LI J, et al. Influenza A virus-encoded NS1 virulence factor protein inhibits innate immune response by targeting IKK. Cell Microbiol, 2012, 14: 1849-1866.

[47] ZENG Y, XU S, WEI Y, et al. The PB1 protein of influenza A virus inhibits the innate immune response by targeting MAVS for NBR1-mediated selective autophagic degradation. PLOS Pathog, 2021, 17: e1009300.

[48] VARGA Z T, RAMOS I, HAI R, et al. The influenza virus protein PB1-F2 inhibits the induction of type I interferon at the level of the MAVS adaptor protein. PLOS Pathog, 2011, 7: e1002067.

[49] REIS A L, MCCAULEY J W. The influenza virus protein PB1-F2 interacts with IKKβ and modulates NF-κB signalling. PLOS ONE, 2013, 8: e63852.

[50] YI C, ZHAO Z, WANG S, et al. Influenza A virus PA antagonizes interferon-b by interacting with interferon regulatory factor 3. Front Immunol, 2017, 8: 1051.

[51] YANG H, DONG Y, BIAN Y, et al. The influenza virus PB2 protein evades antiviral innate immunity by inhibiting JAK1/STAT signalling. Nat Commun, 2022, 13(1): 6288.

[52] XIA C, VIJAYAN M, PRITZL C J, et al. Hemagglutinin of influenza A virus antagonizes type I interferon (IFN) responses by inducing degradation of type I IFN receptor 1. J Virol, 2016, 90: 2403-2417.

[53] XIA C, WOLF J J, VIJAYAN M, STUDSTILL C J, et al. Casein kinase 1 α Mediates the degradation of receptors for type I and type II interferons caused by hemagglutinin of influenza A virus. J Virol, 2018, 92(7): e00006-18.

[54] WOLF J J, XIA C, STUDSTILL C J, et al. Influenza A virus NS1 induces degradation of sphingosine 1-phosphate lyase to obstruct the host innate immune response. Virology, 2021, 558: 67-75.

[55] SEO Y J, BLAKE C, ALEXANDER S, et al. Sphingosine 1-phosphate-metabolizing enzymes control influenza virus propagation and viral cytopathogenicity. J Virol, 2010, 84: 8124-8131.

[56] JOHNSON R, HALDER G. The two faces of Hippo: targeting the Hippo pathway for regenerative medicine and cancer treatment. Nat Rev Drug Discov, 2014, 13: 63-79.

[57] YU F X, ZHAO B, GUAN K L. Hippo pathway in organ size control, tissue homeostasis, and cancer. Cell, 2015, 163: 811-828.

[58] KANGO-SINGH M, NOLO R, TAO C, et al. Shar-pei mediates cell proliferation arrest during imaginal disc growth in Drosophila. Development, 2002, 129: 5719-5730.

[59] HARVEY K F, PflEGER C M, HARIHARAN I K. The Drosophila Mst ortholog, hippo, restricts growth and cell proliferation and promotes apoptosis. Cell, 2003, 114: 457-467.

[60] UDAN R S, KANGO-SINGH M, NOLO R, et al. Hippo promotes proliferation arrest and apoptosis in the Salvador/Warts pathway. Nat Cell Biol, 2003, 5: 914-920.

[61] LAI Z C, WEI X, SHIMIZU T, et al. Control of cell proliferation and apoptosis by mob as tumor suppressor, mats. Cell, 2005, 120: 675-685.

[62] ZHANG L, REN F, ZHANG Q, et al. The TEAD/TEF family of transcription factor Scalloped mediates Hippo signaling in organ size control. Dev Cell, 2008, 14: 377-387.

[63] CHAN E H Y, NOUSIAINEN M, CHALAMALASETTYR B, et al. The Ste20-like kinase Mst2 activates the human large tumor suppressor kinase Lats1. Oncogene, 2005, 24: 2076-2086.

[64] ZHAO B, WEI X, LI W, et al. Inactivation of YAP oncoprotein by the Hippo pathway is involved in cell contact inhibition and tissue growth control. Genes Dev, 2007, 21: 2747-2761.

[65] ZHANG Q, MENG F, CHEN S, et al. Hippo signalling governs cytosolic nucleic acid sensing through YAP/TAZ-mediated TBK1 blockade. Nat Cell Biol, 2017, 19: 362-374.

[66] WANG S, XIE F, CHU F, et al. YAP antagonizes innate antiviral immunity and is targeted for lysosomal degradation through IKKε-mediated phosphorylation. Nat Immunol, 2017, 18: 733-743.

[67] ZHANG Q, ZHANG X, LESSI X, et al. Influenza A virus NS1 protein hijacks YAP/TAZ to suppress TLR3-mediated innate immune response. PLOS Pathog, 2022, 18: e1010505.

[68] CHEN W, CALVO P A, MALIDE D, et al. A novel influenza A virus mitochondrial protein that induces cell death. Nat Med, 2001, 7 :1306-1312.

[69] ZAMARIN D, ORTIGOZAM B, PALESE P. Influenza A virus PB1-F2 protein contributes to viral pathogenesis in mice. J Virol, 2006, 80: 7976-7983.

[70] CONENELLO G M, ZAMARIN D, PERRONE L A, et al. A single mutation in the PB1-F2 of H5N1 (HK/97) and 1918 influenza A viruses contributes to increased virulence. PLOS Pathog, 2007, 3: 1414-1421.

[71] CONENELLO G M, TISONCIK J R, ROSENZWEIG E, et al. A single N66S mutation in the PB1-F2 protein of influenza A virus increases virulence by inhibiting the early interferon response *in vivo*. J Virol, 2011, 85: 652-662.

[72] GARCÍA-SASTRE A, EGOROV A, MATASSOV D, et al. Influenza A virus lacking the NS1 gene replicates in interferon-deicient systems. Virology, 1998, 252: 324-330.

[73] HALE B G, RANDALL R E, ORTÍN J, et al. The multifunctional NS1 protein of influenza A viruses. J Gen Virol, 2008, 89: 2359-2376.

[74] WANG W, RIEDEL K, LYNCH P, et al. RNA binding by the novel helical domain of the influenza virus NS1 protein requires its dimer structure and a small number of specific basic amino acids. RNA, 1999, 5: 195-205.

[75] WANG X, LI M, ZHENG H, et al. Influenza A virus NS1 protein prevents activation of NF-kappaB and induction of alpha/beta interferon. J Virol, 2000, 74: 11566 -11573.

[76] WANG X, CHEN S, WANG D, et al. Synergistic effect of PB2 283M and 526R contributes to enhanced virulence of H5N8 influenza viruses in mice. Vet Res, 2017, 48: 67.

[77] KUMAR K G S, KROLEWSKI J J, FUCHS S Y. Phosphorylation and specific ubiquitin acceptor sites are required for ubiquitination and degradation of the IFNAR1 subunit of type I interferon receptor. J Biol Chem, 2004, 279: 46614-46620.

[78] KUMAR K G S, BARRIERE H, CARBONE C J, et al. Site-specific ubiquitination exposes a linear motif to promote interferon-α receptor endocytosis. J Cell Biol, 2007, 179: 935-950.

[79] ZHENG H, QIAN J, BAKER D P, et al. Tyrosine phosphorylation of protein kinase D2 mediates ligand-inducible elimination of the type 1 interferon receptor. J Biol Chem, 2011, 286: 35733-35741.

[80] ZHENG H, QIAN J, VARGHESE B, et al. Ligand-stimulated downregulation of the alpha interferon receptor: role of protein kinase D2. Mol Cell Biol, 2011, 31: 710-720.

[81] XIA C, WOLF J J, SUN C, et al. PARP1 enhances influenza A virus propagation by facilitating degradation of host type I

interferon receptor. J Virol, 2020, 94(7): e01572-19.

[82] CHAITANYA G V, ALEXANDER J S, BABU P P. PARP-1 cleavage fragments: signatures of cell-death proteases in neurodegeneration. Cell Commun Signal, 2010, 8: 31.

[83] DU Y, YAMAGUCHI H, WEI Y, et al. Blocking c-Met-mediated PARP1 phosphorylation enhances anti-tumor effects of PARP inhibitors. Nat Med, 2016, 22: 194-201.

[84] VERHEUGD P, BÜTEPAGE M, ECKEI L, et al. Players in ADP-ribosylation: readers and erasers. Curr Protein Pept Sci, 2016, 17: 654-667.

[85] ILYUSHINA N A, KHALENKOV A M, SEILER J P, et al. Adaptation of pandemic H1N1 influenza viruses in mice. J Virol, 2010, 84: 8607-8616.

[86] IWAI A, SHIOZAKI T, KAWAI T, et al. Influenza A virus polymerase inhibits type I interferon induction by binding to interferon β Promoter stimulator 1. J Biol Chem, 2010, 285: 32064-32074.

[87] BUSSEY K A, DESMET E A, MATTIACIO J L, et al. PA residues in the 2009 H1N1 pandemic influenza virus enhance avian influenza virus polymerase activity in mammalian cells. J Virol, 2011, 85: 7020-7028.

[88] MEHLE A, DUGAN V G, TAUBENBERGER J K, et al. Reassortment and mutation of the avian influenza virus polymerase PA subunit overcome species barriers. J Virol, 2012, 86: 1750-1757.

[89] GUU T S Y, DONG L, WITTUNG-STAFSHEDE P, et al. Mapping the domain structure of the influenza A virus polymerase acidic protein (PA) and its interaction with the basic protein 1 (PB1) subunit. Virology, 2008, 379: 135-142.

[90] BOIVIN S, CUSACK S, RUIGROK R W H, et al. Influenza A virus polymerase: structural insights into replication and host adaptation mechanisms. J Biol Chem, 2010, 285: 28411-28417.

[91] HARA K, SCHMIDT F I, CROW M, et al. Amino acid residues in the N-terminal region of the PA subunit of influenza A virus RNA polymerase play a critical role in protein stability, endonuclease activity, cap binding, and virion RNA promoter binding. J Virol, 2006, 80: 7789-7798.

[92] WISE H M, FOEGLEIN A, SUN J, et al. A complicated message: identification of a novel PB1-related protein translated from influenza A virus segment 2 mRNA. J Virol, 2009, 83: 8021-8031.

[93] JAGGER B W, WISE H M, KASH J C, et al. An overlapping protein-coding region in influenza A virus segment 3 modulates the host response. Science, 2012, 337: 199-204.

[94] SELMAN M, DANKAR S K, FORBES N E, et al. Adaptive mutation in influenza A virus non-structural gene is linked to host switching and induces a novel protein by alternative splicing. Emerg Microbes Infect, 2012, 1: e42.

[95] WISE H M, HUTCHINSON E C, JAGGER B W, et al. Identification of a novel splice variant form of the influenza A virus M2 ion channel with an antigenically distinct ectodomain. PLOS Pathog, 2012, 8: e1002998.

[96] MURAMOTO Y, NODA T, KAWAKAMI E, et al. Identification of novel influenza A virus proteins translated from PA mRNA. J Virol, 2013, 87: 2455-2462.

Global review of the H5N8 avian influenza virus subtype

Rafique Saba, Rashid Farooq, Mushtaq Sajda, Ali Akbar, Li Meng, Luo Sisi, Xie Liji, and Xie Zhixun

Abstract

Orthomyxoviruses are negative-sense, RNA viruses with segmented genomes that are highly unstable due to reassortment. The highly pathogenic avian influenza (HPAI) subtype H5N8 emerged in wild birds in China. Since its emergence, it has posed a significant threat to poultry and human health. Poultry meat is considered an inexpensive source of protein, but due to outbreaks of HPAI H5N8 from migratory birds in commercial flocks, the poultry meat industry has been facing severe financial crises. This review focuses on occasional epidemics that have damaged food security and poultry production across Europe, Eurasia, the Middle East, Africa, and America. HPAI H5N8 viral sequences have been retrieved from GISAID and analyzed. Virulent HPAI H5N8 belongs to clade 2.3.4.4b, Gs/GD lineage, and has been a threat to the poultry industry and the public in several countries since its first introduction. Continent-wide outbreaks have revealed that this virus is spreading globally. Thus, continuous sero-and viro-surveillance both in commercial and wild birds, and strict biosecurity reduce the risk of the HPAI virus appearing. Furthermore, homologous vaccination practices in commercial poultry need to be introduced to overcome the introduction of emergent strains. This review clearly indicates that HPAI H5N8 is a continuous threat to poultry and people and that further regional epidemiological studies are needed.

Keywords

avian influenza virus, H5N8 subtype, epidemiology, surveillance, control and prevention

Introduction

Avian influenza viruses (AIVs) belong to the Orthomyxoviridae family and contain a segmented genome with eight single-stranded RNA segments and have negative polarity[1]. Hemagglutinin (HA) gene and neuraminidase (NA) gene, two of the envelope proteins of these viruses, are used to classify them into different subtypes[2]. To date, 16 HA and 9 NA subtypes of AIVs have been identified in poultry and wild birds[3].

Low-pathogenic avian influenza (LPAI) viruses are naturally found in wild water birds such as swans, ducks, gulls, geese, swans, shorebirds, and terns[4, 5]. LPAI viruses are transmitted to domestic birds, animals, and even humans from wild water birds. Influenza viruses with H5 HA have been circulating in wild birds and domestic poultry since 1995[6]. The Qinghai Lake-like H5N1 virus was first widely spread by migratory birds and caused huge damage to the poultry industry worldwide, but the origin of the virus remains unclear. The LPAI viruses of the H5 subtype, when infecting poultry, can evolve into HPAI viruses, causing severe mortality[7]. During July and August 2005, HPAI H5 clade 2.2 viruses were detected in poultry farms in Russia and Kazakhstan, where they caused high mortality[8]. These viruses were genetically related to viruses detected in 2005 in Qinghai Lake in China[9]. From July 2005 onward, HPAI H5 viruses were observed to cause outbreaks on poultry farms[8]. The H5N1 virus became endemic in 2003 in southern China, giving rise to several genotypes.

In the mainland of China, the H5N8 virus was detected in poultry between 2009 and 2010, which derived its HA gene from the Asian H5N1 lineage and its neuraminidase (NA), nucleoprotein (NP), and polymerase basic (PB1) genes from unidentified, non-H5N1 viruses. The H5N8 virus is highly pathogenic to chickens and moderately to extremely dangerous to mice[10]. In 2014, a novel reassortant HPAI H5N8 clade 2.3.3.4 virus with the HA gene was identified in Republic of Korea (ROK)[11]. Two types of H5N8 were found during these outbreaks, namely Gochang-like and Buan2-like. The predominant group, Buan2-like, afterward spread to Europe, East Asia, and North America by migratory waterfowl and formed three distinct subgroups[5, 11-14]. In autumn 2016, another High pathogenic AI H5N8 virus of clade 2.3.4.4 spread across different continents[15] and showed sustained prevalence in Africa, Europe, and the Middle East. In early 2020, HAPI H5N8 was continuously reported in Iraq, Kazakhstan, and Russia[16]. Furthermore, in December 2020 in Russia, seven poultry farm workers were infected with a clade 2.3.4.4b H5N8 virus[17]. In June 2021, 2 782 outbreaks of H5N8 were reported, causing the mortality or destruction of approximately 38 million poultry in more than 25 countries[18].

In conclusion, the spread of High pathogenic AI H5N8 viruses has raised serious issues for the security and conservation of animals, poultry, and even public health[19]. All this evidence suggests that H5N8 viruses are likely to spread worldwide; therefore, continuous surveillance and vaccination of poultry are highly recommended. In this review, we describe the emergence of sporadic infection continentally, and the impacts are briefly described.

Intra-and inter-continental transmission patterns of sporadic infection of HPAI H5N8

Asia and Africa

A number of emergence and re-emergence studies of HPAI H5N8 strains have been reported within & across Asia & Africa. One HPAI H5N8 virus (Dkk1203) was isolated from a poultry farm in the mainland of China during 2009-2010. The Dkk1203 isolate derived its HA gene from the Asian H5N1 lineage. Phylogenetic analysis of the HA gene revealed that this isolate was classified into the 2.3.4 clade. Compared to H5N5 viruses that were isolated between December 2008 and January 2009, this strain has longer branches. This strain was distantly related to Eurasian N8 genotype viruses and clustered with three H3N8 viruses with an origin in Eastern Asia. Therefore, the N5 and N8 NA genes of the Dkk1203 isolate are derived from Asian viruses; however, the exact origin is not known[10].

In a breeding duck farm on January 16, 2014, in the Jeonbuk Province of ROK, High-pathogenic AI clinical signs, such as reduced egg production by about 60% and slightly increased mortality rates, were discovered. Moreover, on January 17 of the same year, a farmer was also diagnosed with HPAI from breeder ducks in the Donglim Reservoir[12]. Also, the Donglim Reservoir had 100 Baikal teal carcasses, all of which tested positive for the high pathogenic AI H5N8 virus[12].

A few months later, in April 2014, an outbreak of the HPAI virus with the genotype H5N8, A/chicken/ Kumamoto/1-7/2014, occurred in Japan[20]. The HA clade 2.3.4.4 membership of this virus was also made known. In particular, A/broiler duck/ROK/Buan2/2014 and A/baikal teal/ROK/Donglim3/2014, HPAI H5N8 that were isolated in ROK in January 2014, all eight genomic segments displayed substantial sequence similarity[20]. The experimental work delineated that this isolate from Japan was lethal in chickens when a higher titer of virus was used for infection; however, the chickens were unaffected when challenged with lower viral doses[20].

In the same year (2014), three H5N8 viruses were reported from domestic geese in the mainland of China. The selected strains' sequence analyses revealed that all H5N8 viruses were direct progeny of the K1203 (H5N8) -like viruses discovered in China in 2010 and belonged to the Asian H5N1 HA lineage of clade 2.3.4.4. The recent common clade 2.3.4.4 H5N8 reassortants, which have severely damaged the poultry sector and pose a threat to public health, were created by K1203-like viruses, according to studies[21].

Eight highly pathogenic H5N8 AIVs were discovered in Japan over the winter, particularly in a location where migratory birds overwinter. These isolates were divided into three groups based on genetic analysis, demonstrating that three genetic subgroups of H5N8 HPAIs circulated in these migratory birds. These findings also suggest that the migration of these birds next winter may result in the redistribution of H5N8 HPAI globally[22, 23].

In 2016 in Malard County of the Tehran Province and the Meighan wetland of Arak City, Markazi Province, the HA genes indicated categorization in the 2.3.4.4b subclade. Although being identified as an H5N82.3.4.4b virus, the A/Goose/Iran/180/2016 virus's cluster was split from the A/Chicken/ran/162/2016 virus. This suggests that the entry of these viruses in Iran occurred through more than one window. The most recent HPAI-H5 outbreak in Iran happened in 2015 and was entirely caused by viruses from clade 2.3.2.1c. These findings underscore the necessity to continue proper monitoring activities in the target wild and domestic bird species for early HPAI identification and show that Iran is at high risk of the importation of HPAI H5 of the A/Goose/Guangdong/1/1996 lineage from East Asia. These activities would also allow the study of the genetic and antigenic evolution of H5 HPAI clade 2.3.4.4 viruses in the region and the world[24]. Furthermore, it appears that migrating wild aquatic birds carried these HPAI H5N8 strains into Iran via the West Asia-East African flyway[25].

An H5N8 influenza virus of clade 2.3.4.4 outbreak was reported in 2016 in the Republic of Tyva. The H5N8 clade 2.3.4.4 virus spread over Europe in the fall. The reports provide a clear overview of the viral strains that were discovered in the Russian Federation during the spring and fall of 2016. The strains under investigation were extremely harmful to mice, and several of their antigenic and genetic characteristics were different from an H5N8 strain that was prevalent in Russia in 2014[26].

The newly emerged H5N8 influenza virus was also isolated from green-winged teal ducks. The genomes of the HPAI H5N8 viruses from Egypt were also found to be related to recently identified reassortant H5N8 viruses of clade 2.3.4.4 recovered from several Eurasian nations, according to analyses of the viruses' genomes. The Egyptian H5N8 viruses had a number of genetic shifts that likely allowed for the spread and virulence of these viruses in mammals. Instead of human-like receptors, Egyptian H5N8 viruses prefer to bind to avian-like receptors. Likewise, amantadine and neuraminidase inhibitors had little effect on the Egyptian H5N8 viruses. It is important to continue monitoring waterfowl for avian influenza because it provides early warning of specific dangers to poultry and human health[27]. The presence of this group and clade was also found in Qinghai Lake, China, in 2016, which resulted in the deaths of wild migratory birds[15]. An HPAI H5N8 virus of clade 2.3.4.4b has been detected in Egypt. PA and NP gene replacement identified the strain as A/duck/Egypt/F446/2017. The Russian 2016 HPAI H5N8 virus (A/great crested grebe/Uvs-Nuur Lake/341/2016 (H5N8)) was likely the source of Egyptian H5N8 viruses, according to Bayesian phylogeographic analysis and reassortment most likely took place prior to an incursion into Egypt[28].

In Egypt, multiple introductions of different reassorted strains have been observed. The antigenic sites A and E of the HA gene have two new mutations. With various vaccination seeds, the HA nucleotide sequence identity ranges from 77% to 90%. To determine the main reassorted strain in Egypt, full-genome sequence analysis representing various governorates and sectors has been conducted. All viruses have been shown to be identical to the clade 2.3.4.4b reassorted strain that was discovered in Germany and other nations. Examination of these viruses revealed changes unique to Egyptian strains rather than the original virus identified in 2017 (A/duck/Egypt/F446/2017), and two strains of these viruses had the novel antiviral resistance marker V27A, which indicated amantadine resistance in the M2 protein. The findings showed that circulating H5N8 viruses were more variable than prior viruses analyzed in 2016 and 2017. An early 2017 strain served as the foundation for the main reassorted virus that circulated in 2017 and 2018. To track the development of circulating viruses, it is crucial to keep up this surveillance of AIVs[29]. The Democratic Republic of the Congo strains also belongs to the same clade, 2.3.4.4b. The emergence of this clade in central Africa threatens animal health and food security[30].

The recovered HPAI A (H5N8) viruses in Pakistan during 2018-2019 belonged to clade 2.3.4.4b and were most closely related to the Saudi Arabian A (H5N8) viruses, which were most likely introduced via cross-border transmission from nearby regions about 3 months before the virus was discovered in domestic poultry. It was also found that, prior to the first human A (H5N8) infection in Russian poultry workers in 2020, clade 2.3.4.4b viruses underwent rapid lineage expansion in 2017 and acquired signifcant amino acid mutations, including mutations correlated with increased hemagglutinin affinity to human-2, 6 receptors. Our findings demonstrate the necessity of routine avian influenza surveillance in Pakistan's live bird markets in order to keep an eye out for any potential A (H5Nx) variants that might emerge from poultry populations[31]. Every year, the Indus Flyway, also known as the Green Way, transports between 0.7 and 1.2 million birds from Europe, Central Asian countries, and India to Pakistan.

A thorough investigation was conducted to track the evolution of influenza viruses in poultry during the years 2020-2022 in China. A total of 35 influenza viruses, including 30 H5N8 viruses, 3 H5N1 viruses, and 2 H5N6 viruses, were isolated from chickens, ducks, and geese. The internal genes of H5N1 and H5N6 viruses shared different genetic heterogeneity with H5N8 viruses and had been reassorted with wild bird-origin H5N1 viruses from Europe. All HP H5N8 isolates were derived from clade 2.3.4.4b. The fact that practically all H5N8 viruses in China and ROK showed just one phylogenic cluster with H5N8 viruses of wild bird origin suggests that the H5N8 viruses in China were more stable. We also discovered that the main geographic source for the transmission of these H5N8 viruses to northern and eastern China is ROK. The majority of the co-circulation of H5N8 viruses took place within China, with central China serving as a seeding population during the H5N8 epidemic. Strong statistical evidence supported viral migration from wild birds to chickens and ducks, demonstrating that during 2020-2021, 2.3.4.4b H5N8 viruses with poultry origins were borne by wild birds. Multiple gene segments were also discovered to be involved in the development of severe disease due to H5N8 HPAI viruses, in mallards birds, which explains why no viral gene was found to be solely responsible for reducing the high virulence of an H5N8 virus but the PB2, M and NP segments significantly decreased mortality. Our results give new insights into the dynamics of H5 subtype influenza virus evolution and transmission among poultry following the almost one-year invasion of China by novel H5N8 viruses[32, 33]. In China, the re-emergence of the High Pathogenic H5N8 virus in domestic geese was also reported[34].

The establishment of novel H5N8 strains in China is frequently linked to the migration of migratory birds via the East Asian-Australasian Flyway. This flyway connects Siberia to Australia and includes various stopover spots in China where wild birds gather throughout their annual migration. These locations allow diverse bird species to interact and exchange influenza viruses[32].

During May 2020 in Iraq, H5N8 was reported in poultry. Complete genome sequencing delineated that a novel H5 2.3.3.4b variant had emerged. Furthermore, the long branch lengths for all segments indicated that undetected isolate was circulating for some period and possibly in galliform poultry[16].

After outbreaks in Iraq in July 2020, H5N8 was detected in ducks, geese, and backyard chickens of Chelyabinskaya Oblast (Chelyabinsk), in southern central Russia. During August and September 2020, a total of 11 cases were detected in the Tyumen, Omsk, and Kurgan regions of Russia[16]. Wild birds were described as the cause of the incursion.

Concurrent with the H5N8 outbreak in Russia, the outbreak of H5N8 was also confirmed in several regions of Kazakhstan, including Kostanay, Akmola, and Pavlodar[16]. AI H5N8 diagnosis was confirmed by subtype-specific quantitative RT-PCR[36]. The AI H5N8 virus from Iraq and Kazakhstan shared a lot of genetic similarities, according to genetic analyses[16].

Europe and the Americas

In 2014, European countries, such as Germany, the United Kingdom, the Netherlands, and Italy, reported several outbreaks of H5N8 in poultry. Two different Highly pathogenic viruses, H5N2 and H5N8, were found in the United States in December 2014 in wild birds and later in backyard birds in Washington State. This sparked concerns about potential connections with recent H5N2 outbreaks in Canada and H5N8 in Asia, which is now affecting poultry farms in Europe. The continuous spread of these Eurasian HPAI H5 viruses among wild birds has a significant impact that could arise and the ensuing consequences on American poultry and wildlife rehabilitation facilities. Tundra swans (*C. columbianus*), common teal (*A. crecca*), spot-billed duck (*A. poecilorhyncha*), Eurasian wigeon (*A. penelope*) and mallard, that appeared to be in good health also tested positive for the HPAI H5N8 virus, which raises the possibility that wild birds may be contributing to the spread of this High pathogenic H5 lineage in North America[37].

With a comprehensive review of the spatiotemporal expansion and genetic characteristics of HPAI Gs/GD H5N8 from Poland's 2019/20 epidemic, the highly pathogenic H5 subtype of the Gs/GD lineage repeatedly invaded Poland from 2016 to 2020, posing a major threat to poultry globally. In nine Polish provinces during 2019 and 2020, 35 outbreaks in backyard and commercial poultry holdings as well as one incidence in a wild bird were confirmed. The majority of the outbreaks were found in the meat of ducks and turkeys. All sequenced viruses belonged to a previously unidentified genotype of HPAI H5N8 clade 2.3.4.4b and were closely related to one another. The main methods of HPAI dissemination were found to be human activity and wild birds. A review of current risk assessment techniques is necessary in light of the HPAI virus's unusually delayed emergence[38, 39].

Asia and Europe

A new wave of H5N8 outbreaks in domestic and wild birds was observed in several European nations in October 2020, including the United Kingdom, Denmark, Ireland, Germany, and the Netherlands. In August 2020, several outbreaks of the disease were confirmed from Russia in both domestic and wild birds, and the affected regions spread to Kazakhstan in mid-September. Moreover, H5N8 epidemics in domestic and/or wild birds appeared in East Asia (Japan and ROK) and the Middle East (Israel). A unique variant between clade 2.3.4.4b and Eurasian LPAI viruses in wild birds was described as well as two different forms of HPAI H5N8 variants, one of which only belonged to clade 2.3.4.4b. The geographical areas affected have been steadily expanding, and at least 46 nations have documented highly pathogenic H5N8, with one of the human cases being related to poultry workers during an outbreak in poultry[17].

An influenza A (H5N8) clade 2.3.4.4b strain was recovered from a poultry worker during an outbreak of highly pathogenic H5N8 in chickens at a poultry farm in the Astrakhan region on the Volga River in southern Russia in December 2020, according to a study of a similar nature. Nasopharyngeal swabs were collected from seven poultry workers that tested positive, and two were confirmed by RT-PCR and sequencing. The seven individuals, five of whom were female and two of whom were male, ranged in age from 29 to 60. The HA gene of all five viruses obtained from birds and one from humans shared a significant degree of genetic similarity with other clades. From 2016 to 2021, viruses with the 2.3.4.4b gene were found in wild and domestic birds in Russia. Human influenza A in some cases (H5) 2.3.4.4. A potential public health hazard is infections[17].

H5N8 clade 2.3.4.4b outbreaks were observed in Russia, the Middle East, Central Europe, and Ukraine in 2016. In the southern part of Ukraine, close to areas where migrating waterfowl congregate in large numbers, especially mute swans (*Cygnus olor*), an outbreak of HPAI strains was documented in domestic backyard poultry between 2016 and 2017. Upon sequence analysis, it was found that 2 novel H5N8 HPAI strains were isolated from domestic backyard chickens (*Gallus gallus*) and mallard duck (*Anas platyrhynchos*). HPAI outbreaks in Ukraine underscore the ongoing need for AIV bio-monitoring, genomic sequencing, and mapping of wild bird flyways and their contacts with domestic poultry in Eurasia[40].

Long-distance migratory birds can play a significant role in the global spread of avian influenza viruses, notably through nesting regions in the sub-arctic. The investigation of H5N8 viral sequences, epidemiological studies, waterfowl migration, and chicken trade all revealed that wild birds can spread the virus to poultry via contact with infected water or surfaces. Furthermore, the chicken trade may contribute to the virus's spread. Clade 2.3.4.4 viral hemagglutinin was discovered to be extraordinarily promiscuous, producing reassortants with diverse subtypes and potentially boosting its ability to infect different species of birds and mammals. This promiscuity is likely to have a role in its ability to quickly adapt to various hosts and settings, potentially enhancing its pandemic potential[41].

H5N8 evolution

Whole genome

Gammaviruses are characterized as low pathogenic (LP) viruses or highly pathogenic (HP) viruses based on virulence in chickens. HP viruses may emerge from LP viruses through genetic mutations in wild birds[42]. In this context, AIV subtypes H5 and H7 are characterized as HP viruses. To date, AI viruses have 16 subtypes on the basis of the hemagglutinin gene and 9 due to the neuraminidase gene[1, 42].

The entire genome of HPAI H5N8, is made up of eight single stranded RNA segments. Each segment encodes a distinct gene that is essential for the virus's replication and infection. Polymerase Basic Protein 2 (PB2), which is roughly 2 341 nucleotides long and encodes the PB2 protein, is one of these segments. The polymerase basic protein 1 (PB1) gene is approximately 2 341 nucleotides long and codes for the PB1. The polymerase acidic protein (PA) gene encodes the PA protein and is approximately 2 234 nucleotides long. The hemagglutinin (HA) gene encodes the HA protein and is approximately 1 778 nucleotides long. The nucleoprotein (NP) gene has a length of about 1 565 nucleotides and codes for the NP protein. The neuraminidase (NA) gene encodes the NA protein and is approximately 1 413 nucleotides long. The matrix (M) gene encodes the Ml and M2 proteins and is approximately 1 027 nucleotides long. The non-structural protein

(NS) gene has around 890 nucleotides and encodes the NS1 and NS2 proteins. It is crucial to note that the lengths provided are approximations and may differ slightly across various H5N8 strains or isolates[43].

Hemagglutinin gene (HA)

HA gene sequence analysis was performed, and a phylogenetic tree was constructed by comparing sequences retrieved from the GISAID platform. These HPAI H5 strains belong to different groups and lineages. Sequence analysis was performed by following H5 numbering, which uncovered the genetic diversity during evolution. The cleavage site motif of HPAI H5 includes the polybasic amino acids QGERRRKKR*GLF[44, 45], whereas in the selected isolates reported globally during different years, maximum HPAI H5N8 evolved, and the cleavage site became LREKRRKKR*GLF. Studies have demonstrated that, although HPAI H5N8 attaches to avian-like receptors, it may also attach to human virus-like receptors in the human respiratory tract. HPAI showed more affinity for cats than dogs, which were more susceptible to HPAI. It is suggested that, due to its establishment in ducts, the transmission of HPAI H5N8 viruses may modify the genetic evolution of preexisting avian poultry strains[46].

On the basis of similarity, H5N8 viruses evolved into three groups[21]. Groups I and II contain the isolates belonging to clade 2.3.4.4b and the Eurasian continent, whereas group III contains isolates from the North American lineage, with apparent divergence from those in groups I and II. Moreover, the transmission pattern of this subtype was observed in depth by reviewing the continent wide distribution in Africa (A), Asia (B and C), Europe (D), North America and Oceania (E). In this regard, HA gene sequences of selected HPAI H5N8 viruses were retrieved from the GISAID database. Initially, Bayesian evolutionary analysis was performed using BEAST version 1.10.4, and then FigTree software (v1.4.4) was used for phylogenetic tree construction, as shown (Figure 1-4-1). Moreover, no isolation has been reported from Antarctica or South America. These continent-wide sporadic infection, further clarify that the domestic birds are reassortant hosts for the emergence of novel virus subtypes and are thought to be the reservoir of AIV. The spread of these viruses could endanger the health of both humans and birds.

In addition, asparagine-linked glycosylation sites have been observed among HPAI H5 strains, revealing that some are common during evolution, whereas a number of substitutions and deletions are also seen. Siddique et al. in 2012 reported the same sites along with additional glycosylation sites at the globular head of the HA gene, which is responsible for the prediction of high efficiency of replication[45, 47]. Moreover, the conserved amino acids at positions 222 glutamine and glycine at position 224 of the HA gene are responsible for avian-like receptors at the binding site that is common among all the HPAI H5 proteins selected for analysis, and similar reports are available in this context[45, 48-49].

Furthermore, a number of amino acid mutations have been observed at antigenic sites, including at amino acid position 39, where glutamic acid has been shown to have mutated into glycine, S141P, K169R, D171N, A172T, R178I/R, P197S, R205N/K, and N268Y. These sites have been designated as crucial residues of the antigenic site[50].

In NA, PB1, PB2, PA, NP PA, M, and NS, almost 29 molecular signatures are present that are associated with replication, virulence, transmission, and adaptation in mammals[41, 51-60]. In this regard, a maximum of 20 molecular signatures were present in HPAI H5N1/483, whereas 4-6 were present in HPAI/LPAI H5N8 viruses. The PB2 gene contains the known marker 627 K for mammalian adaptation that has only been shown to be present in 2 HPAI H5 human isolates, HPAI subtype H5N1/483 and H5N6/39715. There are a number of other mutations in the NA gene at the 96A amino acid position and the matrix 2 gene at the S31N site that are

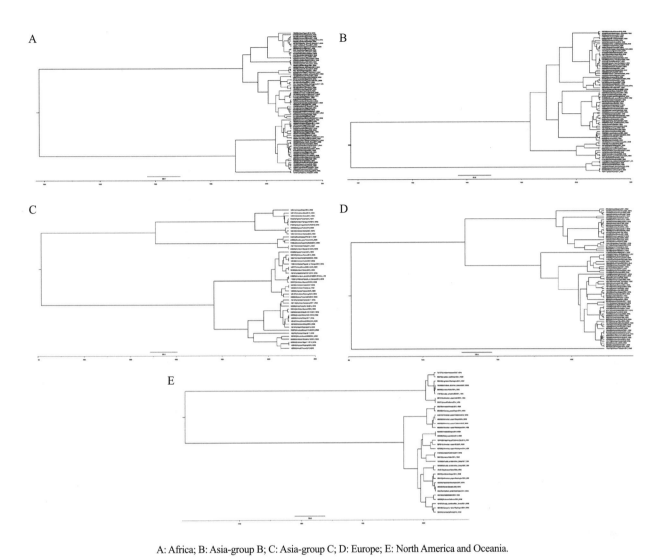

A: Africa; B: Asia-group B; C: Asia-group C; D: Europe; E: North America and Oceania.

Figure 1-4-1   Phylogenetic analysis of the hemagglutinin gene of AIV subtype H5N8 inferred with BEAST software

responsible for dual resistance against antivirals, including oseltamivir and amantadine[61, 62]. However, some other mutations, such as R118K in the NA gene, are associated with additional resistance to zanamivir[63, 64]. Due to these genetic changes, adamantanes and neuraminidase inhibitors may not be able to effectively prevent the replication of these viruses in the host in this situation.

The highly pathogenic AI H5 subtype has been spreading at an unprecedented rate since 2021, which is concerning given the disease's high mortality rate in wild birds and poultry as well as cases that have been observed in mammals and people. This could potentially lead to a future pandemic. Along with causing mass demise in a number of wild mammal species, H5 HPAI has the capacity to switch from infecting avian to mammalian hosts and develop the necessary characteristics for effective transmission from mammal to mammal. Therefore, enhanced surveillance of wild animals, large-scale animal farms, and humans handling them is urgently needed, along with improved biosecurity measures, reduction of poultry farm size and density, vaccination of poultry against HPAI, and avoidance of areas rich in water birds as a location for poultry farms. In addition, the medical sector and society need to prepare for the emergence of the human-to-human spread of H5 HPAI. It is crucial to include the community, communicate about risks, and counter intentional disinformation. The next pandemic, which could result from this AIV, should be prepared for using the lessons learned from the COVID-19 pandemic as a reference[65].

Risk assessment and mitigation strategies

During 2020-2021, in Eurasia, Europe, and Africa, emergent strains were highly pathogenic subtypes of H5N8 belonging to clade 2.3.4.4b and had a significant impact on the poultry industry. In the current scenario, an emergency has been declared for the enhancement of sero-and viro-surveillance across the globe depending on the previous outbreaks in 2005 and 2016[66, 67]. For risk mitigation strategies, an effective risk assessment needs to be performed in terms of tissue/host tropism, pathogenesis, and disease transmission and dissemination. Influenza A virus poses a continuous threat to poultry and the public due to its evolutionary mechanism through reassortment.

HPAI are extremely risky to poultry if not properly vaccinated. The low pathogenic H7N9 virus which emerged in 2013, was converted into high pathogenic due to mutations in early 2017, caused the death of millions of chickens to control the outbreak[68, 69]. The use of H7N9 vaccines effectively controlled the circulation of this virus both in poultry and humans[70]. Since 2004 vaccines are in use against H5 avian viruses in China[70]. Since the emergence of H7N9 in 2017, a bivalent inactivated vaccine against H5/H7 was developed to control both H5 and H7 in poultry in China[69, 71]. Currently, a trivalent vaccine-H5/H7 which contains Re-11, Re-12 and H7-Re3 vaccine seed viruses is in use. This trivalent vaccine was generated by reverse genetics, and HA genes were derived from A/duck/Guizhou/S4184/2017 (H5N6) (DK/GZ/S4184/17) (a clade 2.3.4.4h virus), A/chicken/Liaoning/SD007/2017 (H5N1) (CK/LN/SD007/17) (a clade 2.3.2.1d virus), and A/chicken/Inner Mongolia/SD010/2019 (H7N9) (CK/IM/SD010/2019), respectively[18, 70]. Although the newly emerged H5N8 viruses differ antigenically from currently used vaccines, poultry birds vaccinated in routine with current vaccines still completely protect against H5N8 virus challenge[18]. In another recent study[72], the efficacy of three vaccines was determined against the HPAI A/decoy duck/France/161105a/2016 (H5N8), clade 2.3.3.4b. The first vaccine (Vac1), was derived from HA gene clade 2.3.4.4b A (H5N8) HPAI, the second vaccine (Vac2) used was a commercial bivalent adjuvanted vaccine that contained an expressed HA modified from clade 2.3.2 A (H5N1) HPAI. The third vaccine (Vac3) also incorporated a homologous 2.3.4.4b H5 HA gene. Vac2 partly decreased the respiratory and intestinal excretion of challenge virus, Vac3 completely abolished cloacal shedding while Vac1 abolished oropharyngeal and cloacal shedding to almost undetectable levels. These results provided significant insights in the immunogenicity of recombinant H5 vaccines in mule ducks[72]. Since the H5N8 viruses have been detected in a wide range of wild birds across the globe, it could spread worldwide and can be very lethal to poultry. Therefore, homologous vaccination practices need to be introduced for the control and transmission of the disease, as the exact information on the disease and transmission is still not clear. The Iraqi-like strains are dispersed through poultry or indirect transmission in central Asia. In 2014-2017, there was little evidence of reassortment of HPAI H5N8 and H5N1 viruses in wild birds, as dispersal was unclear, but later, evidence of reassortment was found to be substantive, whereas in Europe in 2020, the emerging HPAI H5N8 strain was clearly a combination of sub-Saharan African viruses with a Eurasian LPAIV origin. Despite the implementation of biosecurity measures, several outbreaks of HPAI H5N8 strains were reported in France during 2016-2017, possibly due to airborne viral transmission. The area around the poultry facilities, almost $50\sim110$ m, is considered contaminated with varied viral concentrations[73]. In case of outbreaks, depopulation methods need to be wisely implemented to further control the airborne contamination of influenza viruses, which could result in instant mass culling.

Conclusions and future perspectives

This study backs up the hypothesis that asymptomatic migrating birds may have assisted viral development and reassortment as well as regional transmission of HPAI subtype H5N8. Another evidence that rapid and active mutation and reassortment of H5 subtypes may occur in these hosts comes from the HPAI subtypes H5N1 and H5N8 coinfecting and cocirculating in migratory ducks. Therefore, intersectoral alliance and coaction for mitigating avian influenza outbreaks based on the One Health approach that is worthwhile and advisable. This review discusses knowledge of the diseases nature, distribution, epidemiology applied surveillance techniques, diagnosis, and control approaches as they related to Sahelian Africa and its surrounding suburbs. Understanding of the influenza virus and its footprint on the well-being of humans and animals would aid in better preparing for the erratic/capricious challenges posed by this infectious disease.

Continuous vigilance, strengthening biosecurity, and intensifying surveillance in wild birds are needed to better manage the risk of HPAI occurrence in the future. Moreover, high-risk countries should vaccinate their poultry birds to prevent further outbreaks of HPAI H5N8. This review clearly indicates that HPAI H5N8 is a threat from a poultry standpoint and public perspective and that continuous surveillance and further epidemiological studies are needed.

References

[1] WEBSTER R G, BEAN W J, GORMAN O T, et al. Evolution and ecology of influenza A viruses. Microbiol Rev. 1992, 56(1): 152-179.

[2] KAWAOKA Y, CHAMBERS T M, SLADEN W L, et al. Is the gene pool of influenza viruses in shorebirds and gulls different from that in wild ducks?. Virology, 1988, 163(1): 247-250.

[3] WANG Y, WANG M, ZHANG H, et al. Emergence, evolution, and biological characteristics of H10N4 and H10N8 avian influenza viruses in migratory wild birds detected in eastern China in 2020. Microbiology spectrum, 2022, 10(2): e0080722.

[4] KRAMMER E, SMITH G J D, FOUCHIER R A M, et al. Influenza. Nat Rev Dis Primers, 2018, 4: 3.

[5] VERHAGEN J H, FOUCHIER R A M, LEWIS N. Highly pathogenic avian influenza viruses at the wild-domestic bird interface in Europe: Future Directions for Research and Surveillance. Viruses, 2021, 13(2): 212.

[6] HARFOOT R, WEBBY R J. H5 influenza, a global update. J Microbiol, 2017, 55(3): 196-203.

[7] ALEXANDER D J, BROWN I H. History of highly pathogenic avian influenza. Rev Sci Tech, 2009, 28: 19-38.

[8] COULOMBIER D, PAGET J, MEIJER A, et al. Highly pathogenic avian influenza reported to be spreading into western Russia. Euro Surveill. 2005, 10(8): e050818.1.

[9] CHEN H, SMITH G J, ZHANG S Y, et al. Avian flu: H5N1 virus outbreak in migratory waterfowl. Nature. 2005, 436(7048):191-192.

[10] ZHAO K, GU M, ZHONG L, et al. Characterization of three H5N5 and one H5N8 highly pathogenic avian influenza viruses in China. Vet Microbiol, 2013, 163(3-4): 351-357.

[11] JEONG J, KANG H M, LEE E K, et al. Highly pathogenic avian influenza virus (H5N8) in domestic poultry and its relationship with migratory birds in South Korea during 2014. Vet Microbiol, 2014, 173(3-4): 249-257.

[12] LEE Y J, KANG H M, LEE E K, et al. Novel reassortant influenza A (H5N8) viruses, South Korea, 2014. Emerg Infect Dis, 2014, 20(6): 1087-1089.

[13] DALBY A R, IQBAL M. The European and Japanese outbreaks of H5N8 derive from a single source population providing evidence for the dispersal along the long distance bird migratory flyways. Peer J, 2015, 3: e934.

[14] LEE D H, TORCHETTI M K, WINKER K, et al. Intercontinental spread of Asian-Origin H5N8 to North America through beringia by migratory birds. J Virol, 2015, 89(12): 6521-6524.

[15] LI M, LIU H, BI Y, et al. Highly pathogenic avian Influenza A (H5N8) Virus in Wild Migratory Birds, Qinghai Lake, China.

Emerg Infect Dis, 2017, 23(4): 637-641.

[16] LEWIS N S, BANYARD A C, WHITTARD E, et al. Emergence and spread of novel H5N8, H5N5 and H5N1 clade 2.3.4.4 highly pathogenic avian influenza in 2020. Emerg Microbes Infect, 2021, 10(1): 148-151.

[17] PYANKOVA O G, SUSLOPAROV I M, MOISEEVA A A, et al. Isolation of clade 2.3.4.4b A(H5N8), a highly pathogenic avian influenza virus, from a worker during an outbreak on a poultry farm, Russia, December 2020. Euro Surveill, 2021, 26(24): 2100439.

[18] CUI P, ZENG X, LI X, et al. Genetic and biological characteristics of the globally circulating H5N8 avian influenza viruses and the protective efficacy offered by the poultry vaccine currently used in China. Sci China Life Sci, 2022, 65(4): 795-808.

[19] SHI W, GAO G F. Emerging H5N8 avian influenza viruses. Science, 2021, 372(6544): 784-786.

[20] KANEHIRA K, UCHIDA Y, TAKEMAE N, et al. Characterization of an H5N8 influenza A virus isolated from chickens during an outbreak of severe avian influenza in Japan in April 2014. Arch Virol, 2015, 160(7): 1629-1643.

[21] LI J, GU M, LIU D, et al. Phylogenetic and biological characterization of three K1203 (H5N8)-like avian influenza A virus reassortants in China in 2014. Arch Virol, 2016, 161(2): 289-302.

[22] OZAWA M, MATSUU A, TOKOROZAKI K, et al. Genetic diversity of highly pathogenic H5N8 avian influenza viruses at a single overwintering site of migratory birds in Japan, 2014/15. Euro Surveill, 2015, 20(20): 21132.

[23] ISODA N, TWABELA A T, BAZARRAGCHAA E, et al. Re-Invasion of H5N8 high pathogenicity avian influenza virus clade 2.3.4.4b in Hokkaido, Japan, 2020. Viruses, 2020, 12(12): 1439.

[24] GHAFOURI S A, GHALYANCHILANGEROUDI A, MAGHSOUDLOO H, et al. Clade 2.3.4.4 avian influenza A (H5N8) outbreak in commercial poultry, Iran, 2016: the first report and update data. Trop Anim Health Prod, 2017, 49(5):1089-1093.

[25] MOTAHHAR M, KEYVANFAR H, SHOUSHTARI A, et al. The arrival of highly pathogenic avian influenza viruses H5N8 in Iran through two windows, 2016. Virus Genes, 2022, 58(6): 527-539.

[26] MARCHENKO V Y, SUSLOPAROV I M, KOMISSAROV A B, et al. Reintroduction of highly pathogenic avian influenza A/H5N8 virus of clade 2.3.4.4. in Russia. Arch Virol, 2017, 162(5): 1381-1385.

[27] KANDEIL A, KAYED A, MOATASIM Y, et al. Genetic characterization of highly pathogenic avian influenza A H5N8 viruses isolated from wild birds in Egypt. J Gen Virol, 2017, 98(7): 1573-1586.

[28] YEHIA N, NAGUIB M M, LI R, et al. Multiple introductions of reassorted highly pathogenic avian influenza viruses (H5N8) clade 2.3.4.4b causing outbreaks in wild birds and poultry in Egypt. Infect Genet Evol, 2018, 58: 56-65.

[29] YEHIA N, HASSAN W M M, SEDEEK A, et al. Genetic variability of avian influenza virus subtype H5N8 in Egypt in 2017 and 2018. Arch Virol, 2020, 165(6): 1357-1366.

[30] TWABELA A T, TSHILENGE G M, SAKODA Y, et al. Highly pathogenic avian influenza A(H5N8) virus, Democratic Republic of the Congo, 2017. Emerg Infect Dis, 2018, 24(7): 1371-1374.

[31] ALI M, YAQUB T, SHAHID M F, et al. Genetic characterization of highly pathogenic avian influenza A(H5N8) virus in pakistani live bird markets reveals rapid diversification of Clade 2.3.4.4b viruses. Viruses, 2021, 13(8): 1633.

[32] LEYSON C M, YOUK S, FERREIRA H L, et al. Multiple gene segments are associated with enhanced virulence of clade 2.3.4.4 H5N8 highly pathogenic avian influenza virus in Mallards. J Virol, 2021, 95(18): e0095521.

[33] YE H, ZHANG J, SANG Y, et al. Divergent reassortment and transmission dynamics of highly pathogenic avian influenza A(H5N8) virus in birds of China during 2021. Front Microbiol, 2022, 13: 913551.

[34] GUO J, YU H, WANG C, et al. Re-emergence of highly pathogenic avian influenza A(H5N8) virus in domestic goose, China. J Infect, 2021, 83(6): 709-737.

[35] LI X, LV X, LI Y, et al. Emergence, prevalence, and evolution of H5N8 avian influenza viruses in central China, 2020. Emerg Microbes Infect, 2022, 11(1): 73-82.

[36] NAGY A, ČERNÍKOVÁ L, KUNTEOVÁ K, et al. A universal RT-qPCR assay for "One Health" detection of influenza A viruses. PLOS ONE. 2021, 16(1): e0244669.

[37] IP H S, TORCHETTI M K, CRESPO R, et al. Novel Eurasian highly pathogenic avian influenza A H5 viruses in wild birds, Washington, USA, 2014. Emerg Infect Dis, 2015, 21(5): 886-890.

[38] SHIN D L, SIEBERT U, LAKEMEYER J, et al. Highly pathogenic avian influenza A(H5N8) virus in Gray Seals, Baltic Sea. Emerg Infect Dis, 2019, 25(12): 2295-2298.

[39] ŚMIETANKA K, ŚWIĘTOŃ E, KOZAK E, et al. Highly pathogenic avian influenza H5N8 in Poland in 2019-2020. J Vet Res, 2020, 64(4): 469-476.

[40] SAPACHOVA M, KOVALENKO G, SUSHKO M, et al. Phylogenetic analysis of H5N8 highly pathogenic avian influenza Viruses in Ukraine, 2016-2017. Vector Borne Zoonotic Dis, 2021, 21(12): 979-988.

[41] LYCETT S J, WARD M J, LEWIS F I, et al. Detection of mammalian virulence determinants in highly pathogenic avian influenza H5N1 viruses: multivariate analysis of published data. J Virol, 2009, 83(19): 9901-9910.

[42] FOUCHIER R A, MUNSTER V, WALLENSTEN A, et al. Characterization of a novel influenza A virus hemagglutinin subtype (H16) obtained from black-headed gulls. J Virol, 2005, 79(5): 2814-2822.

[43] BOUVIER N M, PALESE P. The biology of influenza viruses. Vaccine, 2008, 26 (Suppl 4): D49-D53.

[44] PERDUE M L, GARCÍA M, SENNE D, et al. Virulence-associated sequence duplication at the hemagglutinin cleavage site of avian influenza viruses. Virus Res, 1997, 49(2): 173-186.

[45] SIDDIQUE N, NAEEM K, ABBAS M A, et al. Sequence and phylogenetic analysis of highly pathogenic avian influenza H5N1 viruses isolated during 2006-2008 outbreaks in Pakistan reveals genetic diversity. Virol J, 2012, 9: 300.

[46] KIM Y I, PASCUA P N, KWON H I, et al. Pathobiological features of a novel, highly pathogenic avian influenza A(H5N8) virus. Emerg Microbes Infect, 2014, 3(10): e75.

[47] BENDER C, HALL H, HUANG J, et al. Characterization of the surface proteins of influenza A (H5N1) viruses isolated from humans in 1997-1998. Virology, 1999, 254(1):115-123.

[48] MATROSOVICH M, ZHOU N, KAWAOKA Y, et al. The surface glycoproteins of H5 influenza viruses isolated from humans, chickens, and wild aquatic birds have distinguishable properties. J Virol, 1999, 3(2): 1146-1155.

[49] SMITH G J, NAIPOSPOS T S, NGUYEN T D, et al. Evolution and adaptation of H5N1 influenza virus in avian and human hosts in Indonesia and Vietnam. Virology, 2006, 350(2): 258-268.

[50] KAVERIN N V, RUDNEVA I A, ILYUSHINA N A, et al. Structural differences among hemagglutinins of influenza A virus subtypes are reflected in their antigenic architecture: analysis of H9 escape mutants. J Virol, 2004, 78(1): 240-249.

[51] HIROMOTO Y, YAMAZAKI Y, FUKUSHIMA T, et al. Evolutionary characterization of the six internal genes of H5N1 human influenza A virus. J Gen Virol, 2000, 81(Pt 5): 1293-1303.

[52] SHAW M, COOPER L, XU X, et al. Molecular changes associated with the transmission of avian influenza a H5N1 and H9N2 viruses to humans. J Med Virol, 2002, 66(1): 107-114.

[53] CHEN H, BRIGHT R A, SUBBARAO K, et al. Polygenic virulence factors involved in pathogenesis of 1997 Hong Kong H5N1 influenza viruses in mice. Virus Res, 2007, 128(1-2): 159-163.

[54] GABRIEL G, HERWIG A, KLENK H D. Interaction of polymerase subunit PB2 and NP with importin alpha 1 is a determinant of host range of influenza A virus. PLOS Pathog, 2008, 4(2): e11.

[55] LONG J X, PENG D X, LIU Y L, et al. Virulence of H5N1 avian influenza virus enhanced by a 15-nucleotide deletion in the viral nonstructural gene. Virus Genes, 2008, 36(3): 471-478.

[56] SPESOCK A, MALUR M, HOSSAIN M J, et al. The virulence of 1997 H5N1 influenza viruses in the mouse model is increased by correcting a defect in their NS1 proteins. J Virol, 2011, 85(14): 7048-7058.

[57] HUI K P Y, CHAN L L Y, KUOK D I T, et al. Tropism and innate host responses of influenza A/H5N6 virus: an analysis of *ex vivo* and *in vitro* cultures of the human respiratory tract. Eur Respir J, 2017, 49(3): 1601710.

[58] KAMAL R P, ALYMOVA I V, YORK I A. Evolution and virulence of influenza A virus protein PB1-F2. Int J Mol Sci, 2017, 19(1): 96.

[59] YU Y, ZHANG Z, LI H, et al. Biological characterizations of H5Nx Avian influenza viruses embodying different neuraminidases. Front Microbiol, 2017, 8: 1084.

[60] PULIT-PENALOZA J A, BROCK N, PAPPAS C, et al. Characterization of highly pathogenic avian influenza H5Nx viruses in the ferret model. Sci Rep, 2020, 10(1): 12700.

[61] CHEUNG C L, RAYNER J M, SMITH G J, et al. Distribution of amantadine-resistant H5N1 avian influenza variants in Asia. J Infect Dis, 2006, 193(12): 1626-1629.

[62] ILYUSHINA N A, SEILER J P, REHG J E, et al. Effect of neuraminidase inhibitor-resistant mutations on pathogenicity of clade 2.2 A/Turkey/15/06 (H5N1) influenza virus in ferrets. PLOS Pathog, 2010, 6(5): e1000933.

[63] INTHARATHEP P, LAOHPONGSPAISAN C, RUNGROTMONGKOL T, et al. How amantadine and rimantadine inhibit proton transport in the M2 protein channel. J Mol Graph Model, 2008, 27(3): 342-348.

[64] OROZOVIC G, OROZOVIC K, LENNERSTRAND J, et al. Detection of resistance mutations to antivirals oseltamivir and zanamivir in avian influenza A viruses isolated from wild birds. PLOS ONE, 2011, 6(1): e16028.

[65] KUIKEN T, FOUCHIER R A M, KOOPMANS M P G. Being ready for the next influenza pandemic?. Lancet Infect Dis, 2023, 23(4): 398-399.

[66] ADLHODH C, FUSARO A, KUIKEN T, et al. Avian influenza overview February-May 2020. EFSAJ, 2020, 18: e06194.

[67] ALARCON P, BROUWER A, VENKATESH D, et al. Comparison of 2016-17 and previous epizootics of highly pathogenic avian influenza H5 Guangdong lineage in Europe. Emerging Infectious Diseases; 2018, 24(12): 2270-2283.

[68] SHI J, DENG G, KONG H, et al. H7N9 virulent mutants detected in chickens in China pose an increased threat to humans. Cell Res. 2017, 27(12): 1409-1421.

[69] ZENG X, TIAN G, SHI J, et al. Vaccination of poultry successfully eliminated human infection with H7N9 virus in China. Sci China Life Sci, 2018, 61(12): 1465-1473.

[70] ZENG, X, CHEN, X, MA S, et al. Protective efficacy of an H5/H7 trivalent inactivated vaccine produced from Re-11, Re-12, and H7-Re2 strains against challenge with different H5 and H7 viruses in chickens. J Integr Agric, 2020, 19(9): 2294-2300.

[71] SHI J, DENG G, MA S, et al. Rapid evolution of H7N9 highly pathogenic viruses that emerged in China in 2017. Cell Host Microbe, 2018, 24(4): 558-568.

[72] NIQUEUX É, FLODROPS M, ALLÉE C, et al. Evaluation of three hemagglutinin-based vaccines for the experimental control of a panzootic clade 2.3.4.4b A(H5N8) high pathogenicity avian influenza virus in mule ducks. Vaccine. 2023, 41(1): 145-158.

[73] SCOIZEC A, NIQUEUX E, THOMAS R, et al. Airborne detection of H5N8 highly pathogenic avian influenza virus genome in poultry farms, France. Front Vet Sci, 2018, 5: 15.

Comparative mutational analysis of the Zika virus genome from different geographical locations and its effect on the efficacy of Zika virus-specific neutralizing antibodies

Aziz Abdul, Suleman Muhammad, Shah Abdullah, Ullah Ata, Rashid Farooq, Khan Sikandar, Iqbal Arshad, Luo Sisi, Xie Liji, and Xie Zhixun

Abstract

The Zika virus (ZIKV), which originated in Africa, has become a significant global health threat. It is an RNA virus that continues to mutate and accumulate multiple mutations in its genome. These genetic changes can impact the virus's ability to infect, cause disease, spread, evade the immune system, and drug resistance. In this study genome-wide analysis of 175 ZIKV isolates deposited at the National Center for Biotechnology Information (NCBI), was carried out. The comprehensive mutational analysis of these isolates was carried out by DNASTAR and Clustal W software, which revealed 257 different substitutions at the proteome level in different proteins when compared to the reference sequence (KX369547.1). The substitutions were capsid (17/257), preM (17/257), envelope (44/257), NS1 (34/257), NS2A (30/257), NS2B (11/257), NS3 (37/257), NS4A (6/257), 2K (1/257), NS4B (15/257), and NS5 (56/257). Based on the coexisting mutational analysis, the MN025403.1 isolate from Guinea was identified as having 111 substitutions in proteins and 6 deletions. The effect of coexisting/reoccurring mutations on the structural stability of each protein was also determined by I-mutant and MUpro online servers. Furthermore, molecular docking and simulation results showed that the coexisting mutations (I317V and E393D) in Domain III (DIII) of the envelope protein enhanced the bonding network with ZIKV-specific neutralizing antibodies. This study, therefore, highlighted the rapid accumulation of different substitutions in various ZIKV proteins circulating in different geographical regions of the world. Surveillance of such mutations in the respective proteins will be helpful in the development of effective ZIKV vaccines and neutralizing antibody engineering.

Keywords

mutational analysis, neutralizing antibodies, Guillain-Barré syndrome, MD simulations, Zika virus

Introduction

Zika virus (ZIKV), a mosquito-borne flavivirus, was first isolated in 1947 from a serum sample of rhesus monkeys during an investigation of yellow fever virus (YFV) in the Zika forest of Uganda[1]. Other human pathogenic flaviviruses include dengue virus (DENV), West Nile virus (WNV), and Japanese encephalitis viruses (JEV) [2]. Zika infections usually cause asymptomatic and mild febrile infections associated with vomiting, fever, rash, sweating, pain in the muscles and the back of the eyes, and conjunctivitis, but severe neurological phenotypes such as Guillain-Barré syndrome and congenital Zika syndrome have also been reported[3, 4]. In infected pregnant women and mice, ZIKV can be transmitted to the fetus, leading to severe fetal abnormalities such as microcephaly, spontaneous abortion, and intrauterine growth[5-7]. Sexual and blood-born transmission of ZIKV has also been reported, making the virus a major threat to public health worldwide[8, 9].

The genome of ZIKV consists of positive-sense, single-stranded RNA with 10 794 bases, forming along open reading frame (ORF) flanked by a 5' UTR and 3' UTR. The RNA genome is 5' capped but lacks a 3' poly-A tail. The viral RNA is translated into a large precursor of 3 423 amino acid polyprotein that is co and post-translationally cleaved to yield three structural proteins (the capsid (C), precursor membrane (preM), and envelope protein (E)) and eight non-structural (NS) proteins (NS1, NS2A, NS2B, NS3, NS4A, 2K, NS4B, and NS5) [10, 11]. The 5' terminus has a methylated nucleotide cap, and the 3' terminus has a loop structure that plays a role in cellular translation. Structural proteins are involved in viral particle assembly, while non-structural proteins are responsible for viral replication, assembly, and evasion from the host immune system.

When a virus adapts to a new host, it exploits the host's cellular machinery for its successful entry, replication, and evasion of the host's immune responses[12]. To achieve these goals, the virus must undergo continuous mutations in its genome to modify antigenic epitopes on its proteins. RNA viruses, including ZIKV, have characteristic features of high mutability in their genomes, which in turn increases the virulence and transmission of the viruses[13]. Mutations in the structural protein of ZIKV have been reported to be involved in various biological pathways. For instance, a mutation in the preM protein (S139N) has been shown to increase neurovirulence in neonatal mice[14]. Similarly, three mutations in the E protein of ZIKV, D683E, V763M, and T777M, affect the stability of the E protein, binding the virus particle to its receptor, and are associated with a higher incidence of congenital Zika syndrome[15, 16]. Interestingly, DIII (AA299-403) of the E protein of ZIKV is reported to be targeted by many neutralizing antibodies[17]. Likewise, two mutations (A982V and T233A) have been reported in the NS1 protein of ZIKV. The substitution A982V in NSI not only increased the secretion of NS1 in the circulatory system and enhanced transmission of ZIKV from mice to mosquitos but also reduced the phosphorylation of TBK1 and led to the inhibition of interferon-beta (β-IFN) and facilitated viral replication[18, 19]. Similarly, the substitution T233A in NS1 isolated from the human microcephalic fetus has been reported to destabilize the conformation of NSI and may affect viral replication and pathogenesis[16]. Similarly, a single mutation (A2283S) in NS4B and three mutations (A/T3046I, G/R3107 L, and R/S3167N) in the RNA-dependent RNA polymerase (NS5) may inhibit the interferon regulatory pathway and increase viral replication[15]. As an RNA virus, ZIKV might have a rapid rate of mutation, and the accumulation of coexisting mutations in multiple or single genes might alter its virulence, infectivity, or transmissibility and may also lead to immune evasion and drug resistance. Hence, establishing the coexisting mutation atlas of the ZIKV genome can provide valuable information for assessing the mechanisms linked to pathogenesis, immune modulation, and viral drug resistance.

Since the genomes of ZIKV isolates have been sequenced by many countries and submitted to the National Center for Biotechnology Information (NCBI) database, in this study, we attempted to investigate and identify insertions, deletions and substitutions at the proteome level in the genomes of ZIKV from different geographical regions of the world. We also elaborated the effects of these mutations on the efficacy of already-reported ZIKV-specific neutralizing antibodies using molecular docking and simulation approaches.

Materials and methods

Retrieval of the ZIKV genome sequences

The amino acid sequences of the structural and non-structural proteins of ZIKV deposited in the NCBI repository were downloaded. A total of 175 complete genome sequences of ZIKV have been deposited in the

NCBI by 35 countries and territories. Only the coding and complete gene sequences were used in the present study.

Multiple sequence alignment

Multiple sequence alignment of the three structural (capsid, pre-M and E) and eight non-structural proteins (NS1, NS2A, NS2B, NS3, NS4A, 2K, NS4B, and NS5) of 175 ZIKV isolates was conducted using Lasergene software package (DNASTAR, Madison, WI) (7.1) and Clustal W program of the MEGA software (7.1) concerning already reported prototype KX369547.1 of ZIKV.

Sequence-based structural stability prediction

The I-mutant and MUpro online sequence-based tools were used to analyze the structural stability change of the top reoccurring mutations in both structural and non-structural proteins of ZIKV by keeping all the parameters in default.

Wild structure retrieval and variant modeling

The wild-type structure of the E protein of ZIKV (5JHM) was retrieved from the protein data bank. For the construction of the mutant version of the E protein of ZIKV, Chimera 1.15 software was utilized.

Docking of wild-type and mutant E protein with neutralizing antibodies

The High Ambiguity Driven Protein-protein Docking (HDOCK) online server algorithm was used for interaction of both wild-type and a mutant version of the E protein of ZIKV and previously reported neutralizing antibodies ZV-64 (PDB, 5KVF) and ZV-67 (PDB, 5KVG) to investigate the bonding efficiencies[17].

Molecular dynamics simulation of the E protein and neutralizing antibody complexes

Molecular dynamic simulation (MD) of wild-type E protein with two complexes of neutralizing antibodies (E Wild-type ZV-64, E Wild-type ZV-67) and a mutant version of the E protein with the same neutralizing antibody (E-Mutant-type ZV-64, E-Mutant-type ZV-67) complexes was completed by using the AMBER20 simulation package[20, 21]. The default parameters used by Suleman et al. were employed to complete the 50 nano second (ns) simulation for each complex. Post-simulation analyses, such as Root Mean Square Deviation (RMSD) and Root Mean Square Fluctuation (RMSF), were calculated using CPPTRAJ and PTRAJ modules[22].

Results

Worldwide sequenced genomes of ZIKV deposited in NCBI

In the current study, we have attempted to show the geographic origins of these submitted ZIKV genomes, which were fully sequenced. Only coding regions of the ZIKV genome were examined in this study for mutational analysis at the proteome level. The genomic organization along with the polyprotein (structural and non-structural proteins) of ZIKV is shown in Figure 1-5-1 A. The results of the present study showed that the most abundantly complete genome sequence of the ZIKV originated from Brazil (24%), followed by Thailand (8%), while only one ZIKV genome has been sequenced by Argentina (Figure 1-5-1 B).

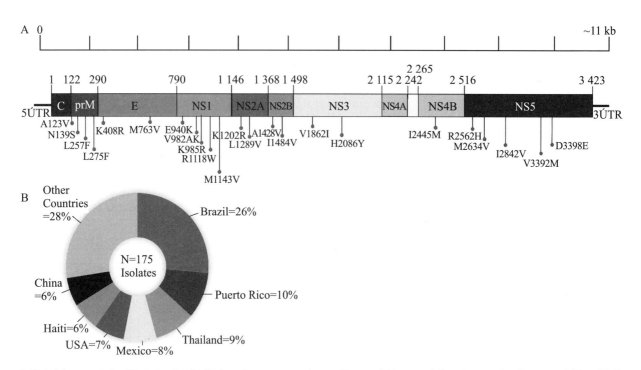

A: Pictorial representation illustrating the distribution of recurrent mutations and hypervariable genomic hot spot mutations in structural (C, preM, E) and non-structural proteins (NS1, NS2A, NS2B, NS3, NS4A, NS4B, and NS5) along with the ZIKV genome. The genomic hotspots are represented by vertical lines. B: Geographic distribution of the 175 complete genome sequences of ZIKV. The pie chart represents the percentage of genomes of the ZIKV genomes sequenced according to their geographic origins. The colors indicate different countries.

Figure 1-5-1　Pictorial representation and geographic distribution of ZIKV genome sequences

Identification of the top ten mutated isolates of ZIKV

In this study, the top ten mutated ZIKV isolates at the protein level were identified. We found that the most mutated isolate was KU963574.2 (Nigeria), harboring 111 substitutions along with six deletions (T446-G447-H448-E449-T450-D451), followed by MN025403.1 (Guinea), which had a total of 103 different substitutions. Similarly, seven isolates were identified from Senegal, where five isolates (AMR39832.1, AMR39833.1, AMR39836.1, KU955591.1, and KU955592.1) harbored 102 substitutions, while two isolates, MF510857.1 and KU955595.1, had 101 and 84 mutations, respectively. Likewise, one isolate from Malaysia ANK57896.1 carried 40 different substitutions in different proteins (Figure 1-5-2 A).

Identification and analysis of mutations in structural proteins of ZIKV isolates circulating in different geographical regions of the world

The three structural proteins of ZIKV, capsid C, preM, and envelope (E), are not only involved in the assembly of mature virions and successful ingress and egress of the host cells, but the latter two (preM and E) are being used as antibody-activating epitopes in ZIKV vaccine development[23]. To identify the mutational profile at the protein level (accumulated through natural selection) in structural proteins of ZIKV, multiple sequence alignments of 175 isolates of ZIKV were carried out from 35 countries and territories that had been deposited in the NCBI repository. We found mutations in all three structural proteins, i.e., Capsid (C), preM, and envelop (E). Collectively, 67 different mutations were observed in these three proteins when compared to the reference strain sequence KX369547.1.

Mutational analysis of the capsid (C) protein

The C protein is 122 amino acids (AAs) in length with a molecular weight of 11 kDa. Multiple copies of the capsid protein encapsulate the genomic RNA of ZIKV and form nucleocapsid core particles of mature ZIKV[24]. The complete CoDing Sequence (CDS) of the capsid protein appears in all 175 isolates of ZIKV in the NCBI database. The present study revealed that among 175 isolates of ZIKV, 67 isolates harbored 17 different amino acid mutations in the capsid protein, with a signature A106T (35%) mutation, which is indicative of widespread mutation in the C protein of ZIKV circulating worldwide in comparison to the reference isolate sequence KX369547.1 (Figure 1-5-2 B, Table 1-5-1). This mutation was predominant (83%) in Cambodian isolates of the ZIKV. The other less common mutations that coexist with A106T in the C protein of ZIKV were K6E, D7E, S8I, G9R, G10R, and F111I in the EU545988.1 (Micronesia) isolate. The second most mutated C protein of the ZIKV isolate was found at KU963574.2 (Nigeria), having seven coexisting mutations, i.e., S25N, F27L, K101R, T108A, V110I, G104S, and A120V when compared to the reference isolate sequence KX369547.1 (Figure 1-5-2 C). Likewise, the most reoccurring mutations observed in the C protein were A106T and D107E, while no insertions or deletions were identified in the C protein (Table 1-5-1). D107E in the capsid protein was previously reported mostly in Colombia, Panama, and Venezuela[25].

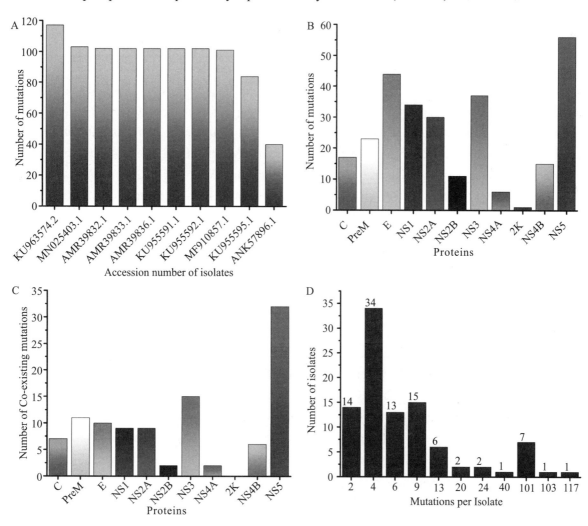

A: Top 10 most mutated ZIKV isolates along with accession number; B: Frequency distribution of substitution in each protein of ZIKV; C: Number of coexisting mutations in each protein of the ZIKV; D: Frequency distribution of ZIKV isolates harboring varying numbers of coexisting mutations.

Figure 1-5-2   Worldwide mutational analysis of the ZIKV genome per isolate

Table 1-5-1　Summary of mutations in each protein of the ZIKV

Protein	No. of amino acids	No. of available isolates	No. of mutated isolates	No. of mutations	Rate of mutation	Indel mutation	Top reoccurring mutations
Capsid	122	175	67	17	0.10		A106T, D107E
preM	168	175	27	23	0.13		A123V, L151M, K242R, L257F, L25F, L275F
Envelop	500	175	110	44	0.25	10 (del)	V313I, K498R, M763V, M777T
NS1	352	175	146	34	0.20		Y916H, E920L, V982A, K985R, M1143V R1118W, M1143V
NS2A	226	175	24	30	0.17		K1202R, L1274P, L1289V
NS2B	130	175	31	11	0.06		A1428V, T1477A, I1484V
NS3	617	175	108	37	0.21		H1857Y, V1862I, H1902N, M207L, H2086Y
NS4A	127	175	76	06	0.10		F2123L
2K	23	175	01	01	0.006		
NS4B	251	175	14	15	0.10		I2295M, I2376M, I2445M
NS5	903	175	136	56	0.32		R2562H, M2634V, I2842V, R3045C, S3162P, T3353A, V3392M, D3398E, M3403V

Mutational analysis of the preM protein

The structural preM protein of the ZIKV is 168 AAs in length with 8 kDa. Initially, after translation, this protein is in an immature form, but later, a cellular furin-like protease cleaves it into prepeptide and M protein. The M protein forms a complex with the envelope protein and protects the E protein[24]. The M protein has been used for vaccine development against ZIKV infection[26]. The present results showed that among 175 isolates of ZIKV, 27 isolates had 23 different substitutions in the preM protein (Figure 1-5-2 B). Among these mutations, the L257F substitution was the most reoccurring mutation in the preM protein. Similarly, other less reoccurring mutations identified in the preM protein were A123V, S130N, N139S, E143K, L151M, K242R, L257F, and L275F (Table 1-5-1). Likewise, 11 coexisting mutations (V125I, N139S, E143K, P148A, M153V, Y157H, I158V, R246K, L257F, A260V, and A262V) were exhibited by two isolates, MN025403.1 (Guinea) and KU963574.2 (Nigeria) (Figure 1-5-2 C). In addition, no deletion or insertion was observed in the preM protein of ZIKV with respect to the reference sequence KX369547.1 (Table 1-5-1).

Mutational analysis of the envelope (E) protein

Among the structural proteins of ZIKV, the E protein is the main and largest surface glycoprotein of virions, having 500 AAs with a molecular weight of 53 kDa[24]. The E protein of ZIKV is one of the major proteins involved in receptor binding and fusion of the virus with host cells. A compact ZIKV particle is composed of 180 copies of the E protein, and each E protein of ZIKV is composed of three domains, namely Domain Ⅰ (DⅠ), which contains the N-terminus of the E protein; Domain Ⅱ (DⅡ), which is involved in dimerization of the E protein; and DⅢ, which mediates attachment to target cells. The E protein is used for vaccine development, and many ZIKV neutralizing antibodies bind to Domain Ⅲ (303 aa～407 aa) of the E protein, thus suggesting that Domain Ⅲ is an important target for neutralizing antibodies[23]. This study reports that among the structural proteins of ZIKV, the E protein was observed to be the most susceptible to mutations, followed by preM and C proteins harboring a total of 44, 23, and 17 different mutations, respectively (Figure 1-5-2B and Table 1-5-1). The most reoccurring mutation in the E protein is M763V, followed by V313I, K408R, and M777T (Table 1-5-1). Furthermore, two segments comprising four AAs and six AAs, N444-D445-T446-G447 and T446-G447-H448-E449-T450-D451, were deleted in isolates KY553111.1 (Republic of Korea) and KU963574.2 (Nigeria), respectively; however, no insertion was observed. Similarly, 10 coexisting mutations (A410T, I459V, F575S, I607V, E683D, A727V, L728F, M763V, M777T, and L785M) were observed in isolate MN025403.1 (Guinea) (Figure 1-5-2).

Identification and analysis of mutations in non-structural proteins of ZIKV strains circulating in different geographical regions of the world

The ZIKV genome encodes eight non-structural proteins, including NS1, NS2A, NS2B, NS3, S4A, 2K, NS4B, and NS5. These non-structural proteins are involved in the viral replication complex inside the infected host cell. In the present study, a total of 190 substitutions were identified in the non-structural proteins of 175 isolates of ZIKV genomes deposited in the NCBI repository from 35 different countries throughout the world. The following are the details of mutations in non-structural proteins of ZIKV.

Non-structural protein 1 (NS1)

The NS1 protein of the ZIKV has 351 AAs with a molecular weight of approximately 48 kDa[24]. It is a secretory protein involved in genome replication as well as in host immune evasion by ZIKV through modulating the host immune mechanism[27, 28]. In the present study, 34 different AAs substitutions at different positions in NSI were identified in 146 isolates (Figure 1-5-2 B, Table 1-5-1). Among these substitutions, nine coexisting mutations (D846E, R863K, S886P, E940K, V956I, V988A, K1007R, I1030V, and M1058V) were observed in isolate MN025403.1 (Guinea) (Figure 1-5-2 C). In NS1, the most recurring mutation was E940L in 44 isolates followed by M1143V in 17 isolates. Other reoccurring mutations include Y916H, V982A, K985R, M1143V, R1118W, and M1143V (Table 1-5-1). Previously, it has been reported that substitution at V982A reduced secretion of NS1 to the host circulatory system, thus leading to reduced uptake by the mosquito and increased interferon inhibition by NS1[18]. Similarly, R1118 and M1143V were first reported in Brazilian isolates[29]. However, the present study identified these two mutations (R1118 and M1143V) in isolates from Haiti, Martinique, and Panama. Furthermore, no insertion or deletion was observed in the NS1 protein (Table 1-5-1).

Non-structural protein (NS2A)

The NS2A protein of the ZIKV has 226 AAs with a molecular weight of 22 kDa. NS2A induces endoplasmic reticulum (ER) membrane rearrangement, interacts with NS3 and NS5 of ZIKV and mediates replication and capsid assembly of ZIKV particles[24]. Among 175 isolates of ZIKV, 24 isolates had 30 different substitutions at different positions (Figure 1-5-2 B and Table 1-5-1), while nine coexisting mutations (I1180M, I1191V, A1204V, I1226V, D1270E, I1275V, V1289A, T1297A, and L1354M) (Figure 1-5-2 C) were identified in isolate MN025403.1 (Guinea). Similarly, two different reoccurring and coexisting mutations, K1202R and L1289V were identified in 15 Brazilian isolates of ZIKV. The most frequently recurring mutations in NS2A were K1202R and L1289V followed by L1274P (Table 1-5-1).

Non-structural protein (NS2B)

The NS2B protein of ZIKV is 130 AAs with a molecular weight of approximately 14 kDa that have been reported to interact with the NS3 protein and be involved in viral replication[24]. The present study revealed that among 175 isolates of ZIKV, only 31 isolates had 11 different substitutions of AAs in the NS2B protein (Figure 1-5-2 B and Table 1-5-1). Among these substitutions, A1428V was observed in 15 Brazilian isolates. Interestingly, the A1428V substitution in NS2B of these isolates coexisted with K1202R and L1298V in the NS2A protein of these isolates. The other reoccurring mutations in different isolates were S1417I, D1461E, and T1477A (Table 1-5-1). Only two coexisting mutations, D1461E and T1477A, were observed in isolate

MN025403.1 (Guinea) (Figure 1-5-2 C), while no insertion or deletion was observed in NS2B during this study (Table 1-5-1).

Non-structural protein (NS3)

The NS3 protein of the ZIKV is 617 AAs, having a molecular weight of approximately 70 kDa. which has been reported to play many important roles in viral RNA replication and methylation processes[24]. This study showed that among 175 isolates of ZIKV, 67 isolates had 37 different substitutions of AAs in the NS3 protein (Figure 1-5-2 B and Table 1-5-1). Among these substitutions, V1862I was observed in 25 isolates, followed by H2086Y in 16 isolates. This V1862I substitution in NS3 coexisted with A1428V of NS2A and K1202R and L1298V of NS2B in 15 Brazilian isolates. Similarly, 15 coexisting mutations (S1558A, H1594L, I1658V, R1671V, K1687R, T1717K, V1722A, T1753I, K1860R, V1862I, H1902N, V1909I, L1974M, R2085K, and H2086Y) were observed in isolate MN025403.1 (Guinea) (Figure 1-5-2 C). In addition, the five most reoccurring mutations were identified, while no deletion or insertion was observed in the NS3 protein during this study (Table 1-5-1).

Non-structural protein (NS4A)

The NS4A protein of the ZIKV is 127 AAs long, having a molecular weight of approximately 16 kDa, which has been reported to play a key role in the localization of the replication complex toward the membrane and the processing of polyproteins of the ZIKV inside the host cell[24]. The present study showed that among 175 isolates of ZIKV, 17 isolates had 6 different substitutions of amino acids in the NS4A protein (Figure 1-5-1 B). Among these substitutions, 2 coexisting mutations, F2123L, and F2127D, were observed in isolate MN025403.1 (Guinea), while no deletion or insertion mutation was observed in the NS4A protein. Similarly, only one mutation, V2259I, was observed in the 2K protein in isolate EU545988.1, while no insertion or deletion was observed in the NS2A and 2K proteins in the present study (Figure 1-5-2 C and Table 1-5-1).

Non-structural protein (NS4B)

The NS4B protein of ZIKV is a small hydrophobic membrane-associated protein of 251 aa with a size of 27 kDa. The NS4B protein along with the NS3 protein plays a significant role in ZIKV genome replication[24]. The present study showed that out of 175 isolates of ZIKV, only 14 isolates have 15 different substitutions of amino acids in the NS4B protein. Among these substitutions, 6 coexisting mutations (L2282, R2289K, A2293T, F2318L, I2453V, S2455L) were observed in isolate MN025403.1 (Guinea). Moreover, the most reoccurring mutations were I2295M, I2367M, and I2445M, while no deletion or insertion mutation was observed in the NS4B protein during the entire study (Figure 1-5-2 C and Table 1-5-1).

Non-structural protein (NS5)

NS5 is the largest protein among all non-structural proteins of ZIKV. It has 903 AAs 103 kDa in size. This has a dual role; it acts as a methyl and guanylyltransferase (MTase and GTase) as well as an RNA-dependent RNA polymerase; thus, it is an important target for drug development against ZIKV infection[24]. In the present study, out of 175 isolates, 136 had 56 different substitutions of AAs in NS5. Among these mutations, 32 coexisting mutations (Y2594H, I2598V, K2621R, A2679T, L2715M, Y2722H, T2749I, V2787A, N2800R, S2807N, I2842V, S2896N, H2909R, E2935V, Q2969H, V3039I, S3034N, I3046A, R3065K, K3080E, I3089V, K3107G, Q3154H, R3161K, S3162P, H3167R, N3172D, D3223S, S3304A, V3333M, T3353K, and N3387D) were observed in isolate MN025403.1 (Guinea) (Figure 1-5-2 C). Similarly, reoccurring mutations were abundant in NS5, while no deletion or insertion was examined during this study when compared to the reference sequence (Table 1-5-1).

Identification of highly mutated proteins of ZIKV

The ZIKV genome encodes three structural and seven non-structural proteins. Among the three structural proteins in all 175 isolates of ZIKV, the E protein was the most mutated protein, having a total of 44 different substitutions, followed by the pre-M protein, having 23 different substitutions. Similarly, among the non-structural proteins, NS5 is the most mutated protein, having 56 different substitutions, followed by the NS3 protein, which harbors 37 different substitutions at various positions. The 2K non-structural protein was found to be the least mutated protein, having only one substitution in only one isolate, EU545988.1 (Micronesia) (Figure 1-5-2 B and Table 1-5-1).

Identification of hypervariable genomic hotspots

Interestingly, among the reoccurring mutations in three structural proteins, pre-M protein has three A123V, N139S, L257F, E protein has two K408R, M763V while no hypervariable hotspot was observed in the C protein. Similarly, among the non-structural proteins, NSP1 (E940K, V982A, K985R, R1118W, M1143V) and NSP5 (R2562H, M2634V, I2842V, V3392M, D3398E) have five hotspots, NS2A (K1202R, L1289V), NS2B (A1428V, I1484V), and NS3 (V1862I, H2086Y) each have two hotspots, NS4B has one (I2445M), and NS4A and K have no hotspots (Figure 1-5-1 B).

Occurrence of coexisting mutations in ZIKV isolates

Analysis of mutational events per isolate of the ZIKV genome revealed a maximum of 111 coexisting substitutions and six (T446-G447-H448-E449-T450-D451) deletions in one isolate KU963574.2 from Nigeria followed by 103 coexisting mutations in the MN025403.1 (Guinea) isolate. Similarly, a maximum of 102 coexisting mutations were observed in five isolates from Senegal (AMR39832.1, AMR39833.1, AMR39836.1, KU955591.1, KU955592.1), while the maximum number of isolates had four mutations in their genome (Figure 1-5-2 D).

Effects of the most frequent reoccurring mutations on the stability of structural and non-structural proteins of the ZIKV isolates

The I-mutant and MUpro online servers revealed only one substitution, H1857Y, in the NS3 protein among non-structural proteins and showed increased stability, while the rest of the substitutions in both structural and non-structural proteins showed decreased stability of their respective proteins.

Mutation modeling and RMSD calculation of the envelope protein (E) of ZIKV

The envelope protein (E), of ZIKV is one of the major proteins involved in receptor binding and fusion of the virus with the host cell, and it also elicits type-specific neutralizing antibodies[30]. The E protein of ZIKV is composed of three domains: Domain I (D I), which contains the N-terminus of the E protein; Domain II (D II), which is involved in dimerization of the E protein; and D III, which mediates attachment to target cells[23]. These three domains in the E protein of ZIKV are arranged in such a way that D I is located in the center, while D II and D II flank the two sides of the D I and form a monomer (Figure 1-5-3 A). Furthermore, each monomer of the E protein interacts with the adjacent monomer of the E protein in an antiparallel manner and forms a dimer (Figure 1-5-3 B) [31]. D III of the E protein is reported to be the main target for neutralizing antibodies. To determine the mechanism of hydrogen bonding strength between the wild-type and mutant ZIKV DⅢ, which are responsible for binding with neutralizing antibodies, we used chimera software to model the shortlisted mutations (I317V, E393D) in the DⅢ of the wild-type structure (Figure 1-5-3 C). Afterward, we superimposed the wild-type and mutant DⅢ to visualize the effect of the generated mutations on the structural confirmation and recorded the RMSD values. The RMSD value for the superimposed structure was 0.8 Å, which showed the perturbation of wild-type structural elements, structural deviations, and changes in the protein conformation (Figure 1-5-3 D).

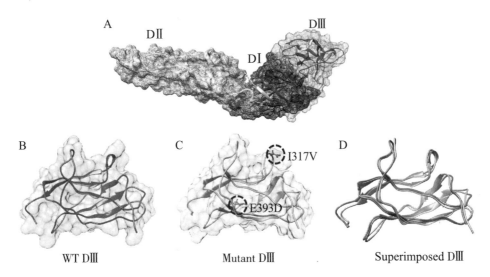

A: Domain mapping of ZIKV E protein; B: Wild-type DⅢ of the E protein; C: Mutant (I317V, E393D) DⅢ; D: Superimposition of wild-type DⅢ on mutant DⅢ to calculate the RMSD (color figure in appendix).

Figure 1-5-3　Domains architecture of ZIKV E protein and mutant modeling

Bonding network analysis using molecular docking of E protein with neutralizing antibodies

Previously, it has been reported that many ZIKV antibodies bind to Domain Ⅲ (303 aa～407 aa) of the E protein of ZIKV, thus suggesting that Domain Ⅲ is an important target for neutralizing antibodies[23]. Structural binding of the wild-type and mutant DⅢ with neutralizing antibodies ZV-64 and ZV-67 was performed to visualize the effect of mutation on the binding efficacy with the antibodies (Figure 1-5-4). Docking of the wild type and mutant by the HDOCK online server revealed key differences in the binding affinity and contact network. The interaction interface analysis of the wild-type DⅢ-ZV-64 complex predicted by HDOCK through PDBsum revealed that there are five hydrogen bonds formed between the DⅢ-ZV-64 complexes. The AAs residues that formed hydrogen bonds in the DⅢ-ZV-64 complex are His400-Tyr96 (double hydrogen bonds), His400-Tyr32, His398-Tyr36, and Asp347-Ser56 (Figure 1-5-4 A). However, the analysis of the interaction interface of the mutant DⅢ-ZV-64 complex revealed that 11 hydrogen bonds exist among the aforementioned complexes. The residues involved in hydrogen bond formation are His400-Tyr94, His398-Tyr36, Tyr386-Trp50, Asp384-Tyr32, Thr397-Glu55, Phe314-Pro44, Thr313-Lys45, and Thr313-Ser43 (Figure 1-5-4 B). Hence, the data revealed that mutations in DⅢ enhanced the binding of ZIKV E protein with the neutralizing antibody ZV-64. To further confirm the above data, we also checked the effect of the identified mutations in the DⅢ of the ZIKV E protein on binding with another neutralizing antibody called ZV-64. After docking the wild-type DⅢ with the ZV-67 antibody, the analysis of the interaction interface identified 3 hydrogen bonds between wild-type DⅢ and the ZV-67 antibody. The residues involved in hydrogen bond formation are Gln349-Trp103, Val346-Gln39, and Phe313-Tyr102 (Figure 1-5-4 C). However, the interaction interface of the mutant DⅢ-ZV-67 complex revealed one salt bridge and four hydrogen bonds. The hydrogen bonds formed between Thr397-Gln39, Val303-Ser168, Ser304-Ser168, and Ser304-Ser167 residues while a salt bridge formed between Lys394-Glu150 residues. These data confirmed that the mutations in the domain (DⅢ) enhanced the binding of ZIKV E protein with the neutralizing antibodies, suggesting that the mutations in DⅢ boosted the hydrogen bonding between the epitopes in DⅢ and neutralizing antibodies ZV-64 and ZV-67. Nevertheless, strong hydrogen bonding between an antigen and antibody is a prerequisite for neutralization of the antigen.

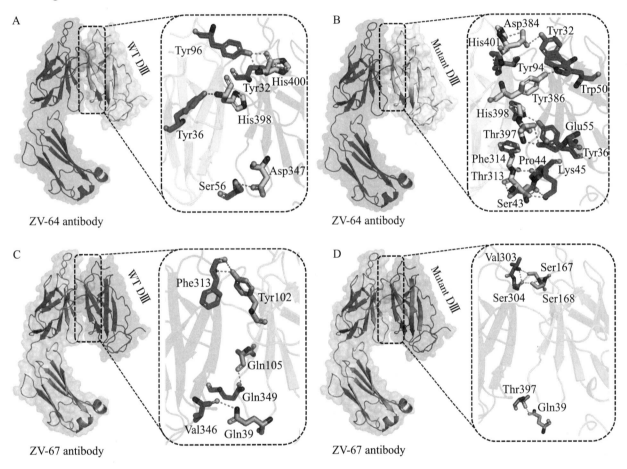

A: Docking of wild-type DⅢ with ZV-64 neutralizing antibody; B: Docking of mutant DⅢ with the ZV-64 neutralizing antibody; C: Docking of wild-type DⅢ with the ZV-67 neutralizing antibody (DⅢ); D: Docking of mutant DⅢ with ZV-67 neutralizing antibody (color figure in appendix).

Figure 1-5-4   Docking of wild-type and mutant DⅢ with ZV-64 and ZV67 neutralizing antibodies

Analysis of dynamics stability and flexibility

A 50 ns simulation was performed to check the dynamic behavior of both wild-type ZV and mutant ZV antibody complexes. Figures 1-5-5 A-D shows that the mutant complexes are more stable than the wild type. Figure 1-5-5 A shows that the wild-type ZV-64 complex gains stability at 9 ns and remains stable until 40 ns, but after this, no major perturbation was recorded in the RMSD value until 50 ns. On the other hand, Figure 1-5-5 B shows that the mutant-ZV-64 complex attained stability at 10 ns and remained stable until 50 ns with an average RMSD of 3 Å. Similarly, the wild-type ZV-67 complex gained stability at 2 ns and remained stable until 15 ns; however, a sudden rise in RMSD was depicted at 15 ns and then gained stability at 30 ns until

50 ns with an average RMSD of 6 Å. In the case of the mutant-ZV-67 complex, the system equilibrated at 3 ns with an average RMSD of 2.5 Å; however, after the system equilibration, no significant fluctuations were seen during the simulation time frame of 50 ns. A lower RMSD represents higher complex stability, whereas a higher RMSD shows lower stability. In summary, the simulation results support the conclusion that the mutant complexes are more stable than the wild type, as indicated by the lack of significant fluctuations in the RMSD values. The RMSD results also suggest that the mutations in the ZIKV enhances the binding affinity with the neutralizing antibodies. Overall, these results provide further evidence that the mutations in ZIKV improves its binding to the neutralizing antibodies, and suggest that the mutant ZIKV may be a more appropriate target for the development of therapeutics.

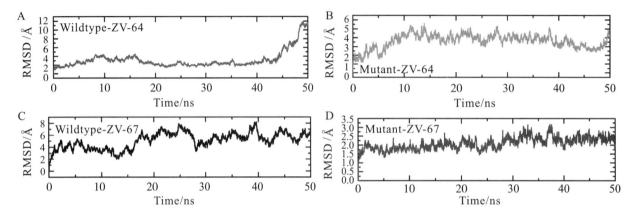

A: Wild-type DⅢ-ZV-64 complex; B: Mutant DⅢ-ZV-64 complex; C: Wild-type DⅢ-ZV-67 complex; D: Mutant DⅢ-ZV-67 complex.

Figure 1-5-5   Calculation of RMSD for all complexes

Afterward, the radius of gyration (Rg) was calculated to check the compactness of the aforementioned complexes (Figures 1-5-6 A-D). A higher Rg value indicates that the complex is unstable, while a lower Rg value represents the stability of the complexes. As shown in Figure 1-5-6, the average Rg values for wild-type ZV-64 and mutant ZV-64 were 27 and 25, respectively, while the average Rg values for wild-type ZV-67 and mutant ZV-67 were 23 and 22, respectively. These results also verified that the mutant complexes have lower residual fluctuations in the interacting residues. The values of Rg for the four complexes are shown in Figure 1-5-6.

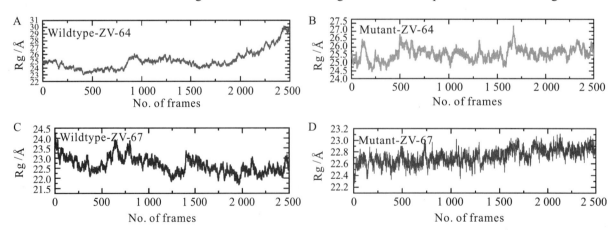

A: Rg value for the wild-type DⅢ-ZV-64 complex; B: Rg value for the mutant DⅢ-ZV-64 complex; C: Rg value for the wild-type DⅢ-ZV-67 complex; D: Rg value for the mutant DⅢ-ZV-67 complex.

Figure 1-5-6   Calculation of Rg for all complexes

Finally, the root-mean-square fluctuation (RMSF) was calculated to check the fluctuation at the residual level. The lower RMSF value shows high stability, while the higher value represents the instability of complexes. As shown in Figure 1-5-7, the average RMSF values of mutant complexes are lower than those of the wild-type complexes, which further confirmed the results of RMSD and Rg. The RMSF values for all complexes are shown in Figure 1-5-7.

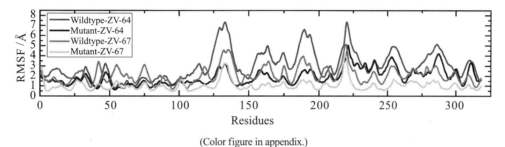

(Color figure in appendix.)

Figure 1-5-7　RMFS calculated for the residual flexibility of all four complexes

Discussion

ZIKV was first isolated from Uganda in 1947 and eventually spread to several countries and territories of the world. Similar to other RNA viruses, ZIKV has undergone several mutations and accumulated various mutations in structural and non-structural proteins. These mutations have enabled it to adapt to various epidemiological settings. In this study, a comprehensive genomewide mutational analysis of 175 ZIKV isolates, available in the NCBI database, deposited by 35 different countries and territories of the world was carried out. This study reports all the reoccurring and coexisting mutations in both structural and non-structural proteins of ZIKV isolates. In addition, the effect of mutations (specifically mutations in domain III of the E protein) on the efficacy of already reported neutralizing antibodies was also evaluated.

Among the structural proteins of ZIKV, the preM and envelope (E) proteins have been used for the development of vaccines in which the E protein is the main target of ZIKV neutralizing antibodies[17, 26]. This study highlighted that among the structural proteins, E protein is the most mutated, harboring 44 different mutations, followed by preM and capsid protein carrying 17 substitutions at different positions. The DIII of the E protein has been reported to be the main target of neutralizing antibodies against ZIKV[17]. Hence, mutations in the E protein are of great importance regarding their involvement in the host immune system and their neutralization. Not all mutations are in the favor of the viruses; hence, mutations sometimes harness the viruses to the antibodies and become easy targets for neutralizing antibodies. It was found that the coexisting mutations (I317 V and E393D) in the DIII domain of ZIKV E protein enhanced binding with the previously reported antibodies ZV64 and ZV67. Similarly, the mutations N139S/S139N play a vital role in the incidence of microcephaly. The 139N residue in the preM protein enhances the tropism of ZIKV for human progenitor cells and increases fetal microcephaly in mice[14, 32]. On the other hand, ZIKV strains circulating in Asian countries carry the 139S residue in the preM protein, which is occasionally reported to be involved in ZIKV-related microcephaly[33]. Based on previous reports[15] and the current study, it was observed that the majority of the isolates of American origin carry the 139N residue in their preM protein; hence, the ratio of ZIKV-related microcephaly is higher (2-12 per 10 000 live births). In contrast, all the isolates of European/Asian origin carry the 139S residue instead of the 139N residue in their preM protein; therefore, the incidence of ZIKV-related microcephaly in these regions is relatively lower (1 in 10 000 live births) [34].

Among the non-structural proteins, one of the biologically important mutations, 892 V/A, in NS1 has been reported to be involved in the secretions of ZIKV to the circulatory system, enhancing uptake of ZIKV by mosquitoes and participating in the phosphorylation of TBK1 and the transmission of the virus, thereby facilitating viral replication in humans[19]. It was found that 892V/A is one of the re-occurring mutations in this study. Another important mutation was T233A (polyprotein T1026A) in the NS1, which has already been reported to destabilize the NS1 dimer and affect viral replication and pathogenesis. One of the re-occurring mutations in this study was M2634V, which was previously identified as neutral having no effect on the function of NS5 mutation. Similarly, mutations such as I3046A/T, K3107G/R, and N3167R/S in NS5 have been reported to inhibit intracellular interferon pathways and increase viral replication[35]. The present study provided Insilco insights into the effect of different mutations on the structural stability and binding affinity of the ZIKV with the human-neutralizing antibodies. However, the limitation of our study is to provide the *in vitro* and *in vivo* experiments and human trials to further confirm the results of our study.

Conclusions

Being an RNA virus, the ZIKV continues to mutate rapidly and accumulate multiple mutations within its genome. In the current study, we reported rapid accumulations of various substitutions of amino acids in both structural and non-structural proteins of ZIKV circulating in different geographical regions of the world. Among the isolates of the ZIKV that were studied, one particular isolate (MN025403.1) from Guinea had the most mutations. This isolate had 111 substitutions and 6 deletions. The molecular docking and simulation approaches in this study verified that the mutated E protein enhanced the binding affinity of the ZIKV to the human-neutralizing antibodies. The study also highlighted the importance of monitoring mutations in the targeted genes of the ZIKV in order to improve the efficacy of vaccines and therapeutic applications. By understanding the impact of mutations on the structure and function of the virus, researchers can develop more effective antiviral drugs and vaccines.

References

[1] DICK G W, KITCHEN S F, HADDOW A J. Zika virus (I). Isolations and Serological Specificity. 1952, 46(5): 509-520.

[2] KUNO G, CHANG G J, TSUCHIYA K R, et al. Phylogeny of the genus Flavivirus. J Virol. 1998, 72(1): 73-83.

[3] CARTEAUX G, MAQUART M, BEDET A, et al. Zika virus associated with meningoencephalitis. N Engl J Med, 2016, 374(16): 1595-1596.

[4] RUBIN E J, GREENE M F, BADEN L R. Zika virus and microcephaly. Mass Med Soc, 2016, 374: 984-985.

[5] BRASIL P, PEREIRA J P JR, MOREIRA M E, et al. Zika virus infection in pregnant women in rio de Janeiro. N Engl J Med, 2016, 375(24): 2321-2334.

[6] CUGOLA F R., FERNANDES I R, RUSSO F B, et al. The brazilian Zika virus strain causes birth defects in experimental models. Nature, 2016, 534(7606): 267-271.

[7] MINER J J, CAO B, GOVERO J, et al. Zika virus infection during pregnancy in mice causes placental damage and fetal demise. Cell. 2016, 165(5): 1081-1091.

[8] MUSSO D, ROCHE C, ROBIN E, et al. Potential sexual transmission of Zika virus. Emerg Infect Dis, 2015, 21(2): 359-361.

[9] GREGORY C J, ODUYEBO T, BRAULT A C, et al. Modes of Transmission of Zika virus. J Infect Dis, 2017, 216(suppl_10): S875-S883.

[10] MUKHOPADHYAY S, KUHN R J, ROSSMANN M G. A structural perspective of the flavivirus life cycle. Nat Rev Microbiol, 2005, 3(1): 13-22.

[11] YE Q, LIU Z Y, HAN J F, et al. Genomic characterization and phylogenetic analysis of Zika virus circulating in the

Americas. Infect Genet Evol, 2016, 43: 43-49.

[12] RAMPERSAD S, PAULA T. Replication and expression strategies of Viruses. Viruses 2018: 55-82.

[13] RASHID F, XIE Z, SULEMAN M, et al. Roles and functions of SARS-CoV-2 proteins in host immune evasion. Front Immunol, 2022, 13: 940756.

[14] YUAN L, HUANG X Y, LIU Z Y, et al. A single mutation in the prM protein of Zika virus contributes to fetal microcephaly. Science, 2017, 358(6365): 933-936.

[15] PETTERSSON J H, ELDHOLM V, SELIGMAN S J, et al. How did Zika virus emerge in the Pacific Islands and Latin America?. MBio. 2016, 7(5): e01239-16.

[16] WEAVER S C. Emergence of Epidemic Zika virus transmission and congenital Zika syndrome: are recently evolved traits to blame?. MBio, 2017, 8(1): e02063-16.

[17] ZHAO H, FERNANDEZ E, DOWD K A, et al. Structural basis of Zika virus-specific antibody protection. Cell, 2016, 166(4): 1016-1027.

[18] LIU J, LIU Y, NIE K, et al. Flavivirus NS1 protein in infected host sera enhances viral acquisition by mosquitoes. Nat Microbiol, 2016, 1(9): 16087.

[19]XIA H, LUO H, SHAN C, MURUATO A E, et al. An evolutionary NS1 mutation enhances Zika virus evasion of host interferon induction. Nat Commun, 2018, 9(1): 414.

[20]CASE D A, CHEATHAM T E III, DARDEN T, et al. The Amber biomolecular simulation programs. J Comput Chem, 2005, 26(16): 1668-1688.

[21] ROE D R, CHEATHAM T E III. PTRAJ and CPPTRAJ: software for processing and analysis of molecular dynamics trajectory data. J Chem Theory Comput, 2013, 9(7): 3084-3095.

[22] SULEMAN M, UMME-I-HANI S, SALMAN M, et al. Sequence-structure functional implications and molecular simulation of high deleterious nonsynonymous substitutions in IDH1 revealed the mechanism of drug resistance in glioma. Front Pharmacol, 2022, 13: 927570.

[23] ROBBIANI D F, BOZZACCO L, KEEFFE J R, et al. recurrent potent human neutralizing antibodies to Zika virus in Brazil and Mexico. Cell, 2017, 169(4): 597-609.

[24] PANWAR U, AND SINGH S K. An overview on Zika virus and the importance of computational drug discovery. J Exp Res Pharmacol 2018, 3: 43-51.

[25] METSKY H C, MATRANGA C B, WOHL S, et al. Zika virus evolution and spread in the Americas. Nature, 2017, 15;546(7658): 411-415.

[26] REICHMUTH A M, OBERLI M A, JAKLENEC A, et al. mRNA vaccine delivery using lipid nanoparticles. Ther Deliv, 2016, 7(5): 319-334.

[27] MACKENZIE J M, JONES M K, YOUNG P R. Immunolocalization of the dengue virus nonstructural glycoprotein NS1 suggests a role in viral RNA replication. Virology, 1996, 220(1): 232-240.

[28] BROWN W C, AKEY D L, KONWERSKI J R, et al. Extended surface for membrane association in Zika virus NS1 structure. Nat Struct Mol Biol, 2016, 23(9): 865-867.

[29] FARIA N R, QUICK J, CLARO I M, et al. Establishment and cryptic transmission of Zika virus in Brazil and the Americas. Nature, 2017, 546(7658):406-410.

[30] WU Y, LI S, DU L, et al. Neutralization of Zika virus by germline-like human monoclonal antibodies targeting cryptic epitopes on envelope domain III. Emerg Microbes Infect, 2017, 6(10): e89.

[31] KOSTYUCHENKO V A, LIM E X, ZHANG S, et al. Structure of the thermally stable Zika virus. Nature, 2016, 533(7603): 425-428.

[32] SHAH A, REHMAT S, ASLAM I, et al. Comparative mutational analysis of SARS-CoV-2 isolates from Pakistan and structural-functional implications using computational modelling and simulation approaches. Comput Biol Med, 2022, 141: 105170.

[33] WONGSURAWAT T, ATHIPANYASILP N, JENJAROENPUN P, et al. Case of microcephaly after congenital infection with

asian lineage Zika virus, Thailand. Emerg Infect Dis, 2018, 24(9): 1758-1761.

[34] NEWS-MEDICAL-LIFE-SCIENCES. We're unable to locate the page you are Looking For. [2022-09-20]. https://www.news-medical.net/health/Microcephaly-Epidemiology.aspx.

[35] WANG L, VALDERRAMOS S G, WU A, et al. From mosquitos to humans: genetic evolution of Zika virus. Cell Host Microbe, 2016, 19(5): 561-565.

Roles and functions of SARS-CoV-2 proteins in host immune evasion

Rashid Farooq, Xie Zhixun, Suleman Muhammad, Shah Abdullah, Khan Suliman, and Luo Sisi

Abstract

Severe acute respiratory syndrome coronavirus 2 (SARS-CoV-2) evades the host immune system through a variety of regulatory mechanisms. The genome of SARS-CoV-2 encodes16 non-structuralproteins (NSPs), four structural proteins, and nine accessory proteins that play indispensable roles to suppress the production and signaling of type Ⅰ and Ⅲ interferons (IFNs). In this review, we discussed the functions and the underlying mechanisms of different proteins of SARS-CoV-2 that evade the host immune system by suppressing the IFN-β production and TANK-binding kinase 1 (TBK1) / interferon regulatory factor 3 (IRF3) / signal transducer and activator of transcription (STAT) 1 and STAT2 phosphorylation. We also described different viral proteins inhibiting the nuclear translocation of IRF3, nuclear factor-κB (NF-κB), and STATs. To date, the following proteins of SARS-CoV-2 including NSP1, NSP6, NSP8, NSP12, NSP13, NSP14, NSP15, open reading frame (ORF) 3a, ORF6, ORF8, ORF9b, ORF10, and Membrane (M) protein have been well studied. However, the detailed mechanisms of immune evasion by NSP5, ORF3b, ORF9c, and Nucleocapsid (N) proteins are not well elucidated. Additionally, we also elaborated the perspectives of SARS-CoV-2 proteins.

Keywords

SARS-CoV-2, immune evasion, structural proteins, non-structural proteins, accessory proteins

Introduction

Coronaviruses (CoVs) are a diverse family of enveloped positive-sense single-stranded RNA viruses[1-3], infecting humans, avian species, and livestock animals, posing a serious threat to public health and economy[4]. CoVs are classified under the order Nidovirales, family Coronaviridae, and subfamily Orthocoronavirinae. The subfamily Orthocoronavirinae is further divided into four genera, i.e., alphacoronavirus (α-CoV), betacoronavirus (β-CoV), gammacoronavirus (γ-CoV), and deltacoronavirus (δ-CoV). The α-and β-CoVs infect only mammals, while the γ-and δ-CoV have a broader host range including avian species[4]. Severe acute respiratory syndrome coronavirus 2 (SARS-CoV-2) has been placed in the subgenus Sarbecovirus under the genus β-CoVs[4, 5]. Human CoVs, such as HCoV-229E (α-CoV) and HCoV-OC43 (β-CoV), HCoV-NL63 (α-CoV) and HCoV-HKU1 (β-CoV), usually cause mild respiratory tract infections associated with symptoms of the "common cold". In contrast, SARS-CoV-2, Middle East respiratory syndrome coronavirus (MERS-CoV), and SARS-CoV have been recognized as highly pathogenic[4].

The genome of SARS-CoV-2 consists of about 30 000 bases[6, 7] (Figure 1-6-1). It contains a 5' cap structure and a 3' poly-A tail. It has a 5' open reading frame (ORF) and a 3' ORF that comprises 2/3 and 1/3 of the complete genome, respectively. After entering the host cell, RNA-dependent RNA polymerase (RdRp) replicates and transcribes the SARS-CoV-2 genome[8]. The 5' ORF (ORF1a/b) is translated into pp1a and pp1ab proteins in the rough endoplasmic reticulum (rER) of the host cell[9]. Proteases cleave these proteins and produce 16 NSPs, ranging from NSP1 to NSP16. The 3' ORF of SARS-CoV-2 has both structural and accessory proteins. There are four structural proteins, i.e., Spike (S), Envelop (E), Nucleocapsid (N), and Membrane (M) proteins (Figure 1-6-1). The structural proteins assemble and help in the budding of new virions at the ER to Golgi compartment that are suggested to exit the infected cells by exocytosis. S protein recognizes and binds to the receptor, angiotensin-converting enzyme 2 (ACE2) of the host cell, mediating the penetration of the virus into the host cell[7, 10]. N protein is multifunctional; its main function is to assemble genomic RNA of the virus into a ribonucleoprotein complex and regulate viral replication[9]. E protein regulates the replication, pathogenicity, and virus dissemination[11, 12], while M protein is responsible for the assembly of viral particles[13, 14]. Interspersed between these structural proteins are nine accessory proteins, i.e., ORF3a, ORF3b, ORF6, ORF7a, ORF7b, ORF8, ORF9b, ORF9c, and ORF10[4] (Figure 1-6-1). The accessory proteins show high variability among CoVs; however, they are conserved within respective viral species to some extent. The accessory proteins do not play roles in virus replication, but they do have important roles in host immune evasion[4].

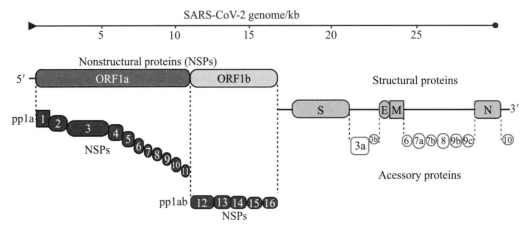

S: Spike protein; E: Envelop protein; M: Membrane protein; N: Nucleocapsid protein. The genome of SARS-CoV-2 consists of 16 non-structural proteins, ranging from NSP1 to NSP16, four structural proteins (S, E, M and N), and interspersed between these structural proteins are nine accessory proteins.

Figure 1-6-1   Genome organization of SARS-CoV-2

Interferons (IFNs) are the first line of defense in hosts against invading viruses[15]. Innate viral recognition triggers a signaling cascade leading to both nuclear factor -κB (NF-κB) mediated induction of pro-inflammatory cytokines(e.g., interleukin (IL) -1, IL-6, tumor necrosis factor-α (TNF-α)) and interferon regulatory factor (IRF) 3-and IRF7 -mediated induction of type I and type III IFNs (IFN-I and IFN-III) [16]. After this, IFN-I and IFN-III responses are activated[17]; IFN-I includes INF-α, IFN-β, IFN-ε, IFN-κ, and IFN-ω, while IFN-III is IFN-λ in humans. IFN-I binds to type I IFN receptor (IFNAR) that is ubiquitously expressed in autocrine and paracrine cells. As a result, hundreds of interferon-stimulated genes (ISGs) are activated, which interfere with every step of viral replication. IFN-III binds to IFN-III receptors (IFNLRs), preferentially expressed on myeloid and epithelial cells, thereby producing ISGs[18]. IFN-I is a key element in providing efficient protection against viral infections including SARS-CoV-2[19]. IFN-I is produced soon after recognition of pathogen-associated molecular patterns (PAMPs), such as viral mRNA[20]. PAMPs are recognized by retinoic acid-inducible gene 1(RIG-I/DExD/H-box helicase 58 (DDX58)) and melanoma differentiation-associated gene 5(MDA5/IFN induced with helicase C domain 1 (IFIH1))[21, 22]. Once activated, the RIG-I and MDA5 interact with the caspase activation and recruitment domain (CARD) domain of mitochondrial antiviral signaling protein (MAVS). The activated MAVS recruits several downstream signaling components to the mitochondria. As a result, an inhibitor of NF-κB kinase ε (IKKε) and TANK-binding kinase 1 (TBK1) are activated that further results in the phosphorylation of IRF3 and IRF7. The phosphorylated IRF3 and IRF7 are dimerized and translocated to the nucleus, where they induce the expression of IFN-I and a subset of ISGs (early ISGs) [23]. The secreted IFN-I expression leads to tyrosine kinase 2 (Tyk2) and Janus kinase 1 (JAK1) activation. After activation, STAT1 and STAT2 are phosphorylated[24, 25]. The phosphorylated STATs form a heterodimer and associate with IRF9, a DNA-binding protein, to form IFN-stimulated growth factor 3 (ISGF3). The ISGF3 complex translocates to the nucleus and binds with interferon-stimulated response elements (ISREs) at ISG promoters and transcribes its downstream genes (Figure 1-6-2).

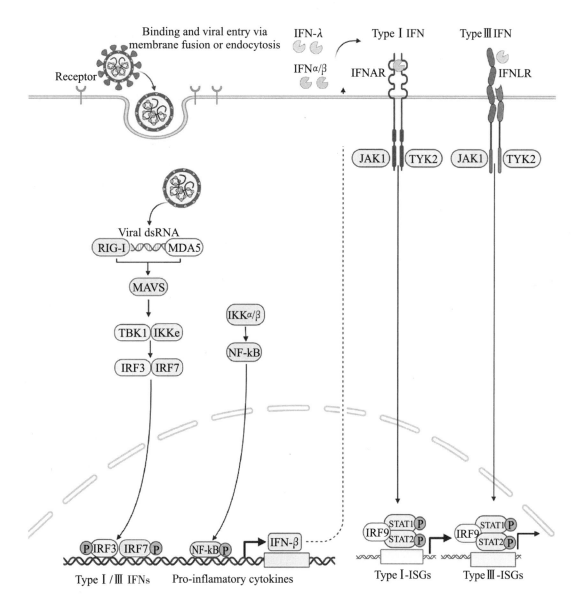

RIG-1: retinoic acid-inducible gene I; MDA5: melanoma differentiation associated gene 5; NF-κB: nuclear factor - κB; IRF3: interferon regulatory factor 3; IRF7: interferon regulatory factor 7. Upon sensing of viruses by RIG-I and MDA5, NF-κB, RF3, and IRF7 stimulate the production of proinlammatory cytokines, IFN-I and IFN-III. IFNs are secreted in an autocrine and paracrine manner to induce the expression of ISGs via the JAK/STAT signaling pathway.

Figure 1-6-2   Innate immune system recognition, IFN signaling, and immune evasion by viruses

Therefore, hundreds of ISG products are expressed at viral infection sites, producing the antiviral state[25]. The ISGs and pro-inflammatory cytokines have diverse functions including inhibition of viral replication, activation, and recruitment of immune cells. IFN-I production is therefore required against viral infections to trigger adaptive immune responses for a longer duration[26, 27].

The transcriptome proiles of different cell types delineate the infection of SARS-CoV-2. The infection triggers a low level of IFN-I and IFN-Ⅲ, and hence, limited ISG response is produced. However, it does induce pro-inflammatory cytokines[28, 29]. Low levels of IFN-I in the serum of coronavirus disease 2019 (COVID-19) patients were detected in the early stages of infection and elevated levels during the advanced stage of infection[29, 30]. Thus, ISG induction requires limited IFN-I production, or IFN-I may be produced in specific immune cells. Compared to SARS-CoV, SARS-CoV-2 induces less IFN-I[30]. IFN-I deiciency in the blood is an indication of -19 severity, which should betaken seriously[31]. SARS-CoV-2 infection also induces the activation of TLR3 and Toll-like receptor 7 (TLR7) RNA sensor pathways in the Clau-3/ Medical research council cell strain (MRC) -5 multicellular spheroids (MTCSs) [32]. TLR3 acts via IRF3 producing IL-1α, IL-1β, IL-4, IL-6, INF-α, and IFN-β. TLR3 also activates the NF-κB transduction pathway. TLR7 acts via NF-κB pathway, inducing IFN-I, IFN-γ, and IFN-λ3. Therefore, TLRs could also be potential targets in controlling the SARS-CoV-2 infection[32]. To have a successful infection, SARS-CoV-2 has evolved several strategies to overcome the host immune system. Since its discovery, SARS-CoV-2 has gained significant genetic diversity in all of its genes that is continuously changing the immune capabilities of its host[33]. To date, several SARS-CoV-2 proteins are known to have helped in immune evasion. In the following sections, we have discussed the host immune evasion by SARS-CoV-2 proteins and the mechanisms through which they help the virus evade the immune system.

NSP1 antagonizes IFN-I signaling by inhibiting STAT1 phosphorylation

Previous work on SARS-CoV has reported several roles for NSP1[34, 35]. It suppresses the translation of host proteins by interacting with the 40S subunit and inhibiting the 80S subunit formation[35]. It also induces endo-nucleolytic cleavage and subsequently degrades host mRNAs, leading to an inhibition of innate immune responses of host cells[34, 35].

The NSP1 of SARS-CoV-2 and SARS-CoV has 84% sequence identity in amino acid residues. Such high conservation suggests similar biological functions and properties[35]. The NSP1 of SARS-CoV contains Lys164 (K164) and His165 (H165) dipeptide motif, which is indispensable for human 40S ribosomal subunit interaction, leading to inhibition of host translation. K164A and H165A substitutions abolish the binding of NSP1 to 40S ribosomal subunit[36]. This dipeptide motif is conserved in SARS-CoV-2 NSP1, and a similar function of protein translation inhibition was observed[37, 38]. When cells were overexpressed with NSP1 followed by Sendai virus (SeV) stimulation, which is an excellent inducer of RIG-I, the endogenous protein levels of IFN-β, IFN-λ, and IL-8 were significantly reduced. However, the transcription of the corresponding genes was induced by NSP1 overexpression[39]. The NSP1 mutant had no inhibitory effects on the protein levels of IFN-β, IFN-λ, and IL-8. Similarly, NSP1 suppressed the luciferase activity driven by ISRE, which is the promoter part of the ISGs[39, 40]. However, autophagy was hardly affected by NSP1 expression, even when induced by rapamycin[38, 41]. These results delineated that NSP1 almost fully inhibits the translation of IFNs, pro-inflammatory cytokines, and ISGs. All of the above studies suggested that NSP1 of SARS-CoV-2 is one of the main immune evasion proteins[35-38, 40, 41]. NSP1, therefore, maybe an attractive therapeutic target against

COVID-19, but further investigation is required to determine whether NSP1 is the best option for vaccine development[42].

Another research also determined that NSP1 can inhibit IFN-β production and suppress 98% IFN-β promoter activity via the proteins that are both upstream and downstream of IRF3[43]. NSP1 significantly inhibits STAT1 phosphorylation, while STAT2 phosphorylation could be marginally inhibited. Moreover, NSP1 also suppresses the nuclear translocation of STAT1, suggesting the role of NSP1 in inhibiting IFN-I signaling (Figure 1-6-3, Table 1-6-1) [43].

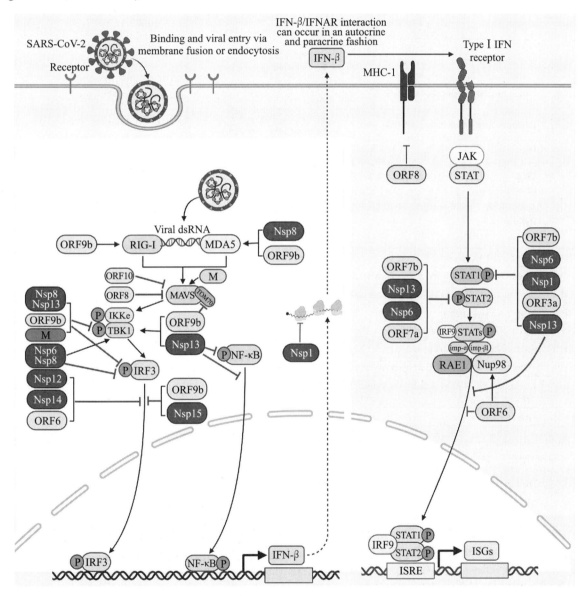

SARS-CoV-2 triggers IFN signaling pathway after being recognized by RIG-1 and MDA5. Different SARS-CoV-2 proteins interfere with these pathways in different ways. NSP8 and NSP13 inhibit TBK1 phosphorylation. NSP6, NSP8, NSP13, OR9b, and M inhibit IRF3 phosphorylation. NSP12, NSP14, NSP15, ORF6, and ORF9b inhibit the nuclear translocation of IRF3. NSP1, NSP6, NSP13, ORF3a, and ORF7b inhibit STAT1 phosphorylation. ORF9b inhibits TOM70. NSP6, NSP13, ORF7a, and ORF7b inhibit STAT2 phosphorylation. NSP13 and ORF6 inhibit the nuclear translocation of STAT1 to antagonize IFN signaling. ORF6 blocks STAT1 nuclear translocation by interacting with the Nup98-RAE1 complex and disrupts the interaction between Nup98 and importin-β1/importin-α1/PY-STAT1 complex, thus preventing the docking of this complex at the nuclear pore. ORF8 inhibits MHC-I to impair antigen-presenting cells. NSP: non-structural protein; ORF: open reading frame; TOM70: translocase of outer membrane 70 KDa Subunit; IFN: interferon; STAT: signal transducer and activator of transcription; IRF3: interferon regulatory factor 3; IRF9: interferon regulatory factor 9; NF-κB; nuclear factor-κB; ISGs: interferon-stimulated genes; MHC-I: major histocompatibility complex I; Nup98-RAE1 complex, nucleoporin 98 (Nup98) -ribonucleic acid export 1 (RAE1).

Figure 1-6-3   Host immune evasion by SARS-CoV-2 proteins

Table 1-6-1 SARS-CoV-2 proteins interfering with IFN induction and signaling

Protein	Mechanism	Experimental approach	Cellular model	References
IFN production inhibition				
NSP1	Multiple targets, may be upstream and downstream of IRF3	Luciferase assay	HEK293T cells	[43]
NSP6	Suppress IRF3 phosphorylation	Western blotting	HEK293T cells	[43]
NSP8	Suppress the phosphorylation of IRF3, TBK1	Western blotting	HEK293T cells	[44]
NSP12	Inhibit nuclear translocation of IRF3	Immunofluorescence assays	HEK293T cells	[45]
NSP13	Physical binds with TBK1, Suppress the phosphorylation of IRF3, TBK1, and NF -κB, Suppress nuclear translocation of NF-κB	Luciferase assay, Western blotting, Immunofluorescence assays	HEK293T, HeLa cells	[43, 46]
NSP14	Inhibit nuclear translocation of IRF3	Luciferase assay, Immunofluorescence assays	293 FT cells	[47]
NSP15	Inhibit nuclear translocation of IRF3	Luciferase assay, Immunofluorescence assays	293 FT cells	[47]
ORF6	Inhibit nuclear translocation of IRF3	Luciferase assay, Western blotting, Immunofluorescence assays	HEK293T	[43, 48-50]
ORF9b	Interact with MDA5, MAVS, TRIF, TBK1, STING, and RIG-1, suppress the phosphorylation of TBK1 and IRF3, suppress nuclear translocation of IRF3	Luciferase assay, Western blotting, Immunofluorescence assays	HEK293T, HeLa cells	[51]
M	Suppress the phosphorylation of IRF3, TBK1, IKKα/β, p65	Luciferase assay, qRT-PCR, Western blotting	HEK293T	[13]
IFN signaling inhibition				
NSP1	Suppress STAT1 phosphorylation	Western blotting	HEK293T cells	[43]
NSP6	Suppress STAT1 and STAT2 phosphorylation	Western blotting	HEK293T cells	[43]
NSP13	Suppress STAT1 and STAT2 phosphorylation	Western blotting	HEK293T cells	[43]
ORF3a	Suppress STAT1 phosphorylation	Western blotting	HEK293T cells	[43]
ORF6	Inhibit STAT1 nuclear translocation	Immunofluorescence assays	HEK293T cells	[43, 48]
ORF7a	Suppress STAT2 phosphorylation	Western blotting	HEK293T cells	[43]
ORF7b	Suppress STAT1 and STAT2 phosphorylation	Western blotting	HEK293T cells	[43]
ORF8	Interact with MHC-I and mediate its down regulation	Western blotting	HEK293T cells	[52]
ORF10	Degrade MAVS	Western blotting	HEK293T cells	[53]

Note: IRF3, Interferon regulatory factor 3; TBK1, TANK binding kinase 1; MDA5, melanoma differentiation-associated gene 5; MAVS, mitochondrial antiviral signaling; TRIF, TIR-domain-containing adapter-inducing interferon-β; STING, stimulator of IFN genes; RIG-1, retinoic acid-inducible gene I; STAT1, signal transducer and activator of transcription 1; STAT2, signal transducer and activator of transcription 2; MHC-I, Major histocompatibility complex I; IKKα/β, inhibitor of nuclear factor -κB (IκB) kinase alpha/beta.

NSP6 inhibits IFN-β production by targeting IRF3 and IFN-I signaling by inhibiting STAT1/STAT2 phosphorylation

NSP6 inhibits IFN-β production and suppresses about 40% luciferase activity[43, 54]. The distinct components of RIG-1 pathway were studied to identify which step of IFN-β production was inhibited by NSP6. The results showed that luciferase activity was suppressed when IFN-β promoter was induced by IKKε, TBK1, or MAVS, suggesting that NSP6 may inhibit IFN-β production by targeting IRF3 (before the activation of IRF3) or other component(s) upstream of IRF3 (that is between IRF3 and TBK1/IKKε) [43].

Similarly, the same study showed that NSP6 could modulate the phosphorylation of TBK1 and IRF3[43]. NSP6 overexpression followed by poly (I:C) transfection inhibited approximately 57% of IRF3 phosphorylation. However, no effect on phosphorylation of TBK1 was observed. Hence, it maybe inferred that

NSP6 binds TBK1, which decreases IRF3 phosphorylation, leading to a reduction of IFN-β production (Figure 1-6-3, Table 1-6-1) [43]. Moreover, NSP6 also antagonizes IFN-I signaling by inhibiting the phosphorylation of STAT1 and STAT2 (Figure 1-6-3, Table 1-6-1). NSP6 could inhibit STAT1 phosphorylation by about 33%~46%, while STAT2 phosphorylation by about 33%~50%[43].

NSP8 interacts with MDA5 to inhibit the phosphorylation of IRF3 and TBK1

NSP8 interacts with MDA5 and antagonizes the phosphorylation of IRF3 and TBK1. MDA5 is the most upstream sensor in the innate immune system and is involved in the recognition of dsRNA. NSP8 interacts with the CARD domain of MDA5 to downregulate the antiviral immune responses[44]. Tertiary structures of NSP8, MDA5 CARD domain, and K63-Ub were determined, and simulation results showed that at the N terminal region of NSP8, there is a short α-helix that covers an area that interacts with K63-Ub. It is already known that the CARD domain of MDA5 undergoes K63-linked polyubiquitination and recruits MAVS to form a signalosome. The structural predictions of the interaction between NSP8 and MDA5 CARD domain showed that NSP8 may interrupt K63-linked polyubiquitination and MAVS recruitment, leading to the inhibition of MDA5 activation[55]. Therefore, NSP8 inhibits K63-linked polyubiquitination of MDA5 by interfering with MDA5-MAVS signalosome.

Clinical data of COVID-19 patients revealed that disease severity is related to cytokine storm[56]. NSP8 overexpression could downregulate the expression of cytokines, including IL-1β, IL-2, IL-5, IL-6, CCL-20, IFN-β, TNF-α, and ISGs IFIT1 and IFIT2[44]. These results indicate that NSP8 strongly impairs the expression of genes involved in antiviral immune and inflammatory responses.

NSP8 expression significantly inhibits the phosphorylation of TBK1, IRF3, IKKα/β, and p65 (Figure 1-6-3, Table 1-6-1). The NF-κB signaling was greatly inhibited, as delineated by decreased p65 phosphorylation. It also inhibits the activation of IRF3 and NF-κB. Moreover, NSP8-downregulated innate immune responses were dependent on MAVS, acting on either MAVS or upstream signals[44]. These results suggest that NSP8 targets the upstream components of IFN-I signaling pathway.

NSP12 inhibits IRF3 nuclear translocation by attenuating IFN-β production

NSP12 inhibits poly (I:C) and SeV-induced IFN-β promoter activation[45, 48]. NSP12 overexpression inhibits IFN-β promoter activation triggered by MAVS, MDA5, RIG-IN, and IRF3-5D. NSP12 does not physically interact with IRF3. However, it decreases the nuclear translocation of IRF3 without impairing its phosphorylation (Figure 1-6-3, Table 1-6-1) [45]. However, another study found that NSP12 is not an IFN-β antagonist[57]. The induction of NSP12 does not affect the production of IFN-β both at mRNA and protein levels. The differences in results of these two studies maybe due to different experimental setups. Therefore, cautions are required while interpreting SARS-CoV-2-related luciferase assays, as different tag proteins and backbones of plasmid could influence the results[57].

NSP13 interacts with TBK1 and antagonizes IFN-I signaling by inhibiting STAT1 and STAT2 phosphorylation

NSP13 inhibits SeV-mediated promoter activity of NF-κB by about 2-fold. It also inhibits the activation and nuclear translocation of NF-κB as it reduces the levels of p-NF-κB when TBK1 and NSP13 co-overexpressed[46]. NSP13 physically binds with TBK1, determined by Co-immunoprecipitation (Co-IP) experiments[14, 43, 46]. In further experiments, it was found that IFN-β and ISRE promoter activities were downregulated after NSP13 overexpression induced by TBK1. This suggests that NSP13 antagonizes IFN response by suppressing IRF3 and TBK1 phosphorylation (Figure 1-6-3, Table 1-6-1) [46]. Another study also suggests that NSP13 inhibits IFN-β production and suppresses about 48% luciferase activity[43]. Furthermore, NSP13 targets IRF3 to inhibit IFN-β production by targeting IRF3 (before the activation of IRF3) or another protein upstream of IRF3 (between IRF3 and TBK1/IKKϵ) [43].

Furthermore, phosphorylation of TBK1 and IRF3 mediated by NSP13 was investigated[43]. NSP13 overexpression followed by poly (I : C) treatment inhibited about 75% IRF3 phosphorylation. NSP13 could also inhibit the phosphorylation of TBK1 in a dose-dependent manner. Hence, it may be deduced that NSP13 interacts with TBK1 and inhibits its phosphorylation, leading to suppression of IRF3 activation and IFN-β production (Figure 1-6-3, Table 1-6-1) [43]. NSP13 significantly suppresses STAT1/2 phosphorylation, suggesting its role in antagonizing IFN-I signaling. Moreover, NSP13 also inhibits the nuclear translocation of STAT1 during IFN-I signaling (Figure 1-6-3, Table 1-6-1) [43].

The recent mutations found in NSP13 make it a stronger IFN antagonist. The different mutations are P77L, Q88H, D260Y, E341D, and M429I, which were observed in different variants of SARS-CoV-2. Structural and biophysical analysis justified the stronger binding of these mutants with TBK1 and helped in more evasion from the host immune system[58]. Therefore, it could be deduced that with the evolution of different variants, the capability of SARS-CoV-2 to evade the host immune system will increase.

NSP14 inhibits IFN-β activation and IRF3 nuclear translocation

Among CoVs, NSP14 is highly conserved and exhibits approximately 99% amino acid similarity with its SARS-CoV counterpart[59]. NSP14 has a guanine-N7-methyltransferase and 3' to 5' exoribonuclease activity[60]. Mutations in the Zinc finger motif and the active site of the exonuclease domain result in a lethal phenotype of this virus[61].

In addition to NSP1, NSP14 also inhibits protein translation in cells[35, 59]. The role of NSP14 in IFN-β and ISG production has been well documented[47, 48, 59]. Overexpression of NSP14 suppresses the production of endogenous ISG proteins but does not affect their mRNA levels. Instead, inhibition of translation was shown to be responsible for the suppression of endogenous expression of ISGs by NSP14[59].

In another study, NSP14 was found to inhibit SeV-mediated IFN-β activation. In addition, NSP14 was able to recapitulate this inhibition when IFN-β promoter activity was induced upon overexpression of RIG-I or MDA5[48]. NSP14 was also found to inhibit IFN production upon RIG-1 activation, and it significantly inhibited SeV-mediated nuclear translocation of IRF3 (Figure 1-6-3, Table 1-6-1) [47].

NSP15 inhibits IFN production and IRF3 nuclear localization

NSP15 was shown to potentially suppress the production and signaling of IFN when the N-terminus RIG-1, an upstream activator of IFN signaling, was used as a potent inducer of IFN production[47]. Furthermore, NSP15 significantly inhibits the nuclear localization of IRF3 upon SeV infection (Figure 1-6-3, Table 1-6-1) [47].

ORF3a antagonizes IFN-I signaling by phosphorylating STAT1

Among all accessory proteins of SARS-CoV-2, ORF3a is the largest, with 275 amino acid residues. It shares approximately 72.7% similarity with the SARS-CoV ORF3a protein[62-64]. ORF3a was shown to suppress more than 40% ISRE promoter activity and significantly suppress IFN-I signaling[43]. It also suppresses STAT1 phosphorylation by 33%~46%; however, STAT2 phosphorylation was only marginally suppressed. Moreover, ORF3a suppresses STAT1 nuclear translocation during IFN-I signaling (Figure 1-6-3, Table 1-6-1) [43]. Furthermore, ORF3a also induces lysosomal damage, necrotic cell death, and cytokine storms[65].

SARS-CoV-2 infection was shown to induce a pro-inflammatory cytokine response through cGAS-STING and NF-κB in human epithelial cells[66, 67]. Inflammatory responses were observed in patients and could be therapeutically targeted to suppress severe disease symptoms. ORF3a has a unique ability to inhibit STING. ORF3a interacts with STING to inhibit NF-κB signaling by blocking the nuclear accumulation of p65. ORF3a therefore can antagonize immune activation induced by the cGAS-STING[67].

ORF6 inhibits IRF3 nuclear translocation and hampers IFN-I signaling by blocking STAT1 nuclear translocation

SARS-CoV-2 ORF6, which contains 61 amino acid residues, shows low similarity to SARS-CoV ORF6[64]. All serbecoviruses, including SARS-CoV and SARS-CoV-2, encode this protein. However, no orthologs of this protein have been found in other β-CoVs, i.e., MERS-CoV, murine hepatitis virus (MHV), and OC43[49]. SARS-CoV ORF6 is known to counteract host antiviral responses at multiple steps of the innate immune pathway[68]. ORF6 of SARS-CoV-2 localizes predominantly in the cytoplasm but can also be found in the Golgi apparatus and ER. ORF6 inhibits IFN-β promoter activation in a dose-dependent manner induced by either poly (I: C) or SeV[43, 48]. Moreover, ORF6 inhibits IFN promoter activation mediated by MAVS, MDA5, RIG-I, and IRF3-5D. ORF6 inhibits IFN-β production at IRF3 levels or downstream of it. Furthermore, ORF6 overexpression inhibited SeV-induced nuclear translocation of IRF3. The amino acid residues 53-61 at the C-terminal tail of ORF6 were important for this antagonistic activity[48].

Furthermore, ORF6 overexpression does not affect IRF3 phosphorylation but significantly blocks its nuclear translocation (Figure 1-6-3, Table 1-6-1). Karyopherin α 1-6 (KPNA1-6) is responsible for the nuclear translocation of IRF3, IRF7, and STAT1[69]. ORF6 binds to KPNA2 but not to other KPNAs; therefore, it was suggested that ORF6 inhibits IFN-β production by interacting with KPNA2 and blocking IRF3 nuclear translocation[43]. Inhibition of IRF3 nuclear translocation was also reported in another study, which suggests that residues E46 and Q56 are essential in providing ORF6 the antagonistic activity. Moreover, it was found that the C-terminal region of ORF6 was responsible for anti-innate immune activity by significantly inhibiting the nuclear translocation of IRF3[49].

ORF6 also significantly suppresses IFN-I signaling[43, 48]and inhibits INF-α-or IFN-β-induced ISRE and ISG56 promoter activities, suggesting that ORF6 antagonizes the downstream IFN signaling[43, 48]. The overexpression ORF6 only marginally suppresses the phosphorylation of STAT1 and STAT2[43]. However, another study showed contradictory results that ORF6 does not suppress STAT1 phosphorylation[48]. These observations indicate that ORF6 might suppress a step downstream of the STAT1/STAT2 phosphorylation[43]. Moreover, ORF6 also suppresses nuclear translocation of STAT1 via ORF6/KPNA2 interaction, thereby inhibiting IFN-I signaling (Figure 1-6-3, Table 1-6-1) [43]. ORF6 also inhibits IFN-β, ISRE, and NF-κB promoter activities in a dose-dependent manner. ORF6 overexpression inhibits the expression of the ISRE promoter, suggesting that different mechanisms are involved to regulate the IFN pathway. Furthermore, ORF6 suppresses SeV-induced mRNA levels of IFN-β, ISG56, and ISG54[50].

ORF6 inhibits nuclear translocation of STAT1 and inhibits IFN signaling yet by a different mechanism as well. It interacts with nucleoporin 98 (Nup98) -ribonucleic acid export 1 (RAE1) (Nup98-RAE1) complex and antagonizes IFN signaling by inhibiting nuclear translocation of STAT1[70] (Figure 1-6-3). The C-terminal region of ORF6 is important for this binding. The binding was impaired and IFN antagonist activity was abolished when Methionine (M) at position 58 was substituted with arginine (R) in ORF6[71]. ORF6 blocks STAT1 nuclear translocation by interacting with the Nup98-RAE1 complex and disrupts the interaction between Nup98 and importin-β1/importin-α1/PY STAT1 complex, thus preventing the binding of this complex at the nuclear pore. Moreover, SARS-CoV ORF6 binds to the Nup98-RAE1 complex, and ORF6s from both viruses share the same binding site on this complex[10]. Nup98 is identified as a critical factor hijacked by SARS-CoV-2 to inhibit IFN signaling. Overexpression of Nup98 successfully rescues the ORF6-mediated inhibition of STAT1 nuclear translocation. Some other studies confirmed the interaction of ORF6 with the Nup98-RAE1 complex, suggesting that ORF6 specifically targets Nup98 to block STAT nuclear import[71-74].

The emergence of new variants of SARS-CoV-2 has enhanced its virulence and human-to-human transmission. The alpha (B.1.1.7) variant, which belongs to variants of concern (VOCs), shows enhanced suppression of innate immune responses in airway epithelial cells compared to first wave isolates[75]. The alpha variants exhibit markedly increased levels of subgenomic RNA and protein levels of ORF6. The resulting increased levels of ORF6 protein inside host cells after infection increase the capability of ORF6 to antagonize the nuclear translocation of IRF3 and STAT1 proteins and hence antagonize IFN signaling more efficiently[75].

ORF7a and ORF7b inhibits IFN-I signaling by suppressing STAT1 and / or STAT2 phosphorylation

ORF7a and ORF7b contain 121 and 43 amino acid residues[64]. ORF7a was shown to inhibit STAT2 phosphorylation by approximately 33%～50% but only marginally suppress STAT1 phosphorylation (Figure 1-6-3, Table 1-6-1). ORF7b, on the other hand, suppressed STAT1 phosphorylation by 33%～46% and suppressed STAT2 phosphorylation by 33%～50%. Both ORF7a and ORF7b were shown to suppress ISRE promoter activity by approximately 40% and STAT1 nuclear translocation during IFN-I signaling (Figure 1-6-3, Table 1-6-1) [43].

ORF8 inhibits IFN-I signaling pathway and downregulates MHC-I

ORF8 is the most puzzling gene of CoVs[76]. It contains 121 amino acid residues with less than 20% sequence identity to SARS-CoV ORF8[77]. It contains a signal sequence for ER import. Antibodies to ORF8 are among the principal biomarkers of SARS-CoV-2 infection[78]. Moreover, several studies have delineated the role of ORF8 in immune evasion[50, 78, 79]. Since the emergence of SARS-CoV-2, several mutations in ORF8 have been recorded. These mutations, which include L84S[80, 81], V62L, S24L[82], and W45L[79], have been observed in different variants of SARS-CoV-2. Different mutants in ORF8 altered their binding efficiency to IRF3 such as W45L mutant was found to bind more stringently to IRF3, indicating its more profound role in immune evasion[79]. Therefore, different variants will have different capabilities for evading the host immune system due to mutations in ORF8.

ORF8 has also been found to interact with major histocompatibility complex I (MHC-I) and mediate its downregulation (Figure 1-6-3, Table 1-6-1). Cells overexpressing MHC-I were found to be targeted for lysosomal degradation by autophagy. ORF8 impairs the activity of antigen-presenting cells; therefore, blocking ORF8 could be used to improve the immune system[52].

The role of ORF8 in IFN antagonism was also described in another study[50]. Cells were cotransfected with NF-κB, IFN-β, or ISRE reporter plasmids and ORF8-overexpressing and respective control plasmids followed by SeV induction. ORF8 significantly inhibited the promoter activity of all three elements (NF-κB, IFN-β, and ISRE). In addition, it was found that ORF8 inhibits the expression of ISRE promoter, suggesting that it may adopt several mechanisms to regulate the host interferon pathway. Moreover, ORF8 significantly suppresses SeV-induced mRNA expression of IFN-β, ISG54, and ISG56[50].

ORF9b antagonizes IFN-I and IFN-III by targeting multiple signaling pathways

ORF9b encodes a protein of 97 amino acid residues and significantly inhibits the production of IFN-I by targeting mitochondria[53, 83]. Antibodies against ORF9b were detected in convalescent sera from SARS-CoV and SARS-CoV-2 patients[84, 85]. Therefore, the role of ORF9b concerning IFN-I production is obvious[86]. To determine which host proteins interact with ORF9b, a biotin-streptavidin affinity purification mass spectrometry approach was used. Translocase of outer membrane 70 KDa Subunit (TOM70) was shown to bind most efficiently to ORF9b. Co-IP experiments further validated these findings. TOM70 is a mitochondrial import receptor that is important for MAVS activation of IRF3 and TBK1[75]. Furthermore, SARS-CoV ORF9b can also bind TOM70. Therefore, ORF9b and TOM70 binding are conserved in SARS-like CoVs. Two domains of TOM70 protein, i.e., the core and the C-terminal are important for this interaction. ORF9b localizes to the outer membrane of mitochondria as TOM70 is also localized at this site. The alpha variants have also markedly increased subgenomic RNA and protein levels of the ORF9b[75]. ORF9b expression alone suppresses the innate immune response by binding to TOM70 (Figure 1-6-3). The binding of ORF9b and TOM70 was regulated by phosphorylation. Mutating ser53 alone or both ser50 and ser53 in ORF9b to phosphomimetic glutamic acid interrupted the binding between ORF9b and TOM70. Therefore, unphosphorylated ORF9b is highly active soon after virus infection to allow effective innate antagonism of the host[75].

Mitochondria and TOM70 play important roles in the IFN-I responses[87]. ORF9b significantly inhibits IFN-β production. Moreover, TOM70 overexpression was shown to largely rescue IFN-β production from ORF9b-mediated inhibition. Therefore, therapeutic agents that inhibit the binding of ORF9b and TOM70 in COVID-19 patients could be developed[86].

Type I and type III IFN responses are inhibited by ORF9b through multiple antiviral pathways[51]. Cells overexpressing ORF9b, followed by poly (I: C) or SeV induction, downregulate the expression of IFN-β, IFN-λ1, ISG56, and CXCL10. To map the step at which ORF9b could exert its inhibitory effects, it was found that ORF9b could antagonize the activities of IFN-β-Luc, IFN-λ 1, ISRE-Luc reporters induced by MAVS, MDA5, TBK1, RIG-IN, but not those induced by IRF3/5D. These results suggest that ORF9b inhibits IFN production at a step upstream of IRF3. ORF9b appeared strongly localized in mitochondria but weakly colocalized with ER and Golgi. Moreover, ORF9b does not interact with IRF3 but it does interact with MDA5, MAVS, TRIF, TBK1, STING, and RIG-1. Therefore, ORF9b targets multiple components of the innate system to inhibit IFN production. Furthermore, it was found that ORF9b suppresses the phosphorylation of TBK1. This phosphorylation suppression is induced by all three important antiviral pathways, i.e., cGAS-STING, TLR3-TRIF, and RIG-I/ MDA5-MAVS. ORF9b also suppresses SeV-induced IRF3 nuclear translocation and phosphorylation[51].

ORF10 suppresses IFN-I signaling by degrading MAVS

The ORF10 protein contains 38 amino acid residues, but the sequence of SARS-CoV-2 ORF10 is different from ORF10s of other CoVs[64]. Since the recent pandemic began, no specific function was attributed to this protein; however, a recent study described its role in the suppression of IFN-I signaling[53]. ORF10 significantly antagonizes IFN-I and ISG expression and degrades MAVS via mitophagy by accumulating LC3 inside mitochondria. ORF10 translocates to mitochondria and induces mitophagy by interacting with Nip3-like protein X (NIX) and LC3B. IFN-I signaling inhibition is blocked when NIX is knocked down. Therefore, ORF10 suppresses the IFN signaling pathway by inhibiting MAVS expression and promoting viral replication (Figure 1-6-3, Table 1-6-1) [53, 88].

Membrane (M) protein interacts with MAVS and inhibits the phosphorylation of TBK1 and IRF3

Membrane (M) is a glycosylated structural protein consisting of 222 amino acid residues. The N-terminal part of this protein contains three membrane spanning domains that are responsible for the assembly of viral particles[13, 14].

M protein inhibits the activation of IFN-β promoter, ISRE, and NF-κB in a dose-dependent manner induced by SeV[13]. Furthermore, stable cell lines expressing M protein inhibit SeV-or poly (I: C) -induced IFNB1, ISG56, CXCL10, and TNF transcription. Stable expression of ACE2 in HEK293 cells (HEK293-ACE2) suppressed SARS-CoV-2-induced transcription of IFN-β 1 and downstream antiviral when overexpressed with M protein. In addition, this protein inhibits IRF3, TBK1, IKKα/β, and p65 phosphorylation in cells[13].

M inhibits ISRE and IFN-β promoter activities mediated by MAVS, RIG-I-CARD, and MDA5 overexpression but not by TBK1. M suppresses NF-κB expression, which is mediated by MDA5, MAVS, and RIG-I-CARD but not by p65. Therefore, it could be deduced that M inhibits innate antiviral signaling at the MAVS level. M protein physically interacts with MAVS (at its transmembrane domain) but not with MDA5, TBK1, or RIG-I. Moreover, M protein disturbs the recruitment of IRF3, TRAF3, and TBK1 to the MAVS complex, and this impairment inhibits the innate antiviral response[13].

Other proteins

In addition to the above mentioned proteins, the following proteins are also found to play a role in IFN antagonism. However, their detailed mechanisms require further investigations.

NSP5 is a protease that cleaves ORF1a and ORF1b into peptides and blocks MAVS-induced IFN-β production. The C145A mutant of NSP5 abrogates the proteolytic activity and fails to inhibit the activation of IFN-β. Therefore, the proteolytic activity is indispensable for NSP5 to suppress IFN-β production[54].

ORF3b, which contains 22 amino acid residues, is considerably shorter than SARS-CoV ORF3b (153 amino acid residues). The results of a luciferase reporter assay suggested that SeV-induced promoter activity of IFN-β was suppressed after overexpression of ORF3b[89].

ORF9c contains 73 amino acid residues and shares 74% sequence identity with SARS-CoV ORF14 and approximately 94% sequence identity with bat SARS-CoV ORF14. It has a putative transmembrane domain that interacts with M protein in various cellular compartments. This interaction disturbs the antiviral process in lung epithelial cells. The expression of this highly unstable protein disturbed IFN signaling, complement signaling, and antigen presentation and induced IL-6 signaling. ORF9c enables evasion of the immune system and coordinates cellular changes that are important in the life cycle of SARS-CoV-2[90].

Nucleocapsid (N) protein inhibits the promoter activities of IFN-β, ISRE, and NF-κB. N protein also suppresses SeV-induced IFN-β, ISG54, and ISG56 mRNA expression levels but does not inhibit the expression from the ISRE promoter[50]. The alpha variants have also markedly increased subgenomic RNA and protein levels of N protein[75]. Therefore, the already IFN antagonistic activity of N against IFN could be highly enhanced in alpha variants.

Perspectives

Innate and adaptive immune responses are considered fundamental elements of host defense against viral infections, but SARS-CoV-2 has devised strategies to evade the immune system. In this article, we summarized the roles and mechanisms of actions of different SARS-CoV-2 proteins playing important roles in the host's immune evasion. SARS-CoV-2 proteins inhibit the production and signaling of IFNs by different mechanisms. For instance, NSP12, NSP14, NSP15, ORF6, and ORF9b inhibit IRF3 nuclear translocation[43, 45, 47, 48, 50, 51], whereas NSP1, NSP6, NSP8, NSP13, ORF3a, ORF7a/b, ORF9b, and Membrane proteins are involved in the phosphorylation inhibition of various components of the immune system[13, 43, 44, 46, 51]. ORF8 downregulates MHC-I to evade the host immune system[52]. NSP8 interacts with MDA5 to impair its K63-linked polyubiquitination and mediate immune evasion[44]. ORF9c interacts with membrane proteins and impairs the antiviral process in lung epithelial cells[90]. These examples demonstrate that each protein of SARS-CoV-2 may perform multiple functions and mediate immune evasion by different mechanisms. Moreover, mutations in the viral genome also play important roles in more aggressive infections. Different variants of interest (VOIs) and VOCs are evolved that favor enhanced human-to-human transmission. Therefore, it is necessary to study host immune evasion in more variants, so that potential therapeutic strategies can be developed. In addition, the inhibitory potential of antagonizing proteins may be different in different experimental setups. Further investigations are required to gain more insights and information about immune responses and COVID-19 interactions. Detailed mechanistic studies of NSP5, ORF3b, ORF9c, and N proteins will further elucidate the pathogenesis of SARS-CoV-2 and therefore need further investigations.

References

[1] PIŠLAR A, MITROVIĆ A, SABOTIČ J, et al. The role of cysteine peptidases in coronavirus cell entry and replication: The therapeutic potential of cathepsin inhibitors. PLOS Pathog, 2020, 16(11): e1009013.

[2] AMARILLA A A, SNG J D J, PARRY R, et al. A versatile reverse genetics platform for SARS-CoV-2 and other positive-strand RNA viruses. Nat Commun, 2021, 12(1): 3431.

[3] SIU Y L, TEOH K T, LO J, et al. The m, e, and n structural proteins of the severe acute respiratory syndrome coronavirus are required for efficient assembly, trafficking, and release of virus-like particles. J Virol, 2008, 82(22): 11318-11330.

[4] V'KOVSKI P, KRATZEL A, STEINER S, et al. Coronavirus biology and replication: implications for SARS-CoV-2. Nat Rev Microbiol, 2021, 19(3): 155-170.

[5] MALIK Y A. Properties of coronavirus and SARS-CoV-2. Malays J Pathol, 2020, 42(1): 3-11.

[6] FINKEL Y, MIZRAHI O, NACHSHON A, et al. The coding capacity of SARS-CoV-2. Nature, 2021, 589 (7840): 125-130.

[7] SHAH A, REHMAT S, ASLAM I, et al. Comparative mutational analysis of SARS-CoV-2 isolates from Pakistan and structural-functional implications using computational modelling and simulation approaches. Comput Biol Med, 2022, 141: 105170.

[8] SANYAL S. How SARS-CoV-2 (COVID-19) spreads within infected hosts-what we know so far. Emerg Top Life Sci, 2020, 4(4): 371-378.

[9] ZHU G, ZHU C, ZHU Y, et al. Minireview of progress in the structural study of SARS-CoV-2 proteins. Curr Res Microb Sci 2020, 1: 53-61.

[10] LI T, WEN Y, GUO H, et al. Molecular mechanism of SARS-CoVs ORF6 targeting the Rae1-Nup98 complex to compete with mRNA nuclear export. Front Mol Biosci, 2021, 8: 813248.

[11] TEOH K T, SIU Y L, CHAN W L, et al. The SARS coronavirus e protein interacts with PALS1 and alters tight junction formation and epithelial morphogenesis. Mol Biol Cell, 2010, 21(22): 3838-3852.

[12] LAZAREVIC I, PRAVICA V, MILJANOVIC D, et al. Immune evasion of SARS- CoV-2 emerging variants: What have we learnt so far?. Viruses, 2021, 13(7): 1192.

[13] FU Y Z, WANG S Y, ZHENG Z Q, et al. SARS-CoV-2 membrane glycoprotein m antagonizes the MAVS-mediated innate antiviral response. Cell Mol Immunol, 2021, 18(3): 613-620.

[14] GORDON D E, JANG G M, BOUHADDOU M, et al. A SARS-CoV-2 protein interaction map reveals targets for drug repurposing. Nature, 2020, 583(7816): 459-468.

[15] PERLMAN S, NETLAND J. Coronaviruses post-SARS: update on replication and pathogenesis. Nat Rev Microbiol, 2009, 7(6): 439-450.

[16] PARK A, IWASAKI A. Type I and type III interferons-induction, signaling, evasion, and application to combat COVID-19. Cell Host Microbe, 2020, 27(6): 870-878.

[17] SCHOGGINS J W, RICE C M. Interferon-stimulated genes and their antiviral effector functions. Curr Opin Virol, 2011, 1(6): 519-525.

[18] KOTENKO S V, RIVERA A, PARKER D, et al. Type III IFNs: Beyond antiviral protection. Semin Immunol, 2019, 43: 101303.

[19] SA RIBERO M, JOUVENET N, DREUX M, et al. Interplay between SARS-CoV-2 and the type I interferon response. PLOS Pathog, 2020, 16(7): e1008737.

[20] STREICHER F, JOUVENET N. Stimulation of innate immunity by host and viral RNAs. Trends Immunol, 2019, 40(12): 1134-1148.

[21] LI J, LIU Y, ZHANG X. Murine coronavirus induces type I interferon in oligodendrocytes through recognition by RIG-I and MDA5. J Virol, 2010, 84 (13): 6472-6482.

[22] ZALINGER Z B, ELLIOTT R, ROSE K M, et al. MDA5 is critical to host defense during infection with murine coronavirus. J Virol, 2015, 89(24): 12330-12340.

[23] LOO Y M, GALE M. Immune signaling by RIG-i-like receptors. Immunity, 2011, 34(5): 680-692.

[24] DARNELL J E, KERR I M, STARK G R. Jak-STAT pathways and transcriptional activation in response to IFNs and other extracellular signaling proteins. Science, 1994, 264(5164): 1415-14121.

[25] SCHINDLER C, LEVY D E, DECKER T. JAK-STAT signaling: from interferons to cytokines. J Biol Chem, 2007, 282(28): 20059-20063.

[26] SCHOGGINSJ W. Interferon-stimulated genes: What do they all do?. Annu Rev Virol, 2019, 6(1): 567-584.

[27] SCHNEIDER W M, CHEVILLOTTE M D, RICE C M. Interferon-stimulated genes: a complex web of host defenses. Annu Rev Immunol, 2010, 32: 513-545.

[28] BLANCO-MELO D, NILSSON-PAYANT B E, LIU W-C, et al. Imbalanced host response to SARS-CoV-2 drives development of COVID-19. Cell, 2020, 181(5): 1036-1045.

[29] CHU H, CHAN J F W, WANG Y, et al. Comparative replication and immune activation profiles of SARS-CoV-2 and SARS-CoV in human lungs: An ex vivo study with implications for the pathogenesis of COVID-19. Clin Infect Dis, 2020, 71(6): 1400-1409.

[30] CHAN J F-W, KOK K-H, ZHU Z, et al. Genomic characterization of the 2019 novel human-pathogenic coronavirus isolated from a patient with atypical pneumonia after visiting wuhan. Emerg Microbes Infect, 2020, 9(1): 221-236.

[31] HADJADJ J, YATIM N, BARNABEI L, et al. Impaired type I interferon activity and inflammatory responses in severe COVID-19 patients. Science, 2020, 369(6504): 718-724.

[32] BORTOLOTTI D, GENTILI V, RIZZO S, et al. TLR3 and TLR7 RNA sensor activation during SARS-CoV-2 infection. Microorganisms, 2021, 9(9): 1820.

[33] BRANT A C, TIAN W, MAJERCIAK V, et al. SARS-CoV-2: from its discovery to genome structure, transcription, and replication. Cell Biosci, 2021, 11(1): 136.

[34] SIMEONI M, CAVINATO T, RODRIGUEZ D, et al. I(nsp1)ecting SARS-CoV-2-ribosome interactions. Commun Biol, 2021, 4(1): 715.

[35] LOKUGAMAGE K G, NARAYANAN K, HUANG C, et al. Severe acute respiratory syndrome coronavirus protein nsp1 is a novel eukaryotic translation inhibitor that represses multiple steps of translation initiation. J Virol, 2012, 86 (24): 13598-13608.

[36] KAMITANI W, HUANG C, NARAYANAN K, et al. A two-pronged strategy to suppress host protein synthesis by SARS coronavirus Nsp1 protein. Nat Struct Mol Biol, 2009, 16(11): 1134-1140.

[37] SCHUBERT K, KAROUSIS E D, JOMAA A, et al. SARS-CoV-2 Nsp1 binds the ribosomal mRNA channel to inhibit translation. Nat Struct Mol Biol, 2020, 27(10): 959-966.

[38] THOMS M, BUSCHAUER R, AMEISMEIER M, et al. Structural basis for translational shutdown and immune evasion by the Nsp1 protein of SARS-CoV-2. Sci, 2020, 369(6508): 1249-1256.

[39] SPARRER K M J, PFALLER C K, CONZELMANN K K. Measles virus c protein interferes with beta interferon transcription in the nucleus. J Virol, 2012, 86 (2): 796-805.

[40] DEVAUX P, VON MESSLING V, SONGSUNGTHONG W, et al. Tyrosine 110 in the measles virus phosphoprotein is required to block STAT1 phosphorylation. Virology, 2007, 360(1): 72-83.

[41] SPARRER K M J, GABLESKE S, ZURENSKI M A, et al. TRIM23 mediates virus-induced autophagy via activation of TBK1. Nat Microbiol, 2017, 2(11): 1543-1557.

[42] VANN K R, TENCER A H, KUTATELADZE T G. Inhibition of translation and immune responses by the virulence factor Nsp1 of SARS-CoV-2. Signal Transduct Target Ther, 2020, 5(1): 234.

[43] XIA H, CAO Z, XIE X, et al. Evasion of type I interferon by SARS-CoV-2. Cell Rep, 2020, 33(1): 108234.

[44] YANG Z, ZHANG X, WANG F, et al. Suppression of MDA5-mediated antiviral immune responses by NSP8 of SARS-CoV-2 (preprint). bioRxiv, 2020.

[45] WANG W, ZHOU Z, XIAO X, et al. SARS-CoV-2 nsp12 attenuates type I interferon production by inhibiting IRF3 nuclear translocation. Cell Mol Immunol, 2021, 18(4): 945-953.

[46] VAZQUEZ C, SWANSON S E, NEGATU S G, et al. SARS-CoV-2 viral proteins NSP1 and NSP13 inhibit interferon activation through distinct mechanisms. PLOS ONE, 2021, 16: 1-15.

[47] YUEN C K, LAM J Y, WONG W M, et al. SARS-CoV-2 nsp13, nsp14, nsp15 and orf6 function as potent interferon antagonists. Emerg Microbes Infect, 2020, 9(1): 1418-1428.

[48] LEI X, DONG X, MA R, et al. Activation and evasion of type I interferon responses by SARS-CoV-2. Nat Commun, 2020, 11(1): 3810.

[49] KIMURA I, KONNO Y, URIU K, et al. Sarbecovirus ORF6 proteins hamper induction of interferon signaling. Cell Rep, 2021, 34(13): 108916.

[50] LI J Y, LIAO C H, WANG Q, et al. The ORF6, ORF8 and nucleocapsid proteins of SARS-CoV-2 inhibit type I interferon signaling pathway. Virus Res, 2020, 286: 198074.

[51] HAN L, ZHUANG M W, DENG J, et al. SARS-CoV-2 ORF9b antagonizes type I and III interferons by targeting multiple components of the RIG-I/MDA-5-MAVS, TLR3-TRIF, and cGAS-STING signaling pathways. J Med Virol, 2021, 93(9): 5376-5389.

[52] ZHANG Y, CHEN Y, LI Y, et al. The ORF8 protein of SARS-CoV-2 mediates immune evasion through down-regulating MHC-i. Proc Natl Acad Sci USA, 2021, 118(23): e2024202118.

[53] LI X, HOU P, MA W, et al. SARS-CoV-2 ORF10 suppresses the antiviral innate immune response by degrading MAVS through mitophagy. Cell Mol Immunol, 2022, 19(1): 67-78.

[54] SHEMESH M, AKTEPE T E, DEERAIN J M, et al. SARS-CoV-2 suppresses IFN β production mediated by NSP1, 5, 6, 15, ORF6 and ORF7b but does not suppress the effects of added interferon. PLOS Pathog, 2021, 17(8): e1009800.

[55] JIANG X, KINCH L N, BRAUTIGAM C A, et al. Ubiquitin-induced oligomerization of the RNA sensors RIG-I and MDA5 activates antiviral innate immune response. Immunity, 2012, 36(6): 959-973.

[56] CHEN Y, LIU Q, GUO D. Emerging coronaviruses: Genome structure, replication, and pathogenesis. JMed Virol, 2020, 92(4): 418-423.

[57] LI A, ZHAO K, ZHANG B, et al. SARS-CoV-2 NSP12 protein is not an interferon- β antagonist. J Virol, 2021, 95(17): e0074721.

[58] RASHID F, SULEMAN M, SHAH A, et al. Structural analysis on the severe acute respiratory syndrome coronavirus 2 non-structural protein 13 mutants revealed altered bonding network with TANK binding kinase 1 to evade host immune system. Front Microbiol, 2021, 12: 789062.

[59] HSU J C C, LAURENT-ROLLE M, PAWLAK J B, et al. Translational shutdown and evasion of the innate immune response by SARS-CoV-2 NSP14 protein. Proc Natl Acad Sci USA, 2021, 118(24): 1-9.

[60] ECKERLE L D, BECKER M M, HALPIN R A, et al. Infidelity of SARS-CoV Nsp14-exonuclease mutant virus replication is revealed by complete genome sequencing. PLOS Pathog, 2010, 6(5): e1000896 .

[61] OGANDO N S, ZEVENHOVEN-DOBBE J C, VAN DER MEER Y, et al. The enzymatic activity of the nsp14 exoribonuclease is critical for replication of MERS-CoV and SARS-CoV-2. J Virol, 2020, 94(23): e01246-20.

[62] ISSA E, MERHI G, PANOSSIAN B, et al. SARS-CoV-2 and ORF3a: Nonsynonymous mutations, functional domains, and viral pathogenesis. mSystems, 2020, 5(3): e00266-20.

[63] FREUNDT E C, YU L, GOLDSMITH C S, et al. The open reading frame 3a protein of severe acute respiratory syndrome-associated coronavirus promotes membrane rearrangement and cell death. J Virol, 2010, 84 (2): 1097-1109.

[64] REDONDO N, ZALDÍVAR-LÓPEZ S, GARRIDO J J, et al. SARS-CoV-2 accessory proteins in viral pathogenesis: Knowns and unknowns. Front Immunol, 2021, 12: 1-8.

[65] SIU K-L, YUEN K-S, CASTAÑO-RODRIGUEZ C, et al. Severe acute respiratory syndrome coronavirus ORF3a protein activates the NLRP3 inflammasome by promoting TRAF3-dependent ubiquitination of ASC. FASEB J, 2019, 33(8): 8865-8877.

[66] NEUFELDT C J, CERIKAN B, CORTESE M, et al. SARS-CoV-2 infection induces a pro-inflammatory cytokine response through cGAS-STING and NF-κB. Commun Biol, 2022, 5(1): 45.

[67] RUI Y, SU J, SHEN S, et al. Unique and complementary suppression of cGAS-STING and RNA sensing-triggered innate immune responses by SARS-CoV-2 proteins. Signal Transduct Target Ther, 2021, 6(1): 123.

[68] FRIEMAN M, YOUNT B, HEISE M, et al. Severe acute respiratory syndrome coronavirus ORF6 antagonizes STAT1 function by sequestering nuclear import factors on the rough endoplasmic reticulum/Golgi membrane. J Virol, 2007, 81(18): 9812-9824.

[69] CHOOK Y M, BLOBEL G. Karyopherins and nuclear import. Curr Opin Struct Biol, 2001, 11(6): 703-715.

[70] SHEN Q, WANG Y E, PALAZZO A F. Crosstalk between nucleocytoplasmic trafficking and the innate immune response to viral infection. J Biol Chem, 2021, 297(1): 100856.

[71] MIORIN L, KEHRER T, SANCHEZ-APARICIO M T, et al. SARS-CoV-2 Orf6 hijacks Nup98 to block STAT nuclear import and antagonize interferon signaling. Proc Natl Acad Sci USA, 2020, 117(45): 28344-28354.

[72] LEE J G, HUANG W, LEE H, et al. Characterization of SARS-CoV-2 proteins reveals Orf6 pathogenicity, subcellular localization, host interactions and attenuation by selinexor. Cell Biosci, 2021, 11(1): 58.

[73] ADDETIA A, LIEBERMAN N A P, PHUNG Q, et al. SARS-CoV-2 ORF6 disrupts bidirectional nucleocytoplasmic transport through interactions with Rae1 and Nup98. MBio, 2021, 12(2): e00065-21.

[74] LI J, GUO M, TIAN X, et al. Virus-host interactome and proteomic survey reveal potential virulence factors influencing SARS-CoV-2 pathogenesis. Med (New York, NY), 2021, 2(1): 99-112.

[75] THORNE L G, BOUHADDOU M, REUSCHL A K, et al. Evolution of enhanced innate immune evasion by SARS-CoV-2. Nature, 2022, 602(7897): 487-495.

[76] NECHES R Y, KYRPIDES N C, OUZOUNIS C A. Atypical divergence of SARS-CoV-2 ORF8 from ORF7a within the coronavirus lineage suggests potential stealthy viral strategies in immune evasion. MBio, 2021, 12(1): 1-12.

[77] SHAH A, RASHID F, AZIZ A, et al. Genetic characterization of structural and open reading fram-8 proteins of SARS-CoV-2 isolates from different countries. Gene Rep, 2020, 21: 100886.

[78] FLOWER T G, BUFFALO C Z, HOOY R M, et al. Structure of SARS-CoV-2 ORF8, a rapidly evolving immune evasion protein. Proc Natl Acad Sci USA, 2021, 118(2): e2021785118.

[79] RASHID F, SULEMAN M, SHAH A, et al. Mutations in SARS-CoV-2 ORF8 altered the bonding network with interferon regulatory factor 3 to evade host immune system. Front Microbiol, 2021, 12: 703145.

[80] CERAOLO C, GIORGI F M. Genomic variance of the 2019-nCoV coronavirus. J Med Virol, 2020, 92(5): 522-528.

[81] TANG X, WU C, LI X, et al. On the origin and continuing evolution of SARS-CoV-2. Natl Sci Rev, 2020, 7(6): 1012-1023.

[82] LAHA S, CHAKRABORTY J, DAS S, et al. Characterizations of SARS-CoV-2 mutational profile, spike protein stability and viral transmission. Infect Genet Evol, 2020, 85: 104445.

[83] SHI C S, QI H Y, BOULARAN C, et al. SARS-coronavirus open reading frame-9b suppresses innate immunity by targeting mitochondria and the MAVS/TRAF3/TRAF6 signalosome. J Immunol, 2010, 193(6): 3080-3089.

[84] GUO J P, PETRIC M, CAMPBELL W, et al. SARS corona virus peptides recognized by antibodies in the sera of convalescent cases. Virology, 2004, 324 (2): 251-256.

[85] JIANG H W, LI Y, ZHANG H N, et al. SARS-CoV-2 proteome microarray for global profiling of COVID-19 specific IgG and IgM responses. Nat Commun, 2020, 11(1): 3581.

[86] JIANG H W, ZHANG H N, MENG Q F, et al. SARS-CoV-2 Orf9b suppresses type I interferon responses by targeting TOM70. Cell Mol Immunol, 2020, 17(9): 998-1000.

[87] LIU X Y, WEI B, SHI H X, et al. Tom70 mediates activation of interferon regulatory factor 3 on mitochondria. Cell Res, 2010, 20(9): 994-1011.

[88] ZANDI M. ORF9c and ORF10 as accessory proteins of SARS-CoV-2 in immune evasion. Nat Rev Immunol, 2022, 22(5): 331.

[89] KONNO Y, KIMURA I, URIU K, et al. SARS-CoV-2 ORF3b is a potent interferon antagonist whose activity is increased by a naturally occurring elongation variant. Cell Rep, 2020, 32(12): 108185.

[90] DOMINGUEZ A, FENG Y, CAMPOSA R, et al. SARS-CoV-2 ORF9c is a membrane-associated protein that suppresses antiviral responses in cells. bioRxiv Prepr Serv Biol, 2020: e256776.

Immunoinformatic-based design of immune-boosting multiepitope subunit vaccines against monkeypox virus and validation through molecular dynamics and immune simulation

Suleman Muhammad, Rashid Farooq, Ali Shahid, Sher Hassan, Luo Sisi, Xie Liji, and Xie Zhixun

Abstract

Monkeypox virus is the causative agent of monkeypox disease, belonging to an orthopoxvirus genus, with a disease pattern similar to that of smallpox. The number of monkeypox cases have robustly increased recently in several countries around the world, potentially causing an international threat. Therefore, serious measures are indispensable to be taken to mitigate the spread of the disease and hence, under these circumstances, vaccination is the best choice to neutralize the monkeypox virus. In the current study, we used immunoinformatic approaches to target the L1R, B5R, and A33R proteins of the monkeypox virus to screen for immunogenic cytotoxic T-lymphocyte (CTL), helper T-lymphocyte (HTL), and B-cell epitopes to construct multiepitope subunit vaccines. Various online tools predicted the best epitope from immunogenic targets (L1R, B5R, and A33R) of monkeypox virus. The predicted epitopes were joined together by different linkers and subjected to 3D structure prediction. Molecular dynamics simulation analysis confirmed the proper folding of the modeled proteins. The strong binding of the constructed vaccines with human TLR-2 was verified by the molecular docking and determination of dissociation constant values. The GC content and codon adaptation index (CAI) values confirmed the high expression of the constructed vaccines in the pET-28a (+) expression vector. The immune response simulation data delineated that the injected vaccines robustly activated the immune system, triggering the production of high titers of IgG and IgM antibodies. In conclusion, this study provided a solid base of concept to develop dynamic and effective vaccines that contain several monkeypox virus-derived highly antigenic and nonallergenic peptides to control the current pandemic of monkeypox virus.

Keywords

monkeypox virus, immunoinformatics, vaccine design, molecular docking, MD simulation

---

Introduction

Monkeypox virus, a double-stranded DNA virus with a genome size of 197 kb, contain more than 197 nonoverlapping open reading frames (ORFs) [1, 2]. Moreover, the virus also contains membrane proteins, structural proteins, and DNA-dependent RNA polymerase[3]. The virus belongs to the family Poxviridae, subfamily Chordopoxvirinae, and genus *Orthopoxvirus* and is the causative agent of a rare zoonotic disease, monkeypox. Monkeypox is similar to smallpox; however, the signs and symptoms are less severe than those of smallpox[2].

The transmission patterns and infection routes of the monkeypox virus are undefined[1]. However, it has been found that the current outbreak of the disease is due to sexual transmission of the virus among men;

nevertheless, this virus could also disseminate through body fluids, scabs, and shared bedding[2].

Monkeypox virus was discovered in monkeys in 1958 at a Danish laboratory, while the first case in humans was reported in 1970 in the Democratic Republic of Congo (DRC), previously Zaire, in a nine months baby[4, 5]. The virus was not reported outside Africa until 2003, but since then, it has spread into central and western African countries[5]. Phylogenetic analyses have revealed that the virus has circulated undetected outside central and western Africa for some time[6]. From 2018 to 2021, monkeypox cases were reported in the United States (US), United Kingdom (UK), Singapore, and Israel[7]. Since May 2022, the virus has spread to more than 50 countries of the world, and the World Health Organization (WHO) on June 23, 2022, declared monkeypox an evolving threat of modern public health concern[6].

Currently, no specific treatments are being devised to treat monkeypox. However, antivirals, vaccinia immune globulin (VIG), and smallpox vaccines are useful to control monkeypox outbreaks. As vaccines are limited, the WHO, on June 14, 2022, issued an interim guidance that mass vaccination is not needed for this disease[8]. Due to the current pandemic situation of monkeypox, the development of vaccines against monkeypox virus is urgently needed. Statistical data showed that vaccination against smallpox with the vaccinia virus was 85% effective against monkeypox[9]. It has been demonstrated that the L1R (intracellular mature virus specific) protein and B5R and A33R (extracellular enveloped specific) proteins are immunogenic in nature and conferred protection against the vaccinia virus[10]. In particular, these proteins are necessary for the virus to bind to, fuse with, and enter target cells[11, 12]. The L1R and B5R immunogens are the target of IMV and EEV neutralizing antibodies while the A33R is the target of complement mediated cytolysis[13, 14]. Moreover, subunit recombinant vaccines of vaccinia virus using LIR, B5R, and A33R have also been shown to be effective against monkeypox[15].

Both cellular and humoral immunity are important for all orthopoxviruses including the monkeypox virus. However, cytotoxic T-cell are of great interest as they provide a better protection. CD8$^+$ T cells protect against the disease, but CD4$^+$ T cell dependent antibody production is indispensable in clearing the virus after acute infection. Moreover, CD4$^+$ cells are more potent in the secretion of IFN-γ compared to CD4$^+$ T cells[16, 17]. Vaccination is one of the best choices to trigger immune responses and neutralize invading pathogens. It has been reported that vaccination is the most reliable and effective strategy against infectious diseases and prevents approximately 2~3 million deaths per year (WHO, 2007). During the recent pandemic of severe acute respiratory syndrome coronavirus 2 (SARS-CoV-2), immunoinformatics based subunit vaccines engineered with adjuvants and conformational linkers against B-cell epitope, helper T-lymphocytes (HTL) and cytotoxic T-lymphocyte (CTL) have been designed[18]. Similarly, in another immunoinformatics study using spike protein of SARS-CoV-2, highly efficient, immunodominant CTL epitopes were predicted that generated specific and robust immune response[19]. Therefore, in the present study, we used an immunoinformatic approach to target the L1R, B5R, and A33R proteins of the monkeypox virus to screen for immunogenic CTL, HTL, and B-cell epitopes to construct multiepitope subunit vaccines (MESVs) against monkeypox virus. The immunoinformatic approach has several useful characteristics over traditional vaccine development; for example, computationally designed vaccines are thermodynamically stable, more effective, low cost, highly specific, and less time consuming to develop. Furthermore, the constructed vaccine was modeled and checked for proper folding and stability through molecular dynamics (MD) simulations. Subsequently, the binding affinity of the designed vaccines with human toll like receptor 2 (TLR-2) was checked by a molecular docking approach. This study will provide cost-effective, highly immunogenic, and nonallergenic MESVs against the monkeypox virus to control the current pandemic.

Materials and methods

Sequence retrieval and immunogenic peptide prediction

The protein sequences of the L1R (Accession ID: Q3I7N2), B5R (Accession ID: Q3I8J3), and A33R (Accession ID: Q3I8L8) proteins of the monkeypox virus were retrieved from UniProt, an online database (accessed on August 5, 2022). First, the aforementioned proteins were submitted to the VaxiJen server to check their antigenicity[20]. The overall steps involved in the present study from epitope prediction to vaccine construction and validation are shown in Figure 1-7-1. For the mentioned purpose, the NetCTL 1.2 online server was used to predict the CTL epitopes sorted based on the combined score and A1 supertype was selected for the NetCTL epitope prediction[21]. However, the IEDB online server was used for the prediction of HTL epitopes specific for human MHC molecules such as HLA-DRB1*03: 01, HLA-DRB1*03: 01, HLA-DRB1*15: 01, HLA-DRB3*01: 01, HLA-DRB3*02: 02, HLA-DRB4*01: 01 and HLA-DRB5*01: 01[22]. The HTL epitopes were selected for vaccine construction based on the lowest percentile ranking. The lower percentile ranking depicts higher binding affinity of epitopes. Finally, the ABCPred server was used to predict the B-cell epitopes that are important for the generation of protective human antibodies[23]. The best epitopes were selected based on high scores, and the cutoff value was set to 0.8.

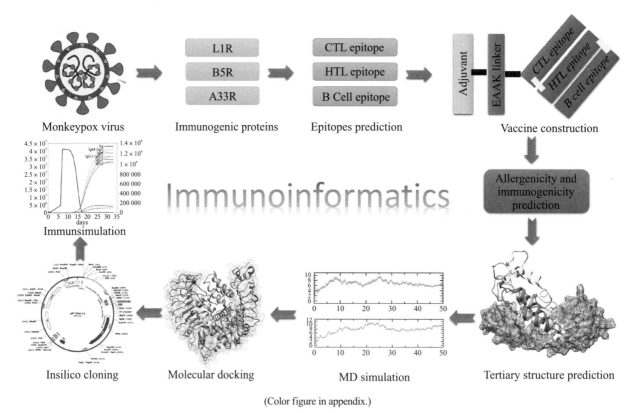

(Color figure in appendix.)

Figure 1-7-1　Overall workflow of the development of a multiepitope subunit vaccine targeting monkeypox virus using the immunoinformatic approach

Construction and characterization of the multiepitope subunit vaccine

After the prediction of immunogenic epitopes for the L1R, B5R, and A33R proteins, we used different linkers to combine the predicted epitopes and finalized the design of the MESVs. The B-cell, HTL and CTL epitopes were combined by using KK, GPGPG and AAY linkers, respectively. These linkers are immunogenic in nature and help the constructed vaccines boost immune responses. Second, these linkers prevent the folding of epitopes, keeping them separated from each other. Human beta (β)-defensin 2 has an important role in innate immunity and generates highly specific immune responses[24]; therefore, to enhance the constancy and immunogenic response of the constructed proteins, we attached human β-defensin 2 to the N-terminal end of the vaccine sequence[25]. Finally, the constructed vaccines for L1R, B5R, A33R, and a proteome-wide construct were used for immunogenicity and allergenicity validation using VaxiJen and Algpred, respectively. We utilized ProtParam tools to check the physiochemical properties of the designed proteins, such as half-life, theoretical PI, molecular weight, and grand average of hydropathy (GRAVY) [26].

Prediction and validation of the 3D structure of the multiple-epitope vaccine

After verification of the immunogenicity and nonallergenicity of the constructed vaccines, the protein sequences of the designed vaccines were submitted to the Robetta server for 3D structural modeling. The Robetta server uses Continuous Automated Model EvaluatiOn (CAMEO) and has been recognized as the most precise and consistent server since 2014[27]. However, the Galaxy Refine sever was used to refine the protein quality[28]. Afterward, to check the quality of the predicted vaccine model, we submitted the 3D structure of the constructed vaccines (L1R, B5R, A33R, and a proteome-wide construct) to the online validation tools ProSa-Web[29]and PROCHECK. The protein structure was analyzed by using the aforementioned servers based on the quality score.

MD simulations of the modeled vaccine structure

MD simulations were performed using the AMBER 20 package to evaluate the protein folding of the constructed vaccine[30]. Proper folding of a protein is necessary to carry out normal biological functions, such as cell-to-cell linkages, and trigger immunogenic responses[31]. Before running the 50-ns simulation, the system was solvated using the FF19SB force field and the TIP3 box (12.0 A°) and then neutralized by using Na^+ counter ions. Afterward, a two-step minimization (3 000 and 6 000 steps) was carried out to remove bad clashes, followed by heating (upto 300K) and equilibrations (1 atm pressure). Finally, the root-mean square deviation (RMSD) and root-mean square fluctuation (RMSF) were calculated by using the CPPTRAJ and PTRAJ packages respectively, which are embedded in AMBER 20, to check the stability, flexibility, and folding of constructed proteins[32].

Molecular docking and dissociation constant (KD) analysis

Binding of the vaccine to the host immune cell receptor is necessary for the induction of human immune responses. Previously, the TLR-2 was recommended for the double-stranded DNA viruses belongs to the family Poxviridae so we selected the TLR-2 human receptor for our study[33]. Therefore, to check the binding of the constructed vaccines (L1R, B5R, A33R, and a proteome-wide construct) with human TLR-2, we utilized the HDOCK sever[34]. The HDOCK server supports the file as an amino acid sequence and then uses a template-based hybrid algorithm to check the interaction between proteins and protein-DNA/RNA, distinguishing this server from other available servers. To perform docking, first, the 3D structure of TLR-2 (PDB ID: 6nig) was retrieved from the UniProt online database and submitted to the HDOCK server along with the designed vaccine proteins. To further verify the strength of vaccine-TLR complexes, we performed dissociation constant

(KD) analysis. Herein, we used the online server PRODIGY to calculate the KD value[35]. To check whether our constructed vaccine can induce the human immune system through multiple TLRs, we also used TLR-3 for molecular docking with the constructed vaccines and followed the same method as for TLR-2.

In silico cloning and codon optimizations

A computer-based genetic method called in silico codon optimization is primarily used to optimize particular amino acid sequences and achieve the highest expression of proteins. Hence, to clone the designed vaccine in the *Escherichia coli* (*E. coli*) K12 expression system, we used the Java codon adaptation tool (JCat) to convert the amino acid sequence of vaccines into a DNA sequence[36]. The GC content and codon adaptation index (CAI) values generated by the JCat server represent the level of expression in the *E. coli* expression system. Finally, SnapGene software was used to clone the constructed vaccines in the pET-28a (+) expression vector.

Immune simulation

To check the response of the human immune system against the constructed L1R, B5R, A33R, and proteome-wide vaccines, we used the online immune simulation server called C-ImmSim[37]. The aforementioned server characterized the constructed vaccines in terms of their antigenicity and ability to trigger immune responses against the injected vaccines. This server estimates the level of immune cells such as helper T-cell 1 (Th1) and helper T-cell 2 (Th2). Moreover, among other immunological reactions, the server also determines the level of interferon, cytokines, and antibodies generated against administered vaccines.

Results and discussion

Monkeypox cases have been reported in multiple countries, such as the US, the UK, Singapore and Israel during 2018-2021[7], and monkeypox has subsequently spread to more than 50 countries since May 2022. Currently, no specific treatments are being devised to treat monkeypox. However, antivirals, VIG, and smallpox vaccines (Dryvax) are useful to control monkeypox outbreaks[2]. However, the Dryvax has adverse side effects such as acquired or congenital defects in the immune system of both the vaccinee and others who are in close contact with the vaccinee. Additionally, the live virus vaccination is contagious and can be shared by the vaccinee with those in their immediate circle, including youngsters and those with compromised immune systems[38, 39]. Due to the current pandemic situation of monkeypox, the development of vaccines against monkeypox virus is urgently needed. The immunoinformatic approach has several advantages over traditional vaccine development. Therefore, in the current study, we used an immunoinformatic approach to target the L1R, B5R, and A33R proteins of the monkeypox virus to screen for immunogenic and non allergenic CTL, HTL, and B-cell epitopes to construct MESVs against monkeypox virus.

Sequence retrieval and characterization

To neutralize invading pathogens and clear them from the human body, acquired immune responses are very important. After the initial attack of the infectious agent, acquired immunity in humans produces memory cells that recognize the pathogens on subsequent attacks. These memory cells provide the basis for the development of a vaccine against specific pathogens[40, 41]. Therefore, in the present study, we used immunoinformatic approaches to develop a MESVs to boost acquired immunity against the monkeypox virus.

As previously verified, plasmid DNA that encodes the monkeypox ortholog L1R, B5R, and A33R proteins induced immunity in rhesus macaques[15]. Therefore, to develop a highly immunogenic and nonallergenic MESVs, we retrieved the protein sequences of L1R, B5R, and A33R from the UniProt database. Then, the protein sequences were screened for antigenicity and allergenicity. The default threshold is 0.4, and the antigenicity scores for L1R, B5R and A33R were 0.737, 0.425 and 0.447, respectively. Furthermore, all three proteins were found to be nonallergenic.

Immunogenic epitope predictions

In acquired immunity, the MHC- II proteins that are present in all nucleated cells support active immunity by presenting epitopes of invading pathogens to CTLs[42]. Due to the importance of CTLs in the clearance of the invading pathogens, we submitted the protein sequences of L1R, B5R, and A33R for prediction of highly immunogenic CTL epitopes. Among the predicted CTL epitopes, 4 epitopes were selected for each of the L1R (TMADDDSKY, SDDSSSDTY, AATTADVRY, and AVGHCYSSY), B5R (LSSINIKEY, CTLLHRCIY, YILDTVNIY, and ILDWSIYLF) and A33R (MLFYMDLSY, KVNNNYNNY, IVTKVNNNY, and YMDLSYHGV) proteins based on high MHC binding affinity and prediction score (Table 1-7-1). Moreover, the CTL epitopes were selected for the proteome-wide vaccine from the aforementioned predicted epitopes (two epitopes from each of L1R, B5R, and A33R) based on high MHC binding affinity and prediction scores (Table 1-7-1).

Table 1-7-1   List of CTL epitopes predicted for the L1R, B5R, and A33R proteins of monkeypox virus

Residue no.	Peptide sequence	MHC binding affinity	Rescaled binding affinity	C-terminal cleavage affinity	Transport affinity	Prediction score	MHC-I binding
L1R							
1	TMADDDSKY	0.352 7	1.497 6	0.976 8	2.990 0	1.793 0	YES
16	SDDSSSDTY	0.211 3	0.897 0	0.950 5	2.648 0	1.172 0	YES
70	AATTADVRY	0.180 8	0.767 5	0.972 1	3.139 0	1.070 3	YES
30	AVGHCYSSY	0.160 7	0.682 1	0.962 6	3.145 0	0.983 8	YES
B5R							
174	LSSINIKEY	0.486 5	2.065 7	0.718 1	2.935 0	2.320 1	YES
11	CTLLHRCIY	0.280 4	1.190 5	0.510 4	2.850 0	1.409 6	YES
147	YILDTVNIY	0.263 9	1.120 3	0.859 2	3.105 0	1.404 5	YES
8	ILDWSIYLF	0.190 1	0.807 1	0.790 2	2.234 0	1.037 3	YES
A33R							
75	MLFYMDLSY	0.290 6	1.233 7	0.967 8	3.139 0	1.535 8	YES
111	KVNNNYNNY	0.257 0	1.091 3	0.855 9	3.086 0	1.374 0	YES
108	IVTKVNNNY	0.247 0	1.048 6	0.929 9	3.060 0	1.341 1	YES
78	YMDLSYHGV	0.267 4	1.135 3	0.806 5	0.135 0	1.263 1	YES
Proteome							
1	TMADDDSKY	0.352 7	1.497 6	0.976 8	2.990 0	1.793 0	YES
16	SDDSSSDTY	0.211 3	0.897 0	0.950 5	2.648 0	1.172 0	YES
174	LSSINIKEY	0.486 5	2.065 7	0.718 1	2.935 0	2.320 1	YES
11	CTLLHRCIY	0.280 4	1.190 5	0.510 4	2.850 0	1.409 6	YES
75	MLFYMDLSY	0.290 6	1.233 7	0.967 8	3.139 0	1.535 8	YES
111	KVNNNYNNY	0.257 0	1.091 3	0.855 9	3.086 0	1.374 0	YES

On the other hand, helper T-cells are also important in generating acquired immunity against invading pathogens including monkeypox virus. Helper-T cells are involved in activating both the B-cell and cytotoxic T-cell pathways. After the entry of monkeypox virus into the human body, macrophages engulf the this virus and present the epitopes to helper T-cells to activate the B-cells to produce neutralizing antibodies. Due to the indispensable role of HTLs in boosting immune responses, we submitted the proteins L1R, B5R, and A33R for prediction of the HTL epitopes that trigger immune responses. Among the predicted epitopes, we selected only four epitopes with the lowest percentile rank for each of the L1R (ADVRYDRRDASNVMD, AATTADVRYDRRDAS, DKKN KVVTDGRVKNK, and KNKVVTDGRVKNKGY), B5R (NRGSFIHNLGLSSIN, FLSYILDTVNIYISI, PSLKIMIPSMIA ITK, and LHIYFIDAANTNIMI) and A33R (YNCYNYDDT FFDDDD, GKVTINDLKMMLFYM, VLGKVTINDLKMMLF, and SGGGIYHDDLVVLGK) proteins. Similarly, two epitopes with the lowest percentile score from each of the above proteins were selected for the construction of a proteome-wide vaccine (Table 1-7-2).

Table 1-7-2   List of Th-cell epitopes predicted for the L1R, B5R, and A33R proteins of monkeypox virus

S. no.	Allele	Start	End	Peptide sequence	Method	Percentile rank
L1R						
1	HLA-DRB1*03: 01	74	88	ADVRYDRRDASNVMD	Consensus (comb.lib./smm/nn)	0.34
2		70	84	AATTADVRYDRRDAS		0.34
3		45	59	DKKNKVVTDGRVKNK		1.90
4		47	61	KNKVVTDGRVKNKGY		1.90
B5R						
1	HLA-DRB3*02: 02	99	113	NRGSFIHNLGLSSIN	NetMHCIIpan	0.01
2	HLA-DRB3*01: 01	144	158	FLSYILDTVNIYISI	Consensus (smm/nn/sturniolo)	0.02
3		400	414	PSLKIMIPSMIAITK		0.06
4		73	87	LHIYFIDAANTNIMI		0.19
A33R						
1	HLA-DRB3*01: 01	128	142	YNCYNYDDTFFDDDD		0.52
2	HLA-DRB1*03: 01	65	79	GKVTINDLKMMLFYM		0.86
3		63	77	VLGKVTINDLKMMLF		0.91
4	HLA-DRB3*01: 01	52	66	SGGGIYHDDLVVLGK		2.40
Proteome						
1	HLA-DRB1*03: 01	74	88	ADVRYDRRDASNVMD		0.34
2		70	84	AATTADVRYDRRDAS		0.34
3	HLA-DRB3*02: 02	99	113	NRGSFIHNLGLSSIN	NetMHCIIpan	0.01
4	HLA-DRB3*01: 01	144	158	FLSYILDTVNIYISI	Consensus (smm/nn/sturniolo)	0.02
5		128	142	YNCYNYDDTFFDDDD		0.52
6	HLA-DRB1*03: 01	65	79	GKVTINDLKMMLFYM		0.86

Differentiated B-cell are the sole cells responsible for producing antibodies in our body and are considered key players in the adaptive and innate immune responses[43]. Furthermore, the design of B-cell epitopes is very important for the generation of immune responses because the antibodies that are produced by B cells induce some protective mechanisms, such as agglutination, neutralization, complement system activation, and cell-mediated cytotoxicity[44]. Therefore, all the selected proteins of the monkeypox virus were submitted for the prediction of B-cell epitopes. Among the predicted epitopes, we selected epitopes with a score greater than 0.8 and used them for designing the final vaccine against the monkeypox virus (Table 1-7-3). Finally, we selected one epitope from each protein with a high score for proteome-wide vaccine construction.

Table 1-7-3   List of B-cell epitopes predicted by the ABCpred server

S. no.	Position	Epitope	Score
L1R			
1	10	DDSKYASDDSSSDTYDSKNK	0.87
2	37	SYRTKTDSDKKNKVVTDGRV	0.86
B5R			
1	60	PNLFHILEYGENILHIYFID	0.82
2	441	AITKHKQHNADLLKMCIKYT	0.82
A33R			
1	53	GGGIYHDDLVVLGKVTINDL	0.91
2	37	FVFYKPKHSTVVKYLSGGGI	0.85
Proteome			
1	10	DDSKYASDDSSSDTYDSKNK	0.87
2	60	PNLFHILEYGENILHIYFID	0.82
3	53	GGGIYHDDLVVLGKVTINDL	0.91

Construction of the MESVs and physiochemical parameter analysis

The computer-aided design of vaccines is a more appropriate replacement for traditional vaccine development, as it is cost-effective, results in the combination of highly immunogenic epitopes from different regions of the protein, is less time consuming and results in the production of no allergenic responses[45]. To construct the final MESVs, we combined all the predicted immunogenicepitopes for L1R, B5R, A33R and the proteome-wide construct by using different linkers. The AAY, GPGPG, and KK linkers were used to combine CTL, HTL, and B-cell epitopes, respectively. These linkers are immunogenic in nature and help the constructed vaccines boost immune responses like the AAY liker enhances the epitope presentation, the GPGPG linker stimulates the HTL responses and the immunogenicity of antibody epitopes while the KK linker enhances the constructed vaccines immunogenicity. Second, these linkers prevent the folding of epitopes, keeping them separated from each other[42, 46]. Finally, the EAAK linker was used to connect the adjuvant to the N-terminal end of the constructed vaccine, which further enhances the efficiency of the vaccine in triggering immune responses. Human β-defensin 2 has an important function in innate immunity and can generate antigen-specific immune responses[24]; therefore, to increase the stability and immunogenic response of the constructed vaccine, we attached human β-defensin 2 as an adjuvant (Figure 1-7-2) [25]. Finally, to provide a highly immunogenic vaccine with no allergenic responses, the constructed vaccines were subjected to allergenicity and antigenicity checks. The constructed vaccines were found to be nonallergenic, with an antigenicity score higher than the threshold value (0.4).

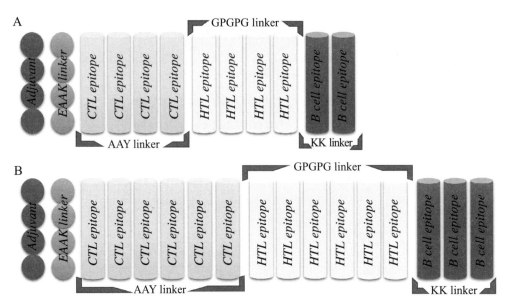

A: The inal vaccine construct of L1R, B5R and A33R; B: The inal proteome-wide vaccine construct.

Figure 1-7-2   Topographical representation of the inal vaccine constructs

Then, the physiochemical properties of the constructed vaccines were calculated, such as half-life, theoretical PI, molecular weight, aliphatic index, instability index, and GRAVY. The calculated molecular weights for L1R, B5R, A33R, and the proteome-wide construct were 31.9, 32.4, 32.1, and 41.7 kDa, respectively, while the half-life of all constructed vaccines was greater than 10 h in *E.coli*. This means that the constructed vaccine can be easily expressed and purified, as the molecular weights are within the normal range. An instability index value of less than 40 indicates that a protein is stable. The instability index values were less than 40, which means that the constructed vaccines are stable (Table 1-7-4). The theoretical PI value, aliphatic index, and GRAVY represent the acidity, thermostability and hydrophobicity of the constructed vaccines, respectively[26]. The values of the physiochemical properties for L1R, B5R, A33R, and the proteome-wide construct are shown (Table 1-7-4).

Table 1-7-4   Designed vaccine physiochemical properties

Vaccine construct	Molecular weight / kDa	Theoretical PI	Half-life in *E.coli*	Instability index	Aliphatic index	GRAVY
L1R	31.9	5.39	>10 h	29.27	65.3	−0.584
B5R	32.4	5.69	>10 h	31.59	108.38	0.263
A33R	32.1	4.98	>10 h	24.46	89.50	0.023
Proteome wide	41.7	4.60	>10 h	37.13	80.34	−0.253

3D structure prediction and validation

The sequences of the predicted vaccines L1R, B5R, A33R, and the proteome-wide construct were submitted to the Robetta server for 3D structure modeling. The Robetta server uses CAMEO and has been recognized as the most precise and consistent server since 2014[27]. The Robetta server generated five models for each of the constructed vaccines, and we selected the best model for each vaccine based on quality analysis (Figure 1-7-3).

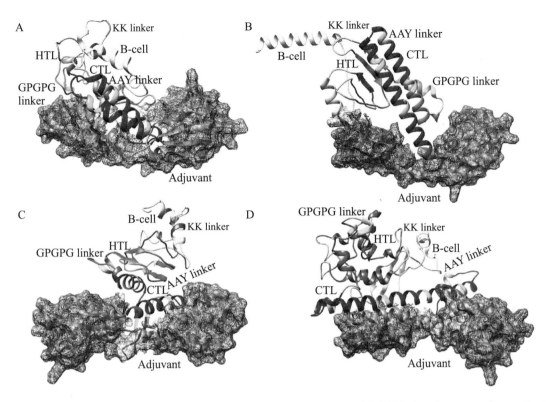

A: The 3D structure of the L1R inal vaccine construct generated by Robetta; B: The 3D structure of the B5R inal vaccine construct generated by Robetta; C: The 3D structure of the A33R inal vaccine construct generated by Robetta; D: The 3D structure of the proteome-wide inal vaccine construct generated by Robetta (color figure in appendix).

Figure 1-7-3   Robetta-generated 3D structures for the inal vaccine constructs of the monkeypox virus

To select the best model for the designed vaccines, we performed ProSA-web and Ramachandran plot analyses. First, we submitted the models for Ramachandran analysis and selected the model with the highest ratio of amino acids present in the favored region and the lowest ratio in the disallowed region. The percentage of residues in the favored region was 87.5%, 92.4%, 88.5%, and 85.4% for L1R, B5R, A33R, and the proteome-wide construct, respectively; however, the percentage of residues present in the outlier region was 0.4%, 0.4%, 1.2%, and 0.9%, respectively. Subsequently, to confirm the quality and errors in the predicted structures, we submitted the models for ProSA-web analysis. The predicted quality Z score was −6.19, −6.17, −8.77, and −8.42 for L1R, B5R, A33R, and the proteome-wide construct, respectively (Figure 1-7-4). The above data indicated that the quality of the selected structures is good, as the Z score lies in the normal range. Furthermore, to check the proper folding of the model structures, we subjected the selected models to molecular dynamics simulation analysis.

Molecular dynamics simulations

The stability investigation of the constructed vaccines by MD simulations revealed the stable behavior of the B5R, A33R and proteome-wide vaccines; however, a steady fluctuation was observed in the L1R vaccine until 50 ns. The B5R protein gained stability at 9 ns and remained stable until 50 ns. However, the protein A33R and proteome-wide gained stability at 20 ns and 30 ns and remained stable until 50 ns. The above data confirmed the stable folding and sustained behavior during the course of simulations (Figure 1-7-5 A-D). The residual fluctuations of the constructed vaccines were also evaluated by calculating the RMSF. The residual fluctuations were found to be minimal; however, little fluctuation was found between 175～200 and 250～300.

The residues present at the ends of proteins show more fluctuations, which might be due to the flexible nature of the N-terminal and C-terminal ends (Figure 1-7-5 E).

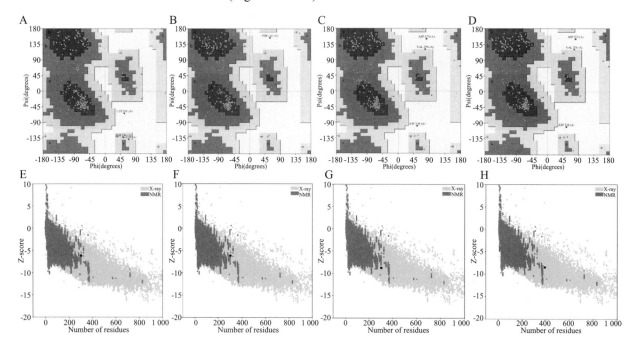

A-D: Structural validation of the L1R, B5R, A33R, and proteome-wide vaccines by Ramachandran plots; E -H: Structural validation of the L1R, B5R, A33R, and proteome-wide vaccines by ProSA -web. Uppercase and lowercase: This is the international farmate for this sever. ProSA-web (protein structure analysis-web) (color figure in appendix).

Figure 1-7-4   Quality analysis of Robetta-generated models by Ramachandran plots and ProSA-web

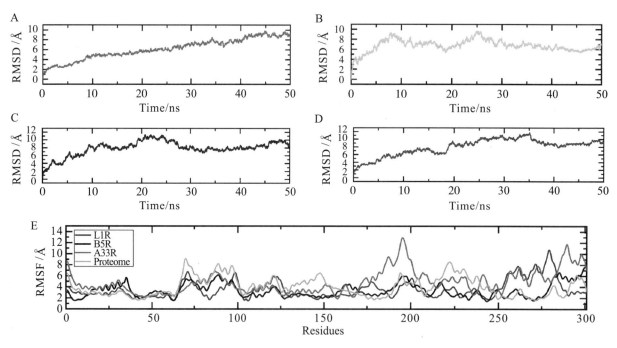

A: RMSD value for the L1R vaccine construct; B: RMSD value for the B5R vaccine construct; C: RMSD value for the A33R vaccine construct; D: RMSD value for the proteome-wide vaccine construct; E: RMSF values for the L1R, B5R, A33R and proteome-wide vaccine constructs (color figure in appendix).

Figure 1-7-5   Molecular dynamics simulations of the constructed vaccines

Molecular docking of constructed vaccines with human TLR-2

Monkeypox virus after entering the host infect the dendritic cells (DCs), macrophages and also inhibit the activation of host innate immune system[47, 48]. TLRs are important regulators of inflammatory pathways that play indispensable roles in mediating immune responses to pathogens. TLRs recognize pathogen-associated molecular patterns (PAMPs), leading to changes in gene expression and triggering intracellular signaling pathways. The innate immune system of the host detects pathogens and responds accordingly via recognition by TLRs[49]. TLRs are important in recognizing the various components of monkeypox viruses, such as nucleic acids and envelope glycoproteins, leading to a series of cascades, including the production of IFN-I, inflammatory cytokines and chemokines[50]. Moreover, adaptive immune responses are also activated when TLRs induce the maturation of dendritic cells (DCs) [51]. Therefore, to check the binding affinity of the constructed vaccines with human TLR-2, we used the HDOCK server. The binding scores predicted by HDOCK for the L1R, B5R, A33R, and proteome-wide vaccines were –266 kcal/mol, –288 kcal/mol, –264 kcal/mol, and –250 kcal/mol, respectively, which shows robust binding of all constructed vaccines with human TLR-2 (Figure 1-7-6). The above data were further verified by interface analysis using the PDBsum server. The binding interface analysis of the L1R-TLR-2 complex revealed 10 hydrogen bonds, 2 salt bridges and 213 non-bonded contacts. However, the B5R-TLR-2 complex formed 1 hydrogen bond, 3 salt bridges and 167 non-bonded contacts. The bonding network of the A33R-TLR-2 complex formed 3 hydrogen bonds, 2 salt bridges and 172 non-bonded contacts. Similarly, the bonding network of the proteome-wide construct-TLR-2 complex formed 8 hydrogen bonds, 5 salt bridges, and 216 non-bonded contacts. To further verify the strength of a vaccine-TLR complex, we performed dissociation constant analysis. The K_D values for L1R-TLR-2, B5R-TLR-2, A33R-TLR-2 and the proteome-wide construct-TLR-2 complex were $4.1E^{-10}$, $5.1E^{-09}$, $3.4E^{-08}$, and $1.5E^{-06}$, respectively. The K_D estimation revealed that the L1R-TLR-2 ($4.1E^{-10}$) complex had a stronger binding

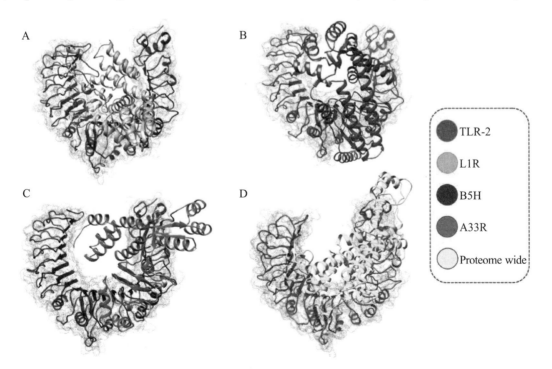

A: L1R-TLR-2 complex; B: B5R-TLR-2 complex; C: A33R-TLR-2 complex; D: proteome-wide construct-TLR-2 complex. The designed vaccines and TLR-2 are shown in different colors (color figure in appendix).

Figure 1-7-6   Complexes of the L1R, B5R, A33R, and proteome-wide vaccine constructs with human TLR-2

affinity than all the other complexes. Furthermore, to check whether our constructed vaccine can induce the human immune responses through multiple TLRs, we docked the constructed vaccines with human TLR-3. The binding scores of L1R, B5R, A33R, and proteome-wide vaccines with human TLR-3 were −241 kcal/mol, −291 kcal/mol, −255 kcal/mol, and −261 kcal/mol, respectively. The analysis of binding interface revealed that the L1R-TLR-3 complex contains 4 salt bridges, 6 hydrogen bonds and 130 non-bonded contacts, while the B5R-TLR-3 complex contains 6 salt bridges, 7 hydrogen bonds and 343 non-bonded contacts. Similarly, the A33R-TLR-3 complex contains 1 hydrogen bond and 118 non-bonded contacts however, the proteome-TLR-3 complex contains 5 salt bridge, 12 hydrogen bonds and 244 non-bonded contacts. The dissociation constant analysis was performed to check the strength of the aforementioned complexes. The K_D values for the LIR-TLR-3, B5R-TLR-3, A33R-TLR-3 and proteome-TLR-3 complexes were $1.6E^{-09}$, $2.2E^{-06}$, $3.8E^{-07}$, and $4.5E^{-12}$, respectively. In summary, the bonding network and dissociation constant analysis revealed that our constructed vaccines can induce the human immune system through multiple TLRs however, the binding affinity of constructed vaccines for TLR-2 is stronger as compared to the TLR-3 except proteome wide vaccine. Secondly, in case of TLR-3 the bonding target of constructed vaccines in not uniform . Therefore, our study verifies the previous recommendation of TLR-2 for the double stranded DNA envelope viruses belongs to the family Poxviridae[33].

In silico cloning of MESVs in the pET-28a (+) expression vector

Codon optimization of the constructed vaccines (L1R, B5R, A33R, and the proteome-wide construct) was carried out for maximum expression in the *E.coli* K12 strain expression system by using the JCat tool. The JCat tool calculates some parameters, such as the CAI value and GC content[36]. The generated CAI values for L1R, B5R, A33R and the proteome-wide vaccine construct were 0.96, 0.95, 0.96, and 0.95, while the GC contents were 69.9%, 65%, 65.7% and 66.1%, respectively. The estimated CAI values for the above constructed vaccines lie in the ideal range for high expression; however, the ideal range of GC content is from 30%~70%. Herein, the above data confirmed the high expression of the constructed vaccines in the *E.coli* expression system. Then, the optimized sequences were cloned into the pET-28a (+) expression vector using the Xho1 and EcoR1 restriction sites (Figure 1-7-7).

Immune response simulation analysis

Vaccination elicits immune responses in the body without causing the disease. Different vaccines stimulate immune responses differently[52]. Protein antigens together with adjuvants trigger local innate immune responses, i.e., proinflammatory cytokine production by macrophages. Dendritic cells or macrophages take up the antigen, and these cells then present the antigen on their surface via MHC molecules. T-cells recognize the MHC/antigen complex via their T-cell receptors, leading to the production of memory T-cells (adaptive immunity). On the other hand, antigens comprising polysaccharides induce antibody responses without T-cell involvement. Polysaccharides bind to mature B-cells, leading to the production of antibody-producing B cells. Such immunological responses do not have memory[53]. Immune response simulation analysis was performed to check the immune boosting efficiency of the constructed vaccines. The immune simulation monitors the response of the immune system in terms of antibody production after the injection of constructed vaccines. After the injection of each vaccine construct, the response of the immune system and antibody production were extremely high (Figure 1-7-7). After injection of each construct, the antigen titer remained low until the 5th day, and then the titer increased until the 13th day; however, the increase in the antibody titer started on the 15th day.

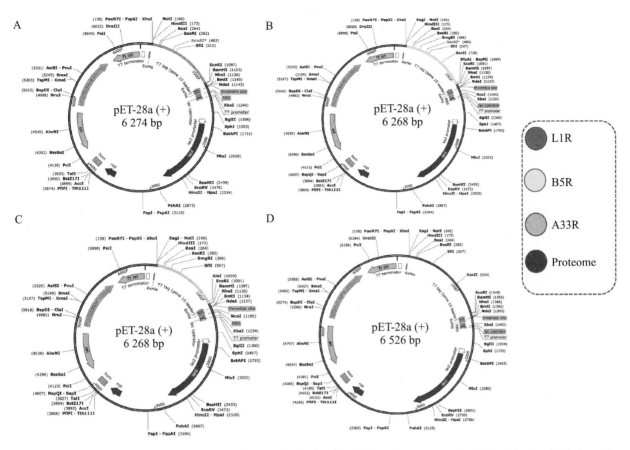

A: Cloning of the L1R vaccine sequence in the pET28a (+) vector; B: Cloning of the B5R vaccine sequence in the pET28a (+) vector; C: Cloning of the A33R vaccine sequence in the pET28a (+) vector; D: Cloning of the proteome-wide vaccine sequence in the pET28a (+) vector.

Figure 1-7-7   Cloning of optimized vaccine sequences in the pET28a (+) expression system

On the 17th day, the total neutralization of antigens was recorded with the induction of the production of other immune system factors. The combined IgM and IgG titer reached 4×10^7 in the case of L1R and B5R, while the combined IgM and IgG titer reached 4.3×10^7 after injection of A33R and the proteome-wide vaccine (Figure 1-7-7). High combined IgG1+IgG2 and IgG1 titers were also observed after injection of the L1R, B5R, A33R, and proteome-wide vaccines. The level of cytokines and interleukins were also analyzed after the injection of constructed vaccines. At 5th day of injection the level of IFN-γ and IL-2 increased slowly and reached to maximum at 18th day. The levels of IFN-γ and IL-2 were higher in case of A33R vaccine as compared to the others. The levels of aforementioned immune factors were significantly higher, which shows the vigorous and steady immune triggering response upon injection. The above data showed that the constructed vaccines can robustly induce immune responses against the invading monkeypox virus (Figure 1-7-8).

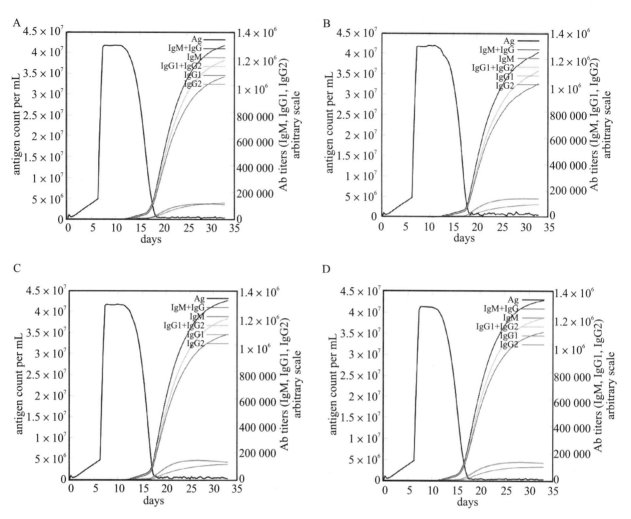

A: Immune responses to the L1R vaccine; B: Immune responses to the B5R vaccine; C: Immune responses to the A33R vaccine; D: Immune responses to the proteome-wide vaccine (color figure in appendix).

Figure 1-7-8　Immune response simulation for a constructed vaccine against monkeypox virus

Conclusions

Computationally designed vaccines are thermodynamically stable, effective, specific and low cost compared to traditional vaccine development. In the current study, we used immunoinformatic approaches to design highly immunogenic, thermostable and nonallergenic MESVs against the monkeypox virus. We used L1R, B5R, and A33R proteins of monkeypox virus to develop highly antigenic and nonallergenic CTL, HTL and B-cell epitopes by using bioinformatics tools. Various bioinformatics tools and approaches confirmed that the 3D structure of each constructed vaccine had a proper folding, a stronger binding with human TLR-2 and high expression of constructed proteins in the *E.coli* expression system. Moreover, the injected vaccines robustly activated the immune system, with high titers of IgG and IgM antibodies. In conclusion, the present study provided dynamic and effective vaccines that are composed of highly antigenic and non-allergenic peptides against the monkeypox virus, demanding further experimental trials.

References

[1] KMIEC D, KIRCHHOFF F. Monkeypox: a new threat?. Int J Mol Sci, 2022, 23 (14): 7866.

[2] RIZK J G, LIPPI G, HENRY B M, et al. Prevention and Treatment of Monkeypox. Drugs, 2022, 82(9): 957-963.

[3] MANES N P, ESTEP R D, MOTTAZ H M, et al. Comparative proteomics of human monkeypox and vaccinia intracellular mature and extracellular enveloped virions. J Proteome Res, 2008, 7(3): 960-968.

[4] MAGNUS P V, ANDERSEN E K, PETERSEN K B, et al. A pox-like disease in cynomolgus monkeys. Acta Pathologica Microbiologica Scandinavica, 1959, 46 (2): 156-176.

[5] BUNGE E M, HOET B, CHEN L, et al. The changing epidemiology of human monkeypox-a potential threat? a systematic review. PLOS Negl Trop Dis, 2022, 16(2): e0010141.

[6] THORNHILL J P, BARKATI S, WALMSLEY S, et al. Monkeypox virus infection in humans across 16 countries-April-June 2022. New Engl J Med, 2022, 387: 679-691.

[7] HAIDER N, GUITIAN J, SIMONS D, et al. Increased outbreaks of monkeypox highlight gaps in actual disease burden in Sub- Saharan Africa and in animal reservoirs. Int J Infect Dis, 2022, 122: 107-111.

[8] GRUBER MF. Current status of monkeypox vaccines. NPJ Vaccines, 2022, 7 (1): 1-3.

[9] FINE P, JEZEK Z, GRAB B, et al. The transmission potential of monkeypox virus in human populations. Int J Epidemiol, 1988, 17(3): 643-650.

[10] FOGG C, LUSTIG S, WHITBECK J C, et al. Protective immunity to vaccinia virus induced by vaccination with multiple recombinant outer membrane proteins of intracellular and extracellular virions. J Virol, 2004, 78 (19): 10230-10237.

[11] ROBERTS K L, SMITH G L. Vaccinia virus morphogenesis and dissemination. Trends Microbiol, 2008, 16(10): 472-479.

[12] MOSS B. Smallpox vaccines: targets of protective immunity. Immunol Rev, 2011, 239(1): 8-26.

[13] RAMÍREZ J C, TAPIA E, ESTEBAN M. Administration to mice of a monoclonal antibody that neutralizes the intracellular mature virus form of vaccinia virus limits virus replication efficiently under prophylactic and therapeutic conditions. J Gen Virol, 2002, 83(5): 1059-1067.

[14] LAW M, SMITH G L. Antibody neutralization of the extracellular enveloped form of vaccinia virus. Virology, 2001, 280(1): 132-142.

[15] HERAUD J M, EDGHILL SMITH Y, AYALA V, et al. Subunit recombinant vaccine protects against monkeypox. J Immunol, 2006, 177(4): 2552-2564.

[16] LANE J M. Immunity to smallpox and vaccinia: the future of smallpox vaccines. Expert Rev Clin Immunol, 2006, 2(3): 325-327.

[17] SHCHELKUNOV S N, YAKUBITSKIY S N, SERGEEV A A, et al. Enhancing the immunogenicity of vaccinia virus. Viruses, 2022, 14(7): 1453.

[18] KUMAR N, SOOD D, CHANDRA R. Design and optimization of a subunit vaccine targeting COVID-19 molecular shreds using an immunoinformatics framework. RSC Adv, 2020, 10(59): 35856-35872.

[19] KUMAR N, ADMANE N, KUMARI A, et al. Cytotoxic T-lymphocyte elicited vaccine against SARS-CoV-2 employing immunoinformatics framework. Sci Rep, 2021, 11(1): 1-14.

[20] DOYTCHINOVA I A, FLOWER D R. VaxiJen: a server for prediction of protective antigens, tumour antigens and subunit vaccines. BMC Bioinf, 2007, 8(1): 1-7.

[21] LARSEN M V, LUNDEGAARD C, LAMBERTH K, et al. Large-Scale validation of methods for cytotoxic T-lymphocyte epitope prediction. BMC Bioinf, 2007, 8(1): 1-12.

[22] VITA R, MAHAJAN S, OVERTON J A, et al. The immune epitope database (IEDB): 2018 update. Nucleic Acids Res, 2019, 47(D1): D339-343.

[23] SAHA S, RAGHAVA G P S. Prediction of continuous b-cell epitopes in an antigen using recurrent neural network. Proteins: Structure Function Bioinf, 2006, 65 (1): 40-48.

[24] KIM J, YANG Y L, JANG S-H, et al. Human β -defensin 2 plays a regulatory role in innate antiviral immunity and is capable of potentiating the induction of antigen-specific immunity. Virol J, 2018, 15(1): 1-12.

[25] SCHRÖDER J-M, HARDER J. Human beta-defensin-2. Int J Biochem Cell Biol, 1999, 31(6): 645-651.

[26] GASTEIGER E, HOOGLAND C, GATTIKER A, et al. Protein identification and analysis tools on the ExPASy server. Proteomics Protoc Handb, 2005, p: 571-607.

[27] KIM D E, CHIVIAN D, BAKER D. Protein structure prediction and analysis using the Robetta server. Nucleic Acids Res, 2004, 32(suppl_2): W526-531.

[28] GIARDINE B, RIEMER C, HARDISON R C, et al. Galaxy: a platform for interactive large-scale genome analysis. Genome Res, 2005, 15(10): 1451 -1455.

[29] WIEDERSTEIN M, SIPPL M J. ProSA-web: interactive web service for the recognition of errors in three-dimensional structures of proteins. Nucleic Acids Res, 2007, 35(suppl_2): W407-410.

[30] Salomon-Ferrer R, Gotz A W, Poole D, et al. Routine microsecond molecular dynamics simulations with AMBER on GPUs. 2. explicit solvent particle mesh ewald. J Chem Theory Comput, 2013, 9(9): 3878-3888.

[31] BHARDWAJ V K, PUROHIT R. Targeting the protein-protein interface pocket of aurora-A-TPX2 complex: rational drug design and validation. J Biomolecular Structure Dynamics, 2021, 39(11): 3882-3891.

[32] ROE D R, CHEATHAM T E III. PTRAJ and CPPTRAJ: software for processing and analysis of molecular dynamics trajectory data. J Chem Theory Comput, 2013, 9(7): 3084-3095.

[33] XAGORARI A, CHLICHLIA K. Toll-like receptors and viruses: induction of innate antiviral immune responses. Open Microbiol J, 2008, 2: 49.

[34] YAN Y, TAO H, HE J, et al. The HDOCK server for integrated protein-protein docking. Nat Protoc, 2020, 15(5): 1829-1852.

[35] XUE L C, RODRIGUES J P, KASTRITIS P L, et al. PRODIGY: A web server for predicting the binding affinity of protein-protein complexes. Bioinformatics, 2016, 32(23): 3676-3678.

[36] GROTE A, HILLER K, SCHEER M, et al. JCat: a novel tool to adapt codon usage of a target gene to its potential expression host. Nucleic Acids Res, 2005, 33(suppl_2): W526-531.

[37] CASTIGLIONE F, CELADA F. Immune system modelling and simulation. Boca Raton: CRC Press, 2015.

[38] ROTZ L D, DOTSON D A, DAMON I K, et al. Vaccinia (smallpox) vaccine: recommendations of the Advisory Committee on Immunization Practices (ACIP), 2001. MMWR Recomm Rep, 2001, 50(RR-10): 1-25; quiz CE1-7.

[39] Centers for Disease Control and Prevention (CDC). Supplemental recommendations on adverse events following smallpox vaccine in the pre-event vaccination program: recommendations of the Advisory Committee on Immunization Practices. MMWR Morb Mortal Wkly Rep, 2003, 52(13): 282-284.

[40] ZHENG B, SULEMAN M, ZAFAR Z, et al. Towards an ensemble vaccine against the pegivirus using computational modelling approaches and its validation through in silico cloning and immune simulation. Vaccines, 2021, 9(8): 818.

[41] SULEMAN M, BALOUCH A R, RANDHAWA A W, et al. Characterization of proteome wide antigenic epitopes to design proteins specific and proteome-wide ensemble vaccines against heartland virus using structural vaccinology and immune simulation approaches. Microbial Pathogenesis, 2022, 168: 105592.

[42] KHAN A, KHAN S, SALEEM S, et al. Immunogenomics guided design of immunomodulatory multi-epitope subunit vaccine against the SARS-CoV-2 new variants, and its validation through in silico cloning and immune simulation. Comput Biol Med, 2021, 133: 104420.

[43] TSAI D Y, HUNG K H, CHANG C W, et al. Regulatory mechanisms of b cell responses and the implication in b cell-related diseases. J Biomed Sci, 2019, 26 (1): 1-13.

[44] SULEMAN M, ASAD U, ARSHAD S, et al. Screening of immune epitope in the proteome of the dabie bandavirus, SFTS, to design a protein-specific and proteome-wide vaccine for immune response instigation using an immunoinformatics approaches. Comput Biol Med, 2022, 148: 105893.

[45] SUNITA, SAJID A, SINGH Y, et al. Computational tools for modern vaccine development. Hum Vaccines Immunotherapeutics, 2020, 16(3): 723-735.

[46] BALDAUF K J, ROYAL J M, HAMORSKY K T, et al. Cholera toxin b: one subunit with many pharmaceutical applications. Toxins, 2015, 7(3): 974-996.

[47] ARNDT W D, COTSMIRE S, TRAINOR K, et al. Evasion of the innate immune type I interferon system by monkeypox virus. J Virol, 2015, 89(20): 10489-10499.

[48] ALAKUNLE E, MOENS U, NCHINDA G, et al. Monkeypox virus in Nigeria: infection biology, epidemiology, and evolution. Viruses, 2020, 12(11): 1257.

[49] KAWAI T, AKIRA S. Pathogen recognition with toll-like receptors. Curr Opin Immunol, 2005, 17(4): 338-344.

[50] KAWAI T, AKIRA S. Innate immune recognition of viral infection. Nat Immunol, 2006, 7(2): 131-137.

[51] PASARE C, MEDZHITOV R. Toll-like receptors: linking innate and adaptive immunity. Mech lymphocyte activation Immune Regul X, 2005, 560: 11-18.

[52] KANG S M, COMPANS R W. Host responses from innate to adaptive immunity after vaccination: molecular and cellular events. Molecules Cells, 2009, 27(1): 5-14.

[53] WEINBERGER B, HERNDLER-BRANDSTETTER D, SCHWANNINGER A, et al. Biology of immune responses to vaccines in elderly persons. Clin Infect Dis, 2008, 46(7): 1078-1084.

Homologous recombination as an evolutionary force in the avian influenza a virus

He Chengqiang, Xie Zhixun, Han Guanzhu, Dong Jianbao, Wang dong, Liu Jiabo, Ma Leyuan, Tang Xiaofei, Liu Xiping, Pang Yaoshan, and Li Guorong

Abstract

Avian influenza A viruses (AIVs), including the H5N1, H9N2, and H7N7 subtypes, have been directly transmitted to humans, raising concerns over the possibility of a new influenza pandemic. To prevent a future avian influenza pandemic, it is very important to fully understand the molecular basis driving the change in AIV virulence and host tropism. Although virulent variants of other viruses have been generated by homologous recombination, the occurrence of homologous recombination within AIV segments is controversial and far from proven. This study reports three circulating H9N2 AIVs with similar mosaic PA genes descended from H9N2 and H5N1. Additionally, many homologous recombinants are also found deposited in GenBank. Recombination events can occur in PB2, PB1, PA, HA, and NP segments and between lineages of the same/different serotype. These results collectively demonstrate that intragenic recombination plays a role in driving the evolution of AIVs, potentially resulting in effects on AIV virulence and host tropism changes.

Keywords

avain influenza virus, homologous recombination, evolution

Introduction

Recently, a highly pathogenic avian influenza A virus (H5N1) has resulted in the deaths of more than 200 people and millions of poultry in Asia, Europe, and Africa[1]. In addition to H5N1, purely avian influenza viruses (AIVs), specifically the H9N2 and H7N7 subtypes, have been directly transmitted to humans, raising concerns over the possibility of a new influenza pandemic among the world's immunologically naive populations[2]. In order to prevent or limit a future avian influenza pandemic, it is very important to fully understand the molecular basis driving the change in AIV virulence and host tropism.

Until now, re-assortment and dynamic gene mutation have been considered the key factors responsible for the evolution of AIV and have thus been studied in detail[3]. It has been previously reported that a re-assortment of gene segments was responsible for creating an entirely novel influenza A virus strain capable of infecting humans[4]. However, phylogenetic analyses of amino acid changes suggested that avian influenza viruses, unlike mammalian strains, show low evolutionary rates[5]; thus, dynamic gene mutation must still play an important role in the virulence change of AIV[6]. In addition, it has been shown that homologous recombination can also play an important role in the evolution of some positive-strand RNA viruses[7-9]. Virulent variants of some other viruses have been generated by homologous recombination[10-12]. Even recombination between a vaccine strain and a persisting pestivirus resulted in a cytopathogenic virus and induction of lethal disease[13]. Although there is evidence that influenza viruses undergo various forms of nonhomologous recombination[14, 15],

112

the occurrence of homologous recombination within segments is highly controversial. Gibbs et al.[16]have proposed that recombination occurred in the HA gene between human and swine influenza viruses. However, the evidence of recombination has been questioned because of the absence of phylogenetic support[17]. Interestingly, it has recently been reported that human influenza viruses do not recombine[18]. Therefore, there is much evidence still needed to demonstrate that homologous recombination occurs within segments.

In this study, we report on three circulating H9N2 AIVs with mosaic gene (s) isolated in China in the year 2 000. To determine whether homologous recombination really shapes the evolution of AIVs and to provide some insights into the recombination itself, we analyzed roughly 9 000 complete segments of AIVs deposited in GenBank and found the evidence of recombination in these genes. Moreover, several mosaic viruses were also found to be able to circulate in the field. These data demonstrate that intragenic recombination can act to drive the evolution of AIVs.

Materials and methods

Viruses

Viruses of the H9N2 subtype, A/chicken/Guangxi/1/00, A/chicken/Guangxi/14/00, and A/chicken/Guangxi/17/00, were isolated from dead chickens exhibiting various clinical symptoms of avian influenza illness in three counties of the Guangxi in 2000. A/chicken/Guangxi/14/00 and A/chicken/Guangxi/17/00 were collected in November, whereas A/chicken/Guangxi/1/00 was collected in March. The serotype of each isolate was determined by hemagglutination inhibition and neuraminidase inhibition tests using polyclonal chicken antisera, standard hemagglutination inhibition antisera (bought from the Harbin Veterinary Research Institute of China: Harbin City, China) and neuraminidase inhibition antisera (a generous gift from Dr Shortridge of Hong Kong University). The three isolates were determined to be of the H9N2 subtype. Before further studies were conducted, the viruses were purified using a plaque-forming method in chicken embryo fibroblast cells. Purified viruses were cultured in 10-day-old, specific pathogen-free embryonated chicken eggs for sequence analysis.

Sequencing

Viral RNA was extracted from allantoic fluid using a protocol described previously[19]. Amplification of the eight full-length genes was carried out by reverse transcription polymerase chain reaction using multipairs of specific primers listed in Table 1-8-1. DNA fragments were cloned into the pMD18-T easy vector (TaKaRa, Dalian, China) and sequenced by TaKaRa. These sequences have been deposited in GenBank (accession number: DQ485205-DQ485228).

Table 1-8-1   The nucleotide sequences of the specific primer pairs used in the study

Gene	Sequence of Primers
PB2	F: 5'-AAAAGCAGGTCAATTATATTC-3'; R: 5'-AAGGTCGTTTTTAAACTATTCA-3'
PBl	F: 5'-AAAAGCAGGCAAACCATTTGA-3'; R: 5'-TTTTCATGAAGGACAAGCTAA-3'
PA	F: 5'-AGCAAAAGCAGGTACTGAT-3'; R: 5'-AGTAGAAACAAGGTACTTTT-3'
HA	F: 5'-AGCAAAAGCAGGGGAATTTCAC-3'; R: 5'-AGTAGAAACAAGGGTGTTTTTGC-3'
NP	F: 5'-GCAGGTAGATAATCACTCACTG-3'; R: 5'-AGTAGAAACAAGGGTATTTTT-3'
NA	F: 5'-AGCAAAAGCAGGAGTAAAAATG-3'; R: 5'-CAAGGAGTTTTTTTTTAAAATTGC-3'

continued

Gene	Sequence of Primers
M	F: 5'-AGCAAAAGCAGGTAGATGTTTAAAG-3'; R: 5'-AGTAGAAACAAGGTAGTTTTTTAC-3'
NS	F: 5'-AAAGCAAGGGTGACAAAGACAT-3'; R: 5'-TAGAAACAAGGGTGTTTTTTATCA-3'

Note: F, forward; R, reverse.

Analysis of recombination

The reference sequences for recombination analysis were downloaded from GenBank. Multialignment was achieved by using ClustalW[20]. Gaps were removed before further analysis was carried out. A Phi test was used to determine whether the homologous recombination event was statistically significant[21]. Maximum-likelihood trees were constructed online using PhyML[22] and displayed using MEGA4[23] to determine the recombination events. Bootstrapping was employed to assess the robustness of a tree with 1 000 replications. The bootstrap values are shown below or above the branch. Phylogenetic trees were also generated using the neighbor-joining (NJ) method with maximum composite likelihood in MEGA4[23]. The scale corresponds to the number of nucleotide substitutions per site. The Shimodaira-Hasegawa test was implemented to determine whether phylogenetic trees estimated from different regions reveal differences that are statistically significant using the Treetest program. Splitstree 4 was employed to find the network of mosaics and their parents[21].

Putative recombinant sequences were identified with the soft package of SimPlot[24] as described in a previous study[25]. Recombination break points were identified by maximization of χ^2 combined with a genetic algorithm[24, 26].

About 9 000 AIV gene sequences were also retrieved from GenBank for scanning for evidence of recombination. These AIV sequences were divided into~100 groups before the recombination analyses were performed. In order to determine the most accurate putative parents, each mosaic was used as a query to perform Blast in GenBank.

Results

We found that three H9N2 AIV isolates collected in different counties of the Guangxi at different times had similar mosaic PA segments. Among the three mosaic isolates, A/chicken/Guangxi/1/00 (H9N2) also had a mosaic PB2. Moreover, we analyzed roughly 9 000 complete segments of AIVs deposited in GenBank (Table 1-8-2) and at least 41 mosaics were found again (Table 1-8-3). Homologous recombination could occur in the same subtype or between different subtypes, such as H5N1 and H5N1, H9N2 and H9N2, and H5N1 and H9N2. The recombination events were located in different genes, including PB2, PB1, PA, HA, NA, and NP. In our analysis, no mosaic was found in NS and MP segments.

Table 1-8-2　The number of complete gene sequences analyzed and the number of mosaic segments found in this study

	H5N1	H9N2	H6N1	H7N1	H7N2	H7N7	H6N2	H7N3	H4N6	H3N8	Total	Mosaics
PB2	382	94	52	21	5	13	50	38	46	43	744	13
PB1	397	95	53	20	5	11	50	39	46	48	764	8
PA	472	113	52	23	4	12	45	40	42	42	845	14
HA	841	156	60	51	26	54	71	62	53	97	1 471	4
NP	496	136	56	29	4	19	54	43	61	53	951	5

continued

	H5N1	H9N2	H6N1	H7N1	H7N2	H7N7	H6N2	H7N3	H4N6	H3N8	Total	Mosaics
NA	839	286	65	31	54	15	69	64	58	78	1 559	1
MP	558	204	55	26	52	21	64	60	125	74	1 239	0
NS	615	292	60	24	57	26	79	58	141	74	1 426	0
Mosaic	Yes	Yes	No	No	No	No	No	No	No	No	No	
Total	4 600	1 376	453	225	207	171	482	404	572	509	8 999	45

Three H9N2 AIVs isolated from Guangxi contain mosaic segment

In the year 2000, avian influenza broke out in different chicken flocks in the Guangxi of China. Approximately 3% of all chickens died from influenza disease in these infected chicken flocks. We isolated and purified three AIVs from three counties in March and November that were determined to be of the H9N2 subtype named A/chicken/Guangxi/1/00(H9N2), A/chicken/Guangxi/14/00(H9N2), and A/chicken/Guangxi/17/00(H9N2). The three viruses were collected from three dead birds. After each segment of the three H9N2 strains was sequenced, we found that they had very high sequence similarity in the PA gene (Figure 1-8-1A): 99.81% (2 138/2 142) between A/ chicken/Guangxi/17/00 and A/chicken/Guangxi/14/00, and 99.49% (2 131/2 142) between A/chicken/Guangxi/17/00 and A/chicken/Guangxi/1/00.

Table 1-8-3   AIV strains with evidence for potential recombination

GenBank number	Mosaic virus strain (subtype) (reference)	Gene	Putative parent lineages (subtype)	Position of break points	Countries
DQ073402	A/tree sparrow/Henan/4/04 (H5N1) J Virol.79 (24): 15460-15466 (2005)	PB2	A/tree sparrow/HN/3/04 (H5N1), A/dk/Hk/293/1978 (H7N2)	1 116, 1 704, 1 812	China
DQ073399	A/tree sparrow/Henan/1/04 (H5N1) JVirol.79 (24): 15460-15466 (2005)	PB2	A/duck/HK/278/1978 (H2N9), A/wild duck/GD/314/04 (H5N1)	636	China
DQ064560	A/chicken/jilin/53/01 (H9N2) Virology.340 (1): 70-83 (2005)	PB2	A/ck/Yokohama/aq144/01 (H9N2), A/chicken/Jilin/hk/2004 (H5N1)	1 925	China
EF124780	A/chicken/Guiyang/1655/06 (H5N1) Proc Natl Acad Sci USA.103: 16936-16941 (2006)	PA	A/chicken/Guiyang/441/06 (H5N1), A/goose Shantou/18442/05 (H5N1)	313, 595	China
AY651625	A/SCk/HK/YU100/2002 (H5N1) Nature.430 (6996): 209-213 (2004)	PA	A/Ck/HK/31.2/2002 (H5N1), A/ chicken/Hubei/327/2004 (H5N1)	859, 1 951	China
AY651512	A/Ck/HK/37.4/2002 (H5N1) Nature.430 (6996): 209-213 (2004)	NP	A/duck/Shanghai/38/2001 (H5N1), A/Ck/HK/31.4/02_ (H5N1)	799	China
AF461526	A/chicken/Tianjing/1/96 (H9N2) Avian Dis.47: 116-127 (2003)	HA	A/chicken/Tianjing/2/96 (H9N2), A/ chicken/Liaoning/1/00 (H9N2)	895	China
DQ485205	A/chicken/Guangxi/1/00 (H9N2) (this study)	PB2	A/duck/Shanghai/13/2001 (H5N1), A/chicken/Jiangsu/1/00 (H9N2)	1 066	China
DQ997093	A/duck/Hubei/wg/2002 (H5N1) (unpublished)	PB2	A/swine/Anhui/ca/2004 (H5N1), A/ ck/Nanchang/4-301/01 (H9N2)	156, 543, 1 192	China
DQ997372	A/chicken/Jilin/hq/2003 (H5N1) (unpublished)	PB2	A/duck/Hubei/wp/2003 (H5N1), A/ black_duck/AUS/4045/80 (H6N5)	1 887	China
DQ997225	A/chicken/Henan/wu/2004 (H5N1) (unpublished)	PB2	A/duck/HK/278/1978 (H2N9), A/ chicken/Jilin/hg/2002 (H5N1)	613, 1 617	China
AY950280	A/chicken/Henan/210/2004 (H5N1) (unpublished)	PB2	A/duck/HK/278/1978 (H2N9), A/ chicken/Jilin/hg/2002 (H5N1)	613, 1 617	China

continued

GenBank number	Mosaic virus strain (subtype) (reference)	Gene	Putative parent lineages (subtype)	Position of break points	Countries
DQ997121	A/chicken/Hubei/wj/1997 (H5N1) (unpublished)	PB2	A/duck/HK/278/1978 (H2N9), A/ck/ HK/NT873.3/01-MB (H5N1)	585	China
DQ351870	A/chicken/Hebei/718/2001 (H5N1) (unpublished)	PB2	A/duck/Guangxi/xa/2001 (H5N1), A/duck/Germany/1215/1973 (H2N3)	783, 979	China
DQ997314	A/chicken/Jilin/hh/2002 (H5N1) (unpublished)	PB2	A/black_duck/AUS/4045/80 (H6N5), A/duck/Hubei/wp/2003 (H5N1)	1 638	China
DQ997101	A/chicken/Hubei/wh/1997 (H5N1) (unpublished)	PB2	A/chicken/Jiangsu/wa/2002 (H9N2), A/chicken/Hubei/wi/1997 (H5N1)	711, 1 323, 1 695	China
DQ997317	A/chicken/Jilin/hj/2003 (H5N1) (unpublished)	PB2	A/duck/Hong Kong/d73/76 (H6N1), A/chicken/Jilin/xv/2002 (H5N1)	1 713	China
DQ997510	A/chicken/Beijing/ne/1999 (H9N2) (unpublished)	PB1	A/chicken/Henan/nd/98 (H9N2), A/ duck/Xuzhou/07/03 (H9N2)	642, 1 566	China
DQ997281	A/goose/Jilin/hb/2003 (H5N1) (unpublished)	PB1	A/chicken/Henan/16/2004 (H5N1), A/chicken/Jiangsu/wa/2002 (H9N2)	1 283, 1 690	China
DQ997084	A/chicken/Hubei/wf/2002 (H5N1) (unpublished)	PB1	A/swine/Anhui/ca/2004 (H5N1), A/ duck/Fujian/19/2000 (H5N1)	541, 6 501 234, 1 963	China
DQ997288	A/chicken/Jilin/hd/2002 (H5N1) (unpublished)	PB1	A/duck/Hubei/wp/2003 (H5N1), A/ swine/shandong/na/2002 (H9N2)	1 182, 1 843	China
DQ351874	A/chicken/Hebei/718/2001 (H5N1)	PB1	AA/chicken/Hebei/1/2002 (H7N2), A/duck/Guangxi/35/2001 (H5N1)	1 638	China
AY653199	A/chicken/Jilin/9/2004 (H5N1) (unpublished)	PB1	A/chicken/Jiangsu/cz1/2002 (H5N1), A/duck/Jiangsu/nf/2003 (H9N2)	1 182, 1 843	China
DQ997297	A/chicken/Jilin/he/2002 (H5N1) (unpublished)	PB1	A/chicken/Jiangsu/cz1/2002 (H5N1), A/swine/shandong/na/2002 (H9N2), A/duck/Hubei/W1/2004 (H9N2)	736, 1 302, 1 791	China
DQ997449	A/chicken/Jiangsu/nf/02 (H9N2) (unpublished)	PB1	A/CK/Yokohama/aq45/02 (H9N2), A/chicken/Jiangsu/wa/2002 (H9N2)	673	China
DQ997137	A/chicken/Hubei/wl/1997 (H5N1) (unpublished)	PA	A/CK/Henan/210/2004_ (H5N1), A/ swine/Anhui/ca/2004 (H5N1), one parent lineage is missing	334, 489, 1 659	China
DQ351867	A/chicken/Hebei/108/02 (H5N1) (unpublished)	PA	A/duck/Fujian/01/2002 (H5N1), A/ quail/Dubai/303/2000 (H9N2)	1 877	China
DQ997107	A/chicken/Hubei/wi/1997 (H5N1) (unpublished)	PA	A/tree sparrow/Henan/2/04 (H5N1), A/goose/Hujian/bb/2003 (H5N1)	1 174, 1 484	China
DQ997274	A/chicken/Jilin/ha/2003 (H5N1) (unpublished)	PA	A/chicken/Jiangsu/wa/2002 (H9N2), A/chicken/Jilin/9/2004 (H5N1)	1 678	China
DQ997280	A/goose/Jilin/hb/2003 (H5N1) (unpublished)	PA	A/chicken/Yamaguchi/7/2004 (H5N1), A/chicken/Jiangsu/cz1/2002 (H5N1), A/quail/Shantou/1461/2001 (H9N2)	674, 1 228, 1 835	China
DQ997414	A/duck/Zhejiang/bj/2002 (H5N1) (unpublished)	PA	A/chicken/Jilin/9/2004 (H5N1), A/ Gf/HK/38/2002 (H5N1)	203, 1 558	China
DQ485223	A/chicken/Guangxi/17/00 (H9N2) (this study)	PA	A/chicken/Jiangsu/cz1/2002 (H5N1), A/chicken/Guangdong/4/00 (H9N2)	1 332	China
DQ485215	A/chicken/Guangxi/14/00 (H9N2) (this study)	PA	A/chicken/Jiangsu/cz1/2002 (H5N1), A/chicken/Guangdong/4/00 (H9N2)	1 332	China
DQ485207	A/chicken/Guangxi/1/00 (H9N2) (this study)	PA	A/chicken/Jiangsu/cz1/2002 (H5N1), A/chicken/Guangdong/4/00 (H9N2)	1 332	China

continued

GenBank number	Mosaic virus strain (subtype) (reference)	Gene	Putative parent lineages (subtype)	Position of break points	Countries
CY014862	A/mallard/Ohio/1801/2005/ (H3N8)	PA	A/mallard/Maryland/1131/05 (H12N5), A/mallard/Maryland/124/05 (H4N6)	1 343	United States
DQ997303	A/chicken/Jilin/hg/02 (H5N1) (unpublished)	PA	A/chicken/Jilin/hl/2004/ (H5N1), A/chicken/Jiangsu/cz1/2002 (H5N1)	661, 1 173	China
DQ997321	A/chicken/Jilin/hj/03 (H5N1) (unpublished)	PA	A/duck/Shandong/093/04 (H5N1), A/chicken/Henan/01/04 (H5N1)	687, 1 173, 1 749	China
AF461512	A/chicken/Gansu/1/99 (H9N2) Avian Dis 47: 116-127 (2003)	HA	A/chicken/Osaka/aq48/97 (H9N2), A/chicken/Shandong/1/98 (H9N2)	911	China
AY664667	A/CK/Hong Kong/NT142/03 (H9N2) (unpublished)	HA	A/CK/HK/NT142/03 (H9N2), A/CK/ Hong Kong/WF120/03 (H9N2)	652, 1 160	China
DQ366322	A/duck/Vietnam/8/05 (H5N1) Arch Virol. 151 (8): 1615-1624 (2006)	HA	A/goose/Vietnam/3/05 (H5N1), the other putative parent is missing	1 086, 1 400	Vietnam
DQ997369	A/chicken/Jilin/hq/2003 (H5N1) (unpublished)	NP	A/chicken/Jilin/hp/2003 (H5N1), A/chicken/Henan/5/98 (H9N2)	486, 903	China
AY653196	A/chicken/Jilin/9/2004 (H5N1) (unpublished)	NP	A/chicken/Jiangsu/cz1/2002 (H5N1), A/chicken/Jilin/hg/2002 (H5N1)	936, 1 135	China
DQ997285	A/chicken/Jilin/hd/2002 (H5N1) (unpublished)	NP	A/chicken/Jiangsu/cz1/2002 (H5N1), A/chicken/Jilin/hg/2002 (H5N1)	936, 1 135	China
DQ997346	A/chicken/Jilin/hn/2003 (H5N1) (unpublished)	NP	A/chicken/Jilin/xv/2002/ (H5N1), A/chicken/Henan/5/98 (H9N2)	612, 1 040	China
DQ349117	A/chicken/Hebei/718/01 (H5N1)	NA	A/starling/England/983/79 (H7N1), A/goose/HK/3014.5/2000 (H5N1)	1 090	China

It is very interesting to note that the three PA segments are mosaic descended from the H9N2 and H5N1 subtypes (Figure 1-8-1 B, C, E, and F). When the mosaics were incorporated into a PA alignment, the evidence for recombination was found (Phi test, $P<0.000\ 1$). Delta distributions from Boni also gave a P value of $10^{-40[27]}$. A single break point was located at position 1 332 (Figure 1-8-1 B and C) using the maximized value of χ^2 combined with a genetic algorithm[24, 26]. A similarity plot constructed using all sites revealed that the PA sequence of the mosaics exhibited greater affinity with one putative parent A/chicken/Jiangsu/CZ1/02 (H5N1) of the H5N1 serotype before the break point (99.32%, 1 323/1 332) (Figure 1-8-1 B). Alternatively, the mosaics shared higher sequence similarity with A/chicken/Guangdong4/00 (H9N2) of the H9N2 serotype after position 1 333 (99.26%, 809/815) (Figure 1-8-1 B). The most compelling evolutionary evidence for recombination is the occurrence of incongruent phylogenetic trees[3]. When maximum-likelihood phylogenetic trees before or after the break point were constructed, incorporating the human lineage, a statistically significant discrepancy between phylogenetic trees was found (Shimodaira-Hasegawa test, $P<0.000\ 1$) (Figure 1-8-1 E and F). In phylogenetic trees, the two regions delimitated by the break point fell in different lineages with 100% bootstrap (1 000 replications); supporting the idea that the three isolates are homologous recombinants originating from the H5N1 and H9N2 lineages (Figure 1-8-1 E and F). In order to see the parent strains network with the mosaics, a split tree of complete PA genes was constructed. The split tree suggested that the PA segment of the three H9N2 isolates descended from A/chicken/Jiangsu/CZ1/02 (H5N1) and A/chicken/Guangdong4/00 (H9N2) lineages (Figure 1-8-1 G).

Before sequencing, the three isolates had been purified using a plaque-forming method to isolate a nearly clonal virus population. Additionally, several clones of each segment from the same isolate were sequenced to avoid artifacts. The mosaics shared very high sequence similarity with their putative parents before and after the break point. The high similarity is reasonable because the recombinants and their parents were contemporary (circulating between 2000 and 2002). Moreover, the major change in sequence cannot be explained by random gene mutation. Therefore, the three repeatable PA mosaics, isolated in different locations and at different times, provided robust evidence for homologous recombination as a natural phenomenon in AIVs.

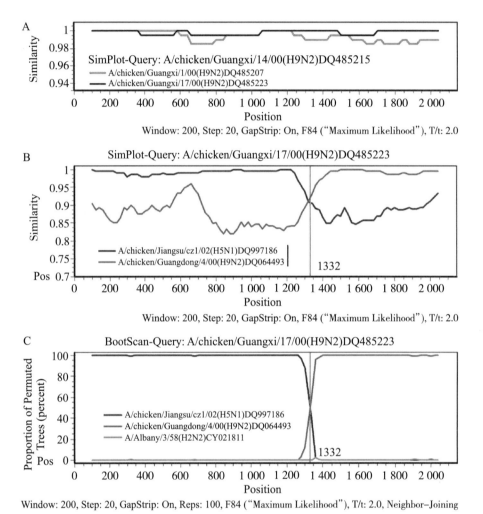

A: A comparison of the PA gene from three AIV isolates: A/chicken/Guangxi/14/00 (H9N2), A/chicken/Guangxi/17/00 (H9N2), and A/chicken/Guangxi/1/00 (H9N2). A/chicken/Guangxi/14/00 (H9N2) is used as the query. The y-axis gives the percentage of identity within a sliding window 200 bp wide centered on the position plotted, with a step size between plots of 20 bp. B: A comparison of PA from A/chicken/Guangxi/17/00 (H9N2) and its putative parents. The red vertical line shows the recombination break point. C: Bootscanning of AIV PA sequences. The y-axis gives the percentage of permutated trees using a sliding window. A human influenza virus isolate A/Albany/3/58 (H2N2) is used as an out-group. The rest is the same as A and B. D, E, and F: The maximum-likelihood trees of the PA segment in regions 1~2 147, 1~1 332, and 1 333~2 147, respectively. G: A split tree inferred from complete PA sequence to show the evolutionary relationship between the mosaics and their parent lineages. The most accurate putative parents are shown in each parent lineage.

Figure 1-8-1　The evidence for recombination in the PA gene of the three H9N2 AIVs isolated from Guangxi in 2000

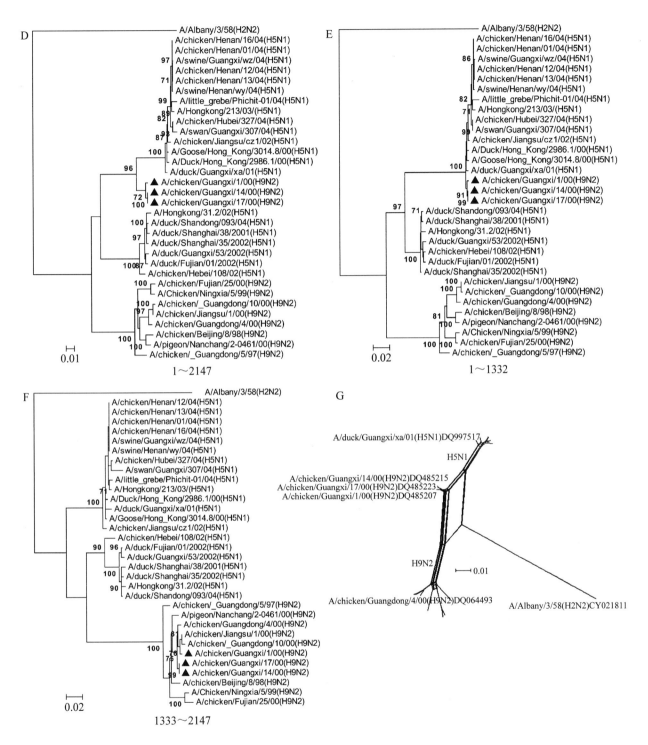

Figure 1-8-1 (Continued)

In addition, PB2 of A/chicken/Guangxi/1/00(H9N2) is also a mosaic descended from the lineages of A/chicken/Guangdong/4/00(H9N2) and A/chicken/Jiangsu/cz1/02(H5N1) (Figure 1-8-2).

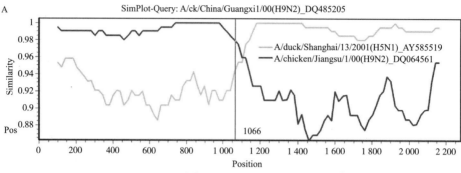

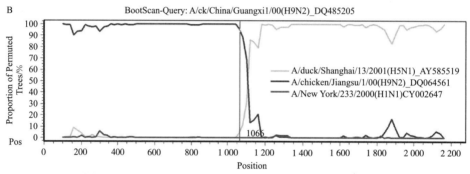

A: Results from SimPlot analysis of A/chicken/Guangxi/1/00 (H9N2) are shown. The y-axis gives the percentage of identity within a sliding window 200 bp wide centered on the position plotted, with a step size between plots of 20 bp. The analysis was carried out using AVchicken/Guangxi/1/00 (H9N2) as a query. The two putative parent lineages are shown with different colors. The red vertical lines show the recombination breakpoints with the maximization of χ^2. B: Bootscanning of AIV PB2 sequences. The y-axis gives the percentage of permutated trees using a sliding window 200 bp wide centered on the position plotted, with a step size between plots of 20 bp. The rest is the same as A. C and D: The ML phylogenetic trees for the recombinant regions from 1 to 1066 and 1067 to 2277, respectively. A human influenza virus isolate A/New York/233/2000(H1N1) is used as the outgroup. E: A split tree inferred from complete PB2 sequence to show the PB2 evolution relation between A/chicken/Guangxi/1/00(H9N2) and their parent lineages. The most accurate putative parents are shown in each parent lineage.

Figure 1-8-2 The evidence for recombination in PB2 segment of A/chicken/Guangxi/1/00 (H9N2)

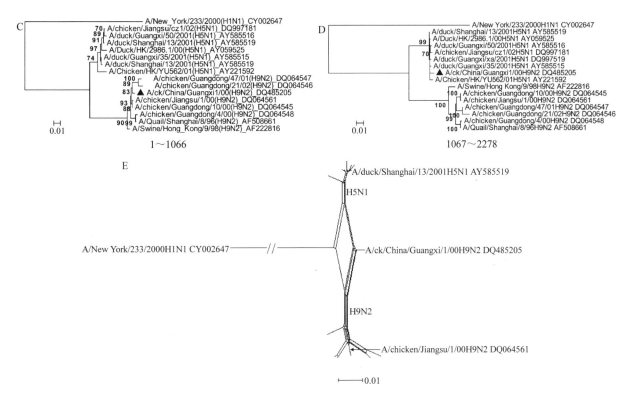

Figure 1-8-2   (Continued)

According to the sequence similarity (Table 1-8-4), A/chicken/Guangdong4/00 (H9N2) lineage could be the putative major parent of A/chicken/Guangxi/1/00 (H9N2), A/chicken/Jiangsu/CZ1/02 (H5N1) lineage might be the putative minor parent, and the PA of A/chicken/Guangxi/14/00 (H9N2) and A/chicken/Guangxi/17/00 (H9N2) seems rearranged from Guangxi1.

Table 1-8-4   Comparison of the putative recombinants isolated from Guangxi and their putative parents

Segment	Similarity					
	GX1/GD4	GX1/JS	GX14/GD4	GX14/JS	GX17/GD4	GX17/JS
HA	99.31	57.6	98.51	57.7	96.84	56.13
PA	92.92	95.48	92.88	95.43	92.97	95.52
PB1	99.18	90.56	89.31	91.25	89.27	91.12
PB2	94.8	95.96	88.97	93	85.66	86.6
NP	98.4	90.12	88.83	92.09	89.22	92.04
NA	98.4	49.14	88.3	50.96	89.2	51.1
M	99.61	92.3	98.44	91.82	99.61	92.31
NS	98.2	90.63	95.17	90.63	94.83	90.29

Note: GX1, A/chicken/Guangxi/1/00(H9N2); GX17, A/chicken/Guangxi/17/00(H9N2); GX14, A/chicken/Guangxi/14/00; JS, A/chicken/Jiangsu/CZ1/02 (H5N1); GD, A/chicken/Guangdong4/00 (H9N2).

Many potential AIV recombinants have been deposited in GenBank

Homologous recombination was thought to be rare in AIVs because no mosaic AIV had been found in a large-scale sequence analysis of avian influenza isolates[28]. To investigate this claim, we analyzed roughly 9 000 AIV sequences deposited in GenBank. Interestingly, in addition to several novel mosaic isolates in this study, we found a series of mosaics in other studies[29-32] (Table 1-8-3). In all, 41 mosaic segments were found among the 9 000 sequences (Table 1-8-3). The whole of these segments are equal to about 1 100 (9 000/8) complete genomes of AIVs, where this means that the rate of recombination is about 3%~4% on the level of the complete AIV genome deposited in GenBank.

These mosaic segments, however, are from public databases, and we cannot ascertain the detailed information of each mosaic. Therefore, these recombinants deposited in GenBank might include some artifact mosaics. There combination rate might be overestimated here. In order to eliminate the artifacts, it is necessary to resequence these mosaic segments.

Similar mosaic segments can be found in different AIV isolates

In addition to the three repeatable PA recombinants isolated here (Figure 1-8-1), several recombinants reported by different research groups at different times shared the same recombination event and near 100% sequence similarity in the mosaic segments. These mosaic segments were PB2 of A/chicken/Henan/wu/2004 (H5N1) and A/chicken/Henan/210/2004 (H5N1), PB2 of A/chicken/Jilin/hj/2003 (H5N1) and A/chicken/Jilin/hh/2002 (H5N1), and PB1 and NP of A/chicken/Jilin/hd/2002 (H5N1) and A/chicken/Jilin/9/2004 (H5N1) . The fact that different AIV strains containing the similar intragenic recombination segments can be isolated in different locations and at different time points by different research groups also supports the hypothesis that recombination can occur in AIVs and exist in the wild.

Discussion

In this study, we report on three AIV recombinants isolated from Guangxi of China. We also identify a series of potential recombinants deposited in GenBank. These recombinants collectively demonstrate that homologous recombination can play a role in shaping genetic diversity of the virus.

The three AIVs isolated in Guangxi contained mosaic PAs sharing very high sequence similarity to each other (Figure 1-8-1 A). The same recombination event and phylogenetic evidence (Figure 1-8-1 B-G) reveal that these PAs descended from the same H9N2/H5N1 recombinant ancestor. Guangxil was collected on 20 March, whereas Guangxi14 and Guangxi17 were collected on 13 November and 20 November, respectively. Therefore, the PAs of Guangxi14 and Guangxi17 might be descended from the Guangxil lineage, which is supported by the phylogenetic relationship of the three mosaics (Figure 1-8-1 D). Similar observations can also be found in PB2 of A/chicken/Jilin/hj/2003(H5N1) and A/chicken/Jilin/hh/2002 (H5N1) and PB2 of A/chicken/Henan/wu/2004(H5N1) and A/chicken/Henan/210/2004(H5N1). The fact that different isolates, originating from the same mosaic ancestor, can be found in different locations and at different time points provides robust evidence that intragenic recombination between different AIV serotypes results in novel viruses. Moreover, these viruses containing mosaic segment (s) are stable and can circulate in the field. Therefore, intragenic recombination does act as a force in the evolution of AIVs.

The recombination, especially between different subtypes, will speed up evolution of the virus. After the AIV H9N2 subtype was first isolated in China in 1994, it became responsible for the 93.89% of avian influenza outbreaks in the country from 1996 to 2000[33]. The H5N1 virus is now endemic in poultry in Asia[29], and the outbreaks occur from time to time in China. This indicates that the two subtypes are circulating at the same time in China, which provides the means for recombination to occur between them. We found that three mosaics descended from H5N1 and H9N2 in Guangxi (Figure 1-8-1 and 1-8-2). Similar H9N2/H5N1 recombinants can be also found in GenBank (Table 1-8-3). Both of the subtypes can infect humans, although infection with H9N2 is thought to be nonfatal[9]. It is therefore important to evaluate the effects on H9N2 virulence as a result of recombination between subtypes.

Interestingly, almost all recombinants were isolated in China (Table 1-8-3). It seems that the AIV has become entrenched in its ecological niche where homologous recombination can occur, highlighting a potential longterm pandemic threat to humans. At present, the mechanism of transmission of these viruses from poultry to humans is poorly understood[29]. In evolutionary terms, and from the perspective of the pathogen, the host species barrier for infection can be thought of as a fitness valley lying between two distinct fitness peaks, representing donor and recipient hosts, respectively[34]. The mutation in polymerase genes was thought to be responsible for the AIV transmission to human populations in the 1918 influenza pandemic[35]. Reassortment and recombination processes will allow some viruses to acquire many of the key adaptive mutations in single step and thus make a major leap in fitness, which might result in a change of host tropism[34]. A reassortment of gene segments between pig and human influenza has resulted in a novel virus to which humans were immunologically naive[36]. Similarly, the intragenic recombination between AIV and human influenza virus is also capable of resulting in a novel virus, adaptable to humans. In fact, intragenic recombination has been found to occur between polymerase members in this analysis, which are known to be involved in many aspects of viral replication and to interact with host factors, thus having a role in host specificity[35]. PB1 is associated with avian-to-human transmission of the PB1 gene of influenza A viruses in the 1957 and 1968 pandemics[36]. Although we did not find the evidence that recombination occurred between human and avian influenza viruses, an AIV strain isolated from swine seems to be able to act as a putative parent (Table 1-8-3). We have also found that the recombination can occur between human and swine influenza viruses[38]. Therefore, proper attention should be given to the potential of homologous recombination events triggering pandemics by altering gene structure or function and/or permitting the highly virulent virus to switch hosts, from birds to humans.

Recently, homologous recombination has also been reported in several negative-strand RNA viruses, such as ambisense arenaviruses[39, 40], the human respiratory syncytial virus[41], and the Newcastle disease virus[42, 44]. Because there is evidence to show that influenza viruses undergo various forms of nonhomologous recombination[14, 15], there should be opportunities for intragenic recombination in the evolution of AIVs. However, the molecular mechanism underlying this recombination is unclear in AIVs. For positive RNA viruses, template switching has been reported as the mechanism of intragenic recombination and as the main mechanism in many cases[44]. For AIVs, however, it is known that the synthesis of RNA strands of the genome is particle associated, and free double-stranded or minus-strand RNA has never been found in infected cells. This seems to reduce the probability of both homologous recombination and template switching occurring in the replication process of the virus. In fact, we also found that the rate of recombination was not as high as that in a positive-sense RNA virus, for example, the foot-and-mouth disease virus in which 10%~20% of viral genomes undergo recombination during a single replication cycle[45]. In studies on intertypic poliovirus, recombination occurred in RNA regions where RNA could potentially form a secondary structure and play an activator role in recombination[46]. This allows the two parental RNAs of different origins to form a complex and thereby force recombination to occur[47, 48]. Therefore, analysis of the characteristics of breakpoints can provide clues about the recombination mechanism in AIVs. When the recombination break point sequences were analyzed, a potential secondary structure of the region around the break point was also found in some mosaics (data not shown), which could provide a molecular basis for the recombination to occur.

In conclusion, this study provides evidence demonstrating that recombination can occur in AIVs and plays a role in the evolution of the virus. It will be important to evaluate its influence on virulence and host tropism and to study the mechanism of AIV recombination.

References

[1] CHANG S, ZHANG J, LIAO X, et al. Influenza Virus Database (IVDB): an integrated information resource and analysis platform for influenza virus research. Nucleic Acids Res, 2007, 35 (Database issue): D376-D380.

[2] HORIMOTO T, KAWAOKA Y. Influenza: lessons from past pandemics, warnings from current incidents. Nat Rev Microbiol, 2005, 3 (8): 591-600.

[3] PETROVA V N, RUSSELL C A. The evolution of seasonal influenza viruses. Nat Rev Microbiol, 2018, 16 (1): 47-60.

[4] STEINHAUER D A, SKEHEL J J. Genetics of influenza viruses. Annu Rev Genet, 2002, 36: 305-332.

[5] GORMAN O T, BEAN W J, KAWAOKA Y, et al. Evolution of the nucleoprotein gene of influenza A virus. J Virol, 1990, 64 (4): 1487-1497.

[6] HATTA M, GAO P, HALFMANN P, et al. Molecular basis for high virulence of Hong Kong H5N1 influenza A viruses. Science, 2001, 293 (5536): 1840-1842.

[7] KIRKEGAARD K, BALTIMORE D. The mechanism of RNA recombination in poliovirus. Cell, 1986, 47 (3): 433-443.

[8] NAGY P D, SIMON A E. New insights into the mechanisms of RNA recombination. Virology, 1997, 235 (1): 1-9.

[9] PEIRIS M, YUEN K Y, LEUNG C W, et al. Human infection with influenza H9N2. Lancet, 1999, 354 (9182): 916-917.

[10] WOROBEY M, RAMBAUT A, PYBUS O G, et al. Questioning the evidence for genetic recombination in the 1918 "Spanish flu" virus. Science, 2002, 296 (5566): 211 discussion 211.

[11] ANDERSON J P, RODRIGO A G, LEARN G H, et al. Testing the hypothesis of a recombinant origin of human immunodeficiency virus type 1 subtype E. J Virol, 2000, 74 (22): 10752-10765.

[12] PITA J S, FONDONG V N, SANGAR A, et al. Recombination, pseudorecombination and synergism of geminiviruses are determinant keys to the epidemic of severe cassava mosaic disease in Uganda. J Gen Virol, 2001, 82 (Pt 3): 655-665.

[13] BECHER P, ORLICH M, THIEL H J. RNA recombination between persisting pestivirus and a vaccine strain: generation of cytopathogenic virus and induction of lethal disease. J Virol, 2001, 75 (14): 6256-6264.

[14] KHATCHIKIAN D, ORLICH M, ROTT R. Increased viral pathogenicity after insertion of a 28S ribosomal RNA sequence into the haemagglutinin gene of an influenza virus. Nature, 1989, 340 (6229): 156-157.

[15] ORLICH M, GOTTWALD H, ROTT R. Nonhomologous recombination between the hemagglutinin gene and the nucleoprotein gene of an influenza virus. Virology, 1994, 204 (1): 462-465.

[16] GIBBS M J, ARMSTRONG J S, GIBBS A J. Recombination in the hemagglutinin gene of the 1918 "Spanish flu". Science, 2001, 293 (5536): 1842-1845.

[17] WOROBEY M, RAMBAUT A, HOLMES E C. Widespread intra-serotype recombination in natural populations of dengue virus. Proc Natl Acad Sci U S A, 1999, 96 (13): 7352-7357.

[18] BONI M F, ZHOU Y, TAUBENBERGER J K, et al. Homologous recombination is very rare or absent in human influenza A virus. J Virol, 2008, 82 (10): 4807-4811.

[19] XIE Z, FADL A A, GIRSHICK T, et al. Amplification of avian reovirus RNA using the reverse transcriptase-polymerase chain reaction. Avian Dis, 1997, 41 (3): 654-660.

[20] THOMPSON J D, GIBSON T J, PLEWNIAK F, et al. The CLUSTAL_X windows interface: flexible strategies for multiple sequence alignment aided by quality analysis tools. Nucleic Acids Res, 1997, 25 (24): 4876-4882.

[21] HUSON D H, BRYANT D. Application of phylogenetic networks in evolutionary studies. Mol Biol Evol, 2006, 23 (2): 254-267.

[22] GUINDON S, GASCUEL O. A simple, fast, and accurate algorithm to estimate large phylogenies by maximum likelihood. Syst Biol, 2003, 52 (5): 696-704.

[23] TAMURA K, DUDLEY J, NEI M, et al. MEGA4: Molecular Evolutionary Genetics Analysis (MEGA) software version 4.0. Mol Biol Evol, 2007, 24 (8): 1596-1599.

[24] LOLE K S, BOLLINGER R C, PARANJAPE R S, et al. Full-length human immunodeficiency virus type 1 genomes from subtype C-infected seroconverters in India, with evidence of intersubtype recombination. J Virol, 1999, 73 (1): 152-160.

[25] HE C Q, DING N Z, FAN W, et al. Identification of chicken anemia virus putative intergenotype recombinants. Virology, 2007, 366 (1): 1-7.

[26] KOSAKOVSKY POND S L, POSADA D, GRAVENOR M B, et al. Automated phylogenetic detection of recombination using a genetic algorithm. Mol Biol Evol, 2006, 23 (10): 1891-1901.

[27] BONI M F, POSADA D, FELDMAN M W. An exact nonparametric method for inferring mosaic structure in sequence triplets. Genetics, 2007, 176 (2): 1035-1047.

[28] OBENAUER J C, DENSON J, MEHTA P K, et al. Large-scale sequence analysis of avian influenza isolates. Science, 2006, 311 (5767): 1576-1580.

[29] LI K S, GUAN Y, WANG J, et al. Genesis of a highly pathogenic and potentially pandemic H5N1 influenza virus in eastern Asia. Nature, 2004, 430 (6996): 209-313.

[30] KOU Z, LEI F M, YU J, et al. New genotype of avian influenza H5N1 viruses isolated from tree sparrows in China. J Virol, 2005, 79 (24): 15460-15466.

[31] LEE M S, CHANG P C, SHIEN J H, et al. Genetic and pathogenic characterization of H6N1 avian influenza viruses isolated in Taiwan between 1972 and 2005. Avian Dis, 2006, 50 (4): 561-571.

[32] SMITH G J, FAN X H, WANG J, et al. Emergence and predominance of an H5N1 influenza variant in China. Proc Natl Acad Sci U S A, 2006, 103 (45): 16936-16941.

[33] GUO Y J, KRAUSS S, SENNE D A, et al. Characterization of the pathogenicity of members of the newly established H9N2 influenza virus lineages in Asia. Virology, 2000, 267 (2): 279-288.

[34] KUIKEN T, HOLMES E C, MCCAULEY J, et al. Host species barriers to influenza virus infections. Science, 2006, 312 (5772): 394-397.

[35] TAUBENBERGER J K, REID A H, LOURENS R M, et al. Characterization of the 1918 influenza virus polymerase genes. Nature, 2005, 437 (7060): 889-893.

[36] RUSSELL C J, WEBSTER R G. The genesis of a pandemic influenza virus. Cell, 2005, 123 (3): 368-371.

[37] KAWAOKA Y, KRAUSS S, WEBSTER R G. Avian-to-human transmission of the PB1 gene of influenza A viruses in the 1957 and 1968 pandemics. J Virol, 1989, 63 (11): 4603-4608.

[38] HE C Q, HAN G Z, WANG D, et al. Homologous recombination evidence in human and swine influenza A viruses. Virology, 2008, 380 (1): 12-20.

[39] CHARREL R N, DE LAMBALLERIE X, FULHORST C F. The Whitewater Arroyo virus: natural evidence for genetic recombination among Tacaribe serocomplex viruses (family Arenaviridae). Virology, 2001, 283 (2): 161-166.

[40] ARCHER A M, RICO-HESSE R. High genetic divergence and recombination in Arenaviruses from the Americas. Virology, 2002, 304 (2): 274-281.

[41] SPANN K M, COLLINS P L, TENG M N. Genetic recombination during coinfection of two mutants of human respiratory syncytial virus. J Virol, 2003, 77 (20): 11201-11211.

[42] HAN G Z, HE C Q, DING N Z, et al. Identification of a natural multi-recombinant of Newcastle disease virus. Virology, 2008, 371 (1): 54-60.

[43] QIN Z, SUN L, MA B, et al. F gene recombination between genotype II and VII Newcastle disease virus. Virus Res, 2008, 131 (2): 299-303.

[44] LAI M M. RNA recombination in animal and plant viruses. Microbiol Rev, 1992, 56 (1): 61-79.

[45] ALEJSKA M, KURZYŃSKA-KOKORNIAK A, BRODA M, et al. How RNA viruses exchange their genetic material. Acta Biochim Pol, 2001, 48 (2): 391-407.

[46] NAGY P D, OGIELA C, BUJARSKI J J. Mapping sequences active in homologous RNA recombination in brome mosaic virus: prediction of recombination hot spots. Virology, 1999, 254 (1): 92-104.

[47] ROMANOVA L I, BLINOV V M, TOLSKAYA E A, et al. The primary structure of crossover regions of intertypic poliovirus recombinants: a model of recombination between RNA genomes. Virology, 1986, 155 (1): 202-213.

[48] TOLSKAYA E A, ROMANOVA L I, BLINOV V M, et al. Studies on the recombination between RNA genomes of poliovirus: the primary structure and nonrandom distribution of crossover regions in the genomes of intertypic poliovirus recombinants. Virology, 1987, 161 (1): 54-61.

Identification of three H1N1 influenza virus groups with natural recombinant genes circulating from 1918 to 2009

He Chengqiang, Ding Naizheng, Mou Xue, Xie Zhixun, Si Hongli, Qiu Rong, Ni Shuai, Zhao Heng, Lu Yan, Yan Hongyan, Gao Yingxue, Chen Linlin, Shen Xiuhuan, Cao Runnan

Abstract

In this study, we identify a recombinant PB1 gene, a recombinant MP segment and a recombinant PA segment. The PB1 gene is recombined from two Eurasia swine H1N1 influenza virus lineages. It belongs to a H1N1 swine clade circulating in Europe and Asia from 1999 to 2009. The mosaic MP segment descends from H7 avian and H1N1 human virus lineages and pertains to a large human H1N1 virus family circulating in Asia, Europe and America from 1918 to 2007. The recombinant PA segment originated from two swine H1N1 lineages is found in a swine H1N1 group prevailing in Asia and Europe from 1999 to 2003. These results collectively falsify the hypothesis that influenza virus do not evolve by homologous recombination. Since recombination not only leads to virus genome diversity but also can alter its host adaptation and pathogenecity; the genetic mechanism should not be neglected in influenza virus surveillance.

Keywords

H1N1 influenza virus, homologous recombination, genetic mechanism

---

Introduction

H1N1 influenza A virus respectively brought severe global pandemic in 1918 and 2009[1, 2]. Although it is unknown in detail why the virus becomes capable of crossing host species barriers, it should be acceptable that genetic diversity plays a key role in its crossing host infection[3]. Therefore, it is very important to fully understand the molecular mechanism that drives the change of its genetic diversity.

Homologous recombination is a key genetic mechanism for rapid evolution. For influenza A virus, although reassortment between segments is also known to be a major factor inducing the rapid change of its genetic diversity, it is still highly controversial whether homologous recombination can naturally occur in the field. Gibbs et al. proposed that recombination could occur in HA gene between human and swine influenza viruses[4]. However, the theory was questioned since the absence of phylogenetic evidence[5]. Boni et al. suggested that homologous recombination could not take place in human influenza virus after they analyzed human influenza sequences from GenBank[6]. Although lots of homologous recombinant influenza A viruses have been found in the public database, they were simply attributed to artifact[7]. We reported several homologous recombinants derived from human and swine influenza viruses[8]. We also found that three H9N2 avian influenza virus isolates contain the same recombinant PA gene[9]. Recently, it was reported that homologous recombination in influenza viruses tends to emerge between strains sharing high sequence similarity[10]. Nevertheless, these findings were still dubious and unacceptable for some researchers[11]. As a result, a number of apparent homologous recombinants are still neglected in influenza surveillances at present[1, 12-28]. Therefore, more comprehensive evidence is necessary to have the genetic mechanism cognized and accepted in influenza A virus.

According to guidelines for identifying homologous recombination events in influenza A virus[11], the most forceful proof for recombination should be to provide a recombinant clade with sequences isolated from different laboratories and animals. In this study, we report a swine H1N1 clade having the same recombinant PB1 segment (viruses collected from 1999 to 2009 in different districts of Europe and Asia by six research groups), a large human H1N1 family possessing similar recombinant segment 7 (more than 100 strains isolated from 1918 to 2007 in America, Europe and Asia), and a swine H1N1 clade sharing the same recombinant PA segment (8 strains collected from 1999 to 2009 in Asia and Europe) to demonstrate that homologous recombination is an important genetic mechanism driving the rapid evolution of H1N1 influenza virus.

Materials and methods

All PB1, PA and MP segments were retrieved from GenBank. These sequences were aligned with CLUSTALW[29]. The alignment of these sequences is available online. Phylogenetic Neighbor-Joining (NJ) trees were built by MEGA5[30], with the Maximum Composite Likelihood model employed. Maximum-Likelihood (ML) trees were constructed employing PhyML[31]or MEGA5 with GTR nucleotide substitution mode. Bootstrap-ping was used to assess the robustness of a tree with 1000 replicates. Shimodaira-Hasegawa test, which was implemented in Treetest program, was employed to prove whether phylogenetic trees estimated from different regions were significantly different.

The aligned sequences were analyzed by operating RDP software package (V.3.34) to search potential recombination events[32]. SimPlot software package[33]was run to determine each recombinant as previous study[9, 34]. Recombination breakpoints were analyzed by maximization of χ^2 employing SimPlot combined with genetic algorithms using GARD[35]. Split tree was constructed and Phi test for recombination was performed utilizing Split Trees software package (version 4.9) [36]. Informative sites were analyzed to differentiate recombination from convergent evolution resulted from mutation through Findsites program implemented in SimPlot software package.

Results and discussion

Referring to previous studies[9, 34], we detected the potential mosaic PB1 segment of H1N1 influenza A virus deposited in the public database employing recombination detecting programs (RDP) [32]. Apart from several single recombinant isolates (data not shown), we found a H1N1 swine influenza virus group having the same mosaic PB1 gene listed in Figure 1-9-1.

In order to demonstrate the recombination event, A/swine/CDA/1488/99 from previous reports[26]was considered as a representative of the PB1 recombinant group for deep recombination analysis since these mosaic PB1 shared at least 97% sequence similarity with each other. Two isolates A/swine/HK/8512/2001[1] and A/swine/Schwerin/103/89[37]were also used as the representatives of the two parental branches (it should be reminded that A/swine/HK/8512/2001 and A/swine/Schwerin/103/89 were not the factual parent strain of the recombinants.). What is more, several 1976 classical swine H1N1 viruses obtained from Influenza Genome Sequencing Project (IGSP) were used as the control of outgroup since they had relative distant relates of the Eurasia swine H1N1 lineage (European avian-like swine H1N1 lineage); moreover, IGSP sequence was thought to have rigorous quality controlling[11].

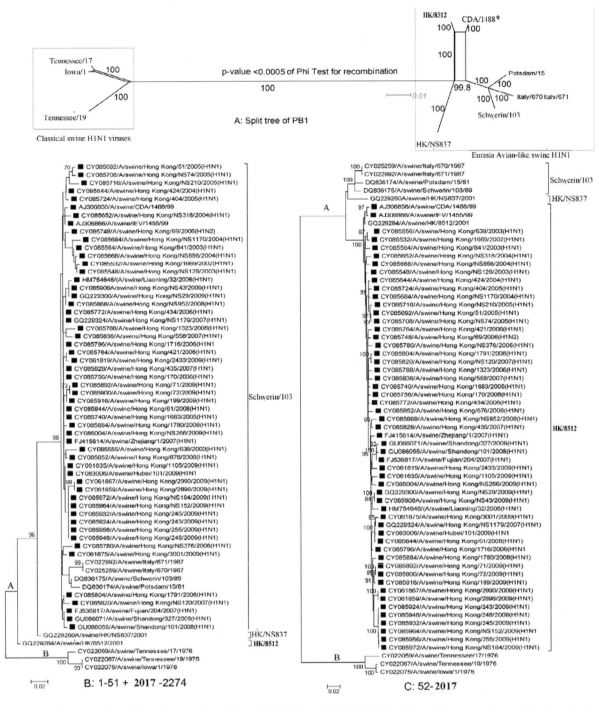

A shows a split phylogenic tree shows the network evolution of PB1 of A/swine/CDA/1488/99 and its putative parent lineages. The three classical H1N1 influenza viruses from USA were used as outgroups. The split tree was constructed employing Splitstree 4[36]. Bootstrap value of each branch (1 000 replicates) is shown near the branches. p-Value of Phi test for recombination is also showed on the tree. The recombinant PB1 is indicated with an asterisk. Please refer B or C for complete name of each isolate. B and C show the incongruous evolution relationships of different regions delimited by the breakpoints. The evolutionary history was inferred by using the Maximum Likelihood method based on the General Time Reversible model. The bootstrap consensus tree inferred from 1 000 replicates is taken to represent the evolutionary history of the taxa analyzed. The percentage (> 70%) of replicate trees in which the associated taxa clustered together in the bootstrap test (1 000 replicates) are shown next to the branches. A discrete Gamma distribution was used to model evolutionary rate differences among sites (4 categories). The rate variation model allowed for some sites to be evolutionarily invariable. The tree is drawn to scale, with branch lengths measured in the number of substitutions per site. ML trees were constructed employing MEGA5[38SS]. Three H1N1 swine influenza virus strains isolated from the US in 1970s were used as an out-group, and a published strain from China GQ229260/A/swine/HK/NS837/2001 was used as an inner-group. B shows the phylogenic tree inferred from positions 1 to 51 and 2 016 to 2 274 of PB1 ORF region. C shows the phylogenic tree inferred from positions 52 to 2 016 of PB1 ORF region. Each recombinant PB1 was marked with "■". A: Eurasia avian-like swine H1N1 viruses; B: Classical swine H1N1 viruses.

Figure 1-9-1　Phylogenic evidence of PB1 recombination

When the representative isolates of A/swine/CDA/1488/99 and its parent lineage were analyzed employing different programs implemented in SimPlot software package[33], significant recombination probability was found ($P < 1.6 \times 10^{-10}$). Combining SimPlot with GARD results[33, 35], two potential breakpoints were located at positions 51 ($\chi^2_{max} = 24.5$) and 2017 ($\chi^2_{max} = 41.8$) of PB1 open reading frame (ORF). Between the two breakpoints, A/swine/CDA/1488/99 and A/swine/HK/8512/2001 (of the major parent lineage) shared higher sequence identity (99.4%). On the contrary, the recombinant had only 93% sequence similarity with A/swine/HK/8512/2001 (versus 98.5% of A/swine/Schwerin/103/89) in the complementary regions of PB1 (Figures 1-9-2 A and C). When A/swine/CDA/1488/99 was used as a query in the Bootscan analysis, 95% of its permuted trees fell into A/swine/HK/8512/2001 lineage between the breakpoints, whereas more than 90% appeared in the Schwerin lineage before the first breakpoint and behind the second (Figure 1-9-2 B).

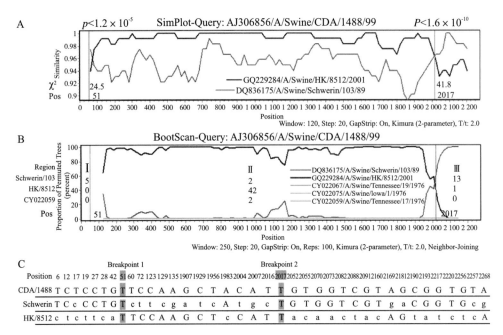

A: PB1 ORF comparison of the recombinant representative of A/swine/CDA/1488/99 with the two putative parent lineage representatives, A/swine/Schwerin/103/89 (Europe) and A/swine/HK/8512/2001 (China). They-axis gave the percentage of similarity within a sliding window of 120 bp wide centered on the position plotted, with a step size between plots of 20 bp. PB1 sequence similarity of each recombinant was nearly parallel on the level of complete PB1 ORF. Vertical lines indicated breakpoints identified by maximization of χ^2. χ^2-value of each breakpoint (lower) and p-values of Fisher's Exact Test (upper) were shown near the vertical lines indicating the breakpoints. B: The result of Bootscan. They-axis gives the percentage of permuted trees using a sliding window of 250 bp width centered on the position plotted, with a step size between plots of 20 bp. CY022067/A/swine/Tennessee/19/1976, CY022075/A/swine/Iowa/1/1976, CY022059/A/swine/Tennessee/17/1976 were used as the out-group and GQ229260A/swine/HK/NS837/2001 as inner-group to determine the recombinant segment. The number of informative sites of different regions is also shown in B. C: Comparison of the variable positions in different regions around the putative breakpoints.

Figure 1-9-2　Recombination signals of PB1 of the recombinant group

PB1 ORF complete sequences of the recombinant and its putative parent lineages were also used to construct a split tree employing Splitstrees 4[36]. In the split tree, the representatives of two parent lineage Schwerin/103 and HK/8512 were the two independent subgenotypes of Eurasia avian-like swine H1N1 virus (Figure 1-9-1 A). And the recombinant PB1 gene of A/swine/CDA/1488/99 and its putative parent lineages had a network relation of evolution (clustered in Schwerin lineage with 99.8% bootstrap value and Hong Kong lineage with 100% bootstrap value) (Figure 1-9-1 A). Phi test provided statistical significant recombination evidence ($P < 0.000\ 5$) when the mosaic gene was implemented in the analyzed data. If the mosaic sequence was removed, the recombination signal would disappear ($P = 0.179$).

According to guidelines for identifying homologous recombination events in influenza A virus, the ideal approach to demonstrate the presence of recombination is a set of statistically incongruent phylogenetic trees[11]. In maximum likelihood (ML) phylogenic trees inferred from different regions of PB1 segment, these recombinants and the Schwerin group were clustered together before the first beakpoint and behind the second (with 98% bootstrap values) (Figure 1-9-1 B). Nonetheless, it turns out that the recombinant group and A/swine/HK/8512/2001 lineage emerged in the same phylogenic branch between the crossover sites (with 100% bootstrap values) (Figure 1-9-1 C). Topology of the two trees was also proven to have statistically significant difference by running the Treetest program (P-value of Shimodaira-Hasegawa test＜0.01) [38].

Although the members of the recombinant group were respectively isolated from 1999 to 2009 in different districts of Europe and Asia by six research groups, they contained the PB1 segment having identical recombination event. Thus, they might be the progenies of the same recombinant ancestor descending from two Eurasia H1N1 swine influenza A virus branches. It was noticed that the two parent lineages were isolated from two geographically distinct regions and 12 years apart. Considering the existence of a global reservoir of influenza virus and the international trade of live pigs, the Europe lineage circulating in 1989 might spread to Asia during 10 years if recombination had happened around 1999. Therefore, it should be reasonable that 1.5% (5/310) diversity of A/swine/Schwerin/103/89 was accumulated in the factual parents train (which has not been isolated) when the recombination occurred around 1999.

We also analyzed segment 7 of H1N1 employing RDP since it is one of the shortest segments in the virus. An old and large human H1N1 family (more than 100 members circulated in America, Europe and Asia from 1918 to 2007) was found to contain the same mosaic segment 7 associated with H7-like avian influenza and classical human H1N1 influenza virus (Figure 1-9-3). The recombinant segment 7 was found in H9N2 and H3N2 swine influenza viruses (Figure 1-9-3 D and E) as well, suggesting that the mosaic segment had been reassorted into other subtype viruses.

Comparing the representative isolate CY045757/A/United_Kingdom/1/1933 with its parent lineages (human and avian branches), we discovered that they had conflicting sequence similarity in different regions delimited by the breakpoints (Figure 1-9-3 A and C). Between the two potential breakpoints, the mosaic sequence and the human branch had higher identity than the avian branch; however, the mosaic sequence shared higher identity with the avian branch in other complement regions. Moreover, statistically significant recombination evidence was found employing recombination detection programs implemented in RDP software package (RDP $P<0.005$; Bootscan $P<0.000\,5$; MaxChi $P<0.000\,05$; Chimaera $P<0.000\,05$; Siscan $P<9 \times 10^{-12}$; 3SEQ $P<0.000\,05$). The Bootscan result was shown in Figure 1-9-3 B.

Employing the Split Trees program, the recombination group and their parent lineages have network evolution relation in the split evolutionary tree. Phi test provided statistically significant recombination evidence ($P=0.011$) when the mosaic genes was implemented in the analyzed data.

Moreover, these recombinant isolates had incongruent phylogenetic relations based on sequence of different regions (Figure 1-9-3 D and E). Between the two putative breakpoints (positions 109 and 665), the recombinants fell into the human branch of H1N1 subtype (Figure 1-9-3 E). However, they appeared in the avian branch before the position 109 and behind the position 665 with more than 90% bootstrap value (Figure 1-9-3 D).

Interestingly, although the oldest MP mosaic segment of A/Brevig_Mission/1/1918 (H1N1) was found in 1918, parent lineages of it were not isolated until 2000. An explanation might be that the parent lineage of human H1N1 virus is represented by isolates from the year 2000 and later, but earlier members of the lineage have not been isolated. And it was one of these earlier viruses of the lineage that donated a gene fragment to the recombinant ancestor in early 20th century. Thus, it was logical that the earlier virus of the human H1N1

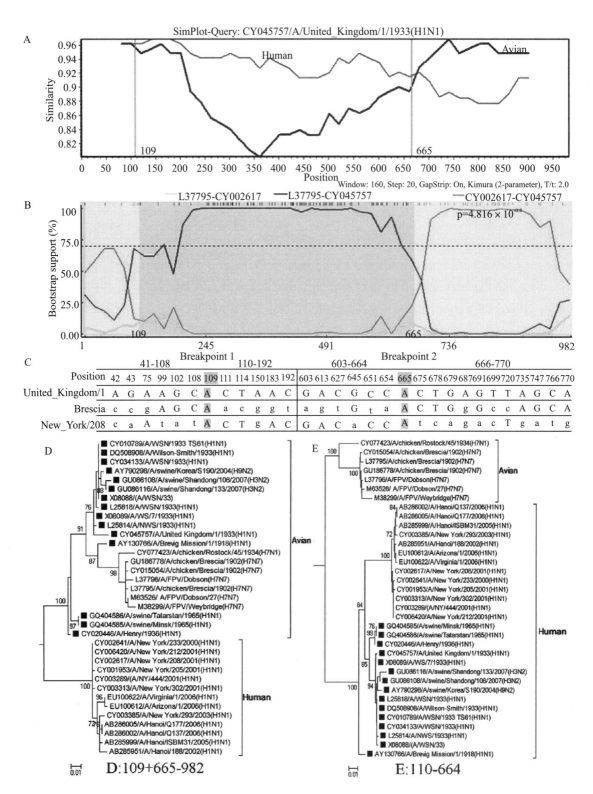

A: MP ORF comparison of the recombinant representative CY045757/A/United_Kingdom/1/1933 (H1N1) with their two putative parent lineage representatives: avian (L37795/A/chicken/Brescia/1902 (H7N7)) and Human (CY002617/A/New_York/208/2001 (H1N1). B: The result of Bootscan. They-axis gave the percentage of Bootstrap value. P-value for recombination was also shown. The result was given by Bootscan program implemented in RDP software package since it was dificult to ind a reasonable out-group in the public database. C: Comparison of the variable positions in different regions around the putative breakpoints. D: The phylogenetic tree inferred from positions 1 to 109 and 665 to 982 of M2 ORF region. E: The phylogenic tree inferred from positions 110 to 665 of M2 ORF region. Recombinant genes were marked with "■". The evolutionary history was inferred by using the Maximum Likelihood method employing Mega 5. Please refer to Figure 1-9-1 for a detailed description.64 C.-Q. He et al. / Virology 427 (2012) 60-66.

Figure 1-9-3   Recombination evidence of Segment 7

lineage played a role in the parent of the recombinant group. Therefore, recombination can also occur between different subtypes.

Similarly, a swine H1N1 group (including 8 isolates) containing the same mosaic PA segment was also detected in this study (Figure 1-9-4). The group circulated in Eurasia countries from 1999 to 2003. And their parent lineages were isolated between 2002 and 2007.

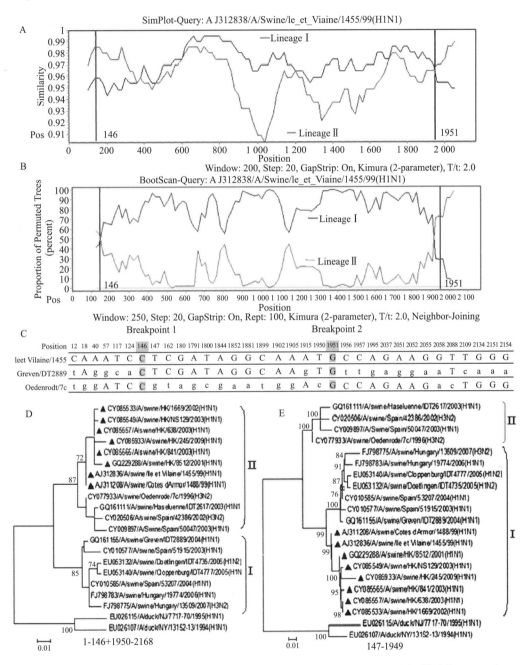

A: PA ORF comparison of the recombinant representative AJ312836/A/swine/Ile_et_Vilaine/1455/99 (H1N1) with their two putative parent lineage representatives: I, GQ161155/A/swine/Greven/IDT2889/2004 (H1N1); and Ⅱ, CY077933/A/swine/Oedenrode/7c/1996 (H3N2). B: The result of Bootscan. They-axis gives the percentage of permutated trees using a sliding window of 250 bp width centered on the position plotted, with a step size between plots of 20 bp. EU026115/A/duck/NJ/ 7717-70/1995 (H1N1) and EU026107/A/duck/NY/13152-13/1994 (H1N1) were used as the out-group. C: Comparison of the variable positions in different regions around the pu-tative breakpoints. D: The phylogenic tree inferred from positions 1 to 146 and 1 950 to 2 168 of PA ORF region. E: The phylogenic tree inferred from positions 147 to 1 949 of PA ORF region. Recombinant genes were marked with "▲" in each tree. The evolutionary history was inferred by using the Maximum Likelihood method employing Mega 5. Please refer to Figure 1-9-1 for a detailed description.

Figure 1-9-4　Recombination evidence of PA Segment

At first, statistically significant recombination evidence was also found in these PA recombinants according to RDP package: RDP $P<0.000\,5$; Bootscan $P<0.000\,5$; MaxChi $P<0.000\,1$; Siscan $P<1\times10^{-10}$; 3SEQ $P<0.000\,5$. Two breakpoints were located at positions 146 and 1951 employing SimPlot and GARD software (Figure 1-9-4 A and B). Comparing the PB1 sequences of the representatives of mosaics and their parent lineages, the mosaics shared higher identity with their parent lineage Ⅰ representative A/swine/Greven/IDT2889/2004 (H1N1) than lineage Ⅱ between the two breakpoint (Figure 1-9-4 A and C); however, the isolate A/swine/Oedenrode/7c/1996 (H3N2) of parent lineage Ⅱ and the mosaics had higher identity in other regions. This observation consisted with the result of Bootscan analysis (Figure 1-9-4 B).

In the split tree of PB1, the recombinant group and its two parent lineages also had a network evolutionary relation. The phi test did ind statistically significant evidence for recombination ($P<0.000\,1$). Between the two putative breakpoints, the recombinants and the parent lineage Ⅰ appeared in a monophyletic group with 84% bootstrap value (Figure 1-9-4 E). However, the recombinant group and the parent lineage Ⅱ had a common ancestor with 99% bootstrap value before position 146 and after position 1 949 (Figure 1-9-4 D).

The three recombinant families provide suficient evidence to safely conclude that homologous recombination occurred in some influenza A virus lineages. It should be noticed that a recombination event detected in a virus sequence does not mean that the virus has undergone the recombination event. The sequences in this case may be (1) a descendant of a recombinant or (2) a descendant of a parent of the recombinant or (3) a descendent of a lineage that was in no way involved in the recombination. Therefore, the three recombinant lineages do not suggest high recombination rate in influenza virus. Nevertheless, they demonstrate that some mosaic segments are stable and can be passed on to progeny.

In conclusion, this study provides three H1N1 clades with different recombinant segments to support that homologous recombination cannot only naturally occur between H1N1 influenza A viruses in the field, but also brought prevailing lineages with new hereditary feature. Therefore, homologous recombination is a potential mechanism driving the rapid evolution of H1N1 influenza A virus. Considering that recombination not only leads to genome diversity but also complicates host adaptation and pathogenecity of virus, those homologous recombinants should not be neglected in influenza virus surveillance.

References

[1] SMITH G J, VIJAYKRISHNA D, BAHL J, et al. Origins and evolutionary genomics of the 2009 swine-origin H1N1 influenza A epidemic. Nature, 2009, 459 (7250): 1122-1125.

[2] TRILLA A, TRILLA G, DAER C. The 1918 "Spanish flu" in Spain. Clin Infect Dis, 2008, 47 (5): 668-673.

[3] PEPIN K M, LASS S, PULLIAM J R, et al. Identifying genetic markers of adaptation for surveillance of viral host jumps. Nat Rev Microbiol, 2010, 8 (11): 802-813.

[4] GIBBS M J, ARMSTRONG J S, GIBBS A J. Recombination in the hemagglutinin gene of the 1918 "Spanish flu". Science, 2001, 293 (5536): 1842-1845.

[5] WOROBEY M, RAMBAUT A, PYBUS O G, et al. Questioning the evidence for genetic recombination in the 1918 "Spanish flu" virus. Science, 2002, 296 (5566): 211 discussion 211.

[6] BONI M F, ZHOU Y, TAUBENBERGER J K, et al. Homologous recombination is very rare or absent in human influenza A virus. J Virol, 2008, 82 (10): 4807-4811.

[7] KRASNITZ M, LEVINE A J, RABADAN R. Anomalies in the influenza virus genome database: new biology or laboratory errors?. J Virol, 2008, 82 (17): 8947-8950.

[8] HE C Q, HAN G Z, WANG D, et al. Homologous recombination evidence in human and swine influenza A viruses. Virology, 2008, 380 (1): 12-20.

[9] HE C Q, XIE Z X, HAN G Z, et al. Homologous recombination as an evolutionary force in the avian influenza A virus. Mol Biol Evol, 2009, 26 (1): 177-187.

[10] HAO W. Evidence of intra-segmental homologous recombination in influenza A virus. Gene, 2011, 481 (2): 57-64.

[11] BONI M F, DE JONG M D, VAN DOORN H R, et al. Guidelines for identifying homologous recombination events in influenza A virus. PLOS ONE, 2010, 5 (5): e10434.

[12] BARANOVICH T, SAITO R, SUZUKI Y, et al. Emergence of H274Y oseltamivir-resistant A(H1N1) influenza viruses in Japan during the 2008-2009 season. J Clin Virol, 2010, 47 (1): 23-28.

[13] CHEN H, SMITH G J, LI K S, et al. Establishment of multiple sublineages of H5N1 influenza virus in Asia: implications for pandemic control. Proc Natl Acad Sci U S A, 2006, 103 (8): 2845-2850.

[14] POONSUK S, SANGTHONG P, PETCHARAT N, et al. Genesis and genetic constellations of swine influenza viruses in Thailand. Vet Microbiol, 2013, 167 (3-4): 314-326.

[15] DUNHAM E J, DUGAN V G, KASER E K, et al. Different evolutionary trajectories of European avian-like and classical swine H1N1 influenza A viruses. J Virol, 2009, 83 (11): 5485-5494.

[16] GUAN Y, SHORTRIDGE K F, KRAUSS S, et al. Molecular characterization of H9N2 influenza viruses: were they the donors of the "internal" genes of H5N1 viruses in Hong Kong?. Proc Natl Acad Sci U S A, 1999, 96 (16): 9363-9367.

[17] KARASIN A I, LANDGRAF J, SWENSON S, et al. Genetic characterization of H1N2 influenza A viruses isolated from pigs throughout the United States. J Clin Microbiol, 2002, 40 (3): 1073-1079.

[18] KARASIN A I, CARMAN S, OLSEN C W. Identification of human H1N2 and human-swine reassortant H1N2 and H1N1 influenza A viruses among pigs in Ontario, Canada (2003 to 2005). J Clin Microbiol, 2006, 44 (3): 1123-1126.

[19] KOU Z, LEI F M, YU J, et al. New genotype of avian influenza H5N1 viruses isolated from tree sparrows in China. J Virol, 2005, 79 (24): 15460-15466.

[20] LI K S, GUAN Y, WANG J, et al. Genesis of a highly pathogenic and potentially pandemic H5N1 influenza virus in eastern Asia. Nature, 2004, 430 (6996): 209-213.

[21] LI Y, LIN Z, SHI J, et al. Detection of Hong Kong 97-like H5N1 influenza viruses from eggs of Vietnamese waterfowl. Arch Virol, 2006, 151 (8): 1615-1624.

[22] LIPATOV A S, KRAUSS S, GUAN Y, et al. Neurovirulence in mice of H5N1 influenza virus genotypes isolated from Hong Kong poultry in 2001. J Virol, 2003, 77 (6): 3816-3823.

[23] LIU H, LIU X, CHENG J, et al. Phylogenetic analysis of the hemagglutinin genes of twenty-six avian influenza viruses of subtype H9N2 isolated from chickens in China during 1996-2001. Avian Dis, 2003, 47 (1): 116-127.

[24] LIU J, BI Y, QIN K, et al. Emergence of European avian influenza virus-like H1N1 swine influenza A viruses in China. J Clin Microbiol, 2009, 47 (8): 2643-2646.

[25] MAROZIN S, GREGORY V, CAMERON K, et al. Antigenic and genetic diversity among swine influenza A H1N1 and H1N2 viruses in Europe. J Gen Virol, 2002, 83 (Pt 4): 735-745.

[26] SMITH G J, FAN X H, WANG J, et al. Emergence and predominance of an H5N1 influenza variant in China. Proc Natl Acad Sci U S A, 2006, 103 (45): 16936-16941.

[27] XU K M, LI K S, SMITH G J, et al. Evolution and molecular epidemiology of H9N2 influenza A viruses from quail in southern China, 2000 to 2005. J Virol, 2007, 81 (6): 2635-2645.

[28] YU H, ZHANG P C, ZHOU Y J, et al. Isolation and genetic characterization of avian-like H1N1 and novel ressortant H1N2 influenza viruses from pigs in China. Biochem Biophys Res Commun, 2009, 386 (2): 278-283.

[29] THOMPSON W W, SHAY D K, WEINTRAUB E, et al. Mortality associated with influenza and respiratory syncytial virus in the United States. JAMA, 2003, 289 (2): 179-186.

[30] TAMURA K, DUDLEY J, NEI M, et al. MEGA4: Molecular evolutionary genetics analysis (MEGA) software version 4.0. Mol Biol Evol, 2007, 24 (8): 1596-1599.

[31] GUINDON S, GASCUEL O. A simple, fast, and accurate algorithm to estimate large phylogenies by maximum likelihood. Syst Biol, 2003, 52 (5): 696-704.

[32] MARTIN D P, WILLIAMSON C, POSADA D. RDP2: recombination detection and analysis from sequence alignments. Bioinformatics, 2005, 21 (2): 260-262.

[33] LOLE K S, BOLLINGER R C, PARANJAPE R S, et al. Full-length human immunodeficiency virus type 1 genomes from subtype C-infected seroconverters in India, with evidence of intersubtype recombination. J Virol, 1999, 73 (1): 152-160.

[34] HE C Q, DING N Z, HE M, et al. Intragenic recombination as a mechanism of genetic diversity in bluetongue virus. J Virol, 2010, 84 (21): 11487-11495.

[35] KOSAKOVSKY POND S L, POSADA D, GRAVENOR M B, et al. Automated phylogenetic detection of recombination using a genetic algorithm. Mol Biol Evol, 2006, 23 (10): 1891-1901.

[36] HUSON D H, BRYANT D. Application of phylogenetic networks in evolutionary studies. Mol Biol Evol, 2006, 23 (2): 254-267.

[37] ZELL R, KRUMBHOLZ A, EITNER A, et al. Prevalence of PB1-F2 of influenza A viruses. J Gen Virol, 2007, 88 (Pt 2): 536-546.

[38] ARIS-BROSOU S. Least and most powerful phylogenetic tests to elucidate the origin of the seed plants in the presence of conflicting signals under misspecified models. Syst Biol, 2003, 52 (6): 781-793.

An overview of avian influenza A H10N8 subtype viruses

Tan Wei, Li Meng, and Xie Zhixun

Abstract

The first avian influenza A subtype H10N8 virus able to infect humans and induce mortality was identified in China at the end of 2013. No such case was observed in other parts of the world. Similar to H7N9, H10N8 viruses are less pathogenic in poultry but could induce severe illness or even death in humans. H10N8 infection poses a potential threat to public health. There is currently no effective means to prevent or treat H10N8 infections. Moreover, it remains unclear how H10N8 viruses are transmitted to humans, and substantial public concerns have been raised about whether H10N8 infections will lead to an outbreak in humans. This article briefly describes the current research on H10N8 subtype viruses.

Keywords

avian influenza virus, gene reassortment, H10N8, H5N1, H7N9, receptor-binding preference

Introduction

Avian influenza virus (AIV) belongs to the Orthomyxoviridae family and is a member of type A influenza viruses. AIV is an enveloped, negative-sense, single-stranded RNA virus. The viral genome consists of eight segmented genes, encoding viral structural and non-structural proteins (NS), respectively, including haemagglutinin (HA), neuraminidase (NA), matrix protein M1, ion channel protein M2, nucleoprotein (NP), and RNA polymerase complex proteins (PB1, PB2, and PA) [1-5].

Currently, sixteen HA subtypes (H1-H16) and nine NA subtypes (N1-N9) of influenza A viruses have been identified in wild aquatic birds, the natural reservoir of low pathogenic avian influenza viruses (LPAIV). LPAIV primarily circulate in wild birds and are typically considered species-specific[6]. However, gene mutations or reassortment might generate a novel AIV strain acquiring the ability to cross species barriers to directly infect humans or other mammals[7, 8]. Prior to 2013, human cases of infection with AIVs were identified, including infections with H5N1, H7N2, H7N3, H7N7, H9N2 and H10N7, respectively[9-11]. Since 2013, in addition to H10N8 virus, three other subtypes of AIVs (H7N9, H6N1 and H5N6) were found to be able to infect humans[12-14].

The first human case of H10N8 infection in China in december 2013

The human-infecting H5N1 virus was first isolated in Hong Kong in 1997[15]. Since then, other subtypes of human-infecting avian influenza viruses (AIVs) have been documented[16]. In December 2013, Chen and colleagues reported the first human case of H10N8 infection in Nanchang City, Jiangxi Province, China[17, 18]. The infected patient was a 73-year-old woman who visited a live poultry market (LPM) four days before the onset of illness. The clinical manifestations included fever, cough and influenza-like symptoms, and she was admitted to a hospital on November 30, 2013. Unfortunately, the patient conditions progressively deteriorated, and the woman eventually died of multiple organ failure on December 6, 2013[17, 18].

Prior to infection, the patient had severe underlying medical conditions, such as hypertension, coronary heart disease, and myasthenia gravis. She underwent a thymectomy in the previous year[17], and she might be susceptible to pathogen infections. On the 4th day of admission, tracheal aspirate specimens were collected from the patient. The specimens were tested at the Nanchang Center for Disease Control and Prevention, and the virus was isolated and identified as an avian influenza A virus[17]. The typed virus was subsequently sent to the Chinese National Influenza Centre in Beijing and subtyped as a novel influenza A H10N8 virus. This virus was subsequently named A/Jiangxi-Donghu/346/2013(H10N8) or JX346 for short[17].

On January 19 and February 9, 2014, a 55-year-old woman and a 75-year-old man infected with H10N8 viruses, respectively, were identified in Nanchang city, Jiangxi, China[20]. The 55-year-old woman recovered from the illness, whereas the 75-year-old man died of multiple organ dysfunctions. All the three individuals infected with H10N8 viruses visited the LPM prior to the onset of clinical symptoms[19, 20].

Origin of influenza A H10N8 subtype viruses

An avian H10N8 subtype virus was first isolated in quails in Italy in 1965[21], and thereafter, avian H10N8 viruses were detected from wild birds or poultry in Australia, Sweden, Canada, US, Republic of Korea and Japan[16]. Viral genome sequencing and phylogenetic analysis revealed that the human-infecting H10N8 (JX346) subtype virus isolated in China represents a novel reassortment among different AIV subtype viruses[17]. Interestingly, all the genes of the JX346 virus are of avian origin. A previous study suggested that the HA gene of H10N8 (JX346) was likely derived from an H10N3 virus isolated from Hunan ducks in China in 2012, whereas the NA gene of JX346 might have derived from an H10N8 virus isolated from a mallard in Republic of Korea in 2010 or an H10N8 virus isolated in wild birds in Japan in 2008[17]. Interestingly, Qi et al. 2014 suggested that the NA gene of H10N8 (JX346) might have derived from an H3N8 virus in duck in Vietnam in 2012.

It has been suggested that the H10 and N8 subtype viruses initially infected wild birds, generating a novel H10N8 virus strain[17]. Thereafter, the H10N8 virus might infect poultry and reassort with an H9N2 virus present in infected birds, thereby resulting in a novel reassortant H10N8 virus with an acquired ability of infecting humans. Epidemiological surveillance revealed that H9N2 viruses circulated in poultry and occasionally infected humans[22]. Similar to H5N1 and H7N9, the six internal genes of the H10N8 (JX346) virus also originated from H9N2 viruses[16, 17, 19]. Previous study showed that an H10N8 virus (A/Chicken/Jiangxi/102/2013/(H10N8)) was isolated in chicken from the LPM, where the 73-year-old patient visited a few days prior to the illness[23]. The human H10N8 virus (JX346) and chicken H10N8 virus (Jiangxi/102) share the same genetic origins in six genes (HA, NA, PA, M, NP and NS genes) but differ in PB1 and PB2 genes[23]. Xu et al., 2015[24] reported that H10N8 viruses were isolated from the LPM A where the 73-year-old woman visited before the onset of her illness. Detailed analysis revealed that the nucleotide sequences of the H10N8 viruses isolated from LPM A are more than 99% identical in sequence to those of the human-infecting H10N8 isolate[24]. These results indicate that the LPM is the most likely source of H10N8 infection for the 73-year-old female patient.

However, currently, the precise origin of the H10N8 virus cannot be definitively determined, reflecting a lack of sufficient epidemiological data on H10N8 infection in poultry and wild birds. Whether the human-infecting H10N8 isolates were generated prior to introduction into the LPM or arose among the poultry in the LPM remains unclear. Before the human H10N8 (JX346) virus was isolated, only two avian H10N8 virus strains were detected in China[25, 26]. One virus, named A/Environment/Dongting Lake/ Hunan/3-9/2007 (H10N8), was isolated from a water sample in the wetlands of the Dongting Lake in Hunan Province in 2007[25]. The other virus, designated A/Duck/Guangdong/E1/2012(H10N8), was isolated from a duck in a LPM in Guangdong Province in 2012[26].

A recent study reported that eight H10N8 viruses were isolated from ducks and chickens in a LPM in China during the period from 2009 to 2013[27]. Ma et al.[28] also reported that they isolated 124 H10N8 or H10N6 viruses from chickens in South China (main in Jiangxi Province) during 2002 to 2014. The sequences of HA gene segments from these H10N8 viruses shared highly homologous to those of the H10N8 (JX346) isolate (Figure 1-10-1). Interestingly, Phylogenetic analysis of H10 and related viruses showed that the six internal genes of the H10N8 viruses were generated through multiple reassortment derived from H9N2 viruses[27, 28]. In 2014, an AIV H10N8 strain was isolated from feral dogs exposed to poultry in the LPM in Guangdong Province, China[29], suggesting that poultry and some mammals should be actively monitored for H10N8 infections. Currently, most H10N8 viruses reported after December 2013 were isolated either from individuals who visited the LPM or poultry and animals in or near the LPM. Increasing evidence suggests that H10N8 viruses primarily circulate in the LPMs[24, 29]; however, the infected birds typically have no apparent signs of illness or mortality. Moreover, the detection of H10N8 viruses on poultry farms seems uncommon[30].

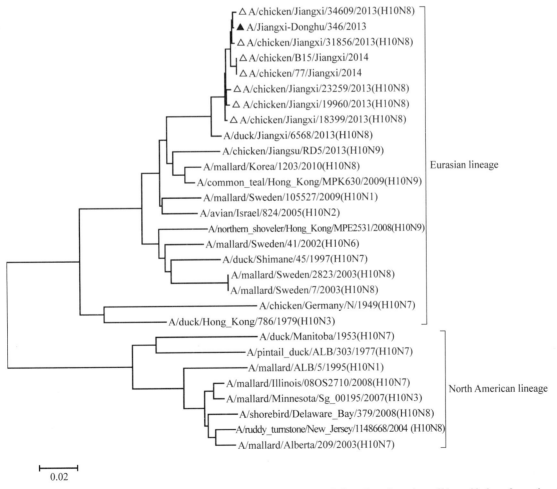

Phylogenetic tree of hemagglutinin gene segments of influenza A virus isolates (H10N8) from Jiangxi province, China, with those from other closely related subtypes of influenza A viruses. The human-origin H10N8 virus isolate, A/Jiangxi/346/2014(H10N8), is marked with ante-black triangle, whereas the rest of the H10N8 isolates marked with ante-white triangles.

Figure 1-10-1 Phylogenetic tree of H10N8

Zhang et al.[31] reported that two H10N8 viruses were isolated from apparently healthy poultry in several LPMs in Nanchang City, during January 2014. Liu et al.[20] also reported that eight H10N8 viruses were isolated in samples collected from LPMs in Nanchang City, Jiangxi Province between December 2013 and February 2014. These authors observed that the sequences of H10N8 virus (A/chicken/Jiangxi/77/2014, A/chicken/Jiangxi/B15/2014 and Ev/JX/03489/2013) were highly homologous to the sequences of H10N8 (JX346) virus. For example, the HA gene from the the H10N8 virus (A/chicken/Jiangxi/77/2014 and A/chicken/Jiangxi/B15/2014) and the NA protein sequence from the H10N8 virus (Ev/JX/03489/2013) shared 99.5% and 100% identity with that from the human-infecting H10N8 virus (JX346), respectively. These results suggest that LPM was the likely source of human infections with H10N8 viruses.

The HA and NA genes of AIVs can be classified as either Eurasian or North American lineages[16]. The eight genes of the H10N8 Hunan strain belong to the Eurasian lineage[25]. Interestingly, although the HA gene of the H10N8 (JX346) virus belongs to the Eurasian lineage, the NA gene of the JX346 belongs to the North American N8 lineage[17]. The human H10N8 (JX346) isolate and the duck H10N8 Hunan strain differ in HA, NA and six internal genes. The HA and NA genes of JX346 were derived from H10N3 and H10N8 viruses, respectively, whereas those of the H10N8 Hunan strain were derived from H10N5 and H6N8 viruses, respectively[17, 25]. The six internal genes of the H10N8 (JX346) virus were all derived from H9N2 viruses[17, 19], whereas those of the H10N8 Hunan strain were primarily derived from H5 or H7 subtype viruses; only the NS1 gene was likely derived from a duck infected with H9N2 viruses[25].

Amino acid substitutions in the viral proteins of human-infecting H10N8 (JX346) virus

Viral genome sequencing and phylogenetic analysis revealed that amino acid substitutions (A135T and S138A) occurred in the H10N8 (JX346) HA protein, and these changes facilitate the adaptation of an AIV to mammalian cells[32]. The NA protein of an AIV can be divided into four regions: head, stalk, membrane-spanning region and cytoplasmic tail[33]. The length of the NA stalk might affect the virulence of an AIV[4]. The H9N2 virus contains a shorter NA stalk and has an increased virulence in chickens[9, 33-34], whereas the H5N1 virus has a shortened NA stalk region and an enhanced virulence in mammalian hosts[35]. Although a deletion in the NA stalk was detected from H5N1 and H7N9 viruses[36, 37], however, such a deletion has not been detected in the H10N8 (JX346) or the H10N8 Guangdong strains[16, 26]. Similar to H7N9, a PDZ domain at the C-terminus of NS1 protein in H10N8 (JX346) virus was deleted, generating a truncated NS1 protein[17]. In contrast, the H10N8 Guangdong strain does not have any amino acid deletions in the NS1, NA or HA proteins[26].

The NS1 protein of H10N8 (JX346) contains a P42S substitution, whereas the M1 protein has two amino acid substitutions (N30D, T215A), all of which might increase viral virulence in mice[38, 39]. The H10N8 (JX346) PB1 protein has L473V and L598P substitutions, responsible for efficient viral adaptation and replication in mammalian cells[17]. The H10N8 (JX346) PB2 protein contains an E627K substitution (PB2-E627K), which has been observed in the PB2 proteins of H1N1, H5N1, H7N7, H7N9, and H9N2 viruses[10, 40-43]. This PB2 E627K substitution has been associated with the enhanced adaptation of viruses to mammalian cells, increased virulence and improved spread of viruses between mammals[17, 44]. The M2 protein of the H10N8 (JX346) virus contains an S31N substitution, suggesting that the JX346 virus might be resistant to M2 ion channel inhibitors[17]. The NA protein of the H10N8 (JX346) virus does not contain H247Y and R292K substitutions, and in vitro assays have shown that the JX346 is sensitive to neuraminidase inhibitors[16, 17].

Mammals infected with H10 or N8 subtype viruses

The first H10 subtype AIV (H10N7) was isolated from chickens in Germany in 1949[45]. One of the features of avian influenza H10 subtype viruses is that these viruses infected wild birds, domestic poultry and certain mammals, and occasionally infected humans[16]. For example, three human cases of H10N8 infection were observed in Nanchang City, Jiangxi Province, China[17, 19]. The H10N7 subtype viruses infected minks and caused mild damage to the lungs[46]. The first H10N7-infecting human case occurred in Egypt in 2004[47], and the second case occurred in Australia in 2010[48]. H10N5 viruses could infect pigs[49], whereas H10N4 viruses infected minks, causing severe lung disease[50]. The cleavage site in the HA precursor of the H10N4 virus does not contain multiple basic amino acid residues[51]; hence, H10N4 is classified as a LPAIV. To date, human infections with H10N4 or H10N5 subtype virus have not been reported.

Some N8 subtype viruses could also infect mammals. For example, H4N8 virus could infect pigs[52], whereas H3N8 virus could infect not only pigs but also horses, dogs, and seals[16, 53]. It should be noted that the H3N8 subtype virus was implicated in the 1889 "Russian Flu" [54]. Prior to the end of 2013, not a single case of human infection with H10N8 subtype virus was reported. Currently, patients infected with an H10N8 subtype virus were only observed in China[17, 19].

A person infected with an H6N1 subtype virus was observed in Taiwan, China, in June 2013[55]. In May 2014, the World Health Organization reported the first human case of an avian H5N6 infection[11]. Therefore, in addition to H10N8, other AIV subtype viruses that infected humans include H5N1, H5N6, H6N1, H7N2, H7N3, H7N7, H7N9, H9N2 and H10N7[9, 11, 17]. Among these, H6N1, H7N2, H7N3, H7N7, H9N2 and H10N7 only caused mild clinical symptoms with some exceptions. For example, the H7N7 subtype resulted in one death among 89 infected individuals in the Netherlands[56], and the H9N2 virus caused relatively severe symptoms only in a bone-marrow transplant patient[22]. Interestingly, during one year period of 2013, it was reported that three AIV subtypes (H7N9, H10N8 and H6N1) infect humans, respectively[16, 55, 57].

Virulence and receptor-binding preference

Virulence of H10N8 viruses

The presence of a multibasic cleavage site in an HA protein is a critical parameter determining the virulence of an AIV in poultry[58]. The cleavage site of the H10N8 (JX346) HA precursor contains only a single basic amino acid, arginine (PELIER*G; *denotes the cleavage site in the H10N8 HA precursor) [17], implying that H10N8 (JX346) is a LPAIV[16, 17, 51]. Interestingly, the cleavage site in the HA precursor of the H10N8 Guangdong strain also contains only a single arginine (PEIVQER*G), and the H10N8 Guangdong isolate is less pathogenic in chickens[16, 26].

One common feature between the H10N8 and H7N9 subtype viruses is that both strains are classified as LPAIVs. Although H10N8 and H7N9 viruses are less pathogenic in poultry, their infection of humans could result in death[30], suggesting that an LPAIV becomes more pathogenic when it crosses host-species barriers to infect a new host. Indeed, after several passages of an avian H10N8 strain in the lungs of mice, the passaged H10N8 viruses became more virulent in mice, leading to death of the infected animals[25]. Chen et al.[10] demonstrated that the human H10N8 virus efficiently replicated in the lungs of a mouse model, inducing acute and persistent cytokine expression and resulting in high pathogenicity in mice, whereas the chicken H10N8 isolate (A/Chicken/Jiangxi/102/2013) was less pathogenic in the same mouse model[10].

Until recently, some human-infecting AIVs, including H5N1, H7N7, H7N9 and H10N8 subtypes, have resulted in death of patients. The six internal genes of the H5N1, H7N9 and H10N8 were all derived from H9N2 viruses[17, 19]. The H9N2 viruses evolve continuously and diversely in poultry displaying both genetic and antigenic variations[59], suggesting that the internal genes of H9N2 viruses might influence the adaptation of an AIV to human cells, and that these genes are likely associated with the enhanced virulence of viruses.

Receptor-binding preference of H10N8 viruses

Two types of receptors bind to the HA proteins of influenza A viruses. One is the avian-type α2, 3 galactose-linked sialic acid (α2, 3-SA) receptor[60], and the other is the human-type α2, 6-SA receptor. High levels of α2, 6-SA receptor and lower levels of α2, 3-SA receptor are present on the surfaces of epithelial cells in the human upper respiratory tract. In contrast, the α2, 3-SA receptor is abundantly present on the surfaces of epithelial cells in the intestinal tract of birds and is also located in human lower respiratory tract[61, 62].

Receptor-binding specificity is primarily determined through a receptor-binding site in an HA protein. Mutations located in or near the receptor-binding site of an HA protein might affect binding to avian-type or human-type receptors[63, 64]. The H10 protein from H10N8 (JX346) binds strongly to avian-type receptors, but weakly to human-type receptors[65, 66], suggesting that the HA protein of the human H10N8 isolate is poorly adapted to bind to the human-type α2, 6-SA receptor, and additional amino-acid substitutions are required to shift to human-type receptor preference. Wang et al.[65] further showed that the arginine residue at position 137 (R137) located within the HA receptor-binding site of H10N8 (JX346) is critical for preferential binding to avian-type receptors[65], whereas Skehel and colleagues demonstrated that a lysine (K) residue at position 137 (K137) in the H10 protein from avian H10N2 virus has some human receptor-binding capacity[64]. Taken together, these studies suggested that the H10 protein of H10N8 viruses is not efficiently adapted for the human type α2, 6-SA receptor, and additional mutations in the H10 protein are required to switch receptor-binding preference to human receptors.

Based on the currently available experimental data, it is clear that the human-infecting H10N8 (JX346) virus possesses preferential binding to avian-type receptors, indicating that JX346 virus behaves like a typical AIV. The H10N8 (JX346) isolate and H5N1 (Hong Kong 1997 isolate) virus are similar in that both viruses bind strongly to avian receptors, but weakly to human receptors[65]. In contrast, the H7N9 virus (A/Anhui/1/2013) has dual receptor-binding capacities because the virus binds strongly to both avian and human receptors[65]. Similar to the H10 protein of the human H10N8 isolate, the H10 proteins of eight H10N8 viruses isolated from chickens and ducks, respectively, bound strongly to avian-type α2, 3-SA receptors, but weakly to human-type α2, 6-SA receptors[27]. Thus, it is important to identify mutations in the HA protein of avian H10N8 virus that could simultaneously increase binding to the α2, 6-SA receptor and decrease binding to the α2, 3-SA receptor or completely switch receptor-binding preference from avian to humans.

Clinical manifestations and laboratory examinations

Clinical manifestations

The incubation period for human infections with H10N8 viruses was approximately four days[17], similar to those of the H5N1 and H7N9 infections[67]. Early in H10N8 infection, patients might show mild symptoms of infection, such as a fever, cough, sputum, sore throat, shortness of breath or poor appetite[17]. Although antiviral and antibacterial drugs and glucocorticoids have been used to treat H10N8-infected patients, the conditions of these patients deteriorated progressively and rapidly. Examining the blood of the infected patients revealed severe lymphocytopenia, slightly increased numbers of neutrophils and leukocytes, and elevated levels of C-reactive protein, creatinine, and lactate dehydrogenase, but reduced levels of total IgG and complement C3[17, 20]. The levels of cytokines and chemokines were substantially increased in the sera of the infected individuals, particularly in fatal cases of infection[20].

Two patients had severe pneumonia, acute respiratory distress syndrome, septic shock, acute kidney failure, and eventually died of multi-organ failure[17, 20]. However, seventeen individuals who had close contact with the patients did not show flu-like symptoms. Virus isolation and antibodies against H10N8 tested negative for all close contacts examined, suggesting that H10N8 viruses have not required the ability to efficiently spread among humans. Co-infections with bacteria, fungi or other subtypes of AIVs were not detected in blood culture tests[17].

Two out of the three patients infected with H10N8 died[17], and the mortality rate was 67%, much higher than the 36% mortality rate in the case of human H7N9 infections[68]. However, this figure might not reflect the actual status because there are currently no sufficient numbers of H10N8-infected human cases to allow reliable statistical estimations. It is likely that some people infected with H10N8 viruses might only exhibit mild symptoms. If those infected individuals did not visit a hospital for an examination, they may have been missed. Therefore, screening for antibodies against H10N8 in individuals within the affected area might facilitate a better evaluation of the prevalence of H10N8 infections in human populations.

Laboratory examinations

It is difficult to distinguish the respiratory symptoms caused by an H10N8 infection from those induced by other respiratory pathogens[16]. Therefore, diagnosing an H10N8 infection requires specific laboratory tests, such as virus isolation, detection of viral RNAs or proteins, sequencing viral gene segments or whole genome, and measuring antibody responses against H10N8 viral proteins[2].

Real-time reverse transcriptase polymerase chain reaction (RT-PCR) is typically applied as a quick diagnostic method for detecting AIVs, including H10N8. Virus isolation and the detection of antibodies against H10N8 viruses can be subsequently performed to verify cases of H10N8 infection. Quantitative RT-PCR is a sensitive, specific assay for H10N8 viral RNAs[2]. Reverse-transcription loop-mediated isothermal amplification assays for detecting H10N8 subtype viruses have recently been developed[69, 70], and these assays can be useful for the surveillance of H10N8 virus infection in live poultry, environment and LPMs.

Similar to H5N1 and H7N9, H10N8 viruses infected humans through the respiratory tract, causing severe symptoms in the lower respiratory tract. The H5N1 virus was isolated through throat swabs[71], whereas H7N9 viruses were detected in the sputa or aspirates from the lower respiratory tract[37]. The collection of sputum or tracheal aspirate specimens from the trachea has been recommended for the detection of human H10N8 infections. Because human infections with H5N1, H7N9 and H10N8 viruses lead to death, the enhanced surveillance of AIV infection in poultry, wild birds and humans is required to monitor the mutations and evolution trends of AIVs. The first case of a human infected with an H6N1 virus was observed in Taiwan, China, in June 2013[55], therefore, the screening of H6N1 infection should also be included as a part of surveillance programmes.

Conclusions

Cases of human infections with AIVs occurred occasionally in the past. However, more cases of human infections with AIVs have been reported since 2013. Between 2013 and 2014, Chinese researchers reported the first fatal case of human infection with an avian H10N8 virus. Subsequent surveillance revealed that the viruses did not acquire the ability for a sustained human-to-human transmission. Currently, the exact origin of human infections with H10N8 viruses remains unclear, although LPMs are the most likely source of human infections. There are currently no efficacious vaccines and antiviral drugs for prevention and treatment of human infections with H10N8 virus.

It is highly important to identify genetic and phenotypic determinants of AIVs that affect virus virulence, reassortment, host adaptation, fitness and transmission between humans, and to examine common biological features of AIVs contributing to cross-species transmission, particularly poultry-to-human transmission. Much effort should be made to determine the mutations or combinations of mutations in HA protein required for switching receptor-binding preferences from avian-to human-type receptors, and to understand the structural and functional basis for receptor-binding specificity of human or avian HA proteins.

Considering the fact that H10N8, H5N1 and H7N9 viruses all possess internal genes from H9N2 viruses, researchers should focus on determining the mechanisms by which the internal genes contribute to the enhanced adaptation of avian influenza viruses to humans. Whether an H10N8 virus can be mutated or reassorted to generate a new virus with a potential to cause a pandemic influenza warrants further investigation. Research should be conducted to explore the origin of human infections with H10N8, the routes of the infections, and the interactions between H10N8 viruses and host factors. These studies will ultimately facilitate the development of novel avenues for global surveillance, diagnosis, prevention and treatment of human infections with AIVs.

References

[1] LAMB R A, KRUG R M. Orthomyxoviridae: The viruses and their replication//KNIPE D M, HOWLEYIN P M. Fields Virology, 4th ed. Philadelphia, USA: Lippincott, Williams and Wilkins, 2001: 1487-1532.

[2] EL ZOWALATY M E, BUSTIN S A, HUSSEINY M I, et al. Avian influenza: virology, diagnosis and surveillance. Future Microbiol, 2013, 8 (9): 1209-1227.

[3] SARACHAI C, SASIPREEYAJAN J, CHANSIRIPORNCHAI N. Characterization of avian influenza H5N1 virosome. Pak Vet J, 2014, 34:201-204.

[4] SUN L S, WANG L F, NING Z Y, et al. Lack of evidence of avian-to-cat transmission of avian H5 subtype influenza virus among cats in southern China. Pak Vet J, 2014, 34:535-537.

[5] RAFIQUE S, SIDDIQUE N, QAYYUM M, et al. In Ovo vaccination against avian influenza virus subtype H9N2. Pak Vet J, 2015, 35:299-302.

[6] PENG Y, XIE Z X, LIU J B, et al. Epidemiological surveillance of low pathogenic avian influenza virus (LPAIV) from poultry in Guangxi Province*, southern China. PLOS ONE, 2013, 8 (10): e77132.

[7] HE C Q, XIE Z X, HAN G Z, et al. Homologous recombination as an evolutionary force in the avian influenza A virus. Mol Biol Evol, 2009, 26 (1): 177-187.

[8] HE C Q, DING N Z, MOU X, et al. Identification of three H1N1 influenza virus groups with natural recombinant genes circulating from 1918 to 2009. Virology, 2012, 427 (1): 60-66.

[9] CHAN J F, TO K K, TSE H, et al. Interspecies transmission and emergence of novel viruses: lessons from bats and birds. Trends Microbiol, 2013, 21 (10): 544-555.

[10] CHEN H, HUANG L, LI H, et al. High Pathogenicity of Influenza A (H10N8) Virus in Mice. Am J Trop Med Hyg, 2015, 93

---

* "Guangxi Province" refers to Guangxi Zhuang Autonomous Region. The same applies to other occurrences in the book.

(6): 1360-1363.

[11] WHO. WHO China Statement on H5N6. (2014-05-07) [2015-11-14]. https://www.who.int/hongkongchina/news/detail/07-05-2014-who-china-statement-on-h5n6.

[12] Tan W, Xie Z X. Research progresson avian influenza avirus subtype H10N8. China Anim Husbandry Vet Med, 2014, 41:236-241.

[13] TAN W, XIE Z. Progresson avian influenza H7N9 Virus. Progress in Veterinary Medicine, 2015, 34:135-139.

[14] LI M, XIE Z, XIE Z, et al. Simultaneous detection of four different neuraminidase types of avian influenza A H5 viruses by multiplex reverse transcription PCR using a GeXP analyser. Influenza Other Respir Viruses, 2016, 10 (2): 141-149.

[15] GUAN Y, SMITH G J. The emergence and diversification of panzootic H5N1 influenza viruses. Virus Res, 2013, 178 (1): 35-43.

[16] TO K K, TSANG A K, CHAN J F, et al. Emergence in China of human disease due to avian influenza A(H10N8)—cause for concern?. J Infect, 2014, 68 (3): 205-215.

[17] CHEN H, YUAN H, GAO R, et al. Clinical and epidemiological characteristics of a fatal case of avian influenza A H10N8 virus infection: a descriptive study. Lancet, 2014, 383 (9918): 714-721.

[18] WAN J, ZHANG J, TAO W, et al. A report of first fetal case of H10N8 avian influenza virus pneumonia in the world. Chinese Critical Care Medicine, 2014, 26:120-122.

[19] GARC A-SASTRE A, SCHMOLKE M. Avian influenza A H10N8—a virus on the verge? Lancet, 2014, 383 (9918): 676-677.

[20] LIU M, LI X, YUAN H, et al. Genetic diversity of avian influenza A (H10N8) virus in live poultry markets and its association with human infections in China. Sci Rep, 2015, 5: 7632.

[21] DE MARCO M A, CAMPITELLI L, FONI E, et al. Influenza surveillance in birds in Italian wetlands (1992-1998): is there a host restricted circulation of influenza viruses in sympatric ducks and coots?. Vet Microbiol, 2004, 98 (3-4): 197-208.

[22] CHENG V C, CHAN J F, WEN X, et al. Infection of immunocompromised patients by avian H9N2 influenza A virus. J Infect, 2011, 62 (5): 394-399.

[23] QI W, ZHOU X, SHI W, et al. Genesis of the novel human-infecting influenza A(H10N8) virus and potential genetic diversity of the virus in poultry, China. Euro Surveill, 2014, 19 (25) : 20841.

[24] XU Y, CAO H, LIU H, et al. Identification of the source of A (H10N8) virus causing human infection. Infect Genet Evol, 2015, 30: 159-163.

[25] ZHANG H, XU B, CHEN Q, et al. Characterization of an H10N8 influenza virus isolated from Dongting lake wetland. Virol J, 2011, 8: 42.

[26] JIAO P, CAO L, YUAN R, et al. Complete genome sequence of an H10N8 avian influenza virus isolated from a live bird market in Southern China. J Virol, 2012, 86 (14): 7716.

[27] DENG G, SHI J, WANG J, et al. Genetics, Receptor Binding, and Virulence in Mice of H10N8 Influenza Viruses Isolated from Ducks and Chickens in Live Poultry Markets in China. J Virol, 2015, 89 (12): 6506-6510.

[28] MA C, LAM T T, CHAI Y, et al. Emergence and evolution of H10 subtype influenza viruses in poultry in China. J Virol, 2015, 89 (7): 3534-3541.

[29] SU S, QI W, ZHOU P, et al. First evidence of H10N8 avian influenza virus infections among feral dogs in live poultry markets in Guangdong province, China. Clin Infect Dis, 2014, 59 (5): 748-750.

[30] SU S, BI Y, WONG G, et al. Epidemiology, evolution, and recent outbreaks of avian influenza virus in China. J Virol, 2015, 89 (17): 8671-8676.

[31] ZHANG T, BI Y, TIAN H, et al. Human infection with influenza virus A(H10N8) from live poultry markets, China, 2014. Emerg Infect Dis, 2014, 20 (12): 2076-2079.

[32] DE WIT E, MUNSTER V J, VAN RIEL D, et al. Molecular determinants of adaptation of highly pathogenic avian influenza H7N7 viruses to efficient replication in the human host. J Virol, 2010, 84 (3): 1597-1606.

[33] SHTYRYA Y A, MOCHALOVA L V, BOVIN N V. Influenza virus neuraminidase: structure and function. Acta Naturae, 2009, 1 (2): 26-32.

[34] SUN Y, TAN Y, WEI K, et al. Amino acid 316 of hemagglutinin and the neuraminidase stalk length influence virulence of H9N2 influenza virus in chickens and mice. J Virol, 2013, 87 (5): 2963-2968.

[35] MATSUOKA Y, SWAYNE D E, THOMAS C, et al. Neuraminidase stalk length and additional glycosylation of the hemagglutinin influence the virulence of influenza H5N1 viruses for mice. J Virol, 2009, 83 (9): 4704-4708.

[36] MATROSOVICH M, ZHOU N, KAWAOKA Y, et al. The surface glycoproteins of H5 influenza viruses isolated from humans, chickens, and wild aquatic birds have distinguishable properties. J Virol, 1999, 73 (2): 1146-1155.

[37] GAO R, CAO B, HU Y, et al. Human infection with a novel avian-origin influenza A (H7N9) virus. N Engl J Med, 2013, 368:1888-1897.

[38] JIAO P, TIAN G, LI Y, et al. A single-amino-acid substitution in the NS1 protein changes the pathogenicity of H5N1 avian influenza viruses in mice. J Virol, 2008, 82 (3): 1146-1154.

[39] FAN S, DENG G, SONG J, et al. Two amino acid residues in the matrix protein M1 contribute to the virulence difference of H5N1 avian influenza viruses in mice. Virology, 2009, 384 (1): 28-32.

[40] TAUBENBERGER J K, REID A H, LOURENS R M, et al. Characterization of the 1918 influenza virus polymerase genes. Nature, 2005, 437 (7060): 889-893.

[41] DE WIT E, FOUCHIER R A. Emerging influenza. J Clin Virol, 2008, 41 (1): 1-6.

[42] TO K K, CHAN J F, CHEN H, et al. The emergence of influenza A H7N9 in human beings 16 years after influenza A H5N1: a tale of two cities. Lancet Infect Dis, 2013, 13 (9): 809-821.

[43] SCHRAUWEN E J, FOUCHIER R A. Host adaptation and transmission of influenza A viruses in mammals. Emerg Microbes Infect, 2014, 3 (2): e9.

[44] HATTA M, GAO P, HALFMANN P, et al. Molecular basis for high virulence of Hong Kong H5N1 influenza A viruses. Science, 2001, 293 (5536): 1840-1842.

[45] FELDMANN H, KRETZSCHMAR E, KLINGEBORN B, et al. The structure of serotype H10 hemagglutinin of influenza A virus: comparison of an apathogenic avian and a mammalian strain pathogenic for mink. Virology, 1988, 165 (2): 428-437.

[46] ENGLUND L. Studies on influenza viruses H10N4 and H10N7 of avian origin in mink. Vet Microbiol, 2000, 74 (1-2): 101-107.

[47] Pan American Health Organization. Avian influenza virus A (H10N7) circulating among humans in Egypt. [2015-11-14]. http://new.paho.org/hq/dmdocuments/2010/Avian_Influenza_Egypt_070503.

[48] ARZEY G G, KIRKLAND P D, ARZEY K E, et al. Influenza virus A (H10N7) in chickens and poultry abattoir workers, Australia. Emerg Infect Dis, 2012, 18 (5): 814-816.

[49] WANG N, ZOU W, YANG Y, et al. Complete genome sequence of an H10N5 avian influenza virus isolated from pigs in central China. J Virol, 2012, 86 (24): 13865-13866.

[50] KLINGEBORN B, ENGLUND L, ROTT R, et al. An avian influenza A virus killing a mammalian species—the mink. Arch Virol, 1985, 86 (3-4): 347-351.

[51] ZOHARI S, METREVELI G, KISS I, et al. Full genome comparison and characterization of avian H10 viruses with different pathogenicity in Mink (Mustela vison) reveals genetic and functional differences in the non-structural gene. Virol J, 2010, 7: 145.

[52] SU S, QI W B, CHEN J D, et al. Complete genome sequence of an avian-like H4N8 swine influenza virus discovered in southern China. J Virol, 2012, 86 (17): 9542.

[53] WEBSTER R G, GERACI J, PETURSSON G, et al. Conjunctivitis in human beings caused by influenza A virus of seals. N Engl J Med, 1981, 304 (15): 911.

[54] TROMBETTA C, PICCIRELLA S, PERINI D, et al. Emerging Influenza Strains in the Last Two Decades: A Threat of a New Pandemic?. Vaccines (Basel), 2015, 3 (1): 172-185.

[55] YUAN J, ZHANG L, KAN X, et al. Origin and molecular characteristics of a novel 2013 avian influenza A(H6N1) virus causing human infection in Taiwan. Clin Infect Dis, 2013, 57 (9): 1367-1368.

[56] FOUCHIER R A, SCHNEEBERGER P M, ROZENDAAL F W, et al. Avian influenza A virus (H7N7) associated with human conjunctivitis and a fatal case of acute respiratory distress syndrome. Proc Natl Acad Sci U S A, 2004, 101 (5): 1356-1361.

[57] MORENS D M, TAUBENBERGER J K, FAUCI A S. H7N9 avian influenza A virus and the perpetual challenge of potential human pandemicity. MBio, 2013, 4 (4) : e00445-13.

[58] SENNE D A, PANIGRAHY B, KAWAOKA Y, et al. Survey of the hemagglutinin (HA) cleavage site sequence of H5 and H7 avian influenza viruses: amino acid sequence at the HA cleavage site as a marker of pathogenicity potential. Avian Dis, 1996, 40 (2): 425-437.

[59] CHOI Y K, OZAKI H, WEBBY R J, et al. Continuing evolution of H9N2 influenza viruses in southeastern China. J Virol, 2004, 78 (16): 8609-8614.

[60] VAN RIEL D, MUNSTER V J, DE WIT E, et al. H5N1 virus attachment to lower respiratory tract. Science, 2006, 312 (5772): 399.

[61] SHINYA K, EBINA M, YAMADA S, et al. Avian flu: influenza virus receptors in the human airway. Nature, 2006, 440 (7083): 435-436.

[62] CHENG V C, TO K K, TSE H, et al. Two years after pandemic influenza A/2009/H1N1: what have we learned? Clin Microbiol Rev, 2012, 25 (2): 223-263.

[63] MATROSOVICH M, TUZIKOV A, BOVIN N, et al. Early alterations of the receptor-binding properties of H1, H2, and H3 avian influenza virus hemagglutinins after their introduction into mammals. J Virol, 2000, 74 (18): 8502-8512.

[64] VACHIERI S G, XIONG X, COLLINS P J, et al. Receptor binding by H10 influenza viruses. Nature, 2014, 511 (7510): 475-477.

[65] WANG M, ZHANG W, QI J, et al. Structural basis for preferential avian receptor binding by the human-infecting H10N8 avian influenza virus. Nat Commun, 2015, 6: 5600.

[66] ZHANG H, DE VRIES R P, TZARUM N, et al. A human-infecting H10N8 influenza virus retains a strong preference for avian-type receptors. Cell Host Microbe, 2015, 17 (3): 377-384.

[67] COWLING B J, JIN L, LAU E H, et al. Comparative epidemiology of human infections with avian influenza A H7N9 and H5N1 viruses in China: a population-based study of laboratory-confirmed cases. Lancet, 2013, 382 (9887): 129-137.

[68] YU H, COWLING B J, FENG L, et al. Human infection with avian influenza A H7N9 virus: an assessment of clinical severity. Lancet, 2013, 382 (9887): 138-145.

[69] BAO H, FENG X, MA Y, et al. Rapid detection of subtype H10N8 influenza virus by one-step reverse transcription-loop-mediated isothermal amplification Methods. J Clin Microbiol, 2015, 53 (12): 3884-3887.

[70] LUO S, XIE Z, XIE L, et al. Reverse-transcription, loop-mediated isothermal amplification assay for the sensitive and rapid detection of H10 subtype avian influenza viruses. Virol J, 2015, 12: 145.

[71] DE JONG M D, SIMMONS C P, THANH T T, et al. Fatal outcome of human influenza A (H5N1) is associated with high viral load and hypercytokinemia. Nat Med, 2006, 12 (10): 1203-1207.

Avian infectious bronchitis virus (AIBV) review by continent

Rafique Saba, Jabeen Zohra, Pervaiz Treeza, Rashid Farooq, Luo Sisi, Xie Liji and Xie Zhixun

Abstract

Infectious bronchitis virus (IBV) is a positive-sense, single-stranded, enveloped RNA virus responsible for substantial economic losses to the poultry industry worldwide by causing a highly contagious respiratory disease. The virus can spread quickly through contact, contaminated equipment, aerosols, and personal-to-person contact. We highlight the prevalence and geographic distribution of all nine genotypes, as well as the relevant symptoms and economic impact, by extensively analyzing the current literature. Moreover, phylogenetic analysis was performed using Molecular Evolutionary Genetics Analysis (MEGA-6), which provided insights into the global molecular diversity and evolution of IBV strains. This review highlights that IBV genotype I (GI) is prevalent worldwide because sporadic cases have been found on many continents. Conversely, G II was identified as a European strain that subsequently dispersed throughout Europe and South America. G III and GV are predominant in Australia, with very few reports from Asia. GIV, GVIII, and GIX originate from North America. GIV was found to circulate in Asia, and GVII was identified in Europe and China. Geographically, the GVI-1 lineage is thought to be restricted to Asia. This review highlights that IBV still often arises in commercial chicken flocks despite immunization and biosecurity measures because of the ongoing introduction of novel IBV variants and inadequate cross-protection provided by the presently available vaccines. Consequently, IB consistently jeopardizes the ability of the poultry industry to grow and prosper. Identifying these domains will aid in discerning the pathogenicity and prevalence of IBV genotypes, potentially enhancing disease prevention and management tactics.

Keywords

AIBV, distribution, continents, prevalence, genotype

Introduction

Infectious bronchitis, which mostly affects poultry, is a viral disease. It causes substantial economic loss and morbidity in poultry industries worldwide[1, 2]. The etiological agents are avian infectious bronchitis virus (AIBV) and avian coronavirus[3], which are members of the genus Gamma coronavirus (γCoV), family Coronaviridae and order Nidovirales[4-6]. The Coronaviridae family is divided into four subfamilies, with four genera belonging to the Coronavirinae subfamily: alpha coronavirus (αCoV), beta coronavirus (βCoV), gam macoronavirus (γCoV) and delta coronavirus (δCoV). Avian species, including chickens, land fowl, and pheasants, are specifically infected by γCoV and δCoV[7].

The genome of IBV is composed of linear and positive single-stranded RNA strands approximately $27 \sim 28$ kb long that encode a variety of structural and nonstructural proteins (NSPs). The IBV genome is organized as 5' UTR-ORF 1a/1b-S-3a-3b-E-M-4b-4c-5a-5b-N-6b-3' UTR, with an a-1 frame shift at the junction of ORF 1a/1b, resulting in the synthesis of the 1a and 1b polyproteins, which are subsequently processed to produce individual NSPs responsible for genome replication and transcription (Figure 1-11-1) [8]. Furthermore, these viruses have important structural components, such as the spike (S), membrane (M), envelope (E), and nucleocapsid (N) proteins[9]. Notably, the spike protein, with a molecular weight of 200 kDa, is the largest glycoprotein among these proteins. Previous research has also identified two accessory genes, ORF3 and ORF5, encoding proteins 3a, 3b, 5a, and 5b[10, 11]. The ORFs, la and lab are approximately 20 kb in length and constitute approximately two-thirds of the genome. NSPs are a collection of 16 proteins that play indispensable roles in viral replication, RNA synthesis, transcription, RNA proofreading, RNA capping, and host immune response modulation[12-16]. The genes for the structural proteins and accessory proteins are ordered in the viral genome as follows.

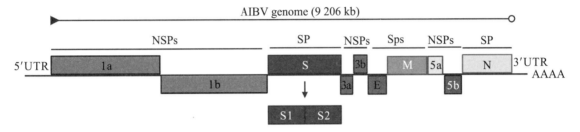

Genome organization of avian infectious bronchitis virus. The total genome of AIBV is 9 206 kb, showing structural proteins (SPs) and non-structural proteins (NSPs). The figure was created with BioRender.com.

Figure 1-11-1

Gene 2 encodes S protein and is the longest gene in this region and allows the virus to attach to host cells and mediates the fusion of viral and cellular membranes, thus contributing to viral tropism[17, 18]. Gene 3 is a polycistronic gene that encodes three proteins, 3a, 3b, and 3c, and is positioned downstream of gene 2. The 3c protein, generally referred to as the E protein, is required for viral particle assembly[19], but the 3a and 3b peptides have yet to be identified[20]. Gene 4 is another polycistronic gene that encodes three distinct peptides. The first ORF encodes an M protein that is a significant component of the viral envelope. The functions of the two small accessory 4b and 4c peptides are unknown. The M protein contributes to virion structure and may activate a humoral immune response in chickens, but this response is insuficient to prevent clinical signs from appearing[21]. Gene 5 encodes two tiny accessory peptides, 5a and 5b[22, 23]. Gene 6 encodes the N protein and is the last ORF of the IBV genome. This protein is a nonspecific RNA-binding protein and has numerous roles in the virus's life cycle[16, 24].

IBV is traditionally thought to be a host-specific respiratory pathogen in poultry, and it replicates at the tracheal mucosa[25]. However, the discovery of new IBV genotypes or serotypes has revealed a wide range of tissue tropisms, including in the bursa of Fabricius[26, 27], urinary tract[28, 29], reproductive tract[30] and gastrointestinal tract[27, 31, 32]. IBV is known to replicate in layers of the reproductive tract epithelium, resulting in decreased egg production and defective eggshell formation[33]. Early infection with the IBV JX181 (GVI-1) strain is highly harmful to laying hens, causing major respiratory problems, irreversible oviduct damage, and growth retardation in the reproductive system[34]. False layer syndrome, which is associated with cystic oviduct development, cystic left oviducts, signs of vent pecking, ovarian regression, and yolk coelomitis, occurs due to IBV infection in young birds aged 25~28 weeks[35]. Cockerel testes can also be infected with IBV[36].

Among all avian species, pheasants are considered the second most common natural host for IBV after poultry[37]. Antibodies against IBV have been found in quail, turkeys and free-ranging rockhopper penguins, though it is unknown whether IBV can be isolated from these animals[38]. In addition, IBV infects a variety of bird species, particularly those raised near domestic poultry, such as ducks, teals, geese, pigeons, partridges, guinea fowl, peafowl and domestic fowl[39]. The genome of IBV shares many similarities among various hosts. One virus found in teal and peafowl, for instance, shows 90% to 99%similarity in sequence to IBV[40]. Incredibly similar nucleotide sequences were found in viruses collected from turkeys, ducks, pheasants, and whooper swans[41]. IB was first detected in the United States in 1931[42]. The causative agent was not identified, which frequently caused confusion with other viral and bacterial pathogens that can cause upper respiratory infections in chickens. Until the first report of the Connecticut isolate (Conn) (GI-1), the Massachusetts isolate (Mass) (GI-1) was the sole serotype discovered. The Conn (GI-1) causes identical symptoms but exhibits antigenic variations from the Massachusetts isolate (GI-1) [43]. Notably, this viral pathogen has demonstrated a remarkable capacity for diversification and adaptation, resulting in the identification of numerous unique serotypes in various parts of the globe[44].

IBV poses no known threat to human health. Neutralizing antibodies have been found in employees who work on commercial chicken farms, though their impact is uncertain[45]. Furthermore, there is currently little evidence to support the idea that IBV may replicate inside a human host; avians can simply transmit IBV infection to chicks mechanically[46]. The continuing processes of genetic recombination and mutation can be attributed to the ongoing emergence of novel IBV variants. Many of these variants show noticeable antigenic differences, which are carefully displayed by cross-neutralization experiments or monoclonal antibody tests. Nonstructural 1ab and accessory proteins, including 3a, 3b, 5a, and 5b, have been reported to be potential tissue tropism regulators as well as pathogenicity determinants. Moreover, the interaction between the S1 and S2 spike subunits of the virus may control the host range and attachment site of IBV. The host immune response and infection route are additional variables that may contribute to tissue tropism[47].

Diverse IBV strains have been found and documented in various geographical locations. Notably, the Georgia 08 (GA08) strain Genotype 1 and lineage 27 (GI-27) of IBV are the most common and prevalent in the United States[48]. Furthermore, reports of its presence outside the United States have been documented. Similarly, GII-1 (D1466) occurs infrequently outside western Europe[49]. However, the GI-13 (793B) and GI-19 (QX) variants are largely distributed in Europe, Africa, and Asia but have not been found in the United States and Australia[50]. The S1 spike gene sequence was used to categorize IBV genotypes, resulting in the identification of nine genotypes spanning dozens of lineages.

Two conditions must be met for classification as a new IBV lineage on the basis of the S1 gene: first, well-supported statistical evidence may usually be ascertained using bootstrap or posterior probability values; second, monophyletic clusters consisting of at least three viruses collected from at least two distinct outbreaks may be present, as defined by[51]. There must be 13% or greater uncorrected pairwise distances in the nucleotide sequences for IBV lineages to be defined by the abovementioned criteria. Various mutations, including insertions, deletions, point mutations, and recombinations, exist across different strains. As a result of these events, several S1 variants have emerged and evolved[52]. As reported, recombination often occurs during IBV replication, resulting in the formation of chimeric viruses made up of genetic sequences from several viruses[5]. Thus, the antigenic diversity of IBV is influenced by many factors.

Genotypic prevalence on continents

The sporadic emergence of different genotypes of IBV across continents can be explained individually. The genetic classification of the reported genotypes and lineages was reproduced by using the S1 gene sequence of all IBV genotypes (Figure 1-11-2). The overall distribution of these genotypes from 1930 to 2022 is also represented using the Ghantt chart (Figure 1-11-3) and (Table 1-11-1) to clarify the reported patterns of the genotypic distribution of this virus. The reported patterns are explained in detail in terms of genotype.

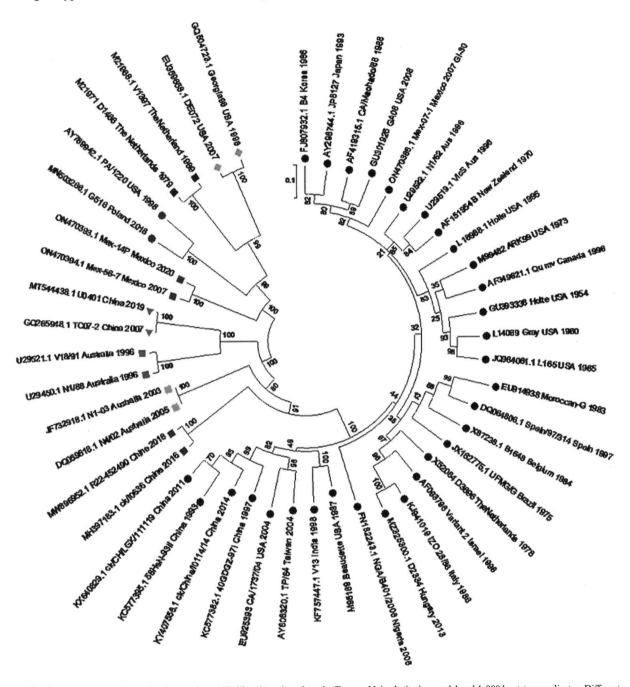

A Circular tree was created by using the maximum likelihood tree based on the Tamura-Nei substitution model and 1 000 bootstrap replicates. Different genotypes are highlighted with different shapes and colors. Black circles: G I lineages (1-31), Meroon square: G II, Blue square: G III, Green rhombus: GIN, Yellow square: GV , Grey inverted-triangle: GVI, Purple square: GVII, Red circle: GVIII, Green square: GIX (color figure in appendix).

Figure 1-11-2   Molecular phylogenetic analysis of the S1 gene of the reference IBV genotypes

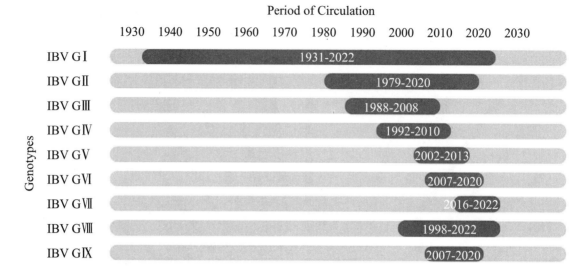

Figure 1-11-3   The Gantt chart represents the circulation of the avian infectious bronchitis virus (GⅠ-GⅨ) genotype throughout the period from 1931-2022.

Table 1-11-1   Genotypes of Infectious Bronchitis Virus distributed across continents

Continent	Reference Strain Name	Prevalent Genotype	References
America	Beudette	GⅠ-1	[47], [51], [60].[63], [67], [68], [69], [71], [72]
	Holte	GⅠ-2	
	Gray	GⅠ-3	
	Holte	GⅠ-4	
	LI65	GⅠ-8	
	ARK99	GⅠ-9	
	UFMG/G	GⅠ-11	
	B1648	GⅠ-14	
	IZO 28/86	GⅠ-16	
	DMV-1639/11	GⅠ-17	
	Qu_mv	GⅠ-20	
	Varient 2	GⅠ-23	
	CA/1737/04	GⅠ-25	
	GA-08	GⅠ-27	
	Mex-07-1	GⅠ-30	
	DE072	GⅡ	[69]
	GA98	GⅣ-1	[73], [146], [148]
	DE072	GⅣ-1	
	MEX-12	GⅧ-1	[73]
	Mex-56-7	GⅨ-1	[73]

continued

Continent	Reference Strain Name	Prevalent Genotype	References
Europe	Beudette	GI-1	[51], [79], [80], [82], [84], [85]
	B	GI-10	
	D274	GI-12	
	793B	GI-13	
	B1648	GI-14	
	IZO 28/86	GI-16	
	QX	GI-19	
	Spain/97/314	GI-21	
	Variant 2	GI-23	
	D1466	GII-1	[141], [142]
	D181	GII-2	
	PA/1220/98	GVIII-1	[157]
Africa	Beaudette	GI-1	[51], [88], [90], [91], [92], [93], [95], [96]
	D3896	GI-12	
	Moroccan-G/83	GI-13	
	B1648	GI-14	
	IZO 28/86	GI-16	
	QX	GI-19	
	Variant 2	GI-23	
	NGA/B401/2006	GI-26	
	D2334/11/2/13/CI	GI-31	
Asia	Beaudette	GI-1	[44], [51], [104], [105], [106], [109], [110], [111], [114], [124], [126], [177]
	TP/64	GI-7	
	D274	GI-12	
	PAK 973	GI-13	
	B4	GI-15	
	JP8127	GI-18	
	QX	GI-19	
	40GDGZ-971	GI-22	
	Variant 2	GI-23	
	V13	GI-24	
	LGX/111119	GI-28	
	I0111/14	GI-29	
	DI466	GII-1	[143]
	K069-01	GIII	[145]
	T07/02	GIV	[145]
	TC07-2	GVI-1	[145], [151], [152], [154], [156]
	IO636/1	GVII-1	[156]

continued

Continent	Reference Strain Name	Prevalent Genotype	References
Australia	N1/62	GI-5	[51], [133]
	VicS	GI-6	
	N1/88	GIII-1	[21], [51], [134]
	23/88	GIII-2	
	NI/03	GV-1	[178]

Genotype I

North America

The first case of IBV was reported in central North American states in the spring of the early 1930s; this is an acute, deadly respiratory condition that appeared to be limited to only chicks aged 2 to 20 days. The chicks developed resistance with age, and those older than three weeks were immune. By the early 1970s, multiple distinct IBV variants had been identified, primarily through serological reports[54-57]. In North America, the GI-9 and GI-27 and GIV-1 genotypes have been linked to continent-wide disease transmission and persistent viral infections[51, 58].

Previously, Arkansas (Ark) (GI-9) was the most prevalent IBV strain known in North America, but recently GA08 (GI-27) became a dominant genotype in this region due to high usage of this strain in vaccines[48, 59]. Another significant IBV strain in the United States is DMV/1639/11 (GI-17), which was first documented in the United States in 2011 and has significantly affected broilers, breeders and layers from 2014-2015; since then, it has prevailed in certain regions of the US[48].

The IBV 4/91 strain IBV/Ck/Can/17 038913 (GI-13) was identified from chicken flocks in eastern Canada that had reduced egg production and egg quality from 2011 to 2018. The isolate was assigned to the GI-13 genotype based on comparative genomics and phylogenetic analysis of the S1 gene, and it exhibited varying similarity with different open reading frames (ORFs) of several reference strains. Despite having wide tissue tropism inlaying hens, IBV/Ck/Can/17 038913 (GI-13) did not affect egg quality or production at the infection dose used in the current investigation. However, the isolate caused considerable macrophage and T-cell migration along with tissue tropism in the kidney[60]. Although Ark (lineage GI-9), Holte/Iowa-97 (lineage GI-3), Mass-type (lineage GI-1), and Q1-like serotypes have been identified within commercial flocks in Mexico, a new study indicated that in 2019-2021, 793B (GI-13) and California variant CAV (GI-17) -like strains were found[61].

The Delmarva (DMV) /1639 variant of IBV, which is a member of GI-17, was linked to numerous nephropathogenic IBV outbreaks that occurred in the Delmarva Peninsula, United States of America (USA), in 2011. IBV strains genetically similar to the DMV/1639 (GI-17) type have been more prevalent in the poultry industry in eastern Canada since 2015. The Canadian DMV/1639 (GI-17) strain has been proven to cause noticeable pathological lesions and tissue tropism in many bodily systems, including the respiratory, reproductive and renal systems[62]. Another study provided a dear illustration of the distribution of DMV strains, in which five different IBVs were isolated from Ontario, Canada. S gene analysis of these strains revealed high similarity to that of the Delmarva (DMV/1639) (GI-17) strain, and the initial detection of this virus occurred during an IBV epidemic on the Delmarva Peninsula in the USA in 2011. However, whole-genome analysis revelated its relevance to the Conn (GI-1) -type vaccine strain. These isolates were grouped according to the GI-17 genotype[63]. Furthermore, a study illustrated the importance of the GI-17 lineage in North American regions; specifically, a virus involved in the IBV epidemic that occurred in Costa Rica from May 2016 through mid-2017 was identified and categorized as a Georgia 13-like on the basis of the S1 region of the S gene. The entire genome sequence has a sequence identity with that of the strain from Georgia, USA, of 94.03%, and the least similar sequence identity with that of the strain from China is 86.03%. The Costa Rican isolate, which shows 96.89% similarity with the S1 subunit of the Georgian strain, was identified as belonging to genotype I and lineage 17 (GI-17) [64].

South America

In the 1950s, IBV spread to South America, with Brazil having the first mass (GI-1) serotype isolate[65]. IBV was recognized as a serious threat to commercial chicken populations by the mid-1980s. New mutations were discovered alongside the Conn (GI-1) and Mass (GI-1) serotypes in both layer and broiler birds[66].

In South America, the GI genotype has exhibited two distinct lineages, GI-11 and GI-16, which have been in extensive circulation for many years. An entirely South American lineage known as GI-11 arose in Brazil, Uruguay, and Argentina in the 1960s. Conversely, the GI-16 lineage is ubiquitous and consists of two genetic groupings that were previously thought to be independent: Q1 (GI-16) and 624/I (GI-16). The 624/I variant was initially detected in 1993 in Italy and subsequently in Slovenia, Poland, and Russia. Emerging in the late 1970s, the GI-16 lineage first appeared and is now found in the majority of South American states. Retrospective research has shown that it has been prevalent in Italy since the early 1960s, with a substantial drop in disease incidence since the 1990s. Q1 (GI-16) was also found in birds with proventriculitis in China in the 1990s and subsequently in other Asian nations, South America, Europe and the Middle East. Recent sequence analysis of the whole genomes of the Italian 624/I (GI-16) and Q1 (GI-16) strains support the idea that these two genetic lineages have a common ancestor[67]. The IBV strain known as VFAR-047 was discovered in 2014 in the broiler flock and was referred to as GI-16 in Lima, Peru's northern area. Examination of the S gene and genome sequencing allowed for the identification of this isolate[68]. According to a recent Peruvian study, the circulation of GI-16 has been determined along with the GI-1 (vaccine derived) lineage in the poultry industry of Peru. GI-1 is prevalent in tropical areas, and GI-16 is prevalent in coastal areas[69].

According to a report from Trinidad and Tobago, the circulating IBV strain shows more than 20% nucleotide differences from the vaccine strain M41 (GI-1) but 16.7% from the GI-14 lineage. Since the Trinidad and Tobago strains do not cluster with other lineages, they are distinct strains of a completely new lineage[70]. Apart from the prevalence of the GI-11 and GI-16 genotypes in South America, the first report of the IBV GI-23 variant was recorded in Brazil in 2022. In 1998, for the first time, the GI-23 variant was found to cause disease in poultry in India. Immediately after, the virus quickly spread to other continents. Among IBV variants, GI-23 strains are significantly pathogenic to the respiratory tract of embryos and specific pathogen-free (SPF) chicks[71]. Another study from Colombia, the country where IBV was initially discovered in 1963, reported the prevalence of GI-1[72].

In Mexico, during 2007 and 2021, 17 avian infectious bronchitis virus strains were isolated from diseased chickens. Different strains were grouped into different lineages; however, six strains in genotype 1 did not belong to any of the identified lineages. Therefore, after clustering in a well-supported clade, the cells were designated as a new lineage (GI-30) [73].

Europe

The Mass (GI-1) serotype of IBVs was considered the only disease-causing IBV until the late 1970s. Dedson & Gough announced in 1971 that other IBV variants had been found in the UK[74]; subsequently, employees at the Doorn Institute discovered IBVs in mass-vaccinated commercial flocks that belonged to not less than four distinct IBV serotypes linked to disease occurrence[75]. According to a surveillance study conducted in Europe, the most prevalent IBV strains found were Massachusetts (GI-1), QX-like (GI-19), 793B (GI-13), D274 (GI-12), and D1466 (GII-1) [76, 77].

The GI-19 genotype was initially reported in China in the late 1990s and was formerly known as the QX (GI-19) genotype; its high prevalence has made it one of the most common IBV[40], after which it spread to other Asian, Middle Eastern and European countries in waves. The GI-19 lineage attracted increased amounts of attention when it was first discovered in the Netherlands between 2003 and 2004. This viral strain was detected in the kidneys and oviducts of infected poultry and was linked to a significant prevalence of false layer syndrome. In Italy, sequencing analysis indicated that the S1 gene shared 99% similarity with the QX (GI-19) strain. In fact, the alignment of the sequence of the hypervariable region of the S1 gene of QX (GI-19) with that of the mass (GI-1) and 4/91 (GI-13) strains was 77.1% and 81%, respectively[78]. Later, in the 2000s, the Chinese QX (GI-19) strain was detected for the first time in the United Kingdom (UK). Sequence analysis revealed 96.6% similarity to the original Chinese QX (GI-19) strain and 98.1% similarity to the common European QX-type (GI-19) IBV known as L-1148[79].

Research carried out in Spain and Italy revealed the simultaneous spread of various IBV genotypes. The genotypes that were most prevalent in the two countries in recent years were 793B (GI-13) and QX (GI-19) [80]. However, a new IBV variant linked to the XDN variant has emerged[81]. The Spanish IBV isolates and the XDN strains were found to form a homogenous clade, and both belonged to the QX (GI-19) genotype, while the Italian strains exhibit greater variation in their genomes. The large and homogeneous clade identified in Spain was thought to have descended from a single recombinant parent and then dispersed across the entire nation. In contrast, nine Italian recombinants, which were distinguished by three distinct recombination events and resulted from independent recombination events between the QX (GI-19) and 793B (GI-13) IBV variants, were considered less viable and less suited strains[82].

The GI-23 lineage IBV variant known as Israel strain 2 was initially identified in Israel in 1998[83]. The first incidence of IBV VAR2 from the GI-23 genotype was found in Poland in December 2015. The whole-genome analysis of the Polish GI-23 lineage revealed that the mosaic pattern of the viral genome from other IBV strains descended from the QX-like (GI-19), 793B-like (GI-13), and mass-like (GI-1) serotypes. The exact origin of the virus that infects the Polish chicken population is still unknown[84].

However, a German broiler farm also reported the IBV Israel strain 2. This was the first time that the GI-23 lineage of Middle Eastern origin was isolated in Germany, adding to the increasing concern about the spread of IBV variants across Europe[85]. The birds exhibited respiratory and nephropathogenic symptoms. Phylogenetic analysis suggested that the current isolates are members of the GI-23 lineage, indicating that they are most genetically related to Polish variants and the IBV VAR206 (GI-23) vaccine. The possibility of recombination events between Polish GI-23 and German field or vaccine strains is evident given the potential of GI-23 strains to undergo recombination[86].

Africa

IBV emerged in southern Africa in the early 1980s, resulting in severe disease conditions such as swollen head syndrome[87]. It was first identified in Mediterranean basin countries. Although IBV is frequently found in both symptomatic and asymptomatic poultry, its effect on the African continent is unknown[88].

Although several lineages of IBV have been recovered in Africa, only GI-26 is considered indigenous to the continent and is predominantly composed of strains from North and West Africa[51]. Other Eurasian-originating African IBV strains include those from the lineages GI-1 (linked with sporadic IBV outbreaks in many countries), GI-12 (Nigeria), GI-13 (South Africa, Morocco, Ethiopia, Sudan, and Algeria), GI-14 (Nigeria & Cameroon), GI-16 (Nigeria & Cted'Ivoire), GI-19 (Ghana, Nigeria, South Africa, Zimbabwe, and Algeria), and GI-23 (Nigeria & Egypt), as well as various lineages from Algeria, Tunisia, Libya, and Ethiopia[51, 89, 90].

IBV has not been reported from Kenya or the East or Central African regions up until this point, except Cameroon in Central Africa[86]. It is possible that the Kenyan strains A374/17 and A376/17 are gastroenteric. S1 gene analysis of this strain revealed its resemblance to GI-23 viruses rather than GI-16 viruses. The four lineages GI-1, GI-13, GI-14, and GI-19 are used to categorize other strains of African IBV. In contrast, the Kenyan isolate A376/17 resembles turkey coronaviruses (TCoVs) from Asia and France more closely than those from North America. However, unlike all other TCoVs that have been studied, TCoVs exhibit distinct differences[91].

In Ethiopia, sequencing analysis confirmed the presence of 793B genotypes (GI-13). Due to the high frequency of 793B (GI-13) live vaccine administration, these Ethiopian IBV isolates exhibited substantial genetic variation, indicating that they are most likely to be field strains. Only backyard poultry showed positive results, indicating that this type of setting is more conducive to IBV circulation[50].

There are several different IBV genotypes in Egypt, comprising GI-13, GI-23, GI-1, and GI-16, each with unique genetic and pathogenic characteristics. Based on phylogenetic analysis, the circulating IBV strains in Egypt can be divided into two major groups. The first group consists of strains from the GI-1 lineage that have developed through genetic drift from live attenuated traditional vaccines[92]. The second group consisted of strains from the GI-23 lineage, including circulating Egyptian variants. According to recombination analysis, a minimum of three distinct IBV strains, 4/91 (GI-13), H120 (GI-1), and QX-type (GI-19), tend to undergo recombination events[2].

In North African nations such as Tunisia, Egypt, and Morocco, the combination of endemic and classic IBV variants prevails, contributing to nearly all the information on IBV strains in Africa[65]. A QX-like (GI-19) IBV strain was recently discovered in Zimbabwe. The IBV strain QX L-1148 (GI-19) from China shows the maximum nucleotide sequence similarity (98.6%) to the S1 gene. Notably, this is the first reported instance of QX-like (GI-19) IBV circulation in Africa[93].

This GI-19 lineage was introduced in 2009, likely as a result of a single introduction event from Germany[94]. In Sudan, two IBV strains were identified from commercial broiler farms that exhibited respiratory symptoms and increased mortality. The Chinese IBV strain, the Italian IBV strain, and the 4/91 (GI-13) vaccine strain had the highest nucleotide sequence similarity for the whole genome (93%). The IBV/Ck/Sudan/AR251-15/2014 strain was genetically distinct from formerly defined IBV strains in Africa and worldwide on the basis of S1 gene analysis and was clustered with viruses of the GI-19 lineage[95]. Greece has a high incidence of IBV field strains, with 20 or more of the GI-19 lineage[94].

In a study in Africa, the complete genome sequences of five IBV strains from the sub-Saharan region were analyzed. Based on phylogenetic analysis, three (GI-14, GI-16, and GI-19) lineages prevailed in that region. However, it has also been observed that a strain isolated from the Ivory Coast, D2334/11/2/13/CI, belongs to a distinct lineage within the GI-31 genotype[88].

A novel IBV genotype known as "IBADAN" has recently been reported. Analysis of the S1 gene sequence indicated that NGA/ A116E7/2006 had 24%~25% amino acid and nucleotide variation from the nearest strain of a different cluster, whereas ITA/90254/ 2005 shows 14% nucleotide and 16% amino acid variation from its closest strain. When the whole genomes of NGA/A116E7/2006 and ITA/90254/2005 were analyzed, 10% nucleotide variation was found, with the IBADAN strain exhibiting a high genetic distance from 9.7% to 16.4% when compared to other known IBV sequences. The S1 protein amino acid sequence revealed particular positions shared by NGA/A116E7/2006 and ITA/90254/2005, differentiating them from other strains[96].

Asia

Several lineages, such as GI-7, -15, -18, -22, and -24, and GVI-1, are exclusive to Asia[51]. In Japan, IBV strains have been genotyped via incomplete nucleotide sequence information that included the HVR-1 and HVR-2 regions[97, 98]. The strains were divided into seven genotypes based on this classification: JP-I (GI-18), JP-II (GI-7), JP-III (GI-19), JP-IV (GVI-1), Mass (GI-1), 4/91 (GI-13), and Gray (GI-3). The JP/KH/64 strain (GI-18) was identified from chickens exhibiting respiratory symptoms in 1964 in Japan, and it was the first JP-I genotype to be identified in Japan[98].

The complete genome sequence of the primary genotype JP-I (GI-18) strains was retrieved to further understand the distribution of IBV variations in Japan. Phylogenetic analysis of the full coding sequence of the S1 gene revealed that JP/KH/64 is part of the GI-18 lineage and groups in a single category with the lineage prototype strain JP8127, which was discovered in Japan. Because the JP-I (GI-18) genotype has been reported in China, and Japan[12, 13], studying the evolution of IBV in East Asian countries may be aided by this knowledge of the JP/KH/64 strain[99]. Another study in Japan determined and studied the entire S1 gene of 61 IBV strains. Except for three strains (JP/Nagasaki/2013, JP/ Kochi/2013, and JP/Nagasaki/2016), all strains grouped into the seven genotypes indicated above and were thought to be produced from recombination within the IBV G1-13 and GI-19 S1 genes[100]. IBV was first characterized in Republic of korea (ROK) in 1986 after respiratory signs were detected in poultry. It was initially classified as ROK group I (K-I) and subsequently assigned to the GI-15 lineage[101]. A nephropathogenic form of IBV, currently known as the ROK II (K-II) (GI-19) subgroup KM91-like variant[102], devastated the ROK chicken industry in 1990. The QX (GI-19) genotype, originally discovered in 1993 in China, is now widespread worldwide and was introduced to ROK in 2002-2003. Presently, this particular strain is referred to as the QX-like (GI-19) K-II subgroup variant[103]. Novel variants of the KM91-like (GI-19) and QX-like (GI-19) strains, termed "K40/09-like", appeared in 2009 and have since become the prevalent strains in ROK[104]. According to Valastro et al.[51], by using the IBV S1 gene categorization method, all K-II (GI-19) subgroups (KM91-like, K40/09-like, and QX-like) and QX genotypes belong to the GI-19 lineage. The S1 gene sequences of 60 ROK IBV isolates were analyzed, and it was discovered that in addition to the GI-15 and GI-19 IBV lineages, five other subgroups of GI-19 cocirculated and expanded in ROK[105]. Since its discovery in 1996, Lineage GI-19 (QX-type) has been the most prevalent IBV strain in China 1996[106, 107] Furthermore, GI-7 (TW-type) and GI-13 (4/91-type) lineages have been recognized to be significant IBV strains in China, with rare infections of GI-1 (Mass-type), GI-9 (Ark-type), and GI-28 (LDT3-type) also reported[108, 109]. The latest studies found 19 IBV variants in clinical specimens collected in China amid January 2021 and June 2022; these included 12 variants of GI-19, three variants of GI-7, and one each of GI-1, GI-9, GI-13, and GI-28. The most common IBV lineages in China according to these studies are GI-19 and GI-7. The progression and dissemination of IBV GI-7 were also detected, and it was proposed that the region of Taiwan might be the source of the IBV lineage GI-7 and that south China has a significant role in IBV transmission[110].

Recently, the respiratory type Mass 41 (GI-1) was the most prevalent type of IBV in India[111]. Elankumaran et al.[112] reported the presence of the IBV variant 793/B by serological testing but did not detect the virus. Bayry et al.[113] found a single strain of nephropathogenic IBV (PDRC/Pune/Ind/1/ 00) in India. Later, Gaba et al.[114] and Sumi et al.[115] identified and genotyped viruses, confirming the presence of 793/B, an IBV strain. Current investigations revealed that the Mass 41 (GI-1) genotype was present in field isolates[116-118]. From 2003 to 2011, Raja et al.[119] examined 20 IBV field strains from India, three of which were identified as novel IBV genotypes. These authors confirmed that these strains are similar to the GI-24 genetic lineage[51] but not to formerly reported strains[88].

The IS-1494 Mahed (variant-2; GI-23) nephropathogenic IBV strain was originally discovered in Iran in December 2015. The strains were grouped with QX (GI-19) and 4/91 (GI-13) according to nucleoprotein gene analysis. However, phylogenetic analysis revealed it to be a chimeric strain[120].

In Pakistan, there have been few reports of IBV circulation. Ahmed et al.[121] documented the dissemination of several European IBV variants in addition to the Massachusetts type. Many IBV vaccines were not effective locally, particularly those with D-1466 (GII-1), 4/91 (GI-13), D-274 (GI-12), or Mass-41 (GI-1), indicating the existence of unidentified IBV variants in the field. Rafique et al.[44] detected the existence of IBV in a one-year study that monitored the spread of IBV in Pakistan. Their results indicated the presence of mass (GI-1) -and 4/91 (GI-13) -type IBV in 43% and 51%, respectively, of the isolates, but only 5% of the various untyped IBV variants were present. Current analysis also revealed one unique Pak-973 isolate (KX013102_Ck/Rwp/NARC-973/2015_Pakistan). Compared to the 4/91 (GI-13) and mass (GI-1) vaccine strains, the isolate had distinct mutations. Another study by Rafique et al.[122] molecularly characterized the field strain KU145467_ NARC/786_Pakistan_2013 (also known as Pak-786) in the same year. The Pak-786 isolate is a member of the GI-13 lineage and includes both vaccine and field strains formerly ascribed to the 793B group[123]. A recent study revealed that IBV-17 of the GI-24 lineage differs from the vaccine strain GI-23, which has been widely used in Pakistan. These variations can result in changes that allow GI-24 lineage viruses to avoid vaccines designed for the GI-23 lineage. It appears that GI-24 has taken over as the main lineage of Pakistani field IBV isolates throughout the past few years (2017-2020) [124].

A "HFT-IBV" variant was discovered in a layer chicken flock in Israel that was routinely vaccinated against 4/91 (GI-13) and H120 (GI-1) variants. The disease was caused by the Israeli variant IS/1494/06, which belongs to the GI-23 lineage. The HFT-IBV isolate shares 97.7% nucleotide sequence similarity with the IS-Var2-like isolates but less than 90% nucleotide sequence similarity with the M41-related (GI-1) (H120, Ma5, M41) and 4/91 (GI-13) (793/B, Moroccan G/83, and CR88) strains. IS-Var2-like genotypes have been assigned to the GI-23 lineage, whereas M41 was assigned to the GI-1 lineage; 4/91-like viruses were assigned to the GI-13 lineage[125]. The IS var-2 IBV field isolates recovered from commercial broiler flocks in Turkey from 2014 to 2019 were discovered to be on the same branch as the GI-23 genotype, which has 99% similarity and is one of the most common wild-type clusters in the Middle East[126].

In 2014, three infectious bronchitis virus (IBV) strains were isolated and identified from hens suspected of being infected with IBV in Guangxi, China: CoV/ck/China/I0111/14, CoV/ck/ China/I0114/14, and CoV/ck/ China/I0118/14. S1 sequencing and phylogenetic analysis revealed that the three IBV isolates are genetically distinct from other known IBV types, indicating the presence of a novel genotype (GI-29) [127].

Southeast Asia

IBV lineages GI-1, GI-13 and GI-19 are prevalent. According to a study confirmed the presence of various IBV lineages in Thai chicken flocks, along with a novel recombinant IBV variantthatoriginatedfrom the GI-19 and GI-13 lineage viruses[128]. In another study, variant IBV isolates were categorized into four groups in a study conducted on IBV infection in chickens in Thailand between 2014 and 2016: QX-like IBV (GI-19), Massachusetts (GI-1), 4/91 (GI-13), and a novel variant. QX-like was the most prevalent IBV genotype among these groups in Thailand[128]. In a recent study, vaccine strains were compared with local common Malaysian IBV strains with two isolates of ACoVs from guinea fowl and jungle fowl. The two isolates from the sample were categorized as genotype I and placed in the GI-13 lineage alongside three other frequently occurring local vaccine strains, namely, CR88, 793B, and 4/91. Molecular characterization revealed homology to the common IBV vaccine strain 4/91 (GI-13) among ACoV isolates from diagnostic cases of junglefowl (isolate 2015) and guinea fowl (isolate 2016) [129]. In the Vietnamese provinces of Hanoi, Thainguyen, and Haiphong, three strains of IBV, known as VNUA3, VNUA8, and VNUA11, were isolated from sick or infected chickens. The Vietnam isolates belonged to three genotypes: Q1-like (GI-16) (VNUA3), QX-like (GI-19) (VNUA8), and TC07-2-like (GVI-1) (VNUA11). This finding indicates that at least three different IBV genotypes are circulating in North Vietnamese poultry[130]. A study conducted at four farms in Myanmar revealed through phylogenetic analysis of the S1 gene that the IBVs are closely related to the C-78 (GI-18) IBV vaccine strain, while the IBV found at farm 2 was found to be closely related to the GN strain (both of which are categorized as JP-1 (GI-18) types). The isolates from farms 4 and 5 are similar to K446-01 (mass type) (GI-1) and TM86 (JP-2 type (GI-7)), respectively. All of the identified IBV types are extensively distributed, and commercial vaccines targeting their weakened strains are accessible. The JP-1 (GI-18), JP-2 (GI-7), and Mass (GI-1) types of IBV were found to be present in Myanmar poultry farms during this survey[131].

Australia

In this geographically isolated country, IBV evolved distinctly from the rest of the world[132]. Since the early 1960s, several distinct IBV variants have been identified and characterized[133], and several lineages (GIII-1, GI-6, GV-1, GI-5, and GI-10) have been found to be exclusive to Australia and New Zealand[51].

The traditional IBV variants from Australia have been divided into two distinct lineages, GI-5 and GI-6, which can be identified by the VicS and N1/62 vaccine strains. Surprisingly, both the N1/62 and VicS strains were discovered in 1962, albeit in different parts of Victoria and New South Wales[133]. Although their S1 gene sequences are 83% identical, their distinction into distinct lineages indicates either a unique ancestral origin or major divergence from a parental strain at the time of isolation[134-136].

IBV was not prevalent in New Zealand until the 1970s, when the first case of IB emerged[137]. Notably, strains from New Zealand are included in the GI-5 and GI-6 lineages. New Zealand's Strain A clustered in the GI-6 lineage[134-136]. Six indigenous New Zealand viruses, three of which were identified in the 1970s and the remainder in the 2000s, are part of the GI-10 lineage[138]. This IBV strain was identified in the region for the first time in 1967[139]. According to virus neutralization testing, 4 serotypes (A, B, C, and D) of the virus were identified in New Zealand in 1976; these serotypes differ from those found in other countries[137]. Four new IBV variants, K43, T6, K32, and K87, were discovered later in 2008 in clinically infected flocks. The sequence homology between these strains and the previously mentioned B, C, and D strains is more than 99%[138]. Phylogenetic analysis of New Zealand strains confirmed the relevance of these strains to Australian

(Vic S) (G I -6) strains rather than European or North American strains[132].

Genotype II

Europe

In the late 1970s, the GII-1 lineage of IBV (also known as D1466 (GII-1) or D212 variation) was identified initially in the Netherlands as an etiological factor associated with egg loss[65, 75]. Due to differences between its S1 coding region and those of other European IBV strains, the variant was grouped with the Dutch V1397 strain under the GII genotype[51]. Many studies indicate that compared to other IBV strains, this variant has considerably different antigenic and molecular characteristics[140]. For many years, the D1466 (GII-1) strain was only sporadically recognized, but the findings of a genetic investigation carried out between 2005 and 2006 revealed that this variant was causing additional problems in Western European countries[76]. Only a few infections are detected by the D1466 (GII-1) mutation in the UK and France; nevertheless, the disease dynamics are increasing in other European countries. In countries such as Germany, the Netherlands, and Belgium, the prevalence of the D1466-like (GII-1) virus was on average between 3% and 5% in 2005 and increased to 10%, 7%, and 16%, respectively, in 2006. In Poland, the conventional nested RT-PCR technique was used from November 2011 to December 2013. The first D1466 (GII-1) IBV was detected during this time, resulting in 26 positive samples or a prevalence of the variant in 11.7% of the surveyed chicken flocks. Our findings revealed that the prevalence of GII-1 IBV is gradually decreasing in Poland. Certain D1466-positive chicken flocks were declared healthy by healthcare personnel, indicating that the virulence and pathogenicity of the GII-1 strains are not severe[141].

GII-2 (D181) is a novel IBV strain that emerged from an unexpected report in 2017 and became the second most isolated IBV strain among the breeders and layers in the Netherlands in 2018. This strain was also found in Belgian and German samples. According to the entire S1 gene and maximum likelihood analysis, D181 is more closely related to GII-1, commonly known as D1466 (GII-1), than to any other IBV strain, the latter of which shares 90% of the sequence. The remaining 10% are mutations that are distributed throughout the whole S1 gene, and are combination study provided no evidence that the S1 gene resulted from recombination between D1466 (GII-1) and other IBV strains[142].

South America

The IBV genotype GII lineage was identified in Georgia, USA, in 2000, and its whole genome was submitted to GenBank under accession number: AF274435.1[69].

In an investigation, RT-PCR was employed to identify the IBV D1466 strain of genotype GII-1[143]. This strategy was validated by studying the spike protein-encoding area in the proprietary S gene of the GII-1 pedigree (pseudo-D1466 strain), which corresponds to IBV[141].

Genotype III

Australia

The GIII-1 lineage originated in 1988 and was given the name Australian subgroup II [134]. The IBV isolates Q3/88 (GIII-2) and N1/88 (GIII-1) were identified from these outbreaks[134, 144]; they are genetically and antigenically different from all formerly identified classical variants, assigned as new or subgroup 2 variants, and were subsequently categorized as the GIII genotype (GIII-1 and GIII-2 lineages, respectively) [51].

Asia

Phylogenetic analysis of the 27 identified IBV strains revealed that amino acid residues of the S1 glycoprotein align with the H120 (GI-1) vaccine variant. The isolated viruses were divided into three genotypes based on their genetic origins (genotype I, II and III). Among these isolates, Li et al.[143] identified these variants at the genotype III level; these included 3 isolates from 2004, 4 from 2006, 1 from 2007, and 6 from 2008[143].

Genotype IV

North America

GIV lineage 1 (GIV-1) is the only North American lineage with a unique genotype. This category included (n=24) vaccinated and field strains identified during 1992 and 2003. The Delaware variant (DE or DE072) (GIV-1) was one of these variants and first discovered in commercial broiler flocks infected with severe respiratory infections in 1992. This difference was attributed to the distinct genotype and novel serotype of this strain compared to the other strains[146, 147]. The IBV strains that were once known as GA98 (GIV-1) were found to be similar to the DE variants despite having a different serotype and are still considered apart of the same lineage[148]. It has been proposed that the GA98 (GIV-1) variant emerged as a result of immunological selection triggered by the DE072 (GIV-1) attenuated live vaccine that was administered throughout the country during 1993[149]. Additionally, in 2000, this lineage included viruses that were found in layer flocks and led to decreased egg production[147].

Asia

Two previously known Taiwanese IBV isolates evolved into a single genotype (GIV), indicating that the development phenomenon in Taiwan of China was isolated[145].

Genotype V

Australia

The GV-1 lineage, known as Australian subgroup III, was reported in 2002; approximately 14 years later, GIII was identified[132]. Respiratory and endemic Australian diseases (4 and 7 variants, respectively) have been characterized in GV lineages. The variants Q1/13 and V1/07 were isolated from broilers with respiratory symptoms in Victoria and Queensland in 2007 and 2013, respectively. These results are identical to those for N1/03 at the genetic level, indicating increased geographic dissemination of genotype GV strains[150].

Genotype VI

Asia

A GVI-1 strain was initially discovered in Guangxi, China, in 2007. The TC07-2 strain was shown to be significantly evolutionarily distant from six other main genotypes[145]. GVI was subsequently isolated in ROK and Japan, among other Asian nations[151, 152]. The respiratory tract tropism of GVI-1 strains may be attributed to extensive recombination of gene 3 with the S gene[153]. Nonetheless, not all GVI-1 variants descended from a single common ancestor.

A variant of IBV GVI is a newly found strain that is not particularly infectious to poultry, but coinfection with an epidemic variant may occur and harm China's poultry industry. In China, IBV genotype VI (GVI-1) was identified in two separate studies from 2019 to 2020[154]. A comparison of the whole genomes of two IBV variant strains in the present study with those of other genotype variants indicated only minor similarities in the 5a, 5b, M, and N genes, with few previously identified GVI-1 variants, but greater similarities with the GI-19, GI-22, and GVII-1 genotype variants. The S gene of GVI IBV was substantially dissimilar to the S genes of the QX (GI-19) and YN (GI-22) strains[155].

Furthermore, genotype VI was produced by three classical American strains and one Japanese strain (GI-18), and the isolated TC07-2 (GVI-1) and published DE/072/92 (GIV-1) strains had the greatest evolutionary distances to all six major genotypes, but their significance was unclear[145]. GVI-I was isolated from Japan, including Ibaraki/168-1/2009, JP/ Chiba/2010, and JP/Kagoshima-3/2014 and shown to cause clinical symptoms such as a reduction in egg production, nephritis and respiratory problems, respectively[100].

Recently, the circulation of the GVI-1 lineage, which was geographically assumed to be contained in Asia, was detected in research in Colombia[72].

Genotype VII

In addition, another novel genotype emerged from China in the late 2020s and grouped as GVII from the I0636/1 isolate and the GX-NN130021 reference strain because it did not resemble other established lineages. In a study, an in-depth comparison and phylogenetic analysis of 74 complete sequences on the basis of the S gene were carried out that involved 73 representatives from each lineage and genotype along with the I0636/1 strain. Additionally, within the S gene of GI-18, at least two recombination sites are replaced with an unidentified sequence that most likely originated from another IBV strain. As a result, a new serotype with limited respiratory tract tropism in poultry emerged[156].

Genotype VIII

Europe

A novel IBV variant was discovered in Poland. This variant differs from previously discovered viruses and is closely related to the North American isolate PA/1220/98. The variant was identified as a distinct isolate on the basis of the S1 coding region and shows homology to other recognized GVII IBV genotypes. This lineage was categorized as distinct within the novel GVIII genotype using the standard criteria for designating a novel IBV genotype or lineage. The nucleotide identity of this strain with any known IBV genotype ranges from 52.7% to 58.1%, with maximum identity (81.4%) with the North American variant. This novel strain was subsequently identified in three other flocks of chickens with poor egg production. Notably, the virus has not yet been found in broilers[157].

North America

The newly identified genotype GVIII was found in two Mexican samples closely related to unique Mexican strains (UNAM-97/ AF288467) [51]. However, GVIII-1 exhibits very little intra-lineage variation, with only a 2% difference in nucleotides and a 3% difference in amino acids. The evidence supporting their classification as a new genotype includes their distinct isolation in the flock and clustering within the phylogenetic tree; a significant divergence of 28% in nucleotides and 45% in amino acids from the nearest GIV-1; and 30 different amino acid alterations. The nucleotide and amino acid sequences of IBV genotypes vary by more than 29%[73].

Genotype IX

Another genotype that emerged in North America in North America is called GIX-1, and distinct clustering of the two Mexican isolates was observed on the phylogenetic tree. The samples were collected 13 years ago from different parts of the country. With 35% nucleotide variation and almost 50% amino acid variation from the closest genotype, GVII-1, this novel genotype has diverged significantly. The same lineage also exhibits a 6% divergence in nucleotides and a 10% in amino acids. It contains 24 distinct residues and has two distinct amino acid insertions[73].

Pathogenicity of IBV strain

Diverse genotypes are thoroughly documented alongside their respective pathogenicity indexes and tissue tropism in the (Table 1-11-2) provided below. The pathogenicity indexes reveal diversity even within identical genotypes, highlighting the complex nature of strain-specific variations. Furthermore, a careful examination reveals that tissue tropism contradicts a one-to-one link with any specific lineage, highlighting the complexities of host-pathogen interactions. This in-depth investigation sheds light on the complex interactions between genotypes, pathogenicity, and tissue tropism, setting the framework for a more sophisticated understanding of microbial dynamics.

Table 1-11-2　Tissue tropism and pathogenicity of the Infectious Bronchitis Strains

Genotype	Lineage	IBV Strain	Pathogenicity	Tissue Tropism/Viral Distribution	References
GI	GI-1	IBVPRO3/ Mass	N. D. (Pathogenic but level is not defined)	Reproductive, Kidneys	[179]
	GI-5	N1/62	High	Kidneys	[159]
	GI-6	Q1/73	Moderate	Kidneys	[159]
	GI-6	Vic/S	Low	Kidney, Respiratory, Reproductive	[159]
	GI-6	Q1/65	High	Kidneys	[159]
	GI-7	TW-Like	High	Trachea, Lungs, Kidneys, and Bursa of Fabricius	[180]
	GI-17	DMV/1639	High	Respiratory, Kidney, Reproductive	[63]
	GI-19	QX	High	Proventriculus, Respiratory, Kidney, Reproductive	[34]
	GI-22	YN	High	Reproductive	[181]
	GI-23	EG/1212B-2012	High	Respiratory, Kidney	[86]
	GI-28	LGX/111119	High	Proventriculus, Respiratory, Kidney	[177]
	GI-29	I0111/14	N. D.	Respiratory, Kidney	[156]
	GI-31	D2334/11/2/ 13/CI	N. D.	caecal tonsil	[88]
GII	GII-1	D1466	Low-Moderate	Respiratory, Kidney, Reproductive	[146]
GIII	GIII-1	V6/92	Low	N. D.	[159]
	GIII-1	N1/08	Low	Respiratory	[159]
GVI	GVI-1	JX181	High	Respiratory, Reproductive, spleen, Bursa of Fabricius	[34]
GVII	GVII-1	I0636/16	Low	Respiratory, Kidney	[156]
GVIII	GVIII-1	PA/1220/08	N. D.	Respiratory, Kidney	[157]
NOVEL	N. D.	CK/CH/ GX/202109	High	Respiratory, Kidneys, Bursa of Fabricius, Proventriculus, Gizzard, Ileum, Jejunum, and Rectum	[47]

Note: N.D., Not Done.

IBV has been linked to a number of clinical symptoms in its host, the domestic chicken. The virus appears to reach host cells via viropexis after initially replicating in the upper respiratory tract[158]. Tissue tropism varies between strains, although the reason for this variation is unknown. Variations in IBV tissue tropism contribute to differences in clinical symptoms in infected birds. In general, these distinctions allow viruses to be classified as proventriculus, respiratory, reproductive, or nephropathogenic based on the major clinical presentations[159]. On the basis of pathogenicity, IBV characterized as low moderate, and high pathogenic. These conditions can also have turned vice versa based on secondary and opportunistic microorganism infection. Nephropathogenic IBV strains induce nephritis in hens and are the most pathogenic IBVs, having mortality more than respiratory and reproductive strains[160]. Proventriculitis causes ruffled feathers and respiratory symptoms such as tracheal rales, nasal discharge, sneezing, and coughing in birds. Ulcers and hemorrhages were found in the proventriculus papillae later on, and the condition is serious[161].

Prevention and control of avian infectious bronchitis virus

To control infectious diseases on modern poultry farms, biosecurity measures, and a productive management system are essential. For the avian influenza virus, this concept depends solely on having appropriate knowledge concerning the variables that affect viral transmission[162, 163]. Consequently, there is a dire need to thoroughly analyze the epidemiological factors responsible for the transmission of IBV, particularly the determinants of spread[164, 165].

Vaccination is still the most effective way to manage IBV infection, despite its limitations, which include serious adverse effects of vaccination in young birds, the need for periodic vaccine replacement due to viral mutation, and the likelihood of viral recombination[166, 167]. Effective and meticulously administered vaccines can reduce the viral load, rate of infection, and occurrence of clinical symptoms[166]. Several factors influence the strength and period of response to vaccination include chick age, vaccine immunogenicity, field strain virulence, vaccine administration strategy, extent of maternal immunity, and the time frame between vaccination and challenge. Vaccinated chickens may remain immune for several months, and in the case of broilers, this immunity may persist throughout their lifetime[168]. The majority of commercial chicken flocks are currently immunized against IBV. The IBV immunization protocol may change based on the type of vaccine used and the particular circumstances of the poultry farm. To maintain immunity, chicks are vaccinated throughout their life beginning at one day old. Booster agents can be administered at 7~10 d, 3~4 weeks, and then every 5~6 weeks afterward. The idea of a protectotype has become more widely accepted for regulating IBV due to its variants circulating worldwide.

Numerous vaccination strategies have been designed to manufacture IBV vaccines that are effective. However, the complex immune protection process prevents the extensive use of new vaccine approaches, which are still in the laboratory research stage[89, 169, 170]. The process of developing in ovo vaccination is also at the research stage. The vaccine will be based on the type of IBV strain and will not kill the embryos[171]. The market offers a variety of IBV vaccines, which might differ in nature and vaccination strains based on local isolates and recombination in variants isolated from various countries with distinct laws and regulations. Heterologous IBV vaccinations effectively provide immunization against the 793B-type (GI-13) variant, which was previously proven to persist with live attenuated IBV vaccines and to be effective against the QX (GI-19) and Italy 02 (GI-21) strains[49]. In a study in ROK, it was demonstrated that the K2 vaccine may be more potent for preventing and controlling novel IBV recombinants and variants that are spreading[173]. To manage IBV in China, H120 (GI-1) vaccination is frequently used in conjunction with the indigenous FNO-55

(GI-13, 4/91-like), QXL87 (GI-19, QX-like), or LDT3-A (GI-28, YN-like) variants.

For serotypes, including Arkansas (GI-9), Massachusetts (GI-1), and Conn (GI-1), improved live vaccines and killed oil-based emulsions are available in North America. The Georgia 98 (GIV-1) and California (GI-17) strain vaccines were obtained from the USA. The vaccinations, designated "Holland variants" D274 (GI-12) and D1466 (GII-1), are generally produced in Europe. Conversely, vaccines based on the H120 (GI-1) strain are being used throughout Europe.

The degree of immunity may rely on regional sources to produce varying levels of immunity and an unusual ability to cross-protect against a few heterologous IBV strains Florida (GI-1) and JMK (GI-3) in the U. S. Overall, combining the IBV 4/91 (GI-13) and Ma5 (GI-1) variant vaccines may provide excellent protection against heterologous IBV strains. Although QX-type (GI-19) live vaccines have been developed in Europe, their use is restricted[33]. A new generation of IBV vaccines against the regionally dominant D274 (GI-12) variant has been produced for future layer and breeding stocks.

One study demonstrated that the type of tissue, inoculation route and vaccine strain all affect the pattern of IBV replication. The distribution and elimination of the vaccine viruses for Massachusetts (GI-1) and 793B (GI-13) were slower when the viruses were administered by drinking water (DW) than when they were administered via the oculonasal (ON) route. Both vaccines were able to induce similar levels of mucosal immunity when administered via the ON route. Regardless of the vaccination technique, the Mass IBV vaccine induces cellular immunity at comparable levels. The 793B vaccine produced noticeably greater levels of humoral immunity when administered via the ON or DW route[172]. In a study, birds were given bivalent live attenuated IB vaccines containing the Mass and Conn serotypes at intervals of two, five, nine, and fourteen weeks after they were first primed with a monovalent live attenuated IB vaccine (mass serotype) at one day old. There was no apparent difference in the ability of the two vaccination regimens to protect laying hens against mass IBV challenge. These findings suggest that the vaccine strain may have a greater level of protection when faced with homologous IBV strains[174]. The probability of postvaccination challenges is infrequent, but postvaccination challenges may lead to reversion to virulence in immunocompromised or unvaccinated chickens, which eventually causes significant mortality and the intermittent spread of IBV[175]. The inability of birds to generate an adequate immune response after vaccination is the cause of vaccine failure[176]. It is noteworthy that vaccination techniques are subject to change overtime in response to the introduction of novel IBV genotypes and improvements in vaccine technology. Furthermore, local elements influence vaccination programs differently across different regions. These include disease prevalence, farm size, biosecurity measures soon.

This review emphasizes the critical relevance of understanding genotypic variations to implement effective control measures. The identification of region-specific genotypes provides poultry stakeholders with customized vaccination, biosecurity, and management measures. With the ever-changing IBV landscape, the incorporation of genotypic information into control systems has emerged as a critical tool. This knowledge synthesis not only improves our understanding of IBV epidemiology but also allows for the creation of customized therapies to reduce the impact of this economically significant poultry virus. This review, in essence, serves as a foundation for expanding the understanding and control of IBV, hence encouraging sustainable and resilient poultry production systems worldwide.

References

[1] CAVANAGH D. Coronavirus avian infectious bronchitis virus. Veterinary Research (Paris), 2007, 38(2): 281-297.

[2] ABOZEID H H, PALDURAI A, KHATTAR S K, et al. Complete genome sequences of two avian infectious bronchitis viruses isolated in Egypt: Evidence for genetic drift and genetic recombination in the circulating viruses. Infection, Genetics and Evolution, 2017, 53: 7-14.

[3] CARSTENS E B, BALL L A. Ratification vote on taxonomic proposals to the International Committee on Taxonomy of Viruses. Archives of Virology, 2009(154): 1181-1188.

[4] CARSTENS E B. Ratification vote on taxonomic proposals to the International Committee on Taxonomy of Viruses (2009). Archives of Virology, 2010, 155(1): 133-146.

[5] ABRO S H, RENSTRÖM L H M, ULLMAN K, et al. Characterization and analysis of the full-length genome of a strain of the European QX-like genotype of infectious bronchitis virus. Archives of Virology, 2012, 157(6): 1211-1215.

[6] DURÃES-CARVALHO R, CASERTA L C, BARNABÉ A C S, et al. Phylogenetic and phylogeographic mapping of the avian coronavirus spike protein-encoding gene in wild and synanthropic birds. Virus Research, 2015, 201: 101-112.

[7] SIDDELL S, SNIJDER E J. An introduction to nidoviruses//PERLMAN S, GALLAGHER T, SNIGDER E J. Nidoviruses. Washington, DC: ASM Press, 2014: 1-13.

[8] BHUIYAN M S A, AMIN Z, BAKAR A M S A, et al. Factor influences for diagnosis and vaccination of avian infectious bronchitis virus (Gammacoronavirus) in chickens. Veterinary Sciences, 2021, 8(3): 47.

[9] LI W, SHI Z, YU M, et al. Bats are natural reservoirs of SARS-like coronaviruses. Science, 2005, 310(5748): 676-679.

[10] LAI M M, CAVANAGH D. The molecular biology of coronaviruses. Advances in Virus Research, 1997, 48: 1-100.

[11] CASAIS R, DAVIES M, CAVANAGH D, et al. Gene 5 of the avian coronavirus infectious bronchitis virus is not essential for replication. Journal of Virology, 2005, 79(13): 8065-8078.

[12] ZIEBUHR J. The coronavirus replicase. Current Topics in Microbiology and Immunology, 2005, 287: 57-94.

[13] CHEN Y, CAI H, PAN J, et al. Functional screen reveals SARS coronavirus nonstructural protein nsp14 as a novel cap N7 methyltransferase. Proceedings of the National Acadmy of Sciences of the United State of America, 2009, 106(9): 3484-3489.

[14] SMITH E C, DENISON M R. Implications of altered replication fidelity on the evolution and pathogenesis of coronaviruses. Current Opinion in Virology, 2012, 2(5): 519-524.

[15] TE VELTHUIS A J W, VAN DEN WORM S H E, SNIJDER E J, et al. The SARS-coronavirus nsp7+nsp8 complex is a unique multimeric RNA polymerase capable of both de novo initiation and primer extension. Nucleic Acids Research, 2012, 40(4): 1737-1747.

[16] KINT J, LANGEREIS M A, MAIER H J, et al. Infectious bronchitis coronavirus limits interferon production by inducing a host shutoff that requires accessory protein 5b. Journal of Virology, 2016, 90(16): 7519-7528.

[17] KANT A, KOCH G, VAN ROOZELAAR D J, et al. Location of antigenic sites defined by neutralizing monoclonal antibodies on the S1 avian infectious bronchitis virus glycopolypeptide. Journal of General Virology, 1992, 73(3): 591-596.

[18] SHAN D, FANG S, HAN Z, et al. Effects of hypervariable regions in spike protein on pathogenicity, tropism, and serotypes of infectious bronchitis virus. Virus Research, 2018, 250: 104-113.

[19] LIU D X, CAVANAGH D, GREEN P, et al. A polycistronic mRNA specified by the coronavirus infectious bronchitis virus. Virology, 1991, 184(2): 531-544.

[20] FISCHER F, STEGEN C F, MASTERS P S, et al. Analysis of constructed E gene mutants of mouse hepatitis virus confirms a pivotal role for E protein in coronavirus assembly. Journal of Virology, 1998, 72(10): 7885-7894.

[21] IGNJATOVIC J, GALLI L. The S1 glycoprotein but not the N or M proteins of avian infectious bronchitis virus induces protection in vaccinated chickens. Archives of Virology, 1994, 138(1-2): 117-134.

[22] YOUN S, LEIBOWITZ J L, COLLISSON E W. In vitro assembled, recombinant infectious bronchitis viruses demonstrate that the 5a open reading frame is not essential for replication. Virology (New York, N Y), 2005, 332(1): 206-215.

[23] LACONI A, VAN BEURDEN S J, BERENDS A J, et al. Deletion of accessory genes 3a, 3b, 5a or 5b from avian coronavirus infectious bronchitis virus induces an attenuated phenotype both in vitro and in vivo. Journal of General Virology, 2018, 99(10): 1381-1390.

[24] ZUNIGA S, CRUZ J L, SOLA I, et al. Coronavirus nucleocapsid protein facilitates template switching and is required for efficient transcription. Journal of Virology, 2010, 84(4): 2169-2175.

[25] AMARASINGHE A, DE SILVA S U, ABDUL-CADER M S, et al. Comparative features of infections of two Massachusetts (Mass) infectious bronchitis virus (IBV) variants isolated from Western Canadian layer flocks. BMC Veterinary Research, 2018, 14(1): 391.

[26] MACDONALD J W, MCMARTIN D A. Observations on the effects of the H52 and H120 vaccine strains of infectious bronchitis virus in the domestic fowl. Avian Pathology, 1976, 5(3): 157-173.

[27] AMBALI A G, JONES R C. Early pathogenesis in chicks of infection with an enterotropic strain of infectious bronchitis virus. Avian Disease, 1990, 34(4): 809-817.

[28] CUMMING R B. The control of avian infectious bronchitis/nephrosis in Australia. Australian Veterinary Journal, 1969, 45(4): 200-203.

[29] FRANCA M, WOOLCOCK P R, YU M, et al. Nephritis associated with infectious bronchitis virus Cal99 variant in game chickens. Avian Disease, 2011, 55(3): 422-428.

[30] BOROOMAND Z, ASASI K, MOHAMMADI A, et al. Pathogenesis and tissue distribution of avian infectious bronchitis virus isolate IRFIBV32 (793/B Serotype) in experimentally infected broiler Chickens. The Scientific World, 2012: 402536-402537.

[31] EL-HOUADFI M, JONES R C, COOK J K A, et al. isolation and characterisation of six avian infectious bronchitis viruses isolated in Morocco. Avian Pathology, 1986, 15(1): 93-105.

[32] LIU S, CHEN J, HAN Z. et al. Infectious bronchitis virus: S1 gene characteristics of vaccines used in China and efficacy of vaccination against heterologous strains from China. Avian Pathology, 2006, 5(35): 394-399.

[33] COOK J K, JACKWOOD M, JONES R C. The long view: 40 years of infectious bronchitis research. Avian Pathology, 2012, 3(41): 239-250.

[34] BO Z, CHEN S, ZHANG C, et al. Pathogenicity evaluation of GVI-1 lineage infectious bronchitis virus and its long-term effects on reproductive system development in SPF hens. Frontiers in Microbiology, 2022, 13: 1049287.

[35] RAMSUBEIK S, STOUTE S, GALLARDO R A, et al. Infectious bronchitis virus california variant CA1737 isolated from a commercial layer flock with cystic oviducts and poor external egg quality. Avian Disease, 2023, 67(2): 212-218.

[36] BOLTZ D A, NAKAI M, BAHRA J M. Avian infectious bronchitis virus: a possible cause of reduced fertility in the rooster. Avian Disease, 2004, 48(4): 909-915.

[37] GOUGH R E, COX W J, GUTIERREZ E, et al. Isolation of "variant" strains of infectious bronchitis virus from vaccinated chickens in Great Britain. Veterinary Record, 1996, 139(22): 552.

[38] KARESH W B, UHART M M, FRERE E, et al. Health evaluation of free-ranging rockhopper penguins (*Eudyptes chrysocomes*) in Argentina. Journal of Zoo and Wildlife Medicine, 1999, 30(1): 25-31.

[39] CAVANAGH D, PICAULT J P, GROUGH R E, et al. Variation in the spike protein of the 793/B type of infectious bronchitis virus, in the field and during alternate passage in chickens and embryonated eggs. Avian Pathology, 2005, 34(1): 20-25.

[40] LIU S, CHEN J, KONG X, et al. Isolation of avian infectious bronchitis coronavirus from domestic peafowl (Pavo cristatus) and teal (Anas). Journal of General Virology, 2005, 86(3): 719-725.

[41] HUGHES L A, SAVAGE C, NAYLOR C, et al. Genetically diverse coronaviruses in wild bird populations of northern England. Emerging Infectious Disease, 2009, 15(7): 1091-1094.

[42] SCHALK A F, HAWN A C. An apparently new respiratory disease of baby chicks. Journal of the American Veterinary Medical Association, 1931(78): 413-423.

[43] JUNGHERR E L, CHOMIAK T W, LUGINBUHL R E. Immunologic differences in strains of infectious bronchitis virus// PROC. 60th Ann Meet U S Livest. Chicago: Sanit Assoc, 1956.

[44] RAFIQUE S, SIDDIQUE N, ABBAS M A, et al. Isolation and molecular characterization of infectious bronchitis virus (IBV) variants circulating in commercial poultry in Pakistan. Pakistan Veterinary Journal, 2018, 38(4): 365-370.

[45] MILLER L T, YATES V J. Neutralization of infectious bronchitis virus human sera. American Journal of Epidemiology, 1968, 88(3): 406-409.

[46] KAPIKIAN A Z, JAMES H D, KELLY S J, et al. Isolation from man of "avian infectious bronchitis virus-like" viruses (Coronaviruses) similar to 229E virus, with some epidemiological observations. The Journal of Infectious Diseases, 1969, 119(3): 282-290.

[47] WANG C, HOU B. A pathogenic and recombinant infectious bronchitis virus variant (CK/CH/GX/202109) with multiorgan tropism. Veterinary Research (Paris), 2023, 54(1): 54.

[48] JACKWOOD M W, JORDAN B J. Molecular Evolution of Infectious Bronchitis Virus and the Emergence of Variant Viruses Circulating in the United States. Avian Disease, 2021, 65(4): 631-636.

[49] COOK J K, ORBELL S J, WOODS M A, et al. Breadth of protection of the respiratory tract provided by different live-attenuated infectious bronchitis vaccines against challenge with infectious bronchitis viruses of heterologous serotypes. Avian Pathology, 1999, 28(5): 477-485.

[50] TEGEGNE D, DENEKE Y, SORI T, et al. Molecular epidemiology and genotyping of infectious bronchitis virus and avian metapneumovirus in backyard and commercial chickens in Jimma Zone, southwestern Ethiopia. Veterinary Sciences, 2020, 7(4): 187.

[51] VALASTRO V, HOLMES E C, BRITTON P, et al. S1 gene-based phylogeny of infectious bronchitis virus: an attempt to harmonize virus classification. Infection, Genetics and Evolution, 2016, 39: 349-364.

[52] JACKWOOD M W. Review of infectious bronchitis virus around the world. Avian Disease, 2012, 56(4): 634-641.

[53] HEWSON K A, NOORMOHAMMADI A H, DEVLIN J M, et al. Evaluation of a novel strain of infectious bronchitis virus emerged as a result of spike gene recombination between two highly diverged parent strains. Avian Pathology, 2014, 43(3): 249-257.

[54] HITCHNER S B, WINTERFIELD R W, APPLETON G S. Infectious bronchitis virus types: incidence in the United States. Avian Disease, 1966, 10(1): 98-102.

[55] HOPKINS S R. Serological comparisons of strains of infectious bronchitis virus using plaque-purified isolants. Avian Disease, 1974, 18(2): 231-239.

[56] COWEN B S, HITCHNER S B. Serotyping of avian infectious bronchitis viruses by the virus-neutralization test[Chickens]. Avian Disease, 1975, 19(3): 583-595.

[57] JOHNSON R B, MARQUARDT W W. Strains of infectious bronchitis virus on the Delmarva peninsula and in Arkansas. Avian Disease, 1976, 20(2): 382-386.

[58] BANDE F, ARSHAD S S, OMAR A R, et al. Global distributions and strain diversity of avian infectious bronchitis virus: a review. Anim Health Research Reviews, 2017, 18(1): 70-83.

[59] FIELDS D B. Arkansas 99, a new infectious bronchitis serotype. Avian Disease, 1973, 17(3): 659-661.

[60] NAJIMUDEEN S M, HASSAN M S H, GOLDSMITH D, et al. Molecular characterization of 4/91 infectious bronchitis virus leading to studies of pathogenesis and host responses in laying hens. Pathogens, 2021, 10(5): 624.

[61] KARIITHI H M, VOLKENING J D, LEYSON C M, et al. Genome sequence variations of infectious bronchitis virus serotypes from commercial chickens in Mexico. Frontiers in Veterinary Science, 2022, 9: 931272.

[62] ALI A, OJKIC D, ELSHAFIEE E A, et al. Genotyping and in silico analysis of delmarva (DMV/1639) infectious bronchitis virus (IBV) spike 1 (S1) glycoprotein. Genes (Basel), 2022, 13(9):1617.

[63] HASSAN M S H, OJKIC D, COFFN C S, et al. Delmarva (DMV/1639) infectious bronchitis virus (IBV) variants isolated in eastern Canada show evidence of recombination. Viruses, 2019, 11(11): 1054.

[64] VILLALOBOS-AGÜERO R A, RAMÍREZ-CARVAJAL L, ZAMORA-SANABRIA R, et al. Molecular characterization of an avian GA13-like infectious bronchitis virus full-length genome from Costa Rica. Virus Disease, 2021, 32(2): 347-353.

[65] SJAAK D W J, COOK J K A, VAN DER HEIJDEN H M. Infectious bronchitis virus variants: a review of the history, current

situation and control measures. Avian Pathology, 2011, 40(3): 223-235.

[66] CUBILLOS A, ULLOA J, CUBILLOS V, et al. Characterisation of strains of infectious bronchitis virus isolated in Chile. Avian Pathology, 1991, 20(1): 85-99.

[67] MARANDINO A, VAGNOZZI A, TOMÁS G, et al. Origin of new lineages by recombination and mutation in avian infectious bronchitis virus from south America. Viruses, 2022, 14(10): 2095.

[68] TATAJE-LAVANDA L, IZQUIERDO-LARA R, ORMEÑO-VÁSQUEZ P, et al. Near-complete genome sequence of infectious bronchitis virus strain VFAR-047 (GI-16 Lineage), isolated in Peru. Microbiology Resource Announcements, 2019, 8(5): e01555-18.

[69] ICOCHEA E, GONZÁLEZ R, CASTRO-SANGUINETTI G, et al. Genetic analysis of infectious bronchitis virus S1 gene reveals novel amino acid changes in the GI-16 lineage in Peru. Microorganisms, 2023, 11(3): 691.

[70] BROWN JORDAN A, FUSARO A, BLAKE L, et al. Characterization of novel, pathogenic field strains of infectious bronchitis virus (IBV) in poultry in Trinidad and Tobago. Transboundary and Emerging Diseases, 2020, 67(6): 2775-2788.

[71] TREVISOL I M, CARON L, MORES M A Z, et al. Pathogenicity of GI-23 avian infectious bronchitis virus strain isolated in Brazil. Viruses, 2023, 15(5): 1200.

[72] RAMIREZ-NIETO G, MIR D, ALMANSA-VILLA D, et al. New insights into avian infectious bronchitis virus in colombia from whole-genome Analysis. Viruses, 2022, 14(11): 2562.

[73] MENDOZA-GONZÁLEZ L, MARANDINO A, PANZERA Y, et al. Research note: high genetic diversity of infectious bronchitis virus from Mexico. Poultry Science, 2022, 101(10): 102076.

[74] DAWSON P S, GOUGH R E. Antigenic variation in strains of avian infectious bronchitis virus. Arch Gesamte Virusforsch, 1971, 34(1): 32-39.

[75] DAVELAAR F G, KOUWENHOVEN B, BURGER A G. Occurrence and significance of infectious bronchitis virus variant strains in egg and broiler production in the Netherlands. Veterinary Quarterly, 1984, 6(3): 114-120.

[76] WORTHINGTON K J, CURRIE R J W, JONES R C. A reverse transcriptase-polymerase chain reaction survey of infectious bronchitis virus genotypes in Western Europe from 2002 to 2006. Avian Pathology, 2008, 37(3): 247-257.

[77] POHJOLA L K, EK-KOMMONEN S C, TAMMIRANTA N E, et al. Emergence of avian infectious bronchitis in a non-vaccinating country. Avian Pathology, 2014, 43(3): 244-248.

[78] TERREGINO C, TOFFAN A, BEATO M S, et al. Pathogenicity of a QX strain of infectious bronchitis virus in specific pathogen free and commercial broiler chickens, and evaluation of protection induced by a vaccination programme based on the Ma5 and 4/91 serotypes. Avian Pathology, 2008, 37(5): 487-493.

[79] GOUGH R E, COX W J, DE B WELCHMAN D, et al. Chinese QX strain of infectious bronchitis virus isolated in the UK. Veterinary Record, 2008, 162(3): 99-100.

[80] FRANZO G, NAYLOR C J, LUPINI C, et al. Continued use of IBV 793B vaccine needs reassessment after its withdrawal led to the genotype's disappearance. Vaccine, 2014, 32(50): 6765-6767.

[81] JI J, XIE J, CHEN F, et al. Phylogenetic distribution and predominant genotype of the avian infectious bronchitis virus in China during 2008-2009. Virology Journal, 2011(8): 1-9.

[82] MORENO A, FRANZO G, MASSI P, et al. A novel variant of the infectious bronchitis virus resulting from recombination events in Italy and Spain. Avian Pathology, 2017, 46(1): 28-35.

[83] MEIR R, ROSENBLUT E, PERL S, et al. Identification of a novel nephropathogenic infectious bronchitis virus in Israel. Avian Disease, 2004, 48(3): 635-641.

[84] LISOWSKA A, SAJEWICZ-KRUKOWSKA J, FUSARO A, et al. First characterization of a Middle-East GI-23 lineage (Var2-like) of infectious bronchitis virus in Europe. Virus Research, 2017, 242: 43-48.

[85] FISCHER S, KLOSTERHALFEN D, WILMS-SCHULZE KUMP F, et al. Research note: first evidence of infectious bronchitis virus Middle-East GI-23 lineage (Var2-like) in Germany. Poultry Science, 2020, 99(2): 797-800.

[86] ZANATY A, NAGUIB M M, EL-HUSSEINY M H, et al. The sequence of the full spike S1 glycoprotein of infectious bronchitis virus circulating in Egypt reveals evidence of intra-genotypic recombination. Archives of Virology, 2016,

161(12): 3583-3587.

[87] MORLEY A J, THOMSON D K, CAIRO U E F O. Swollen-head syndrome in broiler chickens. Avian Disease, 1984, 28(1): 238-243.

[88] BALI K, KASZAB E, MARTON S, et al. Novel lineage of infectious bronchitis virus from Sub-Saharan Africa identified by random amplification and next-generation sequencing of viral genome. Life (Basel), 2022, 12(4)：475.

[89] BANDE F, ARSHAD S S, BEJO M H, et al. Progress and challenges toward the development of vaccines against avian infectious bronchitis. Journal of Immunology Research, 2015, 2015: 424860.

[90] MOHARAM I, SULTAN H, HASSAN K, et al. Emerging infectious bronchitis virus (IBV) in Egypt: Evidence for an evolutionary advantage of a new S1 variant with a unique gene 3ab constellation. Infection, Genetics and Evolution, 2020, 85: 104433.

[91] KARIITHI H M, VOLKENING J D, GORAICHUK I V, et al. Unique variants of avian coronaviruses from indigenous chickens in kenya. Viruses, 2023, 15(2): 264.

[92] ZANATY A, ARAFA A, HAGAG N, et al. Genotyping and pathotyping of diversified strains of infectious bronchitis viruses circulating in Egypt. World Journal of Virology, 2016, 5(3): 125-134.

[93] TOFFAN A, MONNE I, TERREGINO C, et al. QX-like infectious bronchitis virus in Africa. Veterinary Record, 2011, 169(22): 589.

[94] ANDREOPOULOU M, FRANZO G, TUCCIARONE C M, et al. Molecular epidemiology of infectious bronchitis virus and avian metapneumovirus in Greece. Poultry Science, 2019, 98(11): 5374-5384.

[95] NAGUIB M M, HÖPER D, ARAFA A, et al. Full genome sequence analysis of a newly emerged QX-like infectious bronchitis virus from Sudan reveals distinct spots of recombination. Infection, Genetics and Evolution, 2016, 46: 42-49.

[96] DUCATEZ M F, MARTIN A M, OWOADE A A, et al. Characterization of a new genotype and serotype of infectious bronchitis virus in Western Africa. Journal of General Virology, 2009, 90(11): 2679-2685.

[97] MASE M, TSUKAMOTO K, IMAI K, et al. Phylogenetic analysis of avian infectious bronchitis virus strains isolated in Japan. Archives of Virology, 2004, 149(10): 2069-2078.

[98] MASE M, GOTOU M, INOUE D, et al. Genotyping of infectious bronchitis viruses isolated in Japan during 2008–2019. Journal of Veterinary Medical Science, 2021: 20-620.

[99] MASE M, HIRAMATSU K, WATANABE S, et al. Complete Genome Sequence of Infectious Bronchitis Virus Strain JP/KH/64, Isolated in Japan. Microbiology Resource Announcements, 2021, 10(40): e66521.

[100] MASE M, HIRAMATSU K, WATANABE S, et al. Genetic Analysis of the Complete S1 Gene in Japanese Infectious Bronchitis Virus Strains. Viruses, 2022, 14(4):716.

[101] SONG C S, LEE Y J, KIM J H, et al. Epidemiological classification of infectious bronchitis virus isolated in Korea between 1986 and 1997. Avian Pathology, 1998, 27(4): 409-416.

[102] LEE S K, SUNG H W, KWON H M. S1 glycoprotein gene analysis of infectious bronchitis viruses isolated in Korea. Archives of Virology, 2004, 149(3): 481-494.

[103] LEE E K, JEON W J, LEE Y J, et al. Genetic diversity of avian infectious bronchitis virus isolates in Korea between 2003 and 2006. Avian Disease, 2008, 52(2): 332-337.

[104] LIM T H, LEE H J, LEE D H, et al. An emerging recombinant cluster of nephropathogenic strains of avian infectious bronchitis virus in Korea. Infection, Genetics and Evolution, 2011, 11(3): 678-685.

[105] LEE H C, JEONG S, CHO A Y, et al. Genomic analysis of avian infectious bronchitis viruses recently isolated in South Korea reveals multiple introductions of GI-19 Lineage (QX Genotype). Viruses, 2021, 13(6):1045..

[106] FENG K, WANG F, XUE Y, et al. Epidemiology and characterization of avian infectious bronchitis virus strains circulating in southern China during the period from 2013-2015. Scientific Reports, 2017, 7(1): 6576.

[107] XU L, HAN Z, JIANG L, et al. Genetic diversity of avian infectious bronchitis virus in China in recent years. Infection, Genetics and Evolution, 2018, 66(C): 82-94.

[108] ZHAO Y, ZHANG H, ZHAO J, et al. Evolution of infectious bronchitis virus in China over the past two decades. Journal

of General Virology, 2016, 97(7): 1566-1574.

[109] LIAN J, WANG Z, XU Z, et al. Distribution and molecular characterization of avian infectious bronchitis virus in southern China. Poultry Science, 2021, 100(7): 101169.

[110] CHEN L, JIANG W, WU W, et al. Insights into the epidemiology, phylodynamics, and evolutionary changes of lineage GI-7 infectious bronchitis virus. Transboundary and Emerging Disease, 2023: 1-13.

[111] KUMANAN K, SELVAM N T, RAJ G D, et al. Molecular epizootiology of infectious bronchitis virus isolates in Tamil Nadu indicating the possible involvement of a vaccine strain. The Indian Veterinary Journal, 2004(81): 1307-1312.

[112] ELANKUMARAN S, BALACHANDRAN C, CHANDRAN N D, et al. Serological evidence for a 793/B related avian infectious bronchitis virus in India. Veterinary Record, 1999, 144(11): 299-300.

[113] BAYRY J, GOUDAR M S, NIGHOT P K, et al. Emergence of a nephropathogenic avian infectious bronchitis virus with a novel genotype in India. Journal of Clinical Microbiology, 2005, 43(2): 916-918.

[114] GABA A, HANISH D, PAL J K, et al. Isolation, identification and molecular characterization of IBV variant from out break of visceral gout in commercial broilers. Veterinary World, 2010, 3(8): 375-377.

[115] SUMI V, SINGH S D, DHAMA K, et al. Isolation and molecular characterization of infectious bronchitis virus from recent outbreaks in broiler flocks reveals emergence of novel strain in India. Tropical Animal Health and Production, 2012, 44(7): 1791-1795.

[116] PATEL B H, BHIMANI M P, BHANDERI B B, et al. Isolation and molecular characterization of nephropathic infectious bronchitis virus isolates of Gujarat state, India. VirusDisease, 2015, 26(1-2): 42-47.

[117] PARVEEN R, FAROOQ I, AHANGAR S, et al. Genotyping and phylogenetic analysis of infectious bronchitis virus isolated from broiler chickens in Kashmir. VirusDisease, 2017, 28(4): 434-438.

[118] JAKHESARA S J, NATH B, PAL J K, et al. Emergence of a genotype I variant of avian infectious bronchitis virus from Northern part of India. Acta Tropica, 2018, 183: 57-60.

[119] RAJA A, DHINAKAR RAJ G, KUMANAN K. Emergence of variant avian infectious bronchitis virus in India. Iranian Journal of Veterinary Research, 2020, 21(1): 33-39.

[120] MOUSAVI F S, GHALYANCHILANGEROUDI A, HOSSEINI H, et al. Complete genome analysis of Iranian IS-1494 like avian infectious bronchitis virus. VirusDisease, 2018, 29(3): 390-394.

[121] AHMED Z, NAEEM K, HAMEED A. Detection and seroprevalence of infectious bronchitis virus strains in commercial poultry in Pakistan. Poultry Science, 2007, 86(7): 1329-1335.

[122] RAFIQUE S, NAEEM K, SIDDIQUE N, et al. Determination of genetic variability in avian infectious bronchitis virus (AIBV) isolated from Pakistan. Pakistan Journal of Zoology, 2018(50): 695-701.

[123] RAFIQUE S. Seroprevalence and molecular characterization of infectious bronchitis virus variants from poultry in Pakistan. Islamabad, Pakistan: Department of Microbiology, Quaid-i-Azam University, 2018.

[124] SALEEM W, VEREECKE N, ZAMAN M G, et al. Genotyping and phylogeography of Infectious Bronchitis Virus isolates from Pakistan show unique linkage to GI-24 lineage. Poultry Science, 2024, 103(1): 103236.

[125] ÖNGÖR H, TIMURKAAN N, ÇÖVEN F, et al. Detection of Israel variant 2 (IS/1494/06) genotype of infectious bronchitis virus in a layer chicken flock. Ankara Üniversitesi Veteriner Fakültesi Dergisi, 2021(68): 167-172.

[126] MÜŞTAK B, MÜŞTAK H K, BILGEN N. S1 gene based phylogeny of israel variant-2 infectious bronchitis virus isolated in Turkey in a five year period. Polish Journal of Veterinary Sciences, 2022: 45-50.

[127] JIANG L, ZHAO W, HAN Z, et al. Genome characterization, antigenicity and pathogenicity of a novel infectious bronchitis virus type isolated from south China. Infection, Genetics and Evolution, 2017, 54: 437-446.

[128] MUNYAHONGSE S, POHUANG T, NONTHABENJAWAN N, et al. Genetic characterization of infectious bronchitis viruses in Thailand, 2014-2016: identification of a novel recombinant variant. Poultry Science, 2020, 99(4): 1888-1895.

[129] BESAR S A, ARSHAD S S, RAMANOON S Z, et al. Isolation and characterization of avian coronavirus from diagnostic cases of selected bird species in Malaysia. Tropical Agricutural Science, 2023(46): 503-516.

[130] Le T B, Lee H J, Le V P, et al. Multiple genotypes of avian infectious bronchitis virus circulating in Vietnam. Korean

Journal of Poultry Science, 2019(46): 127-136.

[131] FUJISAWA S, MURATA S, TAKEHARA M, et al. Molecular detection and genetic characterization of Mycoplasma gallisepticum, Mycoplama synoviae, and infectious bronchitis virus in poultry in Myanmar. BMC Veterinary Research, 2019, 15(1): 261.

[132] IGNJATOVIC J, GOULD G, SAPATS S. Isolation of a variant infectious bronchitis virus in Australia that further illustrates diversity among emerging strains. Archives of Virology, 2006, 151(8): 1567-1585.

[133] CUMMING R B. Infectious avian nephrosis (Uraemia) in Australia. Australian Veterinary Journal, 1963(39): 145-147.

[134] SAPATS S I, ASHTON F, WRIGHT P J, et al. Sequence analysis of the S1 glycoprotein of infectious bronchitis viruses: identification of a novel genotypic group in Australia. Journal of General Virology, 1996, 77(3): 413-418.

[135] MARDANI K, NOORMOHAMMADI A H, HOOPER P, et al. Infectious Bronchitis Viruses with a Novel Genomic Organization. Journal of Virology, 2008, 82(4): 2013-2024.

[136] QUINTEROS J A, MARKHAM P F, LEE S W, et al. Analysis of the complete genomic sequences of two virus subpopulations of the Australian infectious bronchitis virus vaccine VicS. Avian Pathol, 2015, 44(3): 182-191.

[137] LOHR J E. Serologic differences between strains of infectious bronchitis virus from New Zealand, Australia, and the United States. Avian Disease, 1976, 20(3): 478-482.

[138] MCFARLANE R, VERMA R. Sequence analysis of the gene coding for the S1 glycoprotein of infectious bronchitis virus (IBV) strains from New Zealand. Virus Genes, 2008, 37(3): 351-357.

[139] POHL R M. Infectious bronchitis in chickens. New Zealand Veterinary Journal, 1967: 151.

[140] LIN S Y, CHEN H W. Infectious bronchitis virus variants: molecular analysis and pathogenicity investigation. International Journal of Molecular Sciences, 2017, 18(10): 2030.

[141] DOMANSKA-BLICHARZ K, LISOWSKA A, PIKUŁA A, et al. Specific detection of GII-1 lineage of infectious bronchitis virus. Letters in Applied Microbiology, 2017, 65(2): 141-146.

[142] MOLENAAR R J, DIJKMAN R, WIT J J. Characterization of infectious bronchitis virus D181, a new serotype (GII-2). Avian Pathology, 2020, 49(3): 243-250.

[143] KHALTABADI F R, GHALYANCHILANGEROUDI A, FALLAH M M H, et al. Molecular monitoring of D1466 genotype of avian infectious bronchitis virus in Iran: a retrospective study in 2013-2017. Archives of Razi Institute, 2020, 75(2): 163-168.

[144] IGNJATOVIC J, MCWATERS P G. Monoclonal antibodies to three structural proteins of avian infectious bronchitis virus: characterization of epitopes and antigenic differentiation of australian strains. Journal of General Virology, 1991, 72(12): 2915-2922.

[145] LI L, XUE C, CHEN F, et al. Isolation and genetic analysis revealed no predominant new strains of avian infectious bronchitis virus circulating in south China during 2004-2008. Vet Microbiol, 2010, 143(2-4): 145-154.

[146] GELB J, KEELER C L, NIX W A, et al. Antigenic and S-1 genomic characterization of the Delaware variant serotype of infectious bronchitis virus. Avian Disease, 1997, 41(3): 661-669.

[147] MONDAL S P, LUCIO-MARTÍNEZ B, NAQI S A. Isolation and characterization of a novel antigenic subtype of infectious bronchitis virus serotype DE072. Avian Disease, 2001, 45(4): 1054-1059.

[148] LEE C W, HILT D A, JACKWOOD M W. Identification and analysis of the Georgia 98 serotype, a new serotype of infectious bronchitis virus. Avian Disease, 2001, 45(1): 164-172.

[149] LEE C W, JACKWOOD M W. Origin and evolution of Georgia 98 (GA98), a new serotype of avian infectious bronchitis virus. Virus Research, 2001, 80(1): 33-39.

[150] HEWSON K, NOORMOHAMMADI A H, DEVLIN J M, et al. Rapid detection and non-subjective characterisation of infectious bronchitis virus isolates using high-resolution melt curve analysis and a mathematical model. Archives of Virology, 2009, 154(4): 649-660.

[151] MASE M, KAWANISHI N, OOTANI Y, et al. A novel genotype of avian infectious bronchitis virus isolated in Japan in 2009. Journal of Veterinary Medical Science, 2010, 72(10): 1265-1268.

[152] JANG I, LEE H J, BAE Y C, et al. Genetic and pathologic characterization of a novel recombinant TC07-2-Type avian infectious bronchitis virus. Avian Disease, 2018, 62(1): 109-113.

[153] REN M, SHENG J, MA T, et al. Molecular and biological characteristics of the infectious bronchitis virus TC07-2/GVI-1 lineage isolated in China. Infection, Genetics and Evolution, 2019, 75: 103942.

[154] WANG C Y, LUO Z B, SHAO G Q, et al. Genetic and pathogenic characteristics of a novel infectious bronchitis virus strain in genogroup VI (CK/CH/FJ/202005). Veterinary Microbiology, 2022, 266: 109352.

[155] SUN L, TANG X, QI J, et al. Two newly isolated GVI lineage infectious bronchitis viruses in China show unique molecular and pathogenicity characteristics. Infection, Genetics and Evolution, 2021, 94: 105006.

[156] MA T, XU L, REN M, et al. Novel genotype of infectious bronchitis virus isolated in China. Veterinary Microbiology, 2019, 230: 178-186.

[157] DOMANSKA-BLICHARZ K, SAJEWICZ-KRUKOWSKA J, LISOWSKA A. New PA/1220/98-like variant of infectious bronchitis virus in Poland. Avian Pathology, 2020, 49(4): 380-388.

[158] PATTERSON S, BINGHAM R W. Electron microscope observations on the entry of avian infectious bronchitis virus into susceptible cells. Archives of Virology, 1976, 52(3): 191-200.

[159] QUINTEROS J A, NOORMOHAMMADI A H, LEE S W, et al. Genomics and pathogenesis of the avian coronavirus infectious bronchitis virus. Australian Veterinary Journal, 2022, 100(10): 496-512.

[160] CHEN B Y, HOSI S, NUNOYA T, et al. Histopathology and immunohistochemistry of renal lesions due to infectious bronchitis virus in chicks. Avian Pathology, 1996, 25(2): 269-283.

[161] YU L, JIANG Y, LOW S, et al. Characterization of three infectious bronchitis virus isolates from China associated with proventriculus in vaccinated chickens. Avian Disease, 2001, 45(2): 416-424.

[162] BOENDER G J, HAGENAARS T J, BOUMA A, et al. Risk maps for the spread of highly pathogenic avian influenza in poultry. PLOS Computational Biology, 2007, 3(4): e71.

[163] SINGH M, TORIBIO J A, SCOTT A B, et al. Assessing the probability of introduction and spread of avian influenza (AI) virus in commercial Australian poultry operations using an expert opinion elicitation. PLOS ONE, 2018, 13(3): e193730.

[164] FRANZO G, TUCCIARONE C M, MORENO A, et al. Phylodynamic analysis and evaluation of the balance between anthropic and environmental factors affecting IBV spreading among Italian poultry farms. Scientific Reports, 2020, 10(1): 7289.

[165] NAJIMUDEEN S M, HASSAN M S H, CORK S C, et al. Infectious bronchitis coronavirus infection in chickens: multiple system disease with immune suppression. Pathogens (Basel), 2020, 9(10): 779.

[166] FRANZO G, TUCCIARONE C M, BLANCO A, et al. Effect of different vaccination strategies on IBV QX population dynamics and clinical outbreaks. Vaccine, 2016, 34(46): 5670-5676.

[167] JI W, LI X, ZAI J, et al. Response to comments on "cross-species transmission of the newly identified coronavirus 2019-nCoV" and "codon bias analysis may be insufficient for identifying host(s) of a novel virus. Journal of Medical Virology, 2020, 92(9): 1440.

[168] BIJLENGA G, COOK J K A, GELB J, et al. Development and use of the H strain of avian infectious bronchitis virus from the Netherlands as a vaccine: a review. Avian Pathology, 2004, 33(6): 550-557.

[169] KAPCZYNSKI D R, HILT D A, SHAPIRO D, et al. Protection of chickens from infectious bronchitis by in ovo and intramuscular vaccination with a DNA vaccine expressing the S1 glycoprotein. Avian Disease, 2003, 47(2): 272-285.

[170] YANG T, WANG H N, WANG X, et al. The protective immune response against infectious bronchitis virus induced by multi-epitope based peptide vaccines. Bioscience, Biotechnology, and Biochemistry, 2009, 73(7): 1500-1504.

[171] TARPEY I, ORBELL S J, BRITTON P, et al. Safety and efficacy of an infectious bronchitis virus used for chicken embryo vaccination. Vaccine, 2006, 24(47-48): 6830-6838.

[172] AL-RASHEED M, BALL C, GANAPATHY K. Route of infectious bronchitis virus vaccination determines the type and magnitude of immune responses in table egg laying hens. Veterinary Research (Paris), 2021, 52(1): 1-139.

[173] SHIRVANI E, SAMAL S K. Comparative protective efficacies of novel avian paramyxovirus-vectored vaccines against

virulent infectious bronchitis virus in chickens. Viruses, 2020, 12(7): 697.

[174] ALI A, HASSAN M S H, NAJIMUDEEN S M, et al. Efficacy of two vaccination strategies against infectious bronchitis in laying hens. Vaccines (Basel), 2023, 11(2):338.

[175] HISCOX J A, WURM T, WILSON L, et al. The coronavirus infectious bronchitis virus nucleoprotein localizes to the nucleolus. Journal of Virology, 2001, 75(1): 506-512.

[176] BOSHA J, NONGO N. Common breaches in poultry vaccine handling and administration in makurdi metropolis: A recurrent phenomenon in the tropics. Vom Journal of Veterinary Sciences, 2012(9): 11-16.

[177] CHEN Y, JIANG L, ZHAO W, et al. Identification and molecular characterization of a novel serotype infectious bronchitis virus (GI-28) in China. Veterinary Microbiology, 2017, 198: 108-115.

[178] QUINTEROS J A, IGNJATOVIC J, CHOUSALKAR K K, et al. Infectious bronchitis virus in Australia: a model of coronavirus evolution - a review. Avian Pathology, 2021, 50(4): 295-310.

[179] PEREIRA N A, ALESSI A C, MONTASSIER H J, et al. Gonadal pathogenicity of an infectious bronchitis virus strain from the Massachusetts genotype. Brazilian Journal of Microbiology, 2019, 50(1): 313-320.

[180] YAN S, CHEN Y, ZHAO J, et al. Pathogenicity of a TW-like strain of infectious bronchitis virus and evaluation of the protection induced against it by a QX-Like strain. Frontiers in Microbiology, 2016, 7: 1653.

[181] ZHONG Q, HU Y X, JIN J H, et al. Pathogenicity of virulent infectious bronchitis virus isolate YN on hen ovary and oviduct. Veterinary Microbiology, 2016, 193: 100-105.

Chapter Two
Studies on Molecular Epidemiology

Epidemiological surveillance of low pathogenic avian influenza virus (LPAIV) from poultry in Guangxi, southern China

Peng Yi, Xie Zhixun, Liu Jiabo, Pang Yaoshan, Deng Xianwen, Xie Zhiqin, Xie Liji, Fan Qing, and Luo Sisi

Abstract

Low pathogenic avian influenza virus (LPAIV) usually causes mild disease or asymptomatic infection in poultry. However, some LPAIV strains can be transmitted to humans and cause severe infection. Genetic rearrangement and recombination of even low pathogenic influenza may generate a novel virus with increased virulence, posing a substantial risk to public health. Southern China is regarded as the world "influenza epicenter", due to a rash of outbreaks of influenza in recent years. In this study, we conducted an epidemiological survey of LPAIV at different live bird markets (LBMs) in Guangxi, southern China. From January 2009 to December 2011, we collected 3 121 cotton swab samples of larynx, trachea and cloaca from the poultry at LBMs in Guangxi. Virus isolation, hemagglutination inhibition (HI) assay, and RT-PCR were used to detect and subtype LPAIV in the collected samples. Of the 3 121 samples, 336 samples (10.8%) were LPAIV positive, including 54 (1.7%) in chicken and 282 (9.1%) in duck. The identified LPAIV were H3N1, H3N2, H6N1, H6N2, H6N5, H6N6, H6N8, and H9N2, which are combinations of seven HA subtypes (H1, H3, H4, H6, H9, H10 and H11) and five NA subtypes (N1, N2, N5, N6 and N8). The H3 and H9 subtypes are predominant in the identified LPAIVs. Among the 336 cases, 29 types of mixed infection of different HA subtypes were identified in 87 of the cases (25.9%). The mixed infections may provide opportunities for genetic recombination. Our results suggest that the LPAIV epidemiology in poultry in the Guangxi in southern China is complicated and highlights the need for further epidemiological and genetic studies of LPAIV in this area.

Keywords

avian influenza, epidemiological surveillance

Introduction

Avian influenza (AI) is caused by specified avian influenza viruses (AIVs) that belong to the genus *Influenza A* virus, of the family Orthomyxoviridae. In terms of antigen variations of the surface glycoprotein hemagglutinin (HA) and neuraminidase (NA), AIVs are further divided into different subtypes. So far, 16 HA subtypes and 9 NA subtypes of AIVs have been identified and described[1-3]. Among these subtypes, H5N1 and H7N2 cause high pathogenic avian influenza (HPAI), which is characterized by systemic infections, high mortality and morbidity. Low pathogenic avian influenza (LPAI) can be caused by all HA subtypes (H1-H16)[4, 5]. Poultry infected with LPAI usually have mild symptoms or are asymptomatic carriers, but it is possible to transmit certain viruses to humans, which are capable of causing severe illness. For example, H7N9 are a LPAIV that has been the cause of worldwide outbreaks of avian influenza.

It has been confirmed that genetic recombination in LPAIV has led to deleterious gene mutations, novel phenotypes and increased virulence[6]. Previous studies have shown that seven genes of the H5N1 virus isolated in the Hong Kong outbreak in 1997 have high sequence similarity to the LPAIV virus H6N1[7]. Some LPAIV, such as H9N2, can break species barriers and provide genes to other influenza virus, which could present a risk for severe human infection. It has been suggested that eight of the genes from the H1N1 subtype that caused the 1918 pandemic were originally from a poultry virus[8]. It has also been suggested that this outbreak was transmitted to humans via a pig intermediary[9, 10]. The HA and PB1 genes of the H3N2 subtype that caused the outbreak in Hong Kong in 1968 were originally from the H3 subtype of AIV[11, 12]. Since 1998, the H9N2 subtype has caused several cases of human infections in the mainland of China and Hong Kong. These subtypes posed great threats to public health[13, 14]. Thus, epidemiological surveys of HA subtypes of LPAIV in southern China are critically important.

Southern China is regarded as an "influenza epicenter", due to the recent influenza outbreaks in Hong-Kong, China and southern Asian[15, 16]. This area is on the migratory path of many bird species and is highly populated with poultry, animals and humans. There are several large-scale live bird markets (LBMs) in this area and a large number of small-scale poultry farms. Close contacts between the poultry, animals and humans may facilitate transmission of the influenza virus. In addition, the warm, humid climate in this region may prompt long-term survival and proliferation of the virus. Furthermore, the local people usually use poultry manure to feed their pigs and then the pig manure is used to feed fish. These activities may facilitate the transmission of influenza virus among animals, birds and poultry via water, feed, feces and urine, providing opportunities for genetic recombination of the influenza virus. Recently, H7N9 influenza viruses have caused severe infection in humans in China and transmission by respiratory droplets has been shown in ferrets[17, 18]. In this study, we conducted epidemiological surveillance of LPAIV at LBMs in Guangxi in southern China. The epidemiological data from this region will be useful to plan strategies for preventing LPAIV from causing human infections. Additionally, the results from this surveillance may be helpful in influencing worldwide surveillance of LPAIV, given that this region is on the route of bird immigration.

Materials and methods

Field work

The epidemiological surveillance of LPAIV in poultry was conducted in 15 randomly selected LBMs in Guangxi of southern China during the period from January 2009 to December 2011. Biological samples were collected from the cloaca, larynx and trachea of healthy chickens and ducks using cotton swabs. The cotton swabs were then suspended in 1 mL of storage medium that was transported in 1.5 mL finger tubes on ice. According to the protocol of the world organization for animal health (OIE), the storage medium contained Penicillin (10 000 unit/mL), Streptomycin (10 mg/mL), Gentamycin (10 000 unit/mL), Kanamycin (10 000 unit/mL) and 5% fetal calf serum in sterile PBS (pH = 7.2).

Isolation of virus

The samples were centrifuged at 3 000 r/min for 10 min at 4 ℃, while still in the storage medium. Supernatants were collected and stored at −70 ℃ for virus isolation. The specific pathogen-free (SPF) eggs were purchased from the Beijing Merial Biology Company (Beijing, China), SPF chicken embryos (9～10 days of age) were used for virus inoculation. Each sample (supernatant) was inoculated into three SPF chicken embryos (0.2 mL/embryo), via the allantoic cavity. The inoculated chicken embryos were incubated at 35 ℃ and observed daily. The allantoic fluids were collected from the embryos that died within 120 h of inoculation. The chicken embryos that survived longer than 120 h, post inoculation, were sacrificed at 4 ℃ and the allantoic fluids were collected. The collected allantoic fluids were then tested for hemagglutination (HA). The positive allantoic fluids were stocked at −70 ℃ for identification of virus.

Identification of virus

The hemagglutination inhibition (HI) assay was conducted to determine the HA subtypes of LPAIV. Briefly, the HA-positive allantoic fluids were tested with the serum of different HA subtypes, anti-Newcastle disease virus (NDV) and anti-hemagglutinating adenovirus (EDS76) to determine the HA subtypes. The viruses with determined HA subtypes were further inoculated and grown in Madin-Darby canine kidney (MDCK) cells. The titers of viruses were determined by plaque assay.

Viral RNA was extracted using a Body Fluid Viral DNA/RNA Miniprep Kit (Axygen Biosciences, Hangzhou, China). Based on the protocol, RT-PCR assay was conducted to obtain the full length of the neuraminidase (NA) gene, using primers designed in previous studies[19]. The NA subtype was determined by further sequencing of the amplified NA genes and a BLAST search in the GenBank nr database, which contains the nucleotide sequences of all subtypes of the NA gene.

Statistical analysis

The viral isolation rates were compared using the Fisher exact test with the Epi Info 6.0 software (Centers for Disease Control and Prevention, Atlanta, GA, USA). A difference was considered statistically significant when $P<0.01$.

Results

Isolation rates of LPAIV in different poultry species

From January 2009 to December 2011, a total of 3 121 swab samples were collected from the cloaca, larynx and trachea of healthy chickens and ducks at different LBMs in Guangxi. Among the 3 121 samples, 336 (10.8%) were positive for LPAIV. The LPAIV isolation rate in ducks was significantly higher than in chickens (Figure 2-1-1) ($P<0.01$), which was 9.1%, whereas the LPAIV isolation rate in chickens was 1.7%. The LPAIV isolation rate in chickens in 2010 showed a slight decline, compared to the isolation rate in chickens in 2009, but increased in 2011 when compared to rates from 2009 and 2010 (Figure 2-1-1). The LPAIV isolation rate in ducks, however, showed a continual increase from 2009 to 2011 (Figure 2-1-1). In general, the seasonal pattern of the LPAIV isolation rate in ducks and chickens is highly similar (Figure 2-1-2). In February 2010, the LPAIV isolation rate reached a peak in both ducks and chickens. No significant association between the LPAIV isolation rate and season was observed, however ($P>0.05$).

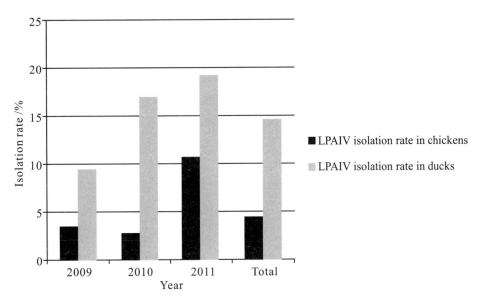

Figure 2-1-1   Annual isolation rates of AIV in poultry at different LBMs in Guangxi, southern China

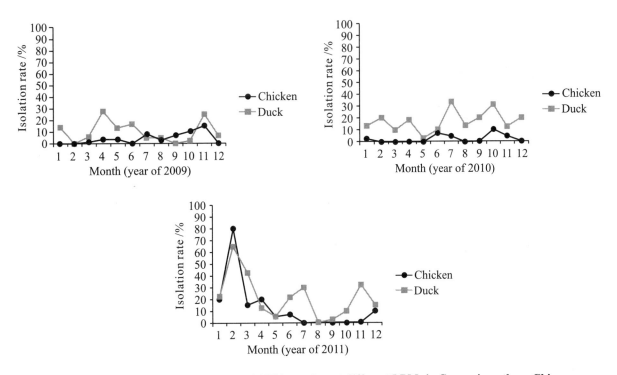

Figure 2-1-2   Monthly isolation rate of AIV in poultry at different LBMs in Guangxi, southern China

HA and NA subtypes identified in chickens and ducks

More than seven HA subtype (Hl, H3, H4, H6, H9, H10, H11 and some which were unknown) were identified in ducks and chickens (Figure 2-1-3). The H3 subtype was dominant in ducks and the H9 subtype was dominant in chickens (Figure 2-1-3). The distribution of different HA subtypes in chickens and ducks was slightly different. Based on RT-PCR, DNA sequencing and a BLAST search in the NCBI nr database, five NA subtypes, including N1, N2, N5, N6 and N8, were identified. The combination of HA and NA subtypes identified in this study are H3N1, H3N2, H6N1, H6N2, H6N5, H6N6, H6N8 and H9N2. The sequences of identified HA subtypes were submitted to GenBank (accession numbers: JN003630, JX304754, JX297583, JX293559, JX304770, JX304762, and KC608159).

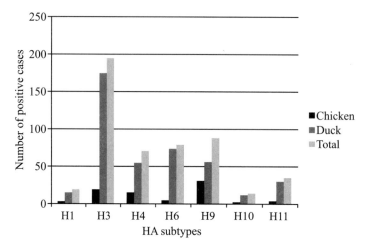

Figure 2-1-3   Distribution of HA subtypes and NA subtypes in poultry at different LBMs in Guangxi, southern China

Mixed infection of different HA subtypes

Of the 336 samples positive for LPAIV, 87 (25.9%) (17 in chickens and 70 in ducks) showed mixed infections of different HA subtypes of LPAIV (Table 2-1-1). As shown in Table 2-1-2, 13 and 23 types of mixed infection of HA subtypes were identified in both chickens and ducks, respectively. Among these concomitant infections, 13 were mixes of two different HA subtypes, 10 were mixes of three different HA subtypes, four were mixes of four different HA subtypes and two were mixes of five different HA subtypes (Table 2-1-2). Mixed infection of H3 and H4 subtypes was the most common mixed infection (Figure 2-1-4).

Table 2-1-1   LPAIV isolated from chickens and ducks in live bird markets (LBMs) in Guangxi, southern China during the period from January 2009 to December 2011

Date	The number of total samples		The number of LPAIV positive samples		LPAIV isolation rate / %		The number of co-infection samples		Co-infection rate / % [a]	
	Chicken	Duck	Chicken	Duck	Chicken	Duck	Chicken	Duck	Chicken	Duck
1/2009	28	14	0	2	0	14.3	0	0	0	0
2/2009	72	39	0	0	0	0	0	0	0	0
3/2009	70	51	1	3	1.4	5.9	1	2	100	66.7
4/2009	81	80	3	22	3.7	27.5	2	17	66.7	77.3
5/2009	57	52	2	7	3.5	13.5	0	3	0	42.9
6/2009	60	54	0	9	0	16.7	0	5	0	55.5
7/2009	75	57	6	3	8	5.3	0	1	0	33.3
8/2009	30	165	1	8	3.3	4.8	1	5	100	62.5
9/2009	30	78	2	0	6.7	0	0	0	0	0
10/2009	20	40	2	1	10	2.5	0	1	0	1

continued

Date	The number of total samples		The number of LPAIV positive samples		LPAIV isolation rate / %		The number of co-infection samples		Co-infection rate / % [a]	
	Chicken	Duck	Chicken	Duck	Chicken	Duck	Chicken	Duck	Chicken	Duck
11/2009	20	40	3	10	15	25	3	4	100	40
12/2009	30	58	0	4	0	6.9	0	2	0	50
Total in year of 2009	573	728	20	69	3.5	9.5	7	40	35	58
1/2010	40	80	1	11	2.5	13.75	0	0	0	0
2/2010	10	20	0	4	0	20	0	0	0	0
3/2010	40	80	0	8	0	10	0	0	0	0
4/2010	50	97	0	18	0	18.6	0	5	0	27.8
5/2010	30	60	0	2	0	3.3	0	0	0	0
6/2010	40	80	3	8	7.5	10	0	0	0	0
7/2010	40	80	2	27	5	33.8	0	10	0	37
8/2010	40	78	0	11	0	14.1	0	0	0	0
9/2010	20	40	0	8	0	20	0	0	0	0
10/2010	30	60	3	19	10	31.7	0	1	0	5.3
11/2010	40	80	2	10	5	12.5	1	0	50	0
12/2010	20	40	0	8	0	20	0	2	0	25
Total in year of 2010	400	795	11	134	2.8	16.9	1	18	9	13.4
1/2011	20	40	4	9	20	22.5	4	1	100	11.1
2/2011	10	20	8	13	80	65	1	3	12.5	23.1
3/2011	20	40	3	17	15	42.5	0	1	0	5.9
4/2011	20	40	4	5	20	12.5	1	0	25	0
5/2011	20	40	1	2	5	5	1	0	100	0
6/2011	30	60	2	13	6.7	21.7	2	2	100	15.4
7/2011	10	20	0	6	0	30	0	3	0	50
8/2011	20	40	0	0	0	0	0	0	0	0
9/2011	14	39	0	1	0	2.6	0	0	0	0
10/2011	20	30	0	3	0	10	0	1	0	33.3
11/2011	20	22	0	7	0	31.8	0	0	0	0
12/2011	10	20	1	3	10	15	0	1	0	33.3
Total in year of 2011	214	411	23	79	10.7	19.2	9	12	39.1	15.2
Total of three years	1 187	1 934	54	282	4.5	14.6	17	70	31	25

Note: [a] The ratio of co-infection is number of samples vs LPAIV positive samples.

Table 2-1-2   Mixed infections of different of HA subtypes of LPAIV in chickens and ducks

Type of mixed infection	Type of mixed infection	Number of case	
		Chicken	Duck
1	H1+H3	0	1
2	H1+H9	1	0
3	H1+H4	0	3
4	H3+H4	4	19
5	H3+H6	0	3
6	H3+H11	0	1
7	H3+H9	0	1
8	H4+H6	1	1
9	H4+H9	1	4
10	H4+H11	1	0
11	H6+H11	0	3
12	H6+H9	1	0
13	H9+H11	0	1
14	H1+H3+H6	1	1
15	H1+H4+H9	2	0
16	H1+H6+H9	1	0
17	H3+H4+H6	0	2
18	H3+H4+H10	0	1
19	H3+H4+H9	0	1
20	H3+H6+H11	0	3
21	H4+H6+H9	0	1
22	H4+H6+H11	1	1
23	H6+H9+H11	1	1
24	H3+H6+H10+H11	0	1
25	H3+H6+H9+H11	0	4
26	H3+H9+H10+H11	1	0
27	H4+H6+H9+H11	0	1
28	H3+H4+H6+H9+H11	0	7
29	H3+H6+H9+H10+H11	1	9

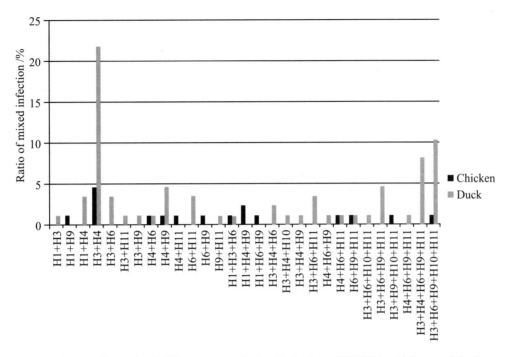

Figure 2-1-4   The ratio of different types of mixed infection of LPAIV in chickens and ducks

Discussion

With periodically emerging novel viral strains, the *Influenza A* virus has caused devastating pandemics and has been identified as a major threat to public health, worldwide. Guangxi is located in southern China and is currently one of the most active areas for epidemic influenza in the world. Guangxi is adjacent to Vietnam, where avian influenza is also endemic and has a complicated epidemiology of human influenza.

Poultry farming is a well-developed industry in this area. There are several large-scale poultry farms, as well as a large number of small-scale farms and villages in the country. The booming poultry farming industry in this area poses a great public health risk for avian influenza. Nevertheless, epidemiological surveillance of LPAIV in this area is largely unknown. In this study, we conducted three years (2009-2011) of epidemiological surveillance of different HA subtype of LPAIV targeting the chickens and ducks sold at LBMs in Guangxi. Our results suggested a high prevalence of LPAIV in the poultry at these markets in southern China. We identified at least 7 HA and 5 NA subtypes of LPAIV in chickens and ducks from this area. In addition, LPAIV were isolated in all seasons, as this area has a particularly warm and humid environment, which may benefit the survival, growth and transmission of LPAIV. We also observed that the isolation rates of LPAIV in ducks were significantly higher than in chickens, which further supports the hypothesis that ducks are the major natural reservoirs of AIVs[4, 20]. It has been reported that the colonic epithelial cells in chickens express both a sialic acid α2-3 galactosidase (Siaα2-3Gal) receptor that binds to AIVs and a sialic acid α2-6 galactosidase (Siaα2-6Gal) receptor that binds to the human influenza virus[21-26]. The human influenza virus receptor, the Siaα2-6Gal, is the dominant receptor. This suggests that chickens may serve as the intermediate host and thus may be the source of transmission of the influenza virus to humans[23]. Therefore, the role of the chicken in the evolution and ecology of the influenza virus needs to be investigated further.

The genome of the influenza virus contains eight RNA fragments. Mixed infections with multiple virus types could lead to reassortment[20, 27]. In addition, concomitant infections of the fungal pathogen *Cryptococcus neoformans* may be associated with enhanced virulence[28]. Genetic reassortment and recombination can occur during the process of viral proliferation and assembly in the host cells, which is a highly efficient way for AIVs to mutate and then to generate a novel virus with new phenotypes. It is thought that the natural reservoir of AIV is the wild bird population. AIVs are capable of switching hosts and causing outbreaks in new species. In nature, the high prevalence of mixed infections in chickens and ducks, as found in our study, suggest that genome reassortment may occur and result in antigenic shift. For example, the H3N2 influenza virus that caused the flu pandemic in Hong Kong in 1968 was generated by gene rearrangement of HA and PBl from the H3 subtype of AIV[8, 11, 12]. The H5N1 subtype of avian influenza virus that caused human infection and death in Hong Kong in 1997 also originated from genetic rearrangement[29]. The HA gene from H5N1 originated from A/goose/Guangdong/1/96 (H5N1), isolated from a goose in Guangdong province in 1996[30]. Subsequently, gene rearrangement of H5N1 with other AIVs led to the novel virus that caused the chicken avian influenza in Hong Kong in 2001[31]. As has been previously reported in eastern China, we discovered mixed infections of HA subtypes in Guangxi. Concurrent infection was more frequent in ducks than in chickens, with concomitant infections of up to five different HA subtypes of LPAIV and 23 different kinds of mixed infections. Our study supported the hypothesis that ducks are the main LPAIV reservoir and promote mixed infections of different HA subtypes of LPAIV[20]. The warm, humid climate in Guangxi may facilitate the survival, growth and transmission of LPAIV, as well as the occurrence of mixed infections. Further investigation focusing on whether genome rearrangement occurs during mixed infection is needed.

Interestingly, the distribution and the isolation rate of the H3 subtype of LPAIV in ducks was significantly higher than other HA subtypes of LPAIV ($P<0.01$) (Figure 2-1-3). The isolation rate of the H3 subtype in chickens is also high, and is only slightly lower than the H9 subtype LPAIV (Figure 2-1-3). The H3 subtype causes the seasonal influenza widely found in humans[32], which highlights the need for further investigation and epidemiological surveillance, as well as genetic and evolutionary studies of the H3 subtypes of the influenza virus. Unlike ducks, the H9 subtype was the most commonly isolated subtype of LPAIV in chickens. The H9 subtype of LPAIV has not only caused economic losses for the poultry industry, but is also capable of infecting humans by directly crossing hosts, which poses a serious threat to human health[13, 14, 33]. It is not clear whether the H9 subtype of LPAIV mutated, or experienced antigenic drift, or whether genetic recombination occurs between the H9 subtype and the H5 subtype of LPAIV, which would improve pathogenicity.

In conclusion, we investigated the epidemiology of LPAIV in poultry from LBMs in Guangxi of southern China, a hotbed for the avian flu. Our study demonstrates a high prevalence of LPAIV in the poultry in this area and highlights the significant need for further investigation of the genetics and evolutionary of LPAIV.

References

[1] WEBSTER R G, BEAN W J, GORMAN O T, et al. Evolution and ecology of influenza A viruses. Microbiological Reviews, 1992, 56(1): 152-179.

[2] FOUCHIER R A, MUNSTER V, WALLENSTEN A, et al. Characterization of a novel influenza A virus hemagglutinin subtype (H16) obtained from black-headed gulls. Journal of Virology, 2005, 79(5): 2814-2822.

[3] WEBSTER R G, AIR G M, METZGER D W, et al. Antigenic structure and variation in an influenza virus N9 neuraminidase. Journal of Virology, 1987, 61(9): 2910-2916.

[4] ALEXANDER D J. A review of avian influenza in different bird species. Veterinary Microbiology, 2000, 74(1-2): 3-13.

[5] EDWARDS S. OIE laboratory standards for avian influenza. Developments in Biologicals, 2006, 124: 159-162.

[6] HE C Q, XIE Z X, HAN G Z, et al. Homologous recombination as an evolutionary force in the avian influenza A virus.

Molecular Biology and Evolution, 2009, 26(1): 177-187.

[7] SHORTRIDGE K F, ZHOU N N, GUAN Y, et al. Characterization of avian H5N1 influenza viruses from poultry in Hong Kong. Virology, 1998, 252(2): 331-342.

[8] REID A H, TAUBENBERGER J K. The origin of the 1918 pandemic influenza virus: a continuing enigma. The Journal of General Virology, 2003, 84(Pt 9): 2285-2292.

[9] KHIABANIAN H, TRIFONOV V, RABADAN R. Reassortment patterns in Swine influenza viruses. PLOS ONE, 2009, 4(10): e7366.

[10] MA W, KAHN R E, RICHT J A. The pig as a mixing vessel for influenza viruses: Human and veterinary implications. Journal of Molecular and Genetic Medicine, 2008, 3(1): 158-166.

[11] KAWAOKA Y, KRAUSS S, WEBSTER R G. Avian-to-human transmission of the PB1 gene of influenza A viruses in the 1957 and 1968 pandemics. Journal of Virology, 1989, 63(11): 4603-4608.

[12] COCKBURN W C, DELON P J, FERREIRA W. Origin and progress of the 1968-69 Hong Kong influenza epidemic. Bulletin of the World Health Organization, 1969, 41(3): 345-348.

[13] BUTT K M, SMITH G J, CHEN H, et al. Human infection with an avian H9N2 influenza A virus in Hong Kong in 2003. Journal of Clinical Microbiology, 2005, 43(11): 5760-5767.

[14] PEIRIS M, YUEN K Y, LEUNG C W, et al. Human infection with influenza H9N2. Lancet, 1999, 354(9182): 916-917.

[15] SHORTRIDGE K F. Is China an influenza epicentre?. Chinese Medical Journal, 1997, 110(8): 637-641.

[16] SHORTRIDGE K F, STUART-HARRIS C H. An influenza epicentre?. Lancet, 1982, 2(8302): 812-813.

[17] XU C L, HAVERS F, WANG L J, et al. Monitoring avian influenza A (H7N9) virus through national influenza-like illness surveillance, China. Emerging Infectious Diseases, 2013, 19(8): 1289-1292.

[18] ZHANG Q Y, SHI J Z, DENG G H, et al. H7N9 influenza viruses are transmissible in ferrets by respiratory droplet. Science, 2013, 341(6144): 410-414.

[19] HOFFMANN E, STECH J, GUAN Y, et al. Universal primer set for the full-length amplification of all influenza A viruses. Archives of Virology, 2001, 146(12): 2275-2289.

[20] KIM J K, NEGOVETICH N J, FORREST H L, et al. Ducks: the "Trojan horses" of H5N1 influenza. Influenza and Other Respiratory Viruses, 2009, 3(4): 121-128.

[21] BATEMAN A C, KARAMANSKA R, BUSCH M G, et al. Glycan analysis and influenza A virus infection of primary swine respiratory epithelial cells: the importance of NeuAc{alpha}2-6 glycans. The Journal of Biological Chemistry, 2010, 285(44): 34016-34026.

[22] CHEN L M, BLIXT O, STEVENS J, et al. In vitro evolution of H5N1 avian influenza virus toward human-type receptor specificity. Virology, 2012, 422(1): 105-113.

[23] GUO C T, TAKAHASHI N, YAGI H, et al. The quail and chicken intestine have sialyl-galactose sugar chains responsible for the binding of influenza A viruses to human type receptors. Glycobiology, 2007, 17(7): 713-724.

[24] HANASHIMA S, SEEBERGER P H. Total synthesis of sialylated glycans related to avian and human influenza virus infection. Chemistry an Asian Journal, 2007, 2(11): 1447-1459.

[25] KIRKEBY S, MARTEL C J, AASTED B. Infection with human H1N1 influenza virus affects the expression of sialic acids of metaplastic mucous cells in the ferret airways. Virus Research, 2009, 144(1-2): 225-232.

[26] WALTHER T, KARAMANSKA R, CHAN R W, et al. Glycomic analysis of human respiratory tract tissues and correlation with influenza virus infection. PLOS Pathogens, 2013, 9(3): e1003223.

[27] FURUSE Y, SUZUKI A, KISHI M, et al. Occurrence of mixed populations of influenza A viruses that can be maintained through transmission in a single host and potential for reassortment. Journal of Clinical Microbiology, 2010, 48(2): 369-374.

[28] DESNOS-OLLIVIER M, PATEL S, SPAULDING A R, et al. Mixed infections and in vivo evolution in the human fungal pathogen Cryptococcus neoformans. MBio, 2010, 1(1): e00091-10.

[29] CLAAS E C, DE JONG J C, VAN BEEK R, et al. Human influenza virus A/Hong Kong/156/97 (H5N1) infection. Vaccine,

1998, 16(9-10): 977-978.

[30] XU X, SUBBARAO, COX N J, et al. Genetic characterization of the pathogenic influenza A/Goose/Guangdong/1/96 (H5N1) virus: similarity of its hemagglutinin gene to those of H5N1 viruses from the 1997 outbreaks in Hong Kong. Virology, 1999, 261(1): 15-19.

[31] SUAREZ D L, PERDUE M L, COX N, et al. Comparisons of highly virulent H5N1 influenza A viruses isolated from humans and chickens from Hong Kong. Journal of Virology, 1998, 72(8): 6678-6688.

[32] ZHAO C A, YANG Y H. Present status of studies on avian influenza and human infection with avian influenza virus. Chinese Journal of Pediatrics, 2004, 42(4): 310-311.

[33] GAO R B, CAO B, HU Y W, et al. Human infection with a novel avian-origin influenza A (H7N9) virus. The New England Journal of Medicine, 2013, 368(20): 1888-1897.

Surveillance of live poultry markets for low pathogenic avian influenza viruses in Guangxi, southern China, from 2012-2015

Luo Sisi, Xie Zhixun, Xie Zhiqin, Xie Liji, Huang Li, Huang Jiaoling, Deng Xianwen, Zeng Tingting, Wang Sheng, Zhang Yanfang, and Liu Jiabo

Abstract

Infections with low pathogenic avian influenza viruses (LPAIVs) can be mild or asymptomatic in poultry; however, in humans, LPAIVs can cause severe infections and death, as demonstrated by the H7N9 and H10N8 human infection outbreaks in 2013 in China. In this study, we conducted an epidemiological survey of LPAIVs at live poultry markets (LPMs) in Guangxi, southern China, which is near several southeast Asian countries. From January 2012 to December 2015, we collected 3 813 swab samples from poultry at LPMs in Guangxi. Viral isolation, hemagglutination inhibition assay and viral sequencing were utilized to identify LPAIVs in the collected samples. Among the samples, 622 (16.3%) were positive for LPAIVs. Six subtypes (H1, H3, H4, H6, H9 and H11) were individually isolated and identified. Of these subtypes, H3, H6 and H9 were predominant in ducks, geese and chickens, respectively. Among the 622 positive samples, 160 (25.7%) contained more than one subtype, and H8, H10, H12, H13, and H16 were identified among them, which highlights the continuous need for enhanced surveillance of AIVs. These results provide detailed information regarding the epidemic situation of LPAIVs in the area, which can aid efforts to prevent and control AIV transmission in humans and animals.

Keywords

low pathogenic avian influenza, surveillance, live poultry markets

---

Introduction

Avian influenza viruses (AIVs) are enveloped viruses of the genus *Influenza A* virus in the family Orthomyxoviridae, and their genome consists of eight segments of single-stranded, negative-sense RNA. AIVs are classified into distinct subtypes based on the antigenicity of their hemagglutinin (HA) and neuraminidase (NA) proteins. To date, 18 HA subtypes (Hl-H18) and 11 NA subtypes (N1-N11) have been identified. All of these subtypes were initially identified from avian species, with the exception of H17N10 and H18N11, which were recently found in bats[1, 2]. According to Office International Des Epizooties (OIE) standards, AIVs are classified according to their level of virulence as either highly pathogenic AIVs (HPAIVs) or low pathogenic AIVs (LPAIVs). A few H5 and H7 subtypes of AIV, such as H5N1 and H7N7, are virulent HPAIVs, leading to high morbidity and mortality in poultry. These subtypes are a public health security concern and cause severe economic losses in the poultry industry[3-5]. In contrast, LPAIVs strains, including the H1-H16 subtypes, typically cause mild disease or asymptomatic infections in poultry. Thus, LPAIVs have been largely neglected in global disease control programs. However, the genetic rearrangement and recombination of LPAIVs afford considerable opportunities for rapid and substantial viral changes through both antigenic shift and drift. Such changes may generate novel viruses with increased virulence that can pose substantial risks to poultry and human health.

In March 2013, H7N9, a novel *Influenza A* virus causing human infection emerged in the Yangtze River Delta region with a fatality rate of approximately 30%. H7N9 quickly spread to more than 18 provinces and municipalities in China[6]. According to the Disease Outbreak News issued by the World Health Organization (WHO) on September 5, 2017, 1 558 laboratory-confirmed cases of human infection with H7N9 virus have been reported since early 2013. During the same period, another novel emergent virus, H10N8, was detected in Jiangxi province, China, resulting in 3 human infections and 2 deaths in December 2013[7]. Nevertheless, both the H7N9 and H10N8 strains showed few symptoms and caused silent out breaks in poultry and were therefore classified as LPAIVs; however, these strains retain the ability to infect humans and cause death. Unlike HPAIV outbreaks that are direct and obvious, LPAIVs, such as H7N9 and H10N8, are harder to detect in poultry until they cross the species barrier and directly infect humans. When reports of human deaths due to LPAIVs emerge, people stop purchasing live poultry in live poultry markets (LPMs), which results in large numbers of poultry products (e.g., chickens, ducks, geese and eggs) that are unable to be sold, thereby causing serious economic losses in the poultry industry. In addition, the genetic reassortment of LPAIVs is cause for concern. The recently emerged H7N9 and H10N8 AIVs are reassortant viruses, and the H9N2 LPAIV contributed six internal genes to the novel H7N9 and H10N8 viruses[8, 9]. A naturally acquired human infection with the H6N1 virus was first documented in Taiwan, China in June 2013. Sequence analyses reveal that the human isolate (A/Taiwan/2/2013, human-H6N1) from the patient was highly homologous to chicken H6N1 virus isolates in Taiwan, China[10]. LPAIVs are closely related to human health and pose potential threats to public health security.

Southern China has been considered an influenza epicenter due to its favorable breeding grounds for the virus. Domestic poultry farming in southern China occurs in high-density settings and in a free-range manner. Waterfowl, especially ducks and geese, occupy lakes and other abundant water resources in Guangxi, southern China. These settings create environments where migratory birds and waterfowl are in close contact, sharing water, food, and habitat, thus contributing to the geographical spread of AIVs via long-distance migration through flyways. Thus, waterfowl species play an important role in AIV transmission and are regarded as a natural reservoir of AIVs[11]. Guangxi borders Vietnam, where complex epidemics of AIVs, including H5N1, continue to occur[12]. Furthermore, trade in LPMs, a traditional practice in this area, is considered a major source of AIV dissemination. In LPMs, birds from different sources can be housed together for several days, providing opportunities for influenza virus reassortment and the cross-species transfer of AIVs[13, 14]. In LPMs, humans interact closely with poultry and potentially share influenza pathogens, which can result in the emergence of novel AIV variants.

Little is known about the infection landscape of LPAIVs in Guangxi. Our comprehensive investigation of LPAIV infection patterns and distribution in LPMs in this region from 2009 to 2011 revealed that the epidemiology of LPAIVs in poultry in Guangxi is complex, highlighting the need for further epidemiological study[15]. In the present work, we report on continuous surveillance data from January 2012 through December 2015 involving a relatively sizeable collection of swab samples from the oropharyngeal and cloacal regions of chickens, ducks, and geese from LPMs in Guangxi. We examined these samples for the presence of LPAIVs (except for H5 and H7 LPAIVs). The resulting epidemiological data offer novel insights for planning strategies to prevent LPAIV infections in humans and other animals.

Materials and methods

Sample collection

The epidemiological surveillance of LPAIVs in poultry was conducted at 21 randomly selected LPMs in Guangxi, southern China. Among these LPMs, six were sampled continuously during the study period. We performed sampling once every week, and the six markets were sampled in turn. We collected 3 813 swab samples (an oropharyngeal swab and a cloacal swab from the same bird were placed into the same collection tube and counted as a single sample) from LPMs from January 2012 to December 2015. The swabs were maintained in 1 mL of transport medium containing antibiotics at 4 ℃ until arrival at the laboratory. The cotton swabs were repeatedly cleaned, wiped and discarded. The sample solutions were stored at −70 ℃ for viral isolation.

Viral isolation and identification

Samples were thawed and centrifuged for 10 min at 3 000 × g at 4 ℃, and 0.2 mL of the supernatant from each sample was inoculated into the allantoic cavity of 9- to 11-day-old SPF embryonated chicken eggs (Beijing Merial Biology Company, Beijing, China). The inoculated chicken embryos were incubated at 35 ℃ and observed daily. Allantoic fluid was harvested from the embryos that died within 120 h of inoculation. The chicken embryos that survived longer than 120 h post-inoculation were chilled at 4 ℃, and the allantoic fluid was then harvested. The harvested allantoic fluid samples were then tested for hemagglutination (HA) using 1% suspensions of chicken erythrocytes. The hemagglutination inhibition (HI) assay was conducted to determine the HA subtypes of the HA-positive samples. Briefly, the HA-positive allantoic fluid samples were tested with sera of different HA subtypes (Hl-H4, H6, H8-H16), Newcastle disease virus (NDV) and egg drop syndrome (EDS) to determine the HA subtypes. The isolates selected for sequence analysis were biologically cloned by at least three rounds of limiting dilution in embryonated SPF eggs. HA subtypes were identified using the HI test and by sequencing. NA subtypes were determined by direct sequencing. Experiments using LPAIVs were conducted in biosafety level 2 laboratory or negative pressure biosafety laboratory facilities.

Viral sequencing

Viral RNA was extracted from infected allantoic fluid using a Body Fluid Viral DNA/RNA Miniprep Kit (Axygen Biosciences, Hangzhou, China) according to the manufacturer's instructions. cDNA was synthesized from viral RNA by reverse transcription with the 12-bp primer 5'-AGCAAAAGCAGG-3'. PCR was performed using specific primers as described in previous research to obtain the full-length HA and NA genes[16]. The PCR products were purified with the TaKaRa Agarose Gel DNA Purification Kit Ver. 2.0 (TaKaRa, Dalian, China) and sequenced by Invitrogen of Guangdong Co., Ltd. A BLAST search was performed to compare the confirmed sequences against the nucleotide sequences of all known HA and NA gene subtypes in the GenBank database, and the HA and NA subtypes of the isolates were determined and verified.

Results

Isolation rates of LPAIVs in chickens, ducks and geese

We collected 3 813 swabs samples from 21 different LPMs in Guangxi from January 2012 to December

2015. The samples were used for the isolation and identification of LPAIVs. As shown in Table 2-2-1, among the 3 813 samples, 622 were identified as positive for LPAIVs, including 161 from chickens, 402 from ducks and 59 from geese. The total isolation rate was 16.3% (622/3 813, positive/sum), and the isolation rates in chickens, ducks and geese were 12.7% (161/1 267), 17.6% (402/2 280) and 22.2% (59/266), respectively. The rate of LPAIVs isolation was the highest from geese, followed by ducks and chickens, indicating that the isolation rate was higher in waterfowl. The numbers of positive samples and collected samples corresponding to the yearly isolation rates in chickens, ducks and geese from 2012 to 2015 are presented in Table 2-2-1. The isolation rate in ducks was slightly higher than the rate in chickens in 2012-2014 and was much higher than the rate in chickens in 2015. Compared with 2012, the isolation rates in both chickens and ducks increased to more than 20% in 2013 and then decreased to approximately 15% in 2014. The isolation rate of LPAIVs in chickens continued to decrease thereafter, reaching 8% in 2015, while that in ducks increased to 20.8%. The isolation rate in geese was highest (50.0%) in 2012. The isolation rate in geese decreased to 34.8% in 2013 and then to 9.8% in 2014, but it subsequently increased to 19.4% in 2015.

Table 2-2-1   LPAIVs isolated from chickens, ducks and geese from live poultry markets (LPMs) in Guangxi, southern China, from January 2012 to December 2015

Sample year	LPAIV isolation rate /% (Number of positive samples/total samples)		
	Chickens	Ducks	Geese
2012	7.5 (14/187)	9.3 (33/356)	50.0 (18/36)
2013	21.6 (52/241)	22.0 (95/431)	34.8 (16/46)
2014	14.8 (61/413)	15.8 (117/740)	9.8 (11/112)
2015	8.0 (34/426)	20.8 (157/753)	19.4 (14/72)
Total (2012-2015)	12.7 (161/1 267)	17.6 (402/2 280)	22.2 (59/266)
Grand total (all poultry samples 2012-2015)	16.3 (622/3 813)		

The season average isolation rates in poultry, including chickens, ducks and geese, during the four years are shown in Figure 2-2-1 A. Higher isolation rates occurred in the spring and winter, and the isolation rates were relatively low in summer and autumn.

The season isolation rates are shown in Figure 2-2-1 B from 2012 to 2015. In general, the isolation rate in 2013 was higher than the rates in the other three years. The highest isolation rates from 2012 to 2015 were in autumn, spring, winter and summer, respectively. Generally, the isolation rates in spring and winter are higher than in summer and autumn, except for the rather high rates in autumn in 2012 and in summer in 2015.

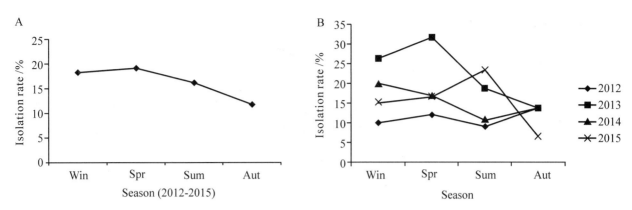

A: Average isolation rates of each season over four years; B: Isolation rates of each season annually.

Figure 2-2-1　Relationship between season and isolation rates of LPAIVs in poultry from LPMs in Guangxi, southern China, from 2012-2015

HA and NA subtypes identified in chickens, ducks and geese

More than six HA subtypes (H1, H3, H4, H6, H9, H11 and some that were unknown) were individually isolated and identified in poultry. The percentages of identified subtypes from poultry in the 622 positive samples and in the 3 813 collected samples are listed in Table 2-2-2. When considering only single infections, H9 was the most commonly isolated subtype; its percentages in all positive samples (24.1%, 150/622) and in collected samples (3.9%, 150/3 813) were the highest for the isolation of any single subtype (Table 2-2-2). The H9, H3 and H6 subtypes were the predominant LPAIVs identified in chickens, ducks and geese, respectively. It is equally important that 78 H4 isolates were identified, which were mainly from ducks.

Table 2-2-2　Number of infections of different HA subtypes of LPAIVs in chickens, ducks and geese

Number	Subtype of infection	Number of cases				Percentages in positive samples /%	Percentages in collected samples /%
		Chickens	Ducks	Geese	Total		
1	H1	0	4	0	4	0.64	0.1
2	H3	7	98	2	107	17.2	2.8
3	H4	5	69	4	78	12.5	2.0
4	H6	10	70	42	122	19.6	3.2
5	H9	99	47	4	150	24.1	3.9
6	H11	0	1	0	1	0.16	0.026
7	Mixed infections	40	113	7	160	25.7	4.2
Total		161	402	59	622		

The distribution of the HA subtypes in poultry is shown in Figure 2-2-2. The H9 subtype was predominant in chickens, representing 61.5% of the isolates, with the remaining three subtypes each representing less than 7%. The duck isolates were diverse, with six identified subtypes. H3 was the most frequent subtype (24.38%), followed by H6 (17.41%), H4 (17.16%) and H9 (11.69%). The least frequent were H1 and H11, which were isolated as single subtypes only in ducks. The goose isolates were less diverse than those of the ducks; the H6 subtype was predominant, with an isolation rate of 71.19%, followed by H4 and H9, at equal rates of 6.78%.

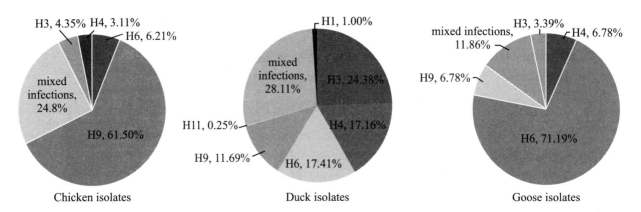

Figure 2-2-2   Distribution of HA subtypes among LPAIVs identified in poultry from LPMs in Guangxi, southern China

The H1 and H11 subtypes were sporadic, while the H3, H4, H6 and H9 subtypes can be isolated in any season. Figure 2-2-3 shows the relationship between the main isolation subtypes and season. The highest isolation rate of the H3 subtype was in winter, and the H6 and H9 subtypes both peaked in spring. The HA and NA genes of the identified LPAIV isolates were sequenced and submitted to a BLAST search of the NCBI database. Three NA subtypes (N2, N6 and N8) were identified. The combinations of HA and NA subtypes identified in this study were H1N2, H3N2, H3N6, H4N2, H4N6, H4N8, H6N2, H6N6, H6N8, H9N2 and H11Nx (x has not been confirmed).

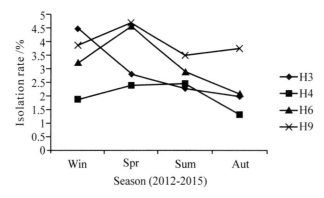

Figure 2-2-3   Relationship between season and isolation rates of the H3, H4, H6 and H9 subtypes in poultry from LPMs in Guangxi, southern China, from 2012-2015

Mixed infections by different HA subtypes

The rate of mixed infections (more than one subtype) among the identified LPAIVs was 25.7% (160/622). Among the 160 samples positive for mixed infections, 40 were from chickens, 113 were from ducks, and 7 were from geese (Table 2-2-2). Among the 161 LPAIV-positive chicken isolates, 40 presented with mixed infections, yielding a rate of 24.84%. The corresponding rates for ducks and geese were 28.11% and 11.86%, respectively (Figure 2-2-2). These results suggest that various mixed infections were widespread, especially in ducks. Ducks likely play an important role in the reassortment events that lead to new AIV variants.

As shown in Tables 2-2-3 and 2-2-4, among the mixed infections, combinations of two HA subtypes, comprising 17 types, were the most common combination type in the three hosts (chickens, ducks and geese), representing 89.4% of the mixed infections (143/160). Combinations of three HA subtypes, comprising 7 types, were observed only in chickens and ducks, and combinations of more than three HA subtypes, comprising 3 types, were observed only in ducks. These findings indicate that mixed infections in ducks were complex.

As shown in Table 2-2-4, the H3+H4 subtype was the most common combination of mixed infection, representing 48.8% (78/160). Among the 78 strains of mixed infections with H3+H4, 57 strains are from 2015, and 21 strains are from 2012-2014; therefore, the isolation of H3+H4 increased sharply in 2015. The different HA subtype combinations yielded 27 different types of mixed infections. Among these 27 types, there were 15 combinations involving the H4 subtype, totaling 127 strains, and 79.4% (127/160) among the mixed infections. Notably, mixed infections of 3~6 HA subtypes all involved the H4 subtype, indicating that the H4 subtype plays a significant role in mixed infections. H9 subtype distributed in the 15 combinations and 56 strains, H3 subtype distributed in the 8 combinations and 88 strains, and H6 subtype distributed in the 8 combinations and 20 strains, suggesting that the H9, H3 and H6 subtypes were also important components in mixed infections. Generally, H4 was the most abundant subtype in mixed infections. The number of combinations and the infection rate of the H9 subtype were both higher than those for H6 but lower than those for the H4 subtype. The rate of mixed infection cases including H3 was higher than that including H9, but only distributed in the 8 combinations, which was lower than those for the H9 (15 combinations). The high rates of cases that included H3 or H4 were mainly due to high proportion of H3+H4 in the mixed infection cases. In addition, H8, H10, H12, H13, and H16 were rare subtypes that were not isolated or identified among the single infections but were identified among the mixed infections.

Table 2-2-3   Number of mixed infections of different HA subtypes of LPAIVs in chickens, ducks and geese

Combinations of mixed infection	Number of cases				Percentages in mixed infections /%	Number of types
	Chickens	Ducks	Geese	Total		
Two different HA subtypes	34	102	7	143	89.4	17
Three different HA subtype	6	6	0	12	7.5	7
Four different HA subtypes	0	4	0	4	2.5	2
Six different HA subtypes	0	1	0	1	0.6	1
Total	40	113	7	160		27

Table 2-2-4   Mixed infections of different HA subtypes of LPAIVs in chickens, ducks and geese

Number	Type of mixed infection (27 combinations)	Number of cases						
		Chickens	Ducks	Geese	Including H9 (15 combinations)	Including H4 (15 combinations)	Including H3 (8 combinations)	Including H6 (8 combinations)
1	H1+H4	0	1	0		1		
2	H1+H9	2	0	0	2			
3	H1+H13	0	1	0				
4	H3+H4	5	69	4		78	78	
5	H3+H9	1	2	0	3		3	
6	H3+H11	0	1	0			1	
7	H4+H6	0	4	0		4		4
8	H4+H9	11	11	0	22	22		
9	H4+H13	0	5	0		5		
10	H6+H9	1	1	0	2			2

continued

Number	Type of mixed infection (27 combinations)	Number of cases						
		Chickens	Ducks	Geese	Including H9 (15 combinations)	Including H4 (15 combinations)	Including H3 (8 combinations)	Including H6 (8 combinations)
11	H6+H11	0	0	1				1
12	H6+H13	2	4	1				7
13	H8+H9	6	0	0	6			
14	H9+H11	0	1	0	1			
15	H9+H12	4	1	1	6			
16	H9+H13	2	0	0	2			
17	H9+H16	0	1	0	1			
18	H3+H4+H6	0	1	0		1	1	1
19	H3+H4+H9	0	1	0	1	1	1	
20	H3+H4+H11	0	1	0		1	1	
21	H4+H6+H9	1	0	0	1	1		1
22	H4+H6+H13	1	1	0		2		2
23	H4+H8+H9	3	2	0	5	5		
24	H4+H9+H16	1	0	0	1	1		
25	H1+H4+H6+H10	0	2	0		2		2
26	H3+H4+H9+H11	0	2	0	2	2	2	
27	H3+H4+H9+ H11+H13+H16	0	1	0	1	1	1	
Total		40	113	7	56	127	88	20
The infection rate					35.0% (56/160)	79.4% (127/160)	55.0% (88/160)	12.5% (20/160)

Discussion

AIVs are distributed worldwide, with a global impact on animal and human health. When a poultry bird, such as a duck, is infected with multiple subtypes of AIV, the eight gene segments from the different HA subtypes may exchange genetic information with each other, which can lead to the emergence of a novel AIV and a potentially unpredictable pathogenic strain that could infect humans. The outbreak of the highly pathogenic H5N1 strain in 1997 was the first evidence that AIVs had the capacity for direct transmission from avian species to humans[3]. The 2009 pandemic H1N1 influenza virus spread to humans in over 215 countries and caused hundreds of thousands of deaths[17]. Previously, it was thought that only the H5 and H7 subtypes posed a significant pandemic risk; however, it is now known that strains of other HA subtypes (e.g., H9, H6 and H10) can infect humans and have pandemic potential[18]. LPAIV surveillance is intrinsically difficult because most LPAIV strains do not cause symptoms or only cause mild symptoms in infected poultry, allowing the viruses to spread silently[19]. To address this obstacle, intensive surveillance must be conducted if newly emerged reassortant viruses are to be detected early.

Livestock breeding is extensive and booming in Guangxi, with many breeding farms and companies due to the subtropical climate. Many large-scale farms are located outside of the city and provide poultry production to meet strong and increasing consumer demand. Some of these products are sold to the adjacent

Guangdong province and Hong Kong. In some rural areas, chickens are freely raised on the mountainside, and ducks and geese range freely in pools, rivers and seaside areas. The China-ASEAN (The Association of Southeast Asian Nations) Exposition has convened in Nanning (the capital of Guangxi) annually since 2004, and people from different countries in Southeast Asia gather to discuss issues related to communication, culture and economy involved with the effective prevention and control of influenza viruses. The prevention and control of AIVs requires further attention and action in this region of the world. In addition, the seaside cities of Beihai and Fangcheng along the Beibu Gulf in Guangxi provide a good resting area for migrating wild birds. This area also has a humid environment, which may benefit the survival, growth and transmission of LPAIVs. To date, thirty cases of human H7N9 infection have been reported in Guangxi since the emergence of the first H7N9 human infection in early 2013. Among the thirty cases, there were three cases in 2014, no cases in 2015 and 2016, and 27 cases in 2017 (as of September 10). As a result, LPMs were closed and sterilized, and people refrained from purchasing poultry products, resulting in severe economic losses to the poultry industry.

In contrast to the results of our previous study[15], the current work revealed that H9, not H3, was the most prevalent subtype, consistent with the current epidemic situation in China[18]. H9 is currently in widespread circulation, occurring mostly in combination with the N2 subtype. However, we did isolate a strain of H9N8 in 2009 (GenBank number: KF768214). H9N2 viruses have been found to possess increased fitness to escape immunization pressure and undergo antigen evolution and drift in chickens, leading to the prevalence of new genotypes[20]. H9N2 has also been reported to reassort with several other subtypes, including H5N1, H6N1 and H6N2[18]. Thus, H9N2 viruses are high on the list of candidates that could cause another human influenza pandemic. Presently, the H9N2 subtype is a primary focus of concern because it is the donor of six internal gene segments to the H7N9 and H10N8 novel reassortment avian influenza strains. It is well known that the H5N1 strain, which was responsible for the 1997 human influenza outbreak in Hong Kong, was found to contain internal genes derived from H9N2[21]. The crucial role of H9N2 viruses at the animal-human interface might be due to its wide host range, adaptation in both poultry and mammals, and extensive gene reassortment[18]. Our epidemiological survey revealed that the H9 subtype was an important component and widely distributed in 15 combinations (total 27 combinations) of the mixed infections. Among the chicken isolates, H9 viruses were the major isolation subtype. In our previous study[22], the results of a receptor-binding analysis revealed that 16 of 17 strains of H9 Guangxi isolates bound to both α-2, 6-linked glycans and α-2, 3-linked glycans; only 1 strain bound solely to α-2, 3-linked glycans. The binding of the human influenza virus receptor, Siaα2-6 Gal, suggests that chickens may serve as an intermediate host and may be the source of transmission of the influenza virus to humans.

In our previous research, H6N1, H6N2, H6N5, H6N6 and H6N8 and combinations of these five subtypes were identified during 2009-2011[15]. These five subtypes also circulated in eastern China from 2002 to 2010[23]; however, in this study, only three H6 subtypes (H6N2, H6N6 and H6N8) were circulating in Guangxi between 2012 and 2015, with H6N2 and H6N6 being predominant. It was reported that the H6N2 and H6N6 viruses coexisted in LPMs in several provinces of southern China between 2008 and 2011, and approximately 34% of the H6 isolates derived from LPMs have acquired the ability to bind to the human-like receptor[24]. The evidence suggests that the H6 subtype was widespread and combined with different NA subtypes. The H6 subtypes were common among the goose and duck isolates in this study, especially the goose isolates, representing much as 71.19% of the identified LPAIVs from geese; the reason for this requires further exploration. Currently, H6 AIVs have worldwide distribution, and strains of the virus have been detected in various animal species, suggesting that H6 has a broader host range than other subtypes[24]. Our previous research[25] suggested that H6 subtype AIVs could be directly transmitted from ducks to pigeons. Many reports have revealed that H6 viruses can infect and be efficiently transmitted among mice and ferrets[26] and that they circulate extensively and reassort frequently[25, 27]. In addition, a serum antibody positive for H6 viruses has been found in poultry workers[28], and a further study indicated that H6 seropositivity in human specimens in southern China was significantly higher than in northern China[29].

Six LPAIV subtypes were individually isolated in ducks, suggesting that ducks are an important host. In the duck isolates, the isolation rate of the H3 subtype AIV was the highest of all the subtypes identified, which emphasizes the importance of enhancing the surveillance of waterfowl-originating AIVs. Most of the H3 isolates matched with N2, while a few matched with N6 or N8. H3, particularly the H3N8 virus, is highly adaptive since it is found in multiple avian and mammal hosts. H3N8, which was first isolated in Miami in 1963, is the major cause of equine influenza; however, H3N8 viruses have not yet been isolated from humans[30]. The H3N8 and H3N2 isolates from LPMs were reported to have a close relationship with the H5N8 HPAIV circulating in Republic of Korea and the United States, suggesting that H3-like AIVs may contribute internal genes to the highly pathogenic H5N8 viruses[31]. In our previous research, a duck isolate named A/duck/ Guangxi/175D12/2014 (H3N6) [32] showed reassortment events between H3 and the H5N6 and H7N2 influenza viruses. Genomic analyses of A/pigeon/Guangxi/020P/2009 (H3N6)[33] and A/goose/Guangxi/020G/2009 (H3N8) [34] suggested that these viruses have undergone extensive reassortment with different AIV subtypes. DEQUAN Y et al. previously reported that H3N2 isolates from LPMs and poultry slaughterhouses in Shanghai had reassorted with other AIVs, especially the H5 and H7 subtypes, probably in pigeons, domestic ducks, and wild birds[35]. An H3N2 isolate from Anhui, which also showed the highest sequence homology to the H7 AIVs, was presumed to be a reassortant of H3 and H7 AIVs[36]. H3 is the most ubiquitous subtype and has a wide host range, including humans, pigs, horses, dogs, cats, seals, poultry, and wild aquatic birds[30]. It was reported that an H3N2 isolate from duck could acquire the potential to infect humans after multiple infections in a pig population[37]. H3 causes the seasonal influenza that is widely found in humans. These findings highlight the need for further investigation, epidemiological surveillance, and genetic and evolutionary studies of H3-subtype AIVs.

H4 viruses also circulate widely throughout the world and are transmitted to mammalian hosts. Our study revealed that the isolation rate of H4 subtypes significantly increased from year to year. In the mixed infections, more than two subtypes included the H4 subtype, suggesting that H4 was highly infectious in combination with other AIV subtypes and had the advantage of increased potential for reassortment compared to other subtypes, thereby leading to the emergence and outbreak of novel subtype influenza viruses, which could threaten public health. We considered the H4 subtype worthy of particular research focus. We found that H4 viruses frequently matched with N2 and N6; however, some H4N8 viruses were also identified in our study. It was reported that H4N8 viruses isolated from shorebird contained a unique PB1 gene and caused severe respiratory disease in mice[38]. We chose some representative isolates of the H4 subtype from this study to experimentally infect specific-pathogen-free (SPF) chickens and did not observe clinical symptoms, which suggests low pathogenicity. Specific antibodies against H4-subtype AIVs were detected in sera from swine and from people working on a chicken farm[39, 40]. Eight genes of H4N6 isolated from a duck farm were closely related to H4N6 viruses from LPMs in Shanghai, indicating a potential correlation between AIVs from LPMs and farms[41]. The H4 LPAIVs infected mice directly without prior adaptation[41, 42]. The extensive reassortment of H4 AIVs is worrisome because it may produce hybrid viruses that can jump to humans and cause major public health issues, as occurred with the newly emerged H7N9 influenza virus[19]. Therefore, further investigations of the mechanisms of H4 AIV mutation and reassortment are important to prepare for potential pandemics.

Not only were the H9, H6, H3 and H4 the main subtypes in the single infections, but they were also important in the mixed infections. H8, H10, H12, H13 and H16 were also involved in mixed infections, but none of them have yet been isolated as a single infection. H8, H12, H13 and H16 were found infrequently compared with the other subtypes but emerged in mixed infections, revealing that they were active and in circulation in the area. These findings suggest a probable increasing trend and the need to monitor these subtypes. The H10 subtype existed in mixed infections and was not isolated alone in this survey; however, H10N8 has caused human deaths in Jiangxi Province in China. Epidemics of mixed infections are wide spread and complex, and the co-infection of different HA subtypes in the same bird presents a great risk of gene rearrangement, which can lead to novel subtypes of AIVs. Furthermore, mixed infections have increased viral diversity through reassortment between viruses from different sources.

In conclusion, we investigated the epidemiology of LPAIVs in poultry from LPMs in Guangxi, southern China, an epicenter of avian flu. Our study demonstrates a high prevalence of LPAIVs in poultry in this area and highlights a need to further investigate the genetics and evolution of LPAIVs.

References

[1] TONG S X, LI Y, RIVAILLER P, et al. A distinct lineage of influenza A virus from bats. Proceedings of the National Academy of Sciences of the United States of America, 2012, 109(11): 4269-4274.

[2] TONG S X, ZHU X Y, LI Y, et al. New world bats harbor diverse influenza A viruses. PLOS Pathogens, 2013, 9(10): e1003657.

[3] CLAAS E C, OSTERHAUS A D, VAN BEEK R, et al. Human influenza A H5N1 virus related to a highly pathogenic avian influenza virus. Lancet, 1998, 351(9101): 472-477.

[4] PUZELLI S, ROSSINI G, FACCHINI M, et al. Human infection with highly pathogenic A (H7N7) avian influenza virus, Italy, 2013. Emerging Infectious Diseases, 2014, 20(10): 1745-1749.

[5] BEVINS S N, DUSEK R J, WHITE C L, et al. Widespread detection of highly pathogenic H5 influenza viruses in wild birds from the pacific flyway of the United States. Scientific Reports, 2016, 6: 28980.

[6] SU S, BI Y H, WONG G, et al. Epidemiology, evolution, and recent outbreaks of avian influenza virus in China. Journal of Virology, 2015, 89(17): 8671-8676.

[7] QI W, ZHOU X, SHI W, et al. Genesis of the novel human-infecting influenza A (H10N8) virus and potential genetic diversity of the virus in poultry, China. European Communicable Disease Bulletin, 2014, 19(25): 20841.

[8] PU J, WANG S G, YIN Y B, et al. Evolution of the H9N2 influenza genotype that facilitated the genesis of the novel H7N9 virus. Proceedings of the National Academy of Sciences of the United States of America, 2015, 112(2): 548-553.

[9] CHEN H Y, YUAN H, GAO R B, et al. Clinical and epidemiological characteristics of a fatal case of avian influenza A H10N8 virus infection: a descriptive study. Lancet, 2014, 383(9918): 714-721.

[10] WEI S H, YANG J R, WU H S, et al. Human infection with avian influenza A H6N1 virus: an epidemiological analysis. The Lancet Respiratory Medicine, 2013, 1(10): 771-778.

[11] WEBSTER R G, BEAN W J, GORMAN O T, et al. Evolution and ecology of influenza A viruses. Microbiological Reviews, 1992, 56(1): 152-179.

[12] MANABE T, YAMAOKA K, TANGO T, et al. Chronological, geographical, and seasonal trends of human cases of avian influenza A (H5N1) in Vietnam, 2003-2014: a spatial analysis. BMC Infectious Diseases, 2016, 16: 64.

[13] YUAN J, LAU E H, LI K, et al. Effect of live poultry market closure on avian influenza A (H7N9) virus activity in Guangzhou, China, 2014. Emerging Infectious Diseases, 2015, 21(10): 1784-1793.

[14] SOARES MAGALHÃES R J, ZHOU X, JIA B, et al. Live poultry trade in southern China provinces and HPAIV H5N1 infection in humans and poultry: the role of Chinese New Year festivities. PLOS ONE, 2012, 7(11): e49712.

[15] PENG Y, XIE Z X, LIU J B, et al. Epidemiological surveillance of low pathogenic avian influenza virus (LPAIV) from poultry in Guangxi Province, southern China. PLOS ONE, 2013, 8(10): e77132.

[16] HOFFMANN E, STECH J, GUAN Y, et al. Universal primer set for the full-length amplification of all influenza A viruses. Archives of Virology, 2001, 146(12): 2275-2289.

[17] NELSON M I, VINCENT A L. Reverse zoonosis of influenza to swine: new perspectives on the human-animal interface. Trends in Microbiology, 2015, 23(3): 142-153.

[18] SUN Y, LIU J. H9N2 influenza virus in China: a cause of concern. Protein & Cell, 2015, 6(1): 18-25.

[19] LIANG L B, DENG G H, SHI J Z, et al. Genetics, receptor binding, replication, and mammalian transmission of H4 avian influenza viruses isolated from live poultry markets in China. Journal of Virology, 2015, 90(3): 1455-1469.

[20] WEI Y D, XU G L, ZHANG G Z, et al. Antigenic evolution of H9N2 chicken influenza viruses isolated in China during 2009-2013 and selection of a candidate vaccine strain with broad cross-reactivity. Veterinary Microbiology, 2016, 182: 1-7.

[21] GUAN Y, SHORTRIDGE K F, KRAUSS S, et al. Molecular characterization of H9N2 influenza viruses: were they the donors of the " internal " genes of H5N1 viruses in Hong Kong?. Proceedings of the National Academy of Sciences of the United States of America, 1999, 96(16): 9363-9367.

[22] QIAN X U, XIE Z X, LUO S S, et al. Isolation, identifcation and biological characteristics of H9 subtype avian infuenza viruses. Progress in Veterinary Medicine, 2016, 37(9): 10-15.

[23] ZHAO G, LU X L, GU X B, et al. Molecular evolution of the H6 subtype influenza A viruses from poultry in eastern China from 2002 to 2010. Virology Journal, 2011, 8: 470.

[24] WANG G J, DENG G H, SHI J Z, et al. H6 influenza viruses pose a potential threat to human health. Journal of Virology, 2014, 88(8): 3953-3964.

[25] LI M, XIE Z X, XIE Z Q, et al. Molecular characteristics of H6N6 influenza virus isolated from pigeons in Guangxi, southern China. Genome Announcements, 2015, 3(6): e01422-15.

[26] GILLIM-ROSS L, SANTOS C, CHEN Z Y, et al. Avian influenza H6 viruses productively infect and cause illness in mice and ferrets. Journal of Virology, 2008, 82(21): 10854-10863.

[27] YUAN R Y, ZOU L R, KANG Y F, et al. Reassortment of avian influenza A/H6N6 viruses from live poultry markets in Guangdong, China. Frontiers in Microbiology, 2016, 7: 65.

[28] YUAN J, ZHANG L, KAN X Z, et al. Origin and molecular characteristics of a novel 2013 avian influenza A (H6N1) virus causing human infection in Taiwan. Clinical infectious diseases : an official publication of the Infectious Diseases Society of America, 2013, 57(9): 1367-1368.

[29] XIN L, BAI T, ZHOU J F, et al. Seropositivity for avian influenza H6 virus among humans, China. Emerging Infectious Diseases, 2015, 21(7): 1267-1269.

[30] SOLÓRZANO A, FONI E, CÓRDOBA L, et al. Cross-species infectivity of H3N8 influenza virus in an experimental infection in swine. Journal of Virology, 2015, 89(22): 11190-11202.

[31] CUI H R, SHI Y, RUAN T, et al. Phylogenetic analysis and pathogenicity of H3 subtype avian influenza viruses isolated from live poultry markets in China. Scientific Reports, 2016, 6: 27360.

[32] LIU T T, XIE Z X, LUO S S, et al. Characterization of the whole-genome sequence of an H3N6 avian influenza virus, isolated from a domestic duck in Guangxi, southern China. Genome Announcements, 2015, 3(5): e01190-15.

[33] LIU T T, XIE Z X, WANG G L, et al. Avian influenza virus with hemagglutinin-neuraminidase combination H3N6, isolated from a domestic pigeon in Guangxi, southern China. Genome Announcements, 2015, 3(1): e01537-14.

[34] LIU T T, XIE Z X, SONG D G, et al. Genetic characterization of a natural reassortant H3N8 avian influenza virus isolated from domestic geese in Guangxi, southern China. Genome Announcements, 2014, 2(4): e00747-14.

[35] YANG D Q, LIU J, JU H B, et al. Genetic analysis of H3N2 avian influenza viruses isolated from live poultry markets and

poultry slaughterhouses in Shanghai, China in 2013. Virus Genes, 2015, 51(1): 25-32.

[36] LI C, YU M, LIU L T, et al. Characterization of a novel H3N2 influenza virus isolated from domestic ducks in China. Virus Genes, 2016, 52(4): 568-572.

[37] SHICHINOHE S, OKAMATSU M, SAKODA Y, et al. Selection of H3 avian influenza viruses with SA α 2, 6Gal receptor specificity in pigs. Virology, 2013, 444(1-2): 404-408.

[38] BUI V N, OGAWA H, XININIGEN, et al. H4N8 subtype avian influenza virus isolated from shorebirds contains a unique PB1 gene and causes severe respiratory disease in mice. Virology, 2012, 423(1): 77-88.

[39] NINOMIYA A, TAKADA A, OKAZAKI K, et al. Seroepidemiological evidence of avian H4, H5, and H9 influenza A virus transmission to pigs in southeastern China. Veterinary Microbiology, 2002, 88(2): 107-114.

[40] KAYALI G, BARBOUR E, DBAIBO G, et al. Evidence of infection with H4 and H11 avian influenza viruses among Lebanese chicken growers. PLOS ONE, 2011, 6(10): e26818.

[41] SHI Y, CUI H R, WANG J H, et al. Characterizations of H4 avian influenza viruses isolated from ducks in live poultry markets and farm in Shanghai. Scientific Reports, 2016, 6: 37843.

[42] KANG H M, CHOI J G, KIM K I, et al. Genetic and antigenic characteristics of H4 subtype avian influenza viruses in Korea and their pathogenicity in quails, domestic ducks and mice. The Journal of General Virology, 2013, 94(Pt1): 30-39.

Survey of low pathogenic avian influenza viruses in live poultry markets in Guangxi, southern China, 2016-2019

Luo Sisi, Xie Zhixun, Li Meng, Li Dan, Xie Liji, Huang Jiaoling, Zhang Minxiu, Zeng Tingting, Wang Sheng, Fan Qing, Zhang Yanfang, Xie Zhiqin, Deng Xianwen, and Liu Jiabo

Abstract

Low pathogenic avian influenza viruses (LPAIVs) have been widespread in poultry and wild birds throughout the world for many decades. LPAIV infections are usually asymptomatic or cause subclinical symptoms. However, the genetic reassortment of LPAIVs may generate novel viruses with increased virulence and cross-species transmission, posing potential risks to public health. To evaluate the epidemic potential and infection landscape of LPAIVs in Guangxi, China, we collected and analyzed throat and cloacal swab samples from chickens, ducks and geese from the live poultry markets on a regular basis from 2016 to 2019. Among the 7 567 samples, 974 (12.87%) were LPAIVs-positive, with 890 single and 84 mixed infections. Higher yearly isolation rates were observed in 2017 and 2018. Additionally, geese had the highest isolation rate, followed by ducks and chickens. Seasonally, spring had the highest isolation rate. Subtype H3, H4, H6 and H9 viruses were detected over prolonged periods, while H1 and H11 viruses were detected transiently. The predominant subtypes in chickens, ducks and geese were H9, H3, and H6, respectively. The 84 mixed infection samples contained 22 combinations. Most mixed infections involved two subtypes, with H3+H4 as the most common combination. Our study provides important epidemiological data regarding the isolation rates, distributions of prevalent subtypes and mixed infections of LPAIVs. These results will improve our knowledge and ability to control epidemics, guide disease management strategies and provide early awareness of newly emerged AIV reassortants with pandemic potential.

Keywords

avian influenza virus, epidemiology, survey, live poultry markets, public health

Introduction

Avian influenza viruses (AIVs) are type A influenza viruses and belong to the Orthomyxoviridae family[1]. AIV is a zoonotic pathogen and can threaten the health of humans and animals[2]. AIV is an enveloped, single-stranded, negative-sense, segmented RNA virus[3]. The genome consists of eight gene segments: basic polymerase 2 (PB2), basic polymerase 1 (PB1), acidic polymerase (PA), hemagglutinin (HA), nucleoprotein (NP), neuraminidase (NA), matrix (M), and nonstructural (NS)[4]. AIVs are subtyped based on the antigenic diversity of two surface glycoproteins: HA and NA. Differences in the antigenicity and phylogenetics of these surface proteins allow characterization of AIV into subtypes H(x)N(y). To date, 16 HA (H1-H16) and 9 NA (N1-N9) subtypes have been recognized as circulating virus strains in poultry or wild birds, whereas H17N10 and H18N11 were newly discovered in bats in recent years[5, 6]. Highly pathogenic AIVs (HPAIVs) and low pathogenic AIVs (LPAIVs) are classified based on virus pathogenicity in poultry[7, 8]. HPAIVs can cause severe disease outbreaks in poultry, resulting in heavy economic losses and posing serious public health concerns[9]. A few H5 and H7 subtypes are virulent HPAIVs and lead to high morbidity and mortality in poultry[10, 11]. Most AIVs are LPAIVs. LPAIV strains, including the H1-H16 subtypes, and their infections are usually asymptomatic or induce subclinical signs of illness, and the infected animals appear to be healthy or exhibit only mild respiratory disease symptoms with low mortality[12]. LPAIVs may be silent in wild birds or poultry, but they could pose a risk to human health. For example, H7N9 emerged in the Yangtze River Delta in the spring of 2013 and caused five waves of infections until 2017. Although the H7N9 AIV had low pathogenicity in poultry at the beginning of the outbreak, it became an apparent public health issue due to increasing human mortality, and severely affected public health and socioeconomic development[13].

The RNA polymerase in AIV lacks proofreading capacity of the eight genes[14]. Genetic reassortment may occur when two or more AIV subtypes coinfect a single cell, leading to the generation of novel AIV viruses with pandemic potential[15]. The virulences of novel viruses are unpredictable and may gain the potential to infect humans or different species. It was also reported in recent years that the H6, H9 and H10 subtypes of LPAIVs could cause zoonotic infections. The first human case of H10N8 infection was confirmed on December 17, 2013, and two other cases were subsequently confirmed, but no further outbreak and occurred, suggesting that H10N8 may be sporadic[16]. The H6N1 virus presented the first case of human infection in Taiwan, China in June 2013[17]. The H9N2 virus provided six internal genes to the emerging H7N9 and H10N8 responsible for human infection in 2013[16, 18, 19]. LPAIVs are usually not undiagnosed, evolve continuously and spread in natural hosts. Monitoring LPAIVs is useful not only for public health to prevent AIVs from spreading to humans, but also to monitor circulation in poultry and virus evolution.

Southern China has been considered an epicenter of influenza virus due to its poultry breeding and trading style. Guangxi is located in the southern part of China. Poultry breeding is extensive and booming in Guangxi; there are many large-scale farms and free range backyards, and more than 1.2 billion birds are raised annually. Chicken, duck and goose comprise the primary species in poultry production in Guangxi, and the output value is more than ￥20 billion. In Guangxi, chicken farming is the most abundant, and yellow meat chicken is widely consumed throughout the country; some chicken farms are located on hillsides, where chickens are raised in a free-range mode. Ducks farming is less common than chicken farming, but still accounts for a large number of birds raised; ducks were raised in rivers, in ponds and on seasides farms. Compared to chickens and ducks, geese production occurs on a smaller scale. Guangxi's seaside cities Beihai and Fangcheng along the Beibu Gulf provide a good resting area for migrating wild birds. In addition, Guangxi neighbors Southeast Asian countries and borders Vietnam. Therefore, epidemiological investigation of LPAIVs in Guangxi is important. At present, comprehensive epidemiological surveillance of LPAIVs in poultry in this region is scarce. Poultry in southern China is mainly traded through live poultry markets (LPMs), where different poultry species are housed together for several days, providing a favorable environment for virus transmission and recombination[20, 21]. There are many LPMs in Guangxi, and the source of poultry in LPMs derived from different farms and free-fenced backyards in the province may vary greatly. Human infections with H5N1, H9N2 and recent H7N9 viruses were associated with exposure to LPMs[22-24]. Continued surveillance of poultry may generate valuable knowledge to support AIV prevention and control. We previously reported on the surveillance of LPAIVs in LPMs in Guangxi from 2009 to 2015[25, 26]. In this study, we conducted an LPAIV surveillance study on chickens, ducks and geese in LPMs in Guangxi from 2016 to 2019, and found that LPAIVs exhibited an infection landscape over four-year periods.

Materials and methods

Sample collection

The epidemiological surveillance of LPAIVs in poultry was conducted at 20 selected LPMs in Guangxi, southern China. Among these LPMs, six representative LPMs in Nanning (the capital of Guangxi) were sampled in turn. Throat and cloacal swab samples from chickens, ducks and geese were collected from January 2016 to December 2019. We collected samples weekly. The sampling buffer was composed of sterile PBS (pH = 7.2) with penicillin (100 unit/mL), streptomycin (10 mg/mL), gentamycin (100 unit/mL) and nystatin (100 unit/mL). We collected throat and cloacal swabs and dipped them in sampling buffer. Swab throat and cloacal samples from the same birds were combined and considered to be one sample. The samples were kept at a low temperature (4 ℃). After sampling, the collected samples were transferred to the laboratory for storage and testing. Swabs were cleaned, squeezed dry and discarded after adequate soaking. The swab solution was stored at -80 ℃ until use.

Virus isolation and hemagglutination inhibition test

The swab solutions were thawed and centrifuged in a sample tube at 3 000 × g for 10 min, and the supernatant was inoculated into 10-day-old specific pathogen-free (SPF) chicken embryos (Beijing Merial Biology Company, Beijing, China) via the allantoic cavity. The chicken embryos were incubated at 35 ℃ after inoculation and were observed daily for the death of chicken embryos. The chicken embryos were chilled if they were still alive 5 days postinoculation, and then the allantoic fluids were collected and identified AIV by hemagglutination (HA) assay, which takes advantage of the tendency of HA protein of AIV to bind to red blood cells (RBCs) causing them to agglutinate. The isolates virus is serially-diluted and incubated with 0.5% RBCs to obtain HA titers. Samples with HA activity were subsequently identified HA subtypes by hemagglutination inhibition (HAI) assay, according to the World Health Organization (WHO) protocol. When antibodies against a specific AIV HA protein bind to the antigenic sites on the HA protein, these sites become blocked and therefore unavailable for binding with RBCs. Briefly, the HAI assay was performed using 96-well microtitre plate, 4 hemagglutination units standardized antigen of isolate was mixed with serially diluted antiserum, and RBCs are then added to assess the degree of binding of the antibody to the HA molecule. The HAI titer was expressed as the reciprocal of the highest serum dilution in which hemagglutination was inhibited. All HI assays were performed in duplicate. A panel of reference antiseras were used against different HA subtypes (Table 2-3-1), Newcastle disease virus (NDV) and eggdrop syndrome (EDS).

Table 2-3-1　List of the primary strains used in the preparation of antisera of HAI assay

Number	Subtype	Strain	HAI titer
1	H1	A/Chicken/Guangxi/GXc-1/2011 (H1N2)	2^6
2	H1	A/Duck/Guangxi/GXd-2/2012 (H1N2)	2^7
3	H2	A/Duck/HK/77/76 (H2N3)	2^7
4	H3	A/Duck/Guangxi/135D20/2013 (H3N2)	2^8
5	H3	A/Chicken/Guangxi/015C10/2009 (H3N2)	2^7
6	H3	A/Goose/Guangxi/139G20/2013 (H3N2)	2^7

continued

Number	Subtype	Strain	HAI titer
7	H4	A/Duck/Guangxi/125D17/2012 (H4N2)	2^7
8	H4	A/Duck/Guangxi/101D18/2011 (H4N6)	2^7
9	H6	A/Duck/Guangxi/GXd-7/2011 (H6N6)	2^8
10	H6	A/Duck/Guangxi/GXd-2/2009 (H6N2)	2^7
11	H6	A/Goose/Guangxi/246G44/2016 (H6N2)	2^6
12	H8	A/Turkey/Ontario/6118/68 (H8N4)	2^7
13	H9	A/Chicken/Guangxi/066C10/2010 (H9N2)	2^8
14	H9	A/Chicken/Guangxi/137C2/2013 (H9N2)	2^7
15	H9	A/Goose/Guangxi/146G30/2013 (H9N2)	2^8
16	H10	A/Duck/HK/876/80 (H10N3)	2^6
17	H11	A/Duck/Guangxi/170/14 (H11N2)	2^7
18	H12	A/Duck/HK/862/80 (H12N5)	2^6
19	H13	A/Gull/Md/704/77 (H13N5)	2^7
20	H14	A/Mallard /Astrakhan/263/82 (H14N5)	2^6
21	H15	A/Shearwater/Western Australia/2576/79 (H15N9)	2^7
22	H16	A/Shorebird/Delaware/168/06 (H16N3)	2^7

Gene sequencing and real-time RT-PCR

The single-infection isolates were further confirmed by plaque-purificationas previously described[27], and then used for subsequent sequencing. Mixed-infection isolations were verified by real-time RT-PCR as previously described[28]. Viral RNA was extracted from viral stock fluid using an EasyPure Viral DNA/RNA kit (TransGen, Beijing, China) according to the manufacturer's manual. cDNA was synthesized from viral RNA by reverse transcription with the 12-bp primer 5'-AGCAAAAGCAGG-3' as previously described[29]. PCR was performed using specific primers as described in previous research to obtain the full-length HA and NA genes[30]. The PCR products were purified with the TaKaRa Agarose Gel DNA Purification Kit Ver. 2.0 (TaKaRa, Dalian, China) and sequenced by Invitrogen of Guangdong Co., Ltd. A BLAST search was performed to compare the sequences against the nucleotide sequences of all known HA and NA genes of AIV in the GenBank database, and the HA and NA subtypes of the isolates were determined and verified.

Statistical analysis

Data statistics were used to determine the differences in isolation rates of LPAIVs from different years, poultry species, seasons and subtypes. The results were analyzed using SPSS 22.0 software (IBM, Chicago, USA), the chi-squared test was used to analyze and compare the differences in these isolation rates, and $P<0.05$ was indicative of a significant difference.

Results

LPAIV isolation rates

We collected a total of 7 567 swab samples from 2 175 chickens, 4 601 ducks and 791 geese in the LPMs of Guangxi from January 2016 to December 2019. Among the 7 567 samples, 974 (12.87%) samples were isolated and tested positive for LPAIVs. The yearly isolation rates were 8.00% (113/1 413), 16.40% (266/1 622), 15.75% (370/2 349) and 10.31% (225/2 183) from 2016 to 2019, respectively. Higher yearly isolation rates were observed in 2017 and 2018 (Figure 2-3-1). Compared to that in 2016, the isolation rate was twice as high in 2017 ($P<0.001$), decreased slightly in 2018, and continued to decrease in 2019 ($P<0.001$).

Among the 974 LPAIV-positive samples, 71 were derived from chickens, 698 were derived from ducks and 205 were derived from geese. The isolation rates of chickens, ducks and geese were 3.26% (71/2 175), 15.17% (698/4 601) and 25.92% (205/791), respectively. Geese had the highest rate of virus isolation, followed by ducks and chickens ($P<0.001$). The annual isolation rates of these poultry species from 2016 to 2019 are exhibited in Figure 2-3-1. The isolation rate of ducks increased annually from 2016 to 2018 but significantly deceased in 2019 ($P<0.001$). The isolation rates of chickens and geese both increased until peaking in 2017 and then decreased annually from 2018 to 2019.

The average seasonal isolation rates for the four years are presented in Figure 2-3-2A. Spring and summer had higher isolation rates, whereas the isolation rates of winter and autumn were relatively low ($P<0.05$). The seasonal isolation rates for each year from 2016 to 2019 are presented in Figure 2-3-2B. Spring had the highest isolation rates except in 2017, when the highest isolation rate was in summer. Generally, the annual isolation rates of 2017 and 2018 were higher than those in 2016 and 2019 ($P<0.001$), so the corresponding seasonal isolation rates were also relatively higher than those in 2016 and 2019.

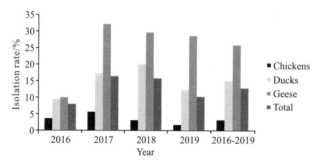

Figure 2-3-1   Yearly isolation rates of LPAIVs in chickens, ducks and geese from 2016 to 2019

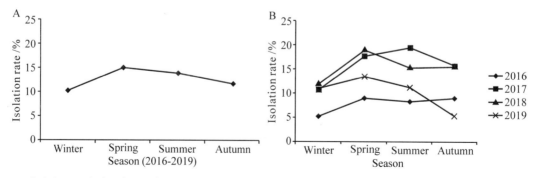

A: The average isolation rates in the winter, spring, summer and autumn for four years from 2016 to 2019; B: The seasonal isolation rates in each year from 2016 to 2019.

Figure 2-3-2   LPAIVs seasonal isolation rates

Distribution and prevalence of HA subtypes among LPAIVs

Among the 7 567 samples, 974 (12.87%) were LPAIV-positive, with 890 single and 84 mixed infections. The numbers of the isolated subtypes and their percentages among species isolates are shown in Table 2-3-2. Of all LPAIV isolates, the H3 subtype accounted for the largest percentage, reaching 46.10%, followed by the H6 subtype (29.47%), the H9 subtype (9.03%), mixed infections (8.62%) and the H4 subtype (5.03%). H3, H4, H6, and H9 were the most prevalent subtypes, whereas H1 and H11 were isolated occasionally. Moreover, Hl and H4 were isolated mainly from ducks. Among the chicken isolates, which included four subtypes (H9, H3, H6 and H1), the H9 subtype was predominant, accounting for nearly half (46.48%), followed by H3 (23.94%). The duck isolates included six subtypes that are highly diversified. The H3 subtype was predominant and accounted for more than half (58.88%) of the isolates, followed by H6 (17.19%). The proportions of the other subtypes and mixed infections were all below 10%. The H11 subtype had only one isolate and was derived from ducks. Among the goose isolates, H6 was the most abundant subtype (77.56%), followed by H3 (10.24%), H9 (5.85%), mixed infection (5.37%) and H4 (0.98%).

Table 2-3-2   Distributions and percentages of the isolated HA subtypes

Number	Isolated subtype	Chickens		Ducks		Geese		Total	
		Number	Percentage in chickens isolations /%	Number	Percentage in ducks isolations /%	Number	Percentage in geese isolations /%	Number	Percentage in all isolations /%
1	H1	1	1.41	15	2.15	0	0	16	1.64
2	H3	17	23.94	411	58.88	21	10.24	449	46.10
3	H4	0	0	47	6.73	2	0.98	49	5.03
4	H6	8	11.27	120	17.19	159	77.56	287	29.47
5	H9	33	46.48	43	6.16	12	5.85	88	9.03
6	H11	0	0	1	0.14	0	0	1	0.10
7	Mixed infections	12	16.90	61	8.74	11	5.37	84	8.62
Total		71		698		205		974	

Figure 2-3-3 also shows the annual distributions of the isolated subtypes from 2016 to 2019. The trend of the annual distribution was similar to that of the four-year total distribution. In the four years, the H3 subtype had the highest percentage among LPAIV-positive samples each year, followed by the H6 subtype. The percentages of the H9 subtype were higher than those of H4 except in 2019. Twenty-four isolates were identified as the H4 subtype in 2019, accounting for 48.98% (24/49) of all H4 isolates. H3, H4, H6 and H9 were the main subtypes and could be identified in all seasons. Figure 2-3-4 reveals their isolation rates in four seasons. The highest isolation rates of H3 and H6 were both observed in the spring, and the H4 subtypes peaked in summer. Higher isolation rates of H9 were observed in winter and summer.

The HA and NA genes of the identified LPAIV isolates were sequenced and submitted to a BLAST search of the NCBI database. H1N2, H1N6, H3N2, H3N6, H3N8, H4N2, H4N3, H4N6, H4N8, H6N2, H6N6, H9N2 and H11N8 have been identified.

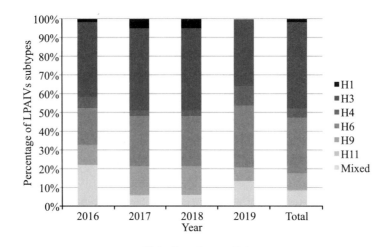

(Color figure in appendix.)

Figure 2-3-3   Percentage of the isolated HA subtypes from 2016 to 2019

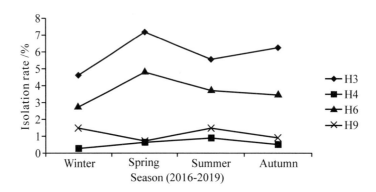

Figure 2-3-4   The isolation rates of the H3, H4, H6 and H9 subtypes in four seasons

The pattern of mixed infections

The percentage of mixed infections among the LPAIV-positive samples was 8.62% (84/974). Among the 84 samples with mixed infections, 12 were derived from chickens, 61 were derived from ducks and 11 were derived from geese (Table 2-3-2). Notably, the percentages of mixed infection cases were relatively high in 2016, decreased to relatively small percentages in 2017 and 2018, and then increased in 2019 (Figure 2-3-3). Among the chicken isolates, mixed infections accounted for 16.90% (12/71). The percentages of mixed infections of ducks and geese corresponded to 8.74% (61/698) and 5.37% (11/205), respectively (Table 2-3-2).

As shown in Table 2-3-3, 84 mixed infection samples contained 22 combinations. Coinfections with two HA subtypes, which were the most common subtypes, comprised 17 combinations and resulted in 78 cases, accounting for 92.86% of the mixed-infection cases (78/84). Coinfections with three HA subtypes comprising 4 combinations and resulting in 5 cases were observed in chickens (2 combinations, 3 cases) and ducks (2 combinations, 2 cases). Coinfection with four HA subtypes comprised only 1 combination in 1 case and was observed in ducks. Among the cases of mixed infection, H3+H4 coinfection was the most common and accounted for 33.33% (28/84) of all mixed infection cases and 27 cases were sourced from ducks.

Table 2-3-3   Mixed infections involving different HA subtypes in chickens, ducks and geese

Number	Types of mixed infections	Chickens	Ducks	Geese	Total
1	H2+H4	1	0	0	1
2	H2+H6	0	0	1	1
3	H3+H4	0	27	1	28
4	H3+H6	0	0	2	2
5	H3+H8	0	1	0	1
6	H3+H9	0	5	0	5
7	H3+H10	0	1	0	1
8	H3+H11	1	9	1	11
9	H4+H6	0	9	1	10
10	H4+H13	0	1	0	1
11	H6+H9	1	1	0	2
12	H6+H11	0	1	5	6
13	H6+H13	0	1	0	1
14	H9+H10	1	0	0	1
15	H9+H12	4	1	0	5
16	H9+H16	0	1	0	1
17	H11+H12	1	0	0	1
18	H1+H2+H10	0	1	0	1
19	H1+H3+H11	0	1	0	1
20	H3+H4+H9	1	0	0	1
21	H4+H8+H9	2	0	0	2
22	H3+H10+H11+H14	0	1	0	1
Total		12	61	11	84

Discussion

The term "highly pathogenicity avian influenza" (HPAI) generally refers to the strains that may induce an "intravenous pathogenicity index" (IVPI) greater than 1.2 or mortality rate over 75% in a defined chicken population during the specified interval of 10 days. Using this definition, all the HPAI strains isolated to date are of the H5 and H7 subtypes. However, viruses of these subtypes can also be of low pathogenicity. The World Organization for Animal Health (OIE) requires notification for all H5 and H7 subtypes, regardless of their pathogenicity, as they have the potential to mutate into HPAI viruses[7]. In this study, non-H5 and non-H7 AIVs are assumed to be LPAIVs. LPAIVs are a potential threat to humans, the prevalence of AIV infections should be monitored, and risks should be evaluated early. Prevention and control of LPAIVs should be continuously conducted in the long run. We analyzed the samples based on the data of three poultry species (chicken, duck, and goose), four years (2016, 2017, 2018 and 2019) and four seasons (winter, spring, summer and autumn) and recorded different subtypes of LPAIVs. We generated epidemiological data regarding the LPAIV isolation rates, the prevalent subtypes, and the percentages and distributions of subtypes and mixed infections. The increased isolation rates in 2017 and 2018 (Figure 2-3-1) were mainly correlated with the increased isolation rates of ducks in the two years. The main reason was that the isolation rate of H3 subtype AIV in ducks significantly increased in 2017 and 2018; in addition, 13 out of 16 isolates of H1 subtype AIV were in ducks in 2017. Duck, as waterfowl, is more likely to interact with wild bird species, increasing the chance of AIV transmission across the wild bird-poultry interface, may lead to increased isolation rates in the two years. In Guangxi, spring had the highest isolation (Figure 2-3-2 A), and warm, humid and rainy conditions in spring may be favorable for the survival, growth and transmission of AIV. Compared to our LPAIV survey from 2012 to 2015[26], mixed infections that decreased may be associated with expanding implementation of the "1110" strategy in live poultry markets. The 1110 strategy involves 1 daily cleaning, 1 weekly disinfection, 1 day of market closure every month and 0 live poultry stock overnight. Implementation of the strategy may decrease the isolation rates and mixed infection cases.

In the present study, the predominant subtypes in chickens, ducks and geese were H9, H3 and H6 (Table 2-3-2), respectively, in agreement with our previous survey results from 2012 to 2015[26]. Waterfowl are well known as a natural reservoir pool for AIV. The water resources of Guangxi are abundant and include many rivers, ponds and lakes etc., which provide good habitat for ducks, geese and other waterfowl. Poultry from LPMs is also derived from backyard farms, in which birds are typically raised as free-range scavengers with minimal to no biosecurity measures. These free-range birds may interact with wild bird species, sharing water, food, and habitat, increasing the chance of AIV transmission across the wild bird-poultry interface. In the present study, it is noteworthy that the proportion of H3 subtypes among the LPAIV samples all increased regardless of poultry species, which was especially obvious in ducks (Table 2-3-2). Domestic ducks serve as an ideal environment for the reassortment of H3 subtype influenza with other subtypes, which plays an important role in the ecology of AIV and may potentially be a threat to human health[31], and in our study coinfection with H3+H4 was the most frequently seen combination (Table 2-3-3). H3 subtype viruses have very diverse host, ranging from birds to various mammalian species, which may be easier to spread to other hosts by waterfowl and wild birds. Some H3N2 viruses of avian-origin were transmitted to dogs in Republic of Korea, causing acute respiratory disease[32]. H3 subtype AIV from LPMs in China may contribute viral internal genes to H5N8 HPAIV[33]. H3 subtype AIV has become an important issue in emerging zoonotic infections and threats to public safety[34, 35]. H6 subtype AIV can cause infection in mammals, for example, an H6N1 virus was isolated from a human with flu-like symptoms in Taiwan, China in 2013[17], and an H6N6 virus was also detected in pigs in Guangdong Province in China in 2010[36]. The increased isolation of H3 in ducks and H6 in geese in our study provides an early warning and suggest that the epidemiolohical surveillance and genetic evolution analysis of H3 and H6 subtypes AIV should be enhanced and to find further information to support the prevention and control of AIV.

Chickens may play a key role in the evolution of the H9 subtype[18]. The H9N2 subtype was first identified from diseased chickens in Guangdong Province, China, in 1994 and became widespread among chickens, causing great economic losses due to reduced egg production and high lethality associated with coinfections with other avian pathogens[37]. Inactivated H9N2 commercial vaccines have been used in chicken flocks since 1998 in China, and the vaccines initially prevented outbreaks and transmission. However, in recent years, genetic alterations in H9N2 virus strains have reduced the effectiveness of vaccines, as H9N2 strains circulate continuously in vaccinated chicken flocks[19, 38]. H9N2 is among the most common subtypes in chickens and can be easily isolated[37]. In our study, the H9 subtype accounted for nearly half of the chicken isolates (Table 2-3-2). Currently, H9 reassortants also comprise H5N2 and H5N6 subtypes[39, 40], in addition to the above mentioned H7N9 and H10N8 subtypes. The genes exchange or recombination of H9 strains with H5 and H7 etc. subtypes AIV may generate novel gene combinations resulting in emergence of HPAIV strains, which seriously threaten the healthy development of poultry industry and public health safety. The H9 reassortants cases continue to emerge, which may pose serious challenges to the prevention and control of AIV. The findings illustrated that the H9 vaccine needs to be updated.

Six subtypes with high diversity and complexity were isolated from ducks (Table 2-3-2). The duck is an important waterfowl species. Therefore, we collected more samples from ducks than from chickens and geese. The H4 subtype was mainly isolated from ducks in our study (Table 2-3-2). It has been reported that H4 combined with various NA subtypes circulates in LPMs in central, eastern and southern China, and H4 and other subtypes (e.g., H3) have undergone complex reassortment events in domestic ducks[4, 41-44]. Our results support the above observations, H4N2, H4N6 and H4N8 circulate, and mixed infections, including H4 are common in Guangxi (Table 2-3-3). The H4 subtype may have acquired the ability to infect, replicate and transmit in mammalian hosts[4, 45]. Our results and other reports suggest that further investigations of the mechanisms of H4 AIV mutation and reassortment are important to prepare for potential pandemics.

Determining the prevalent subtypes of LPAIVs, the distribution and prevalence of these subtypes and the patterns of mixed infections and how they differ among subtypes may provide useful insight and guidance for the prevention and control of LPAIVs. Further epidemiological studies should continue to identify possible risk factors and to better understand the extent of AIV infection as well as potential transmission routes of AIVs. AIVs may occasionally infect humans exposed to infected poultry. Multisectoral coordination at the human-animal-environment interface is critical for zoonotic disease control[46]. Increasing surveillance in human and poultry populations is crucial for predicting and preventing the risk of new outbreaks. Efforts to collect samples not only from chickens, ducks and geese but also from other minor poultry species and humans as well as from the environment should continue in order to help monitor where AIVs circulate. Extensive surveillance of AIV infection and prevalence in LPMs is indispensable for clarifying the epidemic situation and infection landscape.

In conclusion, our study presents a comprehensive analysis of LPAIV infections in Guangxi, southern China from 2016 to 2019. Our results derived from an extensive 4-year effort provide essential surveillance data on the prevalence of different AIV subtypes in chickens, ducks and geese in LPMs, suggesting the need to enhance the monitoring and analysis of the genetic evolution and reassortment of H1, H3, H4, H6 and H9 subtypes AIV, especially for ducks and geese. Meanwhile, biosecurity measures should be improved in farms and LPMs to reduce the transmission of AIV and the occurrence of mixed infection. This information may lay a foundation for reducing the threat of novel or enzootic AIV subtypes of animal and human infections and could provide useful epidemiological data for efforts to effectively prevent and control influenza virus.

References

[1] SU S, BI Y H, WONG G, et al. Epidemiology, evolution, and recent outbreaks of avian influenza virus in China. Journal of Virology, 2015, 89(17): 8671-8676.

[2] TAUBENBERGER J K, REID A H, LOURENS R M, et al. Characterization of the 1918 influenza virus polymerase genes. Nature, 2005, 437(7060): 889-893.

[3] NODA T, SAGARA H, YEN A, et al. Architecture of ribonucleoprotein complexes in influenza A virus particles. Nature, 2006, 439(7075): 490-492.

[4] LIANG L B, DENG G H, SHI J Z, et al. Genetics, receptor binding, replication, and mammalian transmission of H4 avian influenza viruses isolated from live poultry markets in China. Journal of Virology, 2016, 90(3): 1455-1469.

[5] TONG S X, LI Y, RIVAILLER P, et al. A distinct lineage of influenza A virus from bats. Proceedings of the National Academy of Sciences of the United States of America, 2012, 109(11): 4269-4274.

[6] TONG S X, ZHU X Y, LI Y, et al. New world bats harbor diverse influenza A viruses. PLOS Pathogens, 2013, 9(10): e1003657.

[7] CHATZIPRODROMIDOU I P, ARVANITIDOU M, GUITIAN J, et al. Global avian influenza outbreaks 2010-2016: a systematic review of their distribution, avian species and virus subtype. Systematic Reviews, 2018, 7(1): 17.

[8] ŚWIĘTOŃ E, TARASIUK K, ŚMIETANKA K. Low pathogenic avian influenza virus isolates with different levels of defective genome segments vary in pathogenicity and transmission efficiency. Veterinary Research, 2020, 51(1): 108.

[9] WEBSTER R G, BEAN W J, GORMAN O T, et al. Evolution and ecology of influenza A viruses. Microbiological Reviews, 1992, 56(1): 152-179.

[10] HULSE-POST D J, STURM-RAMIREZ K M, HUMBERD J, et al. Role of domestic ducks in the propagation and biological evolution of highly pathogenic H5N1 influenza viruses in Asia. Proceedings of the National Academy of Sciences of the United States of America, 2005, 102(30): 10682-10687.

[11] SWAYNE D E, SUAREZ D L. Highly pathogenic avian influenza. Revue Scientifique et Technique, 2000, 19(2): 463-482.

[12] BERGERVOET S A, GERMERAAD E A, ALDERS M, et al. Susceptibility of chickens to low pathogenic avian influenza (LPAI) viruses of wild bird- and poultry-associated subtypes. Viruses, 2019, 11(11): 1010.

[13] SHEN Y Z, LU H Z. Global concern regarding the fifth epidemic of human infection with avian influenza A (H7N9) virus in China. Bioscience Trends, 2017, 11(1): 120-121.

[14] YOON S W, WEBBY R J, WEBSTER R G. Evolution and ecology of influenza A viruses. Current Topics in Microbiology and Immunology, 2014, 385: 359-375.

[15] ZHANG Y, ZHANG Q Y, KONG H H, et al. H5N1 hybrid viruses bearing 2009/H1N1 virus genes transmit in guinea pigs by respiratory droplet. Science, 2013, 340(6139): 1459-1463.

[16] QI W, ZHOU X, SHI W, et al. Genesis of the novel human-infecting influenza A (H10N8) virus and potential genetic diversity of the virus in poultry, China. Euro Surveillance, 2014, 19(25): 20841.

[17] WEI S H, YANG J R, WU H S, et al. Human infection with avian influenza A H6N1 virus: an epidemiological analysis. The Lancet Respiratory Medicine, 2013, 1(10): 771-778.

[18] LI C, WANG S G, BING G X, et al. Genetic evolution of influenza H9N2 viruses isolated from various hosts in China from 1994 to 2013. Emerging Microbes Infections, 2017, 6(11): e106.

[19] PU J, WANG S G, YIN Y B, et al. Evolution of the H9N2 influenza genotype that facilitated the genesis of the novel H7N9 virus. Proceedings of the National Academy of Sciences of the United States of America, 2015, 112(2): 548-553.

[20] MARTIN V, ZHOU X Y, MARSHALL E, et al. Risk-based surveillance for avian influenza control along poultry market chains in south China: The value of social network analysis. Preventive Veterinary Medicine, 2011, 102(3): 196-205.

[21] OFFEDDU V, COWLING B J, MALIK PEIRIS J S. Interventions in live poultry markets for the control of avian influenza: a systematic review. One Health, 2016, 2: 55-64.

[22] CLAAS E C, OSTERHAUS A D, VAN BEEK R, et al. Human influenza A H5N1 virus related to a highly pathogenic avian influenza virus. Lancet, 1998, 351(9101): 472-477.

[23] PEIRIS M, YUEN K Y, LEUNG C W, et al. Human infection with influenza H9N2. Lancet, 1999, 354(9182): 916-917.

[24] VIRLOGEUX V, FENG L Z, TSANG T K, et al. Evaluation of animal-to-human and human-to-human transmission of influenza A (H7N9) virus in China, 2013-15. Scientific Reports, 2018, 8(1): 552.

[25] PENG Y, XIE Z X, LIU J B, et al. Epidemiological surveillance of low pathogenic avian influenza virus (LPAIV) from poultry in Guangxi Province, Southern China. PLOS ONE, 2013, 8(10): e77132.

[26] LUO S S, XIE Z X, XIE Z Q, et al. Surveillance of live poultry markets for low pathogenic avian influenza viruses in Guangxi Province, southern China, from 2012-2015. Scientific Reports, 2017, 7(1): 17577.

[27] CHEUNG P P, WATSON S J, CHOY K T, et al. Generation and characterization of influenza A viruses with altered

polymerase fidelity. Nature Communications, 2014, 5: 4794.

[28] TSUKAMOTO K, PANEI C J, SHISHIDO M, et al. SYBR green-based real-time reverse transcription-PCR for typing and subtyping of all hemagglutinin and neuraminidase genes of avian influenza viruses and comparison to standard serological subtyping tests. Journal of Clinical Microbiology, 2012, 50(1): 37-45.

[29] LUO S S, XIE Z X, HUANG J L, et al. Simultaneous differentiation of the N1 to N9 neuraminidase subtypes of avian influenza virus by a GeXP analyzer-based multiplex reverse transcription PCR assay. Frontiers in Microbiology, 2019, 10: 1271.

[30] HOFFMANN E, STECH J, GUAN Y, et al. Universal primer set for the full-length amplification of all influenza A viruses. Archives of Virology, 2001, 146(12): 2275-2289.

[31] DENG G H, TAN D, SHI J Z, et al. Complex reassortment of multiple subtypes of avian influenza viruses in domestic ducks at the Dongting Lake region of China. Journal of Virology, 2013, 87(17): 9452-9462.

[32] SONG D, KANG B, LEE C, et al. Transmission of avian influenza virus (H3N2) to dogs. Emerging Infectious Diseases, 2008, 14(5): 741-746.

[33] CUI H R, SHI Y, RUAN T, et al. Phylogenetic analysis and pathogenicity of H3 subtype avian influenza viruses isolated from live poultry markets in China. Scientific Reports, 2016, 6: 27360.

[34] ZOU S M, TANG J, ZHANG Y, et al. Molecular characterization of H3 subtype avian influenza viruses based on poultry-related environmental surveillance in China between 2014 and 2017. Virology, 2020, 542: 8-19.

[35] LUO S S, DENG X W, XIE Z X, et al. Production and identification of monoclonal antibodies and development of a sandwich ELISA for detection of the H3-subtype avian influenza virus antigen. AMB Express, 2020, 10(1): 49.

[36] ZHANG G H, KONG W L, QI W B, et al. Identification of an H6N6 swine influenza virus in southern China. Infection Genetics and Evolution, 2011, 11(5): 1174-1177.

[37] SUN Y, LIU J. H9N2 influenza virus in China: a cause of concern. Protein & Cell, 2015, 6(1): 18-25.

[38] SUN Y P, PU J, FAN L H, et al. Evaluation of the protective efficacy of a commercial vaccine against different antigenic groups of H9N2 influenza viruses in chickens. Veterinary Microbiology, 2012, 156(1-2): 193-199.

[39] ZHAO G, GU X B, LU X L, et al. Novel reassortant highly pathogenic H5N2 avian influenza viruses in poultry in China. PLOS ONE, 2012, 7(9): e46183.

[40] SHEN Y Y, KE C W, LI Q, et al. Novel reassortant avian influenza A (H5N6) viruses in humans, Guangdong, China, 2015. Emerging Infectious Diseases, 2016, 22(8): 1507-1509.

[41] SONG H, QI J X, XIAO H X, et al. Avian-to-human receptor-binding adaptation by influenza A virus hemagglutinin H4. Cell Reports, 2017, 20(5): 1201-1214.

[42] SHI Y, CUI H R, WANG J H, et al. Characterizations of H4 avian influenza viruses isolated from ducks in live poultry markets and farm in Shanghai. Scientific Reports, 2016, 6: 37843.

[43] WU H B, PENG X M, PENG X R, et al. Genetic characterization of natural reassortant H4 subtype avian influenza viruses isolated from domestic ducks in Zhejiang province in China from 2013 to 2014. Virus Genes, 2015, 51(3): 347-355.

[44] YUAN X Y, WANG Y L, YU K X, et al. Isolation and genetic characterization of avian influenza virus H4N6 from ducks in China. Archives of Virology, 2015, 160(1): 55-59.

[45] LI X Y, LIU B T, MA S J, et al. High frequency of reassortment after co-infection of chickens with the H4N6 and H9N2 influenza A viruses and the biological characteristics of the reassortants. Veterinary Microbiology, 2018, 222: 11-17.

[46] PETERS L, GREENE C, AZZIZ-BAUMGARTNER E, et al. Strategies for combating avian influenza in the Asia-Pacific. Western Pacific Surveillance and Response Journal, 2018, 9(5): 8-10.

Genome sequencing and phylogenetic analysis of three avian influenza H9N2 subtypes in Guangxi

Xie Zhixun, Dong Jianbao, Tang Xiaofei, Liu Jiabo, Pang Yaoshan, Deng Xianwen, Xie Zhiqin, Xie Liji, and Khan Mazhar I

Abstract

Three isolates of H9N2 avian influenza viruses (AIV) were isolated from chickens in Guangxi. Eight pairs of specific primers were designed and synthesized according to the sequences of H9N2 at GenBank. Phylogenetic analysis showed a high degree of homology between the Guangxi isolates and isolates from Guangdong and Jiangsu provinces, suggesting that the Guangxi isolates originated from the same source. However, the eight genes of the three isolates from Guangxi were not in the same sublineages in their respective phylogenetic trees, which suggests that they were products of natural reassortment between H9N2 avian influenza viruses from different sublineages. The 9 nucleotides ACAGAGATA which encode amino acids T, G, I were absent between nucleotide 205 and 214 in the open reading frame of the NA gene in the Guangxi isolates. AIV strains that infect human have, in their HA proteins, leucine at position 226. The analysis of deduced amino acid sequence of HA proteins showed that position 226 of these isolates contained glycine instead of leucine, suggesting that these three isolates differ from H9N2 AIV strains isolated from human infections.

Keywords

avian influenza virus (AIV), H9N2 subtype, complete genes, sequencing and analysis

Introduction

Avian influenza virus (AIV) belongs to the Orthomyxoviridae family of RNA viruses. It is an enveloped virus with a helical nucleocapsid and eight segments of single-stranded negative-sense RNA. The envelope contains the haemagglutinin (HA) and neuraminidase (NA) proteins, of which there are currently 16 HA (H1-H16) and 9 NA (N1-N9) subtypes[1, 15]. The H9N2 subtype that has spread worldwide in poultry can infect humans and is prevalent in China[2, 5, 6, 7, 11, 12]. The first outbreak of disease attributed to H9N2 in China happened in Guangdong province in 1994. Other strains of this subtype were subsequently isolated[3, 7, 9, 11].

Studies to characterize these strains were focused on the HA genes for one of the surface viral proteins that plays a crucial role in the pathogenicity[9]. Recent studies characterized the nonstructural[10] and neuraminidase genes[14] of H9N2 viruses in China. Others showed that some novel reassorted H9N2 influenza viruses may possess certain genes from H5 subtype viruses[4, 8, 11]. In this study, we have characterized three strains of H9N2 influenza viruses that were isolated from chicken farms in Guangxi. All eight full-length genes of these isolates were obtained individually by means of RT-PCR. Sequence analysis and phylogenetic study were conducted by comparing the eight genes of each isolate with sequences available in GenBank (H9N2 isolates from 1966-2005).

Materials and methods

Viruses

Three isolates of H9N2 subtype influenza viruses, A/Chicken/Guangxi/1/00 (C/GX/1/00), A/Chicken/Guangxi/14/00 (C/GX/14/00) and A/Chicken/Guangxi/17/00 (C/GX/17/00) were recovered from chickens in different areas of Guangxi during 2000 and 2001. Initial isolation was performed in 10-day-old specific-pathogen-free (SPF) embryonated chicken eggs. Identification of the viruses was determined by standard hemagglutination-inhibition and neuraminidase-inhibition tests[16]. Allantoic fluids were harvested from SPF egg-passaged viruses and used as stock viruses for further analysis.

RNA extraction and RT-PCR

Viral RNA was obtained from allantoic fluid by TRIzol extraction according to the method described by Xie et al.[17]. The concentrations of RNA were determined by spectrophotometry using the UV2501PC (Shimadzu Cooperation, Tokyo, Japan) and then stored at –20 ℃. Reverse transcription was done by using random primers with RNA PCR (AMV) V3.1 kit (TaKaRa Biotechnology, Dalian, China). Amplification of the eight full-length genes was carried out by PCR as described previously by Xie et al[17] using pairs of specific primers as described in Table 2-4-1. The reaction was carried out at 30 ℃ for 5 min followed by 60 min incubation at 37 ℃. PCR was performed in a reaction mixture of 50 μL containing master mix buffer with dNTPs, 50-time Advantage 2 polymerase (Clontech Mountain View, California, USA), 10 μM of each specific primer, and 200 ng cDNA template. The PCR condition for the amplification of PB2, PB1, and HA was 95 ℃ for 2 min denaturation, and 35 cycles of 95 ℃ for 30 s, 62 ℃ for 45 s (annealing), and 68 ℃ for 3 min, followed by 70 ℃ for 10 min final extension. The PCR condition for the amplification of NP, NA, M, and NS genes was the same as above, except that the annealing temperature was reduced to 60 ℃ for NP and NA genes and 58 ℃ for M and NS genes.

Table 2-4-1   The nucleotide sequences of the eight pairs of specific primers for H9N2 subtypes

Gene	Sequence of primers (5' - 3')	
PB2	F: AAAAGCAGGTCAATTATATTC	R: AAGGTCGTTTTTAAACTATTCA
PB1	F: AAAAGCAGGCAAACCATTTGA	R: TTTTCATGAAGGACAAGCTAA
PA	F: AGCAAAAGCAGGTACTGAT	R: AGTAGAAACAAGGTACTTTT
HA	F: AGCAAAAGCAGGGGAATTTCAC	R: AGTAGAAACAAGGGTGTTTTTGC
NP	F: GCAGGTAGATAATCACTCACTG	R: AGTAGAAACAAGGGTATTTTT
NA	F: AGCAAAAGCAGGAGTAAAAATG	R: CAAGGAGTTTTTTTTTAAAATTGC
M	F: AGCAAAAGCAGGTAGATGTTTAAAG	R: AGTAGAAACAAGGTAGTTTTTTAC
NS	F: AAAGCAAGGGTGACAAAGACAT	R: TAGAAACAAGGGTGTTTTTTATCA

Sequence analysis

All PCR products were subjected to electrophoresis in a 1% (w/v) agarose gel. DNA fragments of the expected length were extracted and purified with the DNA Glass-milk Rapid Purification Kit (MK001-2, BioDev, Beijing, China). The purified DNA fragments were cloned into pMD18-T easy vector according to the manufacturer's protocol (TaKaRa Biotechnology Co. Ltd., Dalian, China). Three clones of each of the 8 fragments were sequenced by the TaKaRa Biotechnology Co. Ltd., Dalian, China. DNA sequences of eight genes were compared with the GenBank database. DNA sequences of each cloned gene were repeated twice to confirm the similarity of sequence data. The nucleotide sequences obtained in this study are available in the GenBank database under accession numbers: DQ485205 to DQ485228.

Phylogenetic analysis

The DNA sequences of the ORF nucleotide sequences of NA gene of three Guangxi H9N2 subtypes were compared initially with the Megalign program in the DNASTAR package using the Clustal alignment algorithm (DNASTAR Inc., Madison, Wis, USA) against the twelve H9N2 virus sequences as listed in Table 2-4-3. Pairwise sequence alignments were also performed with the Clustal alignment algorithm in the Megalign program to determine sequence similarity and the phylogenetic relationship of the different H9N2 subtype viruses. The ancestral relationships within eight genes among the seventeen H9N2 subtypes were presented in a phylogenetic tree created by the Megalign program.

Results

Result of sequencing

The lengths of all genes were obtained for the three isolates (C/GX/1/00, C/GX/14/00, and C/GX/17/00): HA 1.7 kb, NA 1.4 kb, NP 1.5 kb, M 1.0 kb, NS 890 bp, PA 2.2 kb, PB1 2.3 kb, PB2 2.3 kb. BLAST software and Megalign programs were used to determine the sequence similarity of the eight genes from the three isolates, as shown in Table 2-4-2.

Table 2-4-2　Length of each gene of AIV strains and virus with the greatest homology to each gene

Gene segment	Strain	Open reading frame	Length of peptides	Virus with greatest homology	Homology /%
HA	C/GX/1/00	34~1 716	560	A/Chicken/Guangdong/4/00	99.3
	C/GX/14/00	34~1 716	560	A/Chicken/Guangdong/4/00	98.5
	C/GX/17/00	34~1 716	560	A/Chicken/Henan/26/00	99.2
NA	C/GX/1/00	20~1 429	469	A/Chicken/Guangdong/4/00	97.0
	C/GX/14/00	20~1 420	466	A/Chicken/Jiangsu/1/00	97.7
	C/GX/17/00	20~1 429	469	A/Chicken/Jiangsu/1/00	97.7
NP	C/GX/1/00	39~1 435	498	A/Chicken/Jiangsu/1/00	98.7
	C/GX/14/00	39~1 435	498	A/quail/Hong Kong/G1/97	98.0
	C/GX/17/00	39~1 435	498	A/quail/Hong Kong/G1/97	98.0
M	C/GX/1/00	26~784	252	A/Chicken/Guangdong/4/00	99.6
	C/GX/14/00	26~784	252	A/Chicken/Guangdong/4/00	98.4
	C/GX/17/00	26~784	252	A/Chicken/Guangdong/4/00	99.6
NS	C/GX/1/00	25~720	231	A/Chicken/Guangdong/47/01	98.9
	C/GX/14/00	66~719	217	A/Chicken/Guangdong/4/00	98.7
	C/GX/17/00	27~680	217	A/Chicken/Guangdong/4/00	98.5
PA	C/GX/1/00	25~2 175	716	A/Chicken/Guangdong/47/01	94.0
	C/GX/14/00	25~2 175	716	A/Duck/Hong Kong/Y439/97	93.8
	C/GX/17/00	25~2 175	716	A/Chicken/Guangdong/4/00	93.0
PB1	C/GX/1/00	22~2 298	758	A/Chicken/Guangdong/4/00	96.2
	C/GX/14/00	22~2 298	758	A/Chicken/Jiangsu/1/00	95.9
	C/GX/17/00	22~2 298	758	A/Chicken/Guangdong/4/00	94.8
PB2	C/GX/1/00	25~2 304	759	A/Chicken/Jilin/53/01	95.5
	C/GX/14/00	25~2 304	759	A/Hong Kong/1047/99	92.4
	C/GX/17/00	25~2 304	759	A/quail/Hong Kong/G1/97	97.9

Comparison of nucleotide and deduced amino acid sequences of eight genes

In this study, the nucleotide sequences obtained from these three H9N2 isolates were compared with those of all the available full-length genomes of H9N2 strains isolated during 1966-2005 and deposited in the GenBank database. Results indicate that nucleotide homology among these isolates with HA genes is between 82.6% to 99.3%; for those with NA genes the similarity is between 91.8% to 97.7%; for NP genes is between 88.0% to 98.7%; for M genes is between 92.7% to 99.6%; for NS genes is between 92.1% to 98.9%; for PA genes is between 86.1% to 94.1%; for PB1 genes is between 86.8% to 96.2%; and for PB2 genes is between 84.3% to 97.9%.

This sequence comparison also shows that 9 nucleotides between nucleotide positions 205 and 214, lying between position 187 to 195 open reading (ORF) in the NA genes, are absent in these three Guangxi isolates (Table 2-4-3) analyzed here. These 9 nucleotides encode for amino acids T, E, I. The deduced amino acid sequences of HA genes of C/GX/1/00, C/GX/14/00, and C/GX/17/00 at the cleavage site of HA contain a PARSSR/GL motif, which denotes the sequence found in low pathogenicity avian influenza (LPAI) viruses, as described previously[5, 9, 13].

Table 2-4-3   Comparison of ORF nucleotide sequences of NA gene

H9N2 subtypes	Site in the ORF
	↓190 nt ↓205 ↓214 ↓230 nt
A-chicken-Guangxi-1-00	AATAGAAAGGAACATA---------GTGCACTTGAATAGCA---------
A-chicken-Guangxi-14-00	AATGGAAAGGAACATA---------GTGTATCTGAATAGCA
A-chicken-Guangxi-17-00	AATGGAAAGGAACATA---------GTGTATCTGAATAGCA
A-chicken-Pakistan-2-99	AATAGAAAGGAACATAAACAGAGATAGTGCATTTGAATAGCA
A-chicken-Shanghai-F-98	AATAGAAAGGAACATA---------GTGCATTTGAATAGTA
A-duck-Hong Kong-Y280-97	AATAGAAAGGAACATA---------GTGCATTTGAATAGTA
A-duck-Hong Kong-Y439-97	AATAGAAAGGAACATAAACAGAGATAGTGCAT T TGAATAATA
A-Quail-Hong Kong-G1-97	AATAGAAAGGAACATAAACAGAGATAGTGCAT T TGAATAATA
A-chicken-Guangdong-47-01	AATAGAAAGGAACATAAACAGAGATAGTGCATTTGAATAGCA
A-chicken-ROK-S1-2003	AATAGAAAAGAACATAAACAAAAATAGTGTATTTGAATAATA
A-chicken-Shenzhen-9-97	AATAGAAAGGAACATAAACAGAGATAGTGCATTTGAATAATA
A-duck-Nanjing-2-97	AATAGAAAGGAGCATA---------GTGCATTTGAATAGTA
A-chicken-Guangdong-4-00	AATAGAAAAGAACATAAACAGAGATAGTGCATTTGAATAGTA
A-chicken-Fujian-25-00	AATAGAAAGGAACATAAACAGAGATAGTGCATTTGAATAA---------
A-chicken-Jiangsu-1-00	AATAGAAAGGAACATAAACAGAGATAGTGCAT T TGAATAGCA

Note: N, nucleotide; ROK, Republic of korea.

Phylogenetic analysis of eight genes

The phylogenetic relationship between each of the eight genes from three isolates, C/GX/1/00, C/GX/14/00, and C/GX/17/00, was analyzed based on their nucleotide sequence (Figure 2-4-1 to Figure 2-4-8). The analysis indicates that there are three sublineages in the Eurasian lineage as described previously[4, 10, 17]. The NP and PB1 genes of C/GX/14/00 and NP, PB1, PB2 genes of C/GX/17/00 were incorporated into the sublineage represented by A/Quail/Hong Kong/G1/97 (Figure 2-4-4, Figure 2-4-7, and Figure 2-4-8). The HA, M, NP, PB1, PB2 genes of C/GX/1/00, the HA and M genes of C/GX/14/00, and HA and M genes of C/GX/17/00 were incorporated into the sublineage represented by A/duck/Hong Kong/Y280/97 (Figure 2-4-1 to Figure 2-4-3). No gene of the three isolates was incorporated into the sublineage represented by A/duck/Hong Kong/Y439/97. The PA, NA, NS genes of C/GX/1/00, NA, PA, NS, PB2 genes of C/GX/14/00, and the NA, NS, PA genes of C/GX/17/00 were not incorporated into any of sublineage of these three isolates (Figure 2-4-4, Figure 2-4-5, Figure 2-4-6, and Figure 2-4-8).

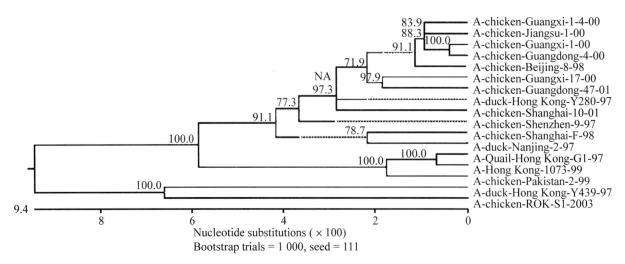

Figure 2-4-1   Phylogenetic tree of HA gene of AIV

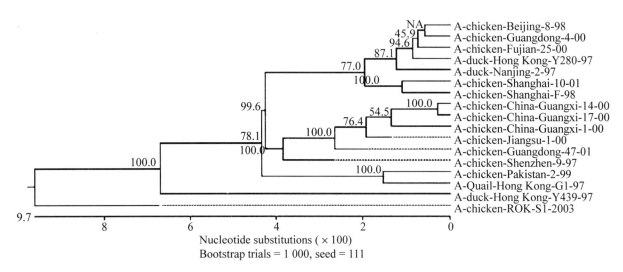

Figure 2-4-2   Phylogenetic tree of NA gene of AIV

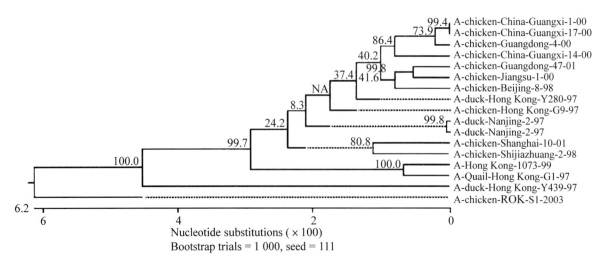

Figure 2-4-3   Phylogenetic tree of M gene of AIV

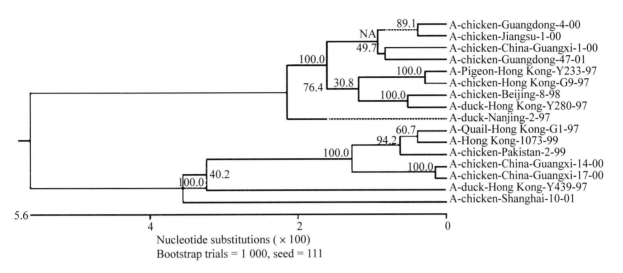

Figure 2-4-4   Phylogenetic tree of NP gene of AIV

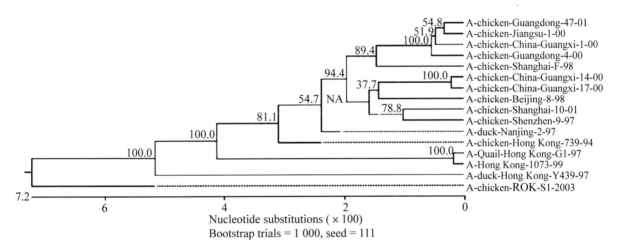

Figure 2-4-5   Phylogenetic tree of NS gene of AIV

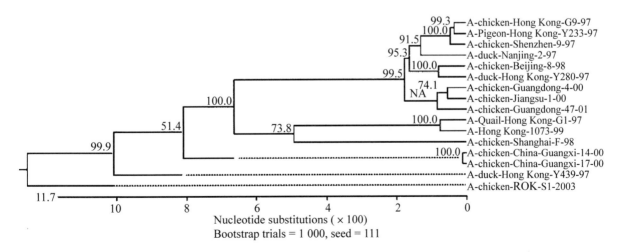

Figure 2-4-6   Phylogenetic tree of PA gene of AIV

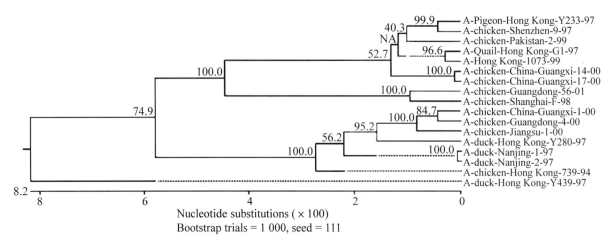

Figure 2-4-7   Phylogenetic tree of PBl gene of AIV

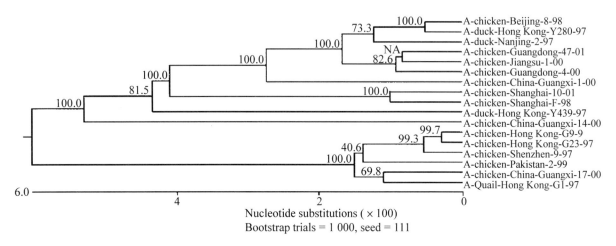

Figure 2-4-8   Phylogenetic tree of PB2 gene of AIV

Discussion

AIV has a segmented negative-strand genome that includes HA, NA, M, NS, NP, PA, PB1, PB2 genes. The results of sequencing and comparisons with the other H9N2 virus sequences showed that we have obtained each of eight genes from three Chinese isolates, C/GX/1/00, C/GX/14/00, and C/GX/17/00 successfully.

Previous studies have defined two distinct lineages of H9N2 influenza viruses: North American and Eurasian lineages. The Eurasian lineage consists of at least three sublineages[4, 8, 11]. Phylogenetic analysis in our study showed similar patterns (Figure 2-4-1 to Figure 2-4-8). Viral genomes of H9N2 viruses analyzed in previous studies had shown a common source of origin from southern China[3, 7, 9]. Sequence comparison and phylogenetic analysis illustrated that NP and PB1 genes of C/GX/14/00, and NP, PB1, and PB2 genes of C/GX/17/00 were incorporated into the sublineage represented by A/Quail/Hong Kong/G1/97. HA, M, NP, PB1, and PB2 genes of C/GX/1/00, HA and M genes of C/GX/14/00, and HA and M genes of C/GX/17/00 were incorporated into the sublineage represented by A/duck/Hong Kong/Y280/97. None of the genes from these three isolates was incorporated into the sublineage represented by A/duck/Hong Kong/Y439/97. PA, NA and NS genes of C/GX/1/00, NA, PA, NS and PB2 genes of C/GX/14/00, and the NA, NS and PA genes of C/GX/17/00 were not incorporated into any of the sublineages. So, it is possible that AIV H9N2 subtype strains C/GX/1/00, C/GX/14/00, and C/GX/17/00 were products of natural reassortment of avian influenza viruses, suggesting that there may be a specific gene pool in China.

Interestingly, gene sequence comparison indicates that there is a 9-nucleotide deletion between nucleotides 205 and 214 (187 position to 195 position of ORF) in the NA genes of the three isolates C/GX/1/00, C/GX/14/00, and C/GX/17/00. These 9 nucleotides encode for amino acids T, E, and I. This deletion did not change the ORF of the NA gene. Noteworthy, about 80% of China isolates carry a similar deletion. Further studies are needed to determine if the lack of 9 nucleotides in NA affects the function of the NA protein.

The HA protein determines the infectious host range of AIV strains. AIV strains infecting humans carry leucine at position 226 of the HA gene[4, 8, 11]. The analysis of deduced amino acid sequence of HA proteins of C/GX/1/00, C/GX/14/00, and C/GX/17/00 isolates showed a glycine at position 226, therefore these three isolates do not belong to H9N2 AIV strains isolated from human infections.

References

[1] ALEXANDER D J. Avian influenza: recent developments. Veterinary Bulletin, 1982, 52(6): 341-359

[2] BUTT K M, SMITH G J, CHEN H L, et al. Human infection with an avian H9N2 influenza A virus in Hong Kong in 2003. Journal of Clinical Microbiology, 2005, 43(11): 5760-5767.

[3] CHOI Y K, OZAKI H, WEBBY R J, et al. Continuing evolution of H9N2 influenza viruses in southeastern China. Journal of Virology, 2004, 78(16): 8609-8614.

[4] GUAN Y, SHORTRIDGE K F, KRAUSS S, et al. H9N2 influenza viruses possessing H5N1-like internal genomes continue to circulate in poultry in southeastern China. Journal of Virology, 2000, 74(20): 9372-9380.

[5] GUO Y J, KRAUSS S, SENNE D A, et al. Characterization of the pathogenicity of members of the newly established H9N2 influenza virus lineages in Asia. Virology, 2000, 267(2): 279-288.

[6] KWON H J, CHO S H, KIM M C, et al. Molecular epizootiology of recurrent low pathogenic avian influenza by H9N2 subtype virus in Korea. Avian Pathology, 2006, 35(4): 309-315.

[7] LI K S, XU K M, PEIRIS J S, et al. Characterization of H9 subtype influenza viruses from the ducks of southern China: a candidate for the next influenza pandemic in humans?. Journal of Virology, 2003, 77(12): 6988-6994.

[8] LIN Y P, SHAW M, GREGORY V, et al. Avian-to-human transmission of H9N2 subtype influenza A viruses: relationship between H9N2 and H5N1 human isolates. Proceedings of the National Academy of Sciences of the United States of America, 2000, 97(17): 9654-9658.

[9] LIU H Q, LIU X F, CHENG J, et al. Phylogenetic analysis of the hemagglutinin genes of twenty-six avian influenza viruses of subtype H9N2 isolated from chickens in China during 1996-2001. Avian Disease, 2003, 47(1): 116-127.

[10] LIU J H, OKAZAKI K, SHI W M, et al. Phylogenetic analysis of neuraminidase gene of H9N2 influenza viruses prevalent in chickens in China during 1995-2002. Virus Genes, 2003, 27(2): 197-202.

[11] MATROSOVICH M N, KRAUSS S, WEBSTER R G. H9N2 influenza A viruses from poultry in Asia have human virus-like receptor specificity. Virology, 2001, 281(2): 156-162.

[12] PEIRIS M, YUEN K Y, LEUNG C W, et al. Human infection with influenza H9N2. Lancet, 1999, 354(9182): 916-917.

[13] STEINHAUER D A. Role of hemagglutinin cleavage for the pathogenicity of influenza virus. Virology, 1999, 258(1): 1-20.

[14] WANG S, SHI W M, MWEENE A, et al. Genetic analysis of the nonstructural (NS) genes of H9N2 chicken influenza viruses isolated in China during 1998-2002. Virus Genes, 2005, 31(3): 329-335.

[15] WEBSTER R G, BEAN W J, GORMAN O T, et al. Evolution and ecology of influenza A viruses. Microbiological Reviews, 1992, 56(1): 152-179.

[16] WORLD HEALTH ORGANIZATION. A revision of the system of nomenclature for influenza viruses: a WHO memorandum. Bulletin of the World Health Organization, 1980, 58(4): 585-591.

[17] XIE Z X, PANG Y S, LIU J B, et al. A multiplex RT-PCR for detection of type A influenza virus and differentiation of avian H5, H7, and H9 hemagglutinin subtypes. Molecular and Cellular Probes, 2006, 20(3): 245-249.

Epidemiological surveillance of parvoviruses in commercial chicken and turkey farms in Guangxi, southern China, during 2014-2019

Zhang Yanfang, Feng Bin, Xie Zhixun, Deng Xianwen, Zhang Minxiu, Xie Zhiqin, Xie Liji, Fan Qing, Luo Sisi, Zeng Tingting, Huang Jiaoling, and Wang Sheng

Abstract

A previously unidentified chicken parvovirus (ChPV) and turkey parvovirus (TuPV) strain, associated with runting-stunting syndrome (RSS) and poultry enteritis and mortality syndrome (PEMS) in turkeys, is now prevalent among chickens in China. In this study, a large-scale surveillance of parvoviruses in chickens and turkeys using conserved PCR assays was performed. We assessed the prevalence of ChPV/TuPV in commercial chicken and turkey farms in China between 2014 and 2019. Parvoviruses were prevalent in 51.73% (1 795/3 470) of commercial chicken and turkey farms in Guangxi, China. The highest frequency of ChPV positive samples tested by PCR occurred in chickens that were broiler chickens 64.18% (1 041/1 622) compared with breeder chickens 38.75% (572/1 476) and layer hens 38.89% (112/288), and TuPV was detected in 83.33% (70/84). Native and exotic chicken species were both prevalent in commercial farms in southern China, and exotic broiler chickens had a higher positive rate with 88.10% (148/168), while native chickens were 50.00% (1 465/2 930). The environmental samples from poultry houses tested positive for ChPV/TuPV were 47.05% (415/874). Samples from open house flocks had higher prevalence rates of ChPV than those of closed house flocks (Table 2-5-5), among which those from the open house showed 84.16% (85/101) positivity, those from litter showed 62.86% (44/70) positivity, and those from drinking water showed 50.00% (56/112) positivity, whereas those from the closed house litter were 53.57% (60/112), those from swabs were 50.18% (138/275), and those from drinking water were 15.69% (32/204). Samples collected during spring were more frequently ChPV/TuPV positive than those collected during other seasons. This study is the first report regarding the epidemiological surveillance of ChPV/TuPV in chicken/turkey flocks in Guangxi, China. Our results suggest that ChPV/TuPV are widely distributed in commercial fowl in Guangxi. These findings highlight the need for further epidemiological and genetic research on ChPV/TuPV in this area.

Keywords

parvovirus, chicken parvovirus, turkey parvovirus, epidemiological surveillance, southern China

Introduction

Vertebrates, both animals and humans can be infected with small, non-enveloped parvoviruses. The genomes of these vertebrate parvoviruses are~5kb in size, and are classified in the subfamily Parvoviridae, including the genus *Parvovirus*[1]. These viruses have been linked to gastrointestinal diseases in human and other mammals[2-4]. Avian parvovirus is one of the most important pathogens causing intestinal diseases in poultry[5]. Chicken parvovirus (ChPV) was detected by electron microscopy for the first time[6, 7]. Since then, ChPV have been identified as the cause of runting and stunting syndrome (RSS) in chickens, which is characterized by significant growth retardation with poor feather development and bone disease[7]. Turkey parvovirus (TuPV) has been reported by Trampel et al.[8], presenting as poult enteritis and mortality syndrome (PEMS) in turkeys, and is considered responsible for the incidence of intestinal diseases and increase in the mortality rate of sick birds. Also, it has been detected in Derzsy's disease in goslings and Muscovy ducks[5, 6, 8, 9].

A non-structural gene (NS) and a structural viral protein (VP) gene[10] are two major genes of parvoviruses. The VP gene is located at the 3' end; the NS gene is at the 5' end, which encodes a small number of replication proteins and appears to be a highly conserved region[10, 11]. ChPV/TuPV genome sequence analysis showed strong similarity between the two, although they are less closely phylogenetically related to geese parvoviruses and Muscovy ducks parvoviruses[12, 13]. Attempts to isolate ChPV/TuPV in tissue cultures or embryonated eggs have remained unsuccessful, except for in a study conducted in Brazil[14].

Economic losses due to increased RSS and PEMS have become a continual worldwide problem that influences the development of the poultry industry. RSS and PEMS are characterized by diarrhea, anorexia, malabsorption, stunting, and poor feed conversion, which lead to immunosuppression[3]. Studies have showed that many viruses have been associated with RSS and PEMS, such as reovirus, coronavirus, astrovirus, rotavirus, and parvoviruses[3, 15-17]. However, none of these viruses has been proved to be the only cause of RSS and PEMS. In addition to a lack of a clear understanding of the cause, vaccines have not been developed for these syndromes. These viruses have been detected and isolated in healthy and diseased birds, which indicates that interaction occurs between the virus and unidentified additional agents[18].

China is mainly based on commercial large-scale chicken farms, and products from commercial poultry farm are an important source of protein in the Chinese population, and ChPV has been a restrictive factor in commercial poultry farms. ChPV shedders among healthy flocks could be one of the main causes of the epidemiological factor in this disease. ChPV/TuPV is emerging and re-emerging worldwide[14, 19-23]. However, research studies on molecular detection and epidemiologic investigations of ChPV/TuPV in China are rarely conducted. The infection statuses of the ChPV/TuPV strains in Guangxi are unknown; thus, we conducted this study to investigate the epidemiology of ChPV/TuPV in Guangxi poultry flocks. Our preliminary findings indicate that newly emerged ChPV/TuPV variants can be detected in commercial poultry in Guangxi, China.

This study could contribute to the design and development of effective disease prevention and control strategies to reduce economic losses due to emerging viruses.

Materials and methods

Sample collection for surveillance

Epidemiological surveillance of ChPV/TuPV in poultry was conducted at 80 randomly selected commercial chicken and turkey farms in Guangxi from October 2014 to November 2019 (Table 2-5-2). Each farm was visited once, and 12 to 120 birds were sampled in each poultry house; 10~30 individual cloacal swabs were collected from each flock. Cloacal samples were collected from 227 commercial chicken (genus *Gallus*) flocks and 6 commercial turkey flocks.

Exotic chickens were imported grandparent stock of specific strains of western chicken breeds, and native chickens were primary breeder stock of miscellaneous breeds that have been kept for several generations in China. Exotic and Chinese native chickens, which included broilers, layers and breeders, were reared intensively at commercial farms. Samples were collected from broiler chickens 1~21 weeks old, breeders 1~54 weeks old and layer hens 3~60 weeks old.

Biological samples were collected from the cloacae of both healthy and diseased chickens and turkeys using cotton swabs. The environmental samples were collected at different points in each poultry house, including litter samples, water samples, and swab samples collected from the walls, floors, feed pads, and drinkers (Table 2-5-2).

According to the World Organization for Animal Health (OIE) protocol, all the samples including cloacal swabs and environmental swabs were collected and stored separately in 1.5 mL of storage medium transported in a 2.0 mL Eppendorf tube (EP) on ice until processing. The storage medium contained 2 000 U/mL of penicillin and 2 mg/mL of streptomyin in sterile PBS (phosphate-buffered saline).

DNA extraction

The cloacal swabs and environmental suspensions were homogenized and centrifuged at 6 000 g for 5 min. In accordance with the manufacturer's instructions for commercial kits (TransGen, Beijing, China), 200 μL of each supernatant was used for DNA extraction.

PCR detection

(1) PCR using a set of specific primers (NS561F and NS561R) targeting the ChPV/TuPV NS gene and amplifying a 561 bp[16] fragment, and (2) nested PCR (nPCR) using 2 sets of specific primers (VP1/VP2 for the first round and VP3/VP4 for the second round) targeting the ChPV/TuPV VP gene and amplifying a 249 bp[24] fragment were carried out to determine whether the samples contained to ChPV/TuPV. The PCR and nPCR primers information is shown in Table 2-5-1.

The PCR assays were conducted as described in published procedures[16]. The nPCRs targeting the VP1/VP2 regions were prepared in the same way, except the annealing temperature reached 56/64 ℃ separately[16, 24]. The PCR products were visualized by 1.2% agarose gel electrophoresis.

Table 2-5-1　Primer information

Primer name	Primer sequence (5'-3')	Product size/bp	Annealing temperature /℃
NS561	F: TTCTAATAACGATATCACTCAAGTTTC	561	55
	R: TTTGCGCTTGCGGTGAAGTCTGGCTCG		
VP1	TGGAATTGTGATACTATATGGG	373	56
VP2	TCYTGATCTGCAAATATTTG		
VP3	CATTGTGTCTGTCTWATGCGTGAC	249	64
VP4	GTTTTCTGGATGACTTGCA		

Results

PCR detection for commercial chicken and turkey flocks

Among the 3 470 tested animals, 283 chickens and 12 turkeys were culled because of exhibiting diarrhea, poor weight gain, malabsorption syndrome, and mortality. Between 2014 and 2019, we collected 3 470 swabs from 233 flocks in 80 commercial poultry farms, and 51.73% (1 795/3 470) were PCR-positive for ChPV/TuPV, including 50.95% of commercial chicken farms and 83.33% of commercial turkey farms (Figure 2-5-1, Table 2-5-2). Compared to in turkeys, natural parvovirus infection was more frequently detected in chickens. However, TuPV was more prevalent than ChPV in the tested flocks. PCR assays showed that there were negative results in one of six examined turkey flocks. These data were similar to the reported 77%～78%

infection rates in the United States commercial chicken and turkey flocks in a survey conducted between 2003 and 2008[16]. The presence of ChPV/TuPV prevalence in Hungarian and Croatian commercial poultry flocks was also reported, but the infection rates in those countries were unknown[19, 20].

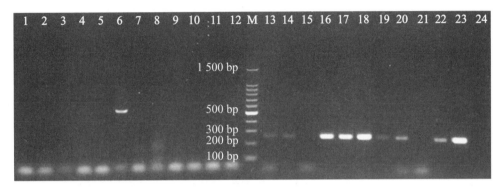

M: 100-bp DNA ladder marker; 1-12 and 13-24: 12 clinical sample of No. 23 farm used PCR and nested PCR for ChPV/TuPV detection, respectively; 6: ChPV/TuPV NS 561 bp product; 13, 14, 16-23: ChPV/TuPV VP 239 bp product.

Figure 2-5-1   The results by some clinical samples used PCR and nested PCR for ChPV/TuPV detection

Table 2-5-2   ChPV/TuPV detection and information on clinical samples

No. farms	Type	No. flocks	No. of the swabs	ChPV/TuPV PCR/%		Environment samples/%		No. of RSS-like/%
				NS1 (PCR)	VP1/VP2 (nested PCR)	Close house	Open house	
2014-1	A	4	60	14 (23.33)	30 (50.00)			
2014-2	B	2	60	6 (10.00)	23 (38.33)			
2014-3	B	7	84	28 (33.33)	40 (47.62)			4/5 (80.00)
	C	3	36	2 (5.56)	2 (5.56)			
2014-4	B	6	48	5 (10.42)	21 (43.75)	10/12 (83.33)		
2014-5	B	3	48	35 (72.92)	47 (97.92)	8/12 (66.67)		
2014-6	C	7	84	59 (70.24)	74 (88.10)	0/2 (0.00)	6/7 (85.71)	18/20 (90.00)
2014-7	B	6	72	32 (53.33)	39 (54.17)	10/15 (66.67)	1/3 (33.33)	
2015-8	B	5	60	24 (40.00)	41 (68.33)	6/12 (50.00)		
2015-9	C	5	60	33 (55.00)	40 (66.67)	2/4 (50.00)	5/8 (62.50)	10/12 (83.33)
2015-10	C	6	90	50 (55.56)	80 (88.89)	5/8 (62.50)	10/16 (62.50)	14/16 (87.5)
2015-11	D	1	12	2 (16.67)	0 (0.00)	0/2 (0.00)	0/4 (0.00)	
2015-12	D	2	24	22 (91.67)	24 (100)	2/4 (50.00)	8/8 (100.00)	11/12 (91.67)
2015-13	C	4	48	4 (8.33)	26 (54.17)	2/4 (50.00)	6/8 (75.00)	8/8 (100.00)
2015-14	B	4	48	24 (50.00)	32 (66.67)	4/12 (33.33)		
2015-15	D	3	48	34 (70.83)	46 (95.83)	3/4 (75.00)	5/8 (62.50)	
2015-16	B	3	36	0 (0.00)	9 (25.00)	2/10 (20.00)		
2015-17	C	2	48	44 (91.67)	46 (95.83)	0/2 (0.00)	5/8 (62.50)	10/10 (100.00)
2015-18	B	4	48	12 (25.00)	30 (62.50)	12/12 (100)		
2015-19	B	4	48	21 (43.75)	25 (52.08)	5/12 (41.67)		
2015-20	E	4	48	32 (66.67)	41 (85.42)	2/4 (50.00)	6/8 (75.00)	11/12 (91.67)
2015-21	E	2	24	3 (12.50)	12 (50.00)	2/12 (16.67)		
2015-22	E	4	48	44 (91.67)	47 (97.92)	1/2 (50.00)	6/7 (85.71)	9/10 (90.00)

continued

No. farms	Type	No. flocks	No. of the swabs	ChPV/TuPV PCR/%		Environment samples/%		No. of RSS-like/%
				NS1 (PCR)	VP1/VP2 (nested PCR)	Close house	Open house	
2015-23	A	5	60	1 (1.67)	10 (16.67)	12/15 (80.00)	4/5 (80.00)	
2015-24	A	4	48	2 (4.17)	16 (33.33)	0/12 (0.00)		
2015-25	B	2	20	7 (35.00)	7 (35.00)	10/30 (33.33)	1/4 (25.00)	
2015-26	E	4	48	40 (83.33)	48 (100)	2/4 (50.00)	3/8 (37.50)	11/12 (91.67)
2015-27	C	2	24	4 (16.67)	11 (45.83)			4/4 (100.00)
2016-28	C	3	24	8 (33.33)	16 (66.67)			3/4 (75.00)
2016-29	C	3	24	4 (16.67)	14 (58.33)			4/4 (100.00)
2016-30	C	4	48	23 (47.92)	33 (68.75)			10/12 (83.33)
2017-31	B	4	48	24 (50.00)	22 (45.83)	4/12 (33.33)		
2017-32	C	4	40	9 (22.50)	22 (55.00)	2/2 (100.00)	6/8 (75.00)	9/10 (90.00)
2017-33	C	2	20	4 (20.00)	10 (50.00)			2/2 (100.00)
2017-34	C	2	20	5 (25.00)	12 (60.00)			5/5 (100.00)
2017-35	B	4	48	0 (0.00)	4 (8.33)	2/12 (16.67)		
2017-36	C	4	48	13 (27.08)	40 (83.33)	2/4 (50.00)	8/8 (100.00)	10/12 (83.33)
2017-37	B	4	48	0 (0.00)	1 (2.08)	1/12 (8.33)		
2017-38	C	4	48	3 (6.25)	26 (54.17)	2/4 (25.00)	7/8 (100.00)	8/8 (100.00)
2017-39	C	4	48	28 (58.33)	39 (81.25)	1/4 (50.00)	6/8 (75.00)	10/10 (100.00)
2017-40	B	4	48	13 (27.08)	37 (77.08)	9/12 (75.00)		
2017-41	B	4	48	11 (22.92)	17 (35.42)	5/12 (41.67)		
2018-42	B	3	60	21 (35.00)	57 (95.00)	11/12 (91.67)		
2018-43	B	3	60	0 (0.00)	6 (10.00)			
2018-44	C	1	20	20 (100.00)	20 (100.00)	4/6 (66.67)	6/6 (100.00)	3/3 (100.00)
2018-45	C	1	20	20 (100.00)	20 (100.00)	3/3 (100.00)	5/5 (100.00)	3/4 (75.00)
2018-46	C	2	40	10 (25.00)	12 (30.00)	0/6 (0.00)	3/10 (30.00)	4/6 (66.67)
2018-47	C	2	40	17 (42.50)	19 (47.50)	1/8 (12.50)	3/8 (37.50)	5/6 (83.33)
2018-48	C	6	120	33 (27.50)	66 (55.00)	3/10 (30.00)	9/12 (75.00)	6/12 (50.00)
2018-49	C	1	20	0 (0.00)	3 (15.00)	0/6 (0.00)	1/6 (16.67)	
2018-50	C	1	20	8 (40.00)	8 (40.00)	0/6 (0.00)	2/6 (33.33)	
2018-51	B	2	40	0 (0.00)	1 (2.50)	1/12 (8.33)		
2018-52	B	2	40	39 (97.50)	39 (97.50)	8/12 (66.67)		
2018-53	B	1	20	9 (45.00)	9 (45.00)			
2018-54	B	1	20	0 (0.00)	0 (0.00)	0/12 (0.00)		
2018-55	C	2	40	9 (22.50)	9 (22.50)	0/6 (0.00)	1/6 (16.67)	
2018-56	C	1	20	20 (100.00)	20 (100.00)	4/4 (100.00)	4/4 (100.00)	6/6 (100.00)
2018-57	C	1	20	14 (70.00)	16 (80.00)	1/2 (50.00)	1/2 (50.00)	4/4 (100.00)
2018-58	C	1	20	3 (15.00)	3 (15.00)			
2018-59	C	1	20	20 (100.00)	20 (100.00)	6/6 (100.00)	6/6 (100.00)	6/6 (100.00)

continued

| No. farms | Type | No. flocks | No. of the swabs | ChPV/TuPV PCR/% | | Environment samples/% | | No. of RSS-like/% |
				NS1 (PCR)	VP1/VP2 (nested PCR)	Close house	Open house	
2018-60	A	1	20	10 (50.00)	10 (50.00)	5/12 (41.67)		
2018-61	A	2	40	15 (37.50)	17 (42.50)	10/20 (50.00)		
2018-62	A	2	40	12 (30.00)	13 (32.50)	3/12 (25.00)		
2018-63	A	1	20	16 (80.00)	16 (80.00)	8/12 (80.00)		
2019-64	C	3	60	32 (53.33)	37 (61.67)	2/16 (12.50)	10/20 (50.00)	4/8 (50.00)
2019-65	B	6	120	2 (1.67)	3 (2.50)	2/12 (16.67)		
2019-66	C	1	20	12 (60.00)	12 (60.00)	1/4 (25.00)	2/4 (50.00)	0/3 (0.00)
2019-67	C	1	20	6 (30.00)	7 (35.00)	0/4 (0.00)	3/4 (75.00)	
2019-68	C	1	20	8 (40.00)	8 (40.00)	1/6 (16.67)	4/6 (66.67)	1/2 (50.00)
2019-69	C	1	20	9 (45.00)	10 (50.00)			1/3 (33.33)
2019-70	C	2	40	30 (75.00)	30 (75.00)	4/6 (66.67)	4/6 (66.67)	3/6 (50.00)
2019-71	C	3	68	45 (66.18)	43 (63.24)	5/12 (41.67)	12/14 (85.71)	7/8 (87.50)
2019-72	B	2	40	0 (0.00)	2 (5.00)	0/12 (0.00)		
2019-73	B	2	40	0 (0.00)	0 (0.00)			
2019-74	B	2	40	0 (0.00)	1 (2.50)			
2019-75	B	2	40	10 (25.00)	11 (27.50)	3/12 (25.00)		
2019-76	B	2	40	0 (0.00)	5 (12.50)	0/12 (0.00)		
2019-77	B	2	40	28 (70.00)	28 (70.00)	8/12 (66.67)		
2019-78	B	2	40	4 (10.00)	3 (7.5)	2/8 (25.00)		
2019-79	C	2	40	21 (52.20)	24 (60.00)	4/12 (33.33)	10/12 (83.33)	5/6 (83.33)
2019-80	C	4	80	18 (22.50)	27 (33.75)	2/16 (12.50)	6/20 (30.00)	10/12 (83.33)
Total 80		233	3 470	1 259 (36.28)	1 795 (51.73)	230/591 (38.92)	185/291 (63.57)	239/295 (81.02)

Note: A, Layer of Chicken farm; B, Breeder of Chicken farm; C, Broiler of Chicken farm; D, Broiler of Turkey; E, Exotic Broiler Chicken farm; Exotic chickens = A+E; Native chickens = B+C.

Epidemiological surveillance of ChPV/TuPV in different commercial chicken and turkey flocks

Of the 283 cloacal swabs collected from the 69 RSS-like flocks, 80.57% (228/283) were PCR-positive, while the 164 healthy chicken flocks had 49.17% (1 567/3 187) prevalence; of the 12 cloacal swabs collected from 1 PEMS-like flocks, 91.67% (11/12) were PCR-positive compared with the 5 healthy turkey flocks, which showed 81.94% (59/72) prevalence. Clinical samples from 227 chicken flocks (50.95%) aged 1~60 weeks and 6 turkey flocks (83.33%) aged 3~47 weeks were collected in 46 different counties in Guangxi. The highest frequency of ChPV positive samples tested by PCR occurred in chickens that were broiler chickens 64.18% (1 041/1 622) compared with breeder chickens 38.75% (572/1 476) and layer hens 38.89% (112/288) (Table 2-5-3). The prevalence of ChPV was higher in broiler chicks aged 1~7 weeks (74.46%) than in birds in the 8~20 weeks (53.56%) and over 21 weeks (50.00%) age groups; in breeders aged 1~7 weeks (74.12%) than in birds in the 8~20 weeks (39.90%) and over 21 weeks (17.28%) age groups; and in layer hens aged

3～28 weeks (46.79%) than in birds in the 29～59 weeks (32.50%) and 60～72 weeks (0.00%) age groups; in turkey, the prevalence of TuPV was higher in the group aged 6～40 weeks (100.00%) than in birds in the 0～5 weeks (91.67%) and over 40 weeks (0.00%) age groups. The highest frequency of ChPV positive samples tested by PCR occurred in chickens that were 1-7 weeks of age, and TuPV was detected in 3～47 weeks old turkeys. All seasons showed ChPV/TuPV circulation, especially from October to March (Figure 2-5-2). The presence of a parvoviral genome in samples was found in healthy poultry and in poultry suffering from RSS/PEMS symptoms. One to three samples from each farm were sequenced and submitted to GenBank[25].

Table 2-5-3　Percentage of positive samples for ChPV/TuPV according to chicken/turkey flock age

Broiler chickens		Breeders		Layer hens		Turkeys	
Weeks	Number of positive samples/%	Weeks	Number of positive samples/%	Weeks	Number of positive samples/%	Weeks	Number of positive samples/%
1～7	624/838[a] (74.46)[b]	1～7	295/398 (74.12)	3～28	73/156 (46.79)	0～5	22/24 (91.67)
8～20	376/702 (53.56)	8～20	160/401 (39.90)	29～59	39/120 (35.00)	6～20	24/24 (100.00)
≥ 21	41/82 (50.00)	≥ 21	117/677 (17.28)	60～72	0/12 (0.00)	20～40	24/24 (100.00)
						≥ 40	0/12 (0.00)
Total	1 041/1 622 (64.18)	572/1 476 (38.75)		112/288 (38.89)		70/84 (83.33)	

Note: [a] Positive/total number of ChPV/TuPV detected; [b] Percentage offocks Positive for ChPV/TuPV.

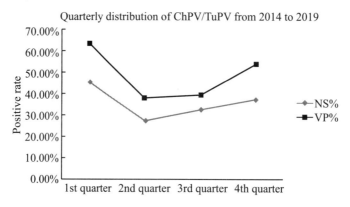

Figure 2-5-2　Quarterly distribution of ChPV/TuPV infections in chicken and turkey flocks from 2014 to 2019

The prevalence rates of ChPV were significantly different in different sources of chickens (Table 2-5-4). All the layer hens belong to exotic chickens. Exotic chickens had a 88.10% (148/168) VP gene positive rate; native chickens had a 50.00% (1 465/2 930) VP gene positive rate.

Poultry farming may introduce pathogens into the environment and food chain. A total of 874 environmental samples were collected from two different systems of bird houses (open house and closed house). A higher prevalence of 59.31% (223/376) ChPV positivity was seen in swab samples (Table 2-5-5) than in other sources of samples, in which litter samples showed 57.14% (104/182) positivity, and drinking water samples showed 27.85% (88/316) positivity. Samples from open house flocks had higher prevalence rates of ChPV than those of closed house flocks (Table 2-5-5), among which those from the open house showed 84.16% (85/101) positivity, those from litter showed 62.86% (44/70) positivity, and those from drinking water showed 50.00% (56/112) positivity, whereas those from the closed house litter were 53.57% (60/112), those from swabs were 50.18% (138/275), and those from drinking water were 15.69% (32/204). Overall, 47.05%

(415/874) of environmental samples from poultry houses tested positive for ChPV and TuPV using PCR-based specific detection.

Forty-six counties belonging to the 10 cities in Guangxi were investigated for the prevalence of ChPV/TuPV in farm, and the top five cities in terms of positive rates of ChPV/TuPV were Liuzhou, Yulin, Beihai, Nanning, and Wuzhou, with positive rates of 100, 67.5, 66.67, 61.09 and 60.42%, respectively.

Table 2-5-4   The results of different ages between exotic and local chickens by ChPV detection

	No. of samples	Weeks	ChPV/TuPV PCR/%
Exotic broiler chickens	14	1~7	94.44[a] (136/144)[b]
	0	8~20	
	24	≥21	50.00 (12/24)
			88.10 (148/168)
Exotic layer hens	156	3~28	46.79 (73/156)
	120	29~59	32.50 (39/120)
	12	60~72	0.00 (0/12)
			38.89 (112/288)
Native chickens	1 092	1~7	73.31 (783/1 092)
	1 103	8~20	48.59 (536/1 103)
	735	≥21	19.86 (146/735)
			50.00 (1 465/2 930)

Note: [a] Percentage of flocks positive for ChPV/TuPV; [b] Positive/total number of ChPV/TuPV detected.

Table 2-5-5   The results of environmental sampling of chicken and turkey houses by ChPV/TuPV detection

	Closed houses/%	Open houses/%	Total/%
Litter A	53.57 (60/112)	62.86[a] (44/70)[b]	57.14 (104/182)
Water B	15.69 (32/204)	50.00 (56/112)	27.85 (88/316)
Swab C	50.18 (138/275)	84.16 (85/101)	59.31 (223/376)
Total	38.92 (230/591)	63.57 (185/291)	47.05 (415/882)

Note: [a] Percentage of flocks positive for ChPV/TuPV; [b] Positive/total number of ChPV/TuPV detected; Litter A, including floor and cleaning poultry wastewater; Water B, including all kinds of drinking water in the henhouse; Swab C, including egg cartons, coop, skip car, and chopping boar.

Discussion

ChPV is a widely prevalent pathogen and has global significance because it adversely affects poultry health and production systems by increasing morbidity, the ratio of feed to meat, and secondary infections. In the present study, specific clinical signs of ChPV were generally observed during the first 2~4 weeks of age, when susceptible chicks were infected by vertical or horizontal routes of transmission. The positive incidence was higher and the age of poultry was lower in all kinds of chickens and turkeys in our study; it was believed that there was age-related development of resistance against ChPV due to immune competence. Exotic broiler chickens showed a higher prevalence of ChPV than Chinese native chickens, indicating the Chinese native chickens may have stronger immune systems. The low ChPV positive rate of exotic Layer hens was due to sick chickens being eliminated in the initial stage. However, subclinical infections usually occur in adult birds and were responsible for production losses and secondary infection. Subclinical infections also act as a source of infection for other young birds and flocks. Additionally, complex factors such as secondary bacterial, fungal, or viral infections may exacerbate the course of parvovirus infection. We used two published PCR assays to confirm the PCR-positive samples, and the results showed that the nPCR (249 bp) has a higher positive rate than the PCR (561 bp), which was used to detect ChPV/TuPV in most countries. It indicated that VP PCR was more sensitive under experimental conditions in our study. This result leads us to ask whether the prevalence of ChPV/TuPV in the world is more severe than previously thought?

The Guangxi Zhuang Autonomous Region is located in southern China. Guangxi borders Vietnam, where avian influenza is also prevalent and the epidemiology of zoonotic diseases is complex. There are several large-scale poultry farms and large numbers of small-scale farms in Guangxi. The booming poultry industry in the region faces a significant risk for RSS/PEMS, and there is little epidemiological surveillance of ChPV/TuPV in the region. The top five cities in terms of positive rates of ChPV/TuPV were Liuzhou, Yulin, Beihai, Nanning, and Wuzhou. Because these are the main regions of the poultry industry of broiler chickens in Guangxi, these findings indicate that the more commercial poultry farms there are in cities, the higher the ChPV/TuPV positivity rate. Moreover, a city's ChPV/TuPV positive rate is also associated with the transportation distance and the clinical sample collection method. All turkey samples were collected around Nanning. In addition, ChPV/TuPV were identified in each season. The region's warm and humid climate may facilitate the survival, growth and transmission of ChPV/TuPV, as well as the occurrence of mixed infections. The selections of poultry farms in this study was designed to cover different geographic areas in Guangxi, but samples from a few areas could not be obtained, meaning that our results were not fully representative. The possibility of some differences in the distribution of viral variants/genotypes due to sampling bias cannot be ruled out, especially for cities where the number of samples was low.

The detection of ChPV in poultry at 1 week of age is not surprising, as published studies suggested the possibility of vertical viral transmission[26, 27]. However, we demonstrated the absence of parvovirus in cloacal swabs from 1-to 5-day-old broiler chickens, which was different from Finkler's report[28], and this finding may be due to the different pathogenicity of the virus strains or the different chicken breeds. In our study, parvovirus was detected mainly in healthy birds. In fact, many chicken and turkey flocks that tested positive for parvovirus showed no signs of disease. This result is consistent with previous work by Zsak et al. although in another study, parvoviruses were detected in poultry with no enteric disorders[16, 20, 29]. In addition, the parvovirus infection rate of turkeys appeared to be higher than that of chickens. The five turkey flocks which were highly positive for parvoviruses included both weak and healthy flocks.

In our study, 47.05% (415/882) of environmental samples from poultry houses tested positive for parvovirus, and a higher prevalence of parvovirus positives was seen in open house flocks' samples than closed house samples, which indicated that the biosecurity measures of farmers who raised the birds in closed houses were applied well. An important factor in RSS/PEMS epidemiology in poultry farms is the introduction of enteric viruses into a flock. Moreover, the persistence of viruses in a flock can influence the course of disease and viral spread or distribution to neighboring farms. Some cases of RSS/PEMS may be attributed to wild birds that are shedding the viruses. These animals begin latent viral excretion before clinical symptoms appear. Birds that have recovered from clinical infection may also be shedders. Birds incubating the virus can spread it to healthy, uninfected birds. Farmers in Guangxi have received little information about RSS and PEMS prevention. Birds move freely, there are no restrictions on human or vehicle movement, and biosecurity measures are not well-applied at many farms. Therefore, it is important to implement biosecurity measures and to control the movements of animals and instruments to minimize the spread of ChPV/TuPV shed by both healthy and sick birds, and, finally but importantly, to improve the nutrition of all the birds in commercial farms.

Considering the emerging status of parvovirus and its wide prevalence, recent advances in diagnosis, vaccinations and therapeutics, along with appropriate disease prevention and control strategies, need to be followed to curtail high losses in the poultry industry due to this economically important pathogen. Further investigations focusing on whether genomic recombination occurred during mixed infection are needed.

In conclusion, this report is the first ChPV/TuPV epidemiological survey to document the presence in Guangxi. Our study determined a prevalence of 51.73% of ChPV/TuPV in apparently healthy and RSS/PEMS-like commercial poultry farms. Our research has published an article comparing the molecular properties of ChPV/TuPV[25]. Commercial poultry farms are very common in China, so we first focused on commercial farms. We are planning to conduct more tests for parvoviruses in various types of chicken farms including small non-commercial flocks for future study. Further extensive epidemiological studies are recommended to determine the true amount of ChPV/TuPV infection in poultry farms and to adopt timely prevention and control strategies, which would help alleviate the economic losses caused by this economically detrimental and emerging virus affecting the poultry industry.

References

[1] LUKASHOV V V, GOUDSMIT J. Evolutionary relationships among parvoviruses: virus-host coevolution among autonomous primate parvoviruses and links between adeno-associated and avian parvoviruses. J Virol, 2001, 75(6): 2729-2740.

[2] MOMOEDA M, WONG S, KAWASE M, et al. A putative nucleoside triphosphate-binding domain in the nonstructural protein of B19 parvovirus is required for cytotoxicity. J Virol, 1994, 68(12): 8443-8446.

[3] GUY J S. Virus infections of the gastrointestinal tract of poultry. Poult Sci, 1998, 77(8): 1166-1175.

[4] SIMPSON A A, HEBERT B, SULLIVAN G M, et al. The structure of porcine parvovirus: comparison with related viruses. J Mol Biol, 2002, 315(5): 1189-1198.

[5] SCHETTLER C H. Virus hepatitis of geese 3 Properties of the causal agent. Avian Pathol, 1973, 2(3): 179-193.

[6] KISARY J, NAGY B, BITAY Z. Presence of parvoviruses in the intestine of chickens showing stunting syndrome. Avian Pathol, 1984, 13(2): 339-343.

[7] KISARY J, AVALOSSE B, MILLER-FAURES A, et al. The genome structure of a new chicken virus identifies it as a parvovirus. J Gen Virol, 1985, 66 (Pt 10): 2259-2263.

[8] TRAMPEL D W, KINDEN D A, SOLORZANO R F, et al. Parvovirus-like enteropathy in Missouri turkeys. Avian Dis, 1983, 27(1): 49-54.

[9] LE GALL-RECULE G, JESTIN V. Biochemical and genomic characterization of muscovy duck parvovirus. Arch Virol, 1994, 139(1-2): 121-131.

[10] GOUGH R E. Parvovirus infections//SAIF Y M, FADLEY A M, GLISSON J R, et al. Diseases of poultry, 12th ed. Iowa: Iowa State University Press, 2008: 397-404.

[11] BRAHAM S, GANDHI J, BEARD S, et al. Evaluation of the Roche LightCycler parvovirus B19 quantification kit for the diagnosis of parvovirus B19 infections. J Clin Virol, 2004, 31(1): 5-10.

[12] ZADORI Z, STEFANCSIK R, RAUCH T, et al. Analysis of the complete nucleotide sequences of goose and muscovy duck parvoviruses indicates common ancestral origin with adeno-associated virus 2. Virology, 1995, 212(2): 562-573.

[13] ZHANG Y, XIE Z, DENG X, et al. Molecular characterization of parvovirus Strain GX-Tu-PV-1, isolated from a Guangxi turkey. Microbiol Resour Announc, 2019, 8(46): e00152-19.

[14] NUNEZ L F, SANTANDER P S, METTIFOGO E, et al. Isolation and molecular characterisation of chicken parvovirus from Brazilian flocks with enteric disorders. Br Poult Sci, 2015, 56(1): 39-47.

[15] PANTIN-JACKWOOD M J, SPACKMAN E, DAY J M, et al. Periodic monitoring of commercial turkeys for enteric viruses indicates continuous presence of astrovirus and rotavirus on the farms. Avian Dis, 2007, 51(3): 674-680.

[16] ZSAK L, STROTHER K O, DAY J M. Development of a polymerase chain reaction procedure for detection of chicken and turkey parvoviruses. Avian Dis, 2009, 53(1): 83-88.

[17] DAY J M, ZSAK L. Determination and analysis of the full-length chicken parvovirus genome. Virology, 2010, 399(1): 59-64.

[18] PANTIN-JACKWOOD M J, DAY J M, JACKWOOD M W, et al. Enteric viruses detected by molecular methods in commercial chicken and turkey flocks in the United States between 2005 and 2006. Avian Dis, 2008, 52(2): 235-244.

[19] BIDIN M, LOJKIC I, BIDIN Z, et al. Identification and phylogenetic diversity of parvovirus circulating in commercial chicken and turkey flocks in Croatia. Avian Dis, 2011, 55(4): 693-696.

[20] PALADE E A, KISARY J, BENYEDA Z, et al. Naturally occurring parvoviral infection in Hungarian broiler flocks. Avian Pathol, 2011, 40(2): 191-197.

[21] DOMANSKA-BLICHARZ K, JACUKOWICZ A, LISOWSKA A, et al. Genetic characterization of parvoviruses circulating in turkey and chicken flocks in Poland. Arch Virol, 2012, 157(12): 2425-2430.

[22] KOO B S, LEE H R, JEON E O, et al. Genetic characterization of three novel chicken parvovirus strains based on analysis of their coding sequences. Avian Pathol, 2015, 44(1): 28-34.

[23] ZSAK L, CHA R M, LI F, et al. Host Specificity and phylogenetic relationships of chicken and turkey parvoviruses. Avian Dis, 2015, 59(1): 157-161.

[24] CARRATALA A, RUSINOL M, HUNDESA A, et al. A novel tool for specific detection and quantification of chicken/turkey parvoviruses to trace poultry fecal contamination in the environment. Appl Environ Microbiol, 2012, 78(20): 7496-7499.

[25] FENG B, XIE Z, DENG X, et al. Genetic and phylogenetic analysis of a novel parvovirus isolated from chickens in Guangxi, China. Arch Virol, 2016, 161(11): 3285-3289.

[26] KISARY J. Experimental infection of chicken embryos and day-old chickens with parvovirus of chicken origin. Avian Pathol, 1985, 14(1): 1-7.

[27] KISARY J, MILLER-FAURES A, ROMMELAERE J. Presence of fowl parvovirus in fibroblast cell cultures prepared from uninoculated White Leghorn chicken embryos. Avian Pathol, 1987, 16(1): 115-121.

[28] FINKLER F, LIMA D A, CERVA C, et al. Chicken parvovirus and its associations with malabsorption syndrome. Res Vet Sci, 2016, 107: 178-181.

[29] PALADE E A, DEMETER Z, HORNYAK A, et al. High prevalence of turkey parvovirus in turkey flocks from Hungary experiencing enteric disease syndromes. Avian Dis, 2011, 55(3): 468-475.

Genetic characterization of fowl aviadenovirus 4 isolates from Guangxi, China, during 2017-2019

Rashid Farooq, Xie Zhixun, Zhang Lei, Luan Yongjiao, Luo Sisi, Deng Xianwen, Xie Liji, Xie Zhiqin, and Fan Qing

Abstract

Hepatitis-hydropericardium syndrome (HHS) is a severe disease that causes 20% to 80% mortality in chickens aged 3 to 6 week. Fowl aviadenovirus serotype 4 (FAdV-4) plays an important role in the etiology of HHS. Since 2015, outbreaks of HHS have been reported in several provinces of China; however, details regarding the FAdV-4 genome properties are lacking. In the present study, the complete genomes of 9 isolates responsible for these outbreaks in Guangxi, China, were sequenced. To investigate the molecular characteristics of these FAdV-4 isolates, we compared their genomes with those of other reported pathogenic and nonpathogenic FAdV-4 isolates. A variable number of GA repeats were observed in the isolates of this study. Each of the isolates GX2017-01, GX2017-02, GX2018-08, and GX2019-09 had 11 GA repeats; GX2017-03, GX2017-04, and GX2017-05 each had 10 GA repeats, while GX2017-06 and GX2018-07 each had 8 GA repeats. We observed several deletions and distinct amino acid mutations in the major structural genes of these isolates when compared with non-Chinese isolates. We found 2 novel putative genetic markers in the hexon protein, one present in GX2017-02, in which aspartic acid (D) was changed to tyrosine (Y), and another present in each of isolates GX2018-08 and GX2019-09, in which serine (S) was changed to arginine (R), when compared with selected Chinese and some non-Chinese isolates. Moreover, the phylogenetic analysis revealed that all the isolates of this study were clustered within FAdV-C. We found that these isolates were closely related to other recently isolated Chinese strains. The data presented in this study will not only increase the understanding of the molecular epidemiology and genetic diversity of FAdV-4 isolates in China but also has an important reference value of the major factors that determine the virulence of FAdV-4 strains.

Keywords

fowl adenovirus 4, genetic characterization, amino acid change, outbreak

Introduction

Fowl aviadenoviruses (FAdV) have been classified into 5 species, A-E, and have 12 serotypes[1]. Fowl aviadenoviruses cause mainly inclusion body hepatitis, hepatitis-hydropericardium syndrome (HHS), and gizzard erosion in chickens[2, 3]. Research has shown that inclusion body hepatitis is caused by all 12 serotypes of FAdV, gizzard erosion is caused by FAdV-1, and HHS is caused by mainly FAdV-4[4].

Hepatitis-hydropericardium syndrome was first reported in Pakistan in 1987[5], and subsequent outbreaks have been recorded in other countries such as India, Japan, Republic of Korea, Iraq, Mexico, and Canada and in those in South and Central America[2, 6-11]. The virus affects mainly 3- to 6-week-old broiler chickens, causing 20% to 80% mortality. The disease is characterized by the accumulation of clear and straw-colored fluid in the pericardial sac. Other signs included an enlarged and discolored liver with hemorrhage or necrotic

233

foci of hemorrhage and necrosis[12].

Several studies have found that recent Chinese FAdV-4 isolates have significant deletions and substitutions in their genome compared with those of FAdV-4 isolates reported from other countries around the world[1, 13]. However, further evaluation of the complete genome of new isolates is needed to better understand the pathogenesis of these viruses.

The important viral structural proteins of FAdV are the penton, hexon, and fiber proteins[14]. The penton protein is responsible for virus internalization during the infectious cycle; the hexon protein contains the main neutralizing epitope and therefore can be used for serotyping, while the fiber protein is responsible for virus-host interactions[15-17]. Moreover, recent findings suggested that hexon and fiber 2 are associated with the virulence of emerging and highly pathogenic FAdV-4 isolates[18].

Since May 2015, outbreaks of HHS have increased in several provinces of China, including Guangxi, causing a massive economic loss to the country. For proper understanding of the disease, molecular characterization of the complete genome of FAdV-4 is indispensable. We isolated, sequenced, and characterized the whole genome of 9 different isolates of FAdV-4 from different broiler flocks in Guangxi, China. To better understand their genome characteristics, we compared them with some of the previously reported pathogenic and nonpathogenic FAdV-4 strains from different countries, including China. Our study presented additional detailed information on the molecular characteristics of FAdV-4 isolates.

Materials and methods

Sample collection and virus isolation

All isolates were collected from commercial broiler farms that were experiencing HHS in Nanning city, Guangxi, China, between 2017 and 2019. The mortality due to FAdV-4 recorded in different commercial broiler farms was from 60% to 80%. The complete information of these isolates is given (Table 2-6-1). The virus was isolated from the livers of infected birds and propagated in primary cultures of chicken embryo liver cells[19]. Medium 199 (Invitrogen, CA) was used, which was supplemented with 10% fetal bovine serum (Invitrogen, CA). Antibiotics, 100 U/mL of penicillin (Sigma-Aldrich, Rockville Pike, Bethesda, MD) and 100 μg/mL of streptomycin (Sigma-Aldrich), were added to the medium and were used for chicken embryo liver cell culture. After isolation, virus dilutions were made for plaque purification in the chicken liver cell line (LMH) (ATCC, Manassas, VA). Each virus was plaque purified. In brief, virus dilutions from 10^{-2} to 10^{-7} were inoculated onto LMH monolayers. 1 h after infection, the cell layer was overlaid with Medium 199 containing 2% agarose (Sigma-Aldrich) and incubated for 3 to 5 D at 37 ℃ in a 5% CO_2 humidified atmosphere. Isolated plaques were then picked and transferred to a new fresh culture of LMH cells[12]. The detection of virus in the samples was performed by PCR using the primers specific for the FAdV hexon gene (GenBank: GU188428), forward: 5'-CAACTACATCGGGGTTCAGGG-3' and reverse: 5'-TGGCGTTTCTCAGCATCA-3'. The PCR profile used was as follows: 95 ℃ for 5 min, followed by 34 cycles at 95 ℃ for 30 s, 58 ℃ for 30 s, and 72 ℃ for 30 s, followed by a final elongation step for 10 min at 72 ℃.

Table 2-6-1　Complete information of all isolates

S. No.	Name of the isolate	GenBank number	Sampling time	Sampling place	Sampling tissue	Age of sampling chicken/d
1	GX2017-01	MN577977	2017.7	Guangxi	Liver	40
2	GX2017-02	MN577978	2017.7	Guangxi	Liver	63

continued

S. No.	Name of the isolate	GenBank number	Sampling time	Sampling place	Sampling tissue	Age of sampling chicken/d
3	GX2017-03	MN577979	2017.11	Guangxi	Liver	56
4	GX2017-04	MN577980	2017.11	Guangxi	Liver	65
5	GX2017-05	MN577981	2017.7	Guangxi	Liver	96
6	GX2017-06	MN577982	2018.5	Guangxi	Liver	75
7	GX2018-07	MN577983	2018.1	Guangxi	Liver	63
8	GX2018-08	MN577984	2018.5	Guangxi	Liver	96
9	GX2019-09	MN577985	2019.4	Guangxi	Liver	50

Complete genome sequencing

The viral DNA of all the isolates used in this study was extracted from the livers of affected broiler chickens using an Easy Pure Viral DNA/RNA kit (cat No.: ER201-01; TransGen, Guangzhou, China) according to the manufacturer's instructions. Specific primers for the complete genome sequencing, which were designed on the basis of the sequence of the FAdV-4 ON1 strain, were used[20]. However, new primers were also designed where necessary to further ensure the sequencing results. PCR products with expected lengths were sequenced directly or cloned into the pEASY-Blunt Cloning vector (TransGen, Beijing, China) according to the manufacturer's instructions. The complete sequence was manually assembled using the Seqman program in the DNAstar software package (version 5.01; Madison, WI).

Sequencing analysis

The nucleotide and deduced amino acid sequences of all the FAdV-4 isolates and certain reference strains used in this study were edited, aligned, and analyzed using the MegAlign program, a part of the Lasergene software package (DNASTAR, Madison, WI). The major structural genes, penton base, hexon, and fiber, were translated into amino acid sequences using EditSeq in the Lasergene software package. The amino acid sequences from these isolates were compared with the amino acid sequences of these proteins of the HNJZ, ON1, and SH95 strains. A phylogenetic tree of the complete genomes of these 9 isolates and additional isolates from China and other parts of the world was produced using the neighbor-joining method and bootstrap test of 1 000 replicates in MEGA 5.0 software (MEGA, Pennsylvania State University, University Park, PA).

Results

Genome sequencing and sequence analysis

The complete genomic sequences of all 9 isolates, FAdV-4-GX2017-001 (GX2017-01), FAdV-4-GX2017-002 (GX2017-02), FAdV-4-GX2017-003 (GX2017-03), FAdV-4-GX2017-004 (GX2017-04), FAdV-4-GX2017-005 (GX2017-05), FAdV-4-GX2017-006 (GX2017-06), FAdV-4-GX2018-007 (GX2018-07), FAdV-4-GX2018-008 (GX2018-08), and FAdV-4-GX2019-009 (GX2019-09), all of which were from Guangxi, China, were deposited in GenBank with accession numbers: MN577977, MN577978, MN577979, MN577980, MN577981, MN577982, MN577983, MN577984, and MN577985, respectively.

The genome sizes of all these isolates were similar to those of previously reported Chinese isolates; however, they differed in size from those of the isolates from other countries, including nonpathogenic isolates[1, 12, 13, 19-21]. The full genome sizes of isolates GX2017-01, GX2017-02, GX2017-03, GX2017-04,

GX2017-05, GX2017-06, GX2018-07, GX2018-08, and GX2019-09 were 43 725, 43 725, 43 723, 43 723, 43 722, 43 718, 43 719, 43 726, and 43 726 bp, respectively. The G+C content was found to be 54.85% for GX2017-02, and for all other isolates, the G+C content was 54.87%.

There was a GA repeat region between the Px and pVI genes. The longest GA repeats were found for GX2017-01, GX2017-02, GX2018-08, and GX2019-09 (11 GA repeats each), followed by those for GX2017-03, GX2017-04, and GX2017-05 (10 GA repeats), whereas the least number of GA repeats were found for GX2017-06 and GX2018-07 (8 GA repeats). However, each of these 9 isolates had a longer GA repeat region than the nonpathogenic ON1 strain which has 5 GA repeats. When comparing the GA repeats of all the 9 isolates of the present study with the previously reported Chinese and Mexico isolates, we found that each of the isolates GX2017-06 and GX2018-07 is 3 GA repeats less than HNJZ and MX-SHP95 (Figure 2-6-1). Moreover, each of the isolates, GX2017-03, GX2017-04, and GX2017-05 is 1 GA repeat less than HNJZ and MX-SHP95 (Figure 2-6-1), whereas GX2017-01, GX2017-02, GX2018-08, and GX2019-09 had the same number of GA repeats as HNJZ and MX-SHP95. All 9 isolates in our study and the previously reported MX-SHP95 and HNJZ isolates had longer TC repeats than the ON1 strain. A deletion of 6 bp at positions 28 784 to 28 789 was also observed in all 9 isolates, as well as in the previous Chinese isolate HNJZ, when compared with the ON1 genome. This deletion was only observed in isolates from China. We found that all 9 isolates had a 1 966 bp deletion in the 3'-end region of the genome when compared with that of ON1. This kind of deletion was also observed in all the Chinese isolates yet studied[13]. The deletion was located at 35 389 bp to 37 354 bp compared with the nucleotide positions within the ON1 genome. Moreover, deletions from 41 675 bp to 41 691 bp, 41 704 bp to 41 710 bp, 41 748 bp to 41 760 bp, and 41 766 bp to 41 809 bp were also observed in all 9 isolates compared with the nucleotide sequence positions within the ON1 genome. These deletions were observed in all pathogenic strains. An insertion of 27 bp in ORF19 A in all 9 isolates at the position 44 376 compared with the nucleotide sequence positions within the ON1 genome was also observed. The same insertion was also found for the HNJZ and MX-SHP95 strains (Figure 2-6-1).

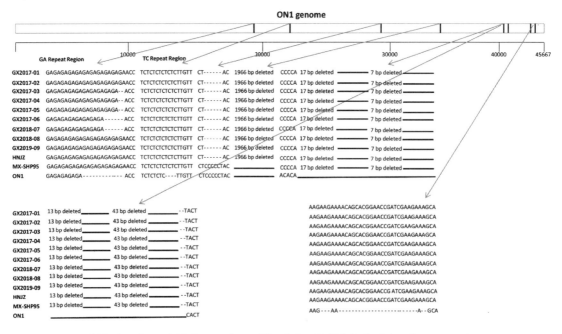

Major insertions and deletions are shown. Abbreviation: FAdV-4, fowl aviadenovirus serotype 4.

Figure 2-6-1　A comparison of the complete genomes of the isolates used in this study and the previously reported sequences of highly pathogenic and nonpathogenic FAdV-4 strains

All of these 9 isolates had high nucleotide similarities with other recently reported Chinese FAdV-4 isolates. After comparison, we found that GX2017-01 has 99.88% to 99.99%, GX2017-02 has 99.85% to 99.96%, GX2017-03 has 99.88% to 99.99%, GX2017-04 and GX2018-07 have 99.89% to 100%, GX2017-05 has 99.99% to 99.88%, GX2017-06 has 99.99% to 99.95%, and GX2018-08 and GX2019-09 have 99.81% to 99.92% similarity with other reported Chinese isolates. However, when we compared the nucleotide identities with non-Chinese isolates, we found that GX2017-01 has 98.46% to 98.73%, GX2017-02 has 98.41% to 98.73%, GX2017-03 has 98.46% to 98.78%, GX2017-04 has 98.46 to 98.78%, GX2017-05 has 98.77% to 98.45%, GX2017-06 has 98.78 to 98.46%, GX2018-07 has 98.47% to 98.79%, and GX2018-08 and GX2019-09 have 98.43% to 98.75% similarity with non-Chinese isolates (Table 2-6-2).

Table 2-6-2　Nucleotide similarity with some Chinese and non-Chinese isolates

Isolates	Similarity with some Chinese isolates at the nucleotide level /%			Similarity with some non-Chinese isolates at the nucleotide level /%		
	Strain	Accession number	Similarity	Strain	Accession number	Similarity
GX2017-01	SD1601	MH006602	99.99	KR5	HE608152	98.78
	AHHQ	MG148334	99.99	SH95	KP295475	98.69
	TCZHP	MG824745	99.99	B1-7	KU342001	98.58
	JSJ13	KM096544	99.88	ON1	GU188428	98.46
GX2017-02	SD1601	MH006602	99.96	KR5	HE608152	98.73
	AHHQ	MG148334	99.96	SH95	KP295475	98.64
	TCZHP	MG824745	99.96	B1-7	KU342001	98.53
	JSJ13	KM096544	99.85	ON1	GU188428	98.41
GX2017-03	SD1601	MH006602	99.99	KR5	HE608152	98.78
	AHHQ	MG148334	99.99	SH95	KP295475	98.69
	TCZHP	MG824745	99.99	B1-7	KU342001	98.58
	JSJ13	KM096544	99.88	ON1	GU188428	98.46
GX2017-04	SD1601	MH006602	100	KR5	HE608152	98.78
	AHHQ	MG148334	100	SH95	KP295475	98.69
	TCZHP	MG824745	100	B1-7	KU342001	98.58
	JSJ13	KM096544	99.89	ON1	GU188428	98.46
GX2017-05	SD1601	MH006602	99.99	KR5	HE608152	98.77
	HB1510	KU587519	99.99	SH95	KP295475	98.69
	TCZHP	MG824745	99.99	B1-7	KU342001	98.57
	JSJ13	KM096544	99.88	ON1	GU188428	98.45
GX2017-06	NIVD2	MG547384	99.99	KR5	HE608152	98.78
	HB1510	KU587519	99.99	SH95	KP295475	98.69
	GX-1	MH45498	99.99	B1-7	KU342001	98.58
	HN/151025	KU245540	99.95	ON1	GU188428	98.46
GX2018-07	NIVD2	MG547384	100	KR5	HE608152	98.79
	AHHQ	MG148334	99.99	SH95	KP295475	98.70
	TCZHP	MG824745	99.98	B1-7	KU342001	98.59
	JSJ13	KM096544	99.89	ON1	GU188428	98.47
GX2018-08	SD1601	MH006602	99.92	KR5	HE608152	98.75
	AHHQ	MG148334	99.92	SH95	KP295475	98.67
	TCZHP	MG824745	99.92	B1-7	KU342001	98.55
	JSJ13	KM096544	99.81	ON1	GU188428	98.43
GX2019-09	SD1601	MH006602	99.92	KR5	HE608152	98.75
	AHHQ	MG148334	99.92	SH95	KP295475	98.68
	TCZHP	MG824745	99.92	B1-7	KU342001	98.57
	JSJ13	KM096544	99.81	ON1	GU188428	98.43

The major structural proteins of FAdV-4

The FAdV capsid consists of 3 main exposed structural proteins, the hexon, the fiber, and the penton base. Hexon is the major capsid protein that contains group-, type-, and subtype-specific antigenic determinants against which antibodies are produced. Moreover, the molecular classification of FAdV is based on the hexon gene loop 1 region and fiber gene[22, 23]. The sizes of penton base, hexon, fiber 1, and fiber 2 genes are 1 578, 2 814, 1 296, and 1 440 bp, respectively. Comparison of the amino acid sequences of each of the penton base, hexon, and fiber proteins revealed the presence of several amino acid mutations and deletions with respect to those of the pathogenic strain MX-SHP95 and nonpathogenic strain ON1. We also compared the amino acid sequences of all 9 isolates with the pathogenic isolate HNJZ recently isolated from China. The size of the penton protein for all the 9 isolates was 525 amino acids, the same as HNJZ, ON1, and MX-SHP95. In the penton base, the substitution from S to P was observed at position 42 in all 9 isolates and in the previously isolates HNJZ and MX-SHP95[1, 19]. In the hexon protein of the isolate GX2017-02, the position 691 amino acid was changed from aspartic acid (D) to tyrosine (Y), and the position 372 amino acid in GX2018-08 and GX2019-09 was changed from serine (S) to arginine (R). The size of the hexon protein of all 9 isolates in this study was 937 amino acids, which is the same as that in HNJZ, ON1, and MX-SHP95. The fiber protein is related to virus neutralization, cellular receptor binding, and virulence variation[24]. Recombinant FAdV-4 fiber 2 protein has been proven to be a protective immunogen against HHS[25]. The fiber gene can also be used to differentiate HHS-inducing FAdV-4 isolates from other FAdV-4 strains[26]. In the present study, the fiber 1 and fiber 2 proteins of all 9 isolates were aligned with those of the HNJZ Chinese isolate, ON1 Canadian isolate, and MX-SHP95 Mexican strain, which enabled us to identify multiple amino acid substitutions. The fiber 1 protein was 431 amino acids long in all 9 isolates, which was the same length as that in HNJZ, but it was 1 amino acid shorter in both the ON1 and MX-SHP95 strains. This 1 amino acid was deleted at position 428. The fiber 2 protein was 479 amino acids long in all 9 isolates, which was the same length as that in HNJZ, but it was 5 amino acids shorter in both the ON1 and MX-SHP95 isolates. These 5 amino acid residues were deleted at position 11. These results indicated that the sizes of the penton, hexon, and fiber proteins were conserved in isolates from China (Table 2-6-3).

Some other substitutions were also observed that were unique to pathogenic strains only. Notably, all 9 isolates and the virulent strains from China, HNJZ, and Mexico, MX-SHP95, had amino acid substitutions at position 188 (I to R) in the hexon protein and at 196 (T to V) and at 431 (S to G) in the fiber 1 protein. In the fiber 2 protein, at position 219 (214 in ON1 and MX-SHP95), the nonpathogenic strain ON1 has a glycine (G), whereas in all 9 isolates, HNJZ and MX-SHP95 have an aspartic acid (D). At position 380 (375 in ON1 and MX-SHP95), the nonpathogenic strain ON1 has an alanine (A), whereas in all 9 isolates, HNJZ and MX-SHP95 have a threonine (T). Moreover, we also observed substitutions at positions 232 (E to Q), 300 (I to T), 305 (S to A), 307 (P to A), 319 (V to I), 329 (V to L), 378 (A to T), 405 (P to S), 435 (T to S), and 453 (S to A) in the fiber 2 protein (Table 2-6-3). Based on the information that we obtained in this study, fiber 2 may serve as a virulence factor protein. Two other putative genetic markers in the hexon protein of the isolates GX2017-02, GX2018-08, and GX2019-09 need to be investigated further. Moreover, to identify critical sites responsible for the high pathogenicity of FAdV-4 isolates, further molecular analysis and reverse genetic systems are required.

Table 2-6-3 Amino acid differences in the major structural genes from FAdV-4 isolates used in this study

Amino acids at position

Genes	Isolates	42	45	193	356	370	426	486
Penton	GX2017-01	P	D	I	V	P	V	T
	GX2017-02	P	D	I	V	P	V	T
	GX2017-03	P	D	I	V	P	V	T
	GX2017-04	P	D	I	V	P	V	T
	GX2017-05	P	D	I	V	P	V	T
	GX2017-06	P	D	I	V	P	V	T
	GX2018-07	P	D	I	V	P	V	T
	GX2018-08	P	D	I	V	P	V	T
	GX2019-09	P	D	I	V	P	V	T
	HNJZ	P	D	I	V	P	V	T
	ON1	S	G	V	A	Q	I	S
	MX-SHP95	P	G	V	A	Q	I	S

Genes	Isolates	164	188	193	195	238	240	243	263	264	372	402	410	574	691	797	842
Hexon	GX2017-01	S	R	R	Q	D	T	N	I	V	S	A	A	I	D	P	A
	GX2017-02	S	R	R	Q	D	T	N	I	V	S	A	A	I	Y	P	A
	GX2017-03	S	R	R	Q	D	T	N	I	V	S	A	A	I	D	P	A
	GX2017-04	S	R	R	Q	D	T	N	I	V	S	A	A	I	D	P	A
	GX2017-05	S	R	R	Q	D	T	N	I	V	S	A	A	I	D	P	A
	GX2017-06	S	R	R	Q	D	T	N	I	V	S	A	A	I	D	P	A
	GX2018-07	S	R	R	Q	D	T	N	I	V	S	A	A	I	D	P	A
	GX2018-08	S	R	R	Q	D	T	N	I	V	R	A	A	I	D	P	A
	GX2019-09	S	R	R	Q	D	T	N	I	V	R	A	A	I	D	P	A
	HNJZ	S	R	R	Q	D	T	N	I	V	S	A	A	I	D	P	A
	ON1	T	I	Q	E	N	A	E	M	I	S	A	T	V	D	A	G
	MX-SHP95	T	R	Q	E	N	A	E	M	I	S	Q	T	V	D	A	G

continued

Fiber 1 — Amino acids at position

Isolates	14	28	44	69	70	119	126	153	186	196	204	251	262	263	310	329	331	374	383	401	428	431
GX2017-01	A	S	R	G	S	N	A	R	D	V	G	L	H	D	H	H	R	S	I	N	-	G
GX2017-02	A	S	R	G	S	N	A	R	D	V	G	L	H	D	H	H	R	S	I	N	-	G
GX2017-03	A	S	R	G	S	N	A	R	D	V	G	L	H	D	H	H	R	S	I	N	-	G
GX2017-04	A	S	R	G	S	N	A	R	D	V	G	L	H	D	H	H	R	S	I	N	-	G
GX2017-05	A	S	R	G	S	N	A	R	D	V	G	L	H	D	H	H	R	S	I	N	-	G
GX2017-06	A	S	R	G	S	N	A	R	D	V	G	L	H	D	H	H	R	S	I	N	-	G
GX2018-07	A	S	R	G	S	N	A	R	D	V	G	L	H	D	H	H	R	S	I	N	-	G
GX2018-08	A	S	R	G	S	N	A	R	D	V	G	L	H	D	H	H	R	S	I	N	-	G
GX2019-09	A	S	R	G	S	N	A	R	D	V	G	L	H	D	H	H	R	S	I	N	-	G
HNJZ	A	S	R	G	S	N	A	R	D	V	G	L	H	D	H	H	R	S	I	N	-	G
ON1	V	I	P	S	G	D	V	H	N	T	G	I	Q	E	R	Q	K	P	L	Y	H	S
MX-SHP95	V	I	P	S	G	D	V	H	N	V	A	I	Q	E	R	Q	K	P	L	Y	N	G

Fiber 2 — Amino acids at position

Isolates	11-15	22	29	114	144	219	232	261	300	305-307	319	324	329	334	338	343-344	346	378	380	391	400	403	405-406	413	435	439	453	459	478
GX2017-01	ENJKP	S	A	D	S	D	Q	T	T	ANA	I	V	L	A	N	LN	A	T	T	T	G	E	SI	S	S	E	A	N	L
GX2017-02	ENJKP	S	A	D	S	D	Q	T	T	ANA	I	V	L	A	N	LN	A	T	T	T	G	E	SI	S	S	E	A	N	L
GX2017-03	ENJKP	S	A	D	S	D	Q	T	T	ANA	I	V	L	A	N	LN	A	T	T	T	G	E	SI	S	S	E	A	N	L
GX2017-04	ENJKP	S	A	D	S	D	Q	T	T	ANA	I	V	L	A	N	LN	A	T	T	T	G	E	SI	S	S	E	A	N	L
GX2017-05	ENJKP	S	A	D	S	D	Q	T	T	ANA	I	V	L	A	N	LN	A	T	T	T	G	E	SI	S	S	E	A	N	L
GX2017-06	ENJKP	S	A	D	S	D	Q	T	T	ANA	I	V	L	A	N	LN	A	T	T	T	G	E	SI	S	S	E	A	N	L
GX2018-07	ENJKP	S	A	D		D	Q	T	T	ANA	I	V	L	A	N	LN	A	T	T	T	G	E	SI	S	S	E	A	N	L
GX2018-08	ENJKP	S	A	D	S	D	Q	T	T	ANA	I	V	L	A	N	LN	A	T	T	T	G	E	SI	S	S	E	A	N	L
GX2019-09	ENJKP	S	A	D	S	D	Q	T	T	ANA	I	V	L	A	N	LN	A	T	T	T	G	E	SI	S	S	E	A	N	L
HNJZ	ENJKP	S	A	D	S	D	Q	T	T	ANA	I	V	L	A	N	LN	A	T	T	T	G	E	SI	S	S	E	A	N	L
ON1	-	S	P	D	S	G	E	S	I	SHP	V	F	V	T	T	NS	A	A	A	S	A	Q	PS	T	T	D	S	A	V
MX-SHP95	-	Y	P	A	A	D	Q	N	T	ANA	I	F	L	T	T	NS	V	T	T	S	A	Q	SS	T	S	D	A	A	V

Note: FAdv-4, fowl aviadenovirus sero type 4.

Pylogenetic analysis

Phylogenetic analysis using nucleotide sequences revealed that the complete genomes of the isolates from the present study clustered within FAdV-C together with other FAdV-4 isolates. The analysis revealed that all 9 isolates in our study were closely related to viruses previously isolated from China. Among these isolates, GX2017-05, GX2017-06, GX2018-07, GX2018-08, and GX2019-09 were closely related among themselves. The isolates GX2017-01 and GX2017-03 were also closely related. The phylogenetic analysis also showed that all the isolates of this study and previous China isolates were in close resemblance with the highly pathogenic MX-SHP95 strain of Mexico (Figure 2-6-2).

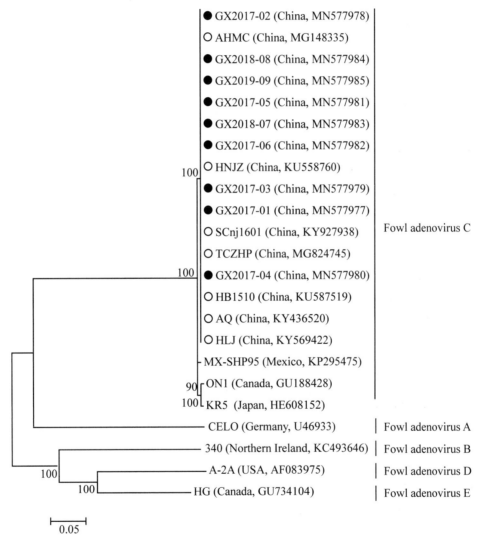

The tree was generated by the neighbor-joining method with a bootstrap test of 1 000 replicates using MEGA 5.0. Circles represent the isolates of the present study. The empty circles represent Chinese isolates from previous studies.

Figure 2-6-2   The results of the phylogenetic analysis based on the nucleotide sequences of selected complete genomes

Discussion

Fowl aviadenoviruses cause different diseases of special importance in broiler chickens; moreover, since 2015, the occurrence of these diseases displayed an increasing trend in China[1]. Fowl adenovirus 4 has been isolated, and its complete genome has been sequenced from different parts of the world and China. To date, the complete genome of 44 isolates of FAdV-4 has been sequenced worldwide; so far, only one isolation has been made from Guangxi, China[12]. Therefore, more knowledge and study was required from this province, so the complete genome of 9 FAdV-4 isolates was sequenced and genetically characterized from this province. Moreover, to fully understand the mechanisms involved in the high pathogenesis of these viruses, information about their complete genome is needed. In this study, the whole-genome sequencing of 9 highly pathogenic FAdV-4 isolates has improved the existing knowledge of FAdV genomes.

The genomes of all 9 isolates varied in length from 43 718 bp to 43 726 bp, which was in accordance with those of previously reported Chinese isolates[1, 27]. However, the genome size of these isolates varied from that of the previously reported highly pathogenic and nonpathogenic FAdV-4 isolates from other parts of the world[19, 20, 28].

Our findings suggested that most of the insertions and deletions occurred on the right-hand side of the genome. All 9 isolates had a deletion of 1 966 bp on the 3'end region of the genome compared with that of non-Chinese isolates (Figure 2-6-1)[19, 20]. However, similar to our findings, all of the viruses previously isolated from China had a similar deletion of nucleotides[1]. It was speculated that this deletion might have a role in the pathogenicity of these isolates, but recent studies using the clustered regularly interspaced short palindromic repeat (CRISPR) and CRISPR associated protein 9 (CRISPR-Cas9) system found that this deletion was dispensable for the high pathogenicity of such isolates[29]. The 6-bp deletion in all 9 isolates at position 28 784, when compared with the ON1 genome, was also present in HNJZ, indicating this deletion was specific to Chinese isolates. Moreover, some characteristics were found that were unique to pathogenic strains only, that is, those with the longer GA repeats; GX2017-01, GX2017-02, GX2018-08, and GX2019-09 have 11 GA repeats each, GX2017-03, GX2017-04, and GX2017-05 have 10 GA repeats each, and the isolates GX2017-06 and GX2018-07 have 8 GA repeats each (Figure 2-6-1). All these isolates have longer GA repeats than the nonpathogenic ON1 strain. It will be interesting to investigate the role of GA repeats in terms of viral pathogenesis. Other characteristics of pathogenic strains are longer TC regions, the deletions between 41 675 and 41 809 and insertions of 27 bp in ORF19 A that were observed in HNJZ, MX-SHP95, and all the 9 isolates of the present study[1, 19].

The penton, hexon, and fiber proteins are the major structural proteins of the FAdV capsid[9, 23]. Therefore, amino acid alignments were used to determine the differences and similarities in these structural proteins among the highly pathogenic and nonpathogenic FAdV isolates. Our findings suggested the presence of several amino acid mutations and deletions when compared with those of the MX-SHP95 and ON1 strains. In the penton protein, a total of 7 substitutions occurred compared with those of the MX-SHP95 and ON1 strains. Interestingly, all these substitutions were similar to those in the previously reported Chinese isolates (Table 2-6-3). The occurrence of the substitution from S to P at position 42 within this protein that is present even in the strain SHP95 may indicate that this mutation might be important for the virulence of this virus. The hexon protein showed unique amino acid substitutions at positions 691 (D to Y) in GX2017-02 and 372 (S to R) within each of GX2018-08 and GX2019-09 isolates with respect to those of the ON1 strain. The role of these 2 new putative genetic markers, each in 3 different isolates, may be further elucidated by using a reverse genetic system, such as that used for the hexon and fiber 2 protein, which were shown to be associated with the virulence of highly pathogenic FAdV[18]. However, all the other substitutions in the hexon protein were similar to those in a previous Chinese isolate[1]. Moreover, in the fiber 1 and fiber 2 proteins of all 9 isolates, all the substitutions that were found were the same as those in previous Chinese isolates[30]. These findings suggested that most of the changes in all the major structural proteins of FAdV found in China were evolutionarily conserved. However, in these 2 fiber proteins, several other amino acid substitutions founded were specific for pathogenic strains only. Moreover, the emergence of highly pathogenic strains of FAdV-4 with some genetic diversity demands additional preventive measures against FAdV-4 infections in poultry farms. The resemblance of all the 9 isolates of the present study and that of previously isolated Chinese isolates with the highly pathogenic MX-SHP95 strain from Mexico revealed a common conserved region which further needs to be investigated.

References

[1] LIU Y, WAN W, GAO D, et al. Genetic characterization of novel fowl aviadenovirus 4 isolates from outbreaks of hepatitis-hydropericardium syndrome in broiler chickens in China. Emerging Microbes & Infection, 2016, 5(11): e117.

[2] ANJUM A D, SABRI M A, IQBAL Z. Hydropericarditis syndrome in broiler chickens in Pakistan. Veterinary Record, 1989, 124(10): 247-248.

[3] ONO M, OKUDA Y, YAZAWA S, et al. Outbreaks of adenoviral gizzard erosion in slaughtered broiler chickens in Japan. Veterinary Record, 2003, 153(25): 775-779.

[4] JIANG Z, LIU M, WANG C, et al. Characterization of fowl adenovirus serotype 4 circulating in chickens in China. Veterinary Microbiology, 2019, 238: 108427.

[5] LI P H, ZHENG P P, ZHANG T F, et al. Fowl adenovirus serotype 4: epidemiology, pathogenesis, diagnostic detection, and vaccine strategies. Poultry Science, 2017, 96(8): 2630-2640.

[6] ABDUL-AZIZ T A, AL-ATTAR M A. New syndrome in Iraqi chicks. Veterinary Record, 1991, 129(12): 272.

[7] ASRANI R K, GUPTA V K, SHARMA S K, et al. Hydropericardium-hepatopathy syndrome in Asian poultry. Veterinary Record, 1997, 141(11): 271-273.

[8] ABE T, NAKAMURA K, TOJO H, et al. Histology, immunohistochemistry, and ultrastructure of hydropericardium syndrome in adult broiler breeders and broiler chicks. Avian Disease, 1998, 42(3): 606-612.

[9] HESS M, RAUE R, PRUSAS C. Epidemiological studies on fowl adenoviruses isolated from cases of infectious hydropericardium. Avian Pathology, 1999, 28(5): 433-439.

[10] TORO H, PRUSAS C, RAUE R, et al. Characterization of fowl adenoviruses from outbreaks of inclusion body hepatitis/hydropericardium syndrome in Chile. Avian Disease, 1999, 43(2): 262-270.

[11] KIM J N, BYUN S H, KIM M J, et al. Outbreaks of hydropericardium syndrome and molecular characterization of Korean fowl adenoviral isolates. Avian Disease, 2008, 52(3): 526-530.

[12] REN G, WANG H, YAN Y, et al. Pathogenicity of a fowl adenovirus serotype 4 isolated from chickens associated with hydropericardium-hepatitis syndrome in China. Poultry Science, 2019, 98(7): 2765-2771.

[13] YE J, LIANG G, ZHANG J, et al. Outbreaks of serotype 4 fowl adenovirus with novel genotype, China. Emerging Microbes & Infection, 2016, 5(5): e50.

[14] ECHAVARRIA M. Adenoviruses in immunocompromised hosts. Clinical Microbiology Reviews, 2008, 21(4): 704-715.

[15] STEER P A, KIRKPATRICK N C, O'ROURKE D, et al. Classification of fowl adenovirus serotypes by use of high-resolution melting-curve analysis of the hexon gene region. Journal of Clinical Microbiology, 2009, 47(2): 311-321.

[16] UUSI-KERTTULA H, HULIN-CURTIS S, DAVIES J, et al. Oncolytic adenovirus: strategies and insights for vector design and immuno-oncolytic applications. Viruses, 2015, 7(11): 6009-6042.

[17] MENG K, YUAN X, YU J, et al. Identification, pathogenicity of novel fowl adenovirus serotype 4 SDJN0105 in Shandong, China and immunoprotective evaluation of the newly developed inactivated oil-emulsion FAdV-4 vaccine. Viruses, 2019, 11(7): 627.

[18] ZHANG Y, LIU R, TIAN K, et al. Fiber-2 and hexon genes are closely associated with the virulence of the emerging and highly pathogenic fowl adenovirus 4. Emerging Microbes & Infection, 2018, 7(1): 199.

[19] VERA-HERNÁNDEZ P F, MORALES-GARZÓN A, CORTÉS-ESPINOSA D V, et al. Clinicopathological characterization and genomic sequence differences observed in a highly virulent fowl Aviadenovirus serotype 4. Avian Pathology, 2016, 45(1): 73-81.

[20] GRIFFIN B D, NAGY E. Coding potential and transcript analysis of fowl adenovirus 4: insight into upstream ORFs as common sequence features in adenoviral transcripts. Journal of General Virology, 2011, 92(Pt 6): 1260-1272.

[21] MO K K, LYU C F, CAO S S, et al. Pathogenicity of an FAdV-4 isolate to chickens and its genomic analysis. Journal of Zhejiang University-Science B, 2019, 20(9): 740-752.

[22] HESS M. Detection and differentiation of avian adenoviruses: a review. Avian Pathology, 2000, 29(3): 195-206.

[23] SCHACHNER A, MAREK A, GRAFL B, et al. Detailed molecular analyses of the hexon loop-1 and fibers of fowl aviadenoviruses reveal new insights into the antigenic relationship and confirm that specific genotypes are involved in field outbreaks of inclusion body hepatitis. Veterinary Microbiology, 2016, 186: 13-20.

[24] PALLISTER J, WRIGHT P J, SHEPPARD M. A single gene encoding the fiber is responsible for variations in virulence in the fowl adenoviruses. Journal of Virology, 1996, 70(8): 5115-5122.

[25] SCHACHNER A, MAREK A, JASKULSKA B, et al. Recombinant FAdV-4 fiber-2 protein protects chickens against hepatitis-hydropericardium syndrome (HHS). Vaccine, 2014, 32(9): 1086-1092.

[26] MASE M, NAKAMURA K, IMADA T. Characterization of fowl adenovirus serotype 4 isolated from chickens with hydropericardium syndrome based on analysis of the short fiber protein gene. Journal of Veterinary Diagnostic Investigation, 2010, 22(2): 218-223.

[27] ZHAO J, ZHONG Q, ZHAO Y, et al. Pathogenicity and complete genome characterization of fowl adenoviruses isolated from chickens associated with inclusion body hepatitis and hydropericardium syndrome in China. PLOS ONE, 2015, 10(7): e133073.

[28] MAREK A, NOLTE V, SCHACHNER A, et al. Two fiber genes of nearly equal lengths are a common and distinctive feature of Fowl adenovirus C members. Veterinary Microbiology, 2012, 156(3-4): 411-417.

[29] PAN Q, WANG J, GAO Y, et al. The natural large genomic deletion is unrelated to the increased virulence of the novel genotype fowl adenovirus 4 recently emerged in China. Viruses, 2018, 10(9): 494.

[30] LI L, WANG J, CHEN P, et al. Pathogenicity and molecular characterization of a fowl adenovirus 4 isolated from chicken associated with IBH and HPS in China. BMC Veterinary Research, 2018, 14(1): 400.

Sequencing and phylogenetic analysis of an avian reovirus genome

Teng Liqiong, Xie Zhixun, Xie Liji, Liu Jiabo, Pang Yaoshan, Deng Xianwen, Xie Zhiqin, Fan Qing, Luo Sisi, Feng Jiaxun, and Khan Mazhar I

Abstract

Avian reovirus infection causes considerable economic loss to the commercial poultry industry. Live-attenuated vaccine strain S1133 (v-S1133, derived from parent strain S1133) is considered the safest and most effective vaccine and is currently used worldwide. To identify the genes responsible for its attenuation, DNA sequences of open reading frames (ORF) of S1133 and its parent strains S1133, 1733, 526, and C78 along with three field isolates (GuangxiR1, GuangxiR2, and GX110058) and one isolate (GX110116) from a vaccinated chicken were performed. The sequence data were compared with available sequences in nucleotide sequence databases of American (AVS-B, 138, 176) and Chinese (C-98 and T-98) origin. Sequence analysis identified that several v-S1133 specific nucleotide substitutions existed in the ORFs of λA, λB, λC, μA, μB, μNS, σA, σB, and σNS genes. The v-S1133 strain could be differentiated from the field-isolated strains based on single nucleotide polymorphisms. Phylogenetic analysis revealed that v-S1133 shared the highest sequence homologies with S1133 and reovirus isolates from China, grouped together in one cluster. Chinese isolates were clearly more distinct from the American reovirus AVS-B strain, which is associated with runting-stunting syndrome in broilers.

Keywords

avian reovirus, attenuated vaccine, single nucleotide polymorphisms, phylogenetic analysis

---

Avian reovirus (ARV) infections are responsible for significant economic losses to the poultry industry worldwide. ARVs have been associated with viral arthritis, stunting syndrome, and tenosynovitis in chickens[1, 2]. The avian orthoreovirus belongs to the genus *Orthoreovirus*, in the Reoviridae family. According to polyacrylamide gel electrophoresis analysis[3], the genome of orthoreovirus contains 10 RNA segments divided into three size classes: large (L1, L2, L3), medium (M1, M2, M3), and small (S1, S2, S3, S4). The proteins encoded by the L-class genes are designated as lambda (λ), those encoded by the M-class as mu (μ) and those encoded by the S-class as sigma (σ) [4, 5]. At least ten structural proteins and five nonstructural proteins encoded by ARV have been described[5]. Protein λA is encoded by the L1 segment and is required for viral RNA polymerase activities. Protein λB, which is encoded by the L2 segment, is a RNA-dependent RNA polymerase (RdRp) enzyme, whereas the L3 segment encoded λC protein is the viral capping enzyme[6, 7]. Protein μA, encoded by the M1 segment, interacts with the λA protein and functions as an NTPase. The M2 segment encodes protein μB, and the M3 segment encodes protein μNS[8-10]. The protein σA, encoded by the S2 segment, is a dsRNA-binding enzyme possessing protein activity. Protein σB, encoded by the S3 segment, is a major component of the reovirion outer capsid, and the S4 segment encodes protein σNS[11-13]. Three encoded proteins (σC, p10, and p17) of the S1 segment have been described, and protein σC was shown to induce apoptosis in cultured cells[14-18].

Live attenuated and inactivated vaccines are readily available for use against ARV infection. The live-attenuated vaccine S1133 (v-S1133, derived from the parent S1133 strain) is considered one of the safest available vaccines and is used widely[19]. Differences in the virulence and molecular makeup of the v-S1133 strain and its parent S1133 stock have not been described in the literature. The complete open reading frames (ORFs) of v-S1133 are not available, and virulent live viruses of the S1133 strain have not been studied.

In the present study, the complete protein coding sequences of v-S1133, S1133, 1733, 526, C78, GuangxiR1, GuangxiR2, GX110058, and GX110116 were sequenced. The v-S1133 was received from a commercial vaccine company (Intervet/Schering-Plough Animal Health), and the S1133 reference strain was acquired from the China Institute of Veterinary Drug Control. Reovirus isolates 1733, 526, and C78 were acquired from the University of Connecticut[20]. Field reovirus isolates GuangxiR1, GuangxiR2, and GX110058 were isolated from Guangxi[21, 22], and the GX110116 strain was isolated from a chicken vaccinated with the v-S1133 commercial vaccine.

Genomic RNA was extracted using a viral RNA/DNA extraction kit (TaKaRa Dalian, China). Genomic RNA was purified and RT-PCR amplification was performed as described by Bányai[23]. The PCR products were purified using agarose gel DNA purification kit (TaKaRa, Dalian, China) and cloned into the pMD18-T vector (TaKaRa, Dalian, China). Two commercial DNA sequence service companies (Invitrogen, Guangzhou China and BGI, Shenzhen, China) performed the DNA sequencing from triplicate clones of one or more PCR product. The nucleotide sequence analysis and sequence alignments were performed using the DNAStar package (Version 7.0, DNASTAR, USA) and MEGA (Version 4.1, MEGA, USA) software. The complete genomic sequences of the three American strains (AVS-B, 138, 176) and two Chinese strains (C-98 and T-98) were retrieved from the GenBank database and used for comparison with the sequence data generated in our laboratory.

Table 2-7-1 presents the single nucleotide polymorphisms (SNPs) that resulted in synonymous and nonsynonymous amino acid substitutions in the ORFs. Nucleotide changes in the ORFs unique to v-S1133, GX110116, and C78 were identified. Reovirus isolates v-S1133, GX110116, and C78 carry the five nonsynonymous nucleotide substitutions (G^{106}, A^{303}, C^{369}, C^{1302}, and T^{3324}), whereas the other sequenced strains have nucleotide substitutions at A^{106} G^{303}, T^{369}, T^{1302}, and C^{3324} in the λA protein gene. Five nonsynonymous nucleotide substitutions (T^{640}, G^{907}, C^{1588} G^{2270}, and C^{3577}) resulted in amino acid changes in the A214S, S303A, M530L, K757R, and I1193L gene segments. Three synonymous nucleotide substitutions (G^{312}, C^{2487}, and C^{2620}) in the λB protein gene and two synonymous nucleotide substitutions (C^{1104}, G^{3172}) and five nonsynonymous nucleotide substitutions (T^{106}, C^{597}, A^{837}, C^{841}, and T^{2859}) in the AC protein gene. The v-S1133, GX110116, and C78 strains are the only one to have nonsynonymous nucleotide substitution of C^{1590} in the μA protein gene. Three nonsynonymous nucleotide substitutions (T^{784}, G^{1168}, and G^{1444}) lead to amino acid changes (T262S, T/S390, and N482D) in the μB protein gene and three nonsynonymous nucleotide substitutions T^{180}, C^{1788} and G^{1890} in the μNS protein gene. Sequences of strains v-S1133, GX110116, and C78 have one synonymous nucleotide substitution (G^{241}) and two nonsynonymous nucleotide substitutions (G^{417} and G^{1110}) in the σA protein gene, two synonymous nucleotide substitutions (C^6 and A^{100}) in the σB protein gene, and one synonymous nucleotide substitution (T^{952}) and one nonsynonymous nucleotide substitution (C^{996}) in the σNS protein gene. Nonsynonymous mutations leading to amino acid substitutions appeared most frequently in the λB and μB genes of v-S1133, GX110116, and C78. Based on the SNPs, the v-S1133 strain differentiated from the field-isolated strains. Substitutions in the nucleotide sequences in AVS-B, 138, and 526 were distributed randomly throughout the complete coding region sequence.

Phylogenetic analysis using nucleotide sequences of the λA, λB, and λC proteins revealed significant genetic diversity. Reovirus v-S1133 was related closely to strains GX110116 and C78. The reovirus strains 526, 138, and AVS-B sequences were clearly more distinct from the other strains (Figure 2-7-1). Phylogenetic analysis of the μA protein gene revealed that the reovirus strain 526 sequence formed a separate lineage, whereas the 138 and AVS-B sequences were most closely related. However, analysis of the μB and μNS gene sequences provided evidence that the AVS-B sequences formed distinct lineages (Figure 2-7-2). Phylogenetic analysis of the σA and σB nucleotide sequences revealed that the reovirus strain 526 formed an isolated lineage, and the 138 and AVS-B strain sequences were most closely related; however, the σNS, the AVS-B, and the 526 sequences formed in a single phylogenetic tree (Figure 2-7-3). Phylogenetic analysis of nucleotide sequences of p10, p17, and σC revealed that the AVS-B sequences formed distinct lineages and were clearly distinct from the sequences of the other strains (Figure 2-7-4). Nucleotide sequences for genes of v-S1133, GX110116, C78, S1133, 1733, 526, GuangxiR1/R2, GX110058, 176, C-98, and T-98 grouped together in a single cluster, while the 138, 526, and AVS-B strain sequences were clearly distinct.

Table 2-7-1   Summary of single nucleotide polymorphisms in the genomes

ORFs Position	λA					λB								λC						
	102	303	369	1302	3324	312	640	907	1588	2270	2487	2620	3577	106	597	837	841	1104	2859	3172
S1133	A	G	T	T	C	A	C	T	A	A	T	T	A	C	T	G	T	A	C	A
v-S1133	G	A	C	C	T	G	T	G	C	G	C	C	C	T	C	A	C	C	T	G
GX110116	G	A	C	C	T	G	T	G	C	G	C	C	C	T	C	A	C	C	T	G
C78	G	A	C	C	T	G	T	G	C	G	C	C	C	T	C	A	C	C	T	G
1733	A	G	T	T	C	A	C	T	A	A	T	T	A	C	T	G	T	A	C	A
526	A	G	T	T	C	A	C	T	A	A	T	T	A	C	T	G	T	A	C	A
GuangxiR1	A	G	T	T	C	A	C	T	A	A	T	T	A	C	T	G	T	A	C	A
GuangxiR2	A	G	T	T	C	A	C	T	A	A	T	T	A	C	T	G	T	A	C	A
GX110058	A	G	T	T	C	A	C	T	A	A	T	T	A	C	T	G	T	A	C	A
AVS-B	A	G	T	T	C	A	C	T	A	A	T	T	A	C	T	G	T	A	C	A
138	A	G	T	T	C	A	C	T	A	A	T	T	A	C	T	G	T	A	C	A
176	A	G	T	T	C	A	C	T	A	A	T	T	A	C	T	G	T	A	C	A
C-98	A	G	T	T	C	A	C	T	A	A	T	T	A	C	T	G	T	A	C	A
T-98	A	G	T	T	C	A	C	T	A	A	T	T	A	C	T	G	T	A	C	A

ORFs Position	μA	μB			μNS				σA			σB		σNS	
	1590	784	1168	1444	172	180	1788	1890	241	417	1110	656	1090	952	996
S1133	C	A	A	A	G	C	T	A	A	A	A	T	G	C	T
v-S1133	T	T	G	G	A	T	C	G	G	G	G	C	A	T	C
GX110116	T	T	G	G	A	T	C	G	G	G	G	C	A	T	C
C78	T	T	G	G	A	T	C	G	G	G	G	C	A	T	C
1733	C	A	A	A	G	C	T	A	A	A	A	T	G	C	T
526	C	A	A	A	G	C	T	A	A	A	A	T	G	C	T
GuangxiR1	C	A	A	A	G	C	T	A	A	A	A	T	G	C	T
GuangxiR2	C	A	A	A	G	C	T	A	A	A	A	T	G	C	T
GX110058	C	A	A	A	G	C	T	A	A	A	A	T	G	C	T
AVS-B	C	A	A	A	G	C	T	A	A	A	A	T	G	C	T
138	C	A	A	A	G	C	T	A	A	A	A	T	G	C	T
176	C	A	A	A	G	C	T	A	A	A	A	T	G	C	T
C-98	C	A	A	A	G	C	T	A	A	A	A	T	G	C	T
T-98	C	A	A	A	G	C	T	A	A	A	A	T	G	C	T

Note: Nucleotide numbering corresponds to that of strain S1133.

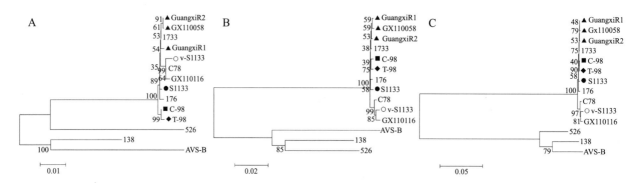

A: λA; B: λB; C: λC. The reference virus v-S1133 is marked by an open circle; S1133 is marked by filled circle; Guangxi isolates are marked by filled triangles; Tianjin isolate is marked by filled diamond, and Inner Mongolia isolate is marked by filled square.

Figure 2-7-1 Phylogenetic trees derived from the nucleotide sequences of the λA, λB, and λC genes of ARV

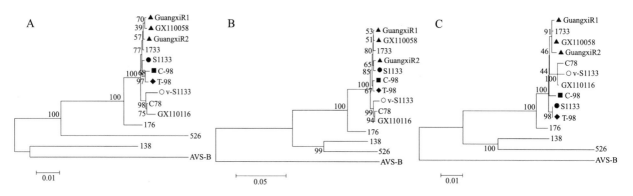

A: μA; B: μB; C:μNS. The reference virus v-S1133 is marked by open circle; S1133 is marked by filled circle; Guangxi isolates are marked by filled triangles; Tianjin isolate is marked by filled diamond, and Inner Mongolia isolate is marked by filled square.

Figure 2-7-2 Phylogenetic trees derived from the nucleotide sequences of the μA, μB, and μNS genes of ARV

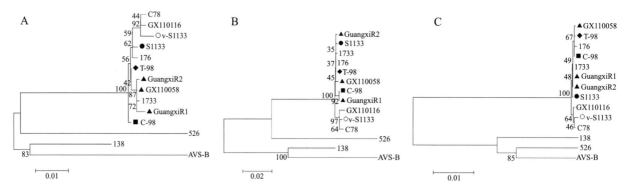

A: σA; B: σB; C: σNS. The reference virus v-S1133 is marked by open circle; S1133 is marked by filled circle; Guangxi isolates are marked by filled triangles; Tianjin isolate is marked by filled diamond, and Inner Mongolia isolate is marked by filled square.

Figure 2-7-3 Phylogenetic trees derived from the nucleotide sequences of the σA, σB, and σNS genes of ARV

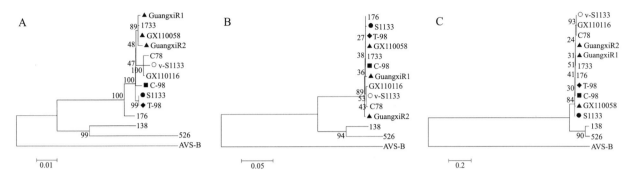

A: p10; B: p17; C: σC. The reference virus v-S1133 is marked by open circle; S1133 is marked by filled circle; Guangxi isolates are marked by filled triangles; Tianjin isolate is marked by filled diamond, and Inner Mongolia isolate is marked by filled square.

Figure 2-7-4   Phylogenetic trees derived from the amino acids sequences of the p10, p17, and σC genes of ARV

The genetic and protein sequence differences of v-S1133 and the field strains of ARVs are not defined clearly, and genomic sequence information is unavailable. The presented data make use of informative polymorphic markers in the complete ORF of the coding region sequence genes to demonstrate the presence of synonymous and nonsynonymous amino acid substitutions in v-S1133, as well as in the parental strain S1133. The v-S1133 λB and μB gene sequences contained a higher frequency of nonsynonymous mutations relative to other genes studied. Based on these SNPs, the v-S1133 strain could be differentiated from S1133 and the other field-isolated strains.

The ARV isolates circulating in Guangxi, Tianjin, and the Inner Mongolia province between 2000 and 2010 were shown to possess a high degree of sequence heterogeneity. However, the American derived strains (AVS-B, 138, and 526) possessed relatively lower similarity to v-S1133. It has been shown that the S-class and M-class gene sequence diversity among strains isolate from Taiwanses(China), as well as American, and Japanese, correlates with the time and geographic location of the isolates[24-28]. Our phylogenetic analyses revealed that the AVS-B, 138, and 526 strains grouped into a single cluster and possessed gene sequences clearly more distinct from the other strains based on λA, λB, λC, and σNS sequences.

References

[1] SU Y P, SU B S, SHIEN J H, et al. The sequence and phylogenetic analysis of avian reovirus genome segments M1, M2, and M3 encoding the minor core protein muA, the major outer capsid protein muB, and the nonstructural protein muNS. Journal of Virological Methods, 2006, 133(2): 146-157.

[2] HEGGEN-PEAY C L, QURESHI M A, EDENS F W, et al. Isolation of a reovirus from poult enteritis and mortality syndrome and its pathogenicity in turkey poults. Avian Dis, 2002, 46(1): 32-47.

[3] SPANDIDOS D A, GRAHAM A F . Physical and chemical characterization of an avian reovirus. Journal of Virology, 1976, 19(3): 968-976.

[4] VARELA R, BENAVENTE J. Protein coding assignment of avian reovirus strain S1133. Journal of Virology, 1994, 68(10): 6775-6777.

[5] BENAVENTE J, MARTÍNEZ-COSTAS J. Avian reovirus: structure and biology. Virus Research, 2007, 123(2): 105-119.

[6] XU W H, COOMBS K M. Avian reovirus L2 genome segment sequences and predicted structure/function of the encoded RNA-dependent RNA polymerase protein. Virology Journal, 2008, 5: 153.

[7] MARTÍNEZ-COSTAS J, VARELA R, BENAVENTE J. Endogenous enzymatic activities of the avian reovirus S1133: identification of the viral capping enzyme. Virology, 1995, 206(2): 1017-1026.

[8] YIN H S, SU Y P, LEE L H. Evidence of nucleotidyl phosphatase activity associated with core protein sigmaA of avian

reovirus S1133. Virology, 2002, 293(2): 379-385.

[9] MARTÍNEZ-COSTAS J, GRANDE A, VARELA R, et al. Protein architecture of avian reovirus S1133 and identification of the cell attachment protein. Journal of Virology, 1997, 71(1): 59-64.

[10] TOURÍS-OTERO F, MARTÍN M C, MARTÍNEZ-COSTAS J, et al. Avian reovirus morphogenesis occurs within viral factories and begins with the selective recruitment of sigmaNS and lambdaA to microNS inclusions. Journal of Molecular Biology, 2004, 341(2): 361-374.

[11] MARTÍNEZ-COSTAS J, GONZÁLEZ-LÓPEZ C, VAKHARIA V N, et al. Possible involvement of the double-stranded RNA-binding core protein sigmaA in the resistance of avian reovirus to interferon. Journal of Virology, 2000, 74 (3): 1124-1131.

[12] VARELA R, MARTÍNEZ-COSTAS J, MALLO M, et al. Intracellular posttranslational modifications of S1133 avian reovirus proteins. Journal of Virology, 1996, 70(5): 2974-2981.

[13] TOURÍS-OTERO F, MARTÍNEZ-COSTAS J, VAKHARIA V N, et al. Characterization of the nucleic acid-binding activity of the avian reovirus non-structural protein sigma NS. The Journal of General Virology, 2005, 86(Pt 4): 1159-1169.

[14] BODELÓN G, LABRADA L, MARTÍNEZ-COSTAS J, et al. The avian reovirus genome segment S1 is a functionally tricistronic gene that expresses one structural and two nonstructural proteins in infected cells. Virology, 2001, 290(2): 181-191.

[15] DAY J M, PANTIN-JACKWOOD M J, SPACKMAN E. Sequence and phylogenetic analysis of the S1 genome segment of turkey-origin reoviruses. Virus Genes, 2007, 35(2): 235-242.

[16] BÁNYAI K, PALYA V, BENKO M, et al. The goose reovirus genome segment encoding the minor outer capsid protein, sigma1/sigmaC, is bicistronic and shares structural similarities with its counterpart in muscovy duck reovirus. Virus Genes, 2005, 31(3): 285-291.

[17] SHMULEVITZ M, YAMEEN Z, DAWE S, et al. Sequential partially overlapping gene arrangement in the tricistronic S1 genome segments of avian reovirus and Nelson Bay reovirus: implications for translation initiation. Journal of Virology, 2002, 76(2): 609-618.

[18] SHIH W L, HSU H W, LIAO M H, et al. Avian reovirus sigmaC protein induces apoptosis in cultured cells. Virology, 2004, 321(1): 65-74.

[19] HEIDE V, KALBAC M, BRUSTOLON M. Development of an attenuated apathogenic reovirus vaccine against viral arthritis/tenosynovitis. Avian Dis, 1983, 27(3): 698-706.

[20] XIE Z X, FADL A A, KHAN M I, et al. Amplification of avian reovirus RNA using the reverse transcriptase-polymerase chain reaction. Avian Dis, 1997, 41(3): 654-660.

[21] LIAO M, XIE Z X, LIU J B, et al. Isolation and identification of avian reovirus. China Poultry, 2002, 24(1): 12-14.

[22] XIE Z X, PENG Y, LUO S S, et al. Development of a reverse transcription loop-mediated isothermal amplification assay for visual detection of avian reovirus. Avian Pathol, 2012, 41(3): 311-316.

[23] BÁNYAI K, DANDÁR E, DORSEY K M, et al. The genomic constellation of a novel avian orthoreovirus strain associated with runting-stunting syndrome in broilers. Virus Genes, 2011, 42(1): 82-89.

[24] LIU H J, HUANG P H. Sequence and phylogenetic analysis of the sigmaA-encoding gene of avian reovirus. Journal of Virological Methods, 2001, 98(2): 99-107.

[25] LIU H J, LEE L H, HSU H W, et al. Molecular evolution of avian reovirus: evidence for genetic diversity and reassortment of the S-class genome segments and multiple cocirculating lineages. Virology, 2003, 314(1): 336-349.

[26] LIU H J, LEE L H, HSU H W, et al. Rapid characterization of avian reoviruses using phylogenetic analysis, reverse transcription-polymerase chain reaction and restriction enzyme fragment length polymorphism. Avian Pathol, 2004, 33(2): 171-180.

[27] HSU H W, SU H Y, HUANG P H, et al. Sequence and phylogenetic analysis of P10- and P17-encoding genes of avian reovirus. Avian Dis, 2005, 49(1): 36-42.

[28] SU Y P, SU B S, SHIEN J H, et al. The sequence and phylogenetic analysis of avian reovirus genome segments M1, M2, and M3 encoding the minor core protein muA, the major outer capsid protein muB, and the nonstructural protein muNS. Journal of Virological Methods, 2006, 133(2): 146-157.

Molecular characterization of chicken anemia virus in Guangxi, southern China, from 2018 to 2020

Zhang Minxiu, Deng Xianwen, Xie Zhixun, Zhang Yanfang, Xie Zhiqin, Xie Liji, Luo Sisi, Fan Qing, Zeng Tingting, Huang Jiaoling, Wang Sheng

Abstract

Background: Chicken anemia virus (CAV) causes chicken infectious anemia, which results in immunosuppression; the virus has spread widely in chicken flocks in China. Objectives: The aim of this study was to understand recent CAV genetic evolution in chicken flocks in Guangxi, southern China. Methods: In total, 350 liver samples were collected from eight commercial broiler chicken farms in Guangxi in southern China from 2018 to 2020. CAV was detected by conventional PCR, and twenty CAV complete genomes were amplified and used for the phylogenetic analysis and recombination analysis. Results: The overall CAV-positive rate was 17.1%. The genetic analysis revealed that 84 CAVs were distributed in groups A, B, C (subgroups C1-C3) and D. In total, 30 of 47 Chinese CAV sequences from 2005-2020 belong to subgroup C3, including 15 CAVs from this study. There were some specific mutation sites among the intergenotypes in the VP1 protein. The amino acids at position 394Q in the VP1 protein of 20 CAV strains were consistent with the characteristics of a highly pathogenic strain. GX1904B was a putative recombinant. Conclusions: Subgroup C3 was the dominant genotype in Guangxi from 2018-2020. The 20 CAV strains in this study might be virulent according to the amino acid residue analysis. These data help improve our understanding of the epidemiological trends of CAV in southern China.

Keywords

chicken anemia virus, genetics, recombination, genome, China

Introduction

Chicken infectious anemia (CIA) is caused by the chicken anemia virus (CAV). CIA is an important immunosuppressive disease characterized by aplastic anemia, bone marrow and lymphoid organ atrophy in young chicks aged less than 2~3 weeks[1]. CAV mainly damages the immune system of chickens and is likely to cause concurrent or secondary infection with other pathogens[2]. For example, CAV coinfection with other immunosuppressive viruses, such as Marek's disease virus (MDV), infectious bursal disease virus (IBDV) and avian leukosis virus (ALV), can not only mutually enhance pathogenicity in chickens, but also has a significant synergistic effect on immunosuppression[3-5]. Therefore, CAV infection can directly or indirectly cause substantial economic losses to the chicken industry.

CAV belongs to the genus *Gyrovirus*, family Anelloviridae[6]. CAV is a nonenveloped, circular, single-stranded DNA virus consisting of a circular genome of 2 298 to 2 319 nucleotides, with three parts or completely overlapping open reading frames (ORFs)[6]. ORF1 encodes the only capsid protein associated

251

with virulence and replication, designated viral protein 1 (VP1)[7-9]. ORF2 encodes viral protein 2 (VP2), a scaffold protein and an auxiliary protein of VP1, which helps VP1 achieve the correct conformation[10]. ORF3 encodes viral protein 3 (VP3), a nonstructural protein that is thought to play a role in the apoptosis of thymic lymphoblasts and primitive haematopoietic cells[11].

CAV has spread widely in chickens worldwide since it was first reported in Japan in 1979[12]. In 1996, CAV was first isolated and identified in chicken flocks in China, and then CAV was subsequently detected in chicken flocks in Guangdong Province, Jiangsu Province, Shandong Province and other places[13-17]. The genetic characterization of CAV in the above provinces of China has been reported[13-17], but the molecular evolution of CAV in Guangxi in southern China has not yet been reported. Guangxi is located on the border in southern China. There are currently thousands of chicken farms in Guangxi, and frequent introduction and transportation pose a risk for the spread and genetic recombination of CAV. Therefore, we report the genetic and recombination characterization of CAVs from Guangxi of southern China from 2018 to 2020 to better understand the genetic evolution of CAV in recent years in Guangxi, southern China.

Materials and methods

Sample collection, DNA/RNA extraction and CAV detection

A total of 350 liver samples were collected from deceased, diseased chickens that were not vaccinated with CAV from chicken farms in four cities in Guangxi in southern China (Table 2-8-1). Among the 350 liver samples, 51 were collected from 5-to 21-day-old chicken flocks, 232 were collected from 22-to 100-day-old flocks, and 67 were collected from 260-to 290-day-old flocks. All samples were stored at −80 ℃ until processing. A commercial TransGen Biotech EasyPure® Genomic DNA/RNA Kit (TransGen, Beijing, China) was used to extract DNA/RNA from the liver samples. CAV and Fowl adenovirus 4 (FAdV-4) DNA from the samples were detected directly by PCR. The primers and PCR procedure of CAV and FAdV-4 were performed according to previously described reports[18, 19]. An isolate CAV GXC060821 (GenBank accession number: JX964755) was preserved in our laboratory and used as a positive control in the PCR. Avian influenza virus (AIV) (H5 and H9 subtypes) and avian leukosis virus subgroup J (ALV-J) RNA from the samples were used for cDNA synthesis, and then PCR for the detection of the viruses was performed according to a previously described procedure[20, 21].

Table 2-8-1 Information on samples collected from commercial chicken farms and 20 CAV genome sequences

Farm name	Breed	Number of chickens	Age of chickens in the farm	City	Positive rate of CAV (positive samples/ total samples)	Age of chickens with positive samples	Accession number	Strain name	Year of collection	Age/ days
Farm A	Ma chicken	8 488	1~300 days	Nanning	13.5% (12/89)	48, 83, 63 and 75 days	MK484614	GX1801	2018	83
							MK484615	GX1804	2018	63
							MN103405	GX1805	2018	75
Farm B	Xiang chicken	11 230	20~110 days	Nanning	11.7% (2/17)	45 and 60 days	MK484616	GX1810	2018	45
Farm C	Xiang chicken	8 825	20~110 days	Baise	26.4% (14/53)	50, 96, 86, 89, 27 and 98 days	MN103402	GX1904A	2019	50
							MN103406	GX1904B	2019	50
Farm D	Three yellow chicken	7 800	20~110 days	Yulin	20.5% (8/39)	40, 60 and 90 days	MN103403	GX1904P	2019	90
							MW554706	GX2020-D1	2020	60
							MW579761	GX2020-D3	2020	60
							MW579762	GX2020-D6	2020	40

continued

Farm name	Breed	Number of chickens	Age of chickens in the farm	City	Positive rate of CAV (positive samples/ total samples)	Age of chickens with positive samples	Accession number	Strain name	Year of collection	Age/ days
Farm E	Xiang chicken	10 002	20~85 days	Nanning	6.8% (4/59)	63 and 70 days	MN103404	GX1905	2019	70
Farm F	Xiang chicken	8 658	20~110 days	Nanning	16.1% (11/68)	40, 45, 48 and 60 days	MN649258	GX1907A	2019	48
							MN649259	GX1907B	2019	48
							OK012319	GXWM201902	2019	40
							OK012320	GXWM201901	2019	40
Farm G	Wu chicken	9 965	20~120 days	Nanning	40.9% (9/22)	24, 46, 53 and 83 days	MN649254	GX1908W1	2019	53
							MN649255	GX1908W3	2019	53
							MN649256	GX1908L2	2019	24
							MN649257	GX1908L3	2019	24
Farm H	Hua chicken	6 590	20~110 days	Qinzhou	33.3% (1/3)	30 days	OK012318	GXQZ202001	2020	30

Note: CAV, chicken anemia virus.

Full-length PCR amplification and sequencing

Twenty CAV-positive samples were randomly selected from CAV-positive samples for complete genome amplification. The primers for CAV complete genome amplification were designed based on the complete genome of the Cux-1 strain (GenBank accession number: M55918). The primers CAV1F 5'-CCGCGCAGGGGCAAGTA-3' and CAV1R 5'-TCGCGGAGGGCAYGTTATTATCTA-3' were applied to amplify an 872 bp fragment. CAV2F 5'-GCCCCATCGCCGGTGAGTTGA-3' and CAV2R 5'-TGCCGGTTACCCA-GTTGCCAVAC-3' were applied to amplify a 907 bp fragment. CAV3F 5'-ATGAGACCCGACGAGCAAC-3' and CAV3R 5'-CCACACAGCGATAGAGTGATTG-3' were applied to amplify a 1 254 bp fragment. The three primer sets covered the entire nucleotide sequence. CAV GXC060821 was used as a positive control in the PCR amplification. The PCR was developed with a 2 × TransTaq-T PCR kit (TransGen, Beijing, China) in a 50 μL volume including 25 μL of 2 × TransTaq-T PCR SuperMix, 1 μmol/L each primer (20 μmol/L), 5 μL of DNA and 15 μL of distilled water. The PCR amplification reaction was performed as follows: 94 ℃ for 5 min; 35 cycles of 94 ℃ for 30 s, 60 ℃ for 1 min, and 72 ℃ for 30 s; and 72 ℃ for 10 min. The purification and cloning of PCR products were performed according to the method used by Teng et al.[22], and the cloned vector containing the target gene was sent for sequencing (TaKaRa, Dalian, China).

Phylogenetic analysis and molecular characterization of CAV

Sixty-four reference CAV sequences (information on CAV sequences is available in Table 2-8-2) were obtained from GenBank. There were few CAV sequences after 2017 in China. Of the 64 reference CAV sequences, 27 were obtained between 2005-2016 in China and the remaining sequences (n=37) were obtained between 1991-2019 in Japan, the USA, Britain, Australia, Italy, Malaysia, Tunisia and Argentina. The sequences in this study obtained by sequencing were assembled, and 20 complete genomes were generated using LaserGene version 7.1 software. A phylogenetic tree based on the complete CAV genomes was constructed using the neighbour-joining algorithm implemented in MEGA version 6.0 with 1 000 bootstrap replications. The nomenclature (Country/GenBank accession number/strain name/year) of CAV strains was used in the phylogenetic tree. Nucleic acid sequences and deduced amino acid sequence alignment of the VP1, VP2 and VP3 genes were performed by comparison with 64 reference CAV sequences using LaserGene version 7.1 software. Alignment reports were exported and analysed.

Table 2-8-2　Reference strains retrieved from the GenBank database

Strain name	Country/Province	GenBank accession number	Reference	Year
Cux-1	Germany	M55918	[6]	1991
C369	Japan	AB046590	[8]	2000
C368	Japan	AB046589	[8]	2000
GD-J-12[a]	China/Guangdong	KF224934	[15]	2012
GD-K-12	China/Hainan	KF224935	[15]	2012
JN1503	China/Shandong	KU641014	[16]	2015
HB1404	China/Hubei	KU645514	[16]	2014
JS1501	China/Jiangsu	KU645518	[16]	2015
SD1510	China/Shandong	KU598851	[16]	2015
CAU269/7	Australia	AF227982	[23]	2000
3711	Australia	EF683159		2007
SD24[a]	China	AY999018	[24]	2005
6	China/Taiwan	KJ728817		2012
Clone 34	Britain	AJ297685	[25]	2001
Clone 33	Britain	AJ297685	[25]	2001
c-CAV	USA	NC001427	[6]	1991
SMSC-1P60	Malaysia	AF390102	[26]	2001
3-1P60	Malaysia	AY040632	[26]	2001
HLJ15108	China/Heilongjiang	KY486137	[27]	2015
HB1517[a]	China/Hubei	KU645516	[27]	2015
HN1504[a]	China/Hunan	KU645512	[27]	2015
GXC060821	China/Guangxi	JX964755		2006
GD-103	China/Guangdong	KU050678		2014
GD-104	China/Guangdong	KU050679		2014
GD-101	China/Guangdong	KU050680		2014
N8	China/Guangdong	MK887164		2016
LN15169	China/Liaoning	KY486154	[27]	2015
SD1505[a]	China/Shandong	KU645523	[27]	2015
HN1405	China/Hunan	KU645520	[27]	2014
SD1508[a]	China/Shandong	KU645519	[27]	2015
GD-1-12	China/Guangdong	JX260426	[28]	2012
TR20	Japan	AB027470		1999
AH4	China/Anhui	DQ124936		2005
CIAV-Shanxi7	China/Shanxi	MH186142		2018
LY-2	China/Shandong	KX447637		2016
SD1513	China/Shandong	KU645517		2015
CAV-18	Argentina	KJ872514	[29]	2007
SMSC-1	Malaysian	AF285882	[30]	2000
SD1403	China/Shandong	KU221054		2014
704	Australia	U65414		1996

continued

Strain name	Country/Province	GenBank accession number	Reference	Year
CAV-EG-14	Egypt	MH001565	[31]	2017
CAV-10	Argentina	KJ872513	[29]	2007
LY-1	China/Shandong	KX447636		2016
CAV-EG-2	Egypt	MH001553	[31]	2017
CAV-EG-11	Egypt	MH001559	[31]	2017
CAV-EG-13[a]	Egypt	MH001560	[31]	2017
CAV-EG-26	Egypt	MH001564	[31]	2017
CIAV/IT/CK/909-06/18	Italy	MT813068	[32]	2018
CIAV/IT/CK/1196/19	Italy	MT813069	[32]	2019
CIAV/IT/CK/1155/19	Italy	MT813070	[32]	2019
CIAV/IT/CK/855/17	Italy	MT813071	[32]	2017
CIAV/IT/CK/1157/19	Italy	MT813072	[32]	2019
CIAV/IT/CK/986-2/18	Italy	MT813073	[32]	2018
CIAV/IT/CK/1099/19	Italy	MT813074	[32]	2019
CIAV/IT/CK/1153-2/19	Italy	MT813075	[32]	2019
CIAV/IT/CK/1188/19	Italy	MT813078	[32]	2019
CIAV/IT/CK/1180/19	Italy	MT813076	[32]	2019
CIAV/IT/CK/1186/19	Italy	MT813077	[32]	2019
CIAV/IT/CK/1188/19	Italy	MT813078	[32]	2019
CIAV_TN_7-1	Tunisia	MZ666088	[33]	2019
CIAV_TN_7-2	Tunisia	MZ666089	[33]	2019
CIAV_TN_7-3	Tunisia	MZ666090	[33]	2019
CIAV_TN_7-4	Tunisia	MZ666091	[33]	2019
CIAV_TN_7-5	Tunisia	MZ666092	[33]	2019
CIAV_TN_7-6	Tunisia	MZ666093	[33]	2019

Note: [a] Recombination events were confirmed by previous reports.

Recombination analysis

Twenty CAV genomes in this study and 64 reference CAV complete genomes of CAV were aligned using MEGA version 6.0 software. Then, the result was imported into recombination detection program 4 (RDP4) software. Seven independent methods (RDP, GENECONV, MaxChi, BootScan, Chimaera, SiScan and 3Seq) were implemented in RDP4 software using default parameters to evaluate the potential recombination sequences[34]. The default settings were used for all methods, and the highest acceptable P value cut-off was set at 0.05. The parental sequences and localization of possible recombination breakpoints were assessed and identified. When a recombination event was supported by more than the above mentioned 6 methods with $P<0.05$ and recombination consensus scores>0.6, the recombination event was regarded as positive. The potential recombinant sequences, parental sequences and CAV sequences from the outgroup were analysed using simplot similarity in SimPlot version 3.5.1[35].

Results

Detection of CAV

The total CAV-positive rate was 17.1% (60/350), and all 60 CAV-positive samples were collected from

22-to 100-day-old chicken flocks (Table 2-8-1). CAVs were not detected in the samples collected from 5-to 21-day-old or 260-to 290-day-old chickens. However, 22 out of the 60 CAV-positive samples (36.7%) were coinfected with one each of FAdV-4 or H5 subtype AIV. CAV coinfection with FAdV-4 was detected in 16 samples, and 6 samples were coinfected with H5 subtype AIV.

Phylogenetic analysis of CAV

Twenty CAV genome sequences in this study were obtained by sequencing, with a total length of 2 298 bp. Detailed information on the 20 CAV genome sequences is provided in Table 2-8-2. Phylogenetic analysis of complete CAV genome sequences (n=84) by the neighbour-joining method separated the CAV strains into 4 distinct genotypes: groups A (n=2), B (n=3), C (n=55) and D (n=24) (Figure 2-8-1). Group C comprised three distinct subgroups (subgroups C1-C3) based on the shape of tree branches (Figure 2-8-1). Phylogenetic analysis suggested that genotyping of CAV in China showed obvious geographical characteristics and no significant correlation with sampling time (Figure 2-8-1). The phylogenetic tree showed that 63.8% (30/47) of Chinese CAV sequences (n=47) from 2005-2020 formed a special branch (subgroup C3) different from CAV sequences from other countries. Subgroup C3 (n=32) was the dominant genotype in China from 2005-2020, and 93.8% (30/32) of subgroup C3 are Chinese strains. 15 complete CAV genome sequences from the present study were classified to subgroup C3 (GX1801, GX1804, GX1805, GX1810, GX1904P, GX1905, GX1907A, GX1907B, GX1908W1, GX1908W3, GX1908L2, GX1908L3, GXQZ202001, GX2020-D6 and GX2020-D3). Five out of 47 Chinese CAV sequences belonged to subgroup C1 (n=12) (Figure 2-8-1). None of the Chinese CAV sequences belonged to subgroup C2 (n=11). Subgroup C2 mainly included British, American, Italian and Egyptian CAV strains. Group D comprised 9 Chinese CAV sequences (including GX2020-D1 and GX1904A) and 15 CAV sequences from Japan, Australia, Italy, Malaysia, Tunisia and Argentina. Group B included three Chinese CAV sequences (including GX1904B). Group A included two Australian CAV sequences.

Molecular characterization

The nucleotide sequences of the 20 CAV genome sequences in this study exhibited 96%~99.8% homology. The nucleotide sequence identities of the complete genome were 94.8%~99.9% between the 20 CAV genome sequences and 64 reference CAV genome sequences from GenBank. GX1804 had the highest identity (99.9%) with strain GD-104 (accession number: KU050679), a Chinese strain isolated in 2014. Strain GX1908W1 had the lowest identity (94.8%) with strain 3711 (accession number: EF683159), an Australian strain isolated in 2007.

All the VP1, VP2 and VP3 amino acid sequences from groups A, B, C (subgroups C1-C3) and D were aligned according to the genotyping from the phylogenetic analysis. The amino acid homology among the 20 CAV VP1 proteins obtained in this study was 97.1%~100%, with a similarity level of 100% among GX1908L2, GX1908L3, GX1805, GX1810 and GX1907A. Amino acid sequence identities among the 20 CAV VP1 and 64 CAV VP1 sequences retrieved from GenBank ranged from 96.4% to 100%. Comparison of the deduced VP2 and VP3 protein amino acid sequences of the 20 CAV strains in this study and 64 reference strains revealed overall predicted amino acid sequence identities ranging from 94.0% to 100%.

The main amino acid variations in VP1 were obtained by aligning the 20 CAV and 64 CAV VP1 amino acid sequences. The alignment results are shown in Figure 2-8-2. The mutational positions were different among the genotypes. Because of the small number of sequences in group A and group B, the mutations of VP1 from group C and group D were analysed. The amino acid residues at position 370 of VP1 exhibited a

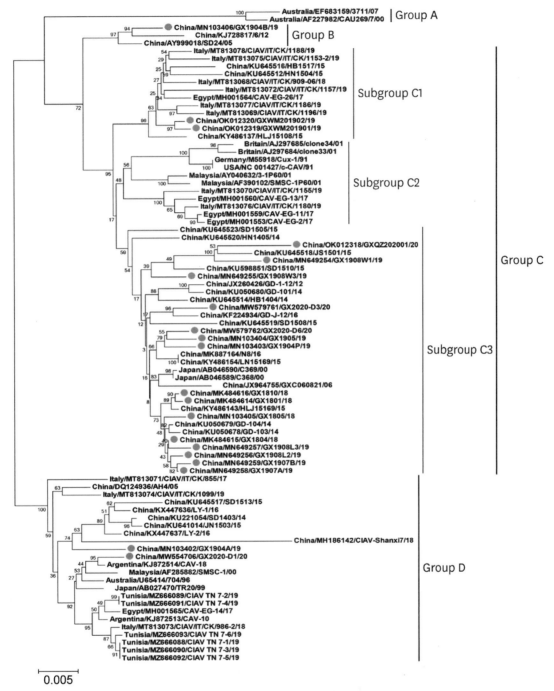

Values≥60 are indicated on the branches (as percentages); the circles represent the new complete CAV genome sequences.

Figure 2-8-1   Phylogenetic analysis of the 20 new complete CAV genome sequences and 64 reference complete CAV genome sequences available in GenBank

relatively high mutation probability (Figure 2-8-2). The mutation at position 370 accounted for approximately 61.9% (52/84) of the total number of VP1 amino acid sequences. The mutation at position 370 mainly manifested in Group B, Subgroup C1 and Group D (Figure 2-8-2). Thirty-two out of 84 VP1amino acid sequences had glycine (Gly) at position 370, and the remaining 52 VP1 amino acid sequences had alanine (Ala), threonine (Thr), serine (Ser) or arginine (Arg) (Figure 2-8-2).

The main amino acid variations in subgroup C3 were mainly concentrated at positions 97

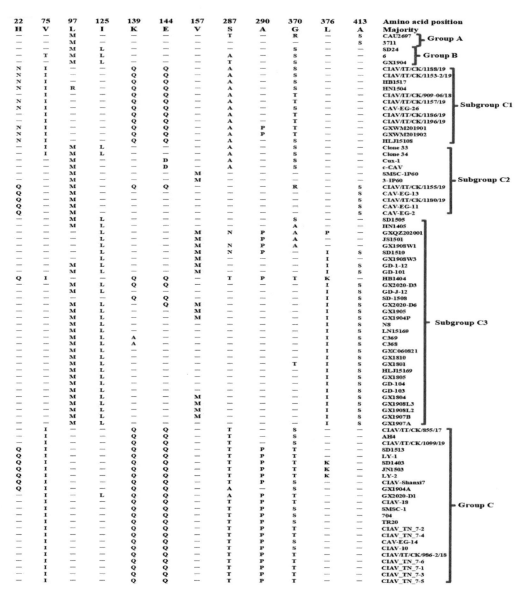

Figure 2-8-2　Analysis of major mutations of amino acid residues derived from VP1 genes of CAVs

(leucine[Leu]→methionine[Met]), 125 (isoleucine[Ile]→Leu), 157 (valine[Val]→Met), 376 (Leu→Ile) and 413 (Ala→Ser). The mutation probability in those positions was approximately 78.1%, 93.8%, 43.8%, 87.5% and 78.1%, respectively, based on the subgroup C3 CAV sequences. The amino acid substitutions in subgroup C2 were mainly concentrated at position 97 (Leu→Met), with a mutation probability of 100% (Figure 2-8-2). The amino acid substitutions in group D were mainly concentrated at positions 75 (Val→Ile), 139 (Lysine[Lys]→glutamine[Gln]), 144 (glutamate[Glu]→Gln), 287 (Ser→Ala/Thr), 290 (Ala→Proline[Pro]) and 370 (Gly→Thrand Ser) (Figure 2-8-2). The amino acid substitutions in subgroup C1 were mainly concentrated at positions 75 (Val→Ile), 139 (Lys→Gln), 144 (Glu) →Gln), 287 (Ser→Ala) and 370 (Gly→Thr and Ser), all with a mutation probability of 100% (Figure 2-8-2). The amino acid substitutions in group D and subgroup C1 had the same amino acid residue at positions 75, 139 and 144, but the residues differed between group D and subgroup C1 at positions 287 and 290 (Figure 2-8-2). The above data reflect the uniqueness of the VP1 amino acid sequences of the group D, subgroup C1 and subgroup C3 genotypes at mutant positions; however, more sequences are needed for alignment. Analysis of the amino acid sequences of VP2 and VP3 showed that there were no special differences in amino acid sequences among the strains (data not shown).

The amino acids at positions 75, 89, 125, 141, 144 and 394 of VP1 related to CAV virulence were analysed by alignment with the virulent (Cux-1 and C368) and attenuated (C369) strains (Table 2-8-3). In this study, it was observed that the amino acids at positions 75, 89, 125, 141, 144 and 394 in the VP1 protein of 14 CAVs were Val, Thr, Leu, Gln, Glu and Gln, respectively, which is consistent with the virulent strain (C368) residues at the same positions. The amino acid at position 394Q in the VP1 protein of the remaining CAV strains from this study were consistent with the virulent strain (Cux-1).

Table 2-8-3   Key amino acid changes in VP1 among the 20 CAVs and Cux-1, C368, and C369

Strain	Amino acid position in VP1					
	75	89	125	141	144	394
Cux-1	V	T	I	Q	D	Q
C369	V	T	L	Q	E	H
C368	V	T	L	Q	E	Q
GX1801	V	T	L	Q	E	Q
GX1804	V	T	L	Q	E	Q
GX1805	V	T	L	Q	E	Q
GX1810	V	T	L	Q	E	Q
GX1904A	I	T	L	Q	Q	Q
GX1904B	V	T	L	Q	E	Q
GX1904P	V	T	L	Q	E	Q
GX1905	V	T	L	Q	E	Q
GX1907A	V	T	L	Q	E	Q
GX1907B	V	T	L	Q	E	Q
GX1908W1	V	T	L	Q	E	Q
GX1908W3	V	T	L	Q	E	Q
GX1908L2	V	T	L	Q	E	Q
GX1908L3	V	T	L	Q	E	Q
GXWM201901	I	T	I	Q	Q	Q
GXWM201902	I	T	I	Q	Q	Q
GX2020-D1	I	T	L	Q	Q	Q
GX2020-D3	V	T	L	Q	Q	Q
GX2020-D6	V	T	L	Q	Q	Q
GXQZ202001	V	T	L	Q	E	Q

Note: CAV, chicken anemia virus.

Recombination analysis

RDP4 software and SimPlot 3.5.1 software were used to detect and confirm the potential recombination events using 20 new CAV genome sequences and 64 reference CAV complete genomes. The two potential recombination events (GX1904B and isolate 6) in CAV sequences are found using the RDP4 software. The results showed that GX1904B of group B was a potential recombinant strain from the major parent isolate 704 (group D) and minor parent isolate HB1404 (group C). The simplot similarity implemented in SimPlot version 3.5.1 software confirmed that the beginning breakpoint and ending breakpoint mapped to positions 689 and 1485, respectively (Figure 2-8-3). The major parent isolate and minor parent isolate of isolate 6 were

GX1908L3 (group C) and GX1904A (group D). The beginning breakpoint and ending breakpoint were mapped to positions 566 and 1584, respectively. SD24 was confirmed to be a recombination event and suspected to be a mosaicism originating from two genotypes by He et al[24]. The phylogenetic analysis classified SD24, isolate 6 and GX1904B to group B, suggesting that group B may be a branch of gene recombination strains. However, additional mosaicisms derived from group C strains and group D strains are needed to verify this conclusion. In this study, six recombination events (the CAV sequences are provided in Table 2-8-2), except for SD24, among 64 reference CAV sequences were used to construct the phylogenetic tree; however, previous reports have confirmed that the sequences were recombinants derived from different strains of the same genotype[15, 27, 31].

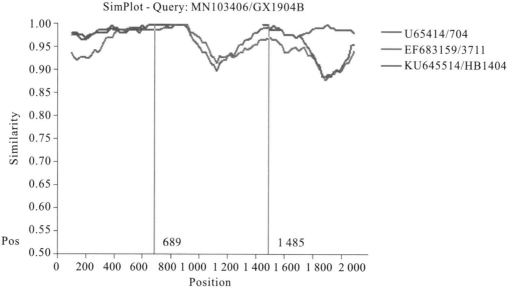

The GX1904B sequence was used as the query; the 3711 strain was included as an outgroup; the y-axis gives the percentage of identity with a sliding window of 200 bp and a step size of 20 bp.

Figure 2-8-3   A comparison of the sequences of three CAVs: 704, HB1404, and GX1904B

Discussion

CAV was first reported in 1996 in China[13], and subsequently, the virus has spread widely in chicken flocks in China. In this study, none of the CAV samples collected from 5-to 21-day-old chickens were positive, which might have been due to the presence of maternal antibodies against CAV in chicks, allowing the chicks to avoid CAV infection. Sixty of the 350 samples were CAV positive, the positive samples were all from 22- to 100-day-old deceased or diseased chickens, and 36.7% of the CAV-positive samples exhibited coinfection of FAdV-4 and H5 subtype AIV. The detection results suggest that chickens aged over 3 weeks infected with CAV might be some cases of latent infection and secondary infections caused by FAdV-4 or other pathogens probably due to the decrease in maternal antibodies. However, because of the small number of samples, the correlation between age and CAV infection rate could not be determined, so it is necessary to collect more clinical samples from chickens of different ages for correlation analysis and to confirm the above conjecture.

In this study, the CAVs were divided into four groups (groups A, B, C and D) according to the CAV classification method of Eltahir et al.[14]. Group C was separated into three subgroups (subgroups C1-C3) in this study. The 20 strains in this study were distributed in group B, subgroup C1, subgroup C3 and D, and 15 out of 20 strains belonged to subgroup C3, indicating that subgroup C3 was the dominant genotype in Guangxi, southern China. In addition, 63.8% (30/47) of Chinese CAV sequences from 2005-2020 were classified to subgroup C3 and 93.8% (30/32) of subgroup C3 are Chinese strains; A total of 19.1% (9/47) of the Chinese CAV sequences were classified to Group D. All eleven CAV sequences from Britain, the USA, Italy and Egypt were in subgroup C2. The CAV sequences from Japan, Australia, Italy, Malaysia, Tunisia and Argentina accounted for approximately 62.5% (15/24) of the total number of CAV sequences in Group D. The results showed that the genotype of CAV from China showed obvious geographic characteristics and no significant correlation with sampling time.

The classification of CAV has been previously described. Zhang et al.[15] reported that CAV was divided into five groups (A, B, C, D and E) based on 54 partial genomic sequences. CAV was separated into four major groups (A-D) by Eltahir using 55 complete VP1, VP2 and VP3 sequences[14]. Li et al.[16] analysed 121 complete CAV genomes and classified them into eight lineages. The reasons for the different CAV classification results obtained by different researchers are as follows: first, different reference strains were analysed by each researcher, and the CAV sequences might not have been complete sequences; second, the methods and software used to construct the phylogenetic tree were different (Zhang and Eltahir used the neighbour-joining statistical method in MEGA software to analyse CAV genetic evolution[14, 15], and, in contrast, maximum likelihood method statistical analysis based on the full-length genomes was implemented in RAxML software by Li et al.[16]); third, the genetic diversity of CAV and the number of CAV sequences used in phylogenetic analysis might explain the clustering of CAV. The fact that different classifications of CAV have been obtained indicates that additional CAV full-length genomes are needed to analyse genetic evolution.

The main amino acid variations in VP1 were obtained by aligning the 20 CAV and 64 CAV VP1 amino acid sequences. There were some special mutation sites among the intergenotypes in the VP1 amino acid sequences. Positions 125, 376 and 413 in the VP1 amino acid sequence of the subgroup C3 genotype with a high mutation probability above 78.1% occurred. The amino acid substitutions of subgroup C2 were mainly concentrated at position 97 with a mutation probability of 100%. The amino acid substitutions of group D and subgroup C1 had the same amino acid residues at positions 75, 139 and 144, but the amino acid residues at positions 287 and 290 differed between group D and subgroup C1. VP1 is the major structural protein of CAV and is believed to be involved in viral replication and virulence[7-9]. Therefore, the mutations at these positions among different genotypes might affect viral virulence and antigenic variation.

The pathogenicity of CAV in chickens is closely related to the amino acid residues of the VP1 protein. The amino acid at position 394 of VP1 has been demonstrated to be crucial for the pathogenicity of CAV, and all the highly virulent cloned strains had Gln at position 394, but all the low-virulence cloned strains had His[8]. Previous studies have shown that all changes at positions 75, 89, 125, 141 and 144 of the VP1 protein caused Cux-1 attenuation (the 394 position was still retained as Gln); however, if only one or four of them was changed, Cux-1 maintained high virulence[9]. The C368 strain was found to be highly virulent by Yamaguchi et al.[8]. The amino acids at positions 75, 89, 125, 141, 144 and 394 in VP1 of 14 strains (GX1801, GX1804, GX1805, GX1810, GX1904P, GX1905, GX1907A, GX1907B, GX1908W1, GX1908W3, GX1908L2, GX1908L3, GX1904B and GXQZ202001) in this study were consistent with those of C368 at the same positions, which indicated that the 14 strains might be highly virulent viruses. The residue at position 394 of the remaining six strains was Gln, which was also consistent with the characteristics of virulent viruses. Therefore, these results indicated that the 20 strains all might be virulent viruses.

Recombination events in CAV occur frequently. Recombination events occur not only among different strains in the same group, but also across different genotype[24, 36]. Previous studies have shown that the gene recombination of CAV can occur in both coding and noncoding regions[24, 36]. The gene recombination event in this study involved the recombination of the coding region of CAV, which might impact the evolution of coding region genes of CAV. In summary, 20 full-length CAV genomes originating from commercial broiler chicken flocks in Guangxi in southern China were characterized. CAV was separated into four major groups A, B, C (subgroup C1-C3) and D in this study, and subgroup C3 was the dominant genotype in Guangxi in southern China. VP1 protein analysis of 20 strains showed that these strains might be virulent, and one CAV recombination event among the 20 full-length CAV genomes was detected. The data in this study are helpful for understanding the recent genetic evolution of CAV in chicken flocks in southern China.

References

[1] TANIGUCHI T, YUASA N, MAEDA M, et al. Chronological observations on hemato-pathological changes in chicks inoculated with chicken anemia agent. Natl Inst Anim Health Q (Tokyo), 1983, 23(1): 1-12.

[2] CHU K S, KANG M S, RIM S H, et al. Coinfected cases with adenovirus, chicken infectious anemia virus and Newcastle disease in broiler chickens. Korean Journal of Veterinary Service, 2010, 33(1): 7-12.

[3] ROSENBERGER J K, CLOUD S S. The effects of age, route of exposure, and coinfection with infectious bursal disease virus on the pathogenicity and transmissibility of chicken anemia agent (CAA). Avian Dis, 1989: 753-759.

[4] MILES A M, REDDY S M, MORGAN R W. Coinfection of specific-pathogen-free chickens with Marek's disease virus (MDV) and chicken infectious anemia virus: effect of MDV pathotype. Avian Dis, 2001, 45(1): 9-18.

[5] CHACON J L, NOGUEIRA E O, BRENTANO L, et al. Detection of chicken anemia virus and infectious bursal disease virus co-infection in broilers. Braz J Vet Res Anim Sci, 2010, 47(4): 293-297.

[6] NOTEBORN M H, DE BOER G F, VAN ROOZELAAR D J, et al. Characterization of cloned chicken anemia virus DNA that contains all elements for the infectious replication cycle. J Virol, 1991, 65(6): 3131-3139.

[7] RENSHAW R W, SOINE C, WEINKLE T, et al. A hypervariable region in VP1 of chicken infectious anemia virus mediates rate of spread and cell tropism in tissue culture. J Virol, 1996, 70(12): 8872-8878.

[8] YAMAGUCHI S, IMADA T, KAJI N, et al. Identification of a genetic determinant of pathogenicity in chicken anaemia virus. J Gen Virol, 2001, 82(Pt 5): 1233-1238.

[9] TODD D, SCOTT A N, BALL N W, et al. Molecular basis of the attenuation exhibited by molecularly cloned highly passaged chicken anemia virus isolates. J Virol, 2002, 76(16): 8472-8474.

[10] DOUGLAS A J, PHENIX K, MAWHINNEY K A, et al. Identification of a 24 kDa protein expressed by chicken anaemia virus. J Gen Virol, 1995, 76 (Pt7): 1557-1562.

[11] NOTEBORN M H, TODD D, VERSCHUEREN C A, et al. A single chicken anemia virus protein induces apoptosis. J Virol, 1994, 68(1): 346-351.

[12] YUASA N, TANIGUCHI T, YOSHIDA I. Isolation and some characteristics of an agent inducing anemia in chicks. Avian Dis, 1979: 366-385.

[13] ZHOU W, SHEN B, YANG B, et al. Isolation and identification of chicken infectious anemia virus in China. Avian Dis, 1997, 41(2): 361-364.

[14] ELTAHIR Y M, QIAN K, JIN W, et al. Molecular epidemiology of chicken anemia virus in commercial farms in China. Virol J, 2011, 8: 145.

[15] ZHANG X, LIU Y, WU B, et al. Phylogenetic and molecular characterization of chicken anemia virus in southern China from 2011 to 2012. Sci Rep, 2013, 3: 3519.

[16] LI Y, FANG L, CUI S, et al. Genomic Characterization of recent chicken anemia virus isolates in China. Front Microbiol, 2017, 8: 401.

[17] DUCATEZ M F, CHEN H, GUAN Y, et al. Molecular epidemiology of chicken anemia virus (CAV) in southeastern Chinese live birds markets. Avian Dis, 2008, 52(1): 68-73.

[18] HUSSEIN H A, YOUSSEF M M, OSMAN A, et al. Immunopathogenesis of attenuated strain of chicken infectious anemia virus in one-day-old specific-pathogen-free chicks. Egypt J Immunol, 2003, 10(1): 89-102.

[19] MASE M, CHUUJOU M, INOUE T, et al. Genetic characterization of fowl adenoviruses isolated from chickens with hydropericardium syndrome in Japan. J Vet Med Sci, 2009, 71(11): 1455-1458.

[20] SMITH L M, BROWN S R, HOWES K, et al. Development and application of polymerase chain reaction (PCR) tests for the detection of subgroup J avian leukosis virus. Virus Res, 1998, 54(1): 87-98.

[21] XIE Z, PANG Y S, LIU J, et al. A multiplex RT-PCR for detection of type A influenza virus and differentiation of avian H5, H7, and H9 hemagglutinin subtypes. Mol Cell Probes, 2006, 20(3-4): 245-249.

[22] TENG L, XIE Z, XIE L, et al. Sequencing and phylogenetic analysis of an avian reovirus genome. Virus Genes, 2014, 48(2): 381-386.

[23] BROWN H K, BROWNING G F, SCOTT P C, et al. Full-length infectious clone of a pathogenic Australian isolate of chicken anaemia virus. Aust Vet J, 2000, 78(9): 637-640.

[24] HE C Q, DING N Z, FAN W, et al. Identification of chicken anemia virus putative intergenotype recombinants. Virology, 2007, 366(1): 1-7.

[25] SCOTT A N, MCNULTY M S, TODD D. Characterisation of a chicken anaemia virus variant population that resists neutralisation with a group-specific monoclonal antibody. Arch Virol, 2001, 146(4): 713-728.

[26] CHOWDHURY S M, OMAR A R, AINI I, et al. Pathogenicity, sequence and phylogenetic analysis of Malaysian chicken anaemia virus obtained after low and high passages in MSB-1 cells. Arch Virol, 2003, 148(12): 2437-2448.

[27] YAO S, TUO T, GAO X, et al. Molecular epidemiology of chicken anaemia virus in sick chickens in China from 2014 to 2015. PLOS ONE, 2019, 14(1): e0210696.

[28] ZHANG X, XIE Q, JI J, et al. Complete genome sequence analysis of a recent chicken anemia virus isolate and comparison with a chicken anemia virus isolate from human fecal samples in China. J Virol, 2012, 86(19): 10896-10897.

[29] RIMONDI A, PINTO S, OLIVERA V, et al. Comparative histopathological and immunological study of two field strains of chicken anemia virus. Vet Res, 2014, 45(1): 102.

[30] CHOWDHURY S M, OMAR A R, AINI I, et al. Isolation, identification and characterization of chicken anaemia virus in Malaysia. J Biochem Mol Biol Biophys, 2002, 6(4): 249-255.

[31] ERFAN A M, SELIM A A, NAGUIB M M. Characterization of full genome sequences of chicken anemia viruses circulating in Egypt reveals distinct genetic diversity and evidence of recombination. Virus Res, 2018, 251: 78-85.

[32] QUAGLIA G, MESCOLINI G, CATELLI E, et al. Genetic heterogeneity among chicken infectious anemia viruses detected in Italian fowl. Animals (Basel), 2021, 11(4): 944.

[33] DI FRANCESCO A, QUAGLIA G, SALVATORE D, et al. Occurrence of chicken infectious anemia virus in industrial and backyard Tunisian broilers: preliminary results. Animals (Basel), 2021, 12(1): 62.

[34] MARTIN D P, MURRELL B, GOLDEN M, et al. RDP4: Detection and analysis of recombination patterns in virus genomes. Virus Evol, 2015, 1(1): vev003.

[35] LOLE K S, BOLLINGER R C, PARANJAPE R S, et al. Full-length human immunodeficiency virus type 1 genomes from subtype C-infected seroconverters in India, with evidence of intersubtype recombination. J Virol, 1999, 73(1): 152-160.

[36] ELTAHIR Y M, QIAN K, JIN W, et al. Analysis of chicken anemia virus genome: evidence of intersubtype recombination. Virol J, 2011, 8: 512.

Genetic and phylogenetic analysis of a novel parvovirus isolated from chickens in Guangxi, China

Feng Bin, Xie Zhixun, Deng Xianwen, Xie Liji, Xie Zhiqin, Huang Li, Fan Qing, Luo Sisi, Huang Jiaoling, Zhang Yanfang, Zeng Tingting, Wang Sheng, and Wang Leyi

Abstract

A previously unidentified chicken parvovirus (ChPV) strain, associated with runting-stunting syndrome (RSS), is now endemic among chickens in China. To explore the genetic diversity of ChPV strains, we determined the first complete genome sequence of a novel ChPV isolate (GX-CH-PV-7) identified in chickens in Guangxi, China, and showed moderate genome sequence similarity to reference strains. Analysis showed that the viral genome sequence is 86.4%~93.9% identical to those of other ChPVs. Genetic and phylogenetic analyses showed that this newly emergent GX-CH-PV-7 is closely related to Gallus gallus enteric parvovirus isolate ChPV 798 from the USA, indicating that they may share a common ancestor. The complete DNA sequence is 4 612 bp long with an A+T content of 56.66%. We determined the first complete genome sequence of a previously unidentified ChPV strain to elucidate its origin and evolutionary status.

Keywords

chicken parvovirus, Guangxi, phylogenetic analysis, gene recombination

Introduction

Parvoviruses (family Parvoviridae) that infect vertebrate hosts make up the subfamily Parvovirinae, while those that infect arthropods make up the subfamily Densovirinae. The traditional classification of parvoviruses was recently replaced by eight genera (*Amdoparvovirus, Aveparvovirus, Bocaparvovirus, Copiparvovirus, Dependoparvovirus, Erythroparvovirus, Protoparvovirus, and Tetraparvovirus*) by the Intermational Committee for Taxonomy of Viruses. Chicken parvovirus (ChPV) and turkey parvovirus (TuPV) are classified as possible members of a different taxon, the new genus *Aveparvovirus*.

ChPV is a non-enveloped, positive-sense, single-stranded DNA virus[1] with a genome of approximately 5 kb containing three open reading frames (ORFs). ORF1 and ORF2 encode a non-structural (NS) protein, which is involved in viral replication, and the capsid proteins (VP1, VP2), respectively, while the function of ORF3 (NP1), located between ORF1 and ORF2, is unclear[2]. ChPV is highly prevalent in poultry flocks affected by runting-stunting syndrome (RSS), but it is also found in apparently healthy flocks[3]. RSS is characterized by stunted growth, diarrhea, immune dysfunction and increased mortality[4]. Economic losses arise from increased production costs due to poor feed conversion and the cost of treatment. The first ChPV sequence was reported by Day and Zsak in 2010[2]. ChPV infection in chickens is endemic to the United States (US), Poland, Hungary, Croatia and Republic of Korea (ROK)[5-9]. However, only 10 complete genomic sequences have been submitted to GenBank so far: one from Hungary, four from the US and four from ROK[2, 10, 11]. In China, the prevalence of

ChPV in chicken flocks is increasing, although the causative strain remains unidentified.

In the present study, we investigated the presence of ChPV strains in Guangxi to establish a genetic database for ChPV genome sequences from chickens and to analyze the evolution of ChPVs. By determining the complete genome sequence of a previously unidentified ChPV strain, we elucidate its origin and evolutionary status and lay the foundation for the development of a new vaccine.

Materials and methods

Viral DNA was extracted from tracheal and cloacal swabs (collected in 2014-2016 from 23 commercial farms in Guangxi, China) using EasyPure viral DNA/RNA kits (TransGen, Beijing, China), according to the manufacturer's instructions. The non-structural (NS) gene was amplified by PCR, using primers (Table 2-9-1) designed to target conserved regions (NS, 561 bp) first[9]. The positive samples were confirmed by partial sequencing of the NS1 gene. BLAST analysis of these samples revealed 98%~100% nt sequence identity to the ChPV ABU-P1 strain (genus *Aveparvovirus*, family Parvoviridae) isolated in Hungary. Based on an alignment of the complete sequences of three ChPV/TuPV isolates deposited in GenBank, three specific primers pairs were then designed to amplify the complete genome of one positive ChPV sample (GX-CH-PV-7) (Table 2-9-1).

Table 2-9-1   Primers used for PCR amplification of the chicken parvovirus genome

Primer[a]	Sequence(5'-3')[b]	Position, nt[c]
F1-1	CTGCTGAGCTGGTAAGATGG	395~414
R1-2	TCTTCCCGACTGACTAGATT	724~743
R1-3	CCCCCATGATACATTTTGCT	1751~1770
F2-1	TTCTAATAACGATATCACTCAAGTTTC	1841~1867
R1-4	AACCAGTATAGGTGGGTTCC	2192~2211
F2-2	GATCCCCAGCGAATCGAGTGG	2345~2365
R1-1	TTTGCGCTTGCGGTGAAGTCTGGCTCG	2375~2401
F3-1	CAAGCCGCCATTGTGTTTGT	3575~3594
R2-1	GTATTGKGTYTGGTTTTCAG	3659~3678
R2-2	AAGTCWAKRTAATTCCATGG	3694~3713
R3-2	GTCCCTGTCAAGTCATTAGAG	3858~3878
F3-2	CGACGAACAATTCAAAATTA	4692~4711
R3-1	TTAATTGGTYYKCGGYRCSCG	5005~5025
PVF1	TTCTAATAACGATATCACTCAAGTTTC	1841~1867
PVR1	TTTGCGCTTGCGGTGAAGTCTGGCTCG	2375~2401

Note: [a] Primers F1-1 to R1-1, F2-1 to R2-1 (or R2-2), and F3-1 to R3-1 were used to amplify the first, second, and third fragments of the near complete chicken parvovirus (GX-CH-PV-7) genome. Primers R1-2, R1-3, R1-4, and R3-2 are also reverse primers. Primers PVF1 to PVR1 were used to amplify partial NS1 sequence. [b] Sequences of primers were designed according to the sequences of three other known ChPV/TuPV strains (GenBank accession no. GU214704, GU214706, NC_024454). [c] Positions of primers located in the complete genome are shown according to the Europe ChPV isolate (Chicken parvovirus ABU-P1).

PCR amplifications were performed in 50 μL (total volume) containing 10 μL of 5 × PrimeSTAR GXL Buffer (Mg^{2+} Plus), 0.5 μL of PrimeSTAR GXL DNA Polymerase, 4 μL of dNTP mixture (2.5 mM each), 0.5 μL of each primer (20 μM), 1 μL of DNA template, and 33.5 μL of RNase-free water. PCR products were separated by electrophoresis on 1.2% agarose gels stained with gel red and visualized using an ultraviolet camera. The products were purified using a TaKaRa MiniBEST Agarose Gel DNA Extraction Kit Ver.4.0 (TaKaRa, Dalian, China). The purified PCR product was cloned into pMD19-T (TaKaRa, Dalian, China) vector. Sanger sequencing was performed on an ABI 3730 DNA Analyzer (TaKaRa, Dalian, China).

Results

The complete genome of GX-CH-PV-7 (GenBank accession number: KU523900) contains 4 612 bp with three ORFs. ORF1 encodes a nonfunctional protein NS1 (nt 1～2 085, 695 aa), ORF3 (nt 2 283～2 587, 101 aa) encodes a putative protein NP1, and ORF2 encodes the minor and major capsid proteins, VP1 (nt 2 585～4 612, 676 aa) and VP2 (nt 3 002～4 612, 537 aa), respectively. BLAST analysis of the whole GX-CH-PV-7 genome revealed 92% nt sequence identity to ChPV ABU-P1, thus confirming the identity of the isolate as a chicken parvovirus.

Analysis of complete and partial genomic sequences using ClustalW and MEGA 6.0.6[12]software indicated that GX-CH-PV-7 shares the highest identity (93.9%) with ChPV 367 (USA) and 89.0%～93.8% identity with other ChPV strains isolated in the USA, Hungary and ROK. Moreover, compared with the regional sequences of 10 other ChPV strains, NS1 of ChPV shared 93.1%～96.1% nt and 87.8%～98.7% aa sequence identity; NP1 shared 72.0%～90.4% nt and 90.2%～95.1% aa sequence identity; VP1 shared 88.9%～92.4% nt and 94.8%～96.8% aa sequence identity, and VP2 shared 83.4%～91.4% nt and 93.5%～97.0% aa sequence identity.

A well-conserved phosphate-binding loop (P-loop) motif (NS1 of ABU-P1 at 392～399 aa) and a NTP-binding motif (NS1 of ABU-P1 at 436～437 aa) were present in all isolates and the ABU-P1 strain[2]. A Z-test of selection revealed that the NS1 genes of GX-CH-PV-7 and ChPV ABU-P1 had undergone purifying selection. Putative nt and aa motifs, including the start codons for VP1 (nt 2 585～2 587) and VP2 (nt 3 002～3 004), the glycine-rich sequence (aa 164～172; VP1), the VP2 leucine residue (aa 291; VP1), and the fivefold cylinder region (aa 273～290; VP1) were identified, which were identified based on multiple alignments of ChPVs (Figure 2-9-1 A). A third start codon with a favorable translation initiation context (GACATGG) was found at position 3 194 in the GX-CH-PV-7 genome, as well as in ChPV ABU-P1, ChPV 736, ChPV 798, and ChPV 367, and this may represent the beginning of a putative VP3 ORF, as has been described in goose and Muscovy duck parvoviruses[13] (Figure 2-9-1 B).

Using SimPlot 3.5.1 software, we observed moderate nucleotide sequence similarity between GX-CH-PV-7 and the other 10 ChPV strains isolated in the USA (ChPV 367, ChPV 798, ChPV 736, ChPV 841), Hungary (ChPV ABU P1) and ROK (ChPV ADL120686, ChPV ADL120019, ChPV ADL120035; ChPV ParvoD62/2013, ChPV ParvoD11/2007) at the whole-genome level, except at the 3'end of ORF2 (encoding VP1 and VP2), which showed less similarity among ChPV/TuPV strains. Genomic similarity was highest at the 5'end of the genome, except in the VP1 region of ChPV ParvoD11/2007 (ROK) and ChPV 841 (USA). We found no evidence of recombination between GX-CH-PV-7 and the other 10 ChPV strains[14], indicating that GX-CH-PV-7 was not generated through recombination from ChPVs (Figure 2-9-1 C).

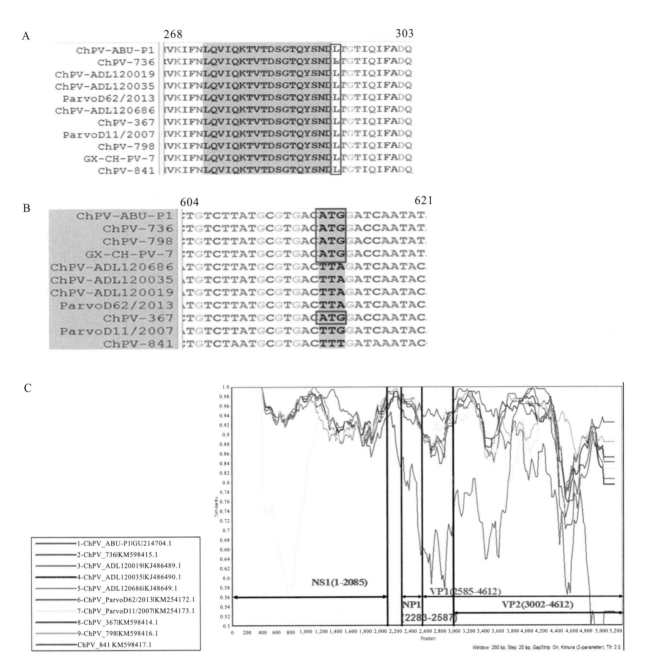

A: Putative VP3 start codons among 11 ChPVs; B: VP2 amino acid sequences comprising the fivefold cylinders in VP2 proteins; C: SimPlot analysis based on full-length nucleotide sequences using GX-CH-PV-7 as the query sequence. Shaded letters indicate the structural motif composed of fivefold cylinder regions. The conserved leucine (shown in a box) represents the structural motif for the constriction of the pore in cylindrical projections at each fivefold axis of symmetry. Both the nucleotide and amino acid positions are from the ABU-P1 strain[2] (color figure in appendix).

Figure 2-9-1　Structural motifs in ORF2 and SimPlot analysis of ChPVs

Neighbor-joining trees were constructed with the substitution model of Tamura and Nei in MEGA 6.0.6; 1 000 bootstrap replicates were used to test the reliability of the tree topology. Phylogenetic analysis of the complete ChPV genome sequence revealed three major clusters. The GX-CH-PV-7 strain segregated into a distinct branch separate from other ChPVs and appeared to have a close relationship to Gallus gallus enteric parvovirus isolate ChPV 798 strain from the USA (GenBank accession number: KM598416), forming a second cluster (Figure 2-9-2).

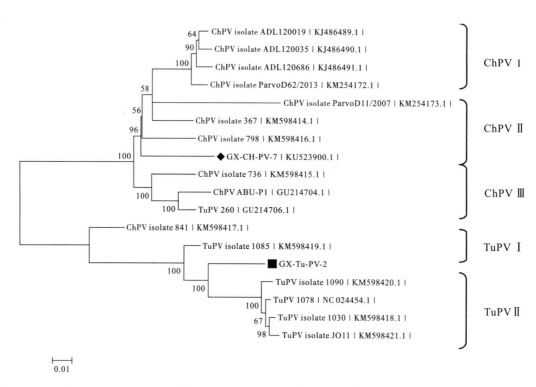

GenBank accession number follow the names of ChPV/TuPV strains. The numbers near the branches indicate the confidence level calculated by bootstrapping ($n = 1\,000$); The length of each pair of branches represents the distance between sequence pairs; Scale bar represents 0.01 nt substitutions per site; "◆" and "■" represent ChPV and TuPV strains, respectively, isolated in China (GX-Tu-PV-2 data not shown).

Figure 2-9-2　Phylogenetic neighbor-joining tree based on near-complete genomic sequences of GX-CH-PV and 17 other ChPV/TuPV isolates

Discussion

Although ChPV strain GX-CH-PV-7 is closely related to Gallus gallus enteric parvovirus isolate ChPV 798 (genus, *Aveparvovirus*) from the USA, it is not known, when and how this virus was introduced into China. Further investigations are required to determine the impact of GX-CH-PV-7 on RSS in commercial chickens. Moreover, whole genome sequence analysis should be performed for other strains from different locations to determine whether the virus was introduced into China by a single entry or multiple entries. Cases of parvovirus infection in chickens have not been reported in China. No apparent full-scale outbreaks of acute or chronic RSS have occurred. Studies on the geographic prevalence and genetic diversity of ChPV and cross-species infection are lacking, and studies on the zoonotic properties of ChPV are incomplete but underway. Knowledge of the diffusion pattern of ChPV around China is also lacking. Given these facts, ChPV surveillance is essential in China.

In conclusion, this study provides the first complete nucleotide sequence of ChPV strain GX-CH-PV-7 from a chicken in China. Genetic and phylogenetic analyses demonstrate that GX-CH-PV-7 is closely related to a ChPV isolate from the USA, suggesting that they may share a common parentage. Given the economic loss caused by ChPV infection in commercial chickens, an extensive epidemiological study of the geographic distribution of ChPV in commercial chicken flocks in China is needed. The data of this study can be referenced to design ChPV diagnostic assays and facilitate further ChPV epidemiological studies in China.

References

[1] KISARY J, AVALOSSE B, MILLER-FAURES A, et al. The genome structure of a new chicken virus identifies it as a parvovirus. J Gen Virol, 1985, 66 (Pt10): 2259-2263.

[2] DAY J M, ZSAK L. Determination and analysis of the full-length chicken parvovirus genome. Virology, 2010, 399(1): 59-64.

[3] PASS D A, ROBERTSON M D, WILCOX G E. Runting syndrome in broiler chickens in Australia. Vet Rec, 1982, 110(16): 386-387.

[4] GUY J S. Virus infections of the gastrointestinal tract of poultry. Poult Sci, 1998, 77(8): 1166-1175.

[5] BIDIN M, LOJKIĆ I, BIDIN Z, et al. Identification and phylogenetic diversity of parvovirus circulating in commercial chicken and turkey flocks in Croatia. Avian Dis, 2011, 55(4): 693-696.

[6] DOMANSKA-BLICHARZ K, JACUKOWICZ A, LISOWSKA A, et al. Genetic characterization of parvoviruses circulating in turkey and chicken flocks in Poland. Arch Virol, 2012, 157(12): 2425-2430.

[7] KOO B S, LEE H R, JEON E O, et al. Molecular survey of enteric viruses in commercial chicken farms in Korea with a history of enteritis. Poult Sci, 2013, 92(11): 2876-2885.

[8] PALADE E A, KISARY J, BENYEDA Z, et al. Naturally occurring parvoviral infection in Hungarian broiler flocks. Avian Pathol, 2011, 40(2): 191-197.

[9] ZSAK L, STROTHER K O, DAY J M. Development of a polymerase chain reaction procedure for detection of chicken and turkey parvoviruses. Avian Dis, 2009, 53(1): 83-88.

[10] KOO B S, LEE H R, JEON E O, et al. Genetic characterization of three novel chicken parvovirus strains based on analysis of their coding sequences. Avian Pathol, 2015, 44(1): 28-34.

[11] ZSAK L, CHA R M, LI F, et al. Host specificity and phylogenetic relationships of chicken and turkey Parvoviruses. Avian Dis, 2015, 59(1): 157-161.

[12] TAMURA K, STECHER G, PETERSON D, et al. MEGA6: Molecular Evolutionary Genetics Analysis version 6.0. Mol Biol Evol, 2013, 30(12): 2725-2729.

[13] ZDORI Z, STEFANCSIK R, RAUCH T, et al. Analysis of the complete nucleotide sequences of goose and muscovy duck parvoviruses indicates common ancestral origin with adeno-associated virus 2. Virology, 1995, 212(2): 562-573.

[14] LOLE K S, BOLLINGER R C, PARANJAPE R S, et al. Full-length human immunodeficiency virus type 1 genomes from subtype C-infected seroconverters in India, with evidence of intersubtype recombination. J Virol, 1999, 73(1): 152-160.

Whole-genome sequence and pathogenicity of a fowl adenovirus 5 isolated from ducks with egg drop syndrome in China

Chen Hao, Li Meng, Liu Siyu, Kong Juan, Li Dan, Feng Jiaxun, and Xie Zhixun

Abstract

Recently, fowl adenovirus (FAdV) infection has become widespread in poultry in China and may be asymptomatic or associated with clinical and other pathological conditions. In 2017, a severe egg drop syndrome outbreak in breeder ducks (45 weeks old) occurred in eastern Shandong province in China. The egg production rate declined from 93% to 41%, finally increasing to 80% (did not reach complete recovery). The presence of the virus was confirmed by FAdV-5 specific PCR assay, and it was designated strain WHRS. Furthermore, next-generation and sanger sequencing of genomic fragments yielded a 45 734 bp genome. Phylogenetic analysis showed that the genomic sequence of the WHRS strain was most homologous- (99.95%) to that of the FAdV-5 17/25 702 and 14/24 408 strain, sharing 32.1%~53.4% similarity with other FAdV strains in the genus *Aviadenovirus*. Infected duck embryos died within 3-5 dpi, but no deaths occurred in the infected ducks. Strain WHRS could cause egg drop syndrome in ducks, accompanied by clinical signs similar to those of natural infections. Overall, strain WHRS is lethal to duck embryos and causes egg drop syndrome in breeder ducks.

Keywords

fowl adenovirus 5, egg drop syndrome, phylogenetic analysis, pathogenicity, duck

---

Introduction

Fowl adenoviruses (FAdVs) belong to the genus *Aviadenovirus* in the family Adenoviridae, which contains five genera (*Mastadenovirus*, *Aviadenovirus*, *Atadenovirus*, *Siadenovirus*, and *Ichtadenovirus*). The genus *Aviadenovirus* is split into five species (FAdV-A to FAdV-E) based on restriction enzyme digestion characteristics[1]. FAdVs have been divided into 12 serotypes (FAdV-1 to 8a and FAdV-8b to 11) based on the results of cross-neutralization tests and restriction enzyme digestion patterns[2, 3]. The 12 serotypes have been grouped into five FAdV species currently recognized as follows: FAdV-A (FAdV-1), FAdV-B (FAdV-5), FAdV-C (FAdV-4 and FAdV-10), FAdV-D (FAdV-2, FAdV-3, FAdV-9 and FAdV-11) and FAdV-E (FAdV-6, FAdV-7, FAdV-8a and FAdV-8b)[4]. FAdVs are nonenveloped icosahedral viruses with a diameter of 70~100 nm. The major structural proteins are hexon, fiber, and the penton base[5, 6]. Due to the complexity of cross-neutralization testing, phylogenetic analyses based on the loop 1 (L1) region of the hexon gene have been applied for the identification and differentiation of at least 12 serotypes within the five species[3, 7].

FAdVs cause different symptoms, including inclusion body hepatitis (IBH), hepatitis-hydropericardium syndrome (HHS), and gizzard erosions in poultry, and some nonpathogenic FAdVs cause no obvious lesions[8-11]. Although the results of postmortem examinations of birds with the three symptoms have varied, several pathological findings were similar, including an enlarged and friable liver with necrotic foci, lymphocytolysis and cyst formation in the spleen and bursa of Fabricius, and atrophy in the thymus. Moreover, there are other clinical symptoms in the FAdV-infected birds, such as severe acute respiratory symptoms, lameness, and swelling of the tarsal joints[12-15]. In addition, members of the genus *Aviadenovirus* have been isolated from ducks, geese, turkeys, falcons, pigeons, and psittacines[16-20] .

FAdV-5 infection causes lameness, swelling of the tarsal joints, and IBH in broilers, and the FAdV-5 strain was isolated from a healthy mallard duck[21]. The chicken antiserum against FAdV-5 reference strain 340 could not neutralize any of the newly isolated viruses[22]. In 2017, an outbreak of severe egg drop syndrome with IBH in breeder ducks (45 weeks old) occurred in eastern Shandong Province in China. The egg production rate declined to 40% of the normal production level and then increased to 80%. Viral nucleotides were detected using (RT-)PCR, and tests for Tembusu virus, EDSV (egg drop syndrome virus), and avian influenza A virus, which could cause a drop in egg production, were negative. In the present study, we confirmed the infectious agent of the disease using virus isolation, next-generation sequencing, phylogenetic analysis and pathogenicity testing.

Materials and methods

Clinical samples

The laying rate decreased from >90% to 40% within 20 days, and hyponoia and anorexia were observed in most breeder ducks. Egg production gradually increased and recovered in the subsequent month. A total of 16/6 000 ducks died during the outbreak. The dead ducks and 20 female ducks with clinical symptoms that had stopped laying eggs were used for aetiological analysis. Nutritional (calcium and phosphorus) and aflatoxin factors were excluded by investigation and testing. Blood agar plates were prepared for sterile isolation of bacteria from the liver, follicles and brain and cultured at 37 ℃ under conventional or facultative anaerobic conditions. The postmortem examinations of the diseased ducks included liver enlargement, congestion, theca folliculi rupture, and even yolk peritonitis in individual diseased ducks. Liver and theca folliculi samples were collected from a breeder duck farm in Shandong province, frozen and transported to our laboratory on dry ice.

Virus propagation

Viral nucleic acid was extracted from samples to detect duck viruses, as assessed by (RT-) PCR assays[23-27]. Matrix genes (a pair of specific primers, F: 5'-TTCTAACCGAGGTCGAAAC-3', R: 5'-AAGCGTCTACGCTGCAGTCC-3') were applied for AIV detection. Tissue homogenates (10%) were freeze-thawed three times and then centrifuged at 8 000 × g for 15 min. Primary cells were collected from 11-day-old duck embryos. The supernatant was filtered through a 0.22 μm pore filter and then added to 80% monolayer duck embryo hepatic (DEH) cell and duck embryo fibroblast (DEF) cell cultures, which were grown in Dulbecco's modified Eagles medium/Ham's F12 medium (DMEM/F12) (HyClone, Waltham, MA, USA) supplemented with 1% fetal bovine serum (FBS) (HyClone). Tests for AIV, TMUV, FAdV, EDSV, DHAV-I, DPV and DRV were negative. The virus was blindly passaged for serial generations on DEH monolayer cells. The cytopathic effects (CPEs) were monitored daily, and the cell culture was harvested when the CPE reached 80%. The viral titres were determined to be $10^{4.0}$ median embryo lethal dose (ELD$_{50}$)/0.2 mL using the Reed and Muench method[28].

Nucleic acid extraction

Based on the appearance of the CPEs, we detected duck reovirus and fowl adenovirus[29, 30]. To obtain high-quality sequencing data, cell culture supernatants were centrifuged at low speed (6 000 × g, 10 min) to remove the precipitate, and then the supernatant was centrifuged at high speed (24 000 × g, 3 h). The pelleted cell-free virions were used for RNA/DNA isolation. RNA was reverse transcribed into cDNA. The DNA concentration was determined with a NanoDrop spectrophotometer.

Whole genome sequencing of the WHRS strain

Whole-genome sequencing was carried out using the HiSeq2500 platform. DNA was disrupted into homogenous fragments by ultrasonication. A paired-end library with a 200 bp insert size was generated according to the manufacturer's protocol (TIANSeq DirectFast DNA Library Prep Kit), and barcodes were used to sequence the virus samples in a single lane. The reads corresponding to different strains were separated based on a perfect match to the barcode sequence. As the virus was passaged in DEH, the genome sequence of *Anas platyrhynchos* should be removed to obtain clean data. Therefore, we initially mapped all reads against the available genome of BGI duck 1.0 and twinkle mtDNA (XM_021279514) and used only unmapped reads for the assembly of the virus genomes.

The next-generation sequencing raw reads were analyzed and assembled into contiguous sequences (contigs) using CLC Genomics Workbench 10.0 software (CLC Bio). All virus-aligned contigs (≥500 bp) were identified and extracted after alignment to the reference genome. The scaffolding and ordering of the contigs for each segment were facilitated by mapping the contigs on the reference virus genome FAdV 340 (KC493646)[21].

Contigs were manually aligned according to the genomic, sequence of the FAdV strain 340. Primers (cF: 5'-TCATAGGGTGAACCTTCTT-3', cR: 5'-TGTCGATGTAGGTTGAGTCA-3') were designed by Oligo 7.0 based on the adjacent contig terminals and utilized to amplify the intermediate sequences. Multiple segments were assembled into two long fragments. The amplified fragments were determined using Sanger dideoxy sequencing (Sangon, Shanghai, China). Finally, the whole genome of the WHRS strain was manually assembled using the Seqman program implemented in Lasergene (version 7.0; DNASTAR Inc., Madison, WI, USA) and deposited into the GenBank database (accession number: OM836676).

Sequence analysis

Prediction of the open reading frames (ORFs) and genes was carried out using GeneMark software, and sequence alignments and analyses were performed using MEGA 5.0[30]. ORFs coding for peptides with ≥30 amino acids were identified as potential protein-coding ORFs. Splice acceptor and donor sites were manually predicted based on previous findings in other FAdV genomes. ORFs and putative proteins were compared to proteins in the NCBI GenBank database using BLASTP.

Phylogenetic analysis

The complete genomic sequence of the FAdV isolate WHRS was aligned with other FAdV genomic sequences available in the GenBank database to determine the nucleotide sequence homologies using ClustalW in MEGA 5.0[31]. Multiple-sequence alignment of deduced amino acid sequences of the hexon and penton proteins was performed using ClustalW. A phylogenetic analysis of complete genome sequences was performed using the maximum-likelihood (ML) method in MEGA 5.0. The phylogenetic trees of the hexon and penton proteins were constructed using the neighbor-joining (NJ) method of MEGA 5.0, with absolute distances based on 1 000 bootstrap replicates.

Detection of FAdV-5 by PCR assay

A pair of FAdV-5-specific primers (dF: 5'-ATTAGCAACATCGCAGAGT-3', dR: 5'-AAGTCGTAGTG GAAGTAGTG-3') was designed using Primer Premier v6.0 to yield a PCR product of 977 bp. The total DNA of the diseased duck liver and the viral stock were extracted using the phenol-chloroform method and used as a template. The amplicons were visualized on a 1% agarose gel.

Experimentally induced infection with the WHRS strain in breeder ducks

To evaluate the pathogenicity in ducks, thirty 400-day-old breeder ducks were subcutaneously inoculated with $10^{4.0}$ ELD_{50} of the WHRS strain, whereas 30 ducks were infected with sterile PBS and served as the negative controls. All animals were free of FAdVs and other egg drop-related viruses, as determined by RT-PCR assays of oral and cloacal swabs. All animals were kept separately in specific pathogen-free chicken isolators. Clinical signs and egg production were monitored daily for 30 days. The cloacal swabs were collected from infected ducks to detect viral shedding by PCR assay using dF/dR.

Results

Clinical features, virus isolation and detection

There were no obvious clinical symptoms, except that egg production declined and then recovered during a 2-month period in the breeder duck flock. Twenty layer ducks, which had stopped laying eggs, were used for postmortem examination and etiology determination. The postmortem examinations revealed severe gross lesions, such as an enlarged and fragile liver with a pale color and steatosis, as well as ruptured, haemorrhagic, and necrotic follicles (Figure 2-10-1 A, C). Mild gross lesions, such as oedema, lung congestion, nephritis, and enteritis, also appeared in diseased ducks. The livers and theca folliculi were homogenized for virus isolation and DNA extraction. (RT-)PCR assay for common duck viruses were negative. The main histopathological changes in the liver of diseased ducks included severe hemorrhage, necrosis and fatty degeneration of hepatocytes, intranuclear inclusions in liver parenchymal cells, and lymphocyte infiltration in the liver (Figure 2-10-1 B). The embryos died at $72 \sim 108$ h post-infection after three serial passages. Severe subcutaneous hemorrhage was observed in the dead embryos and was complicated by hepatitis (Figure 2-10-1 D). The viral titer of the viral stock was $10^{4.7}$ ELD_{50}/mL. A severe CPE was observed in DEH cells in cases of primary infection, but no CPE was observed in DEF cultures inoculated with the WHRS strain (Figure 2-10-2). The field liver sample and viral stock were positive for FAdV-5 according to the PCR assay.

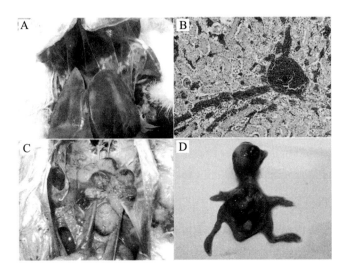

A: Lesions in the liver, swelling, and congestion; B: Necrosis and fatty degeneration of hepatocytes, and intranuclear inclusions in liver parenchymal cells; C: Lesions in the ovary, follicular dysplasia, necrosis, and rupture; D: An infected embryo with subcutaneous hemorrhage (color figure in appendix).

Figure 2-10-1 Gross lesions and histopathologic changes in ducks and embryos

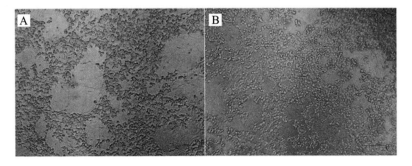

A: DEH cell aggregation and fusion after infection with WHRS; B: Normal DEH cells in the control group. The elongated cells are fibroblasts; the scale bar is 100 μm.

Figure 2-10-2 The cytopathic effect in DEH cells

Genomic sequencing and phylogenetic analyses

After filtering for contaminating duck chromosomal sequences, the average coverage for all sequenced genomes was between 200 and 40 000 reads per nucleotide. *De novo* assembly using 7% of the reads was optimal, resulting in two major contigs of 39 154 bp and 4 244 bp. Based on the reference genomic sequences, PCR amplification and Sanger sequencing of the interval and terminal fragments yielded a final genome of 45 734 bp. The G+C content of the WHRS strain genome was 56.55%. The inverted terminal repeat (ITR) was considered 81 bp long (with a 5 bp deletion compared with a 340 bp deletion). The WHRS strain shared genomic sequence identities of 32.1%~98.6% with other *Aviadenovirus* strains, and it shared a high sequence identity (99.95%) with the FAdV-5 17/25 702 and 14/24 408 strain. WHRS shared low sequence identities (32.1% and 52.6%) with FAdV-4 and FAdV-8b, respectively, which are prevalent in China and have caused IBH and HPS in poultry flocks since 2014. Compared with the deduced proteins of strain 340, the most amino acid deletions/insertions were observed in 100K (9 amino acid insertion, 195D, 234S, 316-318GGG, 332-335NAAR). Moreover, amino acid deletion/insertion occurred in dUTPase, penton, pVI, ORF19A, DNA polymerase, DNA-binding protein, ORF22A, ORF20A and ORF19.

According to the phylogenetic analysis based on the genomic sequences, the WHRS strain was classified into a cluster that included FAdV-5 strain 340 (Figure 2-10-3 A). The ORFs and terminal regions of the genomic sequence were similar to those of FAdV-5 340, including 35 potential ORFs (Figure 2-10-4). To determine the genetic diversity between the WHRS strain and other FAdV strains in the genus *Aviadenovirus*, a phylogenetic tree of the hexon proteins of FAdVs was constructed. The phylogenetic analysis of the hexon proteins showed that the WHRS strain was the most homologous to the FAdV-5 340 strain, sharing an amino acid identity of 97.5%. WHRS shared hexon amino acid sequence identity with other FAdV strains of 74.5%~88.4%. The phylogenetic tree based on the hexon proteins clustered the FAdV strains into five clades, and the FAdV-5 strains WHRS and 340 were in an independent branch (FAdV-B) (Figure 2-10-3 B). The deduced penton protein of WHRS showed 99.8% identity to that of 340. The similarity of penton protein between WHRS and other species of FAdVs ranged from 80.1% to 90.2% (Figure 2-10-3 C).

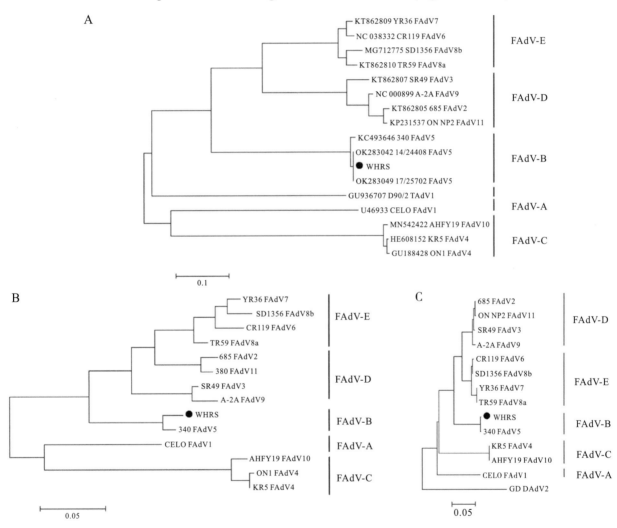

A: Phylogenetic tree constructed using the ML method based on the genomic sequences of the WHRS strain (●) and reference FAdV strains. B: Phylogenetic tree constructed using the NJ method based on the amino acid sequences of hexon proteins of the WHRS strain (●) and other FAdV strains. C: Phylogenetic tree constructed using the NJ method based on the amino acid sequences of the penton proteins of the WHRS strain (●) and other FAdV strains.

Figure 2-10-3   Phylogenetic tree of different methods

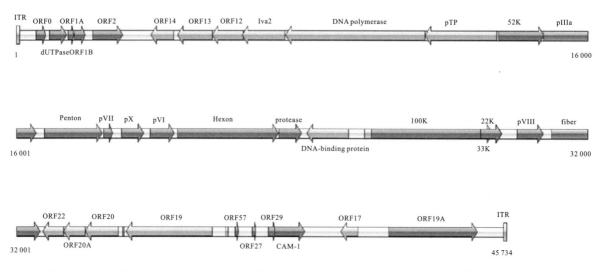

Figure 2-10-4　Schematic representation of genome size and organization of the FAdV-5 strain WHRS

Pathogenicity of WHRS in ducks

There was no mortality in either the infected or control groups during the entire experimental procedure. No clinical signs were observed in the control group. The infected ducks showed loss of appetite, loose feathers and occasional thin stool. The egg production rate of the infected group declined during the entire monitoring period. The number of eggs produced was recorded daily. As shown in Figure 2-10-5, egg production in the infected group decreased to a minimum of 11/30 at 18 dpi but subsequently increased. The egg production rate of the control group was 70.0%~86.7%. Viral nucleic acids were initially detected at 5 dpi and persisted in the feces of the infected group. FAdV-5 DNA was positively detected until 30 dpi. Simultaneously, the tests were continuously negative in the mock animals. The postmortem examination of infected ducks revealed mild liver swelling and necrotizing degenerated follicles that had not yet shed. Compared to those in the field samples, the lesions observed during necropsy of the infected ducks were mild. The livers and follicles were positive for FAdV-5 by PCR, and the FAdV strain could be re-isolated.

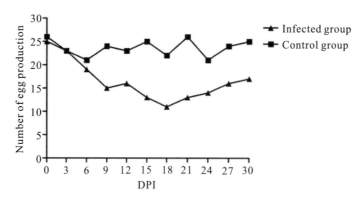

Figure 2-10-5　Daily egg production after the experimental infection

Discussion

In recent years, egg drop syndrome cases in ducks have been reported in breeder duck flocks in China. In addition to stress and nutritional factors that may underlie the disease, many contagious agents, such as Tembusu virus, influenza A virus, and EDSV, have been suggested to be the causative pathogens[24, 25]. Duck hepatitis A virus can also cause egg drop syndrome symptoms[32]. In our study, the egg laying rate gradually declined to 30%~40% at 2 weeks and then increased to >80% but did not completely recover to the initial laying rate. The causative pathogen of duck egg drop syndrome has been suggested to be FAdV-5. Egg drop syndrome might be a serious threat to the duck breeder industry in China. Previous studies have reported that EDSV, which belongs to the genus *Atadenovirus*, can infect healthy hens and quail, leading to a drastic drop in egg production accompanied by the laying of abnormal eggs. EDSV was also isolated from ducks, geese, turkeys, and herring gulls[33-35]. The FAdV-5 340 strain was originally identified in Ireland broiler samples early in the 1970s[21]. FAdV-5 could cause various symptoms in broilers, such as hypoxia, acute cardiac decompensation, incomplete haemostasis, pulmonary oedema, and nephrosis. The necropsy results of the diseased broilers indicated fibrinous airsacculitis, pericarditis, enteritis, and arthritis.

The WHRS strain, considered to belong to the species FAdV-B, was more similar to the Fowl aviadenovirus B isolate 17/25 702 and 14/24 408, which were isolated from dead birds and broilers. Kaján et, al reported that most recent FAdV-B strains were isolated from dead birds during 2014-2018 and indicated the emergence of escape variants in FAdV-B. Till now, pathogenicity of FAdV-B to waterfowl is unclear. The genomic characteristics of the WHRS strain were similar to those of the reference strain 340. However, it shared a low genomic sequence identity with other FAdVs and clustered into a distinct clade/subclade based on the phylogenetic analysis of genomic sequences and hexon and penton proteins. The fibers of FAdVs are considered to play an important role in the infectivity and pathogenicity of FAdVs[36]. The FAdV-A and FAdV-C strains have two fiber genes, and most field isolates were highly pathogenic. The genome of FAdV-5 has a single fiber gene and exhibits lower pathogenicity.

Compared with in chickens, FAdVs show poor pathogenicity and lethality in ducks. A total of 4/16 dead ducks had a secondary bacterial infection, which might be a lethal factor in ducks. In the experimental test, the infected ducks showed only mild clinical symptoms, and no deaths occurred. The results suggest that duck flocks may be resistant to the disease in farms with good feeding conditions. We further investigated whether the WHRS strain was causative and observed a drop in egg production in breeder ducks. The severe drop in egg production of infected ducks was similar to that of naturally infected ducks. The egg production declined at 6 dpi in the two groups, which may have been due to stress. Egg production increased at 18 dpi, but viral DNA was detected. This result indicated that FAdV-5 infection was persistent in duck flocks. The low level of virus replication in ducks might be an important factor for the incomplete recovery of egg production.

In summary, the results of this study indicated that an outbreak of egg drop syndrome in breeder ducks was caused by the FAdV-5 WHRS strain. The disease showed low pathogenicity in the breeder ducks but severely reduced their egg production performance. Epidemiological surveys and pathogenicity tests of FAdV-5 are urgently needed.

Data availability statement

Nucleotide sequence accession numbers have been deposited into the GenBank database under accession number: OM836676.

References

[1] MEULEMANS G, BOSCHMANS M, BERG T P, et al. Polymerase chain reaction combined with restriction enzyme analysis for detection and differentiation of fowl adenoviruses. Avian Pathol, 2001, 30(6): 655-660.

[2] MCFERRAN J B, CLARKE J K, CONNOR T J. Serological classification of avian adenoviruses. Arch Gesamte Virusforsch, 1972, 39(1): 132-139.

[3] MEULEMANS G, COUVREUR B, DECAESSTECKER M, et al. Phylogenetic analysis of fowl adenoviruses. Avian Pathol, 2004, 33(2): 164-170.

[4] HESS M. Detection and differentiation of avian adenoviruses: a review. Avian Pathol, 2000, 29(3): 195-206.

[5] SCHACHNER A, MATOS M, GRAFL B, et al. Fowl adenovirus-induced diseases and strategies for their control-a review on the current global situation. Avian Pathol, 2018, 47(2): 111-126.

[6] VELLINGA J, VAN DER HEIJDT S, HOEBEN R C. The adenovirus capsid: major progress in minor proteins. J Gen Virol, 2005, 86(Pt6): 1581-1588.

[7] MAREK A, GUNES A, SCHULZ E, et al. Classification of fowl adenoviruses by use of phylogenetic analysis and high-resolution melting-curve analysis of the hexon L1 gene region. J Virol Methods, 2010, 170(1-2): 147-154.

[8] DAR A, GOMIS S, SHIRLEY I, et al. Pathotypic and molecular characterization of a fowl adenovirus associated with inclusion body hepatitis in Saskatchewan chickens. Avian Dis, 2012, 56(1): 73-81.

[9] OKUDA Y, ONO M, YAZAWA S, et al. Experimental infection of specific-pathogen-free chickens with serotype-1 fowl adenovirus isolated from a broiler chicken with gizzard erosions. Avian Dis, 2001, 45(1): 19-25.

[10] ZHAO J, ZHONG Q, ZHAO Y, et al. Pathogenicity and complete genome characterization of fowl adenoviruses isolated from chickens associated with inclusion body hepatitis and hydropericardium syndrome in China. PLOS ONE, 2015, 10(7): e0133073.

[11] ABSALON A E, MORALES-GARZON A, VERA-HERNANDEZ P F, et al. Complete genome sequence of a non-pathogenic strain of fowl adenovirus serotype 11: minimal genomic differences between pathogenic and non-pathogenic viruses. Virology, 2017, 501: 63-69.

[12] HESS M. Commensal or pathogen-a challenge to fulfil Koch's Postulates. Br Poult Sci, 2017, 58(1): 1-12.

[13] HOMONNAY Z, JAKAB S, BALI K, et al. Genome sequencing of a novel variant of fowl adenovirus B reveals mosaicism in the pattern of homologous recombination events. Arch Virol, 2021, 166(5): 1477-1480.

[14] KLEINE A, HAFEZ H M, L SCHOW D. Investigations on aviadenoviruses isolated from turkey flocks in Germany. Avian Pathol, 2017, 46(2): 181-187.

[15] MORSHED R, HOSSEINI H, LANGEROUDI A G, et al. Fowl adenoviruses D and E cause inclusion body hepatitis outbreaks in broiler and broiler breeder pullet flocks. Avian Dis, 2017, 61(2): 205-210.

[16] CHEN H, DOU Y, ZHENG X, et al. Hydropericardium hepatitis syndrome emerged in Cherry valley ducks in China. Transbound Emerg Dis, 2017, 64(4): 1262-1267.

[17] KAJAN G L, DAVISON A J, PALYA V, et al. Genome sequence of a waterfowl aviadenovirus, goose adenovirus 4. J Gen Virol, 2012, 93(Pt11): 2457-2465.

[18] MOHAMED M H A, EL-SABAGH I M, ABDELAZIZ A M, et al. Molecular characterization of fowl aviadenoviruses species D and E associated with inclusion body hepatitis in chickens and falcons indicates possible cross-species transmission. Avian Pathol, 2018, 47(4): 384-390.

[19] TESKE L, RUBBENSTROTH D, MEIXNER M, et al. Identification of a novel aviadenovirus, designated pigeon adenovirus 2 in domestic pigeons (Columba livia). Virus Res, 2017, 227: 15-22.

[20] LUSCHOW D, PRUSAS C, LIERZ M, et al. Adenovirus of psittacine birds: investigations on isolation and development of a real-time polymerase chain reaction for specific detection. Avian Pathol, 2007, 36(6): 487-494.

[21] MAREK A, KOSIOL C, HARRACH B, et al. The first whole genome sequence of a Fowl adenovirus B strain enables interspecies comparisons within the genus Aviadenovirus. Vet Microbiol, 2013, 166(1-2): 250-256.

[22] KAJAN G L, SCHACHNER A, GELLERT Á, et al. Species fowl aviadenovirus B consists of a single serotype despite genetic distance of FAdV-5 isolates. Viruses, 2022, 14(2): 248.

[23] TANG Y, GAO X, DIAO Y, et al. Tembusu virus in human, China. Transbound Emerg Dis, 2013, 60(3): 193-196.

[24] KUMAR N S, KATARIA J M, KOTI M, et al. Detection of egg drop syndrome 1976 virus by polymerase chain reaction and study of its persistence in experimentally infected layer birds. Acta Virol, 2003, 47(3): 179-184.

[25] WU R, SUI Z W, ZHANG H B, et al. Characterization of a pathogenic H9N2 influenza A virus isolated from central China

in 2007. Arch Virol, 2008, 153(8): 1549-1555.

[26] FU Y, PAN M, WANG X, et al. Molecular detection and typing of duck hepatitis A virus directly from clinical specimens. Vet Microbiol, 2008, 131(3-4): 247-257.

[27] XIE L, XIE Z, HUANG L, et al. A polymerase chain reaction assay for detection of virulent and attenuated strains of duck plague virus. J Virol Methods, 2017, 249: 66-68.

[28] REED L J, MUENCH H. A simple method of estimating fifty per cent endpoints. Am J Epidemiol, 1938, 27: 493-497.

[29] GANESH K, SURYANARAYANA V, RAGHAVAN R, et al. Nucleotide sequence of L1 and part of P1 of hexon gene of fowl adenovirus associated with hydropericardium hepatitis syndrome differs with the corresponding region of other fowl adenoviruses. Vet Microbiol, 2001, 78(1): 1-11.

[30] LI Z, CAI Y, LIANG G, et al. Detection of Novel duck reovirus (NDRV) using visual reverse transcription loop-mediated isothermal amplification (RT-LAMP). Sci Rep, 2018, 8(1): 14039.

[31] TAMURA K, PETERSON D, PETERSON N, et al. MEGA5: molecular evolutionary genetics analysis using maximum likelihood, evolutionary distance, and maximum parsimony methods. Mol Biol Evol, 2011, 28(10): 2731-2739.

[32] ZHANG R, CHEN J, ZHANG J, et al. Novel duck hepatitis A virus type 1 isolates from adult ducks showing egg drop syndrome. Vet Microbiol, 2018, 221: 33-37.

[33] CHA S Y, KANG M, MOON O K, et al. Respiratory disease due to current egg drop syndrome virus in Pekin ducks. Vet Microbiol, 2013, 165(3-4): 305-311.

[34] KALETA E F, KHALAF S E, SIEGMANN O. Antibodies to egg-drop syndrome 76 virus in wild birds in possible conjunction with egg-shell problems. Avian Pathol, 1980, 9(4): 587-590.

[35] BARTHA A, MESZAROS J, TANYI J. Antibodies against EDS-76 avian adenovirus in bird species before 1975. Avian Pathol, 1982, 11(3): 511-513.

[36] SCHACHNER A, MAREK A, GRAFL B, et al. Detailed molecular analyses of the hexon loop-1 and fibers of fowl aviadenoviruses reveal new insights into the antigenic relationship and confirm that specific genotypes are involved in field outbreaks of inclusion body hepatitis. Vet Microbiol, 2016, 186: 13-20.

Prevalence of *Marteilia* spp. in thirteen shellfish species collected from Chinese coast

Xie Liji, and Xie Zhixun

Abstract

Marteilia spp. is listed as a notifiable parasite by the Office International des Epizooties (OIE), and cause physiological disorders and eventually death of the shellfish. Little is known about the prevalence of *Marteilia* spp. in shellfish of China. Therefore we investigated the prevalence of *Marteilia* spp. in 13 species of shellfish collected in the coastal areas of China between 2006 to 2012. A total of 11 581 individual shellfish comprising 13 different species was collected from seven bay areas in China. The shellfish was tested for the prevalence of *Marteilia* spp. following a PCR protocol recommended by OIE. *Marteilia* spp. was found in the shellfish samples (*Crassostrea rivularis, Crassostrea gigas* and *Crassostrea angulate*), *Meretrix meretrix, Chlamys nobilis, Ruditapes philippinarum, Perna viridis, Tegillarcagranosa*, and *Sinonovacula constricta*. The detection rates of *Marteilia* spp. by PCR method ranged from 0% to 6.67%, with *R. philippinarum* (6.67%) showing the highest detection rate. *Marteilia* spp. sequence of the individual positive shellfish had a high sequence homology (97.4% to 99.6%) with the sequences of other *Marteilia* type "M". Mixed positive of *Marteilia* spp. with other protozoan species (*Haplosporidium nelsoni* and *Perkinsus* spp.) was a common phenomenon (26.28%).

Keywords

shellfish, *Marteilia* spp., PCR detection, Chinese coast

Introduction

In China, the total area of 1 409 000 hectares was used for an annual shellfish production of 12 666 500 tons in the year of 2011, which is up to 60% of the total global output. Oysters are the main shellfish product generated in China. A breeding area of more than 116 000 hectares is employed, and more than 3 600 000 tons are produced annually in China, representing approximately 80% of the global oyster output. In recent years, the annual production of shellfish has been declining due to environmental changes and the occurrence of diseases. *Marteilia* spp. infection poses a serious threat to the coastal ecosystem and the supply of shellfish worldwide[1, 2].

Marteilia spp. is listed as a notifiable parasite by the Office International des Epizooties (OIE). *Marteilia* spp. is a protozoan parasite resulting in physiological disorders and eventually death of the animal[3, 4]. This protozoan parasite is distributed over a wide range of areas, including Oceania and Europe around Albania, Croatia, France, Greece, Italy, Morocco, Portugal, Spain, Sweden, Tunisia, China and the United Kingdom[5-8]. *Marteilia* spp. infection has been identified in many types of mollusks throughout the world, including oyster species[9], mussel species[5, 10] and clam species[11].

However, little has been known about the prevalence of *Marteilia* spp. in shellfish in the coastal areas of China. In this present study, the prevalence of *Marteilia* spp. in 13 shellfish species (*Crassostrea rivularis,*

Crassostrea gigas, Crassostrea angulata, Meretrix meretrix, Chlamys nobilis, Ruditapes philippinarum, Perna viridis, Tegillarca granosa, Sinonovacula constricta, Haliotis discus hannai, Paphia undulate, Tapes dorsatus and *Coelomactra antiquate*) cultured in seven bay areas in China were investigated using a PCR protocol described by Le Roux et al.[10] and this PCR method is also recommended by OIE.

C. rivularis is the major oyster species in Zhanjiang, Haikou and Qin zhou, and *C. gigas* is the major oyster species in Dalian and Qingdao, while *C. angulata* is the major oyster species in Xiamen. We therefore analyzed the prevalence of *Marteilia* spp. in these three different oyster species (*C. rivularis*, *C. gigas* and *C. angulata*) sampled from various locations (Dalian, Qingdao, Wenzhou, Xiamen, Zhanjiang, Haikou and Qinzhou) and in different seasons (spring, summer, autumn and winter).

Materials and methods

Sample collection

Samples of the 13 shellfish species were collected between 2006 and 2012 from seven bay areas in China: Dalian (Bohai Sea), Qingdao (Yellow Sea), Wenzhou (East China Sea), Xiamen (East China Sea), Zhanjiang (South China Sea), Haikou (South China Sea) and Qinzhou (South China Sea) bays. A total of 11 581 shellfish samples were collected for diagnosis, and their specific details are shown in Table 2-25. A total of 5 609 oyster samples were collected. To find out the differentiation of the prevalence of *Marteilia* spp. among different oyster species, at different locations and in different seasons, these 5 609 oyster samples was collected from different oyster species at different locations within four seasons. In terms of oyster species, these 5 609 oyster samples include 3 397 *C. rivularis*, 1 351 *C. gigas* and 861 *C. angulata*. Regarding to the locations, these 5 609 oyster samples include 694 oyster (*C. gigas*) of Dalian, 657 oyster (*C. gigas*) of Qingdao, 597 oyster (122 *C. angulata*+475 *C. rivularis*) of Wenzhou, 739 oyster (*C. angulata*) of Xiamen, 540 oyster (*C. rivularis*) of Zhanjiang, 454 oyster (*C. rivularis*) of Haikou and 1 928 oyster (*C. rivularis*) of Qin zhou. And when counted by different seasons, these 5 609 oyster samples include 1 432 in spring, 1 326 in summer, 1 384 in autumn and 1 467 in winter.

DNA extraction

Genomic DNA was extracted from the excised tissue samples (gill and digestive gland together) using EasyPure® Marine Animal Genomic DNA Kit as described by Xie et al.[12]). The tissue samples were frozen in liquid nitrogen for 3 min and then ground into powder. The powder was then digested overnight with proteinase K at 56 ℃, and the genomic DNA was extracted.

PCR, cloning, and sequencing

PCR is performed with primers targeting the ITS1 (internal transcribed spacer) region (Pr4: 5'-CCGCACACGTTCTTCACTCC-3'and Pr5: 5'-CTCGC GAGTTTCGACAGACG-3'), which were also the OIE recommended primers and expected to amplify a 412 base pairs (bp) DNA fragment. Deionized, nuclease-free water was used as a negative control. The PCR reaction solution consisted of 15 μL of 2 × PCR mix[a], 0.2 μM of each primer, and 3 μL of the purified genomic template DNA. A denaturation step at 94 ℃ for 10 min was followed by 35 cycles consisting of a denaturation step at 94 ℃ for 1 min, an annealing step at 55 ℃ for 1 min and an extension step at 72 ℃ for 1 min. After the final extension at 72 ℃ for 10 min, PCR products were analyzed by 1% agarose gel electrophoresis followed by ethidium bromide staining. The results were visualized under UV light.

Individual positive samples from *C. rivulari*, *C. nobilis*, *R. philippinarum*, *P. viridis*, *T.granosa* and *S. constricta* were selected and again amplified under the conditions similar to those of the screening PCR, except for the composition of the PCR mixture. The total volume of each reaction was increased to 50 μL, including 25 μL of 2 × PCR mix, 0.3 μM of each primer, and 5 μL of the purified genomic DNA as a template. The amplified PCR products were gel-extracted using a Gel Extraction Kit, and were ligated into pMD18-T cloning vector, respectively. Plasmids with the inserted DNA fragment were then transformed into competent *E. coli*. DH5α cells. After an overnight incubation, single colonies were inoculated in liquid culture and incubated at 37 ℃ overnight with shaking. The recombinant plasmids for the individual positive sample were analyzed and verified by the PCR method described above.

Phylogenetic analysis

The recombinant plasmid DNA for the individual positive sample was sequenced, and the sequence data were analyzed using DNAStar software. The sequences were subjected to phylogenetic analysis against the known *Marteilia* spp. sequences deposited in GenBank.

Mixed Positive of *Marteilia* spp. with other protozoans

To investigate if *Marteilia* spp.-positive shellfish were also mixed positive with *Haplosporidium nelsoni*, *Perkinsus* spp., and *Bonamia ostreae*, 156 *Marteilia* spp.-positive shellfish samples were also tested for these three parasites by PCR as previously described[13-15]. The primers used in the PCR amplifications are listed in Table 2-26.

Results

Prevalence of *Marteilia* spp. in 13 shellfish species

A total of 11 581 shellfish was analyzed in this study, of which 156 (1.35%) were *Marteilia* spp. positive by PCR detection.The detection rate in the 13 shellfish species tested ranged from 0%to 6.67% (Table 2-11-1 and Figure 2-11-1). *R.philippinarum* (6.67%) and *S.constricta* (5.14%) showed the highest detection rate, whereas no *Marteilia* spp.positive was detected in *Haliotis discus hamai*, *P.undulate*, *T.dorsatus* or *C.antiquate* samples.The presence of *Marteilia* spp. in the other five shellfish species was lower than 2.17%.

Table 2-11-1　Prevalence of Marteilia spp.in 13 shellfish species in China

Shellfish species	No. of shellfish tested	No. of positive sample	Detection rate/%	No. of mixed positive	Mixed positive rate/%
Oyster (*C.rivulari*)	3 397	14	0.42	5	35.71
Oyster (*C.gigas*)	1 351	2	0.15	0	0
Oyster (*C.angulata*)	861	17	1.97	5	29.41
Oyster (Total)	5 609	33	0.59	10	30.30
M.meretrir	1 026	2	0.19	0	0
C.nobilis	1 012	9	0.89	0	0
R. philippinarum	1 005	67	6.67	23	34.33
P. viridis	513	3	0.58	0	0
T.granosa	506	11	2.17	2	18.18
S.constricta	603	31	5.14	6	19.35

continued

Shellfish species	No. of shellfish tested	No. of positive sample	Detection rate/%	No. of mixed positive	Mixed positive rate/%
Haliotis discus hannai	521	0	0	0	0
P. undulata	569	0	0	0	0
T. dorsatus	107	0	0	0	0
C. aniquate	110	0	0	0	0
Total	11 581	156	1.35	41	26.28

Table 2-11-2   Primers used for PCR detection of pathogens: *Haplosporidium nelsoni*, *Perkinsus* spp.and *Bonamia ostreae*

Parasites	Primers	Sequences (5'- 3')	Length of PCR products/bp	Reference
Haplosporidium nelsoni	MSX-A	CGACTTTGGCATTAGGTTTCAGACC	573	11
	MSX-B	ATGTGTTGGTGACGCTAACCG		
Perkinsus spp.	PerkITS-85 0	CCGCTTTGTTTGGATCCC	703	10
	PerkITS-75	ACATCAGGCCTTCTAATGATG		
Bonamia ostreac	CF	CGGGGGCATAATTCAGGAAC	760	7
	CR	CCATCTGCTGGAGACACAG		

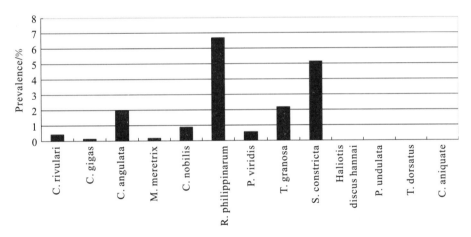

Figure 2-11-1   Prevalence of *Marteilia* spp. in 13 shellfish species

Prevalence of *Marteilia* spp. in three different oyster species

A proportion of *C.rivularis*, *C.gigas* and *C. angulata* was *Marteilia* spp. positive.The detection rate of these three oyster species ranged from 0.15% to 1.97% (Table 2-11-1). *C.angulata* samples showed the highest detection rate (1.97%), whereas *C.gigas* was the least positive oysterspecies.

Prevalence of *Marteilia* spp. in the oysters from different locations and season

The detection rate of oysters from different locations ranged from 0% to 2.30% (Table 2-11-3). *Marteilia* spp.positive samples were not present in the seawaters of Dalian, Wenzhou, Zhanjiang and Haikou bay areas in China.The oysters (*C.angulata*) in Xiamen bay (2.30%) had the highest detection rate of *Marteilia* spp. among the seven tested bay areas. There is no significant difference in the detection rate of *Marreilia* spp.in different seasons (data not shown).

Table 2-11-3 Prevalence of Marteilia spp.in the oysters form different locations

Location (species)	No.of oysters tested	No.of positive samples	Detection rate/%
Dalian (*C. gigas*)	694	0	0
Qingdao (*C. gigas*)	657	2	0.30
Wenzhou (122 *C. angulata*+475 *C. rivularis*)	597	0	0
Xiamen (*C. angulata*)	739	17	2.30
Zhanjiang (*C. rivularis*)	540	0	0
Haikou (*C. rivularis*)	454	0	0
Qinzhou (*C. rivularis*)	1 928	14	0.73
Total	5 609	33	0.59

Phylogenetic analysis

The *Marteilia* spp. sequence of the individual positive sample from *C. rivulari*, *C. nobilis*, *R. philippinarum*, *P. viridis*, *T. granosa and S. constricta* had a high sequence homology (97.4%to 99.6%)with the sequences of other *Marteilia* type "M" from GenBank (Figure 2-11-2).

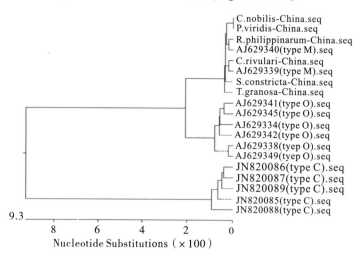

Figure 2-11-2 Phylogenetic sequences analysis of *Marteilia* spp. isolates

Mixed positive of *Marteilia* spp. with other protozoan parasites

Among all 11 581 shellfish samples tested, 156 shellfish were *Marteilia* spp. positive, and 41 samples (10 Oysters, 23 *R. philippinarums*, 6 *S. constrictas* and 2 *T. granosa*) were mixed positive, resulting in a mixed positive rate of 26.28% (41/156, Table 2-11-1). As shown in Table 2-11-4, 30 samples were mixed positive for *Marteilia* spp. with *Perkinsus* spp. (mixed positive rate: 19.23%, 30/156). Three samples were mixed positive for *Marteilia* spp. with *Haplosporidium nelsoni* (mixed positive rate: 1.92%, 3/156). Eight samples were mixed positive for *Marteilia* spp. with *Perkinsus* spp. and *Haplosporidium nelsoni* (mixed positive rate: 5.13%, 8/156). None of the samples were mixed positive for *Marteilia* spp. with *Bonamia ostreae*. Mixed positive rate for *Marteilia* spp. with *Perkinsus* spp. (19.23%) was the highest among all tested samples.

Table 2-11-4   Mixed positive of *Marteilia* spp. with other protozoan parasites

Shellfish species	Number of case			
	Marteilia spp.+*Perkinsus* spp.	*Marteilia* spp.+*Haplosporidium nelsoni*	*Marteilia* spp.+*Haplosporidium nelsoni*+*Perkinsus* spp.	*Marteilia* spp.+*Bonamia ostreae*
C. rivularis	4	1	0	0
C. angulata	5	0	0	0
R. philippinarums	14	1	8	0
S. constrictas	5	1	0	0
T. granosa	2	0	0	0
Total	30	3	8	0
Mixed positive rate/%	19.23	1.92	5.13	0

Discussion

R. philippinarum (6.67%) and *S. constricta* (5.14%) are the highly positive shellfish species among *Marteilia* spp. positive species (*C. rivularis*, *C. gigas* and *C. angulata*, *M. meretrix*, *C. nobilis*, *R. philippinarum*, *P. viridis*, *T. granosa* and *S. constricta*) in the tested seven bay areas in China. To our knowledge, this study is the first report on the prevalence of *Marteilia* spp. in oyster (*C. rivularis*, *C. gigas* and *C. angulata*), *M. meretrix*, *C. nobilis*, *R. philippinarum*, *T. granosa* and *S. constricta* in China. Previous studies have reported the detection rate of *Marteilia* spp. in Mussel in China to be 2.8% (5/180)[5]. Little is known about mortalities or other pathological effects of *Marteilia* spp. infection on oysters, *M. meretrix*, *C. nobilis*, *R. philippinarum*, *P. viridis*, *T. granosa* and *S. constricta* in the seven bay areas in China, as have been documented for infection of *Ostrea edulis* and *Mytilus edulis* and *Mytilus galloprovincialis* in other parts of the world[6, 9 16, 17] Therefore, these need to be studied in the future.

Marteilia spp. prevalence is highly variable up to 98% in *O. edulis*, and higher prevalence is expected depending on farming practices and in the areas that have had more than one year of exposure to infection[1, 3]. Stocking at low density has been shown to be effective on the control and prevention of *Marteilia* spp. Infection[3]. Copepod *Paracartia grani* (Copepoda, Calanoida) could contribute to the transmission of *Marteilia refringens*[18-20]. *Mecocyclops*, *Euchaeta concinna*, *Calanus sinicus* (Copepoda, Calanoida), *Paracalanus parvus* (Copepoda, Calanoida), *Calocalanus pavo* (Copepoda, Calanoida), *Phyllodiaptomus tunguidus* (Copepoda, Calanoida), etc., have been reported in China. However, to our knowledge, no study has been reported on the *Paracartia grani* in China. The general husbandry practices are usually stocking at high density in China, but *Marteilia* spp. has the low prevalence in Chinese coastal bays, which may due to lack of intermediate host *Paracartia grani* (or *Paracartia grani* is not the dominant species) in these seven bay areas of China, so it needs to be further studied.

The detection rates of *Ostrea edulis*, *Mytilus edulis* and *Mytilus gulloprovincicillis* are up to 98%, 40% and 5%[1, 3, 10, 21] respectively. But the detection rates among three oyster species in China are very low, ranging from 0.15% (*C. gigas*) to 1.97% (*C. angulata*), which suggests that the oyster species *Ostrea edulis* and *Mytilusedulis* maybe more suitable for *Marteilia* spp. than other oyster species. The detection rates of oyster also different among the seven tested locations, ranging from 0% (Dalian, Wenzhou, Zhanjiang and Haikou) to 2.30% (Xiamen). *Marteilia* spp. positive was detected in Qingdao (0.30%), Xiamen (2.30%) and Qin zhou (0.73%), and Xiamen (2.30%) had the highest prevalence in the seven bay areas tested. These results suggest that the aquatic conditions in Dalian, Wenzhou, Zhanjiang and Haikou may not be optimal for the survival of *Marteilia* spp., and the aquatic conditions in Xiamen might provide a suitable environment for *Marteilia* spp. to infect oysters. The biological and environmental factors affecting the prevalence of *Marteilia* spp. in oysters should be further investigated.

The *Marteilia* spp. has three types "M", "O" and "C" [22]. The *Marteilia* spp. sequence of the individual positive sample from *C. rivulari*, *C. nobilis*, *R. philippinarum*, *P. viridis*, *T. granosa* and *S. constricta* detected in this study, had a high sequence homology with the sequences of the *Marteilia* type "M". Because the number of the shellfish samples we sequenced was limited, there might be that also *Marteilia* type "O" and "C" in the shellfish of China, indicating further investigation is needed.

In this study, it is the first time to reveal that mixed positive of *Marteilia* spp. *with Perkinsus* spp. or *Haplosporidium nelsoni* was a common phenomenon, and that mixed positive of *Marteilia* spp. with *Perkinsus* spp. was the most common event (19.23%, 30/156). To our knowledge, this is the first report about the mixed positive of *Marteilia* spp. with other protozoan. It is still unknow whether the existence of *Perkinsus* spp. would bring a suitable living condition for *Marteilia* spp.. And more researches are required to figure out whether these mixed positive observations affect the mortality or survival rate of shellfish.

The prevalence of *Marteilia* spp. in 13 species of shellfish in China is based on the PCR detection. In order to confirm the infection of *Marteilia* spp. inside the tissue of the shellfish in China, histological analyses or *in situ* hybridization will be performed in our further study.

References

[1] BERTHE F C J L. Marteiliosis in molluscs: a review. Aquatic, 2004,4(17): 433-448.

[2] VILLALBA A, MOURELLE S G, CARMEN L M. Marteiliasis affecting cultured mussels Mytilus galloprovincialis of Galicia (NW Spain). Diseases of Aquatic Organisms, 1993(16): 16-72.

[3] FEIST S W, HINE P M, BATEMAN K S, et al. *Paramarteilia canceri* sp. n. (Cercozoa) in the European edible crab (*Cancer pagurus*) with a proposal for the revision of the order Paramyxida Chatton, 1911. Folia Parasitol (Praha), 2009,56(2): 73-85.

[4] BERTHE F C J, Le ROUX F, PEYRETAILLADE E, et al. Phylogenetic Analysis of the Small Subunit Ribosomal RNA of Marteilia refringens Validates the Existence of Phylum Paramyxea (Desportes and Perkins, 1990). The Journal of Eukaryotic Microbiology, 2000,47(3): 288-293.

[5] WANG Z, LU X, LIANG Y, et al. Marteilia-Like Parasite in Blue Mussels Mytilus edulis in China. Journal of Aquatic Animal Health, 2012,24(3): 161-164.

[6] LÓPEZ-FLORES I, ROBLES F, VALENCIA J M, et al. Detection of Marteilia refringens using nested PCR and in situ hybridisation in Chamelea gallina from the Balearic Islands (Spain). Diseases of Aquatic Organisms, 2008, 82(1): 79.

[7] AUDEMARD C, SAJUS M, BARNAUD A, et al. Infection dynamics of Marteilia refringens in flat oyster Ostrea edulis and copepod Paracartia grani in a claire pond of Marennes-Oléron Bay. Diseases of Aquatic Organisms, 2004,61(1-2): 103.

[8] KLEEMAN S N, LE R F, BERTHE F, et al. Specificity of PCR and in situ hybridization assays designed for detection of Marteilia sydneyi and M. refringens. Parasitology, 2002,125(2): 131-141.

[9] VILLALBA A, MOURELLE S G, CARBALLAL M J. Effects on infection by the protistan parasite Marteilia refringens on the reproduction of cultured mussels *Mytilus galloprovincialis* in Galicia (NW Spain). Diseases of Aquatic Organisms, 1993,3(17): 205-213.

[10] LE ROUX F, LORENZO G, PEYRET P, et al. Molecular evidence for the existence of two species of Marteilia in Europe. J Eukaryot Microbiol, 2001,48(4): 449-454.

[11] LÓPEZ-FLORES I, GARRIDO-RAMOSB M A, DE LA HERRANB R, et al. Identification of Marteilia refringens infecting the razor clam Solen marginatus by PCR and in situ hybridization. Mol Cell Probes, 2008,22(3): 151-155.

[12] XIE Z, XIE L, FAN Q, et al. A duplex quantitative real-time PCR assay for the detection of Haplosporidium and Perkinsus species in shellfish. Parasitol Res, 2013,112(4): 1597-1606.

[13] CASAS S M, VILLALBA A, REECE K S. Study of perkinsosis in the carpet shell clam Tapes decussatus in Galicia (NW Spain). I. Identification of the aetiological agent and in vitro modulation of zoosporulation by temperature and salinity. Dis Aquat Organ, 2002,50(1): 51-65.

[14] CARNEGIE R B, BARBER B J, CULLOTY S C, et al. Development of a PCR assay for detection of the oyster pathogen Bonamia ostreae and support for its inclusion in the Haplosporidia. Diseases of Aquatic Organisms, 2000,42(3): 199.

[15] DAY J M, FRANKLIN D E, BROWN B L. Use of Competitive PCR to Detect and Quantify Haplosporidium nelsoni Infection (MSX disease) in the Eastern Oyster (*Crassostrea virginica*). Mar Biotechnol (NY), 2000,2(5): 456-465.

[16] ITOH N, MOMOYAMA K, OGAWA K. First report of three protozoan parasites (a haplosporidian, Marteilia sp. and Marteilioides sp.) from the Manila clam, Venerupis (= Ruditapes) philippinarum in Japan. Journal of Invertebrate Pathology, 2005,88(3): 201-206.

[17] FUENTES J, LOPEZ J, MOSQUERA, E. Growth, mortality, pathological conditions and protein expression of Mytilus edulis and *M. galloprovincialis* crosses cultured in the Rıa de Arousa (NW of Spain). Aquaculture, 2002, 1(213): 233-251.

[18] ARZUL I, CHOLLET B, BOYER S, et al. Contribution to the understanding of the cycle of the protozoan parasite Marteilia refringens. Parasitology, 2014,141(2): 227-240.

[19] CARRASCO N, ARZUL I, CHOLLET B, et al. Comparative experimental infection of the copepod Paracartia grani with Marteilia refringens and Marteilia maurini. J Fish Dis, 2008,31(7): 497-504.

[20] AUDEMARD C, LE ROUX F, BARNAUD A, et al. Needle in a haystack: involvement of the copepod Paracartia grani in the life-cycle of the oyster pathogen Marteilia refringens. Parasitology, 2002,124(3): 315-323.

[21] LOPEZ-FLORES I, DE LA HERRÁN R, GARRIDO-RAMOS M A, et al. The molecular diagnosis of Marteilia refringens and differentiation between Marteilia strains infecting oysters and mussels based on the rDNA IGS sequence. Parasitology, 2004, 129(Pt 4): 411-419.

[22] CARRASCO N, ANDREE K B, LACUESTA B, et al. Molecular characterization of the Marteilia parasite infecting the common edible cockle *Cerastoderma edule* in the Spanish Mediterranean coast A new Marteilia species affecting bivalves in Europe?. Aquaculture, 2012, 324(12): 20-26.

Prevalence of *Perkinsus* spp. in selected shellfish species collected off Chinese coast

Xie Liji, and Xie Zhixun

Abstract

We investigated the prevalence of *Perkinsus* spp. in 13 species of shellfish collected from the coastal areas of China between 2006 and 2012. A total of 11 581 shellfish specimens belonging to 13 different species were collected from seven bay areas in China *viz*., Dalian (Bohai Sea), Qingdao (Yellow Sea), Wenzhou (East China Sea), Xiamen (East China Sea), Zhanjiang (South China Sea), Haikou (South China Sea) and Qinzhou (South China Sea). The shellfish samples collected were tested for the prevalence of *Perkinsus* spp. using a PCR protocol. The detection rates of *Perkinsus* spp. by PCR method ranged from 0.38 to 26.37%, with *Ruditapes philippinarum* (26.37%) and *Paratapes undulatus* (=*Paphia undulata*) (20.21%) showing the highest detection rate. The detection rate of *Perkinsus* spp. by PCR in oyster samples were highest in Wenzhou (17.25%), Qinzhou (15.98%) and Zhanjiang (11.67%) bays. Samples taken in autumn showed the highest positive detection rate. Mixed infection by *Perkinsus* spp. and other protozoan species *viz*., *Haplosporidium nelsoni* and *Marteiliia* spp. was found to be a common phenomenon (10.75%).

Keywords

Chinese coast, PCR detection, *Perkinsus* spp., shellfish

As per the report of Ministry of Commerce of the People's Republic of China, a total area of 1 409 000 ha was used for annual shellfish production of 12 666 500 t in the year 2011, which formed more than 60% of the total global output[1]. In recent years, however, the annual production of shellfish has declined due to environmental changes and shellfish diseases. In particular, parasitic diseases caused by *Perkinsus* spp. infection pose a serious threat to the coastal ecosystem and the supply of shellfish worldwide[2, 3].

Perkinsus spp. is listed as a notifiable parasite by the Office International des Epizooties (OIE). It is the major protozoan parasite infecting shellfish, causing mortality among cultured shellfish since 1988[2, 4-8]. However, very little information is available on the prevalence of *Perkinsus* spp. in some species of shellfish in coastal areas of China. The present study investigated prevalence of *Perkinsus* spp. in selected shellfish species occurring along the coast of China.

Samples of shellfishes belonging to 13 different species were collected between 2006 and 2012 from seven bay areas in China *viz*., Dalian (Bohai Sea), Qingdao (Yellow Sea), Wenzhou (East China Sea), Xiamen (East China Sea), Zhanjiang (South China Sea), Haikou (South China Sea) and Qinzhou (South China Sea). A total of 11 581 shellfish samples were collected and the specific details are described in Table 2-12-1.

Genomic DNA was extracted from shellfish samples as described by Xie et al.[9]. Polymerase chain reaction (PCR) was performed with primers PerkITS-85: 5'-CCG-CTT-TGT-TTGGATCCC-3'and PerkITS-750: 5'-ACA-TCA-GGC-CTT-CTA-ATG-ATG-3'[10], yielding an expected amplicon of 703 base pairs (bp). Out of the 11 581 shellfish samples, 1 153 (9.96%) were *Perkinsus* spp. positive by PCR. The detection rate in the 13 shellfish species tested ranged from 0.38 to 26.37% (Table 2-12-1). *Ruditapes philippinarum* (26.37%), *Paratapes undulatus* (=*Paphia undulata*) (20.21%) and *Crassostrea rivularis* (15.16%) showed the highest detection rate. The presence of *Perkinsus* spp. in the other ten shellfish species was lower than 6.63%.

Table 2-12-1   Prevalence of *Perkinsus* spp. in 13 shellfish species in China

Shellfish species	No. of shellfish tested	No. of positive samples	Detection rate /%	No. of mixed positive	Mixed positive detection rate[a]/%
Oyster, *C. rivularis*	3 397	515	15.16	11	2.14
Oyster, *C. gigas*	1 351	70	5.18	28	40
Oyster, *C. angulata*	861	34	3.95	7	20.59
Total (Oysters)	5 609	619	11.04	46	7.43
M. meretrix	1 026	39	3.80	0	0
C. nobilis	1 012	37	3.66	7	18.92
R. philippinarum	1 005	265	26.37	39	14.72
P. viridis	513	19	3.70	0	0
T. granosa	506	12	2.37	3	25
S. constricta	603	40	6.63	11	27.5
H. discus hannai	521	2	0.38	0	0
P. undulatus	569	115	20.21	18	15.65
T. conspersus (= *T. dorsatus*)	107	3	2.80	0	0
M. (= *C.*) *aniquata*	110	2	1.81	0	0
Total	11 581	1 153	9.96	124	10.75

Note: [a] Number of mixed positive / number of positive samples.

The presence of *Perkinsus* spp. infection and the diseases it causes have resulted in population decline of clams since 1980s[6, 8]. Little is known about the pathological effects, including rates of mortality due to *Perkinsus* spp. infection in *Meretrix* spp., *Conus* (=*Eugeniconus*) *nobilis*, *Perna viridis*, *Tegillarca granosa*, *Sinonovacula constricta*, *Haliotis discus hannai*, *P. undulatus*, *Tapes conspersus* (=*dorsatus*) and *Mactra* (=*Coelomactra*) *antiquata* in the seven bay areas in China, although these effects *have been* documented for shellfish species infected with *Perkinsus* spp. in other parts of the world[2, 6, 8]. Therefore, further *studies* are needed in this direction.

Oysters are the main shellfish products in China. In order to find out the difference in the prevalence of *Perkinsus* spp. among different oyster species, at different locations and in different seasons, 5 609 samples of different oyster species were collected from different locations within four seasons which comprised 3 397 *C. rivularis*, 1 351 *Crassostrea gigas* and 861 *Crassostrea angulata*. The detection rates among these three oyster species differed, ranging from 3.95% in *C. angulata* to 15.16% in *C. rivularis*, which suggests that the oyster species *C. rivularis* may be more susceptible to *Perkinsus* spp. than other oyster species. The detection rates of oysters were also found to be different among the seven tested locations, ranging from 1.59% in Dalian Bay to 17.25% in Wenzhou Bay (Table 2-12-2). The areas of high prevalence included Wenzhou Bay (17.25%), Zhanjiang Bay (11.67%) and Qinzhou Bay (15.98%). These results suggest that the waters in Wenzhou, Zhanjiang and Qinzhou bays provided a more conducive environment for *Perkinsus* spp. to infect oysters. The biological and environmental factors that may affect *Perkinsus* spp. infection of oysters in Dalian Bay need to be investigated further.

Table 2-12-2　Prevalence of *Perkinsus* spp. in oysters from different locations

Location (species)	No. of oysters tested	No. of positive samples	Detection rate / %
Dalian (*C. gigas*)	694	11	1.59
Qingdao (*C. gigas*)	657	59	8.98
Wenzhou (122 *C. Angulata*+475 *C. rivularis*)	597	103	17.25
Xiamen (*C. angulata*)	739	31	4.19
Zhanjiang (*C. rivularis*)	540	63	11.67
Haikou (*C. rivularis*)	454	44	9.69
Qinzhou (*C. rivularis*)	1 928	308	15.98
Total	5 609	619	11.04

Highest detection rate (15.39%) of *Perkinsus* spp. in oysters was observed in the bay areas of China in Autumn season and lowest detection rate (3.70%) was noticed in summer (Table 2-12-3). This is consistent with findings showing that infection rates in Korean waters were lowest during summer[11]. The association between season and oyster mortality, however, has not been determined during the present study.

The amplified PCR products (*Perkinsus* spp. positive) were ligated into pMD18-T cloning vector (TaKaRa, Dalian, China) and sequenced. The *Perkinsus* spp. sequences of the individual positive samples from *C. rivularis*, *C gigas* and *Meretrix meretrix* had high sequence homology (98.1 to 99.9%) with the sequences of *Perkinsus atlanticus* strains in GenBank. However the *Perkinsus* spp. sequences of isolates from *C. angulata*, *S. constricta*, *P. undulatus* and *R. philippinarum* showed a high degree of homology (98.5 to 100%) with the sequences of *Perkinsus beihaiensis*. Thus, these isolates were classified as *P. atlanticus* and *P. beihaiensis* respectively, similar to previously observed phylogenetic groupings[12].

To investigate if *Perkinsus* spp. positive shellfish were also mixed infected with *Marteiliia* spp., *Haplosporidium nelsoni* or *Bonamia ostreae*, 1 153 *Perkinsus* spp. positive shellfish samples were also tested for the presence of these three parasites by PCR as previously described[13-15]. Out of 1 153 *Perkinsus* spp. positive samples, 124 (10.75%) were mixed positive for *Perkinsus* spp. along with *H. nelsoni* or *Marteilia refringens* (Table 2-12-1). Eighty-six samples were mixed positive for *Perkinsus* spp.+*H. nelsoni* (86/1153, 7.46%), 30 samples were mixed positive for *Perkinsus* spp.+*M. refringens* (30/1153, 2.60%) and 8 samples were mixed positive for *Perkinsus* spp.+*H. nelsoni* +*Marteilia* spp. (8/1153, 0.69%) (Table 2-12-4). None of the samples were mixed positive for *Perkinsus* spp. +*B. ostreae*. The effects of mixed infections on shellfish mortality or survival rate needs further investigations.

Table 2-12-3　Prevalence of Perkinsus spp. in oysters sampled in different seasons

Season	No. of samples tested	No. of positive samples	Prevalence / %
Spring (February-April)	1 432	175	12.22
Summer (May-July)	1 326	49	3.70
Autumn (August-October)	1 384	213	15.39
Winter (November to January)	1 467	182	12.41
Total	5 609	619	11.04

Table 2-12-4   Mixed positive infections of *Perkinsus* spp. with other protozoan parasites

Shellfish species	Number of cases			
	Perkinsus spp.+ *H. nelsoni*	*Perkinsus* spp.+ *Marteiliia* spp.	*Perkinsus* spp.+*H. nelsoni*+*Marteiliia* spp.	*Perkinsus* spp.+*B. ostreae*
Oysters	37	9	0	0
P. undulatus	18	0	0	0
R. philippinarum	17	14	8	0
C. nobilis	7	0	0	0
S. constricta	6	5	0	0
T. granosa	1	2	0	0
Total	86	30	8	0
Mixed positive detection rate /%	7.46	2.60	0.69	0

References

[1] XIE L, XIE Z. Prevalence of *Marteilia* spp. in thirteen shellfish species collected from China coast. Turk J Fish, 2018,6(18): 753-759.

[2] PARK K, CHOI K. Spatial distribution of the protozoan parasite *Perkinsus* sp. found in the Manila clams, Ruditapes philippinarum, in Korea. Aquaculture, 2001,203(1): 9-22.

[3] FERNÁNDEZ-ROBLEDO J A, LIN Z, VASTA G R. Transfection of the protozoan parasite *Perkinsus marinus*. Molecular and Biochemical Parasitology, 2008,157(1): 44-53.

[4] ANDREWS J D. Epizootiology of the disease caused by the oyster pathogen *Perkinsus marinus* and its effects on the oyster industry. Maryland, USA: Special Publication No. 18, American Fisheries Society, 1988: 47-63.

[5] BURRESON E M, RAGONE C. Epizootiology of *Perkinsus marinus* disease of oysters in Chesapeake.Bay, with emphasis on data since 1985. J. Shellfish Res, 1996(15): 17-34.

[6] HAMAGUCHI M, SUZUKI N, USUKI H, et al. Perkinsus protozoan infection in short-necked clam Tapes (=Ruditapes) philippinarum in Japan. Fish Pathology, 1998,33(5): 473-480.

[7] VILLALBA A, REECE K S. Perkinsosis in molluscs: A review. Aquat. Living Resour. 1998(17): 411-432.

[8] SHIMOKAWA J, YOSHINAGA T, OGAWA K. Experimental evaluation of the pathogenicity of *Perkinsus olseni* in juvenile Manila clams Ruditapes philippinarum. Journal of Invertebrate Pathology, 2010,105(3): 347-351.

[9] XIE Z, XIE L, FAN Q, et al. A duplex quantitative real-time PCR assay for the detection of Haplosporidium and Perkinsus species in shellfish. Parasitol Res, 2013,112(4): 1597-1606.

[10] AUDEMARD C, REECE K S, BURRESON E M. Real-time PCR for detection and quantification of the protistan parasite Perkinsus marinus in environmental waters. Appl Environ Microbiol, 2004,70(11): 6611-6618.

[11] PARK K, NGO T T, CHOI S, et al. Occurrence of *Perkinsus olseni* in the Venus clam *Protothaca jedoensis* in Korean waters. Journal of Invertebrate Pathology, 2006,93(2): 81-87.

[12] MOSS J A, XIAO J, DUNGAN C F, et al. Description of *Perkinsus beihaiensis* n. sp., a new *Perkinsus* sp. Parasite in Oysters of southern China. The Journal of Eukaryotic Microbiology, 2008,55(2): 117-130.

[13] CARNEGIE R B, BARBER B J, CULLOTY S C, et al. Development of a PCR assay for detection of the oyster pathogen Bonamia ostreae and support for its inclusion in the Haplosporidia. Diseases of aquatic organisms, 2000,42(3): 199.

[14] DAY J M, FRANKLIN D E, BROWN B L. Use of competitive PCR to detect and quantify *Haplosporidium nelsoni* infection

(MSX disease) in the Eastern Oyster (*Crassostrea virginica*). Marine Biotechnology (New York, N Y), 2000,2(5): 456-465.

[15] LE ROUX F, LORENZO G, PEYRET P, et al. Molecular evidence for the existence of two species of Marteilia in Europe. J Eukaryot Microbiol, 2001,48(4): 449-454.

Effects of cadmium exposure on intestinal micro-flora of *Cipangopaludina cathayensis*

Jiang Jiaoyun, Li Wenhong, Wu Yangyang, Cheng Chunxing, Ye Quanqing, Feng Jiaxun, and Xie Zhixun

Abstract

As one of the most environmentally toxic heavy metals, cadmium (Cd) has attracted the attention of researchers globally.In particular, Guangxi, a province in southwestern China, has been subjected to severe Cd pollution due to geogenic processes and anthropogenic activities.Cd can be accumulated in aquatic animals and transferred to the human body through the food chain, with potential health risks.The aim of the present study was to explore the effects of waterborne Cd exposure (0.5 mg/L and 1.5 mg/L) on the intestinal microbiota of mudsnail, *Cipangopaludina cathayensis*, which is favored by farmers and consumers in Guangxi. Gut bacterial community composition was investigated using high-throughput sequencing of the V3-V4 segment of the bacterial 16S rRNA gene. Our results indicated that *C. cathayensis* could tolerate low Cd (0.5 mg/L) stress, while Cd exposure at high doses (1.5 mg/L) exerted considerable effects on microbiota composition. At the phylum level, Proteobacteria, Bacteroidetes, and Firmicutes were the dominant phyla in the mudsnail gut microbiota. The relative abundances of Bacteroidetes increased significantly under high Cd exposure (H14) ($P<0.01$), with no significant change in the low Cd exposure (L14) treatment. The dominant genera with significant differences in relative abundance were *Pseudomonas*, *Cloacibacterium*, *Acinetobacter*, *Dechloromonas*, and *Rhodobacter*. In addition, Cd exposure could significantly alter the pathways associated with metabolism, cellular processes, environmental information processing, genetic information processing, human diseases, and organismal systems. Notably, compared to the L14 treatment, some disease-related pathways were enriched, while some xenobiotic and organic compound biodegradation and metabolism pathways were significantly inhibited in the H14 group. Overall, Cd exposure profoundly influenced community structure and function of gut microbiota, which may in turn influence *C. cathayensis* gut homeostasis and health.

Keywords

cadmium, *Cipangopaludina cathayensis*, intestinal microbiota, high-throughput sequencing, microbial diversity

Introduction

In the wake of rapid industrialization, aquatic ecosystem pollution is becoming severe[1]. As ubiquitous hazardous pollutants, heavy metals have attracted the attention or researchers globally due to their environmental toxicity. Cadmium (Cd), a non-essential element, usually exists as Cd (II). As one of the most toxic heavy metals, Cd is released into the environment mainly through anthropogenic activities, indluding electroplating, battery manufacturing, soldering, mining, and agriculture[2]. Cd has numerous negative impacts on aquatic animals, including triggering histopathological changes, inducing oxidative stress, causing metabolic disorders, and altering gut microbial community structure[3-7]. Moreover, Cd is not easily degradable, and can be accumulated in aquatic animals followed by in the human body through the food chain, with potential human health risks[8].

Heavy metal pollution is a major environmental issue in China, and heavy metal pollution in aquatic environment is increasing in severity[8, 9]. Cd has been identified as one of the major soil contaminants in China[10]. In particular, Cd contamination in Guangxi is significantly higher than in other regions in China due to high background geochemical concentrations in the region[11, 12]. In addition, Guangxi is a key non-ferrous metal production area in China, so that Cd pollution is a major challenge in Guangxi. In early January 2012, the Longjiang River of Guangxi was exposed to serious Cd contamination following an accident, with long-term impacts on the regional aquatic ecosystems[13, 14]. Cd is also the primary heavy metal pollutant in Chinese agricultural land, including paddy soils[15, 16]. Indeed, people inhabiting such areas with high levels of Cd pollution may be exposed to Cd toxicity, with potential threats to human health[17].

The mudsnail, *Cipangopaludina cathayensis* (phylum Mollusca, Gastropoda, Prosobranchia, Mesogastropoda, Viviparidae, and *Cipangopaludina*), is a widely distributed species that can be found in Chinese rivers, lakes, ponds, and other water bodies[18]. *C. cathayensis* has high protein and low fat content, is rich in umami amino acid, and has high nutritional value, so that it is highly favored among consumers and farmers in China[19]. Moreover, *C. cathayensis* flesh has been reported to have diverse biological and physiological properties that are beneficial in human disease prevention and treatment[20, 21].

Indeed, *C. cathayensis* is one of the most popular aquatic animals in China. Particularly in Guangxi, the snail family Viviparidae is a source of key components of a famous snack, "snail rice noodle" which represents one of the intangible cultural heritages in China[19]. In recent years, with the continued increase in "snail rice noodle" consumption, demand for mudsnail and its production has been increasing. Paddy field culture is one of the major ways of mudsnail production in Guangxi. However, Cd pollution has been identified as a serious problem in Guangxi paddy soils and aquatic environments[13, 15, 16]. In addition, considering mudsnail is a benthic organism that is closely associated with paddy soil, it could be exposed to high Cd concentrations, with major threats to food safety[22]. At present, only a few studies had explored the adverse impacts of Cd exposure on the snail family Viviparidae. In addition, current studies have largely focused on the oxidative stress caused by Cd exposure[23, 24], so that further investigations on other adverse effects on snails need to be carried out.

The microbiomes associated with aquatic animals, particularly their gut systems, not only participate in digestion but also influence nutrition, growth, reproduction, the immune system, and host vulnerability to disease[3, 6, 25-28]. Cd exposure has been reported to significantly affect the gut microbiota of numerous aquatic organisms[3, 7, 29]. However, the effects of Cd exposure on the intestinal microbiota of *C. cathayensis* remain unclear. To address the knowledge gap, in the present study, *C. cathayensis* individuals were exposed to two doses (0.5 mg/L and 1.5 mg/L) of cadmium chloride (CdCl$_2$·2.5H$_2$O) for > 14 d. The aim of the present study was to investigate the effect of Cd on *C. cathayensis* gut microbiota composition and diversity.

Materials and methods

Experimental snail and treatment

Adult snails (*C. cathayensis*) were obtained from Juhe Agricultural Development Cooperatives (25.75° N, 109.38° E), Sanjiang District, Liuzhou City, Guangxi, China. They were then transferred to the laboratory, and acclimated to the experimental conditions at a temperature 24.0 ± 1.0 ℃ , under a 12-h/12-h light/dark cycle in a 50-L ($65 \times 41 \times 20$ cm) plastic tank for 2 weeks. During the acclimation period, specimens were fed with commercial ground fish food (Tongwei, Chengdu, Sichuan, China) once a day at 0.5% of their body weight. The tank water was changed partially (30%) every day.

After a 2-week acclimation period, 225 snails were divided randomly into three groups and placed in plastic tanks, with three replicates (25 snails per tank) in each treatment. $CdCl_2 \cdot 2.5H_2O$ (Silong, Shantou, Guangdong, China) was dissolved in deionized water to prepare stock solution with a final concentration of 900 mg/L. The 0.5 mg/L and 1.5 mg/L Cd doses were selected according to previous studies[23]. The three treatments in the present study included the control treatment (CK14: with no Cd supplementation), low Cd concentration exposure treatment (L14: 0.5 mg/L), and high Cd concentration exposure treatment (H14: 1.5 mg/L). Other experimental conditions were consistent with those in the acclimation phase. During the experimental period, one-third of the water in the tank was replaced every day by adding fresh water or water with a similar concentration. The experiment lasted 2 weeks, as severe mortality occurred at 14-day in the H14 treatment.

Sample collection

Snail intestine samples were collected on day 14 and used to determine gut microbiota composition and diversity. The guts of three snails were pooled as a single sample, to ensure sample adequacy, with three biological replicates in each treatment. Briefly, the samples were wiped with 75% ethanol before the snails were removed from the shell. Subsequently, the snails were dissected and the guts extracted and rinsed with sterile water three times. The gut samples were flash frozen using liquid nitrogen and stored at −80 ℃ for subsequent analyses.

DNA extraction, bacterial 16S rRNA amplification, and sequencing

Total genomic DNA (gDNA) of the gut microbiota were extracted using a Fast DNA SPIN Extraction Kit (MP Biomedicals, United States) according to the manufacturer's protocol. The V3-V4 regions of the bacterial 16S rRNA genes were amplified by PCR using universal bacterial primers (338F: 50-ACTCCTACGGGAGGGAGCA-30, 806R: 50-GGACTACHVGGGTWTCTAAT-30). The PCR cycle conditions for each sample were as follows: an initial denaturation at 95 ℃ for 5 min; 25 cycles of denaturation at 95 ℃ for 30 s, annealing at 55 ℃ for 30 s, and extension at 72 ℃ for 30 s, with a final extension at 72 ℃ for 5 min. PCR products were purified and quantified using an AxyPrep DNA Gel Extraction Kit (Axygen, Union City, NJ, United States) and a Quant-iT PicoGreen dsDNA Assay Kit (Invitrogen, Waltham, MA, United States), respectively. A TruSeq Nano DNA LT Library Prep Kit (Ilumina, United States) was used to establish the DNA library. The library was sequenced using a MiSeq Reagent Kit v3 (6 000-cycles-PE) (Illumina, United States) on a MiSeq platform by Personal Biotechnology Co, Ltd. (Shanghai, China). The raw reads were deposited into the NCBI Sequence Read Archive database (PRJNA837347).

Sequence processing

The sequencing data were processed using Quantitative Insights Into Microbial Ecology 2 (QIIME2 v2019.4). Briefly, Cutadapt (version 3.7) was used to filter and trim PCR primers from the raw reeds. DADA2 was used for quality control[30], removing chimera sequences, and determining the sequence variants.

Taxonomy was assigned using the DADA2 pipeline, which implements the Naive Bayesian Classifier using the DADA2 default parameters based on the Greengenes database (Release 13.8). Subsequenty, the sequences were rarefied using the feature-table rarefy command in QIIME2.

Data analysis

All sequence analysis steps were performed using QIIME2 and R v3.2.0 (R Foundation for Statistical Computing, Vienna, Austria). The rarefaction curve was generated based on Amplicon Sequence Variants (ASVs) at a 97% similarity cut-off level. For alpha diversity analyses, Chao 1, Observed_species, Shannon, and Simpson indices were calculated using QIIME2 (for calculation methods). Significance between groups was tested using the Kruskal-Wallis H test and the Dunn test. Beta diversity was calculated using weighted Bray-Curtis distance matrix and visualized with Principal Coordinates Analysis (PCoA). Hierarchical clustering using Bray-Curtis distances based on the relative abundances of species was performed to cluster the dataset. A Venn diagram was drawn using the "Venn Diagram" package in R v3.2.0 (R Statistical Foundation). The functional profles of microbial communities were predicted using PICRUSt2 (Phylogenetic Investigation of Communities by Reconstruction of Unobserved States). The predicted genes and their respective functions were annotated using the Kyoto Encyclopedia of Genes and Genomes (KEGG) database. Differences between populations were analyzed using one-way Analysis of Variance. Results were considering statistically significant at $P < 0.05$. The values are expressed as mean $\pm$ SD (Standard deviation).

Results

Relative abundance

After normalization, there were 833 989 sequences across all snail gut contents sampled, with an average of 92 665 sequences per sample. Rarefaction curves indicated that all samples reached the saturation phase. There were 14 951 ASVs derived from all samples; the CK14, L14, and H14 treatments had 4 625, 5 121, and 5 205 ASVs per sample, respectively. Moreover, 859 ASVs were shared among the three treatments, while 2 010, 2 029, and 2 386 ASVs were unique to the CK14, L14, and H14 treatments, respectively (Figure 2-13-1).

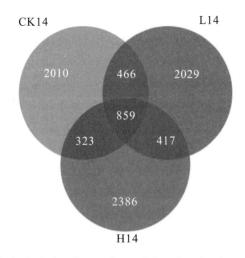

Figure 2-13-1 **Venn diagram analysis depicting the numbers of shared and unique Amplicon Sequence Variants (ASVs) among the control (CK14), 0.5 mg/L (L14), and H14 (1.5 mg/L) treatments**

Intestinal microflora diversity

To compare bacterial community diversity across different groups, alpha-diversity and beta-diversity were evaluated. There were no significant differences in Chao 1 index, Observed_species index, Shannon index, and Simpson index among the three groups ($P > 0.05$) (Figure 2-13-2 A and Supplementary Table S3). In a beta-diversity analysis (PCoA based on Bray-Curtis), the L14 and CK14 treatments were clustered together and could not be distinguished, whereas the H14 group was distinct from the L14 and CK14 groups, with the following main principal component (PC) scores: PC1 = 48.1%, PC2 = 25.7% (Figure 2-13-2 B). In addition, according to the hierarchical clustering tree results, ASVs from *C. cathayensis* in the high Cd exposure group were clustered in one group based on similarity, while the control and low Cd exposure groups clustered into one independent group, excluding one control sample (Figure 2-13-2 C). The results indicate that the high Cd exposure treatment had more severe effects on the diversity of the *C. cathayensis* microbiome than the low Cd exposure treatment.

Gut microbiota community structure

In total, 25 phyla, 50 classes, 115 orders, 185 families, 324 genera, and 90 species were identified. At the phylum level, Proteobacteria was the most abundant phylum across all three treatments (51.9% in CK14, 55.2% in L14, and 38.9% in H14), the other two prevalent phyla were Bacteroidetes and Firmicutes (Figure 2-13-3 A). In addition, Bacteroidetes abundance in the H14 treatment was significantly higher than that in the C14 treatment ($P < 0.01$), although there was no significant difference between the L14 and control treatments (Figure 2-13-3 A). At the genus level, *Pseudomonas*, *Cloacibacterium*, *Acinetobacter*, *Dechloromonas*, *Halomonas*, *Pelomonas*, *Mitochondria*, *Aeromonas*, *Rhodobacter*, and *Aquabacterium* were the dominant (Figure 2-13-3 B). Pseudomonas relative abundance was higher in the Ll4 treatment than in the C14 treatment, although the difference was not significant (Figure 2-13-3 B). Conversely, *Pseudomonas* relative abundance was lower in the H14 treatment than in the C14 treatment, although the difference was not significant (Figure 2-13-3 B). However, *Pseudomonas* relative abundance decreased with an increase in Cd concentration ($P < 0.01$) in the H14. *Acinetobacter* exhibited a similar trend (Figure 2-13-3 B). In addition, *Rhodobacter* relative abundance was significantly lower in the H14 treatment than in the C14 treatment ($P < 0.05$). *Rhodobacter* relative abundance was also lower in the L14 treatment than in the C14 treatment, although the difference was not significant (Figure 2-13-3 B). On the contrary, *Cloacibacterium* and *Dechloromonas* were significantly enriched in the H14 treatment ($P < 0.05$), although there was no significant indifference between the L14 treatment and the CK14 treatment (Figure 2-13-3 B).The results were consistent with beta diversity analysis results (Figure 2-13-2 B, C). Cd exposure at low doses had minimal effect on snail gut microbial diversity, whereas high Cd stress influenced snail gut microbial diversity considerably.

Prediction of microbial community function

PICRUSt functional prediction and KEGG pathway enrichment analysis results showed that the main functional categories included five cellular processes pathways, three environmental information processing pathways, four genetic information processing pathways, five human disease pathways, 11 metabolism pathways, and seven organismal system pathways (Figure 2-13-4). In the L14 treatment, the fluorobenzoate degradation pathway was enriched; on the contrary, carotenoid biosynthesis, steroid biosynthesis, and indole alkaloid biosynthesis pathways were significantly down-regulated in the L14 treatment compared with in the

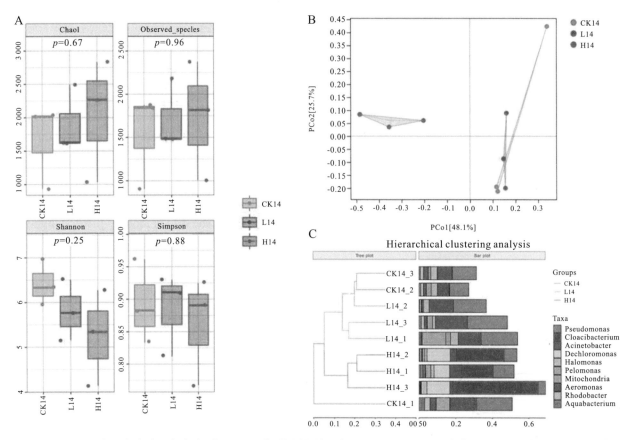

A: α-diversity comparisons in the intestinal microflora among the CK14, L14, and H14 groups; B: Bray-Curtis distances were calculated and visualized through Principal Coordinate Analysis (PCoA) (Ellipses were drawn with 95% confidence intervals); C: Hierarchical cluster analysis of the Bray-Curtis distances generated from taxa tables showed Amplicon Sequence Variant (ASV) similarity across microbial communities among different groups (color figure in appendix).

Figure 2-13-2　Intestinal microbiome diversity in the control (CK14), 0.5 mg/L (L14) and H14 (1.5 mg/L) groups

control treatment (Figure 2-13-5 A, B). In the H14 treatment, five pathways (protein digestion and absorption, apoptosis, lysosome, other glycan degradation, and pathways in cancer) were significantly up-regulated, whereas shigellosis and endocytosis pathways were decreased relative to the control group (Figure 2-13-5 A, C). In addition, notably, compared with in the L14 treatment, some xenobiotic and organic compound biodegradation and metabolism pathways were significantly reduced in the H14 treatment (Figure 2-13-5 A, D).

Discussion

Intestinal microbial diversity

Cadmium is undoubtedly an environmental contaminant. Previous studies have demonstrated that Cd exposure could alter intestinal flora composition in aquatic animals[3, 8, 29]. In the present study, intestinal microbiota in *C. cathayensis* was investigated using high-throughput 16S rRNA gene sequencing. Our results suggested no significant difference in alpha diversity among the three treatments (Figure 2-13-2 A). The results are inconsistent with the findings of previous studies that have reported that Cd exposure altered the alpha diversity of gut microbiota[29, 31], even under relatively low Cd concentration[3]. Furthermore, the PCoA analysis results showed that gut microbial community structure in the high Cd exposure treatment was distinct from that

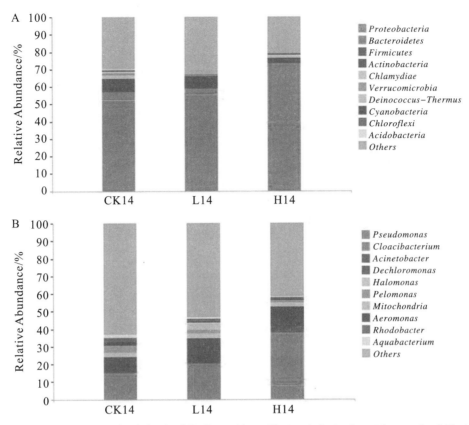

A: Compositions of the intestinal microflora at the phylum level. B: Compositions of the intestinal microflora at the genus level. The top ten abundant genera (higher than 1% in at least one sample) are shown in the figure and the rest are indicated as "Others" (color figure in appendix).

Figure 2-13-3   Compositions of the intestinal microflora among the control (CK14), 0.5 mg/L (L14), and H14 (1.5 mg/L) treatments

in the low Cd exposure and control treatments, whereas the taxonomic groups in the latter two treatments were clustered together (Figure 2-13-2 B). In addition, hierarchical clustering tree construction revealed that the intestinal samples of the three exposure treatments were clustered into two independent groups, excluding one control sample (Figure 2-13-2 C), implying that Cd exposure at high dose (1.5 mg/L) exerted greater effects on the microbiota composition in *C. cathayensis*. The results indicate that *C. cathayensis* could potentially tolerate low Cd stress. Gut microbiome systems of aquatic animals participate in various processes, including nutrition, growth, immunity, and disease resistance[3, 6, 25-28]. In the present study, intestinal microbiota structure was altered in the *C. cathayensis* gut under high Cd exposure when compared with in the control treatment (Figure 2-13-2 B, C). Consequently, alteration of intestinal microbial community structure following Cd exposure could induce adverse effects on *C. cathayensis* health.

Effects of Cd exposure on gut microbial community

In the present study, phylum Proteobacteria was the dominant phylum across all three groups. The results are consistent with the findings of recent studies in other *Cipangopaludina* species[32, 33]. The other two dominant phyla were Bacteroidetes and Firmicutes, which are consistent with the findings of previous studies that have reported that the major bacterial phyla in the gut of aquatic animals, including fish, crustaceans, and mollusks are Proteobacteria, Bacteroidetes, and Firmicutes[3, 4, 7, 29, 33]. However, the two phyla were not dominant in a closely related species, *Cipangopaludina chinensis*[32, 33]. The result implies that although the two

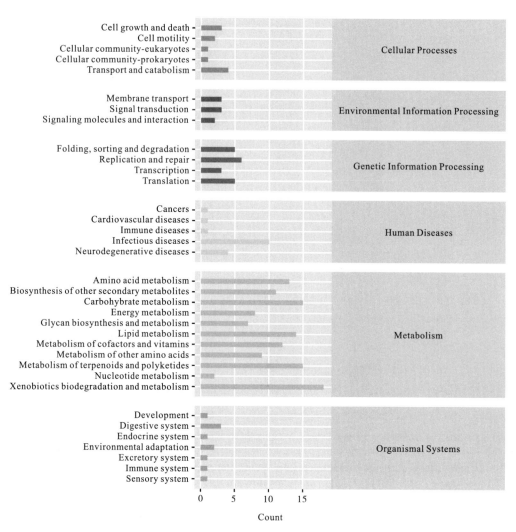

The gene function is showed as color-bars (level 1). The detailed pathways are shown on the left side (level 2).

Figure 2-13-4   Functional annotations and abundance information about the intestinal microbiota at KEGG level 1 and level 2

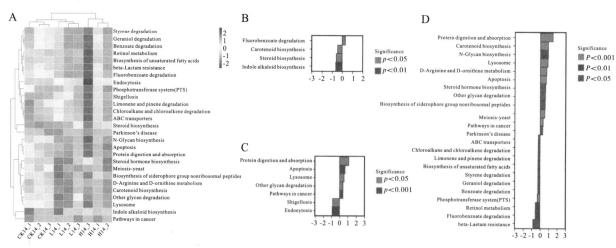

A: Heatmap of the significant differential pathways among the control (CK14), 0.5 mg/L (L14), and H14 (1.5 mg/L) treatments; B: Significant different pathways between the control and L14 treatments; C: Significantly different pathways between the control and H14 treatments; D: Significantly different pathways between the L14 and H14 treatments (color figure in appendix).

Figure 2-13-5   Intestinal microbiota predictive metabolic functions from the KEGG database in all samples

species are closely related, they may have different strategies of responding to Cd stress. Nevertheless, further research is required to investigate the factors responsible for the difference between the two species. Compared to that in the control, the abundance of Bacteroidetes was significantly higher in the H14 treatment, although there was no significant difference between the L14 and control treatments. Bacteroidetes, the largest phylum of Gram-negative bacteria in the human gastrointestinal tract microbiome, has the potential to secrete surface lipopolysaccharides and toxic proteolytic peptides, which can cause inflammation in the gut[34]. The elevated phylum Bacteroidetes abundance in H14 treatment indicated that high Cd exposure has potential adverse effects on *C. cathayensis* health.

Genus *Pseudomonas*, which has been identified as bacterial pathogen in teleosts, exists widely in aquatic environments and in the gut of aquatic animals[28, 35-37]. *Pseudomonas* has also been observed to increase in the guts of different vertebrates, including fish and amphibians, following Cd exposure[3, 5, 31]. Furthermore, *Pseudomonas* outbreaks have been reported in aquacultured animals[35, 36]. However, *Pseudomonas* are also considered probiotics for application in aquaculture[38], that can chelate or oxidize heavy metals, thereby facilitating heavy metal excretion and minimizing the exposure of organisms to heavy metals[27, 39]. In the present study, *Pseudomonas* relative abundance in the L14 treatment was higher than that in the control treatment, although the difference was not significant. However, *Pseudomonas* relative abundance in the H14 treatment was lower than that in the control treatment, although not significant. Notably, *Pseudomonas* abundance decreased with an increase in Cd concentration, suggesting that *Pseudomonas* could play a role in Cd toxicity removal. However, probiotics contents decreased with an increase in Cd concentration, which could adversely affect Cd toxicity tolerance in mudsnail. Indeed, high snail mortality was observed in the H14 treatment but not in the L14 treatment. The results further confirm our postulation above that *C. cathayensis* could acclimate to low Cd concentration, potentially by accumulating *Pseudomonas*. *Acinetobacter* are putative pathogens. Their abundance increased significantly in the gut of Nile tilapia[40] and common carp[3] following Cd exposure, and greatly increased in methyl-mercury (MeHg)-exposed fish[41]. Studies have shown that *Acinetobacter* may exert adverse effects on fish health[42, 43]. Consistent with the previous findings, *Acinetobacter* increased in the L14 treatment, although not significantly. However, when Cd concentration reached 1.5 mg/L, *Acinetobacter* reduced considerably, which may be related to the extremely high Cd content[43]. Similarly, *Rhodobacter* significantly decreased in the H14 treatment. Decreased *Rhodobacter* abundance has been reported to reduce growth[44] and to have adverse effects on fish innate immunity[43], resulting in increased vulnerability to disease[45, 46]. Indeed, *Rhodobacter* is a candidate probiotic for fish[47]. However, some studies have found that higher abundances of such bacteria could be associated with diseased intestines[48], and they could cause neurotoxicity in the hosts[39, 41]. Such findings illustrate the importance of intestinal bacterial community homeostasis in hosts. Decreased *Rhodobacter* abundance in the present study suggest that intestinal function could have been impaired in *C. cathayensis* exposed to Cd, which could result in disease outbreaks under natural conditions[45, 46].

In the present study, *Cloacibacterium*, a key genus in the phylum Bacteroidetes implicated in xenobiotic metabolism and metal removal[27, 49], increased significantly in the H14 treatment. *Cloacibacterium* has been used to detoxify MeHg in MeHg-exposed fish[41]. Enrichment of *Cloacibacterium* has been reported to be an important feature under MeHg-induced neurotoxicity[41]. In addition, in the present study, *Dechloromonas* abundance increased in the H14 treatment. *Dechloromonas*, which belongs to the phylum Proteobacteria and is considered a Cd-resistant microorganism[50], could efficiently degrade polycyclic aromatic hydrocarbons during sludge composting[51, 52], and participate in organic matter degradation in aquaculture pond sediment[53].

Cloacibacterium and *Dechloromonas* enrichment in *C. cathayensis* gut in the present study highlight their potential roles in Cd detoxification, which merit further study.

Intestinal microbiome function

Our function prediction analysis of the gut microbiota showed that most of the genes encoded by the *C. cathayensis* gut microbiota were related to metabolism, followed by organismal systems, cellular processes pathways, human diseases pathways, genetic information processing pathways, and environmental information processing pathways (Figure 2-13-4). The intestines are essential organs involved in the metabolism of nutrients[54]. The results suggest that Cd exposure may alter gut microbial function and host metabolism. In addition, function prediction results showed that, compared with the control treatment, only one pathway related to fluorobenzoate degradation was enriched in the L14 treatment, whereas more pathways were enriched in the H14 treatment (Figure 2-13-5), including protein digestion and absorption, apoptosis, lysosome, other glycan degradation, and pathways in cancer. Fluorobenzoate is the sole carbon and energy source for *Pseudomonas*[55]. The fluorobenzoate degradation pathway was enriched in the L14 treatment, which is consistent with the increasing trends in *Pseudomonas* abundance observed in the treatment group. The cell apoptosis pathway is usually activated following disease infection[56] or exposure to adverse environmental factors[57]. Furthermore, lysosomes not only play a central role in cell decomposition but also participate in metabolism, membrane repair, and cell death[58], and lysosome metabolic pathways are closely related to cell apoptosis[59] In the present study, lysosome pathway and cell apoptosis pathway were both enriched in the H14treatment, which suggests that high Cd exposure may exert more adverse effects on gut microbes of snails than low Cd exposure. Furthermore, pathways in cancer were also enriched in the H14 treatment. The results above partially explain our hypothesis above that *C. cathayensis* has a capacity to acclimate to low Cd stress.

It is also worth noting that pathways associated with xenobiotic and organic compound biodegradation and metabolism, including chloroalkane and chloroalkene degradation, benzoate degradation, fluorobenzoate degradation, styrene degradation, limonene and pinene degradation, geraniol degradation, were significantly inhibited in the H14 treatment in the present study, when compared to in the L14 treatment. Chloroalkane, chloroalkene, benzoate, fluorobenzoate and styrene are xenobiotics found in the environment[60]. However, the pathways associated with the degradation of the xenobiotics were down-regulated in the H14 treatment, which suggested that the capacity of elimination of the compounds decreased following exposure to high Cd doses[60, 61]. Limonene and pinene are considered anti-inflammatory molecules; the down-regulation of the limonene and pinene degradation pathway in the H14 treatment could have increased the levels of limonene and pinene, which could have antagonized the inflammatory response caused by Cd stress[62]. Geraniol is another carbon and energy source for some *Pseudomonas* species[63, 64]. The decline in the geraniol degradation pathway in the present study could be attributed to the decreased contents of the genus *Pseudomonas*. Overall, according to the results of the present study, Cd exposure disrupts gut microbial community structure and their potential functions, and could in turn, adversely influence *C. cathayensis* health.

Conclusions

Our results revealed that Cd exposure could significantly alter the structure and function of intestinal microbial communities, which may in turn influence *C. cathayensis* gut homeostasis and health. To the best of our knowledge, this is the first study to explore the effects of Cd exposure on the intestinal microbiota of *C. cathayensis*. The results obtained in this study provide insights into the mechanisms associated with the response of the intestinal microbiota of *C. cathayensis* to Cd pollution. However, obtaining the 16S rRNA gene sequences through the Illumina HiSeq platform has limitations. In the present study, we did not isolate and identify the putatively pathogenic and putatively beneficial bacteria, which warrants further research.

References

[1] ALI M M, RAHMAN S, ISLAM M S, et al. Distribution of heavy metals in water and sediment of an urban river in a developing country: A probabilistic risk assessment. International Journal of Sediment Research, 2022, 37(2): 173-187.

[2] BURGER J. Assessment and management of risk to wildlife from cadmium. Sci Total Environ, 2008, 389(1): 37-45.

[3] CHANG X L, LI H, FENG J C, et al. Effects of cadmium exposure on the composition and diversity of the intestinal microbial community of common carp (*Cyprinus carpio* L.). Ecotoxicol. Environ. Saf, 2019(171): 92-98.

[4] LIU J, PANG I J, TU Z C, et al. The accumulation, histopathology, and intestinal microorganism effects of waterborne cadmium on *Carassius auratus gibelio*. Fish Physiol Biochem, 2019, 45(1): 231-243.

[5] CHEAIB B, SEGHOUANI H, IJAZ U Z, et al. Community recovery dynamics in yellow perch microbiome after gradual and constant metallic perturbations. Microbiome, 2020, 8(1): 14.

[6] WANG N, JIANG M, ZHANG P J, et al. Amelioration of Cd-induced bioaccumulation, oxidative stress and intestinal microbiota by Bacillus cereus in *Carassius auratus gibelio*. Chemosphere (Oxford), 2020, 245: 125613.

[7] WANG N, GUO Z Y, ZHANG Y L, et al. Effect on intestinal microbiota, bioaccumulation, and oxidative stress of *Carassius auratus gibelio* under waterborne cadmium exposure. Fish Physiol Biochem, 2020, 46(6): 2299-2309.

[8] WANG R, XIA W, EGGLETON M A, et al. Spatial and temporal patterns of heavy metals and potential human impacts in Central Yangtze lakes, China. The Science of the total environment, 2022, 820: 153368.

[9] CHEN Y F, SHI Q G, QU I Y, et al. A pollution risk assessment and source analysis of heavy metals in sediments: A case study of Lake Gehu, China. Chinese Journal of Analytical Chemistry, 2022, 50(5): 100077.

[10] Ministry of Ecology and Environment of the People's Republic of China (2014). National soil pollution survey communique. (2014-04-17)[2022-03-01]. https://www.mee.gov.cn/gkml/sthjbgw/qt/201404/W020140417558995804588.pdf.

[11] ZHAO F J, MA Y, ZHU Y G, et al. Soil contamination in China: current status and mitigation strategies. Environmental Science & Technology, 2015, 49(2): 750-759.

[12] WEN Y B, LI W, YANG Z F, et al. Evaluation of various approaches to predict cadmium bioavailability to rice grown in soils with high geochemical background in the karst region, southwestern China. Environmental Pollution (1987), 2020, 258: 113645.

[13] ZHAO X, YAO L A, MA Q L, et al. Distribution and ecological risk assessment of cadmium in water and sediment in Longjiang River, China: Implication on water quality management after pollution accident. Chemosphere (Oxford), 2018, 194: 107-116.

[14] CUI Y D, WANG B Q, ZHAO Y J, et al. Recovery time of macroinvertebrate community from Cd pollution in Longjiang River, Guangxi, China. Journal of Oceanology and Limnology, 2022, 40(1): 183-194.

[15] SONG B, WANG F P, ZHOU L, et al. Cd Content characteristics and ecological risk assessment of paddy soil in high cadmium anomaly area of Guangxi. Environmental Science, 2019, 40(5): 2443.

[16] YANG Q, YANG Z F, ZHANG Q, et al. Ecological risk assessment of Cd and other heavy metals in soil-rice system in the karst areas with high geochemical background of Guangxi, China. Science China Earth Sciences, 2021, 64(7): 1126-1139.

[17] XU X, NIE S, DING H, et al. Environmental pollution and kidney diseases. Nature reviews. Nephrology, 2018, 14(5): 313-324.

[18] LU H F, DU L N, LI Z Q, et al. Morphological analysis of the Chinese *Cipangopaludina* species (Gastropoda; Caenogastropoda: Viviparidae). Zoo logical Reach, 2014, 35(6): 510-527.

[19] LUO H, CHEN L T, JING T S. Muscle nutrition analysis of four snail species of Viviparidae. J Fish China,2022,46(11).2177-2185.

[20] WANG C, LIU J, HUANG Y, et al. In vitro polysaccharide extraction from *Cipangopaludina cathayensis* and its

pharmacological potential. J Environ Biol, 2016, 37(5 Spec No): 1069-1072.

[21] ZHAO T, XIONG I Q, CHEN W, et al. Purification and characterization of a novel fibrinolytic enzyme from *Cipangopaludina cahayensis*. Iranian Journal of Biotechnology, 2021, 19(1): e2805.

[22] PENG W, HONGPING C, KOPITTKE P M, et al. Cadmium contamination in agricultural soils of China and the impact on food safety. Environmental Pollution (1987), 2019, 249: 1038-1048.

[23] HU R, TANG Z Y. Effect of cadmium and mercury on superoxide dismutase activity of the mudsnail(*Cipangopaludina cahayensis*). J Sichuan Normal Univ, 2012, 35(5): 690-693.

[24] ZHOU Y Y, LUO Z M. Effects of cadmium exposure on the composition and diversity of the intestinal microbial community of common carp (*Cyprinus carpio* L.). Sei Technol Food Ind, 2018(39): 43-47.

[25] TALWAR C, NAGAR S, LAL R, et al. Fish Gut Microbiome: Current Approaches and Future Perspectives. Indian Journal of Microbiology, 2018, 58(4): 397-414.

[26] PAUL J S, SMALL B C. Exposure to environmentally relevant cadmium concentrations negatively impacts early life stages of channel catfish (*Ictalurus punctatus*). Comparative biochemistry and physiology. Toxicology & Pharmacology, 2019, 216: 43-51.

[27] DUAN H, YU L L, TIAN F W, et al. Gut microbiota: A target for heavy metal toxicity and a probiotic protective strategy. The Science of the Total Environment, 2020, 742: 140429.

[28] DIWAN A D, HARKE S N, GOPALKRISHNA, et al. Aquaculture industry prospective from gut microbiome of fish and shellfish: An overview. J Anim Physiol Anim Nutr (Berl), 2022, 106(2): 441-469.

[29] ZHANG Y, LI Z Y, KHOLODKEVICH S, et al. Effects of cadmium on intestinal histology and microbiota in freshwater crayfish (*Procambarus clarkii*). Chemosphere, 2020, 242: 125105.

[30] CALLAHAN B J, MCMURDIE P J, ROSEN M J, et al. DADA2: High-resolution sample inference from Illumina amplicon data. Nature methods, 2016, 13(7): 581-583.

[31] YA J, JU Z Q, WANG H Y, et al. Exposure to cadmium induced gut histopathological damages and microbiota alterations of Chinese toad (*Bufo gargarizans*) larvae. Ecotoxicol Environ Saf, 2019, 180: 449-456.

[32] ZHOU K Q, QIN J Q, PANG H Y, et al. Comparison of the composition and function of gut microbes between adult and juvenile *Cipangopaludina chinensis* in the rice snail system. Peer J (San Francisco, CA), 2022, 10: e13042.

[33] ZHOU Z H, WU H W, LI D H, et al. Comparison of gut microbiome in the Chinese mud snail (*Cipangopaludina chinensis*) and the invasive golden apple snail (*Pomacea canaliculata*). Peer J, 2022, 10: e13245.

[34] LUKIW W I. Bacteroides fragilis Lipopolysaccharide and inflammatory signaling in Alzheimer's disease. Frontiers in Microbiology, 2016, 7: 1544.

[35] LLEWELLYN M S, BOUTIN S, HOSEINIFAR S H, et al. Teleost microbiomes: the state of the art in their characterization, manipulation and importance in aquaculture and fisheries. Front Microbiol, 2014, 5: 207.

[36] XU J, ZENG X H, JIANG N, et al. Pseudomonas alcaligenes infection and mortality in cultured Chinese sturgeon, Acipenser sinensis. Aquaculture, 2015, 446: 37-41.

[37] DEHLER C E, SECOMBES C J, MARTIN S A M. Environmental and physiological factors shape the gut microbiome of Atlantic salmon (*Salmo salar* L). Aquaculture, 2017(467): 149-157.

[38] WANG A R, RAN C, WANG Y B, et al. Use of probiotics in aquaculture of China—a review of the past decade. Fish Shellfish Immunol, 2019, 86: 734-755.

[39] ARUN K B, MADHAVAN A, SINDHU R, et al. Probiotics and gut microbiome—Prospects and challenges in remediating heavy metal toxicity. J Hazard Mater, 2021, 420: 126676.

[40] ZHAI Q X, YU L L, LI T Q, et al. Effect of dietary probiotic supplementation on intestinal microbiota and physiological conditions of Nile tilapia (*Oreochromis niloticus*) under waterborne cadmium exposure. Antonie van Leeuwenhoek, 2017, 110(4): 501-513.

[41] BRIDGES K N, ZHANG Y, CURRAN T E, et al. Alterations to the intestinal microbiome and metabolome of *Pimephales*

promelas and *Mus musculus* following exposure to dietary methylmercury. Environmental Science & Technology, 2018, 52(15): 8774-8784.

[42] WU S G, TIAN J Y, GATESOUPE F J, et al. Intestinal microbiota of gibel carp (*Carassius auratus gibelio*) and its origin as revealed by 454 pyrosequencing. World journal of microbiology & biotechnology, 2013, 29(9): 1585-1595.

[43] WANG X H, HU M H, GU H H, et al. Short-term exposure to norfloxacin induces oxidative stress, neurotoxicity and microbiota alteration in juvenile large yellow croaker *Pseudosciaena crocea*. Environmental pollution (1987), 2020, 267: 115397.

[44] LIU H S, LI X, LEI H J,et al. Dietary seleno-L-methionine alters the microbial communities and causes damage in the gastrointestinal tract of Japanese medaka *Oryzias latipes*. Environ. Sci. Technol, 2021(55): 16515-16525.

[45] SHE R, LI T T, LUO D, et al. Changes in the intestinal microbiota of gibel carp (*Carassius gibelio*) associated with Cyprinid herpesvirus 2 (CyHV-2) infection. Curr Microbiol, 2017, 74(10): 1130-1136.

[46] LIU F P, XU X F, CHAO L, et al. Alteration of the gut microbiome in chronic kidney disease patients and its association with serum free immunoglobulin light chains. Frontiers in Immunology, 2021, 12: 609700.

[47] YE Q, FENG Y Y, WANG Z L, et al. Effects of dietary Gelsemium elegans alkaloids on intestinal morphology, antioxidant status, immune responses and microbiota of *Megalobrama amblycephala*. Fish Shellfish Immunol, 2019, 94: 464-478.

[48] TRAN N T, ZHANG J, XIONG F, et al. Altered gut microbiota associated with intestinal disease in grass carp (*Ctenopharyngodon idellus*). World J Microbiol Biotechnol, 2018, 34(6): 71.

[49] NOUHA K, KUMAR R S, TYAGI R D. Heavy metals removal from wastewater using extracellular polymeric substances produced by *Cloacibacterium normanense* in wastewater sludge supplemented with crude glycerol and study of extracellular polymeric substances extraction by different methods. Bioresour Technol, 2016, 212: 120-129.

[50] ZHANG L Q, FAN J J, NGUYEN H N, et al. Effect of cadmium on the performance of partial nitrification using sequencing batch reactor. Chemosphere (Oxford), 2019, 222: 913-922.

[51] LU Y, ZHENG G Y, ZHOU W B, et al. Bioleaching conditioning increased the bioavailability of polycyclic aromatic hydrocarbons to promote their removal during co-composting of industrial and municipal sewage sludges. Sci Total Environ, 2019, 665: 1073-1082.

[52] CHE I G, BAI Y D, LI X, et al. Linking microbial community structure with molecular composition of dissolved organic matter during an industrial-scale composting. Journal of Hazardous Materials, 2021, 405: 124281.

[53] ZHANG K K, ZHENG X F, HE Z L, et al. Fish growth enhances microbial sulfur cycling in aquaculture pond sediments. Microb Biotechnol, 2020, 13(5): 1597-1610.

[54] LIU Y Y, CHENG J X, XIA Y Q, et al. Response mechanism of gut microbiome and metabolism of European seabass (*Dicentrarchus labrax*) to temperature stress. Sci Total Environ, 2022, 813: 151786.

[55] VORA K A, SINGH C, MODI V V. Degradation of 2-fluorobenzoate by a pseudomonad. Current Microbiology, 1988, 17(5): 249-254.

[56] QIU Y, YIN Y H,RUAN Z Q. Comprehensive transcriptional changes in the liver of kanglang white minnow (*Anabarilius grahami*) in response to the infection of parasite ichthyophthirius multifilis. Animals, 2020(10): 681.

[57] CHEN G, PANG M X, YU X M, et al. Transcriptome sequencing provides insights into the mechanism of hypoxia adaption in bighead carp (*Hypophthalmichthys nobilis*). Comparative biochemistry and physiology. Part D, Genomics & Proteomics, 2021, 40: 100891.

[58] SERRANO-PUEBLA A, BOYA P. Lysosomal membrane permeabilization in cell death: new evidence and implications for health and disease. Annals of the New York Academy of Sciences, 2016, 1371(1): 30-44.

[59] GUO Y S, MA Y, ZHANG Y W, et al. Autophagy-related gene microarray and bioinformatics analysis for ischemic stroke detection. Biochemical and Biophysical Research Communications, 2017, 489(1): 48-55.

[60] VERA A, WILSON F P, CUPPLES A M. Predicted functional genes for the biodegradation of xenobiotics in groundwater and sediment at two contaminated naval sites. Appl Microbiol Biotechnol, 2022, 106(2): 835-853.

[61] GU X I, FU H T, SUN S M, et al. Dietary cholesterol-induced transcriptome differences in the intestine, hepatopancreas, and muscle of Oriental River prawn Macrobrachium nipponense. Comparative biochemistry and physiology. Part D, Genomics & Proteomics, 2017, 23: 39-48.

[62] HAN Y I, KANG L L, LIU X H, et al. Establishment and validation of a logistic regression model for prediction of septic shock severity in children. Hereditas, 2021, 158(1): 45.

[63] VANDENBERGH P A, WRIGHT A M. Plasmid involvement in acyclic isoprenoid metabolism by *Pseudomonas putida*. Appl Environ Microbiol, 1983, 45(6): 1953-1955.

[64] ZHU C S, MILLER M, LUSSKIN N, et al. Snow microbiome functional analyses reveal novel aspects of microbial metabolism of complex organic compounds. Microbiologyopen, 2020, 9(9): e1100.

Pathogenicity and molecular characteristics of fowl adenovirus serotype 4 with moderate virulence in Guangxi, China

Wei You, Xie Zhixun, Fan Qing, Xie Zhiqin, Deng Xianwen, Luo Sisi, Li Xiaofeng, Zhang Yanfang, Zeng Tingting, Huang Jiaoling, Ruan Zhihua, and Wang Sheng

Abstract

The GX2020-019 strain of fowl adenovirus serotype 4 (FAdV-4) was isolated from the liver of chickens with hydropericardium hepatitis syndrome in Guangxi, China, and was purified by plaque assay three times. Pathogenicity studies showed that GX2020-019 can cause typical FAdV-4 pathology, such as hydropericardium syndrome and liver yellowing and swelling. Four-week-old specific pathogen-free (SPF) chickens inoculated with the virus at doses of 10^3 median tissue culture infectious dose ($TCID_{50}$), 10^4 $TCID_{50}$, 10^5 $TCID_{50}$, 10^6 $TCID_{50}$, and 10^7 $TCID_{50}$ had mortality rates of 0%, 20%, 60%, 100%, and 100%, respectively, which were lower than those of chickens inoculated with other highly pathogenic Chinese isolates, indicating that GX2020-019 is a moderately virulent strain. Persistent shedding occurred through the oral and cloacal routes for up to 35 days postinfection. The viral infection caused severe pathological damage to the liver, kidney, lung, bursa of Fabricius, thymus, and spleen. The damage to the liver and immune organs could not be fully restored 21 days after infection, which continued to affect the immune function of chickens. Whole genome analysis indicated that the strain belonged to the FAdV-C group, serotype 4, and had 99.7%~100% homology with recent FAdV-4 strains isolated from China. However, the amino acid sequences encoded by ORF30 and ORF49 are identical to the sequences found in nonpathogenic strains, and none of the 32 amino acid mutation sites that appeared in other Chinese isolates were found. Our research expands understanding of the pathogenicity of FAdV-4 and provides a reference for further studies.

Keywords

fowl adenovirus serotype 4, pathogenicity, molecular characterization, medium virulent, convalescence

Introduction

Fowl adenovirus belongs to the family Adenoviridae and the genus Aviadenovirus. It is a nonenveloped linear double-stranded DNA virus with a genome length of 25~46 kb that can be divided into five species, FAdV-A, B, C, D, E, and 12 serotypes (FAdV-1 to 8a and FAdV-8 to 11)[1, 2]. Hexon, penton, fiber-1, and Fiber-2 are the main structural proteins of the envelope of FAdV-4. The nucleocapsid proteins are primarily pX, pV, pVII, and pVIII, while nonstructural proteins include ElA, E1B, E3, E4, 100k, and 52/55k, among others[3]. Recent research has shown that the hexon protein has the function of neutralizing antigenic sites, contains type-specific antigenic determinant Dlusters, and can be used for serum typing[3]. The Penton protein plays a crucial role in virus entry into cells. The Fiber-1 protein directly triggers infections by pathogenic FAdV-4 through an axial handle-like structure[4]. The region at the top of the fiber-2 protein can bind with host cells, and the strength of binding can be used as a basis for determining the virulence of the virus. Furthermore, the Fiber-

2 protein has good antigenicity, can induce the production of neutralizing antibodies, and can effectively limit FAdV-4 infection[5].

Hydropericardium hepatitis syndrome (HHS) is mainly caused by fowl adenovirus serotype 4 (FAdV-4) and has previously been observed in regions such as Pakistan, Kuwait, Australia, Iraq, South America, Central America, India, Japan, and Republic of Korea[6-8], causing considerable economic losses to the global poultry industry. The virus mainly infects broiler chickens between the ages of 3~6 weeks and has a death rate of up to 80% or more in chicks[9]. However, cases of infection in breeders between 12 and 25 weeks have also been reported in recent years[10]. Poultry begin to show symptoms of FAdV-4 infection from the 3rd to 4th day with a peak in the number of deaths occurring on the 5th to 6th day, followed by a decline[11]. The main gross changes in diseased chickens include the accumulation of a large amount of light yellow fluid in the pericardial sac and the yellowing and noticeable swelling of the liver, and sometimes hemorrhage and necrosis can be observed on the surface of the liver[12]. Previously, FAdV infections in China were primarily asymptomatic secondary infections. The first case of highly pathogenic FAdV-4 in China was reported in June 2015, and the virus then rapidly spread across the country[13]. Outbreaks have occurred in major poultry-rearing provinces, such as Shandong, Henan, Jiangsu, Anhui, Sichuan, Hunan, Guangxi, Guangdong, and others[14-16]. Since 2016, there have been suspected cases of HHS in cities, such as Nanning, Yulin, Guilin, and Qinzhou in Guangxi, seriously threatening the local poultry industry[17, 18].

In this study, an FAdV-4 strain was isolated from chickens suspected to have HHS in Guangxi, China, and subjected to whole-genome sequencing. It was compared with pathogenic and nonpathogenic strains published in GenBank from different regions. The effects of different inoculation doses on viral pathogenicity in chickens and after disease progression were also determined. This study aimed to provide evidence to support the prevalence of FAdV-4 in the Guangxi region and lay the foundation for the research and control of FAdV-4.

Materials and methods

Experimental animal and ethics statement

Specific pathogen-free (SPF) White Leghorn chicken eggs were purchased from Beijing Boehringer Ingelheim Vital Biotechnology Co., Ltd. (Beijing, China). The eggs were incubated for 20~21 d until hatching in an incubator, and the chickens were then raised in a negative pressure SPF isolation unit until 4 weeks of age to test the pathogenicity of the FAdV-4 strain. The animal experiment was approved by the Ethics Committee of Guangxi Veterinary Research Institute. The experimental procedures were conducted in accordance with the regulations of the Animal Ethics Committee of Guangxi Veterinary Research Institute (No.2019c0406).

Sample collection and FAdV-4 PCR detection

In 2020, a commercial broiler flock consisting of approximately 4 000 8-week-old chickens at a poultry farm in Yulin City, Guangxi, south China, exhibited signs of poor health, such as reduced appetite and huddling in corners. The flock experienced a daily mortality rate of approximately 40~60 chickens, accounting for approximately 1% of the total population. Upon autopsy, typical HHS lesions were observed. The virus was isolated from liver tissue specimens showed hepatomegaly, yellow discolouration, and haemorrhagic necrosis. The samples were homogenized in phosphate-buffered saline (PBS), and after three freeze-thaw cycles, the homogenate was centrifuged at 12 000 g for 10 min. The supernatant was used for FAdV-4 PCR detection and virus isolation. A commercial TransGen Biotech EasyPure Genomic DNA/RNA Kit (TransGen, China) was used to extract DNA/RNA from the supernatants.

To detect the presence of viral DNA in the samples, PCR was performed using specific primers for the hexon gene of FAdV-4. The forwards primer sequence was 5'-CGAGGTCTATACCAACACGAGCA-3', and the reverse primer sequence was 5'-TACAGCAGGTTAATGAAGTTATC-3'. The PCR amplification protocol consisted of an initial denaturation step at 95 ℃ for 5 min, followed by 30 cycles of denaturation at 95 ℃ for 30 s, annealing at 56 ℃ for 30 s, extension at 72 ℃ for 30 s, and a final extension step at 72 ℃ for 10 min. To isolate the virus from positive samples, the supernatants were filtered through a 0.22 μm PES membrane filter unit and inoculated into primary cultures of chicken embryo liver (CEL) cells. The culture supernatant was harvested 120 h after virus inoculation or when the cell cytopathic effect reached 7% or more. The harvested supernatant was blindly passaged three times using primary CEL cells.

Virus purification and identification

After the virus was isolated, dilutions of the virus were prepared for plaque purification in CEL cells. Dilutions ranging from 10^{-8} to 10^{-3} were inoculated into CEL cells in six-well cell culture plates. After 1 h of infection, the cells were treated with Dulbecco's Modified Eagle Media/Nutrient Mixture F-12 (DMEM/F12; Gibco, United States) containing 1% low melting point agarose (Promega, United States) and 2% fetal bovine serum (Gibco, United States) and incubated for 5~6 days at 37 ℃ and 5% CO_2. Isolated plaques were then selected and transferred to a new fresh culture of CEL cells. Plaque purifications were performed three times. The purified isolates were named GX2020-019, propagated into seed batches, and stored at −80 ℃ .

The identification of FAdV-4 and determination of its median tissue culture infectious dose ($TCID_{50}$) were performed using an indirect immunofluorescent assay (IFA) to detect virus-infected cells with anti-FAdV-4 monoclonal antibodies prepared in the laboratory[19]. The possibility of contamination was ruled out by PCR detection of avian influenza virus (AIV), Newcastle disease virus (NDV), infectious bronchitis virus (IBV), laryngotracheitis virus (LTV), infectious bursal disease virus (IBDV), reovirus (REV), avian leukosis virus (ALV), reticuloendotheliosis virus (REV), and Mycoplasma[20-26].

Full-length PCR amplification and sequencing

Thirty-nine pairs of primers were synthesized to amplify the full-length DNA segments covering the viral genome of FAdV-4 strain GX2019-010[27]. PCR was performed using PrimeSTAR HS DNA Polymerase (TaKaRa, Japan). Each 50 μL reaction contained 10 μL 5 × PrimeSTAR Buffer, 4 μL dNTP mixture, 0.5 μL polymerase (2.5 U/μL), 2 μL total DNA from FAdV-4 isolates, 1 μL of each primer (10 μmol/L), and nudlease-free water to reach a final volume of 50 μL. The PCR included an initial denaturation at 98 ℃ for 5 min, followed by 30 cycles of 10 s at 98 ℃ , 5 s at 55 ℃ , and 2 min at 72 ℃ and a final extension step for 10 min at 72 ℃ . PCR products were analyzed by 1% agarose gel electrophoresis and visualized by GelRed staining. The PCR products were sequenced directly or cloned into the pMD18-T vector for sequencing. The Seqman program, which is part of the Lasergene software package (version 7.1, DNASTAR, United States), was used to manually assemble the complete sequence.

Phylogenetic analysis and molecular characterization of FAdV-4

The complete nucleotide sequence of GX2020-019 was aligned with 49 reference strains of FAdV-A to E available from the GenBank database using the ClustalW multiple alignment algorithm (shown in Table 2-14-

1). A phylogenetic tree was created by neighbor-joining analysis with 500 replicates for bootstrapping, and evolutionary distances were calculated using the maximum composite likelihood method through MEGA 11 software (version 11, Molecular Evolutionary Genetic Analysis, New Zealand). MegAlign software (version 7.1, DNASTAR, United States) was utilized for a full-genome similarity comparison of GX2020-019 with pathogenic strains (GX2019-004, SD1601, JS07, SCDY, SD1511, and HLJFAd15 strains), nonpathogenic strains (ON1, KR5, and B1-7 strains) of FAdV-C, and reference strains of FAdV-A, B, D, and E. The major structural protein genes and ORFs of GX2020-019 were translated into amino acid sequences using EditSeq software (version 7.1, DNASTAR, United States). These sequences were compared with the sequences of FAdV-4 pathogenic strains (MX-SHP95, GX2019-004, SD1601, JS07, SCDY, SD1511, and HLJFAd15 strains) and nonpathogenic strains (ON1, KR5, and B1-7 strains) using MegAlign software (version 7.1, DNASTAR, United States).

Table 2-14-1　Information of fowl adenovirus reference strains

Strains	GenBank accession number	Species	Serotypes	Country
HLJFAd15	KU991797.1	C	4	China
CH/ SXCZ/2015	K1558762.1	C	4	China
HB1510	KU587519.1	C	4	China
JSJ13	KM096544.1	C	4	China
NIVD2	MG547384	C	4	China
AH-F19	MN781666	C	4	China
CH/ AHMG/2018	MN606303.1	C	4	China
D1910497	MW711380.1	C	4	China
GD616	MW509553.1	C	4	China
S07	KY436519	C	4	China
AQ	KY436520	C	4	China
HN	KY379035	C	4	China
AH712	KY436522	C	4	China
SCDY	MK629523	C	4	China
SD1601	MH006602	C	4	China
GX2017-004	MN577980	C	4	China
GX2019-005	MN577981	C	4	China
GX2019-010	MW439040	C	4	China
SD1511	MF496037	C	4	China
GX2019-011	MW439041	C	4	China
GX2017-001	MN577977	C	4	China
GX2018-007	MN577983	C	4	China
GX2018-008	MN577984	C	4	China
D1910497	MW711380	C	4	USA
MX-SHP95	KP295475	C	4	Mexico
ON1	GU188428.1	C	4	Canada
KR5	HE608152.1	C	4	Austria
B1-7	KU342001.1	C	4	India
CELO	U46933.1	A	1	Austria

continued

Strains	GenBank accession number	Species	Serotypes	Country
61/11z	KX247012.1	A	1	Poland
JM1/1	MF168407	A	1	Japan
11-7, 127	MK572848.1	A	1	Japan
OTE	MK572847.1	A	1	Japan
W-15	KX247011.1	A	1	Poland
340	NC021211.1	B	5	Ireland
WHRS	OM836676.1	B	5	China
19/7209	OK283055.1	B	5	Hungary
LYG	MK757473.1	B	5	China
AF083975	AF0839752	D	9	Canada
SR48	KT862806.1	D	2	Austria
ON-NP2	KP231537.1	D	11	Canada
MX95-S11	KU746335.1	D	11	Mexico
380	KT862812.1	D	11	Britain
685	KT862805.1	D	2	Britain
HG	GU734104.1	E	8	Canada
UPM04217	KU517714.1	E	8b	Malaysia
764	KT862811.1	E	8b	Britain
TR59	KT862810.1	E	8a	Japan
YR36	KT862809.1	E	7	Japan

Pathogenicity assessment of GX2020-019

Four-week-old SPF chickens were used to evaluate the pathogenicity of GX2020-019. The chickens were divided into six groups of 10 birds in each group based on the inoculated dose of GX2020-019 (10^3 $TCID_{50}$, 10^4 $TCID_{50}$, 10^5 $TCID_{50}$, 10^6 $TCID_{50}$ and 10^7 $TCID_{50}$) and were inoculated via intramuscular injection. Control birds received an equal volume of PBS. Each group was housed separately in a negative-pressure SPF chicken isolator. The birds were monitored four times daily for 21 days and scored for clinical signs and mortality. The scale used was as follows: 0 for normal; 1 for precursory symptoms, such as depression and disorganized feathers; 2 for obvious symptoms, such as yellow and green excrement, assumption of the fetal position, and anorexia; and 3 for death. The survival rate curve and clinical signs score curve were plotted using Prism 8.0 software (GraphPad Software Inc., United States).

To assess the impact of GX2020-019 on the body, 30 4-week-old SPF chickens were injected with 10^4 $TCID_{50}$ *via* intramuscular injection, and another five chickens were injected with the same volume of PBS as a control group. Three chickens exhibiting obvious clinical signs, such as green feces, listlessness, weakness, prostration, and decreased appetite, were euthanized on day 5 postinoculation (dpi) and were regarded as the symptomatic group. Three chickens that had recovered from obvious clinical symptoms were euthanized and regarded as the convalescent group at 21 dpi. Additionally, three chickens from the PBS control group were euthanized. Gross lesions were examined and observed. Tissue samples were collected from the liver, heart, spleen, glandular stomach, pancreas, lung, kidney, bursa of Fabricius, thymus, brain, muscle, small intestine, muscular stomach, esophagus, and trachea. A portion of each tissue sample was fixed in 10% neutral formalin buffer solution for histological analysis, while another portion was stored at −80 ℃ for viral load detection in the tissues.

Detection of viral shedding

To determine the duration and concentration of viral shedding in the infected chickens, 20 4-week-old SPF chickens were intramuscularly injected with 10^3 $TCID_{50}$ of GX2020-019, while another five chickens were used as a control group and received the same volume of PBS. Oral and cloacal swabs were collected from 10 birds at 1, 3, 5, 7, 10, 14, 21, 28, and 35 dpi to extract viral DNA and detect FAdV-4 viral loads.

Examination of histopathology

The tissue samples were fixed in a 10% neutral formalin buffer solution at room temperature for at least 48 h and then sent to Guangzhou Maike Biotechnology Co., Ltd. (Guangzhou, China) for tissue sectioning and hematoxylin-eosin (H&E) staining. Lesions associated with FAdV-4 infection were observed under an electron microscope and photographed.

Detection of viral loads

A total of 0.2 g of each tissue sample was homogenized in 1 mL of PBS. The homogenate was subjected to three freeze-thaw cycles, followed by centrifugation at 12 000 g for 10 min. Then, 300 μL of supernatant was collected for DNA extraction using a Universal Genomic DNA Kit (CoWin Biotech Co., Ltd, China) according to the manufacturers instructions. The concentration of the extracted nucleic acid was determined using a NanoDrop ND1000 spectrophotometer (Thermo Scientific, United States) and standardized to 100 μg/mL for use as a template in real-time PCR. Oral and cloacal swabs were suspended in 1 mL of PBS and agitated for 30 sec. The suspension was then allowed to settle at room temperature, and 300 μL of the resulting supernatant was used for DNA extraction with a real-time PCR kit.

All samples were tested in triplicate using the following detection primers: 5'-GCACGAGG CACCTCCAAAGAG-3'and 5'-GTTGTACCCGTCGCAGGAGGATG 3'. Real-time PCR was conducted using a 20 μL total volume containing 2 μL of template DNA, 10 μL of PowerUp TM SYBRTM GreenMasterMix, 1 μL of each primer (10 μmol/L), and 10 μL of deionized water. The thermal cycling program included an initial cycle at 95 ℃ for 120 s, followed by 40 cycles of denaturation at 95 ℃ for 15 s and annealing/extension at 60 ℃ for 60 s[28]. The melting curve was analyzed at the end of the cycling program. Viral load was calculated based on the Ct value of the sample and the standard curve. All samples were analyzed in triplicate.

Statistical analysis

Statistical analysis was performed using one-way ANOVA with Prism 8.0 (GraphPad Software Inc., United States). Results with a p value of less than 0.05 were considered statistically significant. The results are presented as the mean ± SD.

Results

FAdV-4 isolated from chickens with HHS

After DNA extraction from the liver tissue homogenate, a positive band was observed by PCR using the FAdV-4 421 bp detection primer. The supernatant of the homogenate was then inoculated into CEL cells, and significant cytopathic effects were observed in the first generation after inoculation as the cells became round.

The cytopathic effect in the second and third generations reached 70% and remained stable at approximately 72 h, indicating good viral replication in CEL cells. The FAdV-4 isolate was purified by three rounds of plaque purification and named GX2020-019.

An IFA was performed and showed that GX2020-019 infected cells had a significant positive reaction with the anti-fiber-2 monoclonal antibody against FAdV-4. Contamination by AIV, NDV, IBV, LTV, IBDV, RLV, ALV, REV, and Mycoplasma in the viral liquid was ruled out by PCR testing. The $TCID_{50}$ of the seed virus stock was determined to be $10^{8.2}/0.1$ mL.

Pathogenicity of the virus in SPF chickens infected with different doses of GX2020-019

Five different doses were used to infect chickens through the intramuscular route. After 48 h of infection, chickens in the 10^6 $TCID_{50}$ and 10^7 $TCID_{50}$ groups showed clinical symptoms of depression, fluffy feathers, and yellow-green soft stools. At 4 dpi, 100% of the chickens died, and necropsy revealed large amounts of light-yellow effusion in the pericardium, yellowing and fragile liver with obvious blood spots on the surface, as well as congested and enlarged kidneys and spleens (Figure 2-14-1). The average time to death was 79 h in the $10^6 TCID_{50}$ group and 70 h in the 10^7 $TCID_{50}$ group. The time at which the number of deaths peaked in the 10^4 $TCID_{50}$ and $10^5 TCID_{50}$ groups was 4~6 dpi, with a morbidity rate of 90% and 100% and a mortality rate of 20% and 60%, respectively. The average time of death was 86 and 124 h in the two groups, respectively. The symptoms of the chickens that did not die were significantly reduced 6~8 days after onset, appetite returned, and the chickens began to recover. No obvious clinical symptoms or death occurred in the 10^3 $TCID_{50}$ group, and only 30% of chickens showed mental depression from 5 to 8 dpi, which recovered to the levels in the control group. No birds in the control group became sick or died. The clinical symptom score and the survival curve are Figure 2-14-2.

A: Depressed mental state, fluffed feathers, anorexia, and somnolence in the diseased chickens; B: Yellow-green watery stools; C: Presence of effusion in the pericardium, enlarged liver, liver yellowing, and bleeding spots; D: Congested and enlarged kidneys; E: Congested and enlarged spleen (color figure in appendix).

Figure 2-14-1　Clinical symptoms and gross lesions in chickens infected with GX2020-019

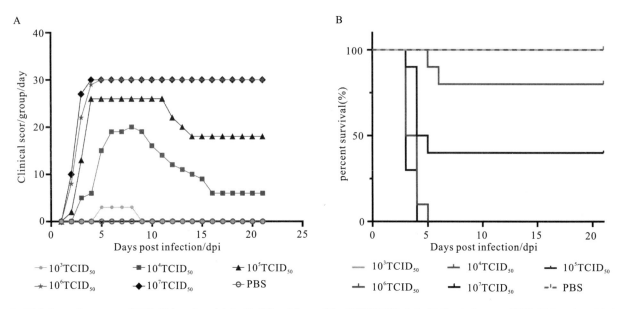

A: Clinical symptom scores of SPF chickens inoculated with different doses of the GX2020-019 strain; B: Survival curve of SPF chickens inoculated with different doses of the GX2020-019 strain.

Figure 2-14-2   Pathogenicity of the GX2020-019 strain in SPF chickens at different doses

Histopathological changes in SPF chickens infected with GX2020-019

During the disease period of chickens infected with GX2020-019, mild to severe degenerative necrosis and widespread inflammatory cell infiltration were found in the liver, bursa of Fabricius, spleen, thymus, lung, and kidney. In particular, inclusion bodies formed within the liver cells, and the lung alveolar cavities were filled with shedding dead cells, red blood cells, and inflammatory cells.

During the convalescence period, the level of liver cord degeneration was reduced, but there was still monocyte infiltration around the hepatic sinusoids, which remained narrow. The congestion and stasis in the lung disappeared, and the exudate in the lung alveoli was absorbed, while the lung alveolar wall cells were the same as those in the control group. The level of atrophy of the lymph follicles in the bursa of Fabricius was reduced, and the distinction between the medulla-cortex boundary and the interfollicular septa became clearer. The degree of necrosis in the thymus cells was reduced, the ratio of cortex to medulla increased, the level of apoptotic lymphocytes decreased, and the abundance of medullary blood vessels and thymic corpuscles increased. Interstitial hemorrhages in the spleen tissue disappeared, and the state of the lymphocytes changed from degenerative necrosis to mild degeneration, with clear distinction in the germinal center. The state of the kidney changed from exhibiting focal necrosis of renal tubular epithelial cells to exhibiting necrosis of individual renal tubular epithelial cells (Figure 2-14-3). The heart, glandular stomach, pancreas, brain, muscle, small intestine, fundus, esophagus, and trachea remained the same as those in the control group during the observation period, with no significant pathological changes.

The viral load of various organs in chickens infected with GX2020-019

The GX2020-019 strain of FAdV-4 is widely tissue tropic in chickens, with high concentrations of the virus detected in all organs and the highest viral load in liver tissue, which is the main target organ; the muscle had the lowest viral load. After entering the regression period, the viral content of each organ significantly decreased ($p < 0.05$), with the thymus and heart having the lowest viral loads and the liver, heart, and thymus showing the highest efficiency of viral clearance (Figure 2-14-4).

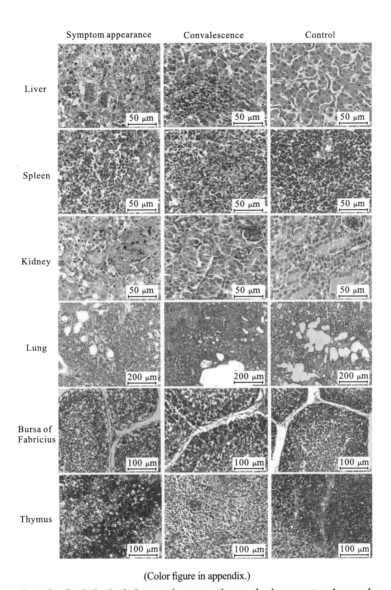

(Color figure in appendix.)

Figure 2-14-3　Pathological changes in organ tissues during onset and convalescence

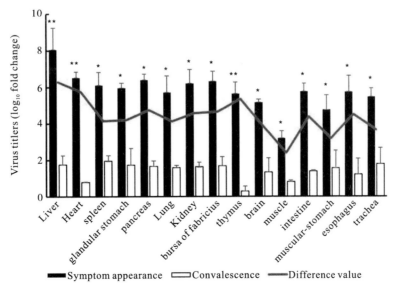

Figure 2-14-4　Organ viral loads in SPF chickens infected with GX2020-019

Viral shedding in SPF chickens after GX2020-019 infection

The level of viral shedding in the oropharyngeal and cloaca was quantified by fluorescence quantitative PCR, as Figure 2-14-5. From 1 to 35 dpi, the level of viral shedding in infected chickens showed a trend of first increasing and then decreasing. The peak level of viral shedding in the cloaca occurred between 5 and 7 dpi, at which time the level of shedding was significantly higher than that in the oropharyngeal area ($P<0.01$). A small amount of viral shedding was still detectable in both the oral cavity and cloaca at 35 dpi (Figure 2-14-6). The results of FAdV-4 shedding detection in the control group were always negative.

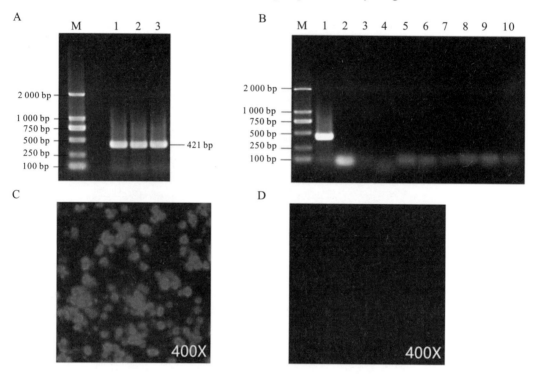

A: FAdV-4 virus was detected using conventional PCR; B: PCR was used to rule out contamination with common mixed infection pathogens, with lanes 1~10 showing detection results for FAdV-4, AIV, NDV, IBV, LTV, IBDV, RLV, ALV, REV, and Mycoplasma; C: GX2020-019 was identified using an anti-FAdV-4 fiber-2 monoclonal antibody by IFA; D: Negative control (color figure in appendix).

Figure 2-14-5　Detection and purity testing of FAdV-4 virus

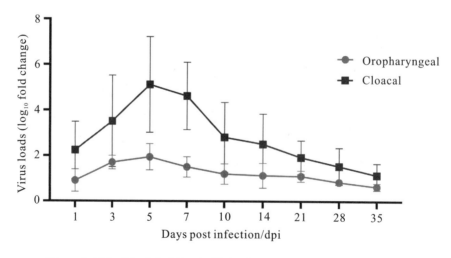

Figure 2-14-6　Viral shedding in SPF chickens infected with GX2020-019

Sequence alignment and phylogenetic analysis

The genome of GX2020-019 is 43 714 nucleotides in length, with A, T, G, and C contents of 23.1%, 22.0%, 27.2%, and 27.7%, respectively, and contains 43 potential protein-coding regions (GenBank accession number: OP378126). Phylogenetic analysis indicated that GX2020-019 belongs to the FAdV-C group along with FAdV-4 strains from China, Canada, India, and Austria (Figure 2-14-7) and is distantly related to FAdV-A, FAdV-B, FAdV-D, and FAdV-E with genome homologies of 54.0%~54.1%, 55.3%~57.7%, 54.7%~55.2%, and 56.4%~56.6%, respectively. The homology between GX2020-019 and FAdV-4 strains from China (GX2017-010, SD1511, HLJFAd15, JS07, SCDY, and SD1601) ranges from 99.7% to 100%, while the homology with strains from outside China (ON1, KR5, and B1-7) ranges from 98.1% to 99.3% (Figure 2-14-8).

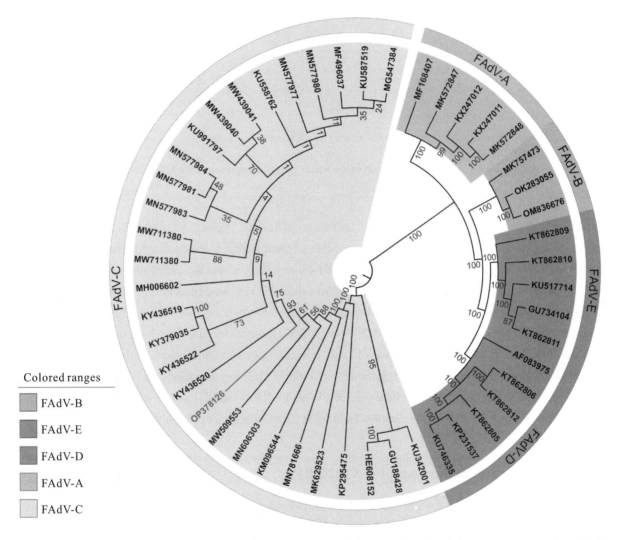

The phylogenetic tree was created by neighbor-joining analysis with 500 replicates for bootstrapping, and evolutionary distances were calculated using the maximum composite likelihood method through MEGA 11 software. The GX2020-019 strain is indicated in red (color figure in appendix).

Figure 2-14-7　Phylogenetic analysis of the whole gene nucleotide sequence of strain GX2020-019 of Guangxi

The insertion/deletion nucleotide sites of GX2020-019

Unlike the oversea strains MX-SHP95, KR5, ON1, and B1-7, GX2020-019 and other FAdV-4 strains reported in China have a 10-base insertion in the tandem repeat sequence (TR-B), three-base insertions in the

Similarity

	1	2	3	4	5	6	7	8	9	10	11	12	13	14	15	16	17	18	19	20	21	22	23	24	25	26	27	28		Strain	Group
1		99.9	99.9	99.9	99.9	99.7	99.9	98.3	98.5	98.2	54.0	54.1	54.1	54.1	54.1	55.3	55.4	57.7	54.7	55.1	54.9	55.0	55.2	56.5	56.6	56.5	56.6	56.4	1	GX2020-019(OP378126)	
2	0.1		100.0	100.0	100.0	99.7	100.0	98.3	98.5	98.2	54.0	54.0	54.1	54.1	54.0	55.3	55.3	57.7	54.7	55.1	54.9	55.0	55.2	56.4	56.5	56.4	56.5	56.4	2	GX2017-004(MN577980)	
3	0.1	0.0		100.0	100.0	99.7	100.0	98.3	98.5	98.2	54.0	54.0	54.1	54.1	54.0	55.3	55.3	57.7	54.7	55.1	54.9	55.0	55.2	56.4	56.5	56.4	56.4	56.4	3	SD1511(MF496037)	
4	0.1	0.0	0.0		100.0	99.7	100.0	98.3	98.5	98.2	54.0	54.0	54.1	54.1	54.0	55.3	55.3	57.7	54.7	55.1	54.9	55.0	55.2	56.4	56.5	56.4	56.5	56.4	4	HLJFAd15(KU991797)	
5	0.1	0.0	0.0	0.0		99.7	100.0	98.3	98.5	98.1	54.0	54.1	54.1	54.1	54.0	55.3	55.3	57.7	54.7	55.1	54.9	55.0	55.2	56.4	56.5	56.4	56.5	56.4	5	JS07(KY436519)	C
6	0.3	0.3	0.3	0.3	0.3		99.7	98.4	98.4	98.1	54.0	54.0	54.1	54.1	54.1	55.3	55.3	57.7	54.6	55.1	54.8	55.0	55.2	56.4	56.5	56.4	56.6	56.4	6	SCDY(MK629523)	
7	0.1	0.0	0.0	0.0	0.0	0.3		98.3	98.5	98.2	54.0	54.0	54.1	54.1	54.0	55.3	55.3	57.7	54.7	55.1	54.9	55.0	55.2	56.4	56.5	56.4	56.4	56.4	7	SD1601(MH006602)	
8	1.7	1.7	1.7	1.7	1.7	1.6	1.7		99.3	98.4	53.2	53.2	53.3	53.3	53.2	54.5	54.5	56.9	54.0	54.5	54.2	54.3	54.6	55.7	55.8	55.7	55.8	55.6	8	ON1(GU188428)	
9	1.5	1.5	1.5	1.5	1.5	1.8	1.5	0.7		98.6	53.2	53.2	53.3	53.3	53.2	54.5	54.5	56.9	54.0	54.5	54.2	54.3	54.6	55.7	55.8	55.7	55.7	55.6	9	KR5(HE608152)	
10	1.8	1.9	1.9	1.9	1.9	1.9	1.6	1.4			53.3	53.3	53.4	53.4	53.3	54.5	54.6	56.9	54.1	54.6	54.3	54.4	54.7	55.7	55.8	55.7	55.8	55.7	10	B1-7(KU342001)	
11	71.8	71.8	71.8	71.8	71.8	71.8	71.8	74.0	74.0	73.7		99.3	99.0	99.2	99.2	55.7	55.7	59.5	55.7	56.6	55.8	56.3	56.7	57.4	57.4	57.4	57.5	57.4	11	CELO(U46933)	
12	71.8	71.8	71.8	71.8	71.8	71.7	71.8	74.0	74.0	73.6	0.7		99.4	99.7	99.7	55.8	55.8	59.6	55.8	56.7	55.8	56.3	56.8	57.5	57.6	57.5	57.6	57.5	12	61-11Z(KX247012)	
13	71.5	71.5	71.5	71.5	71.5	71.5	71.5	73.8	73.8	73.5	1.0	0.7		99.4	99.3	55.8	55.8	59.6	55.9	56.7	55.9	56.3	56.7	57.5	57.5	57.5	57.5	57.5	13	JM1-1(MF168407)	A
14	71.7	71.7	71.7	71.7	71.7	71.6	71.7	73.9	73.9	73.6	0.8	0.3	0.7		100.0	55.8	55.8	59.6	55.9	56.7	55.9	56.4	56.8	57.5	57.5	57.5	57.6	57.5	14	11-7127(MK572848)	
15	71.7	71.7	71.7	71.7	71.8	71.7	71.7	73.9	73.9	73.6	0.8	0.3	0.7	0.0		55.8	55.8	59.6	55.8	56.7	55.8	56.3	56.8	57.5	57.5	57.5	57.6	57.5	15	W-15(KX247011)	
16	68.3	68.4	68.4	68.4	68.4	68.4	68.4	70.5	70.4	70.4	67.5	67.3	67.3	67.3	67.3		99.9	87.4	62.9	63.3	63.1	63.2	63.3	64.2	64.2	64.3	64.4	64.1	16	WHRS(OM836676)	
17	68.2	68.3	68.3	68.3	68.4	68.3	68.3	70.4	70.4	70.3	67.5	67.3	67.3	67.3	67.3	0.1		87.4	62.9	63.3	63.1	63.2	63.3	64.2	64.2	64.3	64.4	64.1	17	19-7209(OK283055)	B
18	62.5	62.6	62.6	62.6	62.7	62.6	62.6	64.5	64.6	64.5	58.5	58.4	58.5	58.4	58.4	13.8	13.8		68.8	68.9	68.5	68.7	68.9	68.5	68.5	68.5	68.6	68.4	18	LYG(MK757473)	
19	70.3	70.3	70.3	70.3	70.3	70.3	70.3	72.0	71.9	71.7	67.6	67.4	67.3	67.3	67.3	51.6	51.6	40.5		96.1	96.3	96.2	96.3	72.4	72.4	72.3	72.3	72.0	19	9(AF083975)	
20	69.1	69.0	69.0	69.0	69.1	69.1	69.0	70.8	70.7	70.5	65.5	65.2	65.2	65.2	65.2	50.9	50.8	40.4	4.0		97.6	97.5	98.3	72.5	72.5	72.5	72.6	72.1	20	SR48(KT862806)	
21	69.7	69.7	69.7	69.7	69.7	69.8	69.7	71.5	71.4	71.1	67.5	67.3	67.3	67.3	67.3	51.3	51.2	41.0	3.8	2.5		100.0	97.7	72.7	72.7	72.6	72.7	72.2	21	ON-NP2(KP231537)	D
22	69.5	69.4	69.4	69.4	69.4	69.5	69.4	71.2	71.1	70.9	66.3	66.1	66.1	66.1	66.1	51.0	51.0	40.7	3.9	2.5	0.0		97.7	72.6	72.6	72.6	72.7	72.2	22	MX95-S11(KU746335)	
23	68.9	68.8	68.8	68.8	68.8	68.8	68.8	70.5	70.4	70.3	65.4	65.1	65.1	65.1	65.1	50.9	50.9	40.4	3.8	1.7	2.3	2.3		72.5	72.5	72.5	72.5	72.1	23	380(KT862812)	
24	65.5	65.7	65.7	65.7	65.7	65.7	65.7	67.5	67.5	67.4	63.6	63.3	63.4	63.4	63.4	49.0	49.0	41.1	34.7	34.4	34.2	34.3	34.4		97.8	98.0	95.4	93.6	24	HG(GU734104)	
25	65.3	65.4	65.4	65.4	65.4	65.4	65.4	67.3	67.3	67.2	63.5	63.2	63.3	63.3	63.3	48.9	48.9	41.1	34.7	34.4	34.2	34.2	34.4	2.2		97.4	95.1	93.3	25	UPM04217(KU517714)	
26	65.6	65.8	65.8	65.8	65.8	65.7	65.8	67.5	67.6	67.5	63.6	63.3	63.4	63.3	63.3	48.8	48.9	41.1	34.8	34.5	34.3	34.3	34.5	2.0	2.7		95.3	93.7	26	764(KT862811)	E
27	65.2	65.4	65.4	65.4	65.4	65.4	65.4	67.3	67.4	67.2	63.3	63.0	63.1	63.1	63.1	48.6	48.7	40.8	34.8	34.4	34.2	34.2	34.4	4.8	5.0	4.9		94.3	27	TR59(KT862810)	
28	65.8	65.9	65.9	65.9	65.9	65.8	65.9	67.7	67.8	67.6	63.5	63.2	63.3	63.3	63.3	49.1	49.1	41.2	35.4	35.1	34.9	34.9	35.1	6.7	7.1	6.6	5.9		28	YR36(KT862809)	
	1	2	3	4	5	6	7	8	9	10	11	12	13	14	15	16	17	18	19	20	21	22	23	24	25	26	27	28			

Divergence

The GX2020-019 strain is indicated in red. Homology analysis between the GX2020-019 strain and reference strains using the MegAlign program within the Lasergene7.0 software package.

Figure 2-14-8　Homology analysis of the whole genomic sequence of the GX2020-019 strain

noncoding regions after the 52/55 kDa protein coding sequence and ORF43, a six-base deletion in the 33 kDa protein coding sequence, long GA sequences in the GA repeat region between the PX and PVI genes, and a deletion of 1 966 bases at ORF19, ORF27, and ORF48, resulting in the almost complete loss of these three ORFs. Unlike nonpathogenic strains, pathogenic FAdV-4 has a three-base insertion in ORF16, resulting in an additional glycine. In addition, unlike the ON1 strain, Chinese isolates including GX2020-019 have a three-base insertion in the coding regions of ORF2, ORF22, and DNApol protein, a 57-base insertion in ORF19A, a 15-base insertion in the coding sequence of fiber-2 resulting in an addition of 5 amino acids (ENGKP), a three-base deletion in the coding region at the end of Fiber-1 resulting in a loss of a histidine residue, a five-base deletion in the noncoding region between ORF17 and ORF30, an 81-base deletion in the tandem repeat sequence (TR-E), and long TC repeat sequences in the TC repeat region between the protease and DBP genes (Figure 2-14-9).

Protein alignment analysis of GX2020-019

The amino acid sequence lengths of the major structural proteins hexon, penton, fiber-1, and fiber-2 in the GX2020-019 strain are 937 amino acids (aa), 525 aa, 431 aa, and 479 aa, respectively. Unlike the amino acid sequences of nonpathogenic strains, they have 10~11, 6~7, 10~20, and 30~32 amino acid mutation sites, respectively. Penton and hexon are relatively conserved, while the Fiber-2 protein has the most mutation sites, and the mutation positions are consistent with previous reports of other isolated strains in China[11,17,29].

Of note, we found that the amino acid sequences encoded by ORF30 and ORF49 in the GX2020-019 strain were the same as those of nonpathogenic strains, and no mutations occurred in the 32 amino acid mutation sites that were mutated in other strains isolated in China (Table 2-14-2). It is inferred that there is a possibility of gene recombination between pathogenic and nonpathogenic FAdV-4 strains.

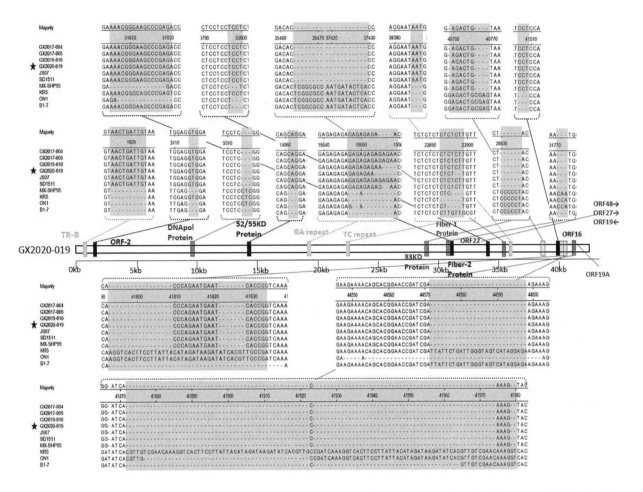

The green region represents the inserted noncoding sequence (CDS) region. The red region represents the inserted coding sequence (CDS) region. The light green region represents the missing noncoding sequence (CDS) region. The light red region represents the missing coding sequence (CDS) region. The blue region represents the observed changes.

Figure 2-14-9   The insertion/deletion nucleotide sites of the whole genomic sequence of the GX2020-019 strain

Table 2-14-2   Differences in the amino acid sequences of ORF30 and ORF 49 between GX2020-019 and the reference strains

CDS	Strain	GenBank	Mutation sites													
			8	23	40	49	51	63	71							
	GX2020-019	OP378126	S	R	W	S	E	E	G							
	B1-7	KU342001.1	●	●	●	●	●	●	●							
	KR5	HE608152	●	●	●	●	●	●	●							
	ON1	GU188428.1	●	●	●	●	●	●	●							
ORF30	MX-SHP95	KP295475	-	Q	●	T	G	V	●							
	GX2017-004	OP378126	-	Q	-	T	G	V	R	-						
	SD1511	MF496037	-	Q	-	T	G	V	R	-						
	HLJFAd15	KU991797	-	Q	-	T	G	V	R	-						
	SCDY	MK629523	-	Q	-	T	G	V	R	-						
	SD1601	MH006602	-	Q	-	T	G	V	R	-						

continued

CDS	Strain	GenBank	Mutation sites													
			26	28	29	30	31	32	33	34	35	36	37	38	39	40
	GX2020-019	OP378126	D	R	S	H	D	C	Y	V	T	E	G	G	A	S
	B1-7	KU342001.1	●	●	●	●	●	●	●	●	●	●	●	●	●	●
	KR5	HE608152.1	●	●	●	●	●	●	●	●	●	●	●	●	●	●
	ON1	GU188428.1	●	●	●	●	●	●	●	●	●	●	●	●	●	●
	MX-SHP95	KP295475	T	-	V	T	-	L	L	R	Q	R	R	R	S	F
	GX2017-004	OP378126	S	-	V	T	-	L	L	R	Q	R	R	R	S	F
	SD1511	MF496037	S	-	V	T	-	L	L	R	Q	R	R	R	S	F
	HLJFAd15	KU991797	S	-	V	T	-	L	L	R	Q	R	R	R	S	F
	SCDY	MK629523	S	-	V	T	-	L	L	R	Q	R	R	R	S	F
	SD1601	MH006602	S	-	V	T	-	L	L	R	Q	R	R	R	S	F
ORF49			41	43	44	46	47	48	49	50	51	52	53			
	GX2020-019	OP378126	D	E	N	S	L	Y	Q	S	L	T	-			
	B1-7	K11342001.1	●	●	●	●	●	●	●	●	●	●	●			
	KR5	HE608152.1	●	●	●	●	●	●	●	●	●	●	●			
	ON1	GU188428.1	●	●	●	●	●	●	●	●	●	●	●			
	MX-SHP95	KP295475	G	R	K	F	A	L	S	I	I	N	L			
	GX2017-004	OP378126	G	R	K	F	A	L	S	I	I	N	L			
	SD1511	MF496037	G	R	K	F	A	L	S	I	I	N	L			
	HLJFAd15	KU991797	G	R	K	F	A	L	S	I	I	N	L			
	SCDY	MK629523	G	R	K	F	A	L	S	I	I	N	L			
	SD1601	MH006602	G	R	K	F	A	L	S	I	I	N	L			

Note: ● indicates that the amino acid at the corresponding position is the same as that in the GX2020-019 strain. - indicates a deletion of the amino acid at the corresponding position in each strain.

Discussion

Since the first report in 2015 of the pathogenic FAdV-4 strain SJ13 isolated from chickens with HHS[13], pathogenic FAdV-4 has spread extensively in China. Prior to the widespread use of commercial fowl adenovirus vaccines, the incidence of HHS in broilers and layers increased annually, resulting in considerable economic losses and raising public awareness of the disease. Consequently, HHS and its causative pathogen FAdV-4 have become a prominent research topic among veterinary researchers in recent years.

In 2020, a strain of the FAdV-4 virus named GX2020-019 was isolated from chickens with symptoms of HHS in a commercial farm in Guangxi. It has been reported that most FAdVs grow well in primary chicken embryo liver cells, primary chicken embryo kidney cells, and chicken liver cancer cell lines (LMH cells)[29]. As the highest viral load is observed in the liver, initially, two purification methods using LMH cells and CEL cells were used in this research. However, the growth rate of plaques formed in LMH cells was slow, and the plaque edges were unclear, making it difficult to select plaques. Therefore, CEL cells were selected for viral purification. After three consecutive rounds of plaque purification in CEL cells, a seed stock with a titer of

$10^{8.2}$ TCID$_{50}$/0.1 mL was obtained.

In the pathogenicity study of GX2020-019 in SPF chickens, we determined its virulence by injecting different doses of the virus into the pectoral muscles of 4-week-old SPF chickens. Each dosage group was kept separate in individual chicken isolators. The course of disease in infected chickens was affected by the inoculation dose, but they all exhibited clinical signs, such as depression, ruffled feathers, decreased appetite, and excretion of yellow-green watery feces. Upon autopsy typical lesions of pathogenic FAdV-4 infection were observed, including $1\sim3$ mL of bright yellow effusion in the pericardium, significant hepatomegaly with yellowing, friable texture, and bleeding points on the surface of the liver.

At present, there is no systematic virulence evaluation method for fowl adenovirus, and only a general classification of pathogenic and nonpathogenic strains based on whether they cause disease in chickens exists. Zhao et al., orally inoculated 3-week-old SPF chickens with $10^{3.5}$ TCID$_{50}$ of the JSJ13 strain, resulting in a mortality rate of 28.6%[13]. Luan et al, intramuscularly injected and orally administered 10^3 TCID$_{50}$ of the GX2019-010 strain to 4-week-old SPF chickens, resulting in a 100% mortality rate[30]. Yuan et al, intramuscularly injected $10^{2.5}$ TCID$_{50}$ of the GD616 strain into 3-week-old SPF chickens, resulting in a mortality rate of 100%[29]. Mo et al. inoculated SPF chickens at 7, 21, and 35 days of age with $3 \times 10^{3.2}$ TCID$_{50}$ of the SD1511 strain through intramuscular injection, resulting in mortality rates of 93%, 80%, and 100%, respectively. When administered through the intranasal route, the mortality rates were 50%, 57.2%, and 50% for the respective age groups[11]. The study showed that the GX2020-019 strain had moderate virulence, as evidenced by mortality rates of 0%, 20%, and 60% when intramuscularly injected at doses of 10^3 TCID$_{50}$, 10^4 TCID$_{50}$ and 10^5 TCID$_{50}$, respectively, in 4-week-old SPF chickens. These mortality rates were much lower than those reported for highly pathogenic isolates in China, indicating that the GX2020-019 strain should be classified as a strain with intermediate virulence.

By conducting pathological histological studies and viral load measurements in 15 organs, including the liver, heart, spleen, proventriculus, pancreas, lung, kidney, bursa of Fabricius, thymus, brain, muscle, small intestine, gizzard, esophagus, and trachea, during both the onset and recovery stages, we found that the liver is the primary target organ of GX2020-019 infection, which can cause liver damage characterized by basophilic nuclear inclusions, lipid degeneration, and multifocal necrosis. Moreover, the liver exhibited the highest viral load during the acute stage. Although the clearance rate of the virus in the liver is high, the hepatic lobular architecture remained unclear during the recovery phase, and the sinusoids were narrowed with inflammatory cell infiltration around the blood vessels. The incomplete recovery of the liver directly affects the functions of oxidative metabolism, glycogen storage, protein synthesis, and bile secretion in the body, leading to reduced weight gain and the decreased economic value of the animals. GX2020-019 displays an affinity for immune tissues that leads to significant cellular degeneration in the bursa of Fabricius, thymus, and spleen during the acute phase of infection. Even during the recovery phase, at 21 dpi, partial lymphofollicular atrophy and blurred follicular-interfollicular boundaries were observed in the bursa of Fabricius, as well as cellular necrosis, increased vascularity in the medulla, and increased size of thymic corpuscles in the thymus. The spleen exhibited only mild lymphocyte degeneration. Damage to immune organs results in immunological impairment, which may decrease the chicken population's resistance to other pathogenic microorganisms. During the recovery phase, only sporadic individual tubular necrosis was observed in the kidneys, which did not affect their urogenital system function.

Although the accumulation of pericardial fluid is a characteristic pathology of HHS, there was no observed tissue damage or infiltration of inflammatory cells in the heart tissue, indicating that the effusion was not a result of apoptosis or inflammation of myocardial cells. These findings are consistent with the results reported by Li et al.[31], and Niu et al.[32], which confirmed that pericardial effusion is derived from vascular exudation based on measurements of components such as total protein, albumin, aspartate aminotransferase, and creatine kinase isoenzymes. Additionally, no pathological changes were observed in the digestive system during the onset period, suggesting that the virus does not cause harm to the digestive system.

Since the outbreak of pathogenic FAdV-4 in China, the factors contributing to its increased virulence have been a hot topic of research. To identify the key virulence factors, pathogenic and nonpathogenic strains were compared using nucleotide and amino acid sequence analysis, and hypothetical nucleotide fragments or amino acid sites were replaced using reverse genetics systems. Recently, studies have shown that the 1 966 bp nucleotide fragment encoding ORF19, ORF27, and ORF48, which are missing in the novel strains, are not the factors that influence virulence[33]; through R188I mutation of the hexon protein, the amino acid residue at position 188 was determined to be a key residue for reducing pathogenicity[34]; through CRISPR/Cas9 gene editing of fiber-1[35, 36] and fiber-2[37], FAdV-4 strains with highly attenuated virulence could be obtained. Additionally, there are differing views that the high pathogenicity factors of FAdV-4 are not related to fiber-2[34, 35] and that the increased virulence of hypervirulent FAdV-4 is independent of fiber-1 and penton[38]. When comparing GX2020-019 with other highly pathogenic Chinese FAdV-4 strains, including GX2019-004, SD1601, JS07, SCDY, SD1511, and HLJFAd15, we found no differences in the main structural proteins hexon, penton, fiber-1, and fiber-2. However, interestingly, we found that the amino acid sequences encoded by ORF30 and ORF49 in the GX2020-019 strain were identical to those of nonpathogenic strains, and the 32 amino acid mutation sites that were present in other isolated strains in China were not present in GX2020-019. Although there are currently no studies on the functional role of the amino acid sequences encoded by ORF30 and ORF49, it is speculated that they may be factors influencing the pathogenicity of FAdV-4.

Our work has expanded the understanding of the pathogenicity and molecular characteristics of the moderately pathogenic FAdV-4 strain GX2020-019 that is prevalent in Guangxi, China, and provided reference materials for further research on this virus.

References

[1] BENKÖ M, HARRACH B. proposal for a new (third) genus within the family Adenoviridae. Archives of Virology, 1998, 143(4): 829-837.

[2] HESS M. Detection and differentiation of avian adenoviruses: a review. Avian Pathol, 2000, 29(3): 195-206.

[3] RUX J J, KUSER P R, BURNETT R M. Structural and phylogenetic analysis of adenovirus hexons by use of high-resolution X-ray crystallographic, molecular modeling, and sequence-based methods. Journal of Virology, 2003, 77(17): 9553-9566.

[4] LU H, GUO Y, XU Z, et al. Fiber-1 of serotype 4 fowl adenovirus mediates superinfection resistance against serotype 8b fowl adenovirus. Frontiers in Microbiology, 2022, 13: 1086383.

[5] WANG X, TANG Q, CHU Z, et al. Immune protection efficacy of FAdV-4 surface proteins fiber-1, fiber-2, hexon and penton base. Virus Res, 2018, 245: 1-6.

[6] ABE T, NAKAMURA K, TOJO H, et al. Histology, immunohistochemistry, and ultrastructure of hydropericardium syndrome in adult broiler breeders and broiler chicks. Avian Dis, 1998, 42(3): 606-612.

[7] DAHIYA S, SRIVASTAVA R N, HESS M, et al. Fowl adenovirus serotype 4 associated with outbreaks of infectious hydropericardium in Haryana, India. Avian Dis, 2002, 46(1): 230-233.

[8] CHOI K S, KYE S J, KIM J Y, et al. Epidemiological investigation of outbreaks of fowl adenovirus infection in commercial chickens in Korea. Poult Sci, 2012, 91(10): 2502-2506.

[9] GRIFFIN B D, NAGY E. Coding potential and transcript analysis of fowl adenovirus 4: insight into upstream ORFs as

common sequence features in adenoviral transcripts. J Gen Virol, 2011, 92(Pt 6): 1260-1272.

[10] CUI J, XU Y, ZHOU Z, et al. Pathogenicity and Molecular Typing of Fowl Adenovirus-Associated With Hepatitis/ Hydropericardium Syndrome in Central China (2015-2018). Front Vet Sci, 2020, 7: 190.

[11] MO K K, LYU C F, CAO S S, et al. Pathogenicity of an FAdV-4 isolate to chickens and its genomic analysis. Journal of Zhejiang University. B. Science, 2019, 20(9): 740-752.

[12] WEI Y, DEGN X W, XIE Z X, et al. Pathological observation of outcome of Hydropericardium hepatitis syndrome. China Poultry, 2023(45): 45-50.

[13] ZHAO J, ZHONG Q, ZHAO Y, et al. Pathogenicity and complete genome characterization of fowl adenoviruses isolated from chickens associated with inclusion body hepatitis and hydropericardium syndrome in China. PLOS ONE, 2015, 10(7): e133073.

[14] NIU Y, SUN Q, ZHANG G, et al. Epidemiological investigation of outbreaks of fowl adenovirus infections in commercial chickens in China. Transbound Emerg Dis, 2018, 65(1): e121-e126.

[15] CHEN L, YIN L, ZHOU Q, et al. Epidemiological investigation of fowl adenovirus infections in poultry in China during 2015-2018. BMC Veterinary Research, 2019, 15(1): 271.

[16] ZHUANG Q Y, WANG S C, ZHANG F Y, et al. Molecular epidemiology analysis of fowl adenovirus detected from apparently healthy birds in eastern China. BMC Veterinary Research, 2023, 19(1): 5.

[17] RASHID E, XIE Z X, ZHANG I, et al. Genetic characterization of fowl aviadenovirus 4 isolates from Guangxi, China, during 2017–2019. Poultry Science, 2020, 99(9): 4166-4173.

[18] ZHANG L, LUAN Y G, XIE Z X, et al. Isolation, identification and analysis of a fowl adenovirus serotype 4. Prog Vet Med, 2020(41): 64-72.

[19] WEI Y, XIE Z X, DENG X W, et al. Expression of the truncated fiber-2 protein of fowl adenovirus serotype 4 and preparation of its monoclonal antibodies. J South Agric, 2022(53): 2341-2349.

[20] FENG B, XIE Z X, Deng X W, et al Development of a triplex PCR assay for detection of Newcastle disease virus, chicken parvovirus and avian. J South Agric, 2019, 50: 2576-2582.

[21] SU W, ZHANG X D, WANG T. Detection of chicken infection Bursal disease virus by polymerase chain reaction. Chin J Vet Sci, 1997(17): 15-17.

[22] WU X Y, ZUANG X, HE H H. Nucleic acid detection, virus isolation and sequence analysis of chicken infectious bronchitis virus in Jiangsu Province. Anim Husband Vet Med, 2021(53): 96-104.

[23] XU Q R, ZHOU Z T, BI D R, et al. PCR detection and TK gene sequence analysis of infectious Laryngotracheitis virus in laying hens. Heilongjiang Anim Sci Vet Med, 2017(47): 123.

[24] YE W C, YU B, LIU Y S, et al. Establishment of universal RT-PCR for detection of Muscovy duck reovirus. Acta Agric Zhejiang, 2020(26): 1453-1456.

[25] BI Y L, ZHUANG J Q, WANG J L, et al. Double PCR detection of poultry birds of reticular endothelial tissue hyperplasia in biological products virus and avian leukosis virus. Anim Husband Vet Med, 2014(46): 74-78.

[26] XIE Z X, DENG X W, TANG X F, et al. The studies and application of PCR kit for Mycoplasma gallisepitcum detection. Anim Husband Vet Med, 2004(40): 3-5.

[27] LUAN Y G, XIE Z X, LUO S S, et al. Whole-genome Sequencing and analysis of fowl adenovirus Serotpye 4 strain GX2019-010 isolate from Guangxi, China. Anim Husb Vet Med, 2020(47): 3793-3804.

[28] LUAN Y G, XIE Z X, LUO S S, et al. Dynamic distribution of infectious fowl adenovirus serotype 4 in different tissues and the virus shedding pattern of SPF chickens. Chin J Vet Sci, 2021(03): 463-468.

[29] YUAN E, SONG H Q, HOU L, et al. Age-dependence of hypervirulent fowl adenovirus type 4 pathogenicity in specific-pathogen-free chickens. Poult Sci, 2021, 100(8): 101238.

[30] LUAN Y G. Whole-genome sequencing analysis, detection method and pathogenicity study of fowl adenovirus serotype 4 isolated from Guangxi. Nanning: University of Guangxi, China, 2021.

[31] LI R, LI G, LIN J, et al. Fowl adenovirus serotype 4 SD0828 infections causes high mortality rate and cytokine levels in

specific pathogen-free chickens compared to ducks. Frontiers in Immunology, 2018, 9: 49.

[32] NIU Y, SUN Q, LIU X, et al. Mechanism of fowl adenovirus serotype 4-induced heart damage and formation of pericardial effusion. Poultry Science, 2019, 98(3): 1134-1145.

[33] PAN Q, WANG J, GAO Y, et al. The natural large genomic deletion is unrelated to the increased virulence of the novel genotype fowl adenovirus 4 recently emerged in China. Viruses, 2018, 10(9).

[34] ZHANG Y, LIU A, WANG Y, et al. A single amino acid at residue 188 of the hexon protein is responsible for the pathogenicity of the emerging novel virus fowl adenovirus 4. Journal of Virology, 2021, 95(17): e60321.

[35] WANG W K, LIU Q, LI T E, et al. Fiber-1, not fiber-2, directly mediates the infection of the pathogenic serotype 4 fowl adenovirus via its shaft and knob domains. Journal of Virology, 2020, 94(17).

[36] MU Y R, XIE Q, WANG W K, et al. A novel Fiber-1-edited and highly attenuated recombinant serotype 4 fowl adenovirus confers efficient protection against lethal challenge. Frontiers in Veterinary Science, 2021, 8: 759418.

[37] XIE Q, CAO S Y, ZHANG W, et al. A novel fiber-2-edited live attenuated vaccine candidate against the highly pathogenic serotype 4 fowl adenovirus. Veterinary Research (Paris), 2021, 52(1): 35.

[38] LIU R, ZHANG Y, GUO H, et al. The increased virulence of hypervirulent fowl adenovirus 4 is independent of fiber-1 and penton. Research in Veterinary Science, 2020, 131: 31-37.

[39] YIN D D, XUE M, YANG K K, et al. Molecular characterization and pathogenicity of highly pathogenic fowl adenovirus serotype 4 isolated from laying flock with hydropericardium-hepatitis syndrome. Microbial Pathogenesis, 2020, 147: 104381.

Molecular characterization of emerging chicken and turkey parvovirus variants and novel strains in Guangxi, China

Zhang Yanfang, Feng Bin, Xie Zhixun, Zhang Minxiu, Fan Qing, Deng Xianwen, Xie Zhiqin, Li Meng, Zeng Tingting, Xie Liji, Luo Sisi, Huang Jiaoling, and Wang Sheng

Abstract

Avian parvoviruses cause several enteric poultry diseases that have been increasingly diagnosed in Guangxi, China, since 2014. In this study, the whole-genome sequences of 32 strains of chicken parvovirus (ChPV) and 3 strains of turkey parvovirus (TuPV) were obtained by traditional PCR techniques. Phylogenetic analyses of 3 genes and full genome sequences were carried out, and 35 of the Guangxi ChPV/TuPV field strains were genetically different from 17 classic ChPV/TuPV reference strains. The nucleotide sequence alignment between ChPVs/TuPVs from Guangxi and other countries revealed 85.2%~99.9% similarity, and the amino acid sequences showed 87.8%~100% identity. The phylogenetic tree of these sequences could be divided into 6 distinct ChPV/TuPV groups. More importantly, 3 novel ChPV/TuPV groups were identified for the first time. Recombination analysis with RDP 5.0 revealed 15 recombinants in 35 ChPV/TuPV isolates. These recombination events were further confirmed by SimPlot 3.5.1 analysis. Phylogenetic analysis based on full genomes showed that Guangxi ChPV/TuPV strains did not cluster according to their geographic origin, and the identified Guangxi ChPV/TuPV strains differed from the reference strains. Overall, whole-genome characterizations of emerging Guangxi ChPV/TuPV field strains will provide more detailed insights into ChPV/TuPV mutations and recombination and their relationships with molecular epidemiological features.

Keywords

molecular characterization, chicken parvovirus, turkey parvovirus phylogenetic analyses

Introduction

According to the description of the International Committee on Viral Taxonomy (ICTV) 2021, the eight genera of parvovirus under the traditional classification have recently been replaced by ten genera. *Aveparvovirus* is one of the genera and includes avian parvoviruses, such as chicken parvoviruses (ChPVs) and turkey parvoviruses (TuPVs)[1]. Parvoviruses are linear, single-stranded DNA viruses with a length of ~5 kb and at least 3 open reading frames (ORFs)[2]. The ORFs include a 5'-ORF, a 3'-ORF and a small ORF located between the other two. The 5'-ORF encodes the nonstructural protein NS1, and the 3'-ORF may encode the capsid proteins VP1, VP2 and VP3. Notably, the small ORF is a hypothetical protein (NP) and remains unknown[3].

Parvoviruses were first identified in chickens via electron microscopy studies and by measurements of their genome sizes[4, 5]. Trampel et al.[6] reported the detection of TuPVs in turkeys and believed that TuPVs increased the incidence of intestinal diseases and bird mortality. The replication efficiency and error correction ability of carnivore parvovirus during the replication process were not strong, so the mutation rate of the virus genome is higher than that of general DNA virus genomes. Therefore, Shackelton[7] believed that single-stranded DNA viruses undergo faster genetic evolution than double-stranded DNA viruses. Additionally, recombination events in parvoviruses have been detected, and these viruses have been found to show high genetic diversity[7, 8].

Avian parvoviruses are similar to parvoviruses in other vertebrates (e.g., cats, dogs, pigs, and cattle)[9-11] and are often associated with gastrointestinal diseases, including runting-stunting syndrome (RSS) in chickens, poult enteritis and mortality syndrome (PEMS) in turkeys, Derzsy's disease in young geese and beak atrophy and dwarfism syndrome (BADS) in different types of ducks[4, 6, 12-14]. Reports[4, 15-17] confirmed that ChPVs were associated with diarrhoea, suggesting that the viruses were important causative agents of intestinal diseases. It has been reported that the occurrence of cerebellar hypoplasia and viral enteritis in commercial chicken flocks is also associated with ChPVs[18, 19].

Recent ChPV/TuPV outbreaks began in the USA in 2008. Research by Zsak et al.[20] suggested that ChPVs and TuPVs diverged from a common ancestor. A similar pattern of ChPV/TuPV infection was observed in chicken and turkey flocks from a Croatian (CRO) farm[21]. ChPV/TuPV infections were identified in intestinal samples from 15 chicken flocks and 2 turkey flocks sampled in Hungary between 2008 and 2010[16]. The prevalence of ChPV/TuPV was examined in individuals of commercial turkeys and flocks at different days of age in Poland from 2008 to 2011[22], and the infection rates of TuPV and ChPV were found to be 29.4% and 22.2%, respectively. In Republic of Korea(ROK), 34 commercial chicken flocks that experienced enteritis outbreaks were investigated for the presence of widespread enteroviruses between 2010 and 2012, and the ChPV positive rate was 26.5%[19]. Recent research by Nunez et al.[23] showed that ChPV was associated with diseases such as enteritis, pancreatitis and pancreatic atrophy. The ChPV/TuPV cases diagnosed in Guangxi, China, from 2014 to 2019 were the first indexed ChPV/TuPV infections in the southern region of China[24-26] and caused enteric disorders and economic losses in the Guangxi poultry industry.

The complete coding regions of only a few classic ChPV/TuPV strains, such as the ChPV ABU-P1 strain and the TuPV 260 and TuPV 1078 strains[3], as well as homologous strains, have been elucidated. The TuPV 260 and TuPV 1078 strains were originally isolated from turkeys with PEMS, and the ChPV ABU-P1 strain was originally isolated from chickens with RSS. As these three classic ChPV/TuPV strains continued to spread in poultry, they may have undergone natural selection and host adaptation to produce newly emerging ChPV/TuPV field strains or variants, as observed for other parvoviruses[27].

Several intestinal disease-related pathogens have been confirmed as pathogens of RSS[16, 28-34]. Nevertheless, the lack of a clear understanding of the complex aetiologies of RSS and PEMS and the existence of numerous virus types related to these syndromes are the main reasons why vaccines for RSS and PEMS have not been developed. Additional studies are needed to demonstrate the role of ChPVs in the aetiology of intestinal diseases. The current report aims to reveal the genetic diversity of ChPV/TuPV strains in China and to determine the phylogenetic relationships between these parvoviruses and highly similar strains to provide a reference for the prevention and treatment of RSS and PEMs.

Materials and methods

Sample collection

The ChPV and TuPV fieldstrains used in this study were obtained from commercial chicken and turkey flocks, including both clinically healthy and suspected RSS/PEMS-affected birds. A total of 1 526 throat and cloacal swab samples were collected from chickens and turkeys from Liuzhou, Guilin, Fangchenggang, Hechi, Chongzuo, Qinzhou, Yulin, Beihai, Nanning and Wuzhou cities in Guangxi, southern China, from 2014 to 2022. All samples were processed according to the protocol of the World Organization for Animal Health (OIE).

DNA extraction, genome-segment amplification and nucleotide sequencing

The presence of ChPV/TuPV in the throat and cloacal swab samples was detected by PCR[17, 20]. The target fragment sizes for the detection are 561 bp (NS1) and 249 bp (VP), repectively. By referring to the complete sequences of 3 prototype ChPV/TuPV strains from GenBank, three specific primer pairs[24] were designed to amplify the complete ChPV/TuPV genomes of 32 positive samples and 3 positive samples, respectively.

Sequence analysis

Sanger sequence assembly and nt sequence translation were performed using DNASTAR Lasergene 7.1. The ORF was predicted on the NCBI website. Sequence similarity was assessed by NCBI BLAST search and using DNAMAN version 10 software (Lynnon Biosoft). Sequence alignment was performed using the ClustalW 2.1 program. Neighbour-joining trees were generated using the MEGA (version 11) program, and bootstrap analysis was performed to verify the tree topology using absolute distances following 1 000 bootstrap replicates[35]. The mVISTA online platform was used for ChPV/TuPV genome-wide comparative analysis. Sequence recombination analysis of the NS1, VP1/VP2 genes of 35 Guangxi ChPV/TuPV strains and 17 reference ChPV/TuPV strains was performed using RDP 5.0 and SimPlot 3.5.1. To ensure the consistency and accuracy of the results, 7 different recombination analysis methods were used for analysis. For example, more than 4 analysis methods showed the presence of recombination events, and at a P value$<10^{-6}$, the recombination event was judged to be credible[36].

Results

PCR confirmation of Guangxi ChPV/TuPV strains

The nonstructural (NS) and VP genes of the positive samples were amplified by PCR using primers targeting the conserved 561 bp NS1 region and 249 bp VP1/VP2 region, respectively. The epidemiological survey results are shown in Table 2-15-1. Table 2-15-1 shows that the total positive rate was 69.72%, while the positive rate of RSS-like cases was as high as 91.86%, and the positive rate of healthy chickens was 66.91%. The positive samples were further confirmed by sequencing the NS1 and VP genes. NCBI BLAST results showed that the samples had 98%~100% homology with the ChPV ABU-P1 strain isolated from Hungary and the TuPV 260 strain isolated from the United States. The full genome sequence was successfully deduced from 32 PCR-positive chicken throat and cloacal swab samples and 3 PCR-positive turkey throat and cloacal swab samples using Sanger sequencing.

Table 2-15-1 Information on the samples and 35 ChPV/TuPV genome sequences

Sampling area	No. flocks	No. swabs	Type	Positive rate		Strain name	Accession number	Age / days	Collection date
				No.swabs / %	No. RSS-like cases / %				
Nanning	4	60	A	30 (50.00)		GX-CH-PV-1	KX084399	300	2014.10.10
Nanning	2	60	B	23 (38.33)		GX-CH-PV-2	KX084400	152	2014.10.10
Nanning	7	84	B	40 (47.62)	4/5 (80.00)	GX-CH-PV-4	KX084401	53	2014.10.31
Nanning	3	48	B	47 (97.92)		GX-CH-PV-5	KX133426	19	2014.12.30
Nanning	7	84	C	74 (88.10)	18/20 (90.00)	GX-CH-PV-6	KX133427	62	2014.12.30
Wuzhou	5	60	B	41 (68.33)		GX-CH-PV-7	KU523900	20	2015.03.20
Yulin	5	60	C	40 (66.67	10/12 (83.33)	GX-CH-PV-8	KX133415	75	2015.07.24
Yulin	6	90	C	80 (88.89)	14/16 (87.50)	GX-CH-PV-9	KX133416	72	2015.07.24
Wuzhou	4	48	C	39 (54.17)	8/8 (100.00	GX-CH-PV-10	KX133417	57	2015.07.24
Wuzhou	4	48	B	32 (66.67)		GX-CH-PV-11	KX133418	43	2015.08.12
Nanning	2	48	C	46 (95.83)	10/10 (100.00)	GX-CH-PV-12	KX133419	20	2015.08.12
Qinzhou	4	48	B	30 (62.50)		GX-CH-PV-13	KX133420	140	2015.08.13
Qinzhou	4	48	B	25 (52.08)		GX-CH-PV-14	KX133421	24	2015.08.13
Nanning	4	48	E	41 (85.42)	11/12 (91.67)	GX-CH-PV-15	KX133422	29	2015.10.09
Nanning	2	24	E	12 (50.00)		GX-CH-PV-16	KX133423	266	2015.10.09
Nanning	4	48	E	47 (97.92)	9/10 (90.00)	GX-CH-PV-17	KX133424	21	2015.10.10
Liuzhou	4	48	E	48 (100.00)	11/12 (91.67)	GX-CH-PV-18	KX133425	22	2015.10.10
Beihai	3	24	C	16 (66.67)		GX-CH-PV-19	MG602509	120	2016.09.30
Beihai	3	24	C	16 (66.67		GX-CH-PV-20	MG602510	130	2016.09.30
Guilin	3	24	C	14 (58.33)	4/4 (100.00)	GX-CH-PV-21	MG602511	26	2016.09.30
Guilin	3	24	C	14 (58.33)	4/4 (100.00)	GX-CH-PV-22	MG602512	16	2016.11.11
Fangchenggang	4	48	C	33 (68.75)		GX-CH-PV-23	MG602513	120	2016.11.11
Fangchenggang	4	40	C	22 (55.00)	9/10 (90.00)	GX-CH-PV-24	MG602514	110	2017.04.21
Fangchenggang	2	20	C	10 (50.00)	2/2 (100.00)	GX-CH-PV-25	MG602515	100	2017.04.21
Nanning	2	20	C	12 (60.00)	5/5 (100.00)	GX-CH-PV-26	MG602516	77	2017.07.18
Nanning	4	48	B	4 (8.33)		GX-CH-PV-27	MG602517	280	2017.07.18
Nanning	4	48	C	40 (83.33)	10/12 (83.33)	GX-CH-PV-28	MG602518	80	2017.07.04
Qinzhou	4	48	C	26 (54.17)	8/8 (100.00)	GX-CH-PV-29	MG602519	40	2017.08.03
Qinzhou	4	48	C	39 (81.25)	10/10 (100.00)	GX-CH-PV-30	MG602520	20	2017.08.03
Nanning		12	D	2 (16.67)		GX-Tu-PV-1	KX084396	20	2015.03.10
Nanning	2	24	D	24 (100.00)	11/12 (91.67)	GX-Tu-PV-2	KX084397	90	2015.04.16
Nanning	3	48	D	46 (95.83)		GX-Tu-PV-3	KX084398	150	2015.04.16
Nanning	3	24	B	17 (70.83)		GX-CH-PV-31	OQ437199	120	2021.10.11
Nanning	3	24	B	18 (75.00)		GX-CH-PV-32	OQ437200	125	2022.05.16
Nanning	3	24	B	16 (66.67)		GX-CH-PV-33	OQ437201	130	2022.09.22
Total	126	1 526		1 064 (69.72)	158/172 (91.86)				

Note: A, layer chicken; B, breeder chicken; C, broiler chicken; D, broiler turkey; E, exotic broiler chicken. exotic chickens=A+E; native chickens=B+C. Each chicken flock had between 8 000 and 12 000 chickens; each turkey flock had between 600 and 1 000 turkeys.

Overall features of the genomes

The genomes of the Guangxi ChPV/TuPV strains ranged from 4 612 to 4 642 bp in length. The approximate GC content of the genomes was 42.88%, and they each contained 3 segments encoding 4 viral proteins. The genomic segments ranged from 305 bp (NP1) to 2085 bp (NS1) in length, and ORF analysis of the nucleotide (nt) sequences indicated that 2 of the 3 genome segments encoded a single ORF, which were all similar to those of the ChPV/TuPV reference strains. The first ORF was predicted to encode 2 putative proteins (NSl on NS1 and NPl on NP1) ranging in size from 101 to 695 amino acids (aa). The 2028-bp VP segment was found to contain two partially overlapping genes encoding VP1 (2 028 bp, 676 aa) and VP2 (1 611 bp, 537 aa).

Comparisons of the similarities between the nt sequences of the Guangxi ChPV/TuPV strains and those of 17 ChPV/TuPV reference strains revealed that all 3 segments identified in the Guangxi ChPV/TuPV strains showed varying degrees of homology with the reference ChPV/TuPV strains. The 35 Guangxi isolates showed 79.4%~99.7% nt identity with each other, and 78.7%~99.7% nt identity with 11 classic ChPV reference strains, including the ChPV ABU-P1, ChPV ADL120686, ChPV ADL120019, ChPV ADL120035, ChPV 367, ChPV 736, ChPV 798, ChPV 841, ChPV ParvoD62/2013, ChPV ParvoD11/2007, and ChPV IPV strains, and 6 classic TuPV reference strains, including the TuPV 260, TuPV 1078, TuPV 1030, TuPV 1085, TuPV 1090 and TuPV JO11 strains.

Nucleotide and amino acid comparisons

Comparing the nt and aa sequences of the NS1 gene revealed high sequence identities between the 35 Guangxi ChPV/TuPV strains and 11 ChPV reference strains and 6 TuPV reference strains. GPV (accession number: NC_001701, from the USA) and DPV (accession number: U22967, from Hungary) were used as outgroups. Homology analysis of the NS1 gene showed that the homologies of the nt and deduced aa sequences of the 35 Guangxi isolates were 88.1%~99.9% and 89.1%~100.0%, respectively. The nt sequence alignment between the ChPV/TuPV strains from Guangxi and those from other countries revealed 85.2%~99.9% similarity, and the aa sequences showed 87.8%~100.0% identities. The sequence identity of the NP1-encoding genes was the highest (>95%); however, the role of this putative protein remains unknown[3].

Compared with the genome fragments encoded by NS1 and NP1, the VP-encoding segments showed higher genetic diversity. For the VPl protein, the Guangxi ChPV/TuPV strains showed similar identities with the ChPV/TuPV reference strains (nt, 73.0%~98.0%; aa, 77.1%~100.0%). Conversely, the VP2 protein shared the lowest identity with the ChPV 367 strain and the highest identity with the TuPV JO11 strain (nt, 72.0%~98.0%; aa, 76.9%~100.0%). For the VP1 gene, the homologies of the nt and deduced aa sequences of the 32 Guangxi isolates were 72.6%~99.9% and 78.0%~99.7%, respectively, and the homologies between the ChPV/TuPV isolates from Guangxi and those from other countries were 72.7%~99.9% and 77.2%~99.6%, respectively. For the VP2 gene, the homologies of the nt and deduced aa sequences of the 32 Guangxi isolates were 70.7%~99.9% and 77.3%~99.6%, respectively, and the homologies between the Guangxi ChPV/TuPV isolates and those from other countries were 71.1%~99.9% and 77.1%~99.4%, respectively.

Sequence analysis

Interestingly, an 8-nt (TTATTTTG) deletion (corresponding to nts 2 778 to 2 785 in the NP1 gene of strain ABU-P1) was observed in all of the ChPV and TuPV strains except for the GX-CH-PV-1 and ChPV 841 strains and the TuPV 1 078, 1 085, 1 090, GX-Tu-PV-1, GX-Tu-PV-2, and GX-Tu-PV-3 strains. Additionally, a 4-nt (CTAA) deletion (corresponding to nt 2 789 to 2 792 in the NP1 gene of strain ABU-P1) was found in all of the ChPV and TuPV strains except for the GX-CH-PV-1 and ChPV 841 strains and the TuPV 1 078, 1 085, 1 090, GX-Tu-PV-1, GX-Tu-PV-2, and GX-Tu-PV-3 strains. Moreover, a 9-nt (TCCATAATG) deletion (corresponding to nt 3 275 to 3 283 in the VP1 gene of strain ABU-P1) was found in all of the ChPV/TuPV strains except for the GX-CH-PV-1 and ChPV 841 strains and the TuPV 1 078, 1 085, 1 090, GX-Tu-PV-1, GX-Tu-PV-2, and GX-Tu-PV-3 strains. Finally, a 3-nt (GAA) deletion (corresponding to nt 3 570 to 3 572 in the VP2 gene of strain ABU-P1) was observed in all of the ChPV/TuPV strains except for the GX-CH-PV-1 and ChPV 841 strains and the TuPV 1 078, 1 085, 1 090, GX-Tu-PV-1, GX-Tu-PV-2, and GX-Tu-PV-3 strains.

Sixteen of the Guangxi ChPV/TuPV strains (ie., GX-CH-PV-4, GX-CH-PV-6, GX-CH-PV-7, GX-CH-PV-13, GX-CH-PV-14, GX-CH-PV-15, GX-CH-PV-17, GX-CH-PV-18, GX-CH-PV-20, GX-CH-PV-21, GX-CH-PV-22, GX-CH-PV-25, GX-CH-PV-27, GX-CH-PV-28, GX-CH-PV-29 and GX-CH-PV-30) sequenced in this study all contained the putative VP3 start codon (spanning nts 3 919 to 3 921 in the ABU-P1 strain), which has also been identified in the ChPV ABU-P1, ChPV 367, ChPV 736, ChPV 798 and TuPV 260 strains. Therefore, the VP3 protein of ChPV is not produced by alternative splicing of ORF2.

A highly conserved phosphate-binding loop (P-loop) motif (aa 392 to 399, GPANTGKT) and NTP binding motif (aa 436 to 437, EE) corresponding to the NS1 gene of strain ABU-P1 were present in all Guangxi ChPV/TuPV strains. The start codons for VP1 (nts 2 998 to 3 000; ABU-P1) and VP2 (nts 3 415 to 3 417; ABU-P1), the leucine residue (L; aa 293, VP1 of ABU-P1; aa 152, VP2 of ABU-P1), and the region of fivefold cylinders (LQVIQKTVTDSGTQYSND; aa 275 to 292, VP1 of ABU-P1; aa 134 to 151, VP2 of ABU-P1) were identified, but a glycine-rich sequence (GGGGGGGGG; aa 164 to 172, VP1 of ABU-P1; aa 23 to 31, VP2 of ABU-P1) was identified in GX-TU-PV-1, GX-TU-PV-2, GX-TU-PV-3, GX-CH-PV-1 and GX-CH-PV-24, while TVGGGGGGG was identified in other Guangxi ChPV isolates.

Phylogenetic analysis

Evolutionary relationships between the Guangxi ChPV/TuPV strains and different members of the *Aveparvovirus* genus, including DPV and GPV, which were used as outgroup controls, were determined by phylogenetic analysis. Based on the nt sequences of the NS1, VP1 and VP2 genome segments and the whole ChPV/TuPV genomes, the neighbour-joining method with 1 000 bootstrap replicates was used to construct the phylogenetic trees (Figure 2-15-1 A-D). All the constructed phylogenetic trees showed marked divergence between the Guangxi ChPV/TuPV strains and the other reference ChPV/TuPV strains. For the 3 genome segments, the vast majority of the ChPV strains formed a host-associated group (except for strains TuPV 260, ChPV 841, GX-Tu-PV-2, and GX-CH-PV-1) that differed from the turkey strains, the FM duck strain, and the virulent B goose strain. Furthermore, the segments encoding the VP1/VP2 proteins exhibited noticeably higher divergence than NS1 in the ChPV/TuPV strains, as indicated by sequence comparisons. A phylogenetic tree based on the VP gene revealed that the 35 Guangxi ChPV/TuPV isolates sequenced in our research clustered into 5 ChPV/TuPV groups designated Groups A, B, C, D, E and F (Figure 2-15-1 D). Genotyping cluster A, which consisted of 12 Guangxi ChPV field strains, included 2 prototype ChPV/TuPV strains (strains ABU-P1 and 260), 1 *Gallus gallus* enteric parvovirus isolate (strain 736) from the USA and 1 prototype ChPV strain (strains ParvoD11-2007) from ROK; genotyping cluster B, which consisted of 4 Guangxi ChPV field strains, included 7 prototype ChPV strains from the USA, ROK and Brazil; and genotyping cluster E which consisted of 3 Guangxi TuPV and 3 ChPV field strains, included 6 prototype ChPV and TuPV strains all from the USA. Eighteen of the 35 ChPV/TuPV isolates were identified as Group A and Group E while 3 Guangxi ChPV and 3 Guangxi TuPV field strains of Group F were field variants and were distinct from the prototype ChPV (strain 841) and TuPV (strains 1085, 1078, 1090, 1030 and JO11) strains, all from the USA. More importantly, 3 novel ChPV/TuPV groups (Groups C, D and E) were identified for the first time, all from Guangxi. Interestingly, ChPV/TuPV whole-genome sequences from chickens with RSS-like symptoms were more concentrated in

Groups C, D and E (Figure 2-15-1 D).

The nt alignment of the full genome of the Guangxi ChPV/TuPV strains and 17 reference ChPV/TuPV strains (Figure 2-15-2 A, B) revealed conserved and divergent regions between the genomes. Visualizing the genomes in this manner supported the results of the phylogenetic study described above.

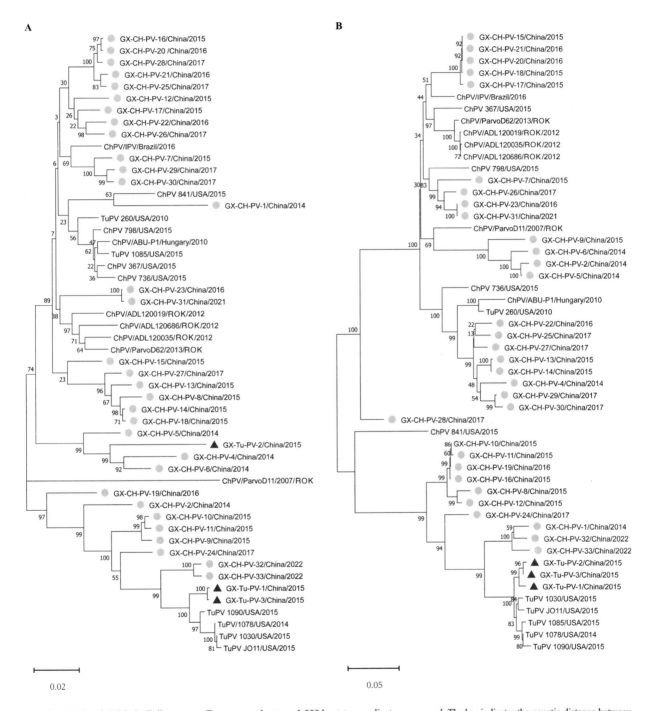

A: NS1; B: VP1; C: VP2; D: Full genomes. To construct the trees, 1 000 bootstrap replicates were used. The bar indicates the genetic distance between sequences, and bootstrap values are shown at the nodes. The circle and triangle represent the Guangxi ChPV strains and TuPV strains, respectively. The ChPV/TuPV in bold black font indicate chickens with RSS-like symptoms.

Figure 2-15-1   Phylogenetic trees constructed using the nucleotide sequences of 3 homologous genome segments and the full genomes of ChPV and TuPV, with DPV (GenBank accession number: U22967) and GPV (GenBank accession number: NC_001701) as outgroups

331

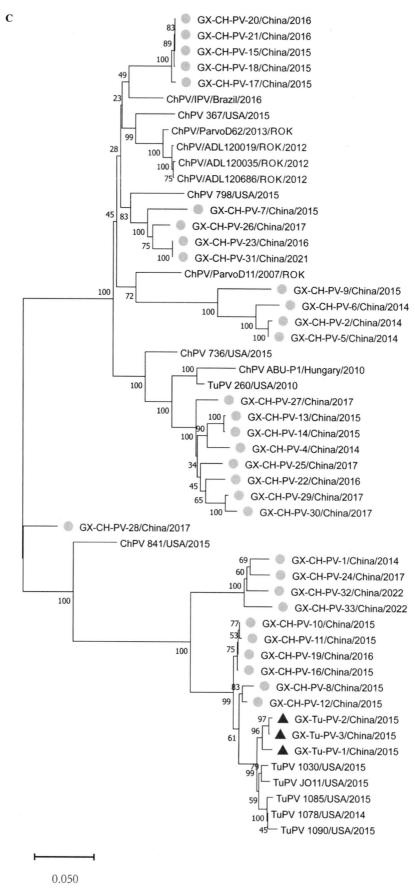

C

0.050

Figure 2-15-1   (continued)

D

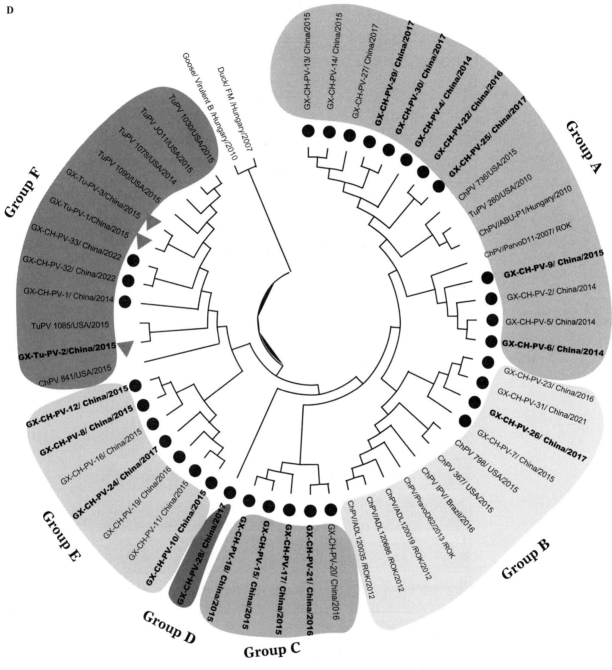

(Color figure in appendix.)

Figure 2-15-1   (continued)

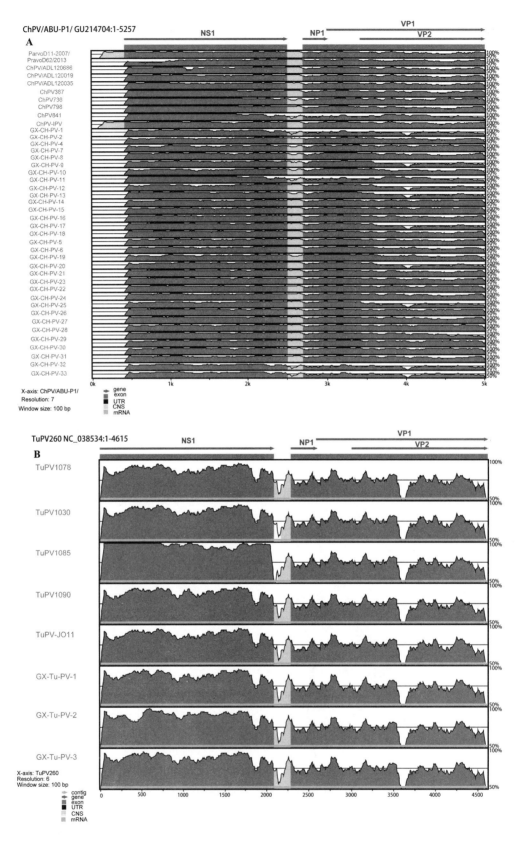

A: Guangxi ChPV field strains; B: Guangxi TuPV field strains. Colour coding: the coloured regions represent a similarity > 90%, and the white regions represent a similarity< 90%. The height of the shaded area at any sampling point is proportional to the genetic relatedness (color figure in appendix).

Figure 2-15-2　mVISTA whole-genome alignments comparing the nucleotide sequences of the Guangxi ChPV and TuPV field strains with representative ABU-P1 and 260, 1078, 1030, 1085, 1090 and JO11 TuPV strains

Recombination analysis

Fifteen recombination events were detected in the NS1, VP1 and VP2 genes of 13 Guangxi strains, as shown in Table 2-15-2. To further verify the recombination events identified by RDP 5.0, SimPlot 3.5.1 software was used to analyse the homology of the recombinant strains. These recombination sequence signals were confirmed by SimPlot analysis.

Table 2-15-2   Information on recombination events detected in the genomes of ChPV/TuPV

Serial no	Strain	Breakpoints		Major parent	Minor parent	Recombinant gene
		Begin	End			
1	GX-CH-PV-16	3 192	4 610	GX-CH-PV-25	GX-Tu-PV-3	VP2
2	GX-Tu-PV-2	4 632	1 530	GX-Tu-PV-3	GX-CH-PV-9	NS1
3	GX-CH-PV-24	1 703	3 046	GX-CH-PV-32	GX-CH-PV-13	NS1, VP1
4	GX-CH-PV-8	3 149	4 604	GX-CH-PV-14	GX-Tu-PV-3	VP2
5		90	1 708	GX-CH-PV-28	GX-CH-PV-18	NS1
6	GX-CH-PV-11	4 610	1 510	GX-CH-PV-14	GX-Tu-PV-1	NS1
7	GX-CH-PV-4	3 174	4 580	GX-CH-PV-18	GX-CH-PV-14	VP2
8		792	1 306	GX-Tu-PV-2	GX-CH-PV-2	NS1
9	GX-CH-PV-18	4 587	3 115	GX-CH-PV-21	GX-CH-PV-14	NS1, VP1
10	GX-CH-PV-28	4 081	4 599	GX-CH-PV-7	GX-CH-PV-24	VP2
11	GX-CH-PV-7	1 552	3 202	GX-CH-PV-23	GX-CH-PV-9	NS1, VP2
12	GX-CH-PV-6	3 235	4 554	CHPV-AUB-P1	GX-CH-PV-2	VP2
13	GX-CH-PV-25	1 772	4 563	GX-CH-PV-20	GX-CH-PV-27	NS1, VP1
14	GX-CH-PV-30	1 503	4 507	GX-CH-PV-7	GX-CH-PV-13	NS1, VP1
15	GX-CH-PV-22	4 591	2 548	GX-CH-PV-27	GX-CH-PV-26	NS1

Note: The "Major parent" is the sequence closely related to that from which the greater part of the recombinants sequence may have been derived; the "Minor parent" is the sequence closely related to that from which sequences in the proposed recombinant region may have been derived. The actual breakpoint position is undetermined. Most likely, it was overprinted by RDP 5.0.

Discussion

In the mid-1980s, ChPVs and TuPVs were identified as the causative agents of a pathogenic poultry disease[4, 6, 10]. Recent genomic characterization studies of the ChPV reference strains ABU-P1, ADL120686, ADL120019, ADL120035, 367, 736, 798, 841, ParvoD62/2013, ParvoD11/2007 and IPV together with the TuPV reference strains 260, 1078, 1030, 1085, 1090 and JO11 have led to an accumulation of genomic sequencing data, providing deeper insights into their molecular features. However, most of the sequence analyses published in the past decade were based on the NS1 and VP genes of ChPVs/TuPVs; few comparative analyses have been based on whole-genome sequences[3, 24, 37]. These reports have facilitated not only the analysis of the overall genetic architecture of ChPVs/TuPVs but also the development of molecular characterization and diagnostic assays.

No full sequence reports of ChPVs/TuPVs in other Chinese provinces have been published, and in this study, the complete nt sequences of Guangxi ChPV/TuPV strains with or without associations with RSS and PEMS were determined and compared with those of other reference ChPV/TuPV strains at the nt and aa levels. We compared the genomes of 32 ChPV strains and 3 TuPV strains isolated in Guangxi, China, with those of reference ChPV/TuPV strains isolated from the USA, Brazil, Hungary, and ROK. The nt and aa sequences of the Guangxi ChPV/TuPV strains showed moderate to low similarity to those of the reference ChPV/TuPV strains, with the C-terminal half of the VP2 protein showing the lowest sequence identity. Sequences of ChPV/TuPV strains were compared with those of classical DPV and GPV isolates and showed rather low identity values. Overall, however, these sequencing data suggest that the Guangxi ChPV/TuPV strains, similar to other ChPV/TuPV strains, belong to the genus *Aveparvovirus*. Comparison of ChPV/TuPV isolates from Guangxi and ChPV ABU-P1 strains revealed evidence of selection for the purification of NS1 and VP genes, suggesting that the Chinese ChPV strains evolved independently from the ABU-P1 strain (Hungary).

In the comparison with other parvoviruses, it was found that all VP protein structures were similar. Glycine enrichment may have implications for antigenicity[38]. A study also showed that the leucine residue in VP1 (aa 293) /VP2 (aa 152) can compress the pores formed by the fivefold cylinder and play an important role in DNA packaging and viral infection[39]. Among the 35 Guangxi ChPVs/TuPVs sequenced in our research, the VP3 start codon was found in 16 strains with reference to the ABU-P1 strain at position 3 919 to 3 921 bp, while the remaining 19 strains (including 3 TuPVs) had no VP3 start codon. Thus, the VP3 protein in ChPVs is not generated from ORF2 by alternative splicing. This conclusion is consistent with that of Koo[37].

Using traditional sequencing methods, we analysed the whole-genome characteristics of 35 ChPV/TuPV strains obtained from Guangxi. Overall, comparing the NS1-, NP1-, and VP-encoding genome segments among the different ChPV/TuPV strains indicated that the regions encoding the outer capsid proteins VP1 (minor capsid protein) and VP2 (major capsid protein) exhibited more variation than the other genes. Specifically, the gene encoding VP2 displayed the greatest sequence divergence, which is reasonable considering that the VP1 and VP2 proteins are components of the outer capsids of ChPVs and TuPVs and may therefore possess several epitopes governing pathogenicity, tissue tropism, and antigenicity[40-43]. Compared with other ChPV/TuPV strains, the Guangxi ChPV/TuPV strains identified in this report may have these properties, and understanding the impact of the nt deletions listed above will require further clarification at the molecular level.

Within genotyping Groups A and B, Guangxi ChPV field strains formed ChPV/TuPV subgroups with the reference ChPV strains, showing high nt similarity to the reference strains. However, the low aa identities (70.6%～88.8%) between the subclusters indicate that the Guangxi ChPV field strains in Groups C, D and E are not identical to the reference ChPV strains in ChPV/TuPV Group A, B and F. Similarly, variations in aa identity were also observed between genotyping ChPV/TuPV Groups C, D and E, in which most of the ChPV/TuPV Guangxi field strains formed their own subgroups, distinguishing them from ChPV/TuPV reference strains detected in other countries (e.g., the USA, Brazil, Hungary and ROK) (Figure 2-15-2). Nonetheless, the novel ChPV/TuPV genotyping Groups C, D and E and emerging variants from the other Guangxi ChPV/TuPV clusters show that ChPV/TuPV has occurred or is continuously undergoing evolutionary mutation or recombination, which should be considered.

Genetic evolution analysis of individual genes and whole genomes based on nucleotide sequences revealed various clustering patterns with reference strains. The topological heterogeneity observed among the phylogenetic trees and sequence alignment indicated that genetic recombination of the VP segment may have occurred between the Guangxi ChPV/TuPV strains and the reference ChPV/TuPV strains. RDP 5.0 and SimPlot recombination analysis detected recombination events in 13 strains (13/35, 37.14%) of ChPVs/TuPVs, occurring in the NS1, VP1and VP2 genes. As shown in the phylogenetic tree of Figure 2-15-1 D and Table 2-15-2, six of the 13 strains were derived from the novel genotyping groups, indicating that recombination between multiple genotypes of ChPVs/TuPVs may accelerate the emergence of new mutants. These findings suggest that genetic recombination between the Guangxi ChPV/TuPV strains and the reference ChPV/TuPV strains may have played a role in the origination of the Guangxi ChPV/TuPV strains, which is consistent with several reports[7, 22] on ChPV/TuPV genes and with the evolutionary strategies observed in most other species within the Parvoviridae family. This observation suggests that ChPVs and TuPVs might have evolved uniquely in Guangxi over time. However, further studies are needed to corroborate this hypothesis. Collectively, the phylogenetic analysis results provided insights into the origins of the unique genetic configurations observed in the novel Guangxi ChPV/TuPV strains detected in China since 2014.

Based on phylogenetic analysis of the NS1 gene, we hypothesized that the parvoviruses detected in turkey flocks were ChPVs adapted to turkey hosts (TuPV 1085, TuPV 260 and GX-Tu-PV-2). For the VP2 gene, the parvoviruses detected in the turkey flock were ChPVs adapted to a turkey host (TuPV 260); for the VP1 gene and the full genome, the parvoviruses detected in the turkey flock were ChPVs adapted to a turkey host (TuPV 260), while those detected in the chicken flock were TuPVs adapted to a chicken host (GX-Tu-PV-1). Genetic evolution analysis showed that the NS gene was more conserved than the VP1gene and VP2 gene, and the VP1 gene sequence had the highest degree of differentiation and the largest degree of variation. Therefore, it was speculated that the VP1 gene could replace the whole gene as a genetic marker for the rapid differentiation and classification of ChPVs and TuPVs. Shackelton et al.[27] also reported that parvovirus has a high atypical mutation rate among DNA viruses, prompting its rapid evolution and host adaptation. Given the epidemiological studies of ChPVs/TuPVs in our laboratory, we suspect that ChPV adaptation to turkeys and TuPV adaptation to chickens are both caused by insufficient disinfection and poor biosafety. In our study, we found that the nt sequences of Guangxi ChPV/TuPV strains showed strong similarity and phylogenetic relationships with the nt sequences of other parvovirus strains isolated from RSS/PEMS cases, which was similar to finding of Zsak et al.[20], who described the similarity between a TuPV isolate and ChPVs. Therefore, it is possible that some regions of the genome were involved in pathogenicity. Additionally, the detection rate of ChPVs/TuPVs in birds with RSS-like symptoms (91.86%) was higher than that in healthy birds (66.91%), and complicating factors such as mixed or secondary infection with other pathogens may exacerbate the process of parvovirus infection. However, the correlation between sequences and RSS-like symptoms remains to be further studied. Our reports have indicated the presence of variations among Guangxi ChPV/TuPV isolates and incidences of emergence of new isolates worldwide. Whole-genome characterizations of newly emerging Guangxi ChPV/TuPV field strains will provide more detailed insights into ChPV/TuPV mutations and recombination and their relationships with molecular epidemiological features. Therefore, the study of ChPVs/TuPVs in Guangxi will be helpful in tracing the source of the viruses causing epidemics at the molecular level and elucidating the potential transmission route and mode, which is of great significance for epidemiological analysis of disease.

Data availability

The datasets generated and analysed during the current study are available in the NCBI genome repository or from the corresponding author upon reasonable request. The accession numbers are as follows: GX-CH-PV-1 (KX084399), GX-CH-PV-2 (KX084400), GX-CH-PV-4 (KX084401), GX-CH-PV-5 (KX133426), GX-CH-PV-6 (KX133427), GX-CH-PV-7 (KU523900), GX-CH-PV-8 (KX133415), GX-CH-PV-9 (KX133416), GX-CH-PV-10 (KX133417), GX-CH-PV-11 (KX133418), GX-CH-PV-12 (KX133419), GX-CH-PV-13 (KX133420), GX-CH-PV-14 (KX133421), GX-CH-PV-15 (KX133422), GX-CH-PV-16 (KX133423), GX-CH-PV-17 (KX133424), GX-CH-PV-18 (KX133425), GX-CH-PV-19 (MG602509), GX-CH-PV-20 (MG602510), GX-CH-PV-21 (MG602511), GX-CH-PV-22 (MG602512), GX-CH-PV-23 (MG602513),

GX-CH-PV-24 (MG602514), GX-CH-PV-25 (MG602515), GX-CH-PV-26 (MG602516), GX-CH-PV-27 (MG602517), GX-CH-PV-28 (MG602518), GX-CH-PV-29 (MG602519), GX-CH-PV-30 (MG602520), GX-CH-PV-31 (OQ437199), GX-CH-PV-32 (OQ437200), GX-CH-PV-33 (OQ437201), GX-Tu-PV-1 (KX084396), GX-Tu-PV-2 (KX084397), GX-Tu-PV-3 (KX084398) and GX-Tu-PV-3 (KX084398).

References

[1] KAPGATE S S, KUMANAN K, VIJAYARANI K, et al. Avian parvovirus: classification, phylogeny, pathogenesis and diagnosis. Avian Pathol, 2018, 47(6): 536-545.

[2] COTMORE S F, AGBANDJE-MCKENNA M, CANUTI M, et al. ICTV virus taxonomy profile: Parvoviridae. J Gen Virol, 2019, 100(3): 367-368.

[3] DAY J M, ZSAK L. Determination and analysis of the full-length chicken parvovirus genome. Virology, 2010, 399(1): 59-64.

[4] KISARY J, NAGY B, BITAY Z. Presence of parvoviruses in the intestine of chickens showing stunting syndrome. Avian Pathol, 1984, 13(2): 339-343.

[5] KISARY J, AVALOSSE B, MILLER-FAURES A, et al. The genome structure of a new chicken virus identifies it as a parvovirus. J Gen Virol, 1985, 66 (Pt10): 2259-2263.

[6] TRAMPEL D W, KINDEN D A, SOLORZANO R F, et al. Parvovirus-like enteropathy in Missouri turkeys. Avian Dis, 1983, 27(1): 49-54.

[7] SHACKELTON L A, HOELZER K, PARRISH C R, et al. Comparative analysis reveals frequent recombination in the parvoviruses. J Gen Virol, 2007, 88(Pt 12): 3294-3301.

[8] LUKASHOV V V, GOUDSMIT J. Evolutionary relationships among parvoviruses: virus-host coevolution among autonomous primate parvoviruses and links between adeno-associated and avian parvoviruses. J Virol, 2001, 75(6): 2729-2740.

[9] MOMOEDA M, WONG S, KAWASE M, et al. A putative nucleoside triphosphate-binding domain in the nonstructural protein of B19 parvovirus is required for cytotoxicity. J Virol, 1994, 68(12): 8443-8446.

[10] GUY J S. Virus infections of the gastrointestinal tract of poultry. Poult Sci, 1998, 77(8): 1166-1175.

[11] SIMPSON A A, H BERT B, SULLIVAN G M, et al. The structure of porcine parvovirus: comparison with related viruses. J Mol Biol, 2002, 315(5): 1189-1198.

[12] LE GALL-RECULE G, JESTIN V. Biochemical and genomic characterization of muscovy duck parvovirus. Arch Virol, 1994, 139(1-2): 121-131.

[13] SCHETTLER C H. Virus hepatitis of geese. 3. Properties of the causal agent. Avian Pathol, 1973, 2(3): 179-193.

[14] CHEN H, DOU Y, TANG Y, et al. Isolation and genomic characterization of a duck-origin GPV-related parvovirus from Cherry valley ducklings in China. PLOS ONE, 2015, 10(10): e0140284.

[15] KISARY J, MILLER-FAURES A, ROMMELAERE J. Presence of fowl parvovirus in fibroblast cell cultures prepared from uninoculated White Leghorn chicken embryos. Avian Pathol, 1987, 16(1): 115-121.

[16] PALADE E A, KISARY J, BENYEDA Z, et al. Naturally occurring parvoviral infection in Hungarian broiler flocks. Avian Pathol, 2011, 40(2): 191-197.

[17] CARRATAL A, RUSINOL M, HUNDESA A, et al. A novel tool for specific detection and quantification of chicken/turkey parvoviruses to trace poultry fecal contamination in the environment. Appl Environ Microbiol, 2012, 78(20): 7496-7499.

[18] MARUSAK R A, GUY J S, ABDUL-AZIZ T A, et al. Parvovirus-associated cerebellar hypoplasia and hydrocephalus in day old broiler chickens. Avian Dis, 2010, 54(1): 156-160.

[19] KOO B S, LEE H R, JEON E O, et al. Molecular survey of enteric viruses in commercial chicken farms in Korea with a history of enteritis. Poult Sci, 2013, 92(11): 2876-2885.

[20] ZSAK L, STROTHER K O, DAY J M. Development of a polymerase chain reaction procedure for detection of chicken and turkey parvoviruses. Avian Dis, 2009, 53(1): 83-88.

[21] BIDIN M, LOJKIĆ I, BIDIN Z, et al. Identification and phylogenetic diversity of parvovirus circulating in commercial

chicken and turkey flocks in Croatia. Avian Dis, 2011, 55(4): 693-696.

[22] DOMANSKA-BLICHARZ K, JACUKOWICZ A, LISOWSKA A, et al. Genetic characterization of parvoviruses circulating in turkey and chicken flocks in Poland. Arch Virol, 2012, 157(12): 2425-2430.

[23] NUNEZ L F, SA L R, PARRA S H, et al. Molecular detection of chicken parvovirus in broilers with enteric disorders presenting curving of duodenal loop, pancreatic atrophy, and mesenteritis. Poult Sci, 2016, 95(4): 802-810.

[24] FENG B, XIE Z, DENG X, et al. Genetic and phylogenetic analysis of a novel parvovirus isolated from chickens in Guangxi, China. Arch Virol, 2016, 161(11): 3285-3289.

[25] ZHANG Y, XIE Z, DENG X, et al. Molecular characterization of parvovirus strain GX-Tu-PV-1, isolated from a Guangxi Turkey. Microbiol Resour Announc, 2019, 8(46):e00152-19.

[26] ZHANG Y, FENG B, XIE Z, et al. Epidemiological surveillance of parvoviruses in commercial chicken and turkey farms in Guangxi, Southern China, During 2014-2019. Front Vet Sci, 2020, 7: 561371.

[27] SHACKELTON L A, PARRISH C R, TRUYEN U, et al. High rate of viral evolution associated with the emergence of carnivore parvovirus. Proc Natl Acad Sci U S A, 2005, 102(2): 379-384.

[28] PANTIN-JACKWOOD M J, DAY J M, JACKWOOD M W, et al. Enteric viruses detected by molecular methods in commercial chicken and turkey flocks in the United States between 2005 and 2006. Avian Dis, 2008, 52(2): 235-244.

[29] CANELLI E, CORDIOLI P, BARBIERI I, et al. Astroviruses as causative agents of poultry enteritis: genetic characterization and longitudinal studies on field conditions. Avian Dis, 2012, 56(1): 173-182.

[30] HEWSON K A, O'ROURKE D, NOORMOHAMMADI A H. Detection of avian nephritis virus in Australian chicken flocks. Avian Dis, 2010, 54(3): 990-993.

[31] OTTO P, LIEBLER-TENORIO E M, ELSCHNER M, et al. Detection of rotaviruses and intestinal lesions in broiler chicks from flocks with runting and stunting syndrome (RSS). Avian Dis, 2006, 50(3): 411-418.

[32] PANTIN-JACKWOOD M J, SPACKMAN E, WOOLCOCK P R. Molecular characterization and typing of chicken and turkey astroviruses circulating in the United States: implications for diagnostics. Avian Dis, 2006, 50(3): 397-404.

[33] YU L, JIANG Y, LOW S, et al. Characterization of three infectious bronchitis virus isolates from China associated with proventriculus in vaccinated chickens. Avian Dis, 2001, 45(2): 416-424.

[34] BANYAI K, DANDAR E, DORSEY K M, et al. The genomic constellation of a novel avian orthoreovirus strain associated with runting-stunting syndrome in broilers. Virus Genes, 2011, 42(1): 82-89.

[35] KOO B S, LEE H R, JEON E O, et al. Genetic characterization of three novel chicken parvovirus strains based on analysis of their coding sequences. Avian Pathol, 2015, 44(1): 28-34.

[36] AGBANDJE-MCKENNA M, LLAMAS-SAIZ A L, WANG F, et al. Functional implications of the structure of the murine parvovirus, minute virus of mice. Structure, 1998, 6(11): 1369-1381.

[37] FARR G A, TATTERSALL P. A conserved leucine that constricts the pore through the capsid fivefold cylinder plays a central role in parvoviral infection. Virology, 2004, 323(2): 243-256.

[38] GOVINDASAMY L, HUEFFER K, PARRISH C R, et al. Structures of host range-controlling regions of the capsids of canine and feline parvoviruses and mutants. J Virol, 2003, 77(22): 12211-12221.

[39] MCKENNA R, OLSON N H, CHIPMAN P R, et al. Three-dimensional structure of Aleutian mink disease parvovirus: implications for disease pathogenicity. J Virol, 1999, 73(8): 6882-6891.

[40] TRUYEN U, GRUENBERG A, CHANG S F, et al. Evolution of the feline-subgroup parvoviruses and the control of canine host range in vivo. J Virol, 1995, 69(8): 4702-4710.

[41] HUEFFER K, PARRISH C R. Parvovirus host range, cell tropism and evolution. Curr Opin Microbiol, 2003, 6(4): 392-398.

[42] TAMURA K, STECHER G, KUMAR S. MEGA11: Molecular Evolutionary Genetics Analysis Version 11. Mol Biol Evol, 2021, 38(7): 3022-3027.

[43] FAN W, CHEN J, ZHANG Y, et al. Phylogenetic and spatiotemporal Analyses of the complete genome sequences of avian coronavirus infectious bronchitis virus in China during 1985-2020: revealing coexistence of multiple transmission chains and the origin of LX4-Type virus. Front Microbiol, 2022, 13: 693196.

Differences in the pathogenicity and molecular characteristics of fowl adenovirus serotype 4 epidemic strains in Guangxi, southern China

Wei You, Xie Zhiqin, Xie Zhixun, Deng Xianwen, Li Xiaofeng, Xie Liji, Fan Qing, Zhang Yanfang, Wang Sheng, Ren Hongyu, Wan Lijun, Luo Sisi, Li Meng

Abstract

Starting in 2015, the widespread prevalence of hydropericardium-hepatitis syndrome (HHS) has led to considerable financial losses within China's poultry farming industry. In this study, pathogenicity assessments, whole-genome sequencing, and analyses were conducted on 10 new isolates of the novel genotype FAdV-4 during a HHS outbreak in Guangxi, China, from 2019 to 2020. The results indicated that strains GX2019-010 to GX2019-013 and GX2019-015 to GX2019-018 were highly virulent, while strain GX2020-019 exhibited moderate virulence. Strain GX2019-014 was characterized as a wild-type strain with low virulence, displaying no pathogenic effects when 0.5 mL containing 10^6 $TCID_{50}$ virus was inoculated into the muscle of specific pathogen-free (SPF) chickens at 4 weeks of age, while 10^7 $TCID_{50}$ and 10^8 $TCID_{50}$ resulted in mortality rates of 80% and 100%, respectively. The whole genomes of strains GX2019-010 to GX2019-013, GX2019-015 to GX2019-018, and GX2020-019 showed high homology with other Chinese newly emerging highly pathogenic FAdV-4 strains, whereas GX2019-014 was closer to nonmutant strains and shared the same residues with known nonpathogenic strains (B1-7, KR5, and ON1) at positions 219AA and 380AA of the Fiber-2 protein. Our work enriches the research on prevalent strains of FAdV-4 in China, expands the knowledge on the virulence diversity of the novel genotype FAdV-4, and provides valuable reference material for further investigations into the key virulence-associated genetic loci of FAdV-4.

Keywords

fowl adenovirus serotype 4 (FAdV-4), wild-type low-virulence strain, pathogenicity difference, molecular characteristics, Guangxi, GX2019-014

Introduction

According to the "ICTV Virus Taxonomy Profile: Adenoviridae 2022" released by the International Committee on Taxonomy of Viruses (ICTV), the Adenoviridae family is presently organized into six acknowledged genera (*Aviadenovirus, Siadenovirus, Atadenovirus, Mastadenovirus, Ichtadenovirus, and Testadenovirus*) based on specific genomic characteristics and molecular phylogenetic analysis. Within the *Aviadenovirus* genus, there are fifteen species (*Fowl Aviadenovirus A-E, Turkey Aviadenovirus B-D, Falcon Aviadenovirus A, Duck Aviadenovirus B, Goose Aviadenovirus A, Pigeon Aviadenovirus A-B and Psittacine Aviadenovirus B-C*) delineated through phylogenetic analysis, genome structure and lack of notable cross-neutralization[1,2]. The five species (A-E) of *Fowl Aviadenoviruses* included 12 serotypes (Serotype 1 to 8a and 8b to 11) [3,4].

Fowl adenovirus serotype 4 (FAdV-4) belongs to the *Aviadenovirus* A species. It is an unenveloped, linear double-stranded DNA virus with a genome spanning 43 to 46 kb in length. The principal structural proteins found on the capsid include hexon, the penton base, fiber-1, and fiber-2. The major core shell proteins include pV, pVII, pVIII, and pX. The known nonstructural proteins include 52/55 k, E1A, E1B, E3, E4, and 100 k.[5]. Recent studies have indicated that hexon creates neutralizing antigenic sites housing clusters of serotype-specific antigenic determinants that are commonly used for phylogenetic analysis[5]. The presence of penton bases influences viral entry into cells [6]. Fiber-1 mediates infection through the shaft-knob structure and is a target of the host cell receptor CAR [7,8]. Fiber-2 is a virulence determinant cluster and exhibits good antigenicity, inducing the production of neutralizing antibodies and effectively combating FAdV-4 infection [9-11].

FAdV-4 has been confirmed as the pathogen causing hydropericardium hepatitis syndrome (HHS) [12]. HHS was initially documented in the Angara region of Pakistan in 1987, with subsequent outbreaks observed in various regions, including Australia, Iraq, Kuwait, India, and Japan. [13-16]. Since 2015, sporadic outbreaks have emerged in China and rapidly spread on a large scale within major poultry farming provinces, including Henan, Shandong, Yunnan, Jilin, Guangdong, and Guangxi [17-19]. HHS has become among the main emerging diseases in chicks in recent years. Chickens typically develop clinical symptoms, such as fluffy feathers, whitish combs, depression, huddling in corners, diarrhea, and sporadic death, 2~4 d after contracting the virus [20]. The onset of evident symptoms and peak mortality occur 3~6 d postinfection; in this phase, chickens pass yellow-green feces, refuse to eat or drink, show cyanotic combs, exhibit lethargy, and eventually die [21,22]. Postmortem examination revealed the accumulation of yellowish fluid in the pericardial sac; hepatomegaly with friable texture, yellow discoloration, and obvious hemorrhagic spots on the liver surface; splenomegaly with congestion; renal hemorrhage; and other typical pathological changes [23], which became more pronounced as the disease progressed. This disease causes mortality and growth retardation in both broiler and layer chickens, with clinical mortality rates reaching 20% to 80%. In addition, the immune organs of surviving birds are damaged, leading to immunosuppression in poultry flocks and substantial financial setbacks for the poultry sector [22]. Therefore, disease prevention and control have become key focuses in poultry farming and clinical settings in recent years.

Since 2016, suspected instances of HHS have been reported in cities such as Nanning, Yulin, Baise, and Qinzhou within Guangxi, China [19]. In this study, 10 strains of FAdV-4 were isolated from commercial poultry farms and individual poultry farms in Guangxi, China, and their pathogenicity in specific pathogen-free (SPF) chickens was assessed. Furthermore, the complete genomes of all 10 isolates were sequenced, and nucleotide and amino acid-level comparisons were conducted with both pathogenic and nonpathogenic strains of FAdV-4 available in GenBank to analyze key virulence-related sites. This study aimed to provide data supporting the epidemiological investigation of FAdV-4 in Guangxi and to provide relevant information for the study of virulence determinant sites of FAdV-4.

Materials and methods

Experimental animals and ethical statement

One thousand SPF chickens utilized to assess the pathogenicity of the FAdV-4 isolates were derived from SPF White Leghorn chicken eggs obtained from Beijing Boehringer Ingelheim Vital Biotechnology Co., Ltd. (Beijing, China) and hatched naturally in an incubator and subsequently reared in negative-pressure SPF isolation facilities until they reached 4 weeks of age. The animal experiments were conducted following approval from the Animal Ethics Committee of the Guangxi Veterinary Research Institute and adhered strictly to their regulations (Approval No. 2019c0406).

Sample collection and PCR detection of FAdV-4

All the isolates were obtained from poultry flocks on chicken farms in the cities of Nanning, Baise, and Yulin in Guangxi, China, during the HHS outbreak from 2019 to 2020. The relevant information regarding the sources of these isolates is provided in Table 2-16-1. The hearts, livers, spleens, kidneys, lungs, bursa of Fabricius, and pancreatic tissues of the suspected chickens were collected, homogenized separately in phosphate-buffered saline (PBS), subjected to three freeze-thaw cycles, and centrifuged at 10 000 × g for 15 minutes to obtain the supernatant. Viral DNA was extracted from the supernatant following the instructions of the TransGen Biotech EasyPure Genomic DNA/RNA Kit (TransGen, China). Common PCR detection of FAdV-4 was conducted using specific primers targeting the hexon gene[24]: forward, 5'-CGAGGTCTATACCAACACGAGCA-3'; reverse, 5'-TACAGCAGGTTAATGAAGTTATC-3'(Sangon Biotech, China). The amplification program comprised an initial denaturation at 95 ℃ for 5 min, followed by 30 amplification cycles, each consisting of denaturation at 95 ℃ for 30 sec, annealing at 55 ℃ for 30 sec, and extension at 72 ℃ for 1 minute, with a final extension step at 72 ℃ for 10 min. Gel electrophoresis was performed, and samples with strongly positive bands were selected for virus isolation.

Table 2-16-1　Information regarding the origins of the isolates

No.	Name of the isolate	Sampling time	Sampling place	Age of the chickens	Sampling tissue
1	GX2019-010	2019.08.07	Guangxi，Nanning City	24	Live
2	GX2019-011	2019.08.07	Guangxi，Nanning City	46	Live
3	GX2019-012	2019.08.07	Guangxi，Nanning City	53	Live
4	GX2019-013	2019.11.07	Guangxi，Baise City	60	Live
5	GX2019-014	2019.11.07	Guangxi，Baise City	50	Kidney
6	GX2019-015	2019.11.07	Guangxi，Baise City	55	Live
7	GX2019-016	2019.11.12	Guangxi，Nanning City	32	Live
8	GX2019-017	2019.04.28	Guangxi，Yulin City	45	Live
9	GX2019-018	2019.08.06	Guangxi，Baise City	56	Live
10	GX2020-019	2020.09.12	Guangxi，Yulin City	42	Live

Primary chicken embryonic liver (CEL) cell preparation and isolation

The chicken embryos were derived from SPF White Leghorn chicken eggs, which were incubated in an incubator until 15 days of age. Liver tissues were dissected from 15-day-old embryos and minced to approximately 2 mm^3. After digestion of liver tissue in 0.25% trypsin-EDTA solution for 10 minutes and gentle pipetting, liver cells were suspended in Dulbecco's modified Eagle's medium/nutrient mixture F-12 (DMEM/F12) (Gibco, USA) supplemented with 10% fetal bovine serum (Gibco, USA). The suspension was filtered through a sterile sieve with a 40-μm aperture, and the cell count was adjusted to 106 CEL cells/mL. The cells were cultured in T-25 flasks for approximately 72 h until they formed a monolayer[22].

The homogenized supernatant from FAdV-4 positive samples was filtered through a sterile syringe filter (PES) with 0.22 μm pores to remove bacteria. The viral suspension was inoculated into CEL at 2% of the volume of the culture medium, cultured for 96～120 hours, and subjected to blind passage for two generations for virus isolation, after which the cell supernatant was collected.

Virus purification and identification

The isolated virus was subjected to plaque purification using CEL cells. The supernatant was diluted from 10^{-3} to 10^{-8} and inoculated onto 6-well cell culture plates containing confluent monolayers of CEL cells. After 1 h of infection, the supernatant was aspirated. The cell surface was then coated with DMEM/F12 supplemented with 2% fetal bovine serum and 1% low melting point agarose (Promega, USA). The cells were incubated at 37 ℃ and 5% CO_2 for 6 days. Isolated, well-defined, medium-sized individual plaques were selected for three rounds of plaque purification.

The purified isolates were named according to the format "Abbreviation of Source Location and Year of Isolation-Purification Order". After purification, the viruses were propagated in CEL cells on a large scale (600 mL/isolate), followed by aliquoting into cryovials at 2 mL/vial, with approximately 300 vials/isolate, to establish a seed bank. Any vial of viral solution from this batch can represent the entire seed bank. The seed bank was stored at –80 ℃ and used for subsequent indirect immunofluorescence assay (IFA) identification, purity testing, and whole-genome sequencing of each isolate. Additionally, the tissue culture infectious dose ($TCID_{50}$) of each strain was determined, and viral dilutions were prepared based on the $TCID_{50}$ results for virulence testing. The isolates were identified using anti-FAdV-4 monoclonal antibodies through IFA [25]. RT-PCR and PCR were used to detect common coinfecting avian viruses, such as Newcastle disease virus (NDV) [26], infectious bronchitis virus (IBV) [27], avian influenza virus (AIV) [28], laryngotracheitis virus (LTV) [29], reovirus (RLV) [30], avian leukosis virus (ALV) [31], reticuloendotheliosis virus (REV) [32], *Mycoplasma* [33], and infectious bursal disease virus (IBDV) [34], to eliminate the possibility of contamination.

$TCID_{50}$ determination for the isolates

The $TCID_{50}$ was measured for each isolated strain separately. The viral stock was diluted in a 10-fold gradient (10^{-1} to 10^{-10}) using DMEM/F12 medium and then inoculated into a 96-well culture plate with a monolayer of CEL cells. Each well received 0.1 mL of the dilution, with 8 replicate wells for each dilution, and blank cell control wells were included. After 1 hour of infection, 0.1 mL of DMEM/F12 culture medium containing 5% FBS was added to each well. The cells were incubated at 37 ℃ with 5% CO_2 for 8 d. Cytopathic effects (CPEs) were observed in each cell well, and the effects in the wells with and without CPEs were recorded. The $TCID_{50}$ was calculated using the Reed-Muench method.

Determination of the pathogenicity of the isolates from SPF chickens

The pathogenicity of 10 isolates of FAdV-4 was evaluated in 4-week-old SPF chickens. The chickens were divided into 9 groups based on the inoculum dose (10^0 $TCID_{50}$, 10^1 $TCID_{50}$, 10^2 $TCID_{50}$, 10^3 $TCID_{50}$, 10^4 $TCID_{50}$, 10^5 $TCID_{50}$, 10^6 $TCID_{50}$, 10^7 $TCID_{50}$, and 10^8 $TCID_{50}$), and 10 birds in each group received intramuscular injections. Additionally, a control group of chickens received injections of an equal volume of PBS solution. Each group of chickens was individually housed in separate SPF chicken isolators. The chickens were monitored for signs of illness or mortality each day for 14 d following inoculation. The incidence and mortality rates were then calculated based on the collected data. Survival curves were plotted using the Prism 8.0 software package (GraphPad Software Inc., San Diego, USA). FAdV-4 was reisolated from the liver samples of deceased chickens in each isolate group.

Whole-genome sequencing and bioinformatics analysis

DNA extracted from the isolates using the EasyPure Viral DNA/RNA Kit (TransGen Biotech) served as the template for PCR amplification, and a total of 56 primer pairs designed according to Luan et al. [35] were used for whole-genome sequencing of FAdV-4. Each position of the entire genome was sequenced independently 2～6 times to ensure the accuracy of the sequencing results. The PCR protocol was the same as that described previously. The PCR products were subjected to 1.2% agarose gel electrophoresis and visualized with GelRed dye. The PCR products were cloned and inserted into the pMD18-T vector and sent to Sangon Biotech Company for sequencing. The complete sequences were manually assembled using the SeqMan program (version 5.01) from the Lasergene software package (DNASTAR, Madison, WI). The entire gene sequence was submitted to the GenBank database, and a sequence accession number was obtained.

The complete nucleotide sequences of the isolates were aligned with sequences from 51 reference strains of FAdV-A to FAdV-E obtained from the GenBank database via the ClustalW alignment method. Neighbor-joining analysis was performed to construct a phylogenetic tree using MEGA 11 software (version 11) for molecular evolutionary genetics analysis. The evolutionary distances were calculated using the maximum composite likelihood method, and the tree was validated with 500 bootstrap replicates. Using MegAlign software (DNASTAR, Madison, WI), we analyzed the similarities/differences in nucleotide and amino acid sequences encoding the major structural proteins hexon, penton base, fiber-1 and fiber-2 from 10 isolates with novel genotype FAdV-4 strains (AH-F19, SD1601, JS07, HLJFAd15, CH/SXCZ/2015, and JSJ13), as well as nonvariant strains (B1-7, ON1, KR5, and MX-SHP95 strains). Characteristic amino acid mutation sites were identified.

To investigate whether strain GX2019-014 is involved in recombination events between the highly pathogenic novel genotype FAdV-4 and nonvariant FAdV-4 strains, we employed the RDP, GENECONV, BootScan, MaxChi, 3Seq, SiScan, Chimaera, and LARD analyses available in the RDP4 software. Analysis was conducted using preset data processing options, and the highest acceptable P value was set to 0.05. Additionally, SimPlot and BootScan analysis methods available in SimPlot software (version 3.5.1) were employed for sequence alignment analysis.

Results

Isolation and identification of FAdV-4

From 2019 to 2020, a total of 10 strains of FAdV-4 were isolated from the liver and kidney tissues of chickens in Guangxi, China. Among these strains, 4 strains were from Nanning City (GX2019-010, GX2019-011, GX2019-012, and GX2019-016), 4 strains were from Baise City (GX2019-013, GX2019-014, GX2019-015, and GX2019-018), and 2 strains were from Yulin City (GX2019-017 and GX2020-019). After three rounds of plaque purification, all 10 isolates replicated well in CEL cells, with observable changes in cell morphology, including rounding and detachment. When CEL cells were inoculated at an MOI of 0.01, the cytopathic effect area reached over 80% between 72 h and 96 h postinoculation. The harvested cell culture supernatant had a $TCID_{50}$ ranging from $10^{8.0}$ to $10^{9.3}$ $TCID_{50}/mL$.

Using indirect immunofluorescence analysis (IFA) with FAdV-4-Fiber-2 monoclonal antibody, all

10 isolates showed positive reactions. PCR analysis was performed to exclude contamination by common coinfecting viruses, such as NDV, IBV, AIV, LTV, RLV, ALV, mycoplasma, REV, and IBDV. The isolation and purification results are shown in Figure 2-16-1.

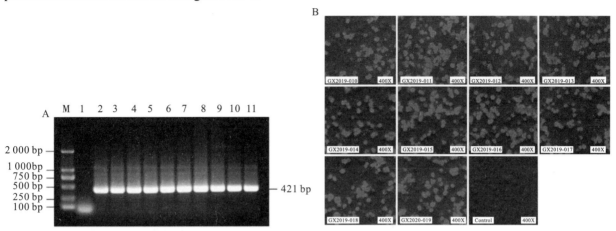

A: Detection of FAdV-4 virus by PCR. M: DL2000 DNA marker; Lane 1- negative control; lanes 2-11: GX2019-010, 011, 012, 013, 014, 015, 016, 017, 018, and GX2020-019 PCR amplification products, respectively. B: Identification of isolates using FAdV-4 fiber-2 monoclonal antibodies via IFA (color figure in appendix).

Figure 2-16-1   Isolation and identification of FAdV-4

Pathogenicity of the isolates on SPF chickens

Different doses of the isolates were intramuscularly injected into 4-week-old SPF chickens to cause infection. In terms of lethality, GX2019-010, GX2019-016, GX2019-017, and GX2019-018 exhibited the greatest virulence, with a dose of 10^2 $TCID_{50}$ resulting in 90% to 100% mortality in SPF chickens. The next highest virulence levels were observed for GX2019-011, GX2019-012, GX2019-013, and GX2019-015, in which a dose of 10^3 $TCID_{50}$ caused 70% to 100% mortality in SPF chickens. The virulence of GX2020-019 was weaker, as a mortality rate of 60% was observed when the chickens were inoculated with 10^5 $TCID_{50}$ of the virus. The virulence of GX2019-014 was the weakest, as experimental chickens inoculated with doses ranging from 10^0 $TCID_{50}$ to 10^6 $TCID_{50}$ showed no mortality, whereas doses of 10^7 $TCID_{50}$ and 10^8 $TCID_{50}$ resulted in mortality rates of 80% and 100%, respectively (Figure 2-16-2 and Table 2-16-2).

Table 2-16-2   Lethality of different doses of isolates on SPF chickens

Strain	Mortality rate（death number/total number）								
	10^8 $TCID_{50}$	10^7 $TCID_{50}$	10^6 $TCID_{50}$	10^5 $TCID_{50}$	10^4 $TCID_{50}$	10^3 $TCID_{50}$	10^2 $TCID_{50}$	10^1 $TCID_{50}$	10^0 $TCID_{50}$
GX2019-010	10/10	10/10	10/10	10/10	10/10	10/10	9/10	3/10	0/10
GX2019-011	10/10	10/10	10/10	10/10	10/10	9/10	6/10	0/10	0/10
GX2019-012	10/10	10/10	10/10	10/10	10/10	10/10	7/10	0/10	0/10
GX2019-013	10/10	10/10	10/10	10/10	10/10	8/10	6/10	0/10	0/10
GX2019-014	10/10	8/10	0/10	0/10	0/10	0/10	0/10	0/10	0/10
GX2019-015	10/10	10/10	10/10	10/10	10/10	7/10	4/10	0/10	0/10
GX2019-016	10/10	10/10	10/10	10/10	10/10	10/10	10/10	6/10	0/10
GX2019-017	10/10	10/10	10/10	10/10	10/10	10/10	10/10	5/10	0/10
GX2019-018	10/10	10/10	10/10	10/10	10/10	10/10	9/10	4/10	0/10
GX2020-019	10/10	10/10	10/10	6/10	2/10	0/10	0/10	0/10	0/10

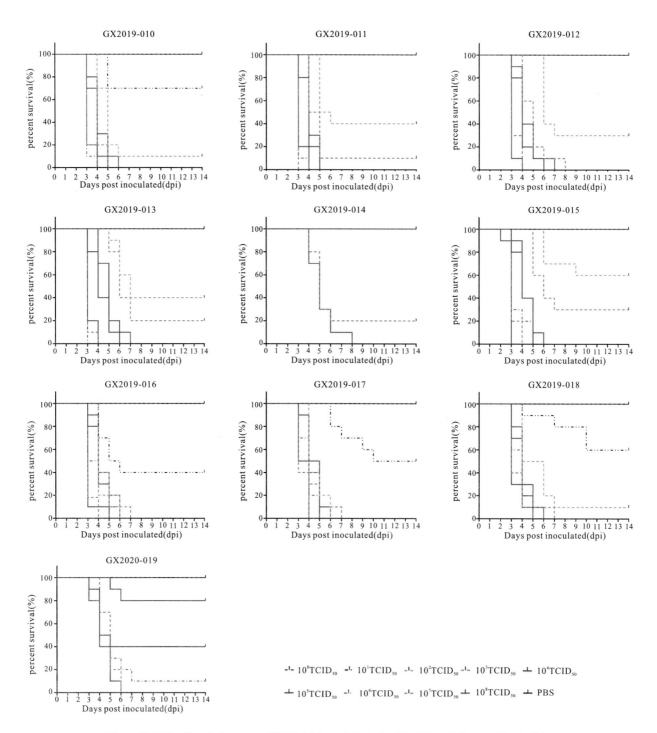

Figure 2-16-2　Survival curves of SPF chickens infected with different doses of the isolates

Autopsies were performed on the deceased chickens and those surviving 14 days postinoculation. All 10 isolates from the dead chickens exhibited typical gross lesions of HHS caused by FAdV-4. These lesions included amber-colored fluid accumulation in the pericardium, significant hepatomegaly, jaundice, and hemorrhagic spots. Up to a dose of 10^6 $TCID_{50}$, no differences were detected between chickens inoculated with the GX2019-014 strain and those in the blank control group (Figure 2-16-3). The results of viral reisolation indicated that the virus isolated from the deceased chickens matched the inoculated isolate.

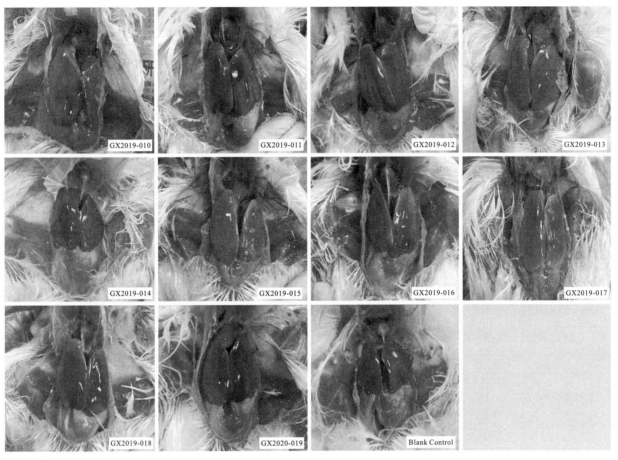

(Color figure in appendix.)

Figure 2-16-3   Gross lesions observed in 4-week-old SPF chickens infected with isolates at a dosage of 10^6 TCID$_{50}$

Phylogenetic Analysis and Molecular Characterization of the FAdV-4 Strains

The complete genome lengths of the 10 FAdV-4 strains obtained in this study ranged from 43 714 to 43 721 bp, with a G+C ratio between 54.87% and 54.98%, and all the genomes encoded 43 open reading frames (ORFs). The complete genome sequences have been uploaded to GenBank and assigned accession numbers, as detailed in Table 2-16-3. Phylogenetic analysis based on the complete genomes indicated that all 10 strains were FAdV-C strains. Their genomic homologies with FAdV-A, FAdV-B, FAdV-D, and FAdV-E ranged from 55.1% to 55.3%, 55.9% to 58.4%, 55.7% to 56.2%, and 56.9% to 57.3%, respectively, suggesting distant relationships. Notably, compared to the newly emerged highly pathogenic FAdV-4 strains in China (such as AH-F19, JS07, SD1601, HLJFAd15, CH/SXCZ/2015, and JSJ13), GX2019-014 shows a closer genetic relationship to virulent strains from other countries (such as MX-SHP95, ON1, KR5, and B1-7), as illustrated in Figure 2-16-4.

Table 2-16-3   Complete genomic information of the isolates and reference strains

Name of the isolate	Genome length	GenBank accession number	GC content	Species	Serotypes	Country
GX2019-010	43 719 bp	MW439040	54.87%	C	4	China
GX2019-011	43 721 bp	MW439041	54.87%	C	4	China

continued

Name of the isolate	Genome length	GenBank accession number	GC content	Species	Serotypes	Country
GX2019-012	43 719 bp	MW439042	54.87%	C	4	China
GX2019-013	43 721 bp	MW439046	54.87%	C	4	China
GX2019-014	43 715 bp	MW448476	54.98%	C	4	China
GX2019-015	43 721 bp	ON665754	54.87%	C	4	China
GX2019-016	43 714 bp	MW439043	54.88%	C	4	China
GX2019-017	43 721 bp	MW439044	54.87%	C	4	China
GX2019-018	43 721 bp	MW439045	54.87%	C	4	China
GX2020-019	43 714 bp	OP378126	54.88%	C	4	China
GX2017-001	43 725 bp	MN577977	54.87%	C	4	China
GX2017-002	43 725 bp	MN577978	54.85%	C	4	China
GX2017-003	43 723 bp	MN577979	54.87%	C	4	China
GX2017-004	43 723 bp	MN577980	54.87%	C	4	China
GX2019-005	43 722 bp	MN577981	54.87%	C	4	China
GX2019-006	43 718 bp	MN577982	54.87%	C	4	China
GX2018-007	43 719 bp	MN577983	54.87%	C	4	China
GX2018-008	43 726 bp	MN577984	54.87%	C	4	China
GX2018-009	43 726 bp	MN577985	54.87%	C	4	China
HLJFAd15	43 720 bp	KU991797.1	54.87%	C	4	China
CH/SXCZ/2015	43 721 bp	KU558762.1	54.87%	C	4	China
HB1510	43 721 bp	KU587519.1	54.87%	C	4	China
JSJ13	43 755 bp	KM096544.1	54.88%	C	4	China
NIVD2	43 719 bp	MG547384	54.87%	C	4	China
AH-F19	43 719 bp	MN781666	54.87%	C	4	China
CH/AHMG/2018	43 721 bp	MN606303.1	54.87%	C	4	China
D1910497	43 717 bp	MW711380.1	54.88%	C	4	China
GD616	43 259 bp	MW509553.1	54.97%	C	4	China
JS07	43 723 bp	KY436519	54.87%	C	4	China
AQ	43 723 bp	KY436520	54.87%	C	4	China
HN	43 724 bp	KY379035	54.87%	C	4	China

continued

Name of the isolate	Genome length	GenBank accession number	GC content	Species	Serotypes	Country
AH712	43 275 bp	KY436522	54.87%	C	4	China
SCDY	43 677 bp	MK629523	54.84%	C	4	China
SD1601	43 723 bp	MH006602	54.87%	C	4	China
SD1511	43 722 bp	MF496037	54.87%	C	4	China
D1910497	43 717 bp	MW711380	54.88%	C	4	USA
MX-SHP95	45 641 bp	KP295475	54.73%	C	4	Mexico
ON1	45 667 bp	GU188428.1	54.63%	C	4	Canada
KR5	45 810 bp	HE608152.1	54.63%	C	4	Austria
B1-7	45 622 bp	KU342001.1	54.64%	C	4	India
CELO	43 804 bp	U46933.1	54.30%	A	1	Austria
61/11z	43 854 bp	KX247012.1	54.30%	A	1	Poland
JM1/1	43 809 bp	MF168407.1	54.31%	A	1	Japan
11-7127	43 795 bp	MK572848.1	54.30%	A	1	Japan
OTE	43 816 bp	MK572847.1	54.25%	A	1	Japan
W-15	43 849 bp	KX247011.1	54.31%	A	1	Poland
340	45 781 bp	NC021211.1	56.52%	B	5	Ireland
WHRS	45 734 bp	OM836676.1	56.55%	B	5	China
19/7209	45 794 bp	OK283055.1	56.66%	B	5	Hungary
LYG	45 781 bp	MK757473.1	57.58%	B	5	China
AF083975	45 063 bp	AF083975.2	53.78%	D	9	Canada
SR48	43 632 bp	KT862806.1	53.33%	D	2	Austria
ON-NP2	45 193 bp	KP231537.1	54.07%	D	11	Canada
MX95-S11	44 326 bp	KU746335.1	53.69%	D	11	Mexico
380	43 347 bp	KT862812.1	53.27%	D	11	Britain
685	44 336 bp	KT862805.1	53.29%	D	2	Britain
HG	44 055 bp	GU734104.1	57.92%	E	8	Canada
UPM04217	44 072 bp	KU517714.1	57.93%	E	8b	Malaysia
764	43 666 bp	KT862811.1	57.81%	E	8b	Britain
TR59	43 287 bp	KT862810.1	57.94%	E	8a	Japan
YR36	43 525 bp	KT862809.1	57.78%	E	7	Japan

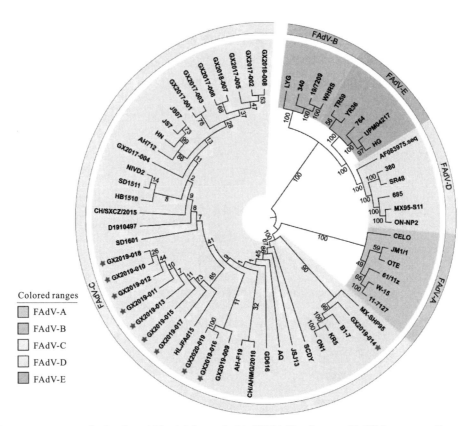

The phylogenetic tree was constructed using the neighbor-joining method in MEGA 11 software, with 500 bootstrap replicates, and evolutionary distances were computed using the maximum composite likelihood method. The strains isolated in this study are indicated by red asterisks.

Figure 2-16-4　Phylogenetic analysis of complete nucleotide sequences based on FAdVs

Compared to the nonvariant FAdV-4 strains B1-7, ON1, KR5, and MX-SHP95, the 10 isolates displayed extended GA sequences within the GA repeat region located between the PX and PVI genes, as well as longer TC repeat sequences within the TC repeat region situated between the protease and DBP genes and 15 to 99 additional base pairs in the ORF19A region. In addition, 3 base pairs in the coding region of the DNApol protein were deleted, and 1 966 base pairs in ORF19, ORF27, and ORF48 were deleted. In contrast to the nonpathogenic ON1 strain, the coding sequence of fiber-2 exhibited an insertion of 15 additional base pairs, leading to the incorporation of 5 extra amino acids (ENGKP). These changes are consistent with the new genotype FAdV-4 isolated in China after 2015. However, other molecular features of the new genotype FAdV-4 in China were observed in only 9 of the isolates (strain GX2019-014 was the exception). These changes included the insertion of 10 base pairs in the tandem repeat sequence (TR-B); the insertion of 3 base pairs each in the coding regions of ORF2, ORF22, and the 52/55K protein; the deletion of 3 base pairs, resulting in the elimination of one histidine residue within the coding region at the end of fiber-1; the deletion of 5 base pairs in the noncoding region between ORF30 and ORF17; and the insertion of 3 base pairs in the noncoding region after ORF43.

The amino acid sequences of the primary structural proteins penton base, hexon, fiber-2, and fiber-1 in the 9 isolates, excluding strain GX2019-014, are highly homologous to those of the highly pathogenic novel genotype strain (99.9%~100%). Compared to those of nonvariant strains, their homologies ranged from 94.8% to 99.7%, with 5 to 7, 12 to 13, 19 to 21, and 22 to 28 amino acid variation sites, respectively. In particular, the Penton base amino acid sequence of strain GX2019-014 has only 2 variation sites, unlike other newly emerged highly pathogenic strains, which exhibit variations at positions 45 (G to D), 356 (A to V), 370 (Q to P), 426 (I to V), and 485 (S to T). A similar pattern was observed for both the fiber-1 and fiber-2 proteins. The new highly pathogenic strains exhibited 7 variation sites in the fiber-1 amino acid sequence at positions 28 (I to S), 122 (D to N), 199 (T to V), 254 (I to L), 265 (Q to H), 266 (E to D), and 311 (R to H) and 15 variation sites in the fiber-2 amino acid sequence at positions 219 (G to D), 261 (S to T), 306 (H to N), 307 (P to A), 319 (V to I), 324 (F to V), 334 (T to A), 343 (N to L), 380 (A to T), 391 (S to T), 400 (A to G), 406 (S to I), 413 (T to S), 439 (D to E), 459 (A to N), and 418 (V to L). Interestingly, these variations are absent in strain GX2019-014 (refer to Figure 2-16-5 and Table 2-16-4). Notably, the amino acids at positions 219 and 380 of the fiber-2 protein in strain GX2019-014 differ from those in the pathogenic FAdV-4 strains but are identical to those in known nonpathogenic strains (B1-7, KR5, and ON1) (refer to the brown area in Table 2-16-4).

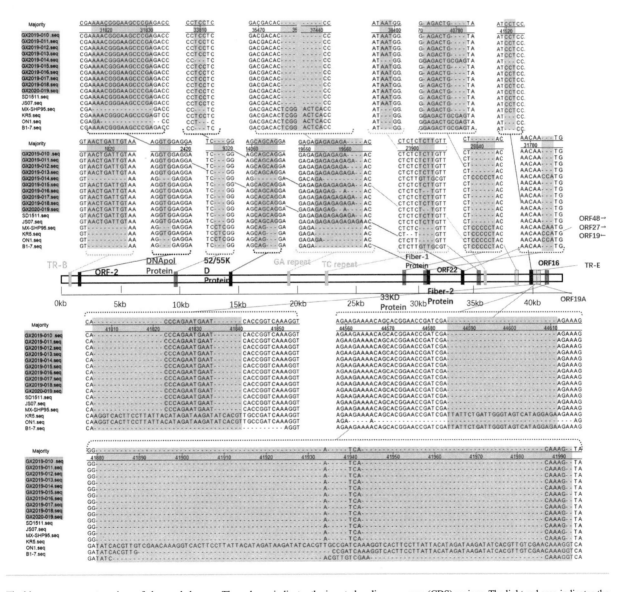

The blue area represents regions of observed changes. The red area indicates the inserted coding sequence (CDS) regions. The light red area indicates the deleted CDS regions. The green area indicates inserted non-CDS regions. The light green area indicates deleted non-CDS regions. RDP4 and SimPlot software analyses did not provide statistical evidence for a recombination event involving strain GX2019-014 between pathogenic and nonpathogenic strains.

Figure 2-16-5   Insertion/deletion nucleotide positions in the whole-genome sequence of the isolates

Table 2-16-4 Amino acid variation sites of hexon, penton base, fiber-1, and fiber-2

Genes	Strains	Mutant sites						
		42	45	193	356	370	426	486
Penton	GX 2019-010	P	D	I	V	P	V	T
	GX 2019-011	•	•	•	•	•	•	•
	GX 2019-012	•	•	•	•	•	•	•
	GX 2019-013	•	•	•	•	•	•	•
	GX 2019-014	•	**G**	•	**A**	**Q**	**I**	**S**
	GX 2019-015	•	•	•	•	•	•	•
	GX 2019-016	•	•	•	•	•	•	•
	GX 2019-017	•	•	•	•	•	•	•
	GX 2019-018	•	•	•	•	•	•	•
	GX 2019-019	•	•	•	•	•	•	•
	SD1511	•	•	•	•	•	•	•
	JSO7	•	•	•	•	•	•	•
	MX-SHP95	•	G	V	A	Q	I	S
	ON1	S	G	V	A	Q	I	S
	KR5	S	G	V	A	Q	I	S
	B1-7	•	G	•	A	Q	I	S

Genes	Strains	Mutant sites														
		164	188	193	195	238	240	243	263	264	402	410	574	795	797	842
Hexon	GX 2019-010	S	R	R	Q	D	T	N	I	V	A	A	I	Q	P	A
	GX 2019-011	•	•	•	•	•	•	•	•	•	•	•	•	•	•	•
	GX 2019-012	•	•	•	•	•	•	•	•	•	•	•	•	•	•	•
	GX 2019-013	•	•	•	•	•	•	•	•	•	•	•	•	•	•	•
	GX 2019-014	•	•	•	•	•	•	•	•	•	•	•	•	•	•	•
	GX 2019-015	•	•	•	•	•	•	•	•	•	•	•	•	•	•	•
	GX 2019-016	•	•	•	•	•	•	•	•	•	•	•	•	•	•	•
	GX 2019-017	•	•	•	•	•	•	•	•	•	•	•	•	•	•	•
	GX 2019-018	•	•	•	•	•	•	•	•	•	•	•	•	•	•	•
	GX 2019-019	•	•	•	•	•	•	•	•	•	•	•	•	•	•	•
	SD1511	•	•	•	•	•	•	•	•	•	•	•	•	•	•	•
	JSO7	•	•	•	•	•	•	•	•	•	•	•	•	•	•	•
	MX-SHP95	T	•	Q	E	N	A	E	M	I	Q	T	V	•	A	G
	ON1	T	I	Q	E	N	A	E	M	I	•	T	V	•	A	G
	KR5	T	I	Q	E	N	A	E	M	I	•	T	•	E	•	G
	B1-7	S	I	Q	E	N	A	E	M	I	•	T	V	•	A	G

Genes	Strains	Mutant sites														
		14	28	44	69	70	80	81	82	98	122	129	138	156	189	199
Fiber-1	GX 2019-010	A	S	R	G	S	--	--	--	D	N	A	A	R	D	V
	GX 2019-011	•	•	•	•	•	--	--	--	•	•	•	•	•	•	•
	GX 2019-012	•	•	•	•	•	--	--	--	•	•	•	•	•	•	•
	GX 2019-013	•	•	•	•	•	--	--	--	•	•	•	•	•	•	•
	GX 2019-014	•	**I**	•	•	•	G	G	G	•	**D**	•	•	•	•	T
	GX 2019-015	•	•	•	•	•	--	--	--	•	•	•	•	•	•	•
	GX 2019-016	•	•	•	•	•	-	--	--	•	•	•	•	•	•	•
	GX 2019-017	•	•	•	•	•	--	--	--	•	•	•	•	•	•	•
	GX 2019-018	•	•	•	•	•	--	--	--	•	•	•	•	•	•	•

continued

Genes	Strains	Mutant sites														
Fiber-1	GX 2019-019	•	•	•	•	•	--	--	--	•		•	•	•	•	•	•
	SD1511	•	•	•	•	•	•	--	--	•		•	•	•	•	•	•
	JSO7	•	•	•	•	•	•	--	--	•		•	•	•	•	•	•
	MX-SHP95	V	I	P	S	G	•	--	--	•		D	V	•	H	N	•
	ON1	V	I	P	S	G	•	--	--	•		D	V	•	H	N	T
	KR5	•	I	P	S	G	G	--	--	N		•	•	S	H	•	•
	B1-7	•	I	•	S	G	G	G	--	N		•	•	S	H	•	•

Genes	Strains	207	254	265	266	313	332	334	377	386		404	431	434
Fiber-1	GX 2019-010	G	L	H	D	H	H	R	S	I		N	--	G
	GX 2019-011	•	•	•	•	•	•	•	•	•		•	•	•
	GX 2019-012	•	•	•	•	•	•	•	•	•		•	•	•
	GX 2019-013	•	•	•	•	•	•	•	•	•		•	•	•
	GX 2019-014	A	**I**	**Q**	**E**	**R**	•	•	•	•		•	H	A
	GX 2019-015	•	•	•	•	•	•	•	•	•		•	•	•
	GX 2019-016	•	•	•	•	•	•	•	•	•		•	•	•
	GX 2019-017	•	•	•	•	•	•	•	•	•		•	•	•
	GX 2019-018	•	•	•	•	•	•	•	•	•		•	•	•
	GX 2019-019	•	•	•	•	•	•	•	•	•		•	•	•
	SD1511	•	•	•	•	•	•	•	•	•		•	•	•
	JSO7	•	•	•	•	•	•	•	•	•		•	•	•
	MX-SHP95	A	I	Q	E	R	Q	K	P	L		Y	N	•
	ON1	•	I	Q	E	R	Q	K	P	L		Y	H	S
	KR5	•	I	Q	E	R	Q	K	P	L		Y	H	S
	B1-7	•	I	•	•	R	•	•	•	L		•	--	•

Genes	Strains	11-15	17	22	29	114	144	219	232	261	300	305	306	307	319	324	329	334	338
Fiber-2	GX 2019-010	ENGKP	T	S	A	D	S	D	Q	T	T	A	N	A	I	V	L	A	N
	GX 2019-011	•	•	•	•	•	•	•	•	•	•	•	•	•	•	•	•	•	•
	GX 2019-012	•	•	•	•	•	•	•	•	•	•	•	•	•	•	•	•	•	•
	GX 2019-013	•	•	•	•	•	•	•	•	•	•	•	•	•	•	•	•	•	•
	GX 2019-014	•	•	•	•	A	A	**G**	•	**N**	•	•	**H**	**P**	**V**	**F**	•	**T**	•
	GX 2019-015	•	•	•	•	•	•	•	•	•	•	•	•	•	•	•	•	•	•
	GX 2019-016	•	•	•	•	•	•	•	•	•	•	•	•	•	•	•	•	•	•
	GX 2019-017	•	•	•	•	•	•	•	•	•	•	•	•	•	•	•	•	•	•
	GX 2019-018	•	•	•	•	•	•	•	•	•	•	•	•	•	•	•	•	•	•
	GX 2019-019	•	•	•	•	•	•	•	•	•	•	•	•	•	•	•	•	•	•
	SD1511	•	•	•	•	•	•	•	•	•	•	•	•	•	•	•	•	•	•
	JSO7	•	•	•	•	•	•	•	•	•	•	•	•	•	•	•	•	•	•
	MX-SHP95	-----	•	Y	P	A	A	•	•	N	•	•	H	•	•	F	•	T	T
	ON1	-----	•	•	P	•	•	G	E	S	I	S	H	P	V	F	V	T	T
	KR5	•	S	•	•	•	•	G	E	S	I	S	H	P	V	F	V	T	T
	B1-7	•	•	•	•	•	•	G	Q	S	•	•	•	•	•	•	•	•	•

Genes	Strains	343	344	346	378	380	391	393	400	403	405	406	413	427	435	439	453	459	478
	GX 201 9-010	L	N	A	T	T	T	P	G	E	S	I	S	I	S	E	A	N	L

continued

Genes	Strains	Mutant sites																	
Fiber-2	GX 201 9-011	•	•	•	•	•	•	•	•	•	•	•	•	•	•	•	•	•	•
	GX 201 9-012	•	•	•	•	•	•	•	•	•	•	•	•	•	•	•	•	•	•
	GX 201 9-013	•	•	•	•	•	•	•	•	•	•	•	•	•	•	•	•	•	•
	GX 201 9-014	W	•	V	•	A	S	•	A	•	•	S	T	V	•	D	•	A	V
	GX 201 9-015	•	•	•	•	•	•	•	•	•	•	•	•	•	•	•	•	•	•
	GX 201 9-016	•	•	•	•	•	•	•	•	•	•	•	•	•	•	•	•	•	•
	GX 201 9-017	•	•	•	•	•	•	•	•	•	•	•	•	•	•	•	•	•	•
	GX 201 9-018	•	•	•	•	•	•	•	•	•	•	•	•	•	•	•	•	•	•
	GX 201 9-019	•	•	•	•	•	•	•	•	•	•	•	•	•	•	•	•	•	•
	SD1511	•	•	•	•	•	•	•	•	•	•	•	•	•	•	•	•	•	•
	JSO7	•	•	•	•	•	•	•	•	•	•	•	•	•	•	•	•	•	•
	MX-SHP95	N	S	V	•	•	S	S	A	Q	•	S	T	V	•	D	•	A	V
	ON1	N	S	•	A	A	S	•	A	Q	P	S	T	•	T	D	S	A	V
	KR5	N	S	•	A	A	S	•	A	Q	•	S	T	•	T	D	S	A	V
	B1-7	•	•	•	A	A	S	•	•	•	•	•	•	V	•	•	S	A	V

Note: • indicates amino acid identity with the GX2019-010 strain at the same locus; -- indicates amino acid deletion at the locus in each strain; bold letters denote amino acid identity with nonvariant strains at the locus in the GX2019-014 strain; grey shading represents potential virulence determinants of FAdV-4 predicted in this study.

Discussion

Since 2015, large-scale outbreaks of HHS have occurred in multiple poultry farming provinces in China. This syndrome is characterized by a significant accumulation of fluid with an amber hue in the pericardium, along with swelling, jaundice, and hemorrhagic spots appearing in the liver. The mortality rate among affected chicken flocks can reach as high as 20% to 80%. The immune organs of surviving chickens are often damaged, leading to immunosuppression, which reduces the resistance of the flock to other pathogenic microorganisms. The significant economic losses have sparked researchers'attention to the pathogenesis and prevention methods of this disease, making HHS and its pathogen FAdV-4 research areas of intense interest in recent years. Researchers have successively isolated and purified the virus from various tissues, such as the liver and kidneys of deceased chickens, identifying the causative agent of this outbreak as a novel genotype of FAdV-4 [15,19,21,36]. The pathogenic identity of these strains was also confirmed through animal regression experiments.

Previous FAdVs typically showed lower virulence and often presented as subclinical infections within chicken populations. The mortality rate among chickens carrying the virus is not typically high, and few laying hens show symptoms [37]. However, the mortality rate among diseased chickens increases during coinfections, particularly coinfections with immunosuppressive viruses such as IBDV and ALV [20]. Investigations into the current HHS outbreak in China revealed that the FAdV-4 strain within the FAdV-C species is significantly predominant rather than the predominant FAdV-D and FAdV-E strains that have been prevalent globally over the past decade; in addition, this FAdV-4 strain exhibits distinctly high pathogenicity [17,38,39]. The novel genotype FAdV-4 strain SD1511, which was inoculated intramuscularly at a dose of $3 \times 10^{3.2}$ TCID$_{50}$ into SPF chickens at 21, 7, and 35 days of age, resulted in mortality rates of 80%, 93%, and 100%, respectively [23]. When administered via intramuscular injection at a dose of $10^{2.5}$ TCID$_{50}$ to SPF chickens at 3 weeks of age, the GD616 strain resulted in a mortality rate of 100% [40]. When administered via intramuscular injection at a dose of 10^6 TCID$_{50}$, the JS07 and HN strains isolated from eastern China exhibited mortality rates of 80% and 100%, respectively [21].

This study was conducted during the HHS outbreak from 2019 to 2020 in the southwestern Chinese province of Guangxi, which includes the major poultry farming cities of Nanning, Yulin, and Baise. Tissue samples from chicks in commercial chicken farms and backyard chicken households were used to detect the pathogenic agents of common avian diseases. Samples that tested positive for FAdV-4 by PCR were selected for viral isolation and purification, resulting in the acquisition of 10 isolates. The tissue samples from which these 10 isolates originated showed no presence of other viruses or mixed infections with other serotypes of fowl adenovirus. Research has shown that the newly emerged highly pathogenic FAdV-4 is an age-dependent virus [23,36], with chickens younger than 60 days being more susceptible and resistance increasing as age progresses. For the study of Chinese newly emerging highly pathogenic FAdV-4 strains, animal models typically utilize SPF chickens aged 3~4 weeks [20,35,41]. We opted for the consistent use of 4-week-old SPF chickens, with the inoculation dosage as the sole variable for virulence testing, aiming to reflect differences in pathogenicity among various strains. This approach facilitates comparisons with other FAdV-4 strains in animal models of similar ages. The results revealed that the mortality rates of strains GX2019-010 to 013 and 015 to 018, which were administered via intramuscular injection at a dose of 10^3 TCID$_{50}$, were 70% and 100%, respectively; these rates were similar to those of other isolates in China. Strain GX2020-019 demonstrated moderate virulence, with mortality rates of 0%, 20% and 60% at 10^3 TCID$_{50}$, 10^4 TCID$_{50}$ and 10^5 TCID$_{50}$, respectively. Of particular significance, the GX2019-014 strain is characterized as a wild-type low-virulence strain. Up to a dose of 10^6 TCID$_{50}$, experimental chickens inoculated with this strain showed no clinical symptoms, and postmortem examination revealed no pericardial effusion, liver swelling or jaundice, consistent with the control group. However, at doses of 10^7 TCID$_{50}$ and 10^8 TCID$_{50}$, the mortality rates were 80% and 100%, respectively.

The novel genotype of FAdV-4 exhibits common molecular characteristics, such as a deletion of 1966 base pairs at ORF19, ORF27, and ORF48; longer GA repeats; and a deletion of three base pairs in the coding region of the DNApol protein [19,22,23,42,43]. These findings have prompted researchers to focus on identifying the key genes responsible for enhancing the virulence of FAdV-4. Using reverse genetic systems, researchers recently reported that deleting the 1 966 bp nucleotide fragment does not affect virulence [42]. Research by Liu et al. [44] ruled out the influence of the hexon protein and penton on the increase in virulence. Through the ExoCET method, Zhang et al. discovered that the high virulence of the strains is associated with the fiber-2 and hexon proteins [45]. Quan et al. [46] used the CRISPR/Cas9 technique to edit the fiber-2 gene, resulting in a decrease in virus virulence. Therefore, it is widely believed that the virulence gene of FAdV-4 resides within the fiber-2 protein.

In this study, with the exception of GX2019-014, the complete genomic sequences of the isolates exhibited high homology with the novel genotype FAdV-4 strains in China. However, GX2019-014 clustered with the nonpathogenic strains KR5, B1-7, and ON1 on the same branch. Analysis of the main structural proteins penton base, hexon, fiber-1, and fiber-2 of the 10 isolates, along with pathogenic and nonpathogenic strains of FAdV-4, revealed that the GX2019-014 strain showed greater homology to nonpathogenic strains. The Chinese isolates of the novel genotype FAdV-4 exhibit molecular characteristics such as a 10-base insertion in TR-B; three-base insertions in the coding regions of ORF2, ORF22, and the 52/55K protein; and a three-base deletion at the end of the coding region of fiber-1, leading to the loss of an amino acid residue.

However, these molecular features were not observed in the GX2019-014 strain. The results of RDP4 and SimPlot software analyses exclude the possibility that strain GX2019-014 is a recombination between nonpathogenic and pathogenic strains. Considering the results of pathogenicity testing, the GX2019-014 strain is more likely a wild-type low-virulence strain. Before the highly pathogenic novel genotype FAdV-4 emerged, studies suggested that residues D219, T300, and T380 in fiber-2 were associated with virulence[46,47]. After more high-pathogenicity novel genotype FAdV-4 whole-genome sequences were deposited in GenBank, Yin et al. conducted comparisons using known pathogenic and nonpathogenic strain proteins, ruling out the impact of T300 on virulence [43]. Zhang et al. [48] previously proposed that amino acid residue 188 of the hexon protein is a key virulence determinant and that the R188I mutation leads to reduced virulence in highly pathogenic strains. However, subsequent validation studies by Wang et al. did not fully support this view[49]. The GX2019-014 strain, a novel low-pathogenicity genotype, provides new data for analyzing key virulence genes through its whole-genome sequence. By comparing the amino acid residues of major structural proteins among known pathogenic strains, nonpathogenic strains, and the GX2019-014 strain, it was found that there are no amino acid residues exclusively present in the highly pathogenic FAdV-4 strains on the hexon, penton, or fiber-1 proteins. The presence of the 188R residue in the hexon protein is consistent with that in other highly pathogenic strains, suggesting that this residue is not a key virulence determinant. However, on the fiber-2 protein, residues D219 and T380 are unique to highly pathogenic FAdV-4 strains, with T380 located within the Knob domain of the fiber-2 protein. This residue may serve as a potential key virulence determinant, consistent with previous research results. Nevertheless, further validation through the establishment of a reverse genetic system is needed.

In summary, we isolated and purified 10 strains of the novel genotype FAdV-4 from Guangxi in southwestern China between 2019 and 2020. We conducted pathogenicity assessments and whole-genome sequencing on these strains. Among these strains, the GX2019-014 strain showed nonpathogenicity in muscle in 4-week-old SPF chickens inoculated with 10^6 TCID$_{50}$ virus, and its whole-genome homology was closer to that of nonpathogenic FAdV-4 strains. This strain is the first reported wild-type low-virulence strain of the novel genotype FAdV-4. Our work has enriched the research data on prevalent strains of FAdV-4 in Guangxi, an important poultry farming region in China. This finding contradicts the previous notion that the Chinese strains isolated from the novel genotype FAdV-4 are highly pathogenic strains, providing reference material for further investigation into the key virulence-associated genetic loci of the novel genotype FAdV-4.

References

[1] BENKŐ, M, AOKI K, ARNBERG N, et al. ICTV virus taxonomy profile: Adenoviridae 2022. J Gen Virol, 2022, 103:001721.

[2] CHEN S, LIN F, JIANG B, et al. Isolation and characterization of a novel strain of duck aviadenovirus B from Muscovy ducklings with acute hepatitis in China. Transboundary and Emerging Diseases, 2022, 69(5): 2769-2778.

[3] MAREK A, KAJÁN, G. L, KOSIOL C, et al. Genetic diversity of species fowl aviadenovirus D and fowl aviadenovirus E. Journal of General Virology, 2016.97(9): 2323-2332.

[4]MCFERRAN, J. B, SMYTH, J. A. Avian adenoviruses. Revue Scientifique et Technique (International Office of Epizootics), 2000, 19(2): 589-601.

[5] RUX J J, KUSER P R, BURNETT R M. Structural and phylogenetic analysis of adenovirus hexons by use of high-resolution x-ray crystallographic, molecular modeling, and sequence-based methods. J Virol, 2003, 77:9553-9566.

[6] FENDER P, BOUSSAID A, MEZIN P, et al. Synthesis, cellular localization, and quantification of penton-dodecahedron in

serotype 3 adenovirus-infected cells. Virology, 2005, 340: 167-173.

[7] WANG W, LIU Q, LI T, et al. Fiber-1, not fiber-2, directly mediates the infection of the pathogenic serotype 4 fowl adenovirus via its shaft and knob domains. J Virol, 2020, 94: e00954-20.

[8] LU H, GUO Y, XU Z, et al. Fiber-1 of serotype 4 fowl adenovirus mediates superinfection resistance against serotype 8b fowl adenovirus. Front Microbiol, 2020, 13:1086383.

[9] SCHACHNER A, MAREK A, JASKULSKA B, et al. Recombinant FAdV-4 fiber-2 protein protects chickens against hepatitis-hydropericardium syndrome (HHS). Vaccine, 2014, 32: 1086-1092.

[10] WANG X, TANG Q, CHU Z, et al. Immune protection efficacy of FAdV-4 surface proteins fiber-1, fiber-2, hexon and penton base. Virus Res, 2018, 245: 1-6.

[11] QUAN X, SHIYA C, WEI Z, et al. A novel fiber-2-edited live attenuated vaccine candidate against the highly pathogenic serotype 4 fowl adenovirus. Vet Res, 2021, 52: 35.

[12] ABE T., NAKAMURA K., TOJO H, et al. Histology, immunohistochemistry, and ultrastructure of hydropericardium syndrome in adult broiler breeders and broiler chicks. Avian Dis, 1998, 42: 606-612.

[13] DAHIYA S, SRIVASTAVA R N, HESS M, et al. Fowl adenovirus serotype 4 associated with outbreaks of infectious hydropericardium in Haryana, India. Avian Dis, 2002, 46: 230-233.

[14] CHOI K S, KYE S J, KIM J Y, et al. Epidemiological investigation of outbreaks of fowl adenovirus infection in commercial chickens in Korea. Poult. Sci.2012. 91:2502-2506.

[15] MIYAOKA Y, YU M, AMIRUL H M, et al. Isolation and molecular characterization of fowl adenovirus and avian reovirus from breeder chickens in Japan in 2019-2021. J Vet Med Sci, 2021, 84: 238-243.

[16]ISHAG H Z A, TERAB A M A, EL TIGANI-ASIL E T A, et al. Pathology and molecular epidemiology of fowl adenovirus serotype 4 outbreaks in broiler chicken in Abu Dhabi Emirate, UAE. Vet Sci, 2022, 9: 154.

[17] NIU Y, SUN Q, ZHANG G, et al. Epidemiological investigation of outbreaks of fowl adenovirus infections in commercial chickens in China. Transbound Emerg Dis, 2018. 65(1): e121-e126.

[18] CHEN L, YIN L, ZHOU Q, et al. Epidemiological investigation of fowl adenovirus infections in poultry in China during 2015-2018. BMC Vet Res, 2019, 15: 271.

[19] RASHID F, XIE Z, ZHANG L, et al. Genetic characterization of fowl aviadenovirus 4 isolates from Guangxi, China, during 2017-2019. Poult Sci, 2022, 99: 4166-4173.

[20] ZHAO J, ZHONG Q, ZHAO Y, et al. Pathogenicity and complete genome characterization of fowl adenoviruses isolated from chickens associated with inclusion body hepatitis and hydropericardium syndrome in China. PLOS ONE, 2015, 10: e0133073.

[21] KAI W, HAIWEI S, YUNZHANG L, et al. Characterization and pathogenicity of fowl adenovirus serotype 4 isolated from eastern China. BMC Vet Res, 2019, 15: 373.

[22] WEI Y, XIE Z, FAN Q, et al. Pathogenicity and molecular characteristics of fowl adenovirus serotype 4 with moderate virulence in Guangxi, China. Front Vet Sci, 2023, 10: 1190126.

[23] MO K K, LYU C F, CAO S S, et al. Pathogenicity of an FAdV-4 isolate to chickens and its genomic analysis. J Zhejiang Univ Sci B, 2019, 20(9): 740-752.

[24] ZHANG L, ZHANG Y, LI Q, et al. Diagnosis and Preventive treatment of Fowl adenovirus-4 infection. Anhui Agri Sci Bull, 2016, 22(10): 116, 122.

[25] WEI Y, XIE Z X, DENG X W, et al. Expression of the truncated fiber-2 protein of fowl adenovirus serotype 4 and preparation of its monoclonal antibodies. J South Agric 2022, 53: 2341-2349.

[26] FENG B, XIE Z X, DENG X W, et al. Development of a triplex PCR assay for detection of Newcastle disease virus, chicken parvovirus and avian influenza virus. Journal of Southern Agriculture, 2019, 50(11): 2576-2582.

[27] WU X Y, ZHUANG X, HE H H, et al. Nucleic acid detection, virus isolation and sequence analysis of chicken infectious bronchitis virus in Jiangsu. Animal Husbandry & Veterinary Medicine, 2021, 53(10): 96-104.

[28] IP D K M, LAU L L H, CHAN K H, et al. The Dynamic Relationship Between Clinical Symptomatology and Viral Shedding

in Naturally Acquired Seasonal and Pandemic Influenza Virus Infections. Clin Infect Dis, 2016, 62(4):431-437.

[29] XU Q R, ZHOU Z T, BI D R, et al. Isolation, PCR Detection and TK Gene Sequence Analysis of Infectious Laryngotracheitis Virus in Laying Hens. Heilongjiang Animal Science and Veterinary Medicine, 2017, 47(24): 122-123.

[30] YE W C, YU B, LIU Y S, et al. Establishment of universal RT-PCR for detection of Muscovy duck reovirus. Acta Agriculturae Zhejiangensis, 2020, 26(6): 1453-1456.

[31] BI Y L, ZHUANG J Q, WANG J L, et al. Double PCR detection of poultry birds of reticular endothelial tissue hyperplasia in biological products virus and avian leukosis virus. Animal Husbandry & Veterinary Medicine, 2014, 46(04): 74-78.

[32] ALY M M, SMITH E J, FADLY A M. Detection of reticuloendotheliosis virus infection using the polymerase chain reaction. Avian Pathol, 1993, 22(3):543-554.

[33] XIE Z X, DENG X W, TANG X, et al. The studies and application of PCR kit for Mycoplasma gallisepitcum detection. Animal Husbandry & Veterinary Medicine 2004, 40(1): 3-5.

[34] SU W, ZHANG X D, WANG T, et al. Detection of Chicken Infection Bursal Disease Virus by Polymerase Chain Reaction. Chinese Journal of Veterinary Science, 1997, 17(1): 15-17.

[35] LUAN Y J, XIE Z X,WANG S, et al. Whole-genome sequencing and analysis of fowl adenovirus serotpye 4 strain GX2019-2010 isolate from Guangxi. China Anim Husb Veterinary Med, 2020. 47(12): 3793-3804.

[36] FENG Y, HUIQI S, LEI H, et al. Age-dependence of hypervirulent fowl adenovirus type 4 pathogenicity in specific-pathogen-free chickens. Poultry Science, 2021, 100(8):101238.

[37] ASTHANA M, CHANDRA R, KUMAR R. Hydropericardium syndrome: current state and future developments. Arch Virol, 2013, 158: 921-931.

[38] QINGYE Z, SUCHUN W, FUYOU Z, et al. Molecular epidemiology analysis of fowl adenovirus detected from apparently healthy birds in eastern China. BMC Vet Res, 2023, 19:5.

[39] JING X, KE-CHANG Y, YUE-YUE L, et al. Isolation and molecular characterization of prevalent fowl adenovirus strains in southwestern China during 2015-2016 for the development of a control strategy. Emerg Microbes Infect, 2017, 6: e103.

[40] YUAN F, HOU L, WEI L, et al. Fowl adenovirus serotype 4 induces hepatic steatosis via activation of liver X receptor-α. J Virol, 2020, 95: e01938-20.

[41] CUI J, XU Y, ZHOU Z, et al. Pathogenicity and Molecular Typing of Fowl Adenovirus-Associated With Hepatitis/Hydropericardium Syndrome in Central China (2015-2018). Front Vet Sci 2020.7, 190.

[42] PAN Q, WANG J, GAO Y, et al. The natural large genomic deletion is unrelated to the increased virulence of the novel genotype fowl adenovirus 4 recently emerged in China. Viruses, 2018, 10: 494.

[43] YIN D, XUE M, YANG K, et al. Molecular characterization and pathogenicity of highly pathogenic fowl adenovirus serotype 4 isolated from laying flock with hydropericardium-hepatitis syndrome. Microb Pathog, 2020, 147: 104381.

[44] LIU R, ZHANG Y, GUO H, et al. The increased virulence of hypervirulent fowl adenovirus 4 is independent of fiber-1 and penton. Res Vet Sci, 2020, 131: 31-37.

[45] ZHANG Y, LIU R, TIAN K, et al. Fiber2 and hexon genes are closely associated with the virulence of the emerging and highly pathogenic fowl adenovirus 4. Emerg Microbes Infect, 2018, 7: 199.

[46] MAREK A, NOLTE V, SCHACHNER A, et al. Two fiber genes of nearly equal lengths are a common and distinctive feature of fowl adenovirus C members. Vet Microbiol, 2012, 156: 411-417.

[47] VERA-HERNÁNDEZ P F, MORALES-GARZÓN A, CORTÉS-ESPINOSA D V, et al. Clinicopathological characterization and genomic sequence differences observed in a highly virulent fowl aviadenovirus serotype 4. Avian Pathol, 2016, 45: 73-81.

[48] ZHANG Y, LIU A, WANG Y, et al.. A single amino acid at residue 188 of the hexon protein is responsible for the pathogenicity of the emerging novel virus fowl adenovirus 4. J Virol, 2021, 95: e0060321.

[49] WANG B, SONG C, YANG P, et al. The role of hexon amino acid 188 varies in fowl adenovirus serotype 4 strains with different virulence. Microbiol Spectr, 2022, 10: e0149322.

Chapter Three
Studies on Molecular Etiology

Complete genome sequence analysis of an H6N1 avian influenza virus isolated from Guangxi Pockmark ducks

Xie Zhixun, Xie Liji, Zhou Chenyu, Liu Jiabo, Pang Yaoshan, Deng Xianwen, Xie Zhiqin, and Fan Qing

Abstract

We report here the complete genomic sequence of a novel H6N1 avian influenzavirus strain, A/Duck/Guangxi/ GXd-5/2010 (H6N1), isolated from Pockmark ducks in Guangxi, southern China. All of the 8 gene segments of A/ Duck/Guangxi/GXd-5/2010 (H6N1) are attributed to the Eurasian lineage; the amino acid motif of the cleavage site between HA1 and HA2 was P-Q-I-E-T-R-G. These are typical characteristics of the low-pathogenicity avian influenza virus. This study will help to understand the epidemiology and molecular characteristics of avian influenzavirus in ducks.

Keywords

avian influenza virus, H6N1 subtype, complete genome sequence

---

The avian influenza virus genome consists of eight separate RNA segments, HA, NA, NP, M, PB1, PB2, PA, and NS, which encode 12 viral structural and nonstructural proteins[6, 7]. Avian influenza viruses are divided into subtypes based on the two major surface glycoproteins, hemagglutinin (HA) and neuraminidase (NA). So far, 16 HA and 9 NA subtypes have been identified[5, 8, 9]. In recent years, there have been several subtypes of avian influenza virus (H5, H6, and H9 subtypes) circulating and evolving in southern China[1, 4, 10], which might provide an opportunity for these virus subtypes to recombine.

In 1997, highly pathogenic avian influenza virus A/Hong Kong/156/97 (H5N1) derived its HA gene from A/Goose/Guangdong/1/96 (H5N1)-like virus and the other seven genes from A/teal/Hong Kong/W312/97 (H6N1)-like virus. It caused disease outbreaks in chickens in Hong Kong[2], consequently resulting in 6 deaths among 18 infected people[3].

In this study, an H6N1 strain, named A/Duck/Guangxi/GXd-5/2010(H6N1), was first isolated from a Pockmark duck in Guangxi, southern China, in 2010. Nucleotide sequences of A/Duck/Guangxi/GXd-5/2010 (H6N1) were amplified through reverse transcription-PCR (RT-PCR). The amplified products were purified and cloned into the pMD18-T vector (TaKaRa, Dalian, China) and then sequenced (TaKaRa, Dalian, China). Sequences were assembled and manually edited to generate the final genome sequence.

The complete genome of the strain consists of eight segments of negative-sense single-stranded RNA molecules, including PB2, PB1, PA, HA, NP, NA, M, and NS. The full lengths of each segment are 2 341, 2 341, 2 233, 1 744, 1 565, 1 463, 1 027, and 838 nucleotides, respectively. The amino acid sequence at the cleavage site in the HA molecule is P-Q-I-E-T-R-G, which is characteristic of low-pathogenicity avian influenza virus.

Sequence analysis showed that five segments (PB2, PB1, PA, NP, and NA) had close relationships with those of the H5 subtype and the others (HA, M, and NS) had the highest homologies with those of the H6 subtype. HA shared the highest sequence homology (97.7%) with A/duck/Shantou/1080/2007 (H6N2) (GenBank accession number: CY109746). NA shared the highest sequence homology (99.1%) with A/Hubei/1/2010 (H5N1) (GenBank accession number: CY098760). PB2 shared the highest sequence homology (99.6%) with A/Hubei/1/2010 (H5N1) (GenBank accession number: CY098755). PA shared the highest sequence homology (96.5%) with A/Guangxi/1/2009 (H5N1) (GenBank accession number: CY098743). M shared the highest sequence homology (98.2%) with A/chicken/eastern China/49/2010 (H6N6) (GenBank accession number: JF965223). NS shared the highest sequence homology (98.5%) with A/duck/Hubei/2/2010 (H6N6) (GenBank accession number: CY110953). The homologies of the PB1 with the sequences of A/Hubei/1/2010 (H5N1), A/Hong Kong/156/97 (H5N1), A/Chicken/Hong Kong/G9/97 (H9N2), and A/Teal/Hong Kong/W312/97 (H6N1) (GenBank accession numbers: CY098756, AF036362, AF156416, and AF250477, respectively), were 95.5%, 90.6%, 90.5%, and 90.7%, respectively. The homologies of the NP with the sequences of A/Hubei/1/2010 (H5N1), A/Hong Kong/156/97 (H5N1), A/Quail/Hong Kong/G1/97 (H9N2), and A/Teal/ Hong Kong/W312/97 (H6N1) (GenBank accession numbers: CY098759, AF036359, AF156407, and AF250480, respectively), were 99.5%, 91.5%, 91.2%, and 91.3%, respectively. These results are useful for analyses of epidemiology and evolutionary characteristics of avian influenza virus.

Nucleotide sequence accession numbers. The GenBank accession numbers of PB2, PB1, PA, HA, NP, NA, M, and NS for A/Duck/Guangxi/GXd-5/2010 (H6N1) are JX304761, JX304760, JX304759, JX304754, JX304757, JX304756, JX304755, and JX304758, respectively.

References

[1] CHEUNG C L, VIJAYKRISHNA D, SMITH G J, et al. Establishment of influenza A virus (H6N1) in minor poultry species in southern China. Journal of Virology, 2007, 81(19): 10402-10412.

[2] CHIN P S, HOFFMANN E, WEBBY R, et al. Molecular evolution of H6 influenza viruses from poultry in southeastern China: prevalence of H6N1 influenza viruses possessing seven A/Hong Kong/156/97 (H5N1)-like genes in poultry. Journal of Virology, 2002, 76(2): 507-516.

[3] CLAAS E C, OSTERHAUS A D, BEEK R V, et al. Human influenza A H5N1 virus related to a highly pathogenic avian influenza virus. Lancet, 1998, 351(9101): 472-477.

[4] DUAN L, BAHL J, SMITH G J D, et al. The development and genetic diversity of H5N1 influenza virus in China, 1996-2006. Virology, 2008, 380(2): 243-254.

[5] FOUCHIER R A M, MUNSTER V, WALLENSTEN A, et al. Characterization of a novel influenza A virus hemagglutinin subtype (H16) obtained from black-headed gulls. Journal of Virology, 2005, 79(5): 2814-2822.

[6] HE C Q, XIE Z X, HAN G Z, et al. Homologous recombination as an evolutionary force in the avian influenza A virus. Molecular Biology and Evolution, 2009, 26(1): 177-187.

[7] JAGGER B W, WISE H M, KASH J C, et al. An overlapping protein-coding region in influenza A virus segment 3 modulates the host response. Science, 2012, 337(6091): 199-204.

[8] SUBBARAO K, JOSEPH T. Scientific barriers to developing vaccines against avian influenza viruses. Nature, 2007, 7(4): 267-278.

[9] WEBSTER R G, BEAN W J, GORMAN O T, et al. Evolution and ecology of influenza A viruses. Microbiology Review, 1992, 56(1): 152-179.

[10] XU K M, SMITH G J, BAHL J, et al. The genesis and evolution of H9N2 influenza viruses in poultry from southern China, 2000 to 2005. Journal of Virology, 2007, 81(19): 10389-10401.

Complete genome sequence analysis of a Newcastle disease virus isolated from a wild egret

Xie Zhixun, Xie Liji, Chen Anli, Liu Jiabo, Pang Yaoshan, Deng Xianwen, Xie Zhiqin, and Fan Qing

Abstract

We report here the complete genomic sequence of a novel Newcastle disease virus (NDV) strain, egret/China/Guangxi/2011, isolated from an egret in Guangxi, southern China. A phylogenetic analysis based on a fusion gene comparison with different NDV strains revealed that egret/China/Guangxi/2011 was phylogenetically close to genotype VIIa NDV, and the deduced amino acid sequence was ^{112}R-R-R-K-R-F^{117} at the fusion protein cleavage site. The whole nucleotide sequence had the highest homology (93.3%) with the sequence of strain chicken/Sukorejo/019/10 (GenBank accession number: HQ697255). This study will help us to understand the epidemiology and molecular characteristics of Newcastle disease virus in a migratory egret.

Keywords

Newcastle disease virus, complete genome sequence, wild egret

---

Newcastle disease virus (NDV), synonymous with *Avian paramyxovirus*-1, is the causative agent of Newcastle disease, which is a highly contagious and fatal viral disease that affects all species of birds[1, 3, 5].

NDV has a nonsegmented, negative-sense RNA genome consisting of six transcriptional units (3'-NP-P-M-F-HN-L-5')[4, 8]. NDV strains have been classified into classes I (9 genotypes) and II (11 genotypes): class I strains are generally avirulent and have been isolated mainly from wild birds, whereas class II strains are virulent and avirulent and have been isolated from wild and domestic birds[2, 6].

In May 2011, NDV was isolated from a wild egret in Guangxi, southern China. The isolate's virus was named egret/China/Guangxi/2011. Nucleotide sequences of egret/China/Guangxi/2011 were amplified through reverse transcription-PCR (RT-PCR). The amplified products were purified and cloned into the pMD18-T vector (TaKaRa, Dalian, China)and then sequenced (TaKaRa, Dalian, China). Sequences were assembled and manually edited to generate the final genome sequence.

Sequence analysis showed that the full genomic length of egret/China/Guangxi/2011 is 15 192 nucleotides (nt). A phylogenetic analysis classified this strain into class II, genotype VIIa. The whole nucleotide sequence had the highest homology (93.3%) with the sequence of strain chicken/Sukorejo/019/10 (GenBank accession number: HQ697255, class II, genotype VIIa). The amino acid sequence identities of the NP, P, M, F, HN, and L proteins between egret/China/Guangxi/2011 and Sukorejo01910 are 97.6%, 96.7%, 97.8%, 98%, 96%, and 97.3%, respectively. The amino acid sequence identities of the NP, P, M, F, HN, and L proteins between egret/China/Guangxi/2011 and strain LaSota (GenBank accession number: JF950510, class II, genotype II) are 84.2%, 81.8%, 83.7%, 84.2%, 81.9%, and 85.8%, respectively.

The homology of the 374-bp partial F gene (positions 4 550 to 4 923) of egret/China/Guangxi/2011 with the sequences of Que-66 (GenBank accession number: M24693, class II, genotype I), LaSota (GenBank accession number: JF950510, class Ⅱ, genotype Ⅱ), Aus-32 (GenBank accession number: M24700, class Ⅱ, genotype Ⅲ), Herts-33 (GenBank accession number: AY741404, class Ⅱ, genotype Ⅳ), CA1085-71 (GenBank accession number: JQ247691, class Ⅱ, genotype Ⅴ), ISreal70-1 (GenBank accession number: AF001111, class Ⅱ, genotype Ⅵ), chicken/Sukorejo/019/10 (GenBank accession number: HQ697255, class Ⅱ, genotype Ⅶa), QH-4-48 (GenBank accession number: AF378252, class Ⅱ, genotype Ⅷ), F48E9 (GenBank accession number: AY508514, class Ⅱ, genotype Ⅸ), US (NJ) (GenBank accession number: EF565065, class Ⅰ) were 79.7%, 75.4%, 80.2%, 81.0%, 81.8%, 84.8%, 96.3%, 82.6%, 78.6%, and 58.3%, respectively.

The sequence at the fusion protein cleavage site is a major determinant of NDV pathogenicity[7]. The cleavage sites of virulent NDV strains contain multiple basic residues, whereas avirulent strains have few basic residues. The egret/China/Guangxi/2011 strain has a virulent fusion protein cleavage site sequence (^{112}R-R-R-K-R-F^{117}), and it accorded with the detection of an intracerebral pathogenicity index of 1.846.

To our knowledge, this is the first report of a phylogenetic analysis of the whole nucleotide sequence of NDV isolated from a wild egret. Thus, it will help us to understand the epidemiology and molecular characteristics of Newcastle disease virus in wild birds.

Nucleotide sequence accession number. The GenBank accession number of egret/China/Guangxi/2011 is JX193074.

References

[1] ALEXANDER D J. Newcastle disease and other avian paramyxoviruses. Revue Scientifique et Technique, 2000, 19(2): 443-462.

[2] CZEGLÉDI A, UJVÁRI D, SOMOGYI E, et al. Third genome size category of avian paramyxovirus serotype 1 (Newcastle disease virus) and evolutionary implications. Virus Research, 2006, 120(1-2): 36-48.

[3] DE LEEUW O S, KOCH G, HARTOG L, et al. Virulence of Newcastle disease virus is determined by the cleavage site of the fusion protein and by both the stem region and globular head of the haemagglutinin-neuraminidase protein. The Journal of General Virology, 2005, 86(Pt 6): 1759-1769.

[4] KNIPED M, LAMB R, PARKS G. Paramyxoviridae: the viruses and their replication//KNIPE D M, LAMB R, PARKS G. Fields Virology, 5th ed. Philadelphia: Lippincott Williams & Wilkins, 2007: 1449-1496.

[5] MAST J, NANBRU C, VAN DEN BERG T, et al. Ultrastructural changes of the tracheal epithelium after vaccination of day-old chickens with the LaSota strain of Newcastle disease virus. Veterinary Pathology, 2005, 42(5): 559-565.

[6] MILLER P J, DECANINI E L, AFONSO C L. Newcastle disease: evolution of genotypes and the related diagnostic challenges. Infection, Genetics and Evolution, 2010, 10(1): 26-35.

[7] PANDA A, HUANG Z, ELANKUMARAN S, et al. Role of fusion protein cleavage site in the virulence of Newcastle disease virus. Microbial Pathogenesis, 2004, 36(1): 1-10.

[8] YAN Y, SAMAL S K. Role of intergenic sequences in Newcastle disease virus RNA transcription and pathogenesis. Journal of Virology, 2008, 82(3): 1323-1331.

Complete genome sequence analysis of a duck circovirus from Guangxi Pockmark ducks

Xie Liji, Xie Zhixun, Zhao Guangyuan, Liu Jiabo, Pang Yaoshan, Deng Xianwen, Xie Zhiqin, and Fan Qing

Abstract

We report here the complete genomic sequence of a novel duck circovirus (DuCV) strain, GX1104, isolated from Guangxi Pockmark ducks in Guangxi, China. The whole nucleotide sequence had the highest homology (97.2%) with the sequence of strain TC/2002 (GenBank accession number: AY394721.1) and had a low homology (76.8% to 78.6%) with the sequences of other strains isolated from China, Germany, and the United States. This report will help to understand the epidemiology and molecular characteristics of Guangxi Pockmark duck circovirus in southern China.

Keywords

Guangxi Pockmark ducks, duck circovirus, complete genome sequence

Duck circovirus (DuCV), isolated from Muscovy duck[3] and Mulard duck[5], is involved in duck diseases, with the main clinical symptoms including immunosuppression and feather disorders in young ducks. The genome of DuCV is circular and 1 996 nucleotides (nt) in size. Two major open reading frames (ORFs) were identified, encoding the replicase (V1) and the capsid protein (C1)[4]. DuCV was first reported in Germany[3] and was also recently reported in China[1, 2, 6]. The whole nucleotide sequences of these strains had a homology of 83.2% to 99.8%.

In April 2011, DuCV was isolated from a commercial Guangxi Pockmark duck farm with an outbreak of an infectious disease whose clinical symptoms manifested with severe immunosuppression and feather disorders in ducklings in Guangxi, southern China. Subsequently, nucleotide sequences of DuCV were amplified through PCR. The amplified products were purified and cloned into the pMD18-T vector (TaKaRa, Dalian, China) and then sequenced (TaKaRa, Dalian, China). Sequences were assembled and manually edited to produce the final genome sequence. The isolate's virus was named duck/Guangxi/04/2011 (GX1104).

Sequence analysis showed that the full genomic length of GX1104 is 1 988 nt. Additionally, the coding region of GX1104 includes two major ORFs, V1 and C1, encoding polypeptides of 292 and 257 amino acids, respectively. Compared with other relative DuCV strains, the V1 and C1 nucleotide sequence homologies were about 85.6% to 97.8% and 74.4% to 96.4%, respectively. The complete genome sequence homology was about 76.8% to 97.2%.

The whole nucleotide sequence had the highest homology (97.2%) with the sequence of strain TC/2002 (GenBank accession number: AY394721.1; origin: Taiwan, China), a medium homology (76.8% to 78.6%) with that of five Chinese strains (GenBank accession numbers: EF451157.1, GQ334371.1, GU131342.1, HM162345.1, and HQ180265.1; origin: mainland of China), strain 33753-52 (GenBank accession number: DQ100076.1; origin: the United States), and a German strain (GenBank accession number: NC005053.1;

origin: Germany), and the lowest homology (54.2%) with that of strain JX1 (Goose circovirus) (GenBank accession number: GU320569.1; origin: Jiangxi, China).

The homologies of the whole nucleotide sequence of GX1104, with the sequences of two Mallard duck circovirus strains (GenBank accession numbers: NC005053 and HQ180266), two Mule duck circovirus strains (GenBank accession numbers: EU344803 and EU499309), two Muscovy duck circovirus (GenBank accession numbers: EF451157 and AY394721), two Perkin duck circovirus (GenBank accession numbers: NC007220 and DQ100076), and two Cherry valley duck circovirus (GenBank accession numbers: GU131342 and HM162352), were 76.8%, 77.1%, 80.4%, 95.5%, 77.9%, 97.1%, 78.6%, 78.6%, 78.6%, and 78.4%, respectively. The homology of the whole nucleotide sequence of GX1104 with the sequence of all 14 Cherry valley duck circovirus in GenBank was less than 80%. The results indicated that Guangxi Pockmark duck circovirus was different from Cherry valley duck circovirus.

Phylogenetic analysis of the whole nucleotide sequence of GX1104 indicates that Guangxi Pockmark duck circovirus has been mutated, resulting in Guangxi Pockmark duck clinical symptoms manifested with severe immunosuppression and feather disorders. Further study is required.

Nucleotide sequence accession number. The GenBank accession number of duck/Guangxi/04/2011 (GX1104) is JX241046.

References

[1] FU G, CHENG L F, SHI S H, et al. Genome cloning and sequence analysis of duck circovirus. Virologica Sinica, 2008, 24(2): 138-143.

[2] FU G, SHI S, HUANG Y, et al. Genetic diversity and genotype analysis of duck circovirus. Avian Dis, 2011, 55(2): 311-318.

[3] HATTERMANN K, SCHMITT C, SOIKE D, et al. Cloning and sequencing of duck circovirus (DuCV). Archives of Virology, 2003, 148(12): 2471-2480.

[4] JOHNE R, FERNÁNDEZ-DE-LUCO D, HÖFLE U, et al. Genome of a novel circovirus of starlings, amplified by multiply primed rolling-circle amplification. The Journal of General Virology, 2006, 87(Pt 5): 1189-1195.

[5] SOIKE D, ALBRECHT K, HATTERMANN K, et al. Novel circovirus in mulard ducks with developmental and feathering disorders. The Veterinary Record, 2004, 154 (25): 792-793.

[6] WAN C H, FU G H, SHI S H, et al. Epidemiological investigation and genome analysis of duck circovirus in southern China. Virologica Sinica, 2011, 26(2): 289-296.

Characterization of an avian influenza virus H9N2 strain isolated from a wild bird in southern China

Xu Qian, Xie Zhixun, Xie Liji, Xie Zhiqin, Deng Xianwen, Liu Jiabo, and Luo Sisi

Abstract

We isolated an avian influenza virus H9N2 strain from a wild bird in the Guangxi of southern China in 2013 named A/turtledove/Guangxi/49B6/2013 (H9N2) (GX49B6). We aimed to understand the genetic characters of the GX49B6 strain by analysis the complete genome sequence. The results showed that our isolated strain has features of low pathogenic avian influenza viruses and viruses that infect humans. The discovery of the complete genome sequence of the GX49B6 strain may be helpful to further the understanding of the epidemiology and surveillance of avian influenza viruses in the field.

Keywords

avian influenza virus, H9N2 subtype, wild bird

Avian influenza A virus (AIV) is a single-stranded negative-sense RNA virus belonging to the family Orthomyxoviridae. It causes a variety of infections in avians and mammals. H9N2 subtype AIVs are widespread in the world and are the most prevalent subtype of avian influenza viruses reported in China over the last decade[1]. Although H9N2 is characterized as a low pathogenic avian influenza virus, occasional infections of humans[2-5] have caused great concerns. So far, H9N2 subtype AIVs are mainly isolated from domestic birds, however, wildfowl and shorebirds are the natural hosts of AIVs[6] and they facilitate the transmission of avian influenza[7]. Guangxi is an autonomous region of major poultry industry and is contiguous to Vietnam where the avian influenza epidemic is complex. Enhancing the surveillance of H9N2 subtype AIVs among wild birds is very important.

In this study, we isolated an H9N2 strain named A/turtledove/Guangxi/49B6/2013(H9N2) (GX49B6) from a wild bird in the Guangxi. We amplified eight genes of the GX49B6 strain by RT-PCR using the universal primers of the influenza A virus[8, 9].

The complete genome of GX49B6 strain consists of PB2, PB1, PA, HA, NP, NA, M, and NS segments. The full lengths of the segments are 2 341 nucleotides (nt), 2 341 nt, 2 233 nt, 1 742 nt, 1 565 nt, 1 459 nt, 1 027 nt, and 890 nt, respectively. The amino acid lengths of the proteins encoded by the eight genes as follows: PB2, 759 aa; PB1, 757 aa; PA, 716 aa; HA, 560 aa; NP, 498 aa; NA, 466 aa; M2, 97 aa; M1, 252 aa; NS2, 121 aa; NSl, 217 aa. The amino acid residue at the cleavage site (335~341) of the HA molecule is RSSR↓GLF, without multiple consecutive basic amino acids, which is characteristic of low pathogenic AIVs. The presence of 158 Glu, 627 Glu, 701 Asp in the amino acid sequence of the PB2 protein, 436 Tyr in the PB1 protein, and 515 Thrin the PA protein, respectively, also provides evidence of low pathogenicity[10, 11]. The GX49B6 strain has L226 and G228 (according to H3 numbering) at the receptor-binding site in the HA protein, which suggests that the GX49B6 strain might have the ability to bind a sialicacid-2, 6-NeuAcGal linkage and might have the potential to infect humans[4, 12].

Sequence analysis revealed that the nucleotide sequences of the HA and NA genes of the GX49B6 strain both belong to the Eurasian lineage. The HA gene shared 96.3% nucleotide homology with the isolate A/Chicken/Guangxi/55/2005(H9N2) that is thought to be are presentative strain in China since 2007[1]. The PB2 and PB1 genes both share a high homology (≥97%) with genes of the H7N9 strains isolated from infected humans in Guangdong and Anhui provinces, respectively. They are also similar to genes of some strains isolated from Vietnam (≥98%).

The data demonstrate that wild birds are involved in the transmission and evolution of AIVs. Our results may be helpful in epidemiological studies on H9N2 subtype AIVs.

Nucleotide sequence accession numbers. The virus genome sequence of the A/turtledove/Guangxi/49B6/2013 (H9N2) strain was deposited in the DDBJ/EMBL/GenBank database under accession numbers KJ725009 through KJ725016.

References

[1] JI K, JIANG W M, LIU S, et al. Characterization of the hemagglutinin gene of subtype H9 avian influenza viruses isolated in 2007-2009 in China. Journal of Virological Methods, 2010, 163(2): 186-189.

[2] ALEXANDER D J, BROWN I H. Recent zoonoses caused by influenza A viruses. Revue Scientifique et Technique, 2000, 19(1): 197-225.

[3] PEIRIS M, YUEN K Y, LEUNG C W, et al. Human infection with influenza H9N2. Lancet, 1999, 354(9182): 916-917.

[4] WAN H, PEREZ D R. Amino acid 226 in the hemagglutinin of H9N2 influenza viruses determines cell tropism and replication in human airway epithelial cells. Journal of Virology, 2007, 81(10): 5181-5191.

[5] XIE Z X, DONG J B, Tang X F, et al. Sequence and phylogenetic analysis of three isolates of avian influenza H9N2 from Chickens in southern China. Scholar Research Exchange, 2008: 1-7.

[6] OLSEN B, MUNSTER V J, WALLENSTEN A, et al. Global patterns of influenza a virus in wild birds. Science, 2006, 312(5772): 384-388.

[7] PENG Y, XIE Z X, LIU J B, et al. Epidemiological surveillance of low pathogenic avian influenza virus (LPAIV) from poultry in Guangxi Province, southern China. PLOS ONE, 2013, 8(20): e77132.

[8] HE C Q, XIE Z X, HAN G Z, et al. Homologous recombination as an evolutionary force in the avian influenza A virus. Molecular Biology and Evolution, 2009, 26(1): 177-187.

[9] HOFFMANN E, STECH J, GUAN Y, et al. Universal primer set for the full-length amplification of all influenza A viruses. Archives of Virology, 2001, 146(12): 2275-2289.

[10] HULSE-POST D I, FRANKS J, BOYD K, et al. Molecular changes in the polymerase genes (PA and PB1) associated with high pathogenicity of H5N1 influenza virus in mallard ducks. Journal of Virology, 2007, 81(16): 8515-8524.

[11] ZHOU B, LI Y, HALPIN R, et al. PB2 residue 158 is a pathogenic determinant of pandemic H1N1 and H5 influenza a viruses in mice. Journal of Virology, 2011, 85(1): 357-365.

[12] MATROSOVICH M, TUZIKOV A, BOVIN N, et al. Early alterations of the receptor-binding properties of H1, H2, and H3 avian influenza virus hemagglutinins after their introduction into mammals. Journal of Virology, 2000, 74(18): 8502-8512.

Complete genome sequence of a novel reassortant avian influenza H1N2 virus isolated from a domestic sparrow in 2012

Xie Zhixun, Guo Jie, Xie Liji, Liu Jiabo, Pang Yaoshan, Deng Xianwen, Xie Zhiqin, Fan Qing, and Luo Sisi

Abstract

We report here the complete genome sequence of a novel H1N2 avian influenza virus strain, A/Sparrow/Guangxi/GXs-1/2012 (H1N2), isolated from a sparrow in the Guangxi of southern China in 2012, All of the 8 gene segments (hemagglutinin(HA), nucleoprotein(NP), matrix(M), polymerase basic 2(PB2), neuraminidase(NA), polymerase acidic(PA), polymerase basic 1(PB1), and nonstructural(NS) genes) of this natural recombinant virus are attributed to the Eurasian lineage, and phylogenetic analysis showed that those genes are derived from H1N2, H3N1, H3N2, H4N6, H6N2, H10N8, H5N1, and H4N6 avian influenza viruses (AIVs), The amino acid motif of the cleavage site of HA is PSIQSR↓GLF, The sequence analysis will help in understanding the molecular characteristics and epidemiology of the H1N2 influenza virus in sparrows.

Keywords

avian influenza virus, H1N2 subtype, complete genome

Avian influenza virus (AIV) is in the influenza A virus family and can cause fatal disease in avian species[1], Hemagglutinin (HA) and neuraminidase (NA) can be used to distinguish different subtypes of AIV, At present, there are at least 17 HA subtypes and 10 NA subtypes of AIV that exist in the world[2-4]. It is notable that among the many subtypes of influenza A virus, influenza A H1 and H3 subtypes are currently circulating among humans, The H1 subtype of AIV is transmitted to pigs, and perhaps through pigs as an intermediate host, through gene rearrangement or mutations, influenza virus is in turn transmitted to humans, causing human influenza outbreaks and epidemics[5-7], Therefore, the H1 subtype of AIVs that circulate in avian species should not be ignored.

In this study, a novel strain named A/Sparrow/Guangxi/GXs-1/2012(H1N2) was first isolated from a domestic sparrow in Guangxi, southern China, in 2012, The nucleotide sequences of this strain were amplified by reverse transcription-PCR (RT-PCR) performed using universal primers[8-10], The amplified products were purified and cloned into the pGEM-T Easy vector (Promega, Madison, USA) and were sequenced (TaKaRa, Dalian, China), Sequences were assembled and manually edited to generate the final genome sequence.

The complete genome of A/Sparrow/Guangxi/GXs-1/2012(H1N2) consists of eight segments, including polymerase (PB2, PB1, PA), HA, nucleoprotein (NP), NA, matrix protein (M), and nonstructural protein (NS) genes, The full lengths of these segments are 2 341, 2 341, 2 233, 1 777, 1 565, 1 466, 1 027, and 890 nucleotides, respectively, The amino acid motif of the cleavage site of HA is PSIQSR↓GLF, which is a typical characteristic of low-pathogenicity AIV, The amino acid residues at the receptor binding site in the HA protein are Q226 and G228, which indicates its AIV-like receptor binding preference, The PB2 protein possesses E627 and D701, which is characteristic of AIV, The PA protein possesses A20, which is antagonistic to antiviral responses, The M2 protein possesses S31, which is not amantadine resistant, E627 and D701 are found in the PB2 sequence, which is characteristic of AIV; A20 in the PA sequence suggests it is antagonistic to antiviral responses; S31 in the M2 protein sequence indicates that the strain has amantadine resistance.

Phylogenetic analysis revealed that the eight genes of A/Sparrow/Guangxi/GXs-1/2012(H1N2) belong to the Eurasian lineage. The HA gene shows the highest sequence homology (96%) to that from A/ostrich/ South Africa/AI2887/2011(H1N2) (GenBank accession number: JX069105), The NP and MP genes show the highest sequence homologies (97%) to the genes from A/chicken/Pakistan/NARC-16945/2010(H3N1) (GenBank accession number: HQ165997) and A/duck/Shanghai/C84/2009(H3N2) (GenBank accession number: JX286592), respectively, The PB2, NA, and PA genes show the highest sequence homologies (98%) to the genes from A/wild duck/Korea/CSM4-28/2010(H4N6) (GenBank accession number: JX454697), A/ duck/Guangxi/GXd-2/2010(H6N2) (GenBank accession number: JX297589), and A/duck/Guangdong/ E1/2012(H10N8) (GenBank accession number: JQ924792), respectively, In addition, the PB1 and NS genes show the highest sequence homologies (99%) to the genes from A/wild duck/Korea/SNU50-5/2009(H5N1) (GenBank accession number: JX497766) and A/wild duck/ Korea/SH5-26/2008 (H4N6) (GenBank accession number: JX454749). These data will help in understanding the molecular characteristics and epidemiology of the Hl subtype influenza virus in avian species in China.

Nucleotide sequence accession numbers, The complete genome sequence of A/Sparrow/Guangxi/GXs-1/2012 (H1N2) was deposited in GenBank under accession number: KF013901 to KF013908.

References

[1] FULLER T L, GILBERT M, MARTIN V, et al. Predicting hotspots for influenza virus reassortment. Emerg Infect Dis, 2013, 19(4): 581-588.

[2] TONG S, LI Y, RIVAILLER P, et al. A distinct lineage of influenza A virus from bats. Proc Natl Acad Sci U S A, 2012, 109(11): 4269-4274.

[3] WEBSTER R G, BEAN W J, GORMAN O T, et al. Evolution and ecology of influenza A viruses. Microbiol Rev, 1992, 56(1): 152-179.

[4] XIE Z X, XIE L J, ZHOU C Y, et al. Complete genome sequence analysis of an H6N1 avian influenza virus isolated from Guangxi Pockmark ducks. J Virol, 2012, 86(24): 13868-13869.

[5] CASTRUCCI M R, DONATELLI I, SIDOLI L, et al. Genetic reassortment between avian and human influenza A viruses in Italian pigs. Virology, 1993, 193(1): 503-506.

[6] GUAN Y, SHORTRIDGE K F, KRAUSS S, et al. Emergence of avian H1N1 influenza viruses in pigs in China. J Virol, 1996, 70(11): 8041-8046.

[7] CHAO L, LIANFEN L, BO W, et al. A review on 2009 influenza A virus. Agric Sci Technol, 2012, 13: 424-427.

[8] HOFFMANN E, STECH J, GUAN Y, et al. Universal primer set for the full-length amplification of all influenza A viruses. Arch Virol, 2001, 146(12): 2275-2289.

[9] HE C Q, DING N Z, MOU X, et al. Identification of three H1N1 influenza virus groups with natural recombinant genes circulating from 1918 to 2009. Virology, 2012, 427(1): 60-66.

[10] HE C Q, XIE Z X, HAN G Z, et al. Homologous recombination as an evolutionary force in the avian influenza A virus. Mol Biol Evol, 2009, 26(1): 177-187.

Avian influenza virus with hemagglutinin-neuraminidase combination H3N6, isolated from a domestic pigeon in Guangxi, southern China

Liu Tingting, Xie Zhixun, Wang Guoli, Song Degui, Huang Li, Xie Zhiqin, Deng Xianweng, Luo Sisi, Huang Jiaoling, and Zeng Tingting

Abstract

The H3 subtype of avian influenza virus can provide genes for human influenza virus through gene reassortment, which has raised great concerns about its potential threat to human health. An H3N6 subtype of avian influenza virus was isolated from Guangxi, China, in 2009. All eight gene segments of the strain were sequenced. The sequence analysis indicated that this H3N6 virus was a nature reassortant virus. The genome sequences now can be used to understand the epidemiological and molecular characteristics of the H3N6 influenza virus in southern China.

Keywords

avian influenza virus, H3N6 subtype, domestic pigeon

Avian influenza virus (AIV) belongs to the type A influenza viruses, which infect many avian species[1]. At present, there are 18 hemagglutinin (HA) and 11 neuraminidase (NA) subtypes of AIV based on the antigenic differences of the HA and NA proteins, which are surface glycoproteins on the viral envelope[2, 3].

The H3 subtype of AIV belongs to the low-pathogenic AIVs (LPAIVs) and is one of predominant subtypes among the LPAIVs[4, 5]. Researchers have shown that the H3 subtype AIV has a high separation rate in poultry, and it may have the ability to cross the species barrier to infect humans through gene reassortment[6-8]. In addition, previous studies have demonstrated that some novel H3N6 subtype viruses were reassortants between highly pathogenic H7 and H5 viruses isolated in Eurasia[9], thus signifying the importance of enhancing the surveillance of the H3N6 subtype AIV.

An H3N6 subtype AIV was isolated from a pigeon in a live poultry market in Guangxi, China, in May 2009 and named A/pigeon/Guangxi/020P/2009(H3N6). In this study, we amplified the full genes by reverse transcription-PCR using AIV universal primers[10, 11]. The amplified products were purified and cloned into the pMD18-T vector (TaKaRa, Dalian, China) and sequenced (TaKaRa Dalian, China). The sequences were assembled using the SeqMan program and manually edited to generate the final full-length genome sequence.

The complete genome of this H3N6 strain consisted of eight gene segments of polymerase basic 2 (PB2), PB1, polymerase acidic (PA), HA, nucleoprotein (NP), NA, matrix (M), and non-structural (NS) genes. The full lengths of these segments were 2 341, 2 341, 2 233, 1 765, 1 565, 1 464, 1 027, and 890 nucleotides, respectively. Those eight genes encoded proteins with the following amino acid lengths: 759 (PB2), 757 (PB1), 716 (PA), 566 (HA), 498 (NP), 470 (NA), 252 (M1), 97 (M2), 230 (NS1), and 121 (NS2). The amino acid sequence at the cleavage site (positions 340 to 348) of the HA molecule was PEKQTR↓GLF, with one basic amino acid, which is characteristic of low-pathogenic AIV. The amino acid residues at the receptor binding site in the HA protein were Q226 and G228, which are different from L226 and S228 in the H3 subtype of human influenza viruses, which preferentially bind to an avian-origin receptor. An analysis of potential glycosylation sites revealed that there were 6 potential N-linked glycosylation sites in the HA protein (positions 24, 38, 54, 181, 301, and 499), while there were 8 in NA (positions 51, 54, 62, 67, 70, 86, 146, and 402). In addition, the PB2 protein identified in this isolate contained E627 and D701, which indicated that the virus was of avian origin[12, 13].

The analysis of the sequence also indicated that the nucleotide sequences of both the HA and NA genes of this H3N6 strain belong to the Eurasian lineage. Also, its other internal genes are closely related to H3N8, H4N6, H6N2, H3N2, and H4N2 subtype AIVs, which suggests that this H3N6 strain went through extensive reassortment with different subtypes of influenza viruses. The genome information of the isolated virus revealed in this study can now be used for conducting an epidemiological investigation on the H3N6 subtype of AIV in China.

Nucleotide sequence accession numbers. The genome sequence of A/pigeon/Guangxi/020P/2009 (H3N6) has been deposited in GenBank under the accession numbers KM186122 to KM186129.

References

[1] XU Q, XIE Z X, XIE L J, et al. Characterization of an avian influenza virus H9N2 strain isolated from a wild bird in southern China. Genome Announc, 2014, 2(3): e00600-e00614.

[2] WU Y, WU Y, TEFSEN B, et al. Bat-derived influenza-like viruses H17N10 and H18N11. Trends Microbiol, 2014, 22(4): 183-191.

[3] TONG S, ZHU X, LI Y, et al. New world bats harbor diverse influenza A viruses. PLOS Pathog, 2013, 9(10): e1003657.

[4] XIE Z X, GUO J, XIE L J, et al. Complete genome sequence of a novel reassortant avian influenza H1N2 virus isolated from a domestic sparrow in 2012. Genome Announc, 2013, 1(4): e00431-e00413.

[5] PENG Y, XIE Z X, LIU J B, et al. Visual detection of H3 subtype avian influenza viruses by reverse transcription loop-mediated isothermal amplification assay. Virol J, 2011, 8: 337.

[6] CAMPITELLI L, FABIANI C, PUZELLI S, et al. H3N2 influenza viruses from domestic chickens in Italy: an increasing role for chickens in the ecology of influenza?. J Gen Virol, 2002, 83(Pt 2): 413-420.

[7] SONG M S, OH T K, MOON H J, et al. Ecology of H3 avian influenza viruses in Korea and assessment of their pathogenic potentials. J Gen Virol, 2008, 89(Pt 4): 949-957.

[8] PENG Y, XIE Z X, LIU J B, et al. Epidemiological surveillance of low pathogenic avian influenza virus (LPAIV) from poultry in Guangxi Province, Southern China. PLOS ONE, 2013, 8(10): e77132.

[9] SIMULUNDU E, MWEENE A S, TOMABECHI D, et al. Characterization of H3N6 avian influenza virus isolated from a wild white pelican in Zambia. Arch Virol, 2009, 154(9): 1517-1522.

[10] HOFFMANN E, STECH J, GUAN Y, et al. Universal primer set for the full-length amplification of all influenza A viruses. Arch Virol, 2001, 146: 2275-2289.

[11] HE C Q, XIE Z X, HAN G Z, et al. Homologous recombination as an evolutionary force in the avian influenza A virus. Mol Biol Evol, 2009, 26: 177-187.

[12] MEHLE A, DOUDNA J A. An inhibitory activity in human cells restricts the function of an avian-like influenza virus polymerase. Cell Host Microbe, 2008, 4(2): 111-122.

[13] LI Z, CHEN H, JIAO P, et al. Molecular basis of replication of duck H5N1 influenza viruses in a mammalian mouse model. J Virol, 2005, 79(18): 12058-12064.

Characterization of the whole-genome sequence of an H3N6 avian influenza virus, isolated from a domestic duck in Guangxi, southern China

Liu Tingting, Xie Zhixun, Luo Sisi, Xie Liji, Deng Xianwen, Xie Zhiqin, Huang Li, Huang Jiaoling, Zhang Yanfang, Zeng Tingting, and Wang Sheng

Abstract

A field strain of H3N6 avian influenza virus (AIV), A/duck/Guangxi/175D12/2014 (H3N6), was isolated from a native duck in Guangxi, southern China, in 2014. All of the eight AIV gene segments were sequenced, and sequence results revealed that there were 11 amino acid deletions at the NA stalk region. The NA, PB2, and NP genes showed highest homology to H5N6 AIV, and the PA gene showed highest homology to H7N2 AIV. Phylogenetic analysis indicated that the eight AIV gene segments belonged to the Eurasian lineage. These findings provide scientific evidence of possible or potential mutations of H3N6 AIV circulating in waterfowl in southern China.

Keywords

avian influenza virus, H3N6 subtype, whole-genome

Avian influenza viruses (AIVs) are members of the family Orthomyxoviridae. To date, a total of 17 hemagglutinin (HA) and 10 neuraminidase (NA) subtypes of AIV have been reported[1-3]. Water fowls are carriers for most of the subtypes as their nature hosts. Domestic ducks play an important role in virus transmission from wild waterfowl species to terrestrial poultry[4]. The H3 subtype AIV is most frequently isolated from feral ducks[5]. Although infected domestic ducks via H3 subtype AIV do not display symptoms, they cary and shield the virus and provide an environment for the virus's reassortment with other subtypes, such as H5 and H7 AIVs, which may cause disease in humans. Therefore, it is important to monitor H3 subtype AIV in ducks and conduct a genomic characterization to detect any potential mutation.

An H3N6 subtype AIV, A/duck/Guangxi/175D12/2014 (H3N6), was isolated from an infected duck in a live poultry market in Guangxi. The eight gene segments of this duck's H3N6 AIV were amplified by RT-PCR using AIV universal primers[6-8]. The amplified PCR products were purified and cloned into the pMD18-T vector and sequenced (TaKaRa, Dalian, China). The genomic sequence was analyzed by MEGA6.0.

The complete genomic segments include PB2, PB1, PA, NP, HA, NA, NS, and M genes, with full lengths of 2 341, 2 341, 2 233, 1 565, 1 765, 1 431, 1 027, and 890 nucleotides (nt), respectively. The NA gene lost 33 nt at positions 194 to 226. There are only six potential N-linked glycosylation sites in the NA protein (positions 51, 59, 75, 135, 190, and 391), which is different from the pigeon H3N6 AIV reported recently[9]. The HA cleavage site possesses only a single basic amino acid (^{340}PEKQTR↓GLF^{348}), which is characteristic of low-pathogenicity AIV[10]. The amino acid residues at the receptor binding site in the HA protein are Q^{226} and G^{228}, which suggests that this duck H3N6 AIV would preferentially bind to alpha[2, 3]-linked sialic acid receptors, which are predominant in avian species[11, 12]. Analysis of potential glycosylation sites reveals that there are 6 potential N-linked glycosylation sites in the HA protein (positions 22, 38, 54, 181, 301, and 499). In addition, this duck's H3N6 AIV possesses E and D at positions 627 and 701 of the PB2 protein, which suggests that this duck's virus is of avian origin[13, 14].

Phylogenetic analysis indicates that all eight gene segments of this duck H3N6 AIV belongs to the Eurasian lineage. The NA gene fragment is closely related to that of H5N6 AIVs isolated in Eurasia in 2013 and 2014, as they share 98.8% to 99.3% nucleotide homology. PB2 and NP gene fragments are most closely related to A/duck/Guangdong/GD01/2014(H5N6) and A/chicken/Dongguan/3363/2013(H5N6), respectively. The PA gene fragment is most closely related to A/duck/Wenzhou/775/2013(H7N2). These data suggest that this duck's H3N6 AIV could share similar original ancestors as those of H7N2 and H5N6 viruses in Eurasia.

In summary, this duck's H3N6 AIV is a novel reassortant AIV, and the findings provide a better understanding of the ecology and epidemiology of H3N6 AIV circulating in southern China.

Nucleotide sequence accession numbers. The genome sequence of A/duck/Guangxi/175D12/2014 (H3N6) was deposited in GenBank under the accession numbers KR919740 to KR919747.

References

[1] WEBSTER R G, BEAN W J, GORMAN O T, et al. Evolution and ecology of influenza A viruses. Microbiol Rev, 1992, 56(1): 152-179.

[2] XIE Z X, XIE L J, ZHOU C Y, et al. Complete genome sequence analysis of an H6N1 avian influenza virus isolated from Guangxi Pockmark ducks. J Virol, 2012, 86(24): 13868-13869.

[3] XIE Z X, GUO J, XIE L J, et al. Complete genome sequence of a novel reassortant avian influenza H1N2 virus isolated from a domestic sparrow in 2012. Genome Announc, 2013, 1(4): e00431-13.

[4] HULSE-POST D J, STURM-RAMIREZ K M, HUMBERD J, et al. Role of domestic ducks in the propagation and biological evolution of highly pathogenic H5N1 influenza viruses in Asia. Proc Natl Acad Sci U S A, 2005, 102(30): 10682-10687.

[5] KALETA E F, HERGARTEN G, YILMAZ A. Avian influenza A viruses in birds-an ecological, ornithological and virological view. Dtsch Tierarztl Wochenschr, 2005, 112(12): 448-456.

[6] HOFFMANN E, STECH J, GUAN Y, et al. Universal primer set for the full-length amplification of all influenza A viruses. Arch Virol, 2001, 146(12): 2275-2289.

[7] HE C Q, XIE Z X, HAN G Z, et al. Homologous recombination as an evolutionary force in the avian influenza A virus. Mol Biol Evol, 2009, 26(1): 177-187.

[8] HE C Q, DING N Z, MOU X, et al. Identification of three H1N1 influenza virus groups with natural recombinant genes circulating from 1918 to 2009. Virology, 2012, 427(1): 60-66.

[9] LIU T T, XIE Z X, WANG G, et al. Avian influenza virus with hemagglutinin-neuraminidase combination H3N6, isolated from a domestic pigeon in Guangxi, southern China. Genome Announc, 2015, 3(1): e01537-14.

[10] TAUBENBERGER J K, REID A H, LOURENS R M, et al. Characterization of the 1918 influenza virus polymerase genes. Nature, 2005, 437(7060): 889-893.

[11] YANG D, LIU J, JU H, et al. Genetic analysis of H3N2 avian influenza viruses isolated from live poultry markets and poultry slaughterhouses in Shanghai, China in 2013. Virus Genes, 2015, 51(1): 25-32.

[12] WU H, WU N, PENG X, et al. Molecular characterization and phylogenetic analysis of H3 subtype avian influenza viruses isolated from domestic ducks in Zhejiang Province in China. Virus Genes, 2014, 49(1): 80-88.

[13] MEHLE A, DOUDNA J A. An inhibitory activity in human cells restricts the function of an avian-like influenza virus polymerase. Cell Host Microbe, 2008, 4(2): 111-122.

[14] LI Z, CHEN H, JIAO P, et al. Molecular basis of replication of duck H5N1 influenza viruses in a mammalian mouse model. J Virol, 2005, 79(18): 12058-12064.

Genetic characterization of a natural reassortant H3N8 avian influenza virus isolated from domestic geese in Guangxi, southern China

Liu Tingting, Xie Zhixun, Song Degui, Luo Sisi, Xie Liji, Li Meng, Xie Zhiqin, and Deng Xianwen

Abstract

A H3N8 subtype of avian influenza virus, A/goose/Guangxi/020G/2009 (H3N8) (GX020G), was isolated from the Guangxi of China in 2009. All eight gene segments of the GX020G strain were sequenced. Sequence analysis indicated that this H3N8 virus is a novel reassortant strain. The genome sequences provide useful information for understanding the epidemiology and molecular characteristics of the H3N8 subtype of influenza virus in southern China.

Keywords

avian influenza virus, H3N8 subtype, genetic characterization, domestic geese

The avian influenza virus (AIV) is a negative-sense, segmented RNA virus that belongs to the genus *influenza A virus* of the family Orthomyxoviridae[1, 2]. There are 17 hemagglutinin (HA) and 10 neuraminidase (NA) subtypes of AIV based on the antigenic differences of the HA and NA proteins, which are surface glycoproteins on the viral envelop[3-5]. This H3 subtype of AIV belongs to low pathogenic AIV (LPAIV), which is one of the predominant subtypes among LPAIVs. Some studies indicated that the H3 subtype has a high separation rate and the seasonal variations in the isolation of the H3 subtype of AIV are consistent with those of human H3 subtype influenza viruses[6, 7]. Research studies also predicts that the H3 subtype of AIV may have the ability to cross the species barrier to infect humans through gene reassortment[8, 9]. Moreover, Hong Kong influenza virus (H3N2) in 1968 was a reassortant with avian (H3) PB1 and HA genes and six other genes from the human (H2N2) virus[10]. Therefore, it is necessary to enhance the surveillance of the H3 subtype of AIV.

In this study, A/goose/Guangxi/020G/2009 (H3N8) was isolated from a goose in a live poultry market in Guangxi, China, in 2009. The eight genes were amplified by real-time PCR using AIV universal primers[11-13]. The amplified products were purified and cloned into the pMD18-T vector (TaKaRa, Dalian, China) and sequenced (TaKaRa, Dalian, China). Sequences were assembled and manually edited to generate the final full-length genome sequence.

The complete genome of the GX020G strain consists of eight gene segments of PB2, PB1, PA, HA, NP, NA, M, and NS genes. The full lengths of these segments are 2 341, 2 341, 2 233, 1 765, 1 565, 1 460, 1 027, and 890 nucleotides, respectively. The amino acid residues at the cleavage site (340~348) of the HA molecule are PEKQTR↓GLF with one basic amino acid, which is characteristic of low pathogenic AIV. The PB2 protein possesses E627and D701, which is characteristic of AIV. The M2 protein possesses S31, which is not amantadine resistant. The amino acid residues at the receptor binding site in the HA protein are Q226 and G228, different than L226 and S228 in the H3 subtype of human influenza viruses, which preferentially bind to an avian-origin receptor.

Sequence analysis indicates that the nucleotide sequences of both HA and NA genes of the GX020G strain both belong to the Eurasian lineage. The HA gene shows the highest homology (95%) to that from A/duck/Beijing/40/2004 (H3N8). The NA gene shows the highest homology (98.1%) to that from A/swine/Guangdong/ K4/2011 (H4N8), which suggests these strains may share similar original ancestors. The other genes show the highest homology (≥97%) to some Eurasia subtypes.

These data indicates that the GX020G is a novel reassortant virus whose genes derived from multiple AIV strains, and the genome information of GX020G is helpful in conducting epidemiology investigation on the H3N8 subtype of AIV in China.

Nucleotide sequence accession numbers. The genome sequence of A/goose/Guangxi/020G/2009 (H3N8) has been deposited at GenBank under the accession no. KJ764713 to KJ764720.

References

[1] PENG Y, XIE Z X, LIU J B, et al. Epidemiological surveillance of low pathogenic avian influenza virus (LPAIV) from poultry in Guangxi Province, southern China. PLOS ONE, 2013, 8(10): e77132.

[2] XU Q, XIE Z X, XIE L J, et al. Characterization of an avian influenza virus H9N2 strain isolated from a wild bird in southern China. Genome Announc, 2014, 2(3): e00600-14.

[3] WEBSTER R G, BEAN W J, GORMAN OT, et al. Evolution and ecology of influenza A viruses. Microbiol Rev, 1992, 56(1): 152-179.

[4] XIE Z X, XIE L J, ZHOU C Y, et al. Complete genome sequence analysis of an H6N1 avian influenza virus isolated from Guangxi Pockmark ducks. J Virol, 2012, 86(24): 13868-13869.

[5] XIE Z X, GUO J, XIE L J, et al. Complete genome sequence of a novel reassortant avian influenza H1N2 virus isolated from a domestic sparrow in 2012. Genome Announc, 2013, 1(4): e00431-13.

[6] PENG Y, ZHANG W, XUE F, et al. Etiological examination on the low pathogenicity avian influenza viruses with different HA subtypes from poultry isolated in eastern China from 2006 to 2008. Chin J Zoonoses, 2009, 25(02): 119-121.

[7] PENG Y, XIE Z X, LIU J B, et al. Visual detection of H3 subtype avian influenza viruses by reverse transcription loop-mediated isothermal amplification assay. Virol J, 2011, 8: 337.

[8] CAMPITELLI L, FABIANI C, PUZELLI S, et al. H3N2 influenza viruses from domestic chickens in Italy: an increasing role for chickens in the ecology of influenza?. J Gen Virol, 2002, 83(Pt 2): 413-420.

[9] SONG M S, OH T K, MOON H J, et al. Ecology of H3 avian influenza viruses in Korea and assessment of their pathogenic potentials. J Gen Virol, 2008, 89(Pt 4): 949-957.

[10] SCHOLTISSEK C, ROHDE W, VON HOYNINGEN V, et al. On the origin of the human influenza virus subtypes H2N2 and H3N2. Virology, 1978, 87(1): 13-20.

[11] HOFFMANN E, STECH J, GUAN Y, et al. Universal primer set for the full-length amplification of all influenza A viruses. Arch Virol, 2001, 146(12): 2275-2289.

[12] HE C Q, XIE Z X, HAN G Z, et al. Homologous recombination as an evolutionary force in the avian influenza A virus. Mol Biol Evol, 2009, 26(1): 177-187.

[13] HE C Q, DING N Z, MOU X, et al. Identification of three H1N1 influenza virus groups with natural recombinant genes circulating from 1918 to 2009. Virology, 2012, 427(1): 60-66.

Full-genome sequence analysis of a natural reassortant H4N2 avian influenza virus isolated from a domestic duck in southern China

Wu Aiqiong, Xie Zhixun, Xie Liji, Xie Zhiqin, Luo Sisi, Deng Xianwen, Huang Li, Huang Jiaoling, and Zeng Tingting

Abstract

We report here the complete genome sequence of a novel reassortant H4N2 avian influenza virus strain, A/duck/Guangxi/125D17/2012 (H4N2) (GX125D17), isolated from a duck in Guangxi, China in 2012. We obtained the complete genome sequence of the GX125D17 virus isolation by PCR, cloning, and sequencing. Sequence analysis revealed that this H4N2 virus strain was a novel reassortant avian influenza virus (AIV). Information about the complete genome sequence of the GX125D17 virus strain will be useful for epidemiological studies.

Keywords

avian influenza virus, H4N2 subtype, full-genome

Avian influenza virus (AIV) is a single-stranded, negative-sense RNA virus belonging to the family *Orthomyxoviridae*[1, 2]. Avian influenza viruses are divided into subtypes based on the two major surface glycoproteins, hemagglutinin (HA), and neuraminidase (NA). So far, 16 HA and 9 NA subtypes have been identified[1, 3-6]. The H4 subtype of AIVs have been circulating and evolving in live poultry markets (LPM) in China[7]. The H4N2 subtype of avian influenza virus has infected migratory water birds[8] and domestic ducks[9]. In addition, the H4 avian influenza virus has infected pigs[10] and poses a threat to mammals. Thus, enhanced the surveillance of H4N2 subtype of AIVs is of great importance.

In this study, an H4N2 strain, named A/Duck/Guangxi/125D17/2012(H4N1), was isolated from a duck in Guangxi, southern China in 2012. The nucleotide sequences of this virus strain were amplified by reverse transcription-PCR (RT-PCR) using universal primers[11-13]. The amplified products were purified and cloned into the pMD18-T vector (TaKaRa, Dalian, China) and then sequenced (Invitrogen, Shanghai, China). Sequences were assembled and manually edited to generate the final genome sequence[14].

The complete genome of GX125D17 strain consists of PB2, PB1, PA, HA, NP, NA, M, and NS segments. The eight genes encoded the following proteins, followed by the deduced amino acid: PB2, 760 aa; PB1, 757 aa; PA, 716 aa; HA, 564 aa; NP, 498 aa; NA, 469 aa; M1, 252 aa; M2, 97 aa; NS1, 230 aa; and NS2, 121 aa. The amino acid sequence at the cleavage site in the HA molecule is EKASR↓GLF, which is characteristic of low-pathogenicity avian influenza virus. Analysis of potential glycosylation sites of surface proteins revealed five potential N-glycosylation sites in HA (18 to 21, 34 to 37, 178 to 181, 310 to 313, and 497 to 500) and seven potential N-glycosylation sites in NA (61 to 64, 69 to 72, 70 to 73, 146 to 149, 200 to 203, 234 to 237, and 402 to 405).

Sequence analysis revealed that the nucleotide sequences of the HA and NA genes of the GX125D17 strain both belong to the Eurasian lineage. The nucleotide homology comparisons revealed that the HA gene of this strain shares 99% homology with the HA gene of a Republic of Korea (ROK) wild bird AIV strain, A/ wild duck/ROK/CSM4-28/2010(H4N6). The NA and NP genes both share the highest homology (≥98%) with those of the H1N2 isolate A/duck/Guangxi/GXd-1/2011(H1N2). The NS gene shares 97% homology with that of the H3N8 isolate A/duck/Chabarovsk/1610/ 1972(H3N8). The PB1 gene shares the highest homology (98%) with that of the H4N6 isolate A/wild duck/ROK/SH5-26/2008(H4N6). The PB2 gene shares the highest homology (98%) with that of the H7N7 isolate A/common teal/Hong Kong/ MPL634/2011(H7N7). The M gene shares 99% homology with the M gene of the H10N8 isolate, A/duck/Guangdong/El/2012(H10N8). The PA gene shares 99% homology with the PA gene of H1N2 isolate, A/wild waterfowl/Dongting/C23831 2012 (H1N2). In conclusion, the GX125D17 virus isolation was proved to be a novel reassortant AIV.

These data will be helpful in epidemiological studies on H4N2 subtype of AIVs in southern China.

Nucleotide sequence accession numbers. The genome sequences of A/Duck/Guangxi/125D17/2012 (H4N2) have been deposited in GenBank under accession numbers KJ881013 to KJ881020.

References

[1] WEBSTER R G, BEAN W J, GORMAN O T, et al. Evolution and ecology of influenza A viruses. Microbiol Rev, 1992, 56(7): 152-179.

[2] XU Q, XIE Z X, XIE L J, et al. Characterization of an Avian influenza virus H9N2 strain isolated from a wild bird in southern China. Genome Announc, 2014, 2(3): e00600-14.

[3] FOUCHIER R A, MUNSTER V, WALLENSTEN A, et al. Characterization of a novel influenza A virus hemagglutinin subtype (H16) obtained from black-headed gulls. J Virol, 2005, 79(5): 2814-2822.

[4] PENG Y, XIE Z X, LIU J B, et al. Epidemiological surveillance of low pathogenic avian influenza virus (LPAIV) from poultry in Guangxi Province, southern China. PLOS ONE, 2013, 8(10): e77132.

[5] SUBBARAO K, JOSEPH T. Scientific barriers to developing vaccines against avian influenza viruses. Nat Rev Immunol, 2007, 7(4): 267-278.

[6] XIE Z, XIE L, ZHOU C, et al. Complete genome sequence analysis of an H6N1 avian influenza virus isolated from Guangxi pockmark ducks. J Virol, 2012, 86(24): 13868-13869.

[7] LIU M, HE S, WALKER D, et al. The influenza virus gene pool in a poultry market in South central china. Virology, 2003, 305(2): 267-275.

[8] BUI V N, OGAWA H, KARIBE K, et al. Surveillance of avian influenza virus in migratory water birds in eastern Hokkaido, Japan. J Vet Med Sci, 2011, 73(2): 209-215.

[9] ZHANG H, CHEN Q, CHEN Z. Characterization of an H4N2 avian influenza virus isolated from domestic duck in Dongting Lake wetland in 2009. Virus Genes, 2012, 44(1): 24-31.

[10] KARASIN A I, BROWN IH, CARMAN S, et al. Isolation and characterization of H4N6 avian influenza viruses from pigs with pneumonia in Canada. J Virol, 2000, 74(19): 9322-9327.

[11] HOFFMANN E, STECH J, GUAN Y, et al. Universal primer set for the full-length amplification of all influenza A viruses. Arch Virol, 2001, 146(12): 2275-2289.

[12] HE C Q, DING N Z, MOU X, et al. Identification of three H1N1 influenza virus groups with natural recombinant genes circulating from 1918 to 2009. Virology, 2012, 427(1): 60-66.

[13] HE C Q, XIE Z X, HAN G Z, et al. Homologous recombination as an evolutionary force in the avian influenza A virus. Mol Biol Evol, 2009, 26(1): 177-187.

[14] XIE Z X, GUO J, XIE L J, et al. Complete genome sequence of a novel reassortant avian influenza H1N2 virus isolated from a domestic sparrow in 2012. Genome Announc, 2013, 1(4): e00431-13.

Genetic characterization of an avian influenza virus H4N6 strain isolated from a Guangxi Pockmark duck

Xie Liji, Xie Zhixun, Wu Aiqiong, Luo Sisi, Zhang Minxiu, Huang Li, Xie Zhiqin, Huang Jiaoling, Zhang Yanfang, Zeng Tingting, and Deng Xianwen

Abstract

An H4N6 subtype avian influenza virus was isolated from a Pockmark duck in southern China in November 2013 and named A/duck/Guangxi/149D24/ 2013 (H4N6). All eight gene segments of the strain were sequenced. Sequence analysis indicated that this H4N6 virus was a natural reassortant virus. This H4N6 virus has two basic amino acids in the cleavage site of hemagglutinin 1 (HA1) and HA2, and the amino acid motif of cleavage site was PEKASRGLF, which is the typical characteristic of the low-pathogenic avian influenza virus. This study will help understand the epidemiology and molecular characteristics of avian influenza virus in Pockmark ducks.

Keywords

avian influenza virus, H4N6 subtype, genetic characterization

---

Avian influenza virus (AIV) is a negative-sense segmented RNA virus that belongs to the genus *influenza A virus* of the family Orthomyxoviridae[1, 2]. At present, there are 18 hemagglutinin (HA) and 11 neuraminidase (NA) subtypes of AIV based on the antigenic differences of the HA and NA proteins, which are surface glycoproteins on the viral envelope[3, 4]. This H4 subtype of AIV belongs to low-pathogenic AIV (LPAIV), which is one of the predominant subtypes among LPAIV. The H4 subtype of AIV has been circulating and evolving in live poultry markets in China[5]. It has been shown that the H4 subtype of AIV has infected migratory water birds and domestic ducks[6, 7]. In addition, the H4 subtype of AIV has infected pigs and poses a threat to mammals[8]. It may have the ability to cross the species barrier to infect humans through gene reassortment, thus signifying the importance of enhancing the surveillance of the H4 subtype of AIV.

An H4N6 subtype AIV was isolated from a Pockmark duck in Guangxi, China, in November 2013 and named A/duck/Guangxi/149D24/2013(H4N6). All eight gene segments were amplified by reverse transcription-PCR using AIV universal primers[9, 10]. The amplified products were gel purified, cloned into the pMD18-T vector (TaKaRa, Dalian, China), and sequenced (TaKaRa, Dalian, China). The sequences were assembled using the SeqMan program and manually edited to generate the final full-length genome sequence.

The complete genome of the A/duck/Guangxi/149D24/2013(H4N6) strain consists of eight gene segments of polymerase basic 2 (PB2), PB1, polymerase acidic (PA), HA, nucleoprotein (NP), NA, matrix (M), and nonstructural (NS) genes. The full lengths of these segments are 2 341, 2 341, 2 233, 1 738, 1 565, 1 464, 1 027, and 890 nucleotides, respectively. The amino acid residues at the cleavage site (positions 338 to 346) of the HA molecule are PEKASRGLF, with two basic amino acids, which is characteristic of low-pathogenic AIV.

Sequence analysis revealed that the nucleotide sequences of the HA and NA genes of the A/duck/ Guangxi/149D24/2013(H4N6) strain both belong to the Eurasian lineage. The nucleotide homology comparisons revealed that the HA gene of this strain shares 99% homology with the HA gene of a Jiangxi AIV strain, A/duck/Jiangxi/32180/2013 (mixed type) (GenBank accession number: KP286940). The NA gene shared the highest sequence homology, at 99%, with A/Duck/Thailand/CU-11825C/2011(H3N6) (GenBank accession number: KJ161948). The PB2 gene shared the highest sequence homology, at 96%, with A/wild duck/Korea/SNU50-5/2009(H5N1) (GenBank accession number: JX497765). The PB1 gene shared the highest sequence homology, at 98%, with A/wild duck/Korea/PSC6-1/2009(H4N6) (GenBank accession number: JX454743). The PA and NP genes shared the highest sequence homology, at 99%, with A/Duck/Vietnam/ LBM48/2011(H3N2) (GenBank accession number: LC028079). The M gene shared the highest sequence homology, at 98%, with A/duck/Guangxi/GXd-1/2011(H1N2) (GenBank accession number: KF013919). The NS gene shared the highest sequence homology, at 99%, with A/duck/Taiwan/WB459/04(H6N5) (GenBank accession number: DQ376795).

These data indicate that the A/duck/Guangxi/149D24/2013(H4N6) strain is a novel reassortant virus whose genes derived from multiple AIV strains, and its genome information is useful for analyses of epidemiology and evolutionary characteristics.

Accession number (s). The genome sequence of A/duck/Guangxi/149D24/2013(H4N6) was deposited in GenBank under the accession numbers MF399054 to MF399061.

References

[1] PENG Y, XIE Z X, LIU J B, et al. Epidemiological surveillance of low pathogenic avian influenza virus (LPAIV) from poultry in Guangxi Province, Southern China. PLOS ONE, 2013, 8(10): e77132.

[2] XU Q, XIE Z X, XIE L J, et al. Characterization of an avian influenza virus H9N2 strain isolated from a wild bird in southern China. Genome Announc, 2014, 2(3): e00600-14.

[3] WU Y, WU Y, TEFSEN B, et al. Bat-derived influenza-like viruses H17N10 and H18N11. Trends Microbiol, 2014, 22(4): 183-191.

[4] TONG S, ZHU X, LI Y, et al. New world bats harbor diverse influenza A viruses. PLOS Pathog, 2013, 9(10): e1003657.

[5] LIU M, HE S, WALKER D, et al. The influenza virus gene pool in a poultry market in south central China. Virology, 2003, 305(2): 267-275.

[6] BUI V N, OGAWA H, KARIBE K, et al. Surveillance of avian influenza virus in migratory water birds in eastern Hokkaido, Japan. J Vet Med Sci, 2011, 73(2): 209-215.

[7] ZHANG H, CHEN Q, CHEN Z. Characterization of an H4N2 avian influenza virus isolated from domestic duck in Dongting Lake wetland in 2009. Virus Genes, 2012, 44(1): 24-31.

[8] KARASIN A I, BROWN I H, CARMAN S, et al. Isolation and characterization of H4N6 avian influenza viruses from pigs with pneumonia in Canada. J Virol, 2000, 74(19): 9322-9327.

[9] HOFFMANN E, STECH J, GUAN Y, et al. Universal primer set for the full-length amplification of all influenza A viruses. Arch Virol, 2001, 146(12): 2275-2289.

[10] HE C Q, XIE Z X, HAN G Z, et al. Homologous recombination as an evolutionary force in the avian influenza A virus. Mol Biol Evol, 2009, 26(1): 177-187.

Molecular characteristics of H6N6 influenza virus isolated from pigeons in Guangxi, southern China

Li Meng, Xie Zhixun, Xie Zhiqin, Luo Sisi, Xie Liji, Huang Li, Deng Xianwen, Huang Jiaoling, Zhang Yanfang, Zeng Tingting, and Wang Sheng

Abstract

Here, we report the complete genome sequence of an H6N6 avian influenza virus (AIV) isolated from a pigeon in Guangxi, southern China, in 2014. The eight RNA segment genes shared a high nucleotide identity (97% to 99%) with H6N6 subtypes of AIV isolated from ducks in the regions around Guangxi. The finding of this study will help us understand the ecology and molecular characteristics of H6 avian influenza virus in wild birds in southern China.

Keywords

avian influenza virus, H6N6 subtype, molecular characteristics

Avian influenza A virus (AIV) is a single-stranded negative-sense RNA virus, which can cause a variety of infections in avian and mammalian species. The H6 subtypes of influenza viruses are the most abundantly detected influenza virus subtype, and they have a broader host range than any other subtype[1]. In the past decade, surveillance studies have revealed the existence of different subtypes of H6 viruses, which showed that most of the strains were isolated from ducks, chickens, and geese[2-6]. Recent reports have detected the H7N9 viruses in healthy pigeons, and these viruses have once again come into the spotlight for their potential role as a bridge among species in the ecology of avian influenza[7, 8]. However, the epidemiology and biological characterization of the H6 virus isolated from pigeons are still unknown in China.

We report here the complete genomic sequence of an H6N6 avian influenza virus strain, A/pigeon/Guangxi/161/2014 (H6N6) (GX161), which was first isolated from pigeons in Guangxi, southern China, in 2014. All eight gene segments were sequenced by a DNA sequencing service company (TaKaRa, Dalian, China). The sequences were assembled and manually edited to generate the final genome sequence, as in a previous study[9].

The full lengths of the polymerase basic 2 (PB2) and PB1, polymerase acidic protein (PA), hemagglutinin (HA), nucleoprotein (NP), neuraminidase (NA), matrix (M), and nonstructural (NS) genes were 2 341, 2 341, 2 233, 1 744, 1 565, 1 464, 1 027, and 890 nucleotides, respectively. Phylogenetic analyses showed that the HA and NA genes belonged to the same clade as H6N6 viruses currently circulating in China, such as A/duck/Guangxi/Gxd-71/2011 and A/duck/Fujian/7818/2007 (up to 97% nucleotide identity with the reference strains). The other six genes were found to be more similar to those of Chinese and Vietnamese H6N6 AIV strains (up to 99% nucleotide identity). Interestingly, all the reference strains were isolated from ducks in the different regions around Guangxi, and this result maybe also suggests that H6 subtype avian influenza viruses can transmit direct from ducks to pigeons.

The amino acid motif of the cleavage site between HA1 and HA2 was PQIETRG; this is a typical characteristic of the low-pathogenicity avian influenza virus[10]. The pigeon GX161 strain has Q226 and G228 (according to H3 numbering) at the receptor-binding site in the HA protein, which was different with S228 detected in both swine and human H6 isolates; this result suggests that the strain has the ability to bind a sialic acid-2, 3-NeuAc Gal linkage[11]. The possession of 627 Glu and 701 Asp in the amino acid sequence of PB2 protein, which still has the characteristics of avian influenza viruses, so it was not be able to replicate in mammalian hosts[12]. Positions 26, 27, 30, 31, and 34, which did not have an amino acid point mutation in the matrix 2 (M2) protein, showed that the GX161 strain is not amantadine resistant[13]. This study will help to understand the molecular characteristics of H6 subtype avian influenza viruses in pigeons.

Nucleotide sequence accession numbers. The genome sequences of A/pigeon/Guangxi/161/2014(H6N6) have been deposited in GenBank under accession numbers KT267019 to KT267026.

References

[1] SPACKMAN E, STALLKNECHT D E, SLEMONS R D, et al. Phylogenetic analyses of type A influenza genes in natural reservoir species in North America reveals genetic variation. Virus Res, 2005, 114(1-2): 89-100.

[2] XIE Z X, XIE L J, ZHOU C Y, et al. Complete genome sequence analysis of an H6N1 avian influenza virus isolated from Guangxi Pockmark ducks. J Virol, 2012, 86(24): 13868-13869.

[3] PENG Y, XIE Z X, LIU J B, et al. Epidemiological surveillance of low pathogenic avian influenza virus (LPAIV) from poultry in Guangxi Province, southern China. PLOS ONE, 2013, 8(10): e77132.

[4] WANG G, DENG G, SHI J, et al. H6 influenza viruses pose a potential threat to human health. J Virol, 2014, 88(8): 3953-3964.

[5] CHEUNG C L, VIJAYKRISHNA D, SMITH G J, et al. Establishment of influenza A virus (H6N1) in minor poultry species in southern China. J Virol, 2007, 81(19): 10402-10412.

[6] HUANG K, ZHU H, FAN X, et al. Establishment and lineage replacement of H6 influenza viruses in domestic ducks in southern China. J Virol, 2012, 86(11): 6075-6083.

[7] SHI J, DENG G, LIU P, et al. Isolation and characterization of H7N9 viruses from live poultry markets—implication of the source of current H7N9 infection in human. Mol Biol Evol, 2009, 26(1): 177-187.

[8] ABOLNIK C. A current review of avian influenza in pigeons and doves (Columbidae). Vet Microbiol, 2014, 170(3-4): 181-196.

[9] XU Q, XIE Z X, XIE L J, et al. Characterization of an avian influenza virus H9N2 strain isolated from a wild bird in southern China. Genome Announc, 2014, 2(3): e00600-14.

[10] ZHAO G, LU X, GU X, et al. Molecular evolution of the H6 subtype influenza A viruses from poultry in eastern China from 2002 to 2010. Virol J, 2011, 8: 470.

[11] MATROSOVICH M, TUZIKOV A, BOVIN N, et al. Early alterations of the receptor-binding properties of H1, H2, and H3 avian influenza virus hemagglutinins after their introduction into mammals. J Virol, 2000, 74(18): 8502-8512.

[12] SUBBARAO E K, LONDON W, MURPHY B R. A single amino acid in the PB2 gene of influenza A virus is a determinant of host range. J Virol, 1993, 67(4): 1761-1764.

[13] FURUSE Y, SUZUKI A, OSHITANI H, et al. Large-scale sequence analysis of M gene of influenza A viruses from different species: mechanisms for emergence and spread of amantadine resistance. Antimicrob Agents Chemother , 2009, 26(1): 177-187.

Genome sequence of an H9N2 avian influenza virus strain with hemagglutinin-neuraminidase combination, isolated from a quail in Guangxi, southern China

Xie Liji, Xie Zhixun, Li Dan, Luo Sisi, Zhang Minxiu, Huang Li, Xie Zhiqin, Huang Jiaoling, Zhang Yanfang, Zeng Tingting, and Deng Xianwen

Abstract

We isolated a strain of H9N2 avian influenza virus from a quail in southern China in May 2015 and named it A/quail/Guangxi/198Q39/2015. All eight gene segments of the strain were sequenced. Sequence analysis indicated that the amino acid motif of the hemagglutinin cleavage site of this H9N2 virus was RSSR↓GLF, which is a typical characteristic of the low pathogenic avian influenza virus. This study will help in better understanding the epidemiology and molecular characteristics of avian influenza virus in wild birds.

Keywords

avian influenza virus, H9N2 subtype, hemagglutinin-neuraminidase combination

The avian influenza virus is a negative-sense, segmented RNA virus that belongs to the genus *influenza A virus* of the family Orthomyxoviridae[1, 2]. At present, there are 18 hemagglutinin (HA) and 11 neuraminidase (NA) subtypes of avian influenza virus based on the antigenic differences of the HA and NA proteins, which are surface glycoproteins on the viral envelope[3, 4].

The H9N2 subtype avian influenza virus is widespread in the world and is the most prevalent subtype of avian influenza viruses reported in China over the past decade[5, 6]. Although H9N2 is characterized as a low-pathogenic avian influenza virus, occasional infections of humans[7-10] have caused great concerns. So far, H9N2 subtype avian influenza viruses are isolated mainly from domestic birds[11]; however, wildfowl and shorebirds are the natural hosts of avian influenza virus and they facilitate the transmission of avian influenza[2, 6].

An H9N2 subtype avian influenza virus was isolated from a quail in Guangxi, China, in May 2015 and named A/quail/Guangxi/198Q39/2015(H9N2). All eight gene segments were amplified by reverse transcription-PCR using avian influenza virus universal primers[12, 13]. The amplified products were gel purified and cloned into the pMD18-T vector (TaKaRa, Dalian, China) and sequenced (TaKaRa, Dalian, China). The sequences were assembled using the SeqMan program and manually edited to generate the final full-length genome sequence.

The complete genome of the A/quail/Guangxi/198Q39/2015 strain consists of eight segments of the HA, NA, NS, M, NP, PA, PB1, and PB2 genes. The full lengths of these segments are 1 742, 1 457, 890, 1 027, 1 565, 2 233, 2 341, and 2 341 nucleotides, respectively. The amino acid (aa) lengths of the proteins encoded by the eight genes are 560 aa (HA), 466 aa (NA), 121 aa (NS2), 217 aa (NS1), 252 aa (M1), 97 aa (M2), 498 aa (NP), 716 aa (PA), 758 aa (PB1), and 259 aa (PB2).

The amino acid residues at the cleavage site (aa 335~341) of the HA molecule are RSSR↓GLF, which is characteristic of low-pathogenic avian influenza virus. Sequence analysis revealed that the nucleotide sequences of the HA and NA genes of the A/quail/Guangxi/198Q39/2015 strain both belong to the Eurasian lineage. The A/quail/Guangxi/198Q39/2015 strain has Leu234 and Gly236 at the receptor-binding site in the HA protein, which suggests that it might have the ability to bind a sialic acid-2, 6-NeuAcGal linkage and might have the potential to infect humans[9].

The nucleotide homology comparisons revealed that the HA gene of this strain shared the highest sequence homology (98%) with the HA gene of a Beijing avian influenza virus strain, A/chicken/Beijing/0309/2013 (GenBank accession number: KM609599). The NA gene shared the highest sequence homology (98%) with A/turtledove/Guangxi/49B6/2013 (GenBank accession number: KJ725014). These results are useful for future analyses of the molecular epidemiology and evolutionary characteristics of avian influenza virus.

Accession number (s). The genome sequence of A/quail/Guangxi/198Q39/2015(H9N2) was deposited in GenBank under the accession numbers MF425642 to MF425649.

References

[1] WEBSTER R G, BEAN W J, GORMAN O T, et al. Evolution and ecology of influenza A viruses. Microbiol Rev, 1992, 56(1): 152-179.

[2] XU Q, XIE Z X, XIE L J, et al. Characterization of an avian influenza virus H9N2 strain isolated from a wild bird in southern China. Genome Announc, 2014, 2(3): e00600-14.

[3] WU Y, WU Y, TEFSEN B, et al. Bat-derived influenza-like viruses H17N10 and H18N11. Trends Microbiol, 2014, 22(4): 183-191.

[4] TONG S, ZHU X, LI Y, et al. New world bats harbor diverse influenza A viruses. PLOS Pathog, 2013, 9(10): e1003657.

[5] JI K, JIANG W M, LIU S, et al. Characterization of the hemagglutinin gene of subtype H9 avian influenza viruses isolated in 2007-2009 in China. J Virol Methods, 2010, 163(2): 186-189.

[6] PENG Y, XIE Z X, LIU J B, et al. Epidemiological surveillance of low pathogenic avian influenza virus (LPAIV) from poultry in Guangxi Province, southern China. PLOS ONE, 2013, 8(10): e77132.

[7] ALEXANDER D J, BROWN I H. Recent zoonoses caused by influenza A viruses. Rev Sci Tech, 2000, 19(1): 197-225.

[8] PEIRIS M, YUEN K Y, LEUNG C W, et al. Human infection with influenza H9N2. Lancet, 1999, 354(9182): 916-917.

[9] WAN H, PEREZ D R. Amino acid 226 in the hemagglutinin of H9N2 influenza viruses determines cell tropism and replication in human airway epithelial cells. J Virol, 2007, 81(10): 5181-5191.

[10] XIE Z X, DONG J B, Tang X F, et al. Sequence and phylogenetic analysis of three isolates of avian influenza H9N2 from chickens in southern China. Sch Res Exch, 2008: 1-7.

[11] OLSEN B, MUNSTER VJ, WALLENSTEN A, et al. Global patterns of influenza a virus in wild birds. Science, 2006, 312(5772): 384-388.

[12] HOFFMANN E, STECH J, GUAN Y, et al. Universal primer set for the full-length amplification of all influenza A viruses. Arch Virol, 2001, 146(12): 2275-2289.

[13] HE C Q, XIE Z X, HAN G Z, et al. Homologous recombination as an evolutionary force in the avian influenza A virus. Mol Biol Evol, 2009, 26(1): 177-187.

Characterization of an avian influenza virus H9N2 strain isolated from dove in southern China

Li Dan, Li Zhengting, Xie Zhixun, Li Meng, Xie Zhiqin, Liu Jiabo, Xie Liji, Deng Xianwen, and Luo Sisi

Abstract

We report here the complete genome sequence of strain H9N2, an avian influenza virus (AIV) isolated from dove in Guangxi, China. Phylogenetic analysis showed that it was a novel reassortant AIV derived from chicken, duck, and wild bird. This finding provides useful information for understanding the H9N2 subtype of AIV circulating in southern China.

Keywords

avian influenza virus, H9N2 subtype, dove

Avian influenza virus (AIV) is a negative-sense RNA virus of the family Orthomyxoviridae. Currently, there are 18 hemagglutinin (HA) and 11 neuraminidase (NA) subtypes of AIV based on the antigenic differences of the HA and NA proteins[1-5]. Although H9N2-subtype AIVs belong to the group of low-pathogenic AIVs, this subtype of influenza virus has spread to many poultry farms in China and is considered endemic[6, 7]. In addition, the spread of H9N2 subtype AIV has resulted in significant economic losses due to reduced egg production and high mortality associated with coinfection with other respiratory pathogens.

In February 2014, an H9N2 subtype AIV was isolated from an infected dove, named A/dove/Guangxi/96B8/2014 (H9N2) (DV/GX/96B8). Dove is a very popular dish in Guangxi, southern China, and doves are raised and sold with other birds and animals in live poultry markets. In this study, eight gene segments of the isolated AIV were amplified by reverse transcription-PCR using the universal primers of the influenza A virus[8]. The amplified products were purified and cloned into the pMD18-T vector and sequenced at TaKaRa (Dalian, China).

The complete genome of this H9N2 strain consisted of eight gene segments of polymerase basic 2 (PB2), polymerase basic 1 (PB1), polymerase acidic (PA), hemagglutinin (HA), nucleoprotein (NP), neuraminidase (NA), matrix (M), and nonstructural (NS) genes. The full lengths of the segments were 2 341 nucleotides (nt), 2 341 nt, 2 233 nt, 1 742 nt, 1 565 nt, 1 458 nt, 1 027 nt, and 890 nt, respectively. The amino acid residues at the cleavage site (nt 335 to 341) of the HA molecule were RSSR↓GLF without basic amino acid, which is characteristic of low-pathogenic AIVs[9]. The PA protein possesses T515, the PB1 protein possesses Y436, and the PB2 protein possesses E158, E627, and D701, providing further evidence of low pathogenicity[10, 11]. The DV/GX/96B8 strain has L226 and G228 (according to H3 numbering) at the receptor-binding site in the HA protein, which suggests that the DV/GX/96B8 strain has the ability to bind with sialic acid-2, 6-NeuAcGal linkage and might have the potential to infect humans[12, 13].

Phylogenetic analysis of the DV/GX/96B8 surface genes HA and NA showed that they belonged to a G1-like virus and that their nucleotide homologies were 98% and 95% compared with the G1-like virus, respectively[14]. The virus in this branch differs from the vaccine strains that are used to immunize chickens against H9N2 subtype AIVs. The internal genes showed that the PB1, NP, PB2, PA, and NS genes belonged to F98-like virus, and nucleotide homologies were all greater than 98% compared with the F98-like virus. The M gene belonged to a G1-like virus and had nucleotide homology greater than 98% compared with the strain A/chicken/Shandong/yt0106/2012 (H9N2)[15]. Thus, we isolated a natural recombinant H9N2 influenza virus from dove that differed from other H9N2 genotype strains. The genomic information of DV/GX/96B8 is crucial for conducting an epidemiological investigation of the H9N2 subtype of AIV in China.

Nucleotide sequence accession number. The complete genome sequence of A/dove/Guangxi/96B8/2014 (H9N2) has been deposited in GenBank under the accession numbers MF465797 to MF465804.

References

[1] ZHU X, YU W, MCBRIDE R, et al. Hemagglutinin homologue from H17N10 bat influenza virus exhibits divergent receptor-binding and pH-dependent fusion activities. Proc Natl Acad Sci U S A, 2013, 110(4): 1458-1463.

[2] XIE L, XIE Z, LI D, et al. Genome sequence of an H9N2 avian influenza virus strain with hemagglutinin-neuraminidase combination, isolated from a quail in Guangxi, southern China. Genome Announc, 2017, 5(38): e00965-17.

[3] WU Y, WU Y, TEFSEN B, et al. Bat-derived influenza-like viruses H17N10 and H18N11. Trends Microbiol, 2014, 22(4): 183-191.

[4] CHAN J F, TO K K, TSE H, et al. Interspecies transmission and emergence of novel viruses: lessons from bats and birds. Trends Microbiol, 2013, 21(10): 544-555.

[5] LUO S S, XIE Z X, XIE Z Q, et al. Surveillance of live poultry markets for low pathogenic avian influenza viruses in Guangxi Province, southern China, from 2012-2015. Sci Rep, 2017, 7(1): 17577.

[6] PENG Y, XIE Z X, LIU J B, et al. Epidemiological surveillance of low pathogenic avian influenza virus (LPAIV) from poultry in Guangxi Province, Southern China. PLOS ONE, 2013, 8(10): e77132.

[7] XU Q, XIE Z X, XIE L J, et al. Characterization of an avian influenza virus H9N2 strain isolated from a wild bird in southern China. Genome Announc, 2014, 2(3): e00600-14.

[8] HOFFMANN E, STECH J, GUAN Y, et al. Universal primer set for the full-length amplification of all influenza A viruses. Arch Virol, 2001, 146(12): 2275-2289.

[9] XIE Z X, XIE L J, ZHOU C Y, et al. Complete genome sequence analysis of an H6N1 avian influenza virus isolated from Guangxi Pockmark ducks. J Virol, 2012, 86(24): 13868-13869.

[10] HULSE-POST D I, FRANKS J, BOYD K, et al. Molecular changes in the polymerase genes (PA and PB1) associated with high pathogenicity of H5N1 influenza virus in mallard ducks. J Virol, 2007, 81(16): 8515-8524.

[11] ZHOU B, LI Y, HALPIN R, et al. PB2 residue 158 is a pathogenic determinant of pandemic H1N1 and H5 influenza a viruses in mice. J Virol, 2011, 85(1): 357-365.

[12] MATROSOVICH M, TUZIKOV A, BOVIN N, et al. Early alterations of the receptor-binding properties of H1, H2, and H3 avian influenza virus hemagglutinins after their introduction into mammals. J Virol, 2000, 74(18): 8502-8512.

[13] LI M, XIE Z X, XIE Z Q, et al. Molecular Characteristics of H6N6 Influenza Virus Isolated from Pigeons in Guangxi, Southern China. Genome Announc, 2015, 3(6): e01422-15.

[14] WANG B, LIU Z, CHEN Q, et al. Genotype diversity of H9N2 viruses isolated from wild birds and chickens in Hunan Province, China. PLOS ONE, 2014, 9(6): e101287.

[15] PU J, WANG S, YIN Y, et al. Evolution of the H9N2 influenza genotype that facilitated the genesis of the novel H7N9 virus. Proc Natl Acad Sci U S A, 2015, 112(2): 548-553.

Identification of a triple-reassortant H1N1 swine influenza virus in a southern China swine

Xie Zhixun, Zhang Minxiu, Xie Liji, Luo Sisi, Liu Jiabo, Deng Xianwen, Xie Zhiqin, Pang Yaoshan, and Fan Qing

Abstract

We report here the complete genome sequence of a triple-reassortant H1N1 swine influenza virus strain, A/swine/Guangxi/BB1/2013(H1N1) (GXBB1), isolated from a swine in the Guangxi of southern China in 2013. We obtained the complete genome sequence of the GXBB1 virus. Sequence analysis demonstrated that this H1N1 virus was a triple-reassortant swine influenza virus (SIV) whose genes originated from avian, human, and swine, respectively. Knowledge regarding the complete genome sequence of the GXBB1 virus will be useful for epidemiological surveillance.

Keywords

swine influenza virus, H1N1 subtype, swine

Swine influenza virus (SIV), a member of the genus *Orthomyxovirus* (family Orthomyxoviridae), is a single-stranded, negative-sense RNA virus that causes an acute and highly contagious respiratory disease in swine. The pig can serve as a mixing vessel for the generation of genetically reassortant viruses because the porcine tracheal cells have receptors for both avian and human influenza viruses[1, 2]. It has been reported that some new variants were generated by reassortment between avian and human viruses that occurred in pigs in nature[3, 4, 5], and the new variants have the potential to pose a serious threat to public health. Several sporadic human infections caused by swine-like viruses were reported[6, 7]. A novel influenza A (H1N1) virus genome which contained human, swine, and avian virus genes emerged in the human population in 2009, and this pandemic strain may be derived from pigs[8, 9]. Thus, it is important to enhance the surveillance of SIVs.

In this study, A/swine/Guangxi/BB1/2013(H1N1) (GXBB1) was isolated from a sick pig in Guangxi, China. The eight genes of GXBB1 were amplified by reverse transcription (RT)-PCR using universal primers[10, 11]. The amplified products were purified and cloned into the pMD18-T vector (TaKaRa, Dalian, China) and sequenced (TaKaRa, Dalian, China)[12]. Sequences were assembled and manually edited to generate the final full-length genome sequence.

The genome of the GXBB1 virus consisted of eight gene segments, which included polymerase (PB2, PB1, and PA), hemagglutinin (HA), nucleoprotein (NP), neuraminidase (NA), matrix protein (M), and nonstructural protein (NS) genes, respectively. The PB2 gene consisted of 2 280 nucleotides (nt), the PB1 gene 2 274 nt, the PA gene 2 150 nt, the HA gene 1 701 nt, the NP gene 1 515 nt, the NA gene 1 410 nt, the M gene 982 nt, and the NS gene 838 nt. The PSIQSR ↓ G in the HA cleavage site and 92D in the NS1 site classified GXBB1 as a low pathogenic influenza virus. 190D, 220R, 225E, 226Q, and 228G (according to H3 numbering) in HA conferred preferential binding to the SAα-2, 6-Gal receptor. The amino acid residues in NA associated with neuraminidase inhibitory drugs were conservative, which implied that GXBB1 might be sensitive to neuraminidase inhibitors, but S31N in M2 suggested that GXBB1 might be resistant to M2 ion channel inhibitors.

Phylogenetic analysis showed that the HA, NA, and M genes belonged to the European avian lineage. The PB2 and PA genes were both from a North American avian lineage. PB1 was a North American triple-reassortant SIV of human origin. The NP and NS genes were from a classical H1N1 swine lineage, and the PB2, PB1, PA, NS, and NP genes were derived from the reassortant H1N2 or H3N2 SIV. Phylogenetic analysis showed that the GXBB1 virus was a "human-swine-avian" triple-reassortant H1N1 virus, which was generated through gene reassortment between triple-reassortant H1N2 SIV or H3N2 SIV and European avian-like H1N1 SIV.

Our results demonstrated that the GXBB1 isolate is a triple-reassortant SIV whose genes were derived from avian, human, and swine origins.

Nucleotide sequence accession numbers. The genome sequence of A/swine/Guangxi/BB1/2013 (H1N1) was deposited in GenBank under the accession numbers KJ174942-KJ174949.

References

[1] CAMPITELLI L, DONATELLI I, FONI E, et al. Continued evolution of H1N1 and H3N2 influenza viruses in pigs in Italy. Virology, 1997, 232(2): 310-318.

[2] ITO T, COUCEIRO JN, KELM S, et al. Molecular basis for the generation in pigs of influenza A viruses with pandemic potential. J Virol, 1998, 72(9): 7367-7373.

[3] BROWN I H, ALEXANDER D J, CHAKRAVERTY P, et al. Isolation of an influenza A virus of unusual subtype (H1N7) from pigs in England, and the subsequent experimental transmission from pig to pig. Vet Microbiol, 1994, 39(1-2): 125-134.

[4] BROWN I H, HARRIS P A, MCCAULEY J W, et al. Multiple genetic reassortment of avian and human influenza A viruses in European pigs, resulting in the emergence of an H1N2 virus of novel genotype. J Gen Virol, 1998, 79(Pt 12): 2947-2955.

[5] CASTRUCCI M R, DONATELLI I, SIDOLI L, et al. Genetic reassortment between avian and human influenza A viruses in Italian pigs. Virology, 1993, 193(1): 503-506.

[6] MYERS K P, OLSEN C W, GRAY G C. Cases of swine influenza in humans: a review of the literature. Clin Infect Dis, 2007, 44(8): 1084-1088.

[7] SHINDE V, BRIDGES C B, UYEKI T M, et al. Triple-reassortant swine influenza A (H1) in humans in the United States, 2005-2009. N Engl J Med, 2009, 360(25): 2616-2625.

[8] SMITH G J, VIJAYKRISHNA D, BAHL J, et al. Origins and evolutionary genomics of the 2009 swine-origin H1N1 influenza A epidemic. Nature, 2009, 459(7250): 1122-1125.

[9] PEIRIS J S, POON L L, GUAN Y. Emergence of a novel swine-origin influenza A virus (S-OIV) H1N1 virus in humans. J Clin Virol, 2009, 45(3): 169-173.

[10] HE C Q, DING N Z, MOU X, et al. Identification of three H1N1 influenza virus groups with natural recombinant genes circulating from 1918 to 2009. Virology, 2012, 427(1): 60-66.

[11] HE C Q, XIE Z X, HAN G Z, et al. Homologous recombination as an evolutionary force in the avian influenza A virus. Mol Biol Evol, 2009, 26(1): 177-187.

[12] XIE Z, XIE L, ZHOU C, et al. Complete genome sequence analysis of an H6N1 avian influenza virus isolated from Guangxi Pockmark ducks. J Virol, 2012, 86(24): 13868-13869.

Identification of a genotype IX Newcastle disease virus in a Guangxi white duck

Xie Zhixun, Xie Liji, Xu Zongli, Liu Jiabo, Pang Yaoshan, Deng Xianwen, Xie Zhiqin, Fan Qing, and Luo Sisi

Abstract

We report the complete genomic sequence of a novel Newcastle disease virus (NDV) strain, duck/China/Guangxi19/2011, isolated from a white duck in Guangxi, southern China. Phylogenetic analysis based on a fusion gene comparison with different NDV strains revealed that duck/China/Guangxi19/2011 is phylogenetically close to genotype IX NDV, and the deduced amino acid sequence of the fusion protein cleavage site was 112R-R-Q-R-R-F117. The whole nucleotide sequence had the highest homology (99.7%) to the sequence of strain F48E8 (GenBank accession number: FJ436302). This study will help us understand the epidemiology and molecular characteristics of genotype IX Newcastle disease virus in ducks.

Keywords

genotype IX, Newcastle disease virus, white duck, identification

---

Newcastle disease virus (NDV) is a single-stranded, negative-stranded RNA virus. The full genome of NDV has three sequence types, with lengths of 15 198 nucleotides, 15 192 nucleotides, and 15 186 nucleotides. NDVs of 15 198 nucleotides belong to class I (containing 9 genotypes), and NDVs of 15 192 nucleo-tides and 15 186 nucleotides belong to class II (containing 11 genotypes)[1-3]. The NDV genome sequence contains six open reading frames, which encode 6 kinds of proteins (3'-NP-P-M-F-HN-L-5'), nucleocapsid proteins, phosphoproteins, matrix proteins, fusion proteins, hemagglutinin-neuraminidase proteins, and large polymerases[4-6].

In December 2011, NDV was isolated from a white duck in Guangxi, southern China. The isolate was named duck/China/Guangxi19/2011. Nucleotide sequences of duck/China/Guangxi19/2011 were amplified by PCR. The amplified products were purified and cloned into the pMD18-T vector (TaKaRa, Dalian, China) and then sequenced (TaKaRa, Dalian, China). Sequences were assembled and manually edited to generate the final genome sequence.

Sequence analysis showed that the full genome sequence of duck/China/Guangxi19/2011 is 15 192 nucleotides and has the highest homology (99.7%) to the sequence of strain F48E8 (GenBank accession number: FJ436302, class II, genotype IX). The amino acid sequence identities of the NP, P, M, F, HN, and L proteins between duck/China/Guangxi19/2011 and F48E8 are 99.6%, 99.0%, 98.6%, 99.3%, 99.1%, and 99.6%, respectively. The amino acid sequence identities of the NP, P, M, F, HN, and L proteins between duck/China/Guangxi/2011 and strain LaSota (GenBank accession number: AF077761, class II, genotype II) are 91.5%, 86.5%, 90.2%, 91.9%, 90.6%, and 93.4%, respectively. The amino acid sequence identities of the NP, P, M, F, HN, and L proteins between duck/China/Guangxi/2011 and the Newcastle disease virus isolate SDWF02 (GenBank accession number: HM188399, class II, genotype VII) are 94.5%, 80.8%, 89.2%, 90.3%, 88.3%, and 93.6%, respectively.

The sequence at the fusion protein cleavage site is a major determinant of NDV pathogenicity[7-9]. The F gene of duck/China/Guangxi19/2011 has the highest sequence homology (99.8%) to strain F48E8, and its virulence fusion protein cleavage site sequence (112R-R-Q-R-R-F117)[10] is in accord with the detected biological characteristics (mean death time, 51.4 h; intracerebral pathogenicity index, 1.80; intravenous pathogenicity index, 2.82).

The first to 21st amino acid sites of the F protein are the signal peptide areas of the N terminus and comprise one of the main variant areas of the F protein[11]. There are four amino acid mutations in duck/China/Guangxi19/2011 compared with strain F48E8. The sites of amino acid mutations are the third amino acid, for which proline (hydrophobic) in F48E8 is mutated to serine (hydrophilic) in duck/China/Guangxi19/2011; the fourth amino acid, for which lysine (alkaline) in F48E8 is mutated to arginine (basic) in duck/China/Guangxi19/2011; the 380th amino acid, for which threonine (hydrophilic) is mutated to alanine (hydrophobic) in duck/China/Guangxi19/2011; and the 553rd amino acid, for which methionine (hydrophobic) is mutated to isoleucine (hydrophobic) in duck/China/Guangxi19/2011. Further study is needed to determine whether these variations affect viral fusion.

This report of the phylogenetic analysis of the whole-genome sequence of genotype IX NDV isolated from a white duck will further understanding of the epidemiology and molecular characteristics of NDV in duck.

Nucleotide sequence accession number. The GenBank accession number for duck/China/Guangxi19/2011 is KC920893.

References

[1] KNIPE D M, LAMB R, PARKS G. Paramyxoviridae: the viruses and their replication//KNIPE D M, HOWLEY P M. Fields Virology 5th ed. Philadelphia : Lippincott Williams &Wilkins, 2007: 1449-1496.

[2] HUANG Y, WAN H Q, LIU H Q, et al. Genomic sequence of an isolate of Newcastle disease virus isolated from an outbreak in geese: a novel six nucleotide insertion in the non-coding region of the nucleoprotein gene. Brief Report. Arch Virol, 2004, 149(7): 1445-1457.

[3] CZEGLÉDI A, UJVÁRI D, SOMOGYI E, et al. Third genome size category of avian paramyxovirus serotype 1 (Newcastle disease virus) and evolutionary implications. Virus Res, 2006, 120(1-2): 36-48.

[4] MAMINIAINA OF, GIL P, BRIAND FX, et al. Newcastle disease virus in Madagascar: identification of an original genotype possibly deriving from a died out ancestor of genotype IV. PLOS ONE, 2010, 5(11): e13987.

[5] RIMA B K, WISHAUPT R G, WELSH M J, et al. The evolution of morbilliviruses: a comparison of nucleocapsid gene sequences including a porpoise morbillivirus. Vet Microbiol, 1995, 44(2-4): 127-134.

[6] GARCÍA-SASTRE A, CABEZAS J A, VILLAR E. Proteins of Newcastle disease virus envelope: interaction between the outer hemagglutinin-neuraminidase glycoprotein and the inner non-glycosylated matrix protein. Biochim Biophys Acta, 1989, 999(2): 171-175.

[7] PANDA A, HUANG Z, ELANKUMARAN S, et al. Role of fusion protein cleavage site in the virulence of Newcastle disease virus. Microb Pathog, 2004, 36(1): 1-10.

[8] XIE Z, XIE L, CHEN A, et al. Complete genome sequence analysis of a Newcastle disease virus isolated from a wild egret. J Virol, 2013, 86(24): 13854-13855.

[9] DE LEEUW O S, KOCH G, HARTOG L, et al. Virulence of Newcastle disease virus is determined by the cleavage site of the fusion protein and by both the stem region and globular head of the haemagglutinin-neuraminidase protein. J Gen Virol, 2005, 86(Pt 6): 1759-1769.

[10] MANIN T B, SHCHERBAKOVA L O, BOCHKOV I U A, et al. Characteristics of field isolates of Newcastle disease virus isolated in the course of outbreaks in the poultry plant in the Leningrad region in 2000. Vopr Virusol, 2002, 47(6): 41-43.

[11] COLLINS M S, STRONG I, ALEXANDER D J. Evaluation of the molecular basis of pathogenicity of the variant Newcastle disease viruses termed "pigeon PMV-1 viruses". Arch Virol, 1994, 134(3-4): 403-411.

Complete genome sequences of an avian orthoreovirus isolated from Guangxi, China

Teng Liqiong, Xie Zhixun, Xie Liji, Liu Jiabo, Pang Yaoshan, Deng Xianwen, Xie Zhiqin, Fan Qing, and Luo Sisi

Abstract

We report the complete genomic sequences of an avian orthoreovirus, strain GuangxiR1, isolated from a chicken flock in Guangxi, southern China, in 2000. Phylogenetic analyses suggest that the strain is closely related to the S1133 strain, which is associated with tenosynovitis, but is far different from strain AVS-B, which is associated with runting-stunting syndrome in broilers.

Keywords

avian orthoreovirus, complete genome characterization, chicken

---

Avian orthoreovirus (ARV) is the etiological agent of viral arthritis, runting-stunting syndrome, and tenosynovitis in chickens, causing significant economic losses in the poultry industry[1, 2]. Avian orthoreovirus belongs to the genus *Orthoreovirus*, family Reoviridae[3]. The genome of 10 RNA segments was divided into three size classes, designated large (Ll, L2, L3), medium (M1, M2, M3), and small (S1, S2, S3, and S4)[4]. These segments encode 10 viral structural proteins (λA, λB, λC, μA, μB, μBC, μBN, σC, σA, and σB) and five nonstructural proteins (μNSC, μNS, σNS, P10, and P17)[5, 6, 7]. The genetic characterization of the complete genomic sequence is limited to a few strains[8].

The avian orthoreovirus strain GuangxiR1, which caused paralysis and hock joints in commercial broilers, was isolated from Guangxi, southern China, in 2000[9, 10]. The nucleotide sequences of GuangxiR1 were amplified through reverse transcription PCR (RT-PCR). The amplified products were purified using a DNA purification kit (TaKaRa, Dalian, China) and cloned into pMD18-T vector (TaKaRa, Dalian, China). The PCR product DNA sequencing was performed by two commercial DNA sequence service companies (Invitrogen, Guangzhou, China; Huada, Shenzhen, China). Nucleotide sequence analysis and alignment were done using the MEGAv.4.1 software package (MEGA Software).

The complete genome of GuangxiR1 is 23 494 bp. The full lengths of the Ll, L2, L3, M1, M2, M3, S1, S2, S3, and S4 segments are 3 959, 3 830, 3 907, 2 283, 2 158, 1 996, 1 643, 1 324, 1 202, and 1 192 nucleotides, respectively. The deduced lengths of the structural and nonstructural proteins λa, λB, λC, μA, μB, μNS, σC, σA, σB, and σNS are 1 293, 1 259, 1 285, 732, 676, 635, 326, 416, 367, and 367 amino acids, respectively. The σC protein, involved in the induction of apoptosis, is coded by the S1 segment. Amino acid comparative analyses indicated that σC shares the highest sequence homology (98.2%) with that from the S1133 strain, associated with tenosynovitis (GenBank accession number: AF330703), but only 48.5% with that from the AVS-B strain, associated with runting-stunting syndrome in broilers (GenBank accession number: FR694197). These data are useful for analyses of the epidemiology and evolutionary characteristics of avian orthoreovirus GuangxiR1.

Nucleotide sequence accession numbers. The complete genome sequences of GuangxiR1 are available in GenBank. The accession number KC183743 to KC183752 correspond to Ll, L2, L3, M1, M2, M3, S1, S2, S3, and S4, respectively.

References

[1] GLASS S E, NAQI S A, HALL C F, et al. Isolation and characterization of a virus associated with arthritis of chickens. Avian Dis, 1973, 17(2): 415-424.

[2] HIERONYMUS D R, VILLEGAS P, KLEVEN S H. Identification and serological differentiation of several reovirus strains isolated from chickens with suspected malabsorption syndrome. Avian Dis, 1983, 27(1): 246-254.

[3] SPANDIDOS D A, GRAHAM A F. Physical and chemical characterization of an avian reovirus. J Virol, 1976, 19(3): 968-976.

[4] VARELA R, BENAVENTE J. Protein coding assignment of avian reovirus strain S1133. J Virol, 1994, 68(10): 6775-6777.

[5] BODELÓN G, LABRADA L, MART NEZ-COSTAS J, et al. The avian reovirus genome segment S1 is a functionally tricistronic gene that expresses one structural and two nonstructural proteins in infected cells. Virology, 2001, 290(2): 181-191.

[6] SHIH W L, HSU H W, LIAO M H, et al. Avian reovirus sigmaC protein induces apoptosis in cultured cells. Virology, 2004, 321(1): 65-74.

[7] BENAVENTE J, MART NEZ-COSTAS J. Avian reovirus: structure and biology. Virus Res, 2007, 123(2): 105-119.

[8] BÁNYAI K, DANDÁR E, DORSEY K M, et al. The genomic constellation of a novel avian orthoreovirus strain associated with runting-stunting syndrome in broilers. Virus Genes, 2011, 42(1): 82-89.

[9] LIAO M, XIE Z X, LIU J B, et al. Isolation and identification of avian reovirus. China Poult, 2002, 24(1): 12-14.

[10] XIE Z X, PENG Y S, LUO S S, et al. Development of a reverse transcription loop-mediated isothermal amplification assay for visual detection of avian reovirus. Avian Pathol, 2012, 41: 311-316.

Full-genome sequence of chicken anemia virus strain GXC060821, isolated from a Guangxi Sanhuang chicken

Xie Zhixun, Deng Xianwen, Xie Liji, Liu Jiabo, Pang Yaoshan, Xie Zhiqin, Fan Qing, and Luo Sisi

Abstract

We report here the complete genomic sequence of a novel chicken anemia virus strain GXC060821, isolated from a Sanhuang chicken in Guangxi of southern China. The complete genome of GXC060821 was sequenced. The full-length of GXC060821 is 2 292 bp and contains three overlapping open reading frames (ORFs). A comparison of the complete sequences and the deduced amino acid sequences of GXC060821 with 31 other published chicken anemia virus sequences showed that the homologies of the nucleotides are 96.1% to 98.5% and the homologies of the deduced amino acid sequences are 89.8% to 94.2%. Phylogenetic tree analysis indicated that GXC060821 is closely related to the two Chinese strains, TJBD40 (accession number: AY843527) and LF4 (accession number: AY839944), and it has a distant relationship with the American isolate 98D06073 (accession number: AF311900). This report will help to understand the epidemiology and molecular characteristics of chicken anemia virus in a Guangxi Sanhuang chicken.

Keywords

chicken anemia virus, full-genome sequence, Sanhuang chicken

Chicken anemia virus (CAV) is a small DNA virus with a circular, covalently linked, single negative-stranded genome. It is the causative agent of chicken infectious anemia (CIA) and classified in the family Circoviridae, genus *Gyrovirus*[1]. CAV is an economically important pathogen with a worldwide distribution. CAV was first isolated in 1979 in Japan and is the major agent responsible for a disease causing severe anemia and immunosuppression[2]. The characteristic symptoms of the disease include aplasia of the bone marrow and the destruction of T-lymphoid tissue, which has been shown histopathologically after CAV infection[3, 4]. Generally, CAV as the causative agent of chicken anemia disease affects 1-day-old chicks that lack maternal antibodies[5]. Mortality rates as high as 55% and morbidity rates as high as 80% have been described when chicks are infected with CAV[6, 7].

In this study, a novel strain of CAV, named GXC060821, was first isolated from a Sanhuang chicken in Guangxi, southern China. The nucleotide sequences of this strain were amplified by PCR. The amplified products were purified and cloned into the pMD18-T vector (TaKaRa, Dalian, China) and then sequenced (Invitrogen, Guangzhou, China)[8]. The sequences were assembled and manually edited to generate the final genome sequence.

Sequence analysis showed that the full-genome sequence of GXC060821 is 2 292 nucleotides and contains 3 overlapping open reading frames (ORFs), including viral protein 1 (VP1), viral protein 2 (VP2), and viral protein 3 (VP3)[7-9]. The full lengths of these ORFs are 1 350, 651, and 366 nucleotides, respectively.

GXC060821 was compared with 31 CAV stains, four American strains, one Australian strain, one German strain, two Japanese strains, six Malaysian strains, 10 Chinese strains, and seven strains from the United

Kingdom. The nucleotide sequence identities of the VP1, VP2, and VP3 genes between GXC060821 and 31 CAV strains are 94.4% to 98.6%, 98.6% to 99.5%, and 98.6% to 99.7%, respectively, and the amino acid sequence identities are 96.7% to 99.6%, 96.8% to 98.6%, and 96.7% to 99.2%, respectively.

Immunogenicity studies have shown that VP1 and VP2 are crucial components for the elicitation of host-produced virus neutralizing antibodies in chickens[10], and the amino acid sequence of GXC060821 with other 31 CAV strains is conservative. Therefore, VP1 and VP2 have previously been thought to be good candidates for use as immunogens when developing subunit vaccines or diagnostic kits[10, 11].

The complete sequences and the deduced amino acid sequences of GXC060821 with 31 other published CAV sequences showed that the homologies of the nucleotides are 96.1% to 98.5% and the homologies of the deduced amino acid sequences are 89.8% to 94.2%. Phylogenetic tree analysis indicated that GXC060821 is closely related to the two Chinese strains, TJBD40 (GenBank accession number: AY843527) and LF4 (GenBank accession number: AY839944), and it has a distant relationship with the American isolate 98D06073 (GenBank accession number: AF311900). This report will help to understand the epidemiology and molecular characteristics of CAV in a Guangxi Sanhuang chicken.

Nucleotide sequence accession number. The complete genomic sequence of GXC060821 was deposited in GenBank under the accession number JX964755.

References

[1] PRINGLE C R. Virus taxonomy at the XIth International Congress of Virology, Sydney, Australia, 1999. Arch Virol, 1999, 144(10): 2065-2070.

[2] YUASA N, TANIGUCHI T, YOSHIDA I, et al. Isolation and some characteristics of an agent inducing anemia in chicks. Avian Dis, 1979, 23(2): 366-385.

[3] YUASA N, IMAI K, WATANABE K, et al. Aetiological examination of an outbreak of haemorrhagic syndrome in a broiler flock in Japan. Avian Pathol, 1987, 16(3): 521-526.

[4] LUCIO B, SCHAT K A, SHIVAPRASAD H L. Identification of the chicken anemia agent, reproduction of the disease, and serological survey in the United States. Avian Dis, 1990, 34(1): 146-153.

[5] YUASA N, TANIGUCHI T, FURUTA K, et al. Maternal antibody and its effect on the susceptibility of chicks to chicken anemia agent. Avian Dis, 1979, 24(1): 197-201.

[6] LEE M S, CHOU Y M, LIEN Y Y, et al. Production and diagnostic application of a purified, E. coli-expressed, serological-specific chicken anaemia virus antigen VP3. Transbound Emerg Dis, 2011, 58(3): 232-239.

[7] LAI G H, LIN M K, LIEN Y Y, et al. Expression and characterization of highly antigenic domains of chicken anemia virus viral VP2 and VP3 subunit proteins in a recombinant E. coli for sero-diagnostic applications. BMC Vet Res, 2013, 9: 161.

[8] XIE Z X, XIE L J, PANG Y S, et al. Development of a real-time multiplex PCR assay for detection of viral pathogens of penaeid shrimp. Arch Virol, 2008, 153(12): 2245-2251.

[9] ZHANG X, LIU Y, WU B, et al. Phylogenetic and molecular characterization of chicken anemia virus in southern China from 2011 to 2012. Sci Rep, 2013, 3: 3519.

[10] NOTEBORN M H, VERSCHUEREN C A, KOCH G, et al. Simultaneous expression of recombinant baculovirus-encoded chicken anaemia virus (CAV) proteins VP1 and VP2 is required for formation of the CAV-specific neutralizing epitope. J Gen Virol, 1998, 79(Pt 12): 3073-3077.

[11] KOCH G, VAN ROOZELAAR D J, VERSCHUEREN C A, et al. Immunogenic and protective properties of chicken anaemia virus proteins expressed by baculovirus. Vaccine, 1995, 13(8): 763-770.

Genome analysis of a Tembusu virus, GX2013H, isolated from a Cheery Valley duck in Guangxi, China

Xie Zhixun, Zeng Tingting, Xie Liji, Deng Xianwen, Xie Zhiqin, Liu Jiabo, Fan Qing, Pang Yaoshan, and Luo Sisi

Abstract

We report here the complete genome sequence of a duck Tembusu virus (DTMUV) strain, GX2013H, isolated from a duck from Cheery Valley in Guangxi of southern China in 2013. We obtained the strain GX2013H from a Cheery Valley duck with severely decreased egg production and neurological signs. The genome of GX2013H is 10 990 nucleotides (nt) in length and contains a single open reading frame encoding a putative polyprotein of 3 425 amino acids (aa). A comparison of the complete sequence and the deduced amino acid sequence of GX2013H with published sequences of 15 other duck Tembusu viruses from China showed that the homologies of the nucleotides are approximately 96.5% to 97.5% and the homologies of the deduced amino acid sequences are approximately 98.9% to 99.3%. This report will help to understand the epidemiology and molecular characteristics of TMUV in Guangxi.

Keywords

Tembusu virus, genome analysis, Cheery Valley duck

In 2010, a novel infectious agent emerged in China and caused extensive epidemics in layer and breeder ducks. The isolated strain was considered to be a new genotype of Tembusu virus (TMUV)[1]. TMUV infects mainly ducks and geese[2, 3], and nearly all duck species can be infected by TMUV, such as Beijing ducks, Muscovy ducks, and Cheery Valley ducks[1, 2, 4]. There was nearly 100% morbidity but 0% to 12% mortality in infected ducks, which were observed in the area of greatest waterfowl production in China. TMUV causes a range of symptoms in infected ducks, including decreased egg production, high fever, loss of appetite, and neurological signs. To date, many TMUVs have been isolated and viral genomes have been sequenced[3, 5-8]. However, genome sequences of strains isolated from Guangxi have rarely been published.

In 2013, breeder ducks showed a decrease in egg production from 90% to almost 0%, as well as neurological signs, in a Cheery Valley duck farm in Guangxi. We collected ovary and brain samples from affected ducks and isolated the virus under the conventional procedure[9]. A strain of TMUV was isolated, and other pathogens that cause similar symptoms were ruled out[10]. Thirteen pairs of primers were designed to amplify the different genomic regions of the strain GX2013H, with an overlapping genome fragment covering each region. The 5' -and 3' -terminal sequences were determined using the SMARTer RACE cDNA amplification kit (Clontech, Dalian, China). The amplified products were purified, cloned into pMD18-T vector (TaKaRa, Dalian, China), and sequenced (Invitrogen, Shanghai, China). The sequences were assembled using the SeqMan program to produce the complete genome sequence of GX2013H[11, 12]. The full-length genome sequence of GX2013H is 10 990 nucleotides (nt) in length, with a typical flavivirus genome organization, and the 5' and 3' untranslated regions (UTR) are 94 and 618 nt, respectively. Additionally, the coding region of GX2013H includes a single open reading frame (ORF) (10 278 nt) that encodes a polypeptide of 3 425 amino acids (aa), three structural proteins (capsid, prM, and envelope), and seven nonstruc-tural proteins (NS1, NS2A, NS2B, NS3, NS4A, NS4B, and NS5).

Compared to the genome sequences of previously isolated TMUVs from different species of ducks and geese in various areas in China, there is 96.5% to 97.5% homology at the nucleotide level and 98.9% to 99.3% homology at the amino acid level. A phylogenetic tree based on the whole polyprotein sequence showed that GX2013H is in a single clade, whereas other strains are in a different clade.

By predicting the potential glycosylation sites, we found 13 glycosylation sites in six viral proteins. The numbers of glycosylation sites in the prM, E, NS1, NS2A, NS4B, and NS5 genes are 2, 1, 3, 1, 3, and 3, respectively.

In conclusion, the study of the whole-genome sequence of TMUV profits further investigation on the epidemiology and evolution of TMUV, and it may help elucidate the mechanisms of virus replication and pathogenesis.

Nucleotide sequence accession number. The complete genome sequence of the duck Tembusu virus isolate has been deposited to GenBank under the accession number KJ700462.

References

[1] CAO Z, ZHANG C, LIU Y, et al. Tembusu virus in ducks, China. Emerg Infect Dis, 2011, 17(10): 1873-1875.

[2] SU J, LI S, HU X, et al. Duck egg-drop syndrome caused by BYD virus, a new Tembusu-related flavivirus. PLOS ONE, 2011, 6(3): e18106.

[3] YUN T, ZHANG D, MA X, et al. Complete genome sequence of a novel flavivirus, duck Tembusu virus, isolated from ducks and geese in China. J Virol, 2012, 86(6): 3406-3407.

[4] YUN T, YE W, NI Z, et al. Identification and molecular characterization of a novel flavivirus isolated from Pekin ducklings in China. Vet Microbiol, 2012, 157(3-4): 311-319.

[5] LIU M, LIU C, LI G, et al. Complete genomic sequence of duck flavivirus from China. J Virol, 2012, 86(6): 3398-3399.

[6] WAN C, HUANG Y, FU G, et al. Complete genome sequence of avian Tembusu-related virus strain WR isolated from White Kaiya ducks in Fujian, China. J Virol, 2012, 86(19): 10912.

[7] ZHU W, CHEN J, WEI C, et al. Complete genome sequence of duck Tembusu virus, isolated from Muscovy ducks in southern China. J Virol, 2012, 86(23): 13119.

[8] TANG Y, DIAO Y, GAO X, et al. Analysis of the complete genome of Tembusu virus, a flavivirus isolated from ducks in China. Transbound Emerg Dis, 2012, 59(4): 336-343.

[9] PENG Y, XIE Z X, LIU J B, et al. Epidemiological surveillance of low pathogenic avian influenza virus (LPAIV) from poultry in Guangxi Province, southern China. PLOS ONE, 2013, 8(10): e77132.

[10] XIE Z X, PANG Y S, LIU J B, et al. A multiplex RT-PCR for detection of type A influenza virus and differentiation of avian H5, H7, and H9 hemagglutinin subtypes. Mol Cell Probes, 2006, 20(3-4): 245-249.

[11] XIE Z X, ZHANG M X, XIE L J, et al. Identification of a triple-reassortant H1N1 swine influenza virus in a southern China Pig. Genome Announc, 2014, 2(2): e00229-14.

[12] TENG L Q, XIE Z X, XIE L J, et al. Sequencing and phylogenetic analysis of an avian reovirus genome. Virus Genes, 2014, 48(2): 381-386.

Identification and whole-genome sequence analysis of Tembusu virus GX2013G, isolated from a Cherry valley duckling in southern China

Zeng Tingting, Xie Zhixun, Xie Liji, Deng Xianwen, Xie Zhiqin, Huang Li, Luo Sisi, and Huang Jiaoling

Abstract

A duck Tembusu virus (DTMUV) was isolated from the brain of a Cherry Valley duckling that showed neurological signs by using a specific-pathogen-free chicken embryo. The isolate was named GX2013G (GenBank accession number: KM275941). The strain GX2013G was identified with reverse transcription-PCR (RT-PCR), and the amplicon was sequenced. The genome that was obtained is 10 990 nucleotides in length and contains a single open reading frame encoding a putative polyprotein of 3 425 amino acids. This study will advance the understanding of the epidemiology and molecular characteristics of Tembusu virus (TMUV) in Guangxi and further studies of the mechanisms of virus replication and pathogenesis.

Keywords

Tembusu virus, whole-genome, Cherry valley duckling

Tembusu virus (TMUV) was isolated and identified for the first time in 2010 in the east of China[1]. It was an extensive epidemic in the densest waterfowl breeding area of China, including Fujian[2], Shandong[3], Zhejiang[4], Jiangsu[5], and Guangxi[6]. Almost all duck breeds were reported as having been infected by TMUV[3, 6-9]; infections in geese and chickens were also reported[10, 11]. Lying ducks showed a major decrease in egg production, high fever, loss of appetite, and neurological signs after infection. Morbidity was nearly 100%, but there was only a 0% to 12% mortality rate for the affected ducks. Ducklings could also be infected by TMUV, but were rarely reported.

In 2013, a flock of ducklings showed neurological signs and loss of appetite in a Cherry valley duck farm in Guangxi. Encephaledema was observed, and the brains were collected to isolate viruses by using specific-pathogen-free (SPF) chicken embryos and identified by reverse transcription-PCR (RT-PCR). A strain of TMUV was isolated, and other pathogens which cause similar symptoms were ruled out[12-14]. Thirteen pairs of primers were used to amplify the different regions of the strain GX2013G with an overlapping genome fragment covering each region. The 5' -and 3' -terminal sequences were determined by using a SMARTer rapid amplification of cDNA ends (RACE) cDNA amplification kit (Clonetech, Dalian, China). The amplified products were purified and cloned into pMD18-T vectors (TaKaRa, Dalian, China) and sequenced (Invitrogen, Shanghai, China). Sequences were assembled by a Seqman program (Lasergene, Madison, USA) to produce the complete genome sequence of GX2013G. The full-length genome sequence of GX2013G was 10 990 nudeotides (nt) in length, with a typical flavivirus genome organization, and the 5' and 3' untranslated regions (UTR) were 94 and 618 nt, respectively. Additionally, the coding region of GX2013G included a single open reading frame (ORF) (10 278 nt) that encoded apolypeptide of 3 425 amino acids (aa), three structural proteins (capsid, prM, and envelope), and seven nonstructural proteins (NS1, NS2A, NS2B, NS3, NS4A, NS4B, and NS5).

Compared to genome sequences of previously isolated TMUVs from different species of ducks, geese, and chickens in various areas of China, there were 96.5% to 97.5% homology at the nucleotide level and 99.7% homology with GX2013H isolated in Guangxi[6]. A phylogenetic tree based on the whole sequence showed that GX2013G was in a single clade with GX2013H, whereas other strains were in a different clade. It was demonstrated that TMUVs in China had conservative nucleic acid levels and the most homology in the area.

By predicting the potential glycosylation sites, we found 13 glycosylation sites in six viral proteins: 2 glycosylation sites in prM, lin E, 3 in NS1, 1 in NS2A, 3 in NS4B, and 3 in NS5. The numbers and positions of glycosylation sites in each protein were the same as GX2013H.

In conclusion, this study of the whole-genome sequence of TMUV provides further information about the epidemiology and evolution of TMUV and may help elucidate mechanisms of virus replication and pathogenesis.

Nucleotide sequence accession number. The complete genome sequence of the duck Tembusu virus isolate has been deposited in GenBank under the accession number KM275941.

References

[1] CAO Z Z, ZHANG C, LIU Y, et al. Tembusu virus in ducks, china. Emerging Infectious Disease, 2011, 17(10): 1873-1875.

[2] WAN C H, HUANG Y, FU G H, et al. Complete genome sequence of avian Tembusu-related virus strain WR isolated from White Kaiya ducks in Fujian, China. Journal of virology, 2012, 86(19): 10912.

[3] TANG Y, DIAO Y X, CHEN H, et al. Isolation and genetic characterization of a tembusu virus strain isolated from mosquitoes in Shandong, China. Transboundary and Emerging Disease, 2013, 62(2): 209-216.

[4] YUN T, ZHANG D, MA X, et al. Complete genome sequence of a novel flavivirus, duck Tembusu virus, isolated from ducks and geese in china. Journal of Virology, 2012, 86(6): 3406-3407.

[5] HAN K, HUANG X, LI Y, et al. Complete genome sequence of goose Tembusu virus, isolated from jiangnan white geese in jiangsu, china. Genome Announcement, 2013, 1(2): e0023612.

[6] XIE Z X, ZENG T T, XIE L J, et al. Genome analysis of a Tembusu virus, GX2013H, isolated from a Cheery valley duck in Guangxi, China. Genome Announcement, 2014, 2(4): e00466-14.

[7] YUN T, YE W, NI Z, et al. Identification and molecular characterization of a novel flavivirus isolated from Pekin ducklings in China. Veterinary Microbiology, 2012, 157(3-4): 311-319.

[8] YU K, SHENG Z Z, HUANG B, et al. Structural, antigenic, and evolutionary characterizations of the envelope protein of newly emerging duck Tembusu virus. PLOS ONE, 2013, 8(8): e71319.

[9] ZHU W, CHEN J, WEI C, et al. Complete genome sequence of duck Tembusu virus, isolated from Muscovy ducks in southern China. Journal of Virology, 2012, 86(23): 13119.

[10] LIU M, CHEN S, CHEN Y, et al. Adapted Tembusu-like virus in chickens and geese in China. Journal of Clinical Microbiology, 2012, 50(8): 2807-2809.

[11] HUANG X, HAN K, ZHAO D, et al. Identification and molecular characterization of a novel flavivirus isolated from geese in China. Research in Veterinary Science, 2013, 94(3): 774-780.

[12] XIE L J, XIE Z X, ZHAO G Y, et al. Complete genome sequence analysis of a duck circovirus from Guangxi Pockmark ducks. Journal of Virology, 2012, 86(23): 13136.

[13] XIE Z X, PANG Y S, LIU J B, et al. A multiplex RT-PCR for detection of type A influenza virus and differentiation of avian H5, H7, and H9 hemagglutinin subtypes. Molecular and Cellular Probes, 2006, 20(3-4): 245-249.

[14] XIE Z X, XIE L J, XU Z, et al. Identification of a genotype IX Newcastle disease virus in a Guangxi White Duck. Genome Announcement, 2013, 1(5): e00836-13.

Molecular characterization of the full Muscovy duck parvovirus, isolated in Guangxi, China

Zhao Hong, Xie Zhixun, Xie Liji, Deng Xianwen, Xie Zhiqin, Luo Sisi, Huang Li, Huang Jiaoling, and Zeng Tingting

Abstract

We report the complete genomic sequence of the full Muscovy duck parvovirus (MDPV) strain, designated GX2011-5, isolated from a Muscovy duck in Guangxi, China. The complete genomic sequence was 5 132 bp in length and contained two major open reading frames encoding a 1 844 nucleotide (nt) nonstructural protein and a 2 199 nt capsid protein. Comparison of the complete sequence of GX2011-5 with other published sequences of Muscovy duck parvovirus revealed that this strain exhibited 90.4% to 95.1% sequence homology. This report will advance our understanding of the epidemiology and molecular characteristics of MDPV in the Muscovy duck population in Guangxi, China.

Keywords

duck parvovirus, MDPV, genome sequence, Muscovy duck

Muscovy duck parvovirus (MDPV) is a single-stranded DNA virus that belongs to the family Parvoviridae[1]. Muscovy ducklings under 3 weeks of age are the most susceptible to contracting MDPV infection. The clinical signs of the disease included watery diarrhea, wheezing, and locomotor dysfunction. Once these signs are observed, death occurs rapidly. Although the mortality is much lower in ducks infected after 3 weeks of age, the production performance of ducks that survive the infection is reduced because of abnormal feathering and growth retardation[2, 3, 4]. To date, many MDPVs have been isolated and sequenced; however, there are few reports of strains isolated from Guangxi[5, 6, 7]. The aim of this study is to obtain the full genomic sequence of an MDPV strain isolated from Muscovy ducks.

In May 2011, MDPV was isolated from a commercial Muscovy duck farm during an outbreak of an infectious disease. The clinical symptoms manifested with severe watery diarrhea and locomotor dysfunction in ducklings in the city of Yulin in Guangxi in southern China. Subsequently, nucleotide sequences of MDPV were amplified through PCR. The amplified products were purified and cloned into the pMD18-T vector (TaKaRa, Dalian, China) and then sequenced (TaKaRa, Dalian, China)[8, 9, 10]. The sequences were assembled and manually edited to produce the final genome sequence. The isolated virus was named duck/ Guangxi/05/2011 (GX2011-5).

The full-length genome sequence of GX2011-5 was 5 132 nudeotides (nt) in length, containing two major open reading frames (ORFs). The left ORF (ORF1, 1 884 nt) encodes pleiotropic regulatory proteins (NS1, NS2) and 627 amino acids (aa), the right ORF (ORF2, 2 199 nt) encodes the viral capsid proteins (VP1, VP2, VP3) and 732 amino acids (aa), Analysis of the potential glycosylation sites of the surface proteins revealed four potential N-glycosylation sites in NS1 (150~153, 225~228, 360~363, and 433~436) and four potential N-glycosylation sites in VP1 (219~222, 331~334, 582~585, and 703~705)[11, 12].

Compared to the genome sequences of isolates of MDPV from different areas, the nucleotide sequence of GX2011-5 shared 95.1% homology with the sequence of the strain SAAS-SHNH (GenBank accession number: KC171936.1; origin, Shanghai, China), it shared 90.4% homology with the sequence of the strain FM (GenBank accession number: NC006147.2; origin, France). The NS1 and VP1 genes of GX2011-5 shared 97.9% to 99.5% and 88.7% to 99.3% homology, respectively, with the NS1 and VP1 sequences of the strains SAAS-SHNH, FM, P (GenBank accession number: JF26997.1; origin, Fujian, China), P1 (GenBank accession number: JF26998.1; origin, Fujian, China), and X75093 (GenBank accession number: X75093.1; origin, France). These results indicate that the NS1 gene is much more conserved than the VP1 gene[13].

In conclusion, the study of the whole-genome sequence of MDPV requires further investigation to assess the epidemiology and evolution of MDPV, and such studies may help to elucidate the mechanisms of virus replication and pathogenesis.

Nucleotide sequence accession number. The complete genome sequence of the Muscovy duck parvovirus isolate described here has been deposited in GenBank under the accession number KM093740.

References

[1] LE GALL-RECULÉ G, JESTIN V. Biochemical and genomic characterization of muscovy duck parvovirus. Arch Virol, 1994, 139(1-2): 121-131.

[2] GLÁVITS R, ZOLNAI A, SZABÓ E, et al. Comparative pathological studies on domestic geese (Anser anser domestica) and Muscovy ducks (Cairina moschata) experimentally infected with parvovirus strains of goose and Muscovy duck origin. Acta Vet Hung, 2005, 53(1): 73-89.

[3] JI J, XIE Q M, CHEN C Y, et al. Molecular detection of Muscovy duck parvovirus by loop-mediated isothermal amplification assay. Poult Sci, 2010, 89(3): 477-483.

[4] BARNES H J. Muscovy duck parvovirus//CALNEK B W, BARNES H J, BEARD C W, et al. Diseases of Poultry, 10th ed. Ames: Iowa State University Press, 1997: 1032-1033.

[5] CHANG P C, SHIEN J H, WANG M S, et al. Phylogenetic analysis of parvoviruses isolated in Taiwan from ducks and geese. Avian Pathol, 2000, 29(1): 45-49.

[6] TAKEHARA K, HYAKUTAKE K, IMAMURA T, et al. Isolation, identification, and plaque titration of parvovirus from Muscovy ducks in Japan. Avian Dis, 1994, 38(4): 810-815.

[7] XIE Z X, ZENG T T, XIE L J, et al. Genome analysis of a Tembusu virus, GX2013H, isolated from a Cheery valley duck in Guangxi, China. Genome Announc, 2014, 2(4): e00466-14.

[8] PENG Y, XIE Z X, LIU J B, et al. Epidemiological surveillance of low pathogenic avian influenza virus (LPAIV) from poultry in Guangxi Province, southern China. PLOS ONE, 2013, 8(10): e77132.

[9] XIE Z X, PANG Y S, LIU J B, et al. A multiplex RT-PCR for detection of type A influenza virus and differentiation of avian H5, H7, and H9 hemagglutinin subtypes. Mol Cell Probes, 2006, 20(3-4): 245-249.

[10] XIE L J, XIE Z X, ZHAO G Y, et al. Complete genome sequence analysis of a duck circovirus from Guangxi Pockmark ducks. J Virol, 2012, 86(23): 13136.

[11] LIU H M, WANG H, TIAN X J, et al. Complete genome sequence of goose parvovirus Y strain isolated from Muscovy ducks in China. Virus Genes, 2014, 48(1): 199-202.

[12] XIE Z X, ZHANG M X, XIE L J, et al. Identification of a triple-reassortant H1N1 swine influenza virus in a southern China Pig. Genome Announc, 2014, 2(2): e00229-14.

[13] POONIA B, DUNN PA, LU H, et al. Isolation and molecular characterization of a new Muscovy duck parvovirus from Muscovy ducks in the USA. Avian Pathol, 2006, 35(6): 435-441.

Molecular characterization of parvovirus strain GX-Tu-PV-1, isolated from a Guangxi turkey

Zhang Yanfang, Xie Zhixun, Deng Xianwen, Xie Zhiqin, Xie Liji, Zhang Minxiu, Luo Sisi, Fan Qing, Huang Jiaoling, Zeng Tingting, and Wang Sheng

Abstract

The aim of the current study was to determine the genomic sequence of parvovirus strain GX-Tu-PV-1, which was isolated from a turkey in Guangxi, south China. The analysis showed that the genome sequence of GX-Tu-PV-1 was 81.3% to 99.3% similar to those of other turkey parvoviruses (TuPVs) and 79.8% to 92.1% related to chicken parvovirus (ChPV). This study will help in understanding the epidemiology and molecular characteristics of parvovirus in turkeys.

Keywords

parvovirus, molecular characterization, turkey

Parvovirus (family Parvoviridae) spp. that infect vertebrate hosts make up the subfamily Parvovirinae, while those that infect arthropods make up the subfamily Densovirinae[1-4]. According to the latest classification by the International Commitee on Taxonomy of Viruses, the Parvovirinae subfamily is divided into eight genera[5-8] (*Amdoparvovirus, Aveparvovirus, Bocaparvovirus, Copiparvovirus, Dependoparvovirus, Erythroparvovirus, Protoparvovirus, and Tetraparvovirus*). Chicken parvovirus (ChPV) and turkey parvovirus (TuPV) are classified as members of the new genus *Aveparvovirus (Galliform aveparvovirus* 1).

In this study, viral DNA was extracted from tracheal and cloacal swabs from 12 turkeys using the EasyPure viral DNA/RNA kit (TransGen, Beijing, China). TuPV was identified by PCR, using primers designed to target the conserved region (nonstructural (NS) gene, 561 bp) first[1]. The two positive samples were confirmed by partial sequencing of the nonstructural 1 (NS1) gene using primers NSF and NSR (Table 3-22-1). BLAST analysis revealed 98.0% to 100% nucleotide (nt) sequence identity to the TuPV 1 030 strain (GenBank accession number: KM598418). Based on an alignment of the sequences of three TuPV isolates deposited in GenBank, the specific pairs of primers (Table 3-22-1) were then designed to amplify the genome of one positive TuPV sample named GX-Tu-PV-1. The gel-electrophoresed PCR products were purified by using the AxyPrep DNA gel extraction kit (Axygen, Hangzhou, China), followed by cloning into the pMD19-T vector (TaKaRa, Dalian, China). Sanger sequencing was performed on a DNA analyzer (Invitrogen, Guangzhou, China). The undetermined 5' and 3' teminal fragments were amplified using the 5' rapid amplification of cDNA ends (RACE) system kit v.2 (TaKaRa, Dalian, China) and reverse transcription-PCR with oligo (dT) primers, respectively. The amplified fragments were cloned into the pMD19-T vector (TaKaRa, Dalian, China), and 4 and 8 clones of the 5' and 3' terminal regions, respectively, were sequenced. The sequences were obtained by assembling overlapping contigs, followed by editing using the EditSeq and MegAlign programs of DNAStar 7.0 green (DNAStar, WI, USA).

Table 3-21-1 Primers used for PCR amplification of the turkey parvovirus genome

Primer [a]	Sequence (5'-3') [b]	Nucleotide positions [c]
F1-1	CTGCTGAGCTGGTAAGATGG	395~414
R1-2	TCTTCCCGACTGACTAGATT	724~743
R1-3	CCCCCATGATACATTTTGCT	1 751~1 770
F2-1	TTCTAATAACGATATCACTCAAGTTTC	1 841~1 867
R1-4	AACCAGTATAGGTGGGTTCC	2 192~2 211
R1-1	TTTGCGCTTGCGGTGAAGTCTGGCTCG	2 375~2 401
F3-1	CAAGCCGCCATTGTGTTTGT	3 575~3 594
R2-2	GTATTGKGTYTGGTTTTCAG	3 659~3 678
R2-1	AAGTCWAKRTAATTCCATGG	3 694~3 713
R3-2	GTCCCTGTCAAGTCATTAGAG	3 858~3 878
R3-1	TTAATTGGTYYKCGGYRCSCG	5 005~5 025
NSF	TTCTAATAACGATATCACTCAAGTTTC	1 841~1 867
NSR	TTTGCGCTTGCGGTGAAGTCTGGCTCG	2 375~2 401

Note: [a] Primers F1-1/R1-1, F1-1/R1-2, F1-1/R1-3, and F1-1/R1-4 were used to amplify the first fragment, F2-1/R1-1 and F2-1/R2-2 were used to amplify the second fragment, and F3-1/R3-1 and F3-1/R3-2 were used to amplify the third fragment of the turkey parvovirus (GX-Tu-PV-1) genome. Primers NSF to NSR were used to amplify the partial NS sequence.
[b] The sequences of the primers were designed according to the sequences of three other known ChPV/TuPV strains (GenBank accession numbers: GU214704, GU214706, and NC_024454).
[c] The positions of primers located in the genome are shown according to the U. S. TuPV isolate (TuPV 1030).

The DNA sequence of the obtained isolate was 4 642 bp long, with an A+T content of 57.1% and G+C content of 42.9%. The entire genome of GX-Tu-PV-1 consists of three open reading frames (ORFs). ORF1 and ORF2 encode a nonstructural (NS) protein, which is involved in viral replication, and the major capsid proteins (VP1 and VP2), respectively. ORF3 encodes a putative protein, NP1, of which the function is unclear[9-13]. The genetic diversity of GX-Tu-PV-1 was explored using phylogenetic analyses using ClustalW and MEGA 7.0[14]. It showed that the genome sequence of GX-Tu-PV-1 was 81.3% to 99.3% related to TuPVs (GenBank accession numbers: NC_038534, EU304809, KM598418, KM598419, and KX084398) and 79.8% to 92.1% related to ChPV strains (GenBank accession numbers: KM598417, GU214704, KM598416, KY069111, KU569162, KJ486489, KX133418, and KM254173) (Figure 3-22-1).

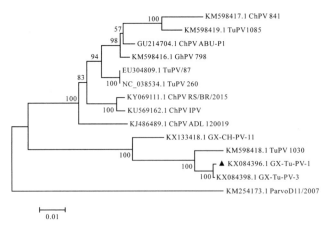

▲: GX-Tu-PV-1. GenBank accession numbers follow the names of the ChPV/TuPV strains. The numbers near the branches indicate the confidence level calculated by bootstrapping (n = 1 000). The length of each pair of branches represents the distance between sequence pairs. The scale bar represents 0.01-nt substitutions per site.

Figure 3-21-1 Neighbor joining (NJ) tree of the genomic sequences of GX-Tu-PV-1 and 13 other ChPV/TuPV isolates

The sequence data of the GX-Tu-PV-1 strain will facilitate research on the epidemiology and evolutionary biology of parvoviruses in China.

Data availability. The genome sequence of GX-Tu-PV-1 was deposited in GenBank under the accession number KX084396.

References

[1] FENG B, XIE Z X, DENG X W, et al. Genetic and phylogenetic analysis of a novel parvovirus isolated from chickens in Guangxi, China. Arch Virol, 2016, 161(11):(11): 3285-3289.

[2] TSAI H J, TSENG C H, CHANG P C, et al. Genetic variation of viral protein 1 genes of field strains of waterfowl parvoviruses and their attenuated derivatives. Avian Dis, 2004, 48(3): 512-521.

[3] ZSAK L, STROTHER K O, DAY J M. Development of a polymerase chain reaction procedure for detection of chicken and turkey parvoviruses. Avian Dis, 2004, 53(1): 83-88.

[4] MURGIA M V, RAUF A, TANG Y, et al. Prevalence of parvoviruses in commercial turkey flocks. Avian Dis, 2012, 56(4): 744-749.

[5] COTMORE S F, AGBANDJE-MCKENNA M, CHIORINI J A, et al. The family Parvoviridae. Arch Virol, 2014, 159(5): 1239-1247.

[6] COTMORE S F, TATTERSALL P. Parvoviruses: small does not mean simple. Annu Rev Virol, 2014, 1(1): 517-537.

[7] KAPGATE S S, KUMANAN K, VIJAYARANI K, et al. Avian parvovirus: classification, phylogeny, pathogenesis and diagnosis. Avian Pathol, 2018, 47(6): 536-545.

[8] DE LA TORRE D I, NUÑEZ L F, ASTOLFI-FERREIRA C S, et al. Enteric virus diversity examined by molecular methods in Brazilian poultry flocks. Vet Sci, 2018, 5(2): 38.

[9] DAY J M, ZSAK L. Recent progress in the characterization of avian enteric viruses. Avian Dis, 2013, 57(3): 573-580.

[10] BIDIN M, LOJKIĆ I, BIDIN Z, et al. Identification and phylogenetic diversity of parvovirus circulating in commercial chicken and turkey flocks in Croatia. Avian Dis, 2011, 55(4): 693-696.

[11] ZSAK L, CHA R M, LI F, et al. Host specificity and phylogenetic relationships of chicken and turkey parvoviruses. Avian Dis, 2015, 59(1): 157-161.

[12] DOMANSKA-BLICHARZ K, JACUKOWICZ A, LISOWSKA A, et al. Genetic characterization of parvoviruses circulating in turkey and chicken flocks in Poland. Arch Virol, 2012, 157(12): 2425-2430.

[13] NUÑEZ L, SANTANDER-PARRA S, CHAIBLE L, et al. Development of a sensitive real-time fast-qPCR based on SYBR green for detection and quantification of chicken parvovirus (ChPV). Vet Sci, 2018, 5(3): 69.

[14] KUMAR S, STECHER G, TAMURA K. MEGA7: molecular evolutionary genetics analysis version 7.0 for bigger datasets. Mol Biol Evol, 2016, 33(7): 1870-1874.

Two subgroups of porcine circovirus 2 appearing among pigs in southern China

Zhang Minxiu, Xie Zhixun, Xie Liji, Deng Xianwen, Xie Zhiqin, Luo Sisi, Huang Li, Huang Jiaoling, and Zeng Tingting

Abstract

We report here two complete genome sequences of porcine circovirus 2. The complete genomes of BH5 and BH6 were amplified and analyzed. Sequence analysis demonstrated that strains BH5 and BH6 belonged to PCV2d and PCV2b, respectively. Knowledge regarding the complete genome sequences of strains BH5 and BH6 will be useful for epidemiological surveillance.

Keywords

porcine circovirus 2, genome sequences, characterization

Porcine circovirus 2 (PCV-2) belongs to the Circoviridae family, which is a small nonenveloped virus and contains a single-stranded circular DNA genome of about 1.76 kb[1, 2]. The genome contains two open reading frames, including ORF1 and ORF2[3]. ORF1 consists of 945 nucleotides (nt), encoding 314 amino acids, called the Rap protein, responsible for viral replication proteins. ORF2 consists of 702 or 705 nt, encoding 234 or 235 amino acids, called the Cap protein[4]. Porcine circovirus 2 (PCV-2) is considered to be the main pathogen that causes postweaning multisystemic wasting syndrome (PMWS), an epidemic on pig farms[5]. PCV-2 has serious economic impact on the pig industry. Thus, it is important to enhance the surveillance of PCV-2.

In this study, various tissue samples (liver, spleen, tonsil, lymph nodes) were collected from pigs at a farm in Guangxi, China. DNA was extracted from various tissue samples with a viral DNA/RNA kit (TransGen, Beijing, China). The complete genomes of PCV-2 strains BH5 and BH6, which were collected from the same pig farm, were amplified by PCR using Primers[6]. The amplified products were purified and cloned into the pMD18-T vector (TaKaRa, Dalian, China) and sequenced (TaKaRa, Dalian, China)[7-9]. Sequences were assembled and manually edited to generate the final full-length genome sequences[10-14].

The genomes of strains BH5 and BH6 consist of 1 767 nt and the ORF1 of strains BH5 and BH6 consists of 945 nt, encoding 314 amino acids. ORF2 of strain BH5 consists of 705 nt, encoding 235 amino acids; ORF2 of strain BH6 consists of 702 nt, encoding 234 amino acids. The genome sequence of strain BH5 presented 96.4% similarity compared with the strain BH6. Genome sequences of strain BH5 shared the highest sequence homology (98.8%) with porcine circovirus 2 isolate Henan (GenBank accession number: AY969004). Genome sequences of strain BH6 shared the highest sequence homology (97.9%) with porcine circovirus type 2 strain TJ (GenBank accession number: AY181946). The homologies of strain BH5 with the sequences of representative porcine circovirus 2 strain BF and France (GenBank accession number: AF381175 and AF055394) were 95.2% and 96.2%, and the homologies of strain BH6 with these strains were 96.2% and 98.7%.

PCV2 can be divided into five subgroups according to ORF2. These five subgroups are PCV2a, PCV2b, PCV2c, PCV2d, and PCV2e. Phylogenetic analysis showed that strains BH5 and BH6 belong to

PCV2d and PCV2b. The results showed that strain BH6 has a close relationship with porcine circovirus 2 isolate Henan, and also has a relationship with the porcine circovirus 2 isolate DK1987PMWSfree (GenBank accession number: EU148504). Strain BH5 has a close relationship with porcine circovirus 2 isolate TJ06 (GenBank accession number: EF524539), and also has a relationship with the porcine circovirus 2 isolate DK1987PMWSfree. These results suggested two subgroups of porcine circovirus 2 exist in the same pig farm and showed some degree of complexity in the epidemic of porcine circovirus 2.

Nucleotide sequence accession numbers. The complete genome sequences of BH5 and BH6 were deposited in GenBank under the accession numbers KJ956689, KJ956690, KJ956691, KJ956692, and KM245558.

References

[1] MEEHAN B M, CREELAN J L, MCNULTY M S, et al. Sequence of porcine circovirus DNA: affinities with plant circoviruses. J Gen Virol, 1997, 78 (Pt 1): 221-227.

[2] TISCHER I, GELDERBLOM H, VETTERMANN W, et al. A very small porcine virus with circular single-stranded DNA. Nature, 1982, 295(5844): 64-66.

[3] LIU J, CHEN I, KWANG J. Characterization of a previously unidentified viral protein in porcine circovirus type 2-infected cells and its role in virus-induced apoptosis. J Virol, 2005, 79(13): 8262-8274.

[4] NAWAGITGUL P, MOROZOV I, BOLIN S R, et al. Open reading frame 2 of porcine circovirus type 2 encodes a major capsid protein. J Gen Virol, 2000, 81(Pt 9): 2281-2287.

[5] HARDING J C. The clinical expression and emergence of porcine circovirus 2. Vet Microbiol, 2004, 98(2): 131-135.

[6] LI B, MA J, LIU Y, et al. Complete genome sequence of a highly prevalent porcine circovirus 2 isolated from piglet stool samples in China. J Virol, 2012, 86(8): 4716.

[7] XIE Z X, XIE L J, ZHOU C Y, et al. Complete genome sequence analysis of an H6N1 avian influenza virus isolated from Guangxi Pockmark ducks. J Virol, 2012, 86(24):13868-13869.

[8] XIE Z X, ZHANG M X, XIE L J, et al. Identification of a Triple-Reassortant H1N1 Swine Influenza Virus in a Southern China Pig . Genome Announc, 2014, 2(2): e00229-14.

[9] XIE Z X, ZENG T T, XIE L J, et al. Genome analysis of a Tembusu virus, GX2013H, isolated from a Cheery valley duck in Guangxi, China. Genome Announc, 2014, 2(4): e00466-14.

[10] HE C Q, XIE Z X, HAN G Z, et al. Homologous recombination as an evolutionary force in the avian influenza A virus. Mol Biol Evol, 2009, 26(1): 177-187.

[11] LIU T T, XIE Z X, SONG D G, et al. Genetic characterization of a natural reassortant H3N8 avian influenza virus isolated from domestic geese in Guangxi, southern China. Genome Announc, 2014, 2(4): e00747-14.

[12] XIE Z X, FAN Q, XIE Z Q, et al. Complete genome sequence of a bovine viral diarrhea virus strain isolated in southern china. Genome Announc, 2014, 2(3): e00512-14.

[13] XU Q, XIE Z X, XIE L J, et al. Characterization of an avian influenza virus H9N2 strain isolated from a wild bird in southern China. Genome Announc, 2014, 2(3): e00600-14.

[14] XIE Z X, GUO J, XIE L J, et al. Complete genome sequence of a novel reassortant avian influenza H1N2 virus isolated from a domestic sparrow in 2012. Genome Announc, 2013, 1(4): e00431-13.

Complete genome sequence of a bovine viral diarrhea virus strain isolated in southern China

Xie Zhixun, Fan Qing, Xie Zhiqin, Liu Jiabo, Pang Yaoshan, Deng Xianwen, Xie Liji, Luo Sisi, and Khan Mazhar I

Abstract

We report here the full-length RNA genomic sequence of the bovine viral diarrhea virus (BVDV) strain GX4, isolated from a cow in southern China. Studies indicate that BVDV GX4 belongs to the BVDV-1b subtype. This report will help in understanding the epidemiology and molecular characteristics of BVDV in southern China cattle.

Keywords

bovine viral diarrhea virus, complete genome sequence, characterization

---

Bovine viral diarrhea (BVD) has a high prevalence rate and low mortality rate, leading to significant economic losses[1]. Bovine viral diarrhea virus (BVDV) is a positive-sense single-stranded RNA virus, with a genome size of approximately 12.5 kb. BVDV is a member of the genus *Pestivirus* in the family Flaviviridae. BVDV is classified into two biotypes, cytopathogenic and noncytopathogenic, based on the presence or absence, respectively, of cytopathogenic effects in cell cultures. In addition, there are two major genotypes, BVDV1 and BVDV2, based on genetic relatedness[2].BVDV1 has at least 15 subtypes, and the complete genomic sequences of BVDV-1a, -1b, -1d, -1e, -1m, 1k, and BVDV-2a and-2b have been reported[3-5].

A strain of BVDV, named GX4, was isolated from a dairy farm in Guangxi in southern China. Fecal swabs were collected from a Holstein cow with typical acute diarrheal disease. The swabs were eluted with phosphate-buffered saline (PBS) (pH7.2), inoculated onto MDBK monolayer cultures, and incubated at 37 ℃ for 4 d. No cytopathic effect was observed after eight serial passages. Viral genomic RNA was extracted from GX4-infected culture supernatant using the TRIzol RNA kit (Invitrogen, Guangzhou, China). The Primer Premier software was used to design 12 pairs of primers, and a BLAST search was performed to verify the oligonucleotide specificities of the primers. Reverse transcription-PCR (RT-PCR) amplifications were performed using a Prime-Script one-step RT-PCR kit (TaKaRa, Dalian, China)[6].The amplified products were purified and cloned into the pMD18-T vector (TaKaRa, Dalian, China) and then sequenced (Invitrogen, Guangzhou, China).The sequences were assembled and manually edited to generate the final genome sequence using the DNAStar MegAlign software.

The genome of GX4 comprises 12 218 nucleotides, with a 3' untranslated region (UTR), a 5' UTR, and one open reading frame (ORF) encompassing 3 898 amino acids. The complete sequences and the deduced amino acid sequences of GX4 were compared with those of published BVDV sequences. The homologies at the nucleotide level are 89.0% to 99.2%, and the homologies at the deduced amino acid level are 88.6% to 97.2% for the BVDV-1b subtype. Immunogenicity studies have shown that the nonstructural 3 (NS3) and envelope 2 (E2) glycoproteins are crucial components for eliciting neutralizing antibodies in cattle[7, 8]. However, the amino acid sequence of E2 of GX4, as with the other 29 BVDV-1b subtypes, is highly variable, whereas that of NS3 is conserved. The NS3 glycoprotein has been thought to be a good candidate for use as an immunogen in the development of subunit vaccines or diagnostic kits[9, 10].

Phylogenetic tree analysis indicated that GX4 is closely related to strain Av69 VEDEVAC (GenBank accession number: KC695814.1) from Lanzhou, China; the major difference between these strains is the presence of 45 nucleotides at positions 4 307 to 4 351 in Av69 VEDEVAC. However, these nucleotides are not present in BVDV-1b strains, indicating that BVDV GX4 belongs to the BVDV-1b subtype. This report will help to understand the epidemiology and molecular characteristics of BVDV in Guangxi cattle.

Nucleotide sequence accession number. The complete genomic sequence of BVDV GX4 has been deposited in GenBank under the accession number KJ689448.

References

[1] RIDPATH J F, BOLIN S R, DUBOVI E J. Segregation of bovine viral diarrhea virus into genotypes. Virology, 1994, 205(1): 66-74.

[2] HOUE H. Epidemiology of bovine viral diarrhea virus. Vet Clin North Am Food Anim Pract, 1995, 11(3): 521-547.

[3] KADIR Y, CHRISTINE F, BARBARA BW, et al. Genetic heterogeneity of bovine viral diarrhoea virus (BVDV) isolates from turkey: identification of a new subgroup in BVDV-1. Vet Microbiol, 2008, 130(3-4): 258-267.

[4] VILCEK S, DURKOVIC B, KOLESÁROVÁ M, et al. Genetic diversity of international bovine viral diarrhoea virus (BVDV) isolates: identification of a new BVDV-1 genetic group. Vet Res, 2004, 35(5):(5): 609-615.

[5] MISCHKALE K, REIMANN I, ZEMKE J, et al. Characterisation of a new infectious full-length cDNA clone of BVDV genotype 2 and generation of virus mutants. Vet Microbiol, 2010, 142(1-2): 3-12.

[6] FAN Q, XIE Z X, XIE L J, et al. Comparative study of three nucleic acid amplification assays for the detection of bovine viral diarrhea virus. Chin Vet Sci , 2012, 42(3): 294-298.

[7] FERRER F, ZOTH S C, CALAMANTE G, et al. Induction of virus-neutralizing antibodies by immunization with Rachiplusia nu per os infected with a recombinant baculovirus expressing the E2 glycoprotein of bovine viral diarrhea virus. J Virol Methods, 2007, 146(1-2): 424-427.

[8] DEREGT D, DUBOVI E J, JOLLEY M E, et al. Mapping of two antigenic domains on the NS3 protein of the pestivirus bovine viral diarrhea virus. Vet Microbiol, 2005, 108(1-2): 13-22.

[9] MAKOSCHEY B, SONNEMANS D, BIELSA JM, et al. Evaluation of the induction of NS3 specific BVDV antibodies using a commercial inactivated BVDV vaccine in immunization and challenge trials. Vaccine, 2007, 25(32): 6140-6145.

[10] KUIJK H, FRANKEN P, MARS M H, et al. Monitoring of BVDV in a vaccinated herd by testing milk for antibodies to NS3 protein. Vet Rec, 2008, 163(16): 482-484.

Chapter Four

Studies on Pathogenesis and Immune Mechanism

Transcriptomic and translatomic analyses reveal insights into the signaling pathways of the innate immune response in the spleens of SPF chickens infected with avian reovirus

Wang Sheng, Huang Tengda, Xie Zhixun, Wan Lijun, Ren Hongyu, Wu Tian, Xie Liji, Luo Sisi, Li Meng, Xie Zhiqin, Fan Qing, Huang Jiaoling, Zeng Tingting, Zhang Yanfang, Zhang Minxiu, and Wei You

Abstract

Avian reovirus (ARV)infection is prevalent in farmed poultry and causes viral arthritis and severe immunosuppression. The spleen plays a very important part in protecting hosts against infectious pathogens. In this research, transcriptome and translatome sequencing technology were combined to investigate the mechanisms of transcriptional and translational regulation in the spleen after ARV infection. On a genome-wide scale, ARV infection can significantly reduce the translation efficiency (TE) of splenic genes. Differentially expressed translational efficiency genes (DTEGs)were identified, including 15 upregulated DTEGs and 396 downregulated DTEGs. These DTEGs were mainly enriched in immune regulation signaling pathways, which indicates that ARV infection reduces the innate immune response in the spleen. In addition, combined analyses revealed that the innate immune response involves the effects of transcriptional and translational regulation. Moreover, we discovered the key gene IL4I1, the most significantly upregulated gene at both the transcriptional and translational levels. Further studies in DF1 cells showed that overexpression of IL4I1 could inhibit the replication of ARV, while inhibiting the expression of endogenous IL4I1 with siRNA promoted the replication of ARV. Overexpression of IL4I1 significantly downregulated the mRNA expression of IFN-β, LGP2, TBK1 and NF-κB; however, the expression of these genes was significantly upregulated after inhibition of IL4I1, suggesting that IL4I1 may be a negative feedback effect of innate immune signaling pathways. In addition, there may be an interaction between IL4I1 and ARV σA protein, and we speculate that the IL4I1 protein plays a regulatory role by interacting with the σA protein. This study not only provides a new perspective on the regulatory mechanisms of the innate immune response after ARV infection but also enriches the knowledge of the host defense mechanisms against ARV invasion and the outcome of ARV evasion of the host's innate immune response.

Keywords

translatomics, Ribo-seq, avian reovirus, spleen, innate immunity, IL4I1

Introduction

Avian reovirus (ARV) is part of the genus *Orthoreovirus* in the Spinareoviridae family, and the main symptoms it causes include viral arthritis, chronic respiratory disease, growth retardation and malabsorption syndrome. In addition, ARV infection can cause severe immunosuppression and predispose patients to other complications or secondary infections. ARV infection is widespread in the global poultry industry; vaccination is mainly used to prevent it in the fowl industry, but it is still not well prevented or controlled. ARV infection causes major economic losses for farmers[1-3].

At present, good progress has been made in the research and development of ARV vaccines and the establishment of detection methods, but research on the pathogenic mechanism of ARV and the antiviral response of host innate immunity is still in the development stage. Furthermore, clarifying the interaction mechanism of ARV is necessary for its prevention, control and treatment.

Innate immunity is the body's first line of defense against ARV invasion. The body activates the production of inflammatory factors and interferons by recognizing pathogenic pattern-related molecules through pattern recognition receptors and activates acquired immunity to trigger a comprehensive immune response[4, 5]. Many studies have been conducted on the regulation of innate immune-related pattern recognition receptors and their effector factors during ARV infection. In the early period of ARV infection, the PI3K/Akt/ NF-κB and STAT3 signaling pathways can be activated to induce an inflammatory response[6, 7]. Lostalé-Seijo et al. found that ARV infection of chicken embryonic fibroblasts induced the expression of interferon and interferon-stimulated genes Mx and double-stranded RNA-dependent protein kinases (PKRs) to exert antiviral effects[8].

In the early stage of ARV infection, the virus can induce the activation of MDA5 signaling pathway-related molecules in chicken peripheral blood lymphocytes, thereby inducing the production of inflammatory factors and interferons[9]. In a previous study, we detected the transcriptional expression level changes in interferon and interferon-stimulated genes (Mx, IFITM3, IFI6 and IFIT5)in several different tissues and organs of ARV-infected specific pathogen-free (SPF) chickens using real-time PCR, and the results suggested that ARV infection can cause significant changes in these effector factors, indicating that the process of ARV infection is closely linked to the recognition of host innate immune-related model receptors and the production of effector factors[10, 11]. However, the limited number of genes that were tested made it impossible to fully characterize the gene regulation of innate immunity during ARV infection.

Next-generation sequencing technology has become an effective tool for studying the interaction between viruses and hosts. However, due to the poor correlations between mRNA abundance and protein abundance, traditional mRNA sequencing (RNA-seq) cannot accurately depict the whole realm of gene expression, particularly in regards to reflecting the actual expression levels of proteins[12]. Ribosome profiling, also known as ribosomal footprint sequencing (Ribo-seq), is a new high-throughput sequencing technology developed in recent years that sequences ribosome-protected mRNA fragments (RPFs)[13]. Ribo-seq can accurately measure translational activity and abundance genome-wide[14], making it possible to investigate the ribosome density profile of the translatome[15, 16]. Association analysis combined with RNA-seq and Ribo-seq methods can be used to study post-transcriptional regulation and translational regulation mechanisms. Ribo-seq has been used to explore the mechanisms of translation regulation in different species, such as humans[17], mice[18], zebrafish[19], Drosophila[20], rice[21], Arabidopsis[22] and maize[23]. However, the translational regulation mechanisms in chickens remain poorly studied.

The spleen is the largest and most important peripheral immune organ in chickens, and it plays a major role in maintaining the balance of immune function and evading the invasion of pathogenic microorganisms. Previous studies have found that ARV infection can cause harm to the spleen, which leads to immunosuppression[24, 25]. The results of our previous study showed that the viral load in the spleen after ARV infection was noticeably above those in the thymus and bursa of Fabricius, suggesting that the spleen is the main immune organ attacked by ARV. In addition, the mRNA expression of various interferon-stimulated genes in the spleen after ARV infection was rapidly upregulated in the early stage of infection, indicating that ARV infection can induce a strong innate immune response in the spleen. Analysis of the pathological changes in the spleen after ARV infection showed that there were no obvious lesions on days 1 to 2, while generalized necrotic degeneration of the lymphocytes and homogeneous red staining of the splenic body were observed on day 3. This pathological injury continued until day 7 and was gradually relieved. Interestingly, the ARV viral load in the spleen remained high for 1 to 3 days after infection and then decreased sharply on day 4[11]. Therefore, it is speculated that the early stage of ARV infection, especially day 3, is a critical period for ARV invasion of the spleen. Transcriptional and translational regulation play major roles in host innate immunity against viral infection. In this research, SPF chickens artificially infected with the ARV S1133 strain were used as subjects. Their splenic tissues were dissected on the third day after infection, and RNA-seq and Ribo-seq analyses were performed to study chicken spleen gene regulation after ARV infection at the transcriptional and translational levels. Finally, functional genes that play an important part in the innate immune response were identified, and their functions were further analyzed. Our results provide a comprehensive understanding of the immune evasion of ARVs in the spleen and of the host immune defense against ARVs.

Materials and Methods

Viral inoculations and animal experiments

The ARV S1133 strain used in the study was purchased from the China Institute of Veterinary Drug Control. "White Leghorn" SPF chicken eggs were purchased from Beijing Boehringer Ingelheim Vital Biotechnology Co., Ltd. (Beijing, China). Incubation was performed using a fully automated incubator, after which chicks were raised in SPF chicken isolators. A total of twenty 7-day-old SPF chickens were randomized into two groups and raised aseptically in an SPF chicken isolator. Group A was the experimental group (ARV), and each chicken was inoculated with 0.1 mL 10^4 $TCID_{50}$/0.1 mL ARV S1133 virus by foot pad injection. Group B was the control group (CON), which was inoculated with the same amount of PBS via foot pad injection. Samples were collected on day 3 after infection. The chickens were taken from the ARV infection group and the control group for dissection and collection of spleens, and then the collected samples were snap-frozen in liquid nitrogen and stored in a –80 ℃ freezer for subsequent analysis. The standard for selecting sequencing samples in this study was to use the two samples whose viral load of spleen was closest to the mean in the group as 2 biological replicates.

RNA extraction and transcriptome sequencing

Total RNA of the spleen in each group was extracted by using TRIzol® RNA extraction reagent (Invitrogen, Carlsbad, CA, USA) according to the manufacturer's instructions. The integrity of RNA was examined by agarose gel electrophoresis, and the concentration of RNA was measured by NanoDrop 2 000 spectrophotometers (Thermo Fisher Scientific, Boston, MA, USA). The mRNA was purified by oligo (dT) magnetic beads and fragmented into short fragments using fragmentation buffer. First-strand cDNA was synthesized with SuperScript Ⅱ Reverse Transcriptase (Invitrogen) using random primers, and second-strand cDNA was synthesized using the synthesized first strand of cDNA as a template. The obtained double-stranded cDNA was purified by a VAHTS® mRNA-seq V3 Library Prep Kit for Illumina (Vazyme, Nanjing, China), end repaired, poly (A) added and then ligated to Illumina sequencing adapters. Sequencing was performed on the Illumina HiSeq-2000 platform for 50 cycles. High-quality reads passed through the Illumina quality filter were retained in fastq. gz format for sequence analysis.

Preparation of ribosome-protected fragments and ribosome profling

RPF extraction and sequencing were performed by a commercial company (Chi-Biotech, Wuhan, China) according to a previous study[17]. Spleen tissue from each group was added to lysis buffer, ground at low temperature and then low concentrations of RNase I were added for digestion. The digested samples were pooled and layered on the surface of 15 mL sucrose buffer (30% sucrose in RB buffer). The ribosomes were pelleted by ultracentrifugation at 42 500 r/min for 5 h at 4 ℃. RPF extraction was then performed using TRIzol, and ribosomal RNA (rRNA) was depleted using the Ribo-off® rRNA Depletion Kit (Vazyme) following the manufacturer's instructions. Sequencing libraries of RPFs were constructed following the VAHTS® Small RNA Library Prep Kit for Illumina (Vazyme). The library was resolved by a 6% polyacrylamide gel. The fraction with an insertion size of ~28 nt was excised and purified from the gel. This fraction was sequenced by an Illumina HiSeq-2000 sequencer for 50 cycles. High-quality reads that passed the Illumina quality filters were kept for sequence analysis.

Sequence analysis

For both mRNA and RPF sequencing data sets, high-quality reads were mapped to the mRNA reference sequence (GRCg6a) through the FANSe2 algorithm[26] with the parameters-E5%-indel-S14. The expression abundance of mRNA and RPFs was normalized by RPKM (reads per kilobase per million reads)[27]. Differentially expressed genes (DEGs) in RNA-seq and Ribo-seq were identified via the edgeR package[28] with $|\log_2$ fold change$|>1$ and false discovery rate (FDR)<0.01. The quotient of RPFs and mRNA expression abundance is translation efficiency (TE)[29, 30]. Differential TE genes (DTEGs) were calculated by a t test with $|\log_2$ fold change$|>1$ and p value<0.05. Bioinformatic analysis was performed using Omicsmart, a real-time, interactive online platform for data analysis (http: //www.omicsmart.com, accessed on 20 November 2023).

Overexpression of IL4I1 protein

The recombinant plasmid pEFIα-Myc-IL4I1 was constructed from the IL4I1 gene sequence (Genbank accession number: NM_001099351.3) from chicken. DF1 cells were cultured in 6-well plates. When the cell confluency reached 70%~80%, the recombinant IL4I1 plasmid was transfected with liposome Lipofectamine™ 3000 (Invitrogen) to overexpress IL4I1 protein. After 24 h of transfection, DF1 cells were infected with the ARV S1133 strain at a multiplicity of infection (MOI) of 1. Then, the sample of cells and medium supernatant were gathered at 24 h postinfection. RNA was extracted from the above cell samples and reverse-transcribed to synthesize cDNA using the GeneJET RNA Purification Kit and Maxima™ H minus cDNA synthesis master mix (Thermo Fisher Scientific). Real-time PCR detected the replication of ARV at the gene level and the expression changes in innate immune signaling pathway-related molecules. The primer sequences of molecules associated with the innate immune signaling pathway and ARV σC gene are shown in Table 4-1-1[31]. In addition, the above medium supernatant was used to infect DF1 cells and the virulence was determined by the Reed-Muench method to detect the replication of ARV.

Table 4-1-1   Primers used in this study

Gene	GenBank accession number	Primer sequences (5'-3')
ARV σC	L39002.1	F: CCACGGGAAATCTCACGGTCACT, R: TACGCACGGTCAAGGAACGAATGT
IL4I1	NM_001099351.3	F: CACGCCGTATCAGTTCACC, R: CCTCACCGCAGCCTTCAT
IFN-α	AB021154.1	F: ATGCCACCTTCTCTCACGAC, R: AGGCGCTGTAATCGTTGTCT
IFN-β	X92479.1	F: ACCAGGATGCCAACTTCT, R: TCACTGGGTGTTGAGACG
MDA5	NM_001193638	F: CAGCCAGTTGCCCTCGCCTCA, R: AACAGCTCCCTTGCACCGTCT
LGP2	MF563595.1	F: CCAGAATGAGCAGCAGGAC, R: AATGTTGCACTCAGGGATGT
MAVS	MF289560.1	F: CCTGACTCAAACAAGGGAAG, R: AATCAGAGCGATGCCAACAG
TRAF3	XM_040672281.1	F: GGACGCACTTGTCGCTGTTT, R: CGGACCCTGATCCATTAGCAT
TRAF6	XM_040673314.1	F: GATGGAGACGCAAAACACTCAC, R: GCATCACAACAGGTCTCTCTTC
IKKε	XM_428036.4	F: TGGATGGGATGGTGTCTGAAC, R: TGCGGAACTGCTTGTAGATG
TBK1	MF159109.1	F: AAGAAGGCACACATCCGAGA, R: GGTAGCGTGCAAATACAGC
IRF7	NM_205372.1	F: CAGTGCTTCTCCAGCACAAA, R: TGCATGTGGTATTGCTCGAT
NF-κB	NM_205129.1	F: CATTGCCAGCATGGCTACTAT, R: TTCCAGTTCCCGTTTCTTCAC
GAPDH	NM_204305.1	F: GCACTGTCAAGGCTGAGAACG, R: GATGATAACACGCTTAGCACCAC

RNA interference assay

According to the sequence of IL4I1 genes, three small interfering RNAs (siRNAs) (Table 4-1-2)were designed and synthesized (GenePharma, Shanghai, China). IL4I1 siRNA inhibitory molecules were transfected into DF1 cells using Lipofectamine® RNAiMAX Reagent (Invitrogen) to inhibit IL4I1 protein expression. After 24 h of transfection, cells were infected with the ARV S1133 strain at a MOI of 1, and after another 24 h, cell samples and medium supernatants were collected to detect replication of ARV virus.

Table 4-1-2   siRNA sequence targeting the IL4I1 gene

siRNA	Sense	Antisense
si108	GCUGCUGAGUAUUGUGAAATT	UUUCACAAUACUCAGCAGCTT
si336	GCUGGUGCGUGAGUUUAUATT	UAUAAACUCACGCACCAGCTT
si1357	CCGUAUCAGUUCACCGAUUTT	AAUCGGUGAACUGAUACGGTT
siNC	UUCUCCGAACGUGUCACGUTT	ACGUGACACGUUCGGAGAATT

Real-time PCR

Real-time PCR was performed using the PowerUp™ SYBR™ Green Master Mix (Thermo Scientific)via the QuantStudio 5 real-time PCR system (Thermo Lifetech ABI, Boston, MA, USA).

Co-immunoprecipitation (Co-IP) assays

The interactions between IL4I1 and ARV σA or σC protein were detected by Co-IP. We designed three experimental groups: the test group was co-transfected with pEFIα-Myc-IL4I1 and pEFIα-HA-σA/σC and the control group was co-transfected with pEFIα-Myc-IL4I1 and pEFIα-HA or pEFIα-Myc and pEFIα-HA-σA/σC. Three biological replicates were assigned to each group. After 24 h of transfection, cells were lysed, protein samples were collected and Co-IP analysis with the Pierce Classic IP Kit (Thermo Scientific) was performed. The forward Co-IP test used rabbit-derived HA antibody (Invitrogen) for fixed adsorption samples, and Western blot analysis used murine Myc antibody and murine HA antibody (Invitrogen) as primary antibodies for detection. The reverse Co-IP used rabbit-derived Myc antibody (Invitrogen) for fixed adsorbed samples, and Western blot analysis used murine HA antibody and murine Myc antibody (Invitrogen) as primary antibodies. Goat anti-murine IgG (Beyotime, Shanghai, China) labeled with alkaline phosphatase was the secondary antibody.

Statistical analysis

At least three valid repeat tests were performed for each treatment, and the results are expressed as the mean$\pm$SD. Graph analysis and statistical comparisons used GraphPad Prism statistical software, version 9.5.0. The unpaired two tailed t-test (for two groups) and one-way ANOVA (for multiple groups) were used to identify the significance of difference. The results' difference were considered statistically significant at $P<0.05$.

Results

Overview of high-throughput sequencing data between the spleens of control group and ARV SPF chickens

To explore genome-wide innate immune response regulation from a translational perspective, we compared the ribosomal maps of the spleens of control group (CON) and ARV SPF chickens using ribosomal footprint sequencing and mRNA sequencing. The Ribo-seq and RNA-seq libraries of the CON and ARV groups were prepared and sequenced on HiSeq-2000 platforms, resulting in 13.1~14.8 million and 16~16.2 million ribosome profiling clean reads for the CON and ARV groups, as well as 17.3~17.8 million and 14.5~16.5 million RNA-seq clean reads for the CON and ARV groups, respectively. RNA-seq identified and quantified 17 721 genes and 17 693 genes in the CON and ARV groups, respectively. Ribo-seq identified and quantified 15 760 genes and 14 776 genes in the CON and ARV groups, respectively, and their expression abundance levels all had a similar normal distribution. The Pearson correlation analysis of RNA-seq and Ribo-seq exhibited similarities and differences between the CON and ARV groups ($R^2>0.92$). These results indicate that subsequent analyses are performed on the basis of reliable data.

Global translatome characteristics

To investigate whether the characteristics of the ribosome-protected fragments change with innate immune response regulation in the spleen, the basic ribosome profiles of RPFs were compared between the CON and ARV groups. The length distribution of RPF peaks at 28 nt for both the CON and ARV groups (Figure 4-1-1 A). Figure 4-1-1 B shows that the distribution patterns of RPFs for the CON and ARV groups were similar, with the vast majority of ribosome footprints located at the CDS of both CON and ARV mRNAs. These results are similar to those in most eukaryotes, suggesting that the translation process is highly conserved[32]. However, compared to the CON group, the RPFs of the ARV group in the CDSs and 5' UTRs decreased to 3.87% and 3.31%, respectively. The ARV RPFs in 3' UTRs increased from 8.02% to 15.19%. These results suggest that the apparent activation of RPFs in 3' UTRs may be caused by ARV infection in chickens. In addition, three-nucleotide periodicity was clearly observed around the start and stop codon regions of RPFs at different read lengths of 28 nt. A vast number of RPFs were enriched in the region approximately from position -12 nt to the annotated start codon (Figure 4-1-1 C), indicating that the initiation stage is the principal rate-limiting stage of translation.

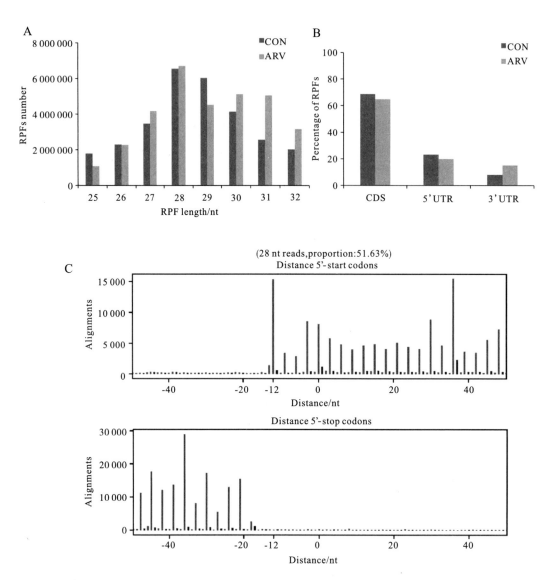

A: Length distribution of RPFs; B: The percentage of RPFs located in CDS, 5' UTR and 3' UTR; C: The total number of RPFs along CDS start and stop codon regions in ARV group. A codon contains three bases, which are represented by red, blue and green (color figure in appendix).

Figure 4-1-1   Characteristics of ribosome profiling data in the CON and ARV groups

Translational efficiency significantly decreased after ARV infection

TE is an important index of translation that reflects the efficiency of mRNA utilization, and the formula is as follows: TE = (RPKM from translatome)/(FPKM from transcriptome)[14, 33]. The average TE of genes across the whole genome decreased significantly after ARV infection ($\log_2$ mean TE of CON =−0.092 97, $\log_2$ mean TE of ARV=−0.642 5; Figure 4-1-2 A). In addition, the ratio of genes with higher TE ($\log_2$TE >1)in the ARV group was lower than that in the CON group (Figure 4-1-2 B). Moreover, there were 15 upregulated, differentially expressed TE genes and 396 downregulated, differentially expressed TE genes after ARV infection compared to the CON group (Figure 4-1-2 C). In regard to the functions of the differentially expressed TE genes, GO and KEGG analyses were conducted. The GO analysis displayed the top 20 terms, which mainly included cell activation, regulation of response to stimulus, regulation of immune system process, etc. (Figure 4-1-2 D). The KEGG pathway enrichment analysis showed the top 20 pathways for which gene expression was enriched. Among the top 20 pathways, 7 pathways belonged to the "immune system" category, including the chemokine signaling pathway, Fc epsilon RI signaling pathway, T-cell receptor signaling pathway, intestinal immune network for IgA production, complement and coagulation cascades, B-cell receptor signaling pathway and hematopoietic cell lineage. The rheumatoid arthritis pathway was also among the top 20 pathways and belonged to the "immune diseases" category (Figure 4-1-2 E). The above results indicate that after infection with ARV, gene translation efficiency is reduced at the overall level, and the translation efficiency of immune-related genes is significantly affected.

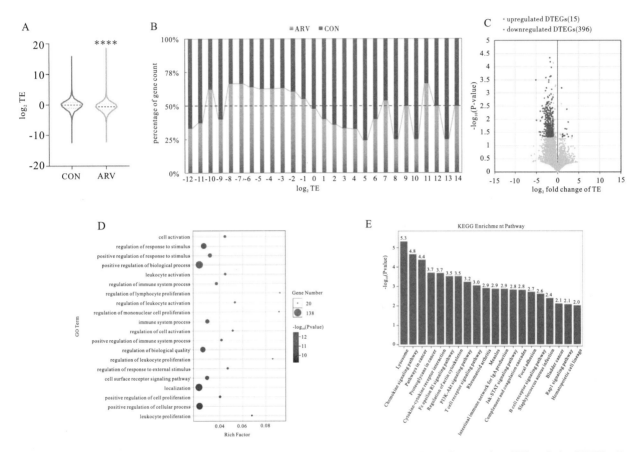

A: Violin plot of TE in the CON and ARV groups; B: Relative TE ratio; C: Volcano map of DTEGs; D: Gene ontology (GO) analysis of DTEGs; E: Kyoto Encyclopedia of Genes and Genomes (KEGG) analysis of DTEGs. Asterisk indicates significant difference (****$P<0.000$ 1). Genes were classified based on their rounded $\log_2$ TE values. The gray dot represents the non-differentially expressed translation efficiency genes (color figure in appendix).

Figure 4-1-2   Translation efficiency analysis

Regulation patterns of the transcriptome and translatome

Based on both ribosome profiling and RNA sequencing data, the transcriptional and translational expression differences between the CON and ARV groups were examined. The CON and ARV groups had high correlations for transcriptome and translatome ($R^2=0.836$ 4 in RNA-seq, $R^2=0.853$ 1 in Ribo-seq), which illustrates that transcriptome analysis and translatome analysis are reliable. The differentially expressed genes in both the RNA-seq and Ribo-seq data sets were filtered based on the criteria of $|\log_2$ fold change $|>$ 1 and FDR<0.01. There were 225 transcriptionally upregulated and 439 downregulated DEGs in the ARV group compared to the CON group, corresponding to 851 upregulated and 1 128 downregulated DEGs at the translational level. The quantities of downregulated genes were much greater than those of the upregulated genes at two levels, suggesting a global decline in gene expression in the ARV group.

To explore the relationships and differences in the regulation of gene expression in the spleen at the transcriptional level and the translation level after ARV infection, we conducted a combined analysis of the transcriptome and the translatome. Figure 4-1-3 A displays the scatter plot of the fold changes in transcriptional and translational expression. The scatter plot was classified into nine categories based on the criteria of $|\log_2$ fold change in RPKM $|>1$ and FDR<0.01. Our results revealed that 81.96% of genes were categorized in the unchanged class (quadrant E), and 14.7% of genes were in the discordant classes (quadrants A, B, D, F, H, I). Notably, 1.31% (154) and 2.02% (238) of genes were located in quadrants C and G, respectively, which meant that the expression of genes changed congruously at the transcriptional and translational levels (upregulation for quadrant C; downregulation for quadrant G). Furthermore, to explore the synergistic functions of transcription and translation, GO analysis of the biological processes enriched for the congruous DEGs (quadrants C and G) was conducted. The results showed that the congruous DEGs were significantly enriched for terms related to innate immunity such as cell surface receptor signaling pathway, immune response, innate immune response, immune system process and regulation of immune system process (Figure 4-1-3 B). These results suggest that the innate immune response of the body to ARV infection involves co-regulation of transcription and translation.

Screening functional genes after ARV infection

The above association analysis showed that genes in quadrants C and G were mainly enriched for biological processes related to immune regulation, indicating that these DEGs may be involved in the innate immune response to ARV infection. We identified 392 DEGs in quadrants C and G. We compared the top 30 DEGs at the transcriptional and translational levels (Figure 4-1-4 A, B). Based on the significance of the DEGs, the most significant DEG at the transcriptional and translational levels was determined to be IL4I1 (FDR in transcriptome $= 1.85 \times 10^{-27}$, FDR in translatome $= 6.2 \times 10^{-57}$). According to the RNA-seq and Ribo-seq data, the expression level of IL4I1 significantly increased after ARV infection (Figure 4-1-4 C). The RT-qPCR results were consistent with the RNA-seq and Ribo-seq results (Figure 4-1-4 D).

IL4I1 expression reduced ARV replication

To verify the effect of IL4I1 overexpression on ARV replication, we transfected the pEF1α-Myc-IL4I1 recombinant plasmid into DF1 cells, then verified IL4I1 overexpression by real-time PCR and Western blotting 24 h later. Real-time PCR showed that IL4I1 gene expression in DF1 cells was significantly upregulated after transfection with the pEFlα-Myc-IL4I1 (Figure 4-1-5 A). Detection at approximately 60 kDa using Myc-tagged antibodies showed that IL4I1 was correctly expressed in DF1 cells transfected with the pEFlα-Myc-IL4I1, while IL4I1 protein expression was not detected in cells transfected with empty vectors (Figure 4-1-5 B). Cells were infected with ARV after 24 h of transfection, and cell samples and supernatants were collected after another 24 h. Real-time PCR detection showed that the expression of the ARV σC gene at the mRNA level was significantly reduced (Figure 4-1-5 C), and the ARV virus titer in the cell supernatant was also significantly reduced (Figure 4-1-5 D). These results suggest that the overexpression of IL4l1 in DF1 cells inhibits the replication of ARV.

Therefore, we speculated that IL4I1 can cause negative feedback in the replication of ARV and that inhibiting the expression of IL4I1 can promote the replication of ARV. We designed and synthesized three siRNAs against the IL4I1 gene to inhibit the expression of IL4I1, of which siRNA1357 was the most ideal

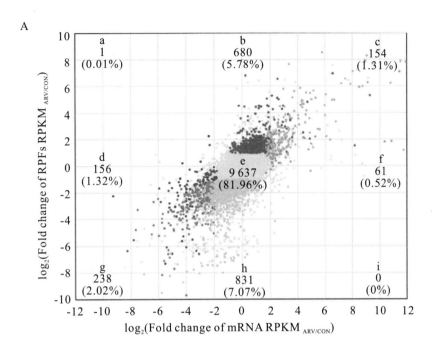

- a: Downregulation for transcription and upregulation for translation;
- b: No change for transcription and upregulation for translation;
- c: Upregulation for both transcription and translation;
- d: Downregulation for transcription and no change for translation;
- e: No change for both transcription and translation;
- f: Upregulation for both transcription and no change for translation;
- g: Downregulation for both transcription and translation;
- h: No change for transcription and downregulation for translation;
- i: Upregulation for transcription and downregulation for translation.

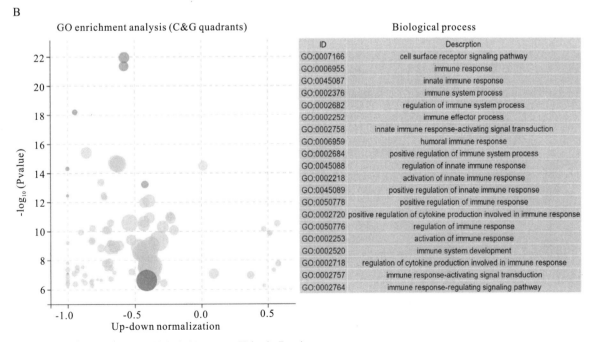

A: Scatter plot of the fold change in the ARV/CON group at the transcriptional and translational levels; B: GO enrichment analysis of genes in quadrants C and G (color figure in appendix).

Figure 4-1-3   Avian reovirus altered gene expression at both the transcriptional and translational levels

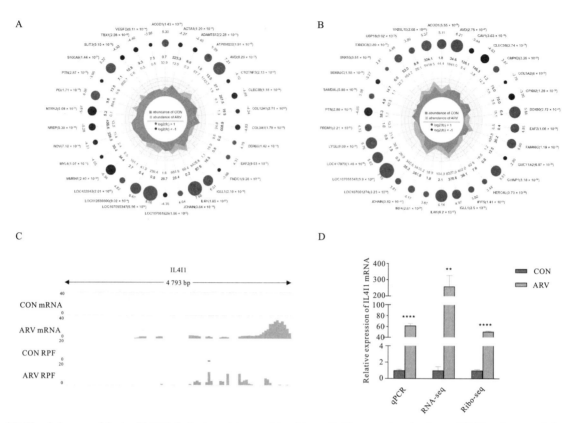

A and B: The circle maps of the top 30 DEGs in the transcriptome and translatome; C: The expression abundance of IL4I1 at the transcriptional and translational levels; D: The relative level of IL4I1 mRNA expression. Asterisks indicate significant differences (**$P<0.01$, ****$P<0.000\ 1$) (color figure in appendix).

Figure 4-1-4   Functional gene screening

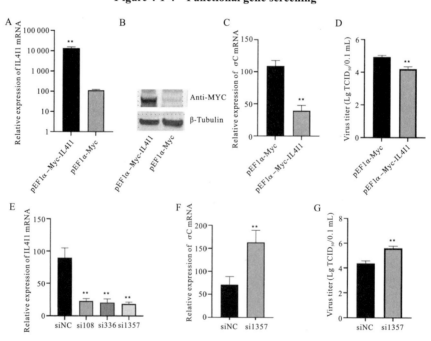

A and B: DF-1 cells were transfected with pEFla-Myc-IL4I1 or pEFla-Myc plasmids, and both real-time PCR and Western blot confirmed high levels of IL4I1 expression in DF-1 cells; C and D: DF-1 cells were transfected with pEFla-Myc-IL4I1 or pEFla-Myc plasmids and infected with the ARV S1133 strain at an MOI of 1, viral replication was detected by real-time PCR and viral titer 24 h postinfection; E: Comparison of the inhibition efficiency of three siRNAs of IL4I1 by real-time PCR; F and G: DF1 cells were transfected with si1357 or siNC and infected with the ARV S1133 strain of virus at an MOI of 1, viral replication was detected by real-time PCR and viral titer 24 h postinfection. Data are represented as the mean ± SD of three independent experiments. Asterisks indicate significant differences (**$P<0.01$).

Figure 4-1-5   IL4I1 inhibits the replication of ARV in DF1 cells

(Figure 4-1-5 E). siRNA1357 was transfected into DF1 cells, and the cells were infected with ARV virus after 24 h of transfection. Then, cell samples and supernatants were collected at 24 h postinfection to detect ARV replication at the gene expression and viral titer levels, and the results were consistent with expectations (Figure 4-1-5 F, G). The above results show that inhibiting the expression of IL4I1 can promote the replication of ARV.

The effect of IL4I1 expression on the innate immune response during ARV infection

To investigate how IL4I1 regulates the innate immune response induced by ARV infection, we overexpressed or inhibited IL4I1 and detected the effect of IL4I1 expression on the expression of innate immune signaling pathway-correlated factors during ARV infection by real-time PCR. The results showed that the mRNA expression of MDA5, TRAF3 and TRAF6 was significantly upregulated after overexpression or inhibition of IL4I1. The mRNA expression of MAVS was upregulated after overexpression of IL4I1 and downregulated after inhibition of IL4I1. The mRNA expression of IKKε did not differ significantly after overexpression of IL4I1 and was significantly upregulated after inhibition of IL4I1. The mRNA expression of IRF7 was significantly upregulated after overexpression of IL4I1, but there was no significant difference after IL4I1 inhibition. There was no significant difference in the mRNA expression of IFN-α after overexpression or inhibition of IL4I1. The mRNA expression of IFN-β, LGP2, TBK1 and NF-κB was significantly downregulated after overexpression of IL4I1 and upregulated after inhibition of IL4I1 (Figure 4-1-6 A, B). These results suggest that IL4I1 may be a negative feedback regulator of innate immune signaling pathways and that IL4I1 expression may reduce IFN-β production by inhibiting the expression of LGP2, TBK1 and NF-κB.

Interaction between IL4I1 and ARV σA/σC proteins

The σA and σC proteins are important structural proteins of ARV and play a significant part in the interaction between ARV and the host. Therefore, we studied the relationship between IL4I1 and ARV σA/σC. We transfected the eukaryotic expression plasmids pEF1α-HA-σA and pEF1α-HA-σC into DF1 cells and overexpressed ARV σA and σC proteins in DF1 cells. Real-time PCR detection showed that the expression of IL4I1 was significantly upregulated after the overexpression of σA and σC proteins in DF1 cells. The overexpression of σA in particular upregulated IL4I1 by a relatively high fold change (Figure 4-1-7 A).

Subsequently, we used Co-IP to determine whether IL4I1 interacted with ARV σA and σC proteins *in vitro*. The Co-IP results of IL4I1 protein and ARV σA protein showed that when Co-IP immobilization used the anti-Myc monoclonal antibody, the σA-HA protein could be identified by Western blot. However, when Co-IP immobilization used the anti-HA monoclonal antibody, the IL4I1-Myc protein was not identified by Western blot (Figure 4-1-7 B).

The Co-IP results of IL4I1 protein and ARV σC protein showed that when Co-IP immobilization used the anti-Myc monoclonal antibody, no σC-HA protein was identified by Western blot. IL4I1-Myc protein was also not identified by Western blot when Co-IP immobilization used the anti-HA monoclonal antibody (Figure 4-1-7 C). Therefore, there may be an interaction between the IL4I1 and ARV σA proteins.

Discussion

Gene expression is closely related to the occurrence and development of various physiological and pathological activities and diseases. High-throughput sequencing technology enables the sequencing and identification of millions of nucleotide molecules simultaneously. To date, it is widely used in the screening of important functional genes and research on animal diseases[34]. Translation regulation is a key element in the

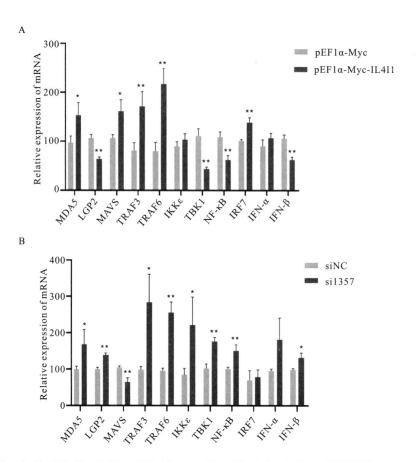

A: DF-1 cells were transfected with pEF1α-Myc-IL4I1 or pEF1α-Myc plasmids and then infected with the ARV S1133 strain at an MOI of 1. Expression changes in genes associated with the innate immune signaling pathway were detected by real-time PCR 24 h postinfection. B: DF1 cells were transfected with si1357 or siNC and infected with the ARV S1133 strain of virus at an MOI of 1. Expression changes in genes associated with the innate immune signaling pathway were detected by real-time PCR 24 h postinfection. Data are represented as the mean±SD of three independent experiments. Asterisks indicate significant differences (*P<0.05, **P<0.01).

Figure 4-1-6   Effect of IL4I1 on the innate immune response during ARV infection

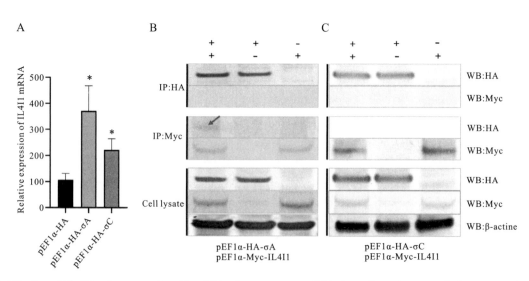

A: PEFla-HA-σA or pEFla-HA-σC were transfected into DF1 cells to overexpress ARV σA or σC proteins, respectively. Determination of IL4I1 expression by real-time PCR. B: Verification of IL4I1 interaction with ARV σA by Co-IP. C: Verification of IL4I1 interaction with ARV σC by Co-IP. Asterisks indicate significant differences (*P<0.05).

Figure 4-1-7   IL4I1 in DF1 cells interacts with ARV σA and σC proteins

regulation of gene expression. Omics studies have shown that translation regulation accounts for more than half of all regulation overseeing gene expression, and the translational differences better reflect the expression changes in the proteome than those of the transcriptome[35, 36].

In this study, Ribo-seq was conducted to reveal the gene expression profile of the spleen after ARV infection. We analyzed the characteristics of RPFs in the spleen; after ARV infection, the abundance of splenic RPFs in the CDS region and the 5' UTR was lower than that in the control group, which also indicated that ARV infection inhibited the process of protein synthesis in the spleen. After ARV infection, the abundance of splenic RPFs in the 3' UTR was higher than that in the CON group. In the eukaryotic translational process, the 5' UTR and the 3' UTR play important roles in post-transcriptional regulation. The 5' UTR mediates post-transcriptional regulation through the main elements present in this region, such as uORFs, secondary structures and RPF-binding motifs, and the 3' UTR contains a large number of regulatory elements, such as microRNA binding sites and protein binding sites[37-39]. After ARV infection, the abundance of splenic RPFs increased significantly in the 3' UTR, suggesting that there may be potential translational regulation of ARV infection in the 3' UTR, which may be related to the innate immune regulation process of the host after ARV infection. These complex regulatory mechanisms still need to be further studied.

Studies have confirmed that genes with higher translation efficiency perform more important biological functions, and in addition, the specific array of genes with high translation efficiency or upregulated translation efficiency reflect the function and phenotype of a particular cell[14]. Our sequencing results showed that ARV infection significantly reduced the overall translation efficiency of splenic genes. Further analysis showed that, compared with the control group, there were 15 significantly upregulated DTEGs and 369 significantly downregulated DTEGs in the ARV infection group. These DTEGs were mainly enriched in signaling pathways related to immune regulation, which led to speculation that ARV infection would reduce the immune response ability of the spleen. Recent studies have confirmed that ARV infection causes immunosuppression[40]. However, the specific mechanism of ARV-induced immunosuppression still needs to be studied in greater depth. After ARV infection, the translation efficiency of immune regulation-related signaling pathways in the spleen is reduced, which we speculate is one of the important reasons for immunosuppression caused by ARV infection.

The transcriptome and translatome association data analysis showed that 392 genes were expressed in common with significant differences at the transcription and translation levels. GO enrichment analysis showed that the 20 most significant GO terms were all enriched in signaling pathways related to immune regulation. This suggests that these DEGs play important roles in the innate immune response to ARV infection. We further analyzed the significance of these 392 congruous DEGs and screened the 30 genes with the highest significance. The most significantly differentially expressed gene was IL4I1 in both the transcriptome and translatome. The expression of IL4I1 was highly induced after ARV infection, suggesting that IL4I1 plays an important part in the response to ARV infection. Interleukin-4-induced-1 (IL4I1) is a less-studied amino acid catabolic enzyme that belongs to the L-amino acid oxidase family. IL4I1 plays an important role in the body's defense against infection, regulation of immune homeostasis and injury response. In recent years, it has been found that IL4I1 is closely related to the regulatory process of human immune metabolism, and it is an important immunosuppressive factor and a key metabolic immune checkpoint[41]. Tumor cells produce large amounts of the IL4I1 metabolic enzyme, which promotes the spread of tumor cells and suppresses the immune system. IL4l1 breaks down tryptophan to form indole metabolites and kynurequinolinic acid, which are agonists of the aryl hydrocarbon receptor (AHR). Indole metabolites and kynurequinolinic acid bind to and activate AHR receptors, thereby mediating the toxic effects of dioxins, which reduces the utilization of essential or semi-essential amino acids. At the same time, toxic metabolites are produced, which cause damage to antitumor T lymphocytes and promote the growth of tumor cells[42].

There are few reports on the role and mechanism of IL4I1 in avian virus infection. Studies by high-throughput sequencing technology have found that a large amount of IL4I1 expression can be induced after a variety of viral infections. Hu et al.found that IL4I1 expression was significantly upregulated in chicken embryo fibroblasts infected by the J subpopulation of avian leukemia virus (ALV-J) by transcriptome sequencing[43]. Feng et al.used transcriptome sequencing analysis to find that the expression of IL4I1 in chicken primary mononuclear macrophages infected with ALV-J was significantly upregulated 3 h and 36 h after infection, and overexpression of IL4I1 at the gene level could facilitate the replication of ALV-J in chicken macrophages[44]. Dong et al.found that the IL4I1 gene was upregulated in chicken spleen tissues infected with Marek virus (MDV) by transcriptome sequencing technology[45]. Conversely, our study showed that IL4I1 inhibits the replication of the ARV virus. This suggests that there may be differences in the roles of IL4I1 in the replication of different viruses. We found that IL4I1 was able to prevent the replication of ARV in DF1 cells by overexpression or inhibition assays. Our previous studies showed that the ARV viral load in the spleen was high on days 1 to 3 after ARV infection and then decreased sharply on day 4[11]. Therefore, we speculate that the rapid upregulation of IL4I1 expression in the spleen after ARV infection is beneficial to inhibiting the proliferation of ARV in the spleen. However, IL4I1 was able to promote the replication of ALV-J in chicken macrophages, suggesting that there may be differences in the roles of IL4I1 in the replication of different avian viruses. We also detected a regulatory role for IL4I1 on innate immune signaling pathway-correlated factors during ARV infection by real-time PCR. The mRNA expression of IFN-β, LGP2, TBK1 and NF-κB was significantly downregulated after overexpression of IL4I1, and the mRNA expression of IFN-β, LGP2, TBK1 and NF-κB was significantly upregulated after inhibiting IL4I1. This suggests that IL4I1 may be a negative feedback effect of innate immune signaling pathways during ARV infection, and IL4I1 may reduce the production of IFN-β by inhibiting the expression of LGP2, TBK1 and NF-κB. The current study has shown that IL4I1 is an important immunosuppressive molecule that plays a key role in the immune evasion of tumors[42]. Previous studies have also found that IL4I1 can inhibit the production of IFN-γ and inflammatory cytokines, limit local Th1 inflammation and inhibit the inflammatory response[46]. ARV is an important avian immunosuppressive disease, and our sequencing analysis found that ARV infection can significantly reduce the translational efficiency of immunomodulatory-related genes in chicken spleen. ARV, ALV-J and MDV infection can cause avian immunosuppressive diseases, and IL4I1 expression was upregulated after each of these three viral infections. Whether the regulation of IL4I1 expression is related to the immunosuppressive processes caused by avian immunosuppressive viruses is an interesting mechanism that deserves more in-depth study.

Virus-encoded proteins play a crucial part in virus interactions with its host. The σA protein of ARV plays a vital role in the pathogenesis of ARV infection. A study found that the ARV σA protein binds irreversibly to viral dsRNA, thereby inhibiting the dsRNA-dependent protein kinases activation and ultimately interfering with the antiviral effects of interferon[47, 48]. In addition, the ARV σA protein can also activate the PI3K/Akt signal transduction pathways in cells, increase the expression of phosphorylated Akt (p-Akt) in cells and, thus, inhibit the apoptosis of infected cells to facilitate ARV infection and replication[49]. It has also been showed that the ARV σA protein affects the replication of ARV in DF1 cells by interacting with the NME2 protein of the host[50]. The ARV σC protein is related to the adsorption and proliferation of virions[51]. The ARV σC protein is able to induce apoptosis by interacting with the host protein EFF1A1[52]. In this study, we found that overexpression of ARV σA and σC proteins in DF1 cells can cause significant upregulation of IL4I1 expression at the transcriptional level. The interaction between IL4I1 and ARV σC proteins was not found by co-immunoprecipitation experiments, while the co-immunoprecipitation analysis of IL4I1 protein and ARV σA proteins uncovered an interesting phenomenon. When we used the anti-Myc antibody to fix IL4I1 protein in the Co-IP experiment, σA protein interacted with IL4I1 protein, but when we fixed σA protein using the anti-HA antibody in the Co-IP experiment, the interaction between σA protein and IL4I1 protein could not be detected. To ensure the rigor of the experimental data, we performed multiple replicates using antibodies and Co-IP kits of different brands, all with the same results. We reviewed the literature and found that the human IL4I1 protein is a glycosylated secreted protein[53]. The structure and function of avian IL4I1 protein have not been reported, and our online software analysis shows that avian IL4I1 protein is also a secreted protein. We hypothesize that the IL4I1 protein failed to be detected by Western blot analysis when the σA protein was immobilized in the Co-IP experiment, which is related to the fact that IL4I1 is secreted extracellularly after synthesis. Therefore, we speculate that there may be an interaction between the IL4I1 protein and the ARV σA protein, and that the IL4I1 protein may actively bind to the σA protein. We speculate that the IL4I1 protein may play a regulatory role by interacting with the ARV σA protein. We further speculate that after ARV infection, IL4I1 is modulated and then transcribed and expressed in large quantities, and the IL4I1 protein competitively binds to the ARV σA protein, thereby affecting the function of the ARV σA protein and inhibiting the replication of ARV. However, due to a lack of avian-derived IL4I1 protein-specific antibodies, it is difficult to further verify the interaction between the IL4I1 and ARV σA proteins. In future studies, we will prepare monoclonal antibodies against avian IL4I1 protein and conduct in-depth research on the role and regulatory mechanism of IL4I1 in the interaction network of ARV or σA protein

Conclusions

In this study, we determined that the spleen produces a strong innate immune response at both the transcriptional and translational levels after ARV infection and that the spleen is an important immune response organ in ARV infection. ARV infection reduces the translation efficiency of innate immunity-related genes, and we speculate that the decrease in translation efficiency is the key cause of immunosuppression caused by ARV infection. ARV infection can significantly upregulate the expression of IL4I1, while the upregulation of IL4I1 helps to inhibit the replication of ARV. IL4I1 may inhibit the innate immune response triggered by ARV infection. In addition, the IL4I1 protein may interact with the viral protein σA of ARV. These results provide new insights into ARV-host interactions and will facilitate the development of new vaccines or other therapeutic agents to control ARV based on the IL4I1 gene in chickens.

References

[1] VAN DER HEIDE L. The history of avian reovirus. Avian Dis, 2000, 44(3): 638-641.

[2] DANDAR E, BALINT A, KECSKEMETI S, et al. Detection and characterization of a divergent avian reovirus strain from a broiler chicken with central nervous system disease. Arch Virol, 2013, 158(12): 2583-2588.

[3] TENG L, XIE Z, XIE L, et al. Sequencing and phylogenetic analysis of an avian reovirus genome. Virus Genes, 2014, 48(2): 381-386.

[4] GREEN A M, BEATTY P R, HADJILAOU A, et al. Innate immunity to dengue virus infection and subversion of antiviral responses. J Mol Biol, 2014, 426(6): 1148-1160.

[5] IWASAKI A, PILLAI P S. Innate immunity to influenza virus infection. Nat Rev Immunol, 2014, 14(5): 315-328.

[6] LIN P, LIU H, LIAO M, et al. Activation of PI 3-kinase/Akt/NF-kappaB and Stat3 signaling by avian reovirus S1133 in the early stages of infection results in an inflammatory response and delayed apoptosis. Virology, 2010, 400(1): 104-114.

[7] NEELIMA S, RAM G C, KATARIA J M, et al. Avian reovirus induces an inhibitory effect on lymphoproliferation in chickens. Vet Res Commun, 2003, 27(1): 73-85.

[8] LOSTALE-SEIJO I, MARTINEZ-COSTAS J, BENAVENTE J. Interferon induction by avian reovirus. Virology, 2016, 487: 104-111.

[9] XIE L, XIE Z, WANG S, et al. Altered gene expression profiles of the MDA5 signaling pathway in peripheral blood lymphocytes of chickens infected with avian reovirus. Arch Virol, 2019, 164(10): 2451-2458.

[10] WANG S, XIE L, XIE Z, et al. Dynamic changes in the expression of interferon-stimulated genes in joints of SPF chickens infected with avian reovirus. Front Vet Sci, 2021, 8: 618124.

[11] WANG S, WAN L, REN H, et al. Screening of interferon-stimulated genes against avian reovirus infection and mechanistic exploration of the antiviral activity of IFIT5. Front Microbiol, 2022, 13: 998505.

[12] HUANG T, YU J, LUO Z, et al. Translatome analysis reveals the regulatory role of betaine in high fat diet (HFD)-induced hepatic steatosis. Biochem Biophys Res Commun, 2021, 575: 20-27.

[13] HUANG T, YU J, MA Z, et al. Translatomics probes into the role of lycopene on improving hepatic steatosis induced by high-fat diet. Frontiers in Nutrition, 2021, 8.

[14] INGOLIA N T, GHAEMMAGHAMI S, NEWMAN J R, et al. Genome-wide analysis in vivo of translation with nucleotide resolution using ribosome profiling. Science, 2009, 324(5924): 218-223.

[15] INGOLIA N T, LAREAU L F, WEISSMAN J S. Ribosome profiling of mouse embryonic stem cells reveals the complexity and dynamics of mammalian proteomes. Cell, 2011, 147(4): 789-802.

[16] INGOLIA N T. Ribosome profiling: new views of translation, from single codons to genome scale. Nat Rev Genet, 2014, 15(3): 205-213.

[17] LIAN X, GUO J, GU W, et al. Genome-wide and experimental resolution of relative translation elongation speed at individual gene level in human cells. PLOS Genetics, 2016, 12(2): e1005901.

[18] HUANG T, YU L, PAN H, et al. Integrated Transcriptomic and Translatomic Inquiry of the Role of Betaine on Lipid Metabolic Dysregulation Induced by a High-Fat Diet. Frontiers in Nutrition, 2021, 8.

[19] BAZZINI A A, LEE M T, GIRALDEZ A J. Ribosome profiling shows that miR-430 reduces translation before causing mRNA decay in zebrafish. Science, 2012, 336(6078): 233-237.

[20] ZHANG H, DOU S, HE F, et al. Genome-wide maps of ribosomal occupancy provide insights into adaptive evolution and regulatory roles of uORFs during Drosophila development. PLOS Biol, 2018, 16(7): e2003903.

[21] XIONG Q, ZHONG L, DU J, et al. Ribosome profiling reveals the effects of nitrogen application translational regulation of yield recovery after abrupt drought-flood alternation in rice. Plant Physiology and Biochemistry, 2020, 155: 42-58.

[22] JUNTAWONG P, GIRKE T, BAZIN J, et al. Translational dynamics revealed by genome-wide profiling of ribosome footprints in Arabidopsis. Proc Natl Acad Sci USA, 2014, 111(1): E203-E212.

[23] LEI L, SHI J, CHEN J, et al. Ribosome profiling reveals dynamic translational landscape in maize seedlings under drought stress. The Plant Journal, 2015, 84(6): 1206-1218.

[24] ROSENBERGER J K, STERNER F J, BOTTS S, et al. In vitro and in vivo characterization of avian reoviruses. I. Pathogenicity and antigenic relatedness of several avian reovirus isolates. Avian Dis, 1989, 33(3): 535-544.

[25] ROESSLER D E, ROSENBERGER J K. In vitro and in vivo characterization of avian reoviruses. III. Host factors affecting virulence and persistence. Avian Dis, 1989, 33(3): 555-565.

[26] XIAO C, MAI Z, LIAN X, et al. FANSe2: a robust and cost-efficient alignment tool for quantitative next-generation sequencing applications. PLOS ONE, 2014, 9(4): e94250.

[27] MORTAZAVI A, WILLIAMS B A, MCCUE K, et al. Mapping and quantifying mammalian transcriptomes by RNA-Seq. Nat Methods, 2008, 5(7): 621-628.

[28] ROBINSON M D, MCCARTHY D J, SMYTH G K. edgeR: a Bioconductor package for differential expression analysis of digital gene expression data. Bioinformatics, 2010, 26(1): 139-140.

[29] WANG T, CUI Y, JIN J, et al. Translating mRNAs strongly correlate to proteins in a multivariate manner and their translation ratios are phenotype specific. Nucleic Acids Research, 2013, 41(9): 4743-4754.

[30] LI G, BURKHARDT D, GROSS C, et al. Quantifying absolute protein synthesis rates reveals principles underlying allocation of cellular resources. Cell, 2014, 157(3): 624-635.

[31] HE Y, XIE Z, DAI J, et al. Responses of the Toll-like receptor and melanoma differentiation-associated protein 5 signaling pathways to avian infectious bronchitis virus infection in chicks. Virol Sin, 2016, 31(1): 57-68.

[32] FUJITA T, KURIHARA Y, IWASAKI S. The plant translatome surveyed by ribosome profiling. Plant Cell Physiol, 2019, 60(9): 1917-1926.

[33] DUNN J G, FOO C K, BELLETIER N G, et al. Ribosome profiling reveals pervasive and regulated stop codon readthrough in *Drosophila melanogaster*. Elife, 2013, 2: e1179.

[34] MAIER T, GUELL M, SERRANO L. Correlation of mRNA and protein in complex biological samples. FEBS Lett, 2009, 583(24): 3966-3973.

[35] INGOLIA N T, BRAR G A, ROUSKIN S, et al. The ribosome profiling strategy for monitoring translation in vivo by deep sequencing of ribosome-protected mRNA fragments. Nat Protoc, 2012, 7(8): 1534-1550.

[36] SCHAFER S, ADAMI E, HEINIG M, et al. Translational regulation shapes the molecular landscape of complex disease phenotypes. Nat Commun, 2015, 6: 7200.

[37] CABRERA-QUIO L E, HERBERG S, PAULI A. Decoding sORF translation - from small proteins to gene regulation. RNA Biol, 2016, 13(11): 1051-1059.

[38] ARAUJO P R, YOON K, KO D, et al. Before it gets started: regulating translation at the 5'UTR. Comp Funct Genomics, 2012, 2012: 475731.

[39] CHAUDHURY A, HUSSEY G S, HOWE P H. 3'-UTR-mediated post-transcriptional regulation of cancer metastasis: beginning at the end. RNA Biol, 2011, 8(4): 595-599.

[40] LIN H, CHUANG S T, CHEN Y, et al. Avian reovirus-induced apoptosis related to tissue injury. Avian Pathol, 2007, 36(2): 155-159.

[41] MASON J M, NAIDU M D, BARCIA M, et al. IL-4-induced gene-1 is a leukocyte L-amino acid oxidase with an unusual acidic pH preference and lysosomal localization. J Immunol, 2004, 173(7): 4561-4567.

[42] SADIK A, SOMARRIBAS P L, OZTURK S, et al. IL4I1 is a metabolic immune checkpoint that activates the AHR and promotes tumor progression. Cell, 2020, 182(5): 1252-1270.

[43] HU X, CHEN S, JIA C, et al. Gene expression profile and long non-coding RNA analysis, using RNA-Seq, in chicken embryonic fibroblast cells infected by avian leukosis virus J. Arch Virol, 2018, 163(3): 639-647.

[44] FENG M, XIE T, LI Y, et al. A balanced game: chicken macrophage response to ALV-J infection. Vet Res, 2019, 50(1): 20.

[45] DONG K, CHANG S, XIE Q, et al. RNA Sequencing revealed differentially expressed genes functionally associated with immunity and tumor suppression during latent phase infection of a vv + MDV in chickens. Sci Rep, 2019, 9(1): 14182.

[46] MARQUET J, LASOUDRIS F, COUSIN C, et al. Dichotomy between factors inducing the immunosuppressive enzyme IL-4-induced gene 1 (IL4I1) in B lymphocytes and mononuclear phagocytes. Eur J Immunol, 2010, 40(9): 2557-2568.

[47] VAZQUEZ-IGLESIAS L, LOSTALE-SEIJO I, MARTINEZ-COSTAS J, et al. Avian reovirus sigmaA localizes to the nucleolus and enters the nucleus by a nonclassical energy- and carrier-independent pathway. J Virol, 2009, 83(19): 10163-10175.

[48] GONZALEZ-LOPEZ C, MARTINEZ-COSTAS J, ESTEBAN M, et al. Evidence that avian reovirus sigmaA protein is an inhibitor of the double-stranded RNA-dependent protein kinase. J Gen Virol, 2003, 84(Pt 6): 1629-1639.

[49] XIE L, XIE Z, HUANG L, et al. Avian reovirus sigmaA and sigmaNS proteins activate the phosphatidylinositol 3-kinase-dependent Akt signalling pathway. Arch Virol, 2016, 161(8): 2243-2248.

[50] XIE L, WANG S, XIE Z, et al. Gallus NME/NM23 nucleoside diphosphate kinase 2 interacts with viral sigmaA and affects the replication of avian reovirus. Vet Microbiol, 2021, 252: 108926.

[51] SHMULEVITZ M, YAMEEN Z, DAWE S, et al. Sequential partially overlapping gene arrangement in the tricistronic S1 genome segments of avian reovirus and Nelson Bay reovirus: implications for translation initiation. J Virol, 2002, 76(2): 609-618.

[52] ZHANG Z, LIN W, LI X, et al. Critical role of eukaryotic elongation factor 1 alpha 1 (EEF1A1) in avian reovirus sigma-C-induced apoptosis and inhibition of viral growth. Arch Virol, 2015, 160(6): 1449-1461.

[53] CASTELLANO F, MOLINIER-FRENKEL V. An overview of l-amino acid oxidase functions from bacteria to mammals: focus on the immunoregulatory phenylalanine oxidase IL4I1. Molecules, 2017, 22(12):2151.

Chicken IFI6 inhibits avian reovirus replication and affects related innate immune signaling pathways

Wan Lijun, Wang Sheng, Xie Zhixun, Ren Hongyu, Xie Liji, Luo Sisi, Li Meng, Xie Zhiqin, Fan Qing, Zeng Tingting, Zhang Yanfang, Zhang Minxiu, Huang Jiaoling, and Wei You

Abstract

Interferon-alpha inducible protein 6 (IFI6) is an important interferon-stimulated gene. To date, research on IFI6 has mainly focused on human malignant tumors, virus-related diseases and autoimmune diseases. Previous studies have shown that IFI6 plays an important role in antiviral, antiapoptotic and tumor-promoting cellular functions, but few studies have focused on the structure or function of avian IFI6. Avian reovirus (ARV) is an important virus that can exert immunosuppressive effects on poultry. Preliminary studies have shown that IFI6 expression is upregulated in various tissues and organs of specific-pathogen-free chickens infected with ARV, suggesting that IFI6 plays an important role in ARV infection. To analyze the function of avian IFI6, particularly in ARV infection, the chicken IFI6 gene was cloned, a bioinformatics analysis was conducted, and the roles of IFI6 in ARV replication and the innate immune response were investigated after the overexpression or knockdown of IFI6 in vitro. The results indicated that the molecular weight of the chicken IFI6 protein was approximately 11 kDa and that its structure was similar to that of the human IFI27L1 protein. A phylogenetic tree analysis of the IFI6 amino acid sequence revealed that the evolution of mammals and birds was clearly divided into two branches. The evolutionary history and homology of chickens are similar to those of other birds. Avian IFI6 localized to the cytoplasm and was abundantly expressed in the chicken lung, intestine, pancreas, liver, spleen, glandular stomach, thymus, bursa of Fabricius and trachea. Further studies demonstrated that IFI6 overexpression in DF-1 cells inhibited ARV replication and that the inhibition of IFI6 expression promoted ARV replication. After ARV infection, IFI6 modulated the expression of various innate immunity-related factors. Notably, the expression patterns of MAVS and IFI6 were similar, and the expression patterns of IRF1 and IFN-β were opposite to those of IFI6. The results of this study further advance the research on avian IFI6 and provide a theoretical basis for further research on the role of IFI6 in avian virus infection and innate immunity.

Keywords

IFI6, structure, function, antiviral, ARV, innate immune signaling pathway

---

Introduction

Innate immunity is the body's first line of immune defense against pathogen invasion. After infection of a host, a virus is sensed by pattern recognition receptors, and this sensing triggers a series of intracellular signaling pathways and induces the expression and secretion of a variety of cytokines, such as proinflammatory factors, chemokines, and interferons (IFNs)[1]. As important cytokines, IFNs mediate an antiviral response and play an extremely important role in innate immunity[2]. IFNs themselves exert no antiviral effects; in contrast,

IFNs induce antiviral activity by inducing the expression of a variety of interferon-stimulated genes (ISGs)[3]. Therefore, in-depth research on the expression of ISGs and their mechanisms of action is expected to increase the understanding of host antiviral immune responses and suggest distinct drug targets in the treatment of specific infectious diseases[1]. Hundreds of ISGs have been identified thus far, and these play important roles in host antiviral infection responses, including immune system regulation. The gene encoding interferon-alpha inducible protein 6 (IFI6), which is also known as G1P3 and IFI6-16, is an important ISG. IFI6 expression can be upregulated by type I IFNs and belongs to the FAM14 family[4, 5]. IFI6 is found only in eukaryotes. Studies have shown that human IFI6 is located on chromosome 1 and comprises 812 bp that encode 130 amino acids with a molecular weight of approximately 13 kDa. The protein structure of IFI6 is sequentially composed of signal peptides in the N-terminus, a hydrophilic region, a transmembrane region, a connecting region, a transmembrane region, and a hydrophilic region[5]. The structure and function of human IFI6 are clearly understood, and studies have shown that IFI6 plays an important role in a variety of human diseases. In recent years, several studies have shown that IFI6 is associated with a variety of malignant diseases; notably, IFI6 is highly expressed in various malignant tumors, virus-related diseases and autoimmune diseases[6, 7]. Further research has revealed that IFI6 plays an important role in many processes, such as antiviral, antiapoptotic and tumor-promoting activities. For example, IFI6 is highly expressed in gastric cancer cell lines and tissues, is enriched mainly in the inner mitochondrial membrane, co-localizes with cytochrome c in mitochondria, and can inhibit caspase-3 activity by inhibiting mitochondrial membrane depolarization and the release of cytochrome c, which results in the inhibition of apoptosis[6]. Other studies have found that the expression of IFI6 is significantly elevated in esophageal squamous cell carcinoma (ESCC) patients and ESCC cell lines cultured *in vitro*. Further in-depth studies have revealed that high expression of IFI6 in ESCC cells exerts an oncogenic effect and that the knockdown of IFI6 expression increases the accumulation of reactive oxygen species, which leads to inhibition of the proliferation of cancer cells and induction of apoptosis[8]. In terms of its antiviral activity, IFI6 confers protection to uninfected cells by blocking yellow fever virus-, West Nile virus-, dengue virus- and Zika virus-induced endoplasmic reticulum membrane invagination; i.e., IFI6 interacts with Bip, which is a companion to endoplasmic reticulum heat shock protein 70 encoded by HSPA5, to prevent virus-induced endoplasmic reticulum membrane invagination and protect uninfected cells and thereby achieves antiviral effects[9]. Studies of influenza A virus, severe acute respiratory syndrome coronavirus 2, and Sendai virus found that IFI6 negatively regulates the innate immune response induced by these viruses by affecting RIG-1 activation[10]. However, little is known about the role of avian IFI6 in avian diseases; therefore, the structure and function of avian IFI6 are worth exploring.

Avian reovirus (ARV), which is a double-stranded RNA virus that exists widely in nature and can infect chickens, ducks, geese, turkeys and other birds, has been detected in some wild birds[11-13]. ARV infection can cause viral arthritis/tenosynovitis, short stature syndrome, and malabsorption syndrome[14]. In addition, ARV infection can lead to immunosuppression, causing cell damage in various immune-related organs, such as the bursa of Fabricius, thymus, and spleen of chickens[15]. ARV infection makes the host more susceptible to infection with other pathogens, resulting in increased mortality due to coinfection and posing serious threats to healthy poultry development and to the poultry industry[16]. Innate immunity plays an important role in host defense against ARV infection, and considerable research has investigated ARV and the innate immunity of hosts. Studies have found that ARV infection activates the expression of many innate immune response-related factors in a host. ARV S1133 infection in specific-pathogen-free chickens can induce changes in the expression of various innate immunity-related factors, including IFN-α, IFN-β, IFN-γ, IL-6, IL-17, IL-18, MX, IFITM3,

PKR, OAS, IFIT5, ISG12, VIPERIN, and IFI6, in chicken peripheral blood lymphocytes and joints[17-19]. In addition, many studies have aimed to understand the intracellular molecular mechanisms underlying the effects of ARV infection. The host protein NME2 affects the replication of ARV in chicken embryo fibroblasts (CEFs) by binding to the structural ARV protein δA[20]; ARV induces apoptosis by activating UPR-related signaling pathways through ATF6[21]; and in CEFs, ARV induces IFN production through caspases[22].

In this study, we cloned the chicken IFI6 gene, performed bioinformatics analysis and subcellular localization assays of IFI6, determined the expression distribution of IFI6 in different chicken tissues, and investigated the effect of IFI6 on ARV replication and the expression of innate immunity signaling pathway-related factors after IFI6 overexpression and knockdown. The results of the study increase the understanding of the function of avian IFI6 and provide a theoretical basis for further studies on the pathogenic mechanism of the effects of ARV and for developing specific drugs to fight ARV infection.

Materials and methods

Chicken, cells, and virus

Specific-pathogen-free White Leghorn chicken embryos were purchased from Merial Vital Laboratory Animal Technology Co., Ltd. (Beijing, China), and hatched using an automatic incubator. After hatching, the chicks were transferred to a specific-pathogen-free chicken incubator for rearing. Three 14-day-old specific-pathogen-free chickens that exhibited good growth were selected, and specimens from 15 tissues were collected: heart, liver, lung, bursa of Fabricius, thymus, spleen, intestine, glandular stomach, gizzard, muscle, trachea, brain, kidney, joint and pancreas. The specimens were snap frozen in liquid nitrogen and stored at −70 ℃. The virus strain ARV S1133 used in this study was purchased from the China Academy of Veterinary Drug Inspection (Beijing, China). DF-1 cells were cultured in Dulbecco's modified Eagle's medium: nutrient mixture F-12 (Gibco, USA) supplemented with 10% fetal bovine serum (Gibco), 100 μg/mL streptomycin sulfate and 100 U/mL penicillin sodium (Beyotime Biotechnology, China).

Cloning and bioinformatics analysis of the IFI6 gene

The gene and amino acid sequences of IFI6 were downloaded from the National Center for Biotechnology Information (NCBI) database. Sequence alignment was performed, and primers (IFI6-U and IFI6-D) were designed (Table 4-2-1). The avian IFI6 gene was cloned using cDNA obtained by reverse transcription of mRNA extracted from DF-1 cells and used as the template. DNAStar, MEGA software and the online tools GOR4 and SWISS-MODEL were used to perform the sequence alignment, phylogenetic tree construction, homology analysis, and secondary and tertiary structure prediction analysis based on the IFI6 gene sequence.

Plasmid construction and transfection

IFI6-U and IFI6-D primers were designed using the pEF1α-Myc and IFI6 gene sequences (Table 4-2-1). The IFI6 gene was cloned and inserted into a pEF1α-Myc plasmid, and its accurate insertion was confirmed by RayBiotech (Guangzhou, China). Then, 2 μg of the correctly sequenced plasmid was transfected into 1×10^6 DF-1 cells using a Lipofectamine™ 3000 Transfection Kit (Thermo Scientific, USA), and protein from DF-1 cells transfected with the pEF1α-Myc plasmid was used as the blank control. The overexpression of IFI6 was verified by Western blotting 48 h after transfection.

Table 4-2-1   Primers used in this study

Primer name	Sequence (5'-3')	Amplified sequence length
IFI6-U	GCGTCGACCATGTCTGACCAGAACGTCCAC	324 bp
IFI6-D	GCGCGGCCGCTCAGCGCCTTCCTCCTTTGCCA	

RNA interference

Using the chicken IFI6 gene sequence, three specific short interfering RNAs (siRNAs) (si54, si156, and si274) and siRNAs directed to an unrelated molecule, siNC, were designed and synthesized by Suzhou GenePharma Co., Ltd. The siRNA sequences are shown in Table 4-2-2. Lipofectamine^{TM} RNAiMAX Reagent (Thermo Scientific) was used for the transfection of 30 pmol of si54, si156, and si274 into DF-1 cells to inhibit IFI6 expression. Using siNC-transfected DF-1 cells as the control cells, we performed real-time fluorescence quantitative PCR to identify the siRNA that exerted the greatest inhibitory effect 24 h after transfection.

Table 4-2-2   Short interfering RNA (siRNA) sequences targeting the IFI6 gene

siRNA	Sequence (5'-3')	Sequence (3'-5')
si54	GCAAGAGGUUCUCUUGCUUTT	AAGCAAGAGAACCUCUUGCTT
si156	GCCAAAGGCUCAACACACUTT	AGUGUGUUGAGCCUUUGGCTT
si274	GCACAACUUCCACUAUCCATT	UGGAUAGUGGAAGUUGUGCTT
siNC	UUCUCCGAACGUGUCACGUTT	ACGUGACACGUUCGGAGAATT

RNA extraction and real-time fluorescence quantitative PCR

Total RNA was extracted from cells and chicken tissue specimens using a Thermo Scientific GeneJET RNA Purification Kit (Thermo Scientific), and cDNA was synthesized by reverse transcription. The relative expression levels of IFI6, GAPDH, MAVS, IRF1, IRF7, STING, TBK1, NF-κB, MDA5, LGP2, IFN-α, IFN-β, and IFN-γ were quantified by real-time fluorescence quantitative PCR with specific primers as previously reported[23]. Real-time fluorescent quantitative PCRs were performed in 96-well plates using PowerUp SYBR Green Master Mix (Thermo Scientific). The reaction conditions were as follows: 94 ℃ for 2 min and 40 cycles of 94 ℃ for 15 s, and 60 ℃ for 30 s. Using the GAPDH housekeeping gene as the reference, the expression levels of target genes were normalized, and the relative expression levels of the genes were calculated using the $2^{-\Delta\Delta CT}$ method.

Protein extraction and western blotting

48 h after transfection, the culture medium was discarded, and the cells were washed three times with 1 × PBS and then lysed in lysis buffer (Beyotime Biotechnology) on ice for 30 min. After lysis, the cell supernatant was obtained by centrifugation at 4 ℃ and processed for use in SDS-PAGE. After electrophoresis, the protein was transferred to a polyvinylidene fluoride membrane, and the membrane was incubated with Western blot blocking solution for 6 h. The membranes were incubated with diluted mouse anti-Myc monoclonal antibody (Abcam, UK) and β-actin monoclonal antibody (Abcam) at 37 ℃ for 2 h, and the

membrane was then washed 3 times (10 min each time) with $1 \times$ TBST buffer (Beyotime Biotechnology), incubated with HRP-labeled IgG (Beyotime Biotechnology) at 37 ℃ for 1 h and washed 3 times with $1 \times$ TBST buffer (Beyotime Biotechnology). Photographs were taken after color development using a 3, 3' -diaminobenzidine (DAB) horseradish peroxidase color development kit (Beyotime Biotechnology).

Confocal microscopy analysis of the subcellular localization of target proteins

The cells were cultured in a special dish designed for use in laser confocal microscopy. 48 h after transfection, the culture medium was discarded, and the cells were fixed with a 4% tissue cell fixative solution for 10 min, permeabilized with 0.1% Triton X-100 (Beyotime Biotechnology) for 15 min, and blocked with 5% bovine serum albumin (Beyotime Biotechnology) for 1 h. The cells were then incubated with diluted mouse anti-Myc monoclonal antibody (Abcam) at 37 ℃ for 2 h, washed three times (5 min each) with $1 \times$ TBST buffer (Beyotime Biotechnology) and incubated with Alexa Fluor 488-labeled goat anti-mouse IgG (H + L) cross-adsorbed secondary antibody (Invitrogen, USA) for 1 h at 37 ℃ . After three washes with $1 \times$ TBST buffer (Beyotime Biotechnology), DAPI (Beyotime Biotechnology) was added to stain the nuclei. The stain was discarded after 10 min, and the cells were washed 3 times (5 min each time) with $1 \times$ PBS buffer (Beyotime Biotechnology). Subsequently, an appropriate amount of $1 \times$ PBS buffer (Beyotime Biotechnology) was added to the cells, and the cells were then observed and photographed using a laser confocal microscope at $63 \times$ magnification and excitation wavelengths of 405 nm and 488 nm.

Statistical analysis

All data represent the results from at least three independent experiments. Statistical analyses were performed by t-test with GraphPad Prism 5.0 software. The results are presented as follows: * indicates $P < 0.05$, ** indicates $P < 0.01$, *** indicates $P < 0.001$, and **** indicates $P < 0.000\ 1$.

Results

Cloning and bioinformatics analysis of IFI6

The IFI6 gene was amplified using cDNA reverse transcribed from mRNA extracted from DF-1 cells as a template and the primers IFI6-U and IFI6-R (Figure 4-2-1). The sequence was uploaded to the NCBI-Blast online website for alignment. The cloned sequence was confirmed to be the full-length sequence encoded by the IFI6 gene of *Gallus gallus* located on chicken chromosome 2, and its coding DNA sequence (CDS) consisted of 324 bases encoding a total of 107 amino acids.

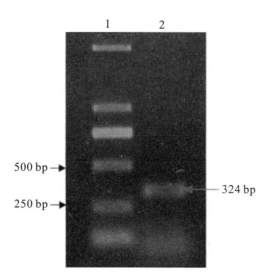

Lane 1: DL2000 DNA marker; lane 2: PCR amplification product, the arrow indicates the amplified IFI6 fragment, which is 324 bp.

Figure 4-2-1   Cloning of the IFI6 gene

Using DNAStar software, sequence homology analysis was performed with the full-length cloned sequence of the chicken IFI6 gene coding region and corresponding sequences from other species (Table 4-2-3). The similarity of the cloned sequence with the corresponding sequences of *Merops nubicus* and *Taeniopygia guttata* birds was generally higher than the similarity of the cloned sequence with the corresponding sequences of mammals.

Table 4-2-3   Sequence homology of the coding region of the IFI6 gene in different species

Species	Accession number	Identity
Merops nubicus	XM_008944065.1	73.6%
Taeniopygia guttata	NM_0011971792	87.4%
Homo sapiens	NM_022873.3	37.9%
Bos taurus	NM_001075588.1	40.0%
Ornithorhynchus anatius	XM_029080218.2	34.8%
Ovis arie	XM_027965432.2	40.0%
Canis lupus familiaris	XM_535344.7	39.8%
Sus scrofa	XM_021095658.1	40.3%

The primary structure analysis of the chicken IFI6 protein showed that the polypeptide chain encoded by the gene consisted of 107 amino acids and had a molecular weight of approximately 10.18 kDa (Figure 4-2-2 A). GOR4 was used to predict the secondary structure of the IFI6 protein (Figure 4-2-2 B): alpha helices, 23.36%; extended strands, 10.28%; beta turns, 19.63%; and random coils, 46.73%. The specific distribution is shown in Figure 4-2-2 C, where blue represents alpha helices, purple represents random coils, red represents extended strands, and green represents beta turns.

SWISS-MODEL was used to construct a three-dimensional structural model of the amino acid sequence of the IFI6 protein (Figures 4-2-2 D-F). The global model quality estimation (GMQE) score for the predicted IFI6 protein structure and interferon alpha-inducible protein 27-like protein 1 (IFI27L1) structure was 0.30 and the coverage rate was 51%.

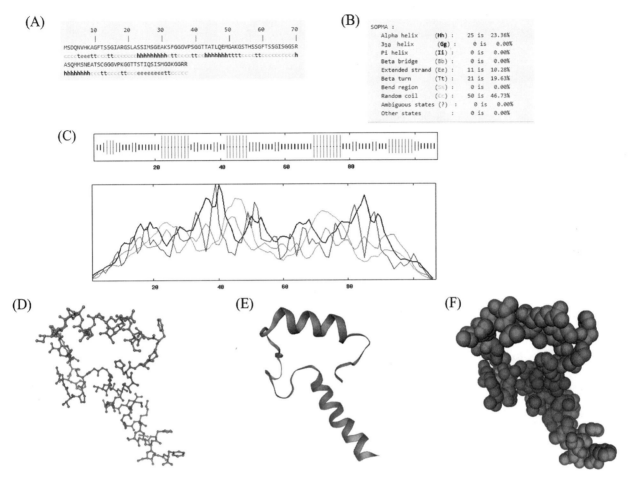

A: Primary structure showing the polypeptide chain encoded by the gene consisting of 107 amino acids; B: Secondary structure; C: Amino acid sequence analysis (blue represents alpha helices, purple represents random coils, red represents extended strands, and green represents beta turns); D-F: Prediction of the three-dimensional structural model of the IFI6 protein: backbone structure (D), ribbon structure (E), and spherical structure (F). The global model quality estimation (GMQE) score for the predicted IFI6 protein structure and interferon alpha-inducible protein 27-like protein 1 (IFI27L1) structure was 0.30 and the coverage was 51% (color figure in appendix).

Figure 4-2-2   Structural prediction of the IFI6 protein

National Center for Biotechnology Information, DNAStar and other software packages were used to analyze the homology between the amino acid sequence encoded by the chicken IFI6 gene and the corresponding amino acid sequences in other species. The homology of the amino acid sequences of chicken IFI6 with that of *Homo sapiens*, *Bos taurus*, *Ornithorhynchus anatinus*, *Ovis aries*, *Canis lupus familiaris*, *Sus scrofa*, *M. nubicus*, and *T. guttata* IFI6 was 29.6%, 26.8%, 21.8%, 26.1%, 26.8%, 26.1%, 67.6%, and 83.1%, respectively.

After multiple alignment of the amino acid sequences of the abovementioned species, a phylogenetic tree of the IFI6 amino acid sequence was constructed using MEGA software with the neighbor-joining (NJ) method and 1 000 bootstrap replicates. The results are shown in Figure 4-2-3. Mammals and birds were clearly clustered in their respective categories, and *O. anatinus* was positioned between the two branches but closer to the mammalian branch.

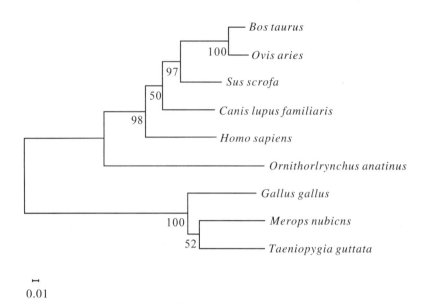

0.01

A phylogenetic tree was constructed using the neighbor-joining (NJ) method with MEGA software. The scale bars represent the branch lengths, and bootstrap confidence values are displayed at the nodes of the tree.

Figure 4-2-3   Evolutionary analysis of the amino acid sequence of IFI6 among different species

Subcellular localization

Alexa Fluor 488 (green)-labeled goat anti-mouse IgG (H + L) cross-adsorbed secondary antibody (Invitrogen) was used to label the IFI6 protein. Nuclei were stained with DAPI (blue) (Beyotime Biotechnology). The subcellular localization of the IFI6 protein was observed by laser confocal microscopy, and the results are shown in Figure 4-2-4. DF-1 cells transfected with a pEF1α-Myc-IFI6 plasmid emitted green fluorescence throughout the cytoplasm, indicating that the IFI6 protein localized to the cytoplasm. After receiving the same treatment, DF-1 cells transfected with an empty plasmid, pEF1α-Myc, emitted no fluorescence.

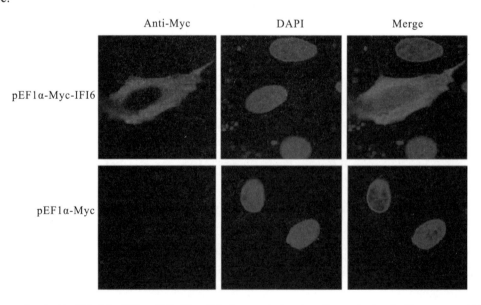

DF-1 cells were transfected with pEF1α-Myc-IFI6 or pEF1α-Myc. 48 h after transfection, the cells were incubated with a mouse anti-Myc monoclonal antibody for 2 h and then incubated with an Alexa Fluor 488 (green)-labeled goat anti-mouse IgG (H+L) cross-adsorbed secondary antibody for 1 h. Nuclei were stained with DAPI (blue), and the fluorescence intensity was detected by confocal microscopy and used to assess the location of expressed IFI6 (color figure in appendix).

Figure 4-2-4   Subcellular localization of the IFI6 protein (63×magnification)

Distribution of IFI6 in different chicken tissues

Real-time fluorescence quantitative PCR was used to detect the distribution of IFI6 in different tissues of 14-day-old specific-pathogen-free chickens. The highest expression of IFI6 was found in the lung, followed by the intestine, pancreas, liver, spleen, glandular stomach, thymus, bursa of Fabricius and trachea; a weak signal was detected in the heart, kidney, brain, gizzard and joints; and the lowest signal was detected in muscle (Figure 4-2-5).

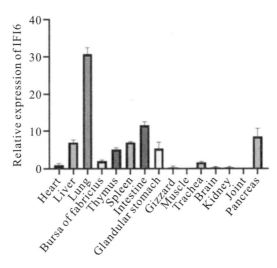

Real-time fluorescence quantitative PCR was used to measure the IFI6 mRNA levels in the heart, liver, lung, bursa of Fabricius, thymus, spleen, glandular stomach, gizzard, muscle, trachea, brain, kidney, joint and pancreas of 14-day-old specific-pathogen-free chickens.

Figure 4-2-5   Expression analysis of chicken IFI6 in various tissues

High expression of IFI6 reduced ARV replication

To examine the effect of IFI6 on ARV replication, the pEF1α-Myc-IFI6 plasmid was transfected into DF-1 cells, and the cells were then infected with ARV. 48 h after infection, the expression of the target protein was verified by Western blotting and real-time fluorescence quantitative PCR. Due to the lack of an IFI6 monocdonal antibody, anti-Myc monodonal antibody tags fused to the N-terminus of IFI6 were used to detect the expression of the target proteins. IFI6 protein expression (approximately 11 kDa) was confirmed in the cells transfected with pEF1α-Myc-IFI6, and target protein expression was not detected in the cells transfected with the empty vector (Vec) (Figure 4-2-6 A). Real-time fluorescence quantitative PCR was used to measure the relative expression of IFI6 24 h after infection, and the expression level was found to be increased by approximately 196-fold in the experimental group compared with the control group (Figure 4-2-6 B). As shown in Figure 4-2-6 C, the viral load, which is represented by the σC gene level of ARV as measured by real-time fluorescence quantitative PCR, was significantly decreased after infection ($P<0.000\ 1$). In addition, the supernatant of the cells transfected with pEF1α-Myc-IFI6 showed a lower virus level than that of the control group cells, as determined by a viral titer assay (Figure 4-2-6 D). Although viral proliferation was observed, the overexpression of IFI6 significantly inhibited ARV replication compared with that observed in control cells.

Therefore, we hypothesized that IFI6 negatively regulates ARV replication and that the inhibition of IFI6 expression promotes ARV replication. Three different siRNAs were used to downregulate IFI6 expression, and of these, si274 exhibited the greatest inhibitory effect (Figure 4-2-6 E). Therefore, si274 was transfected into DF-1 cells, and the cells were then infected with ARV. Both the RNA and virus titers were significantly increased ($P < 0.01$ or $P < 0.05$), demonstrating that the downregulation of IFI6 expression increased ARV amplification in cells (Figures 4-2-6 F, G). Taken together, these results suggest that ARV infection stimulates upregulation of the expression of IFI6 and that IFI6 exhibits anti-ARV activity.

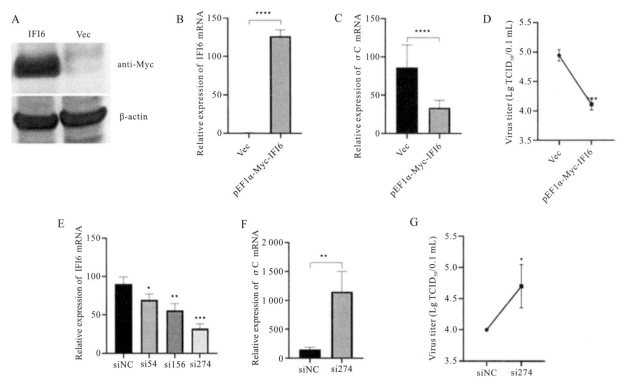

DF-1 cells were transfected with pEF1α-Myc-IFI6 or pEF1α-Myc (Vec). High IFI6 expression in DF-1 cells was confirmed by Western blotting (A) and real-time fluorescence quantitative PCR (B). DF-1 cells were transfected with pEF1α-Myc-IFI6 or pEF1α-Myc (Vec) and then infected with ARV at a multiplicity of infection (MOI) of 1. The number of virions was assessed by real-time fluorescence quantitative PCR (C), and the viral titer (D) was measured 24 h after infection. DF-1 cells were transfected with IFI6-targeting short interfering RNA (siRNA) and untargeted control siRNA (siNC), and the interference efficiencies of the three siRNAs were determined by real-time fluorescence quantitative PCR and compared (E). DF-1 cells were transfected with si274 or siNC and then infected with ARV at an MOI of 1; the number of virions was assessed by real-time fluorescence quantitative PCR (F), and the virus titer (G) was measured 24 h after infection. The data are presented as the mean ± SD deviations of three independent experiments. $*P < 0.05$, $**P < 0.01$, $***P < 0.001$, and $****P < 0.000\ 1$.

Figure 4-2-6　IFI6 impedes ARV replication in DF-1 cells

Effects of IFI6 on innate immune signaling pathway-related factors during ARV infection

The aforementioned results suggest that chicken IFI6 inhibits ARV replication, but the relevant mechanism remains unknown. To explore the mechanism, we transfected the pEF1α-Myc-IFI6 recombinant plasmid and si274 against IFI6 into DF-1 cells, overexpressed or inhibited the expression of the IFI6 gene in these DF-1 cells, infected the cells with ARV 24 h after transfection, and collected infected cell samples 24 h after infection. The effect of IFI6 on the expression of innate immune signaling pathway-related factors after ARV infection was assessed by real-time fluorescence quantitative PCR, and the results are shown in Figure 4-2-7. IFI6 overexpression and knockdown significantly increased the IRF7 mRNA expression level ($P < 0.01$ or $P < 0.000\ 1$) and downregulated the TBK1, LGP2, IFN-α, and IFN-γ mRNA expression levels ($P < 0.05$, $P < 0.01$, $P < 0.001$, or $P < 0.000\ 1$). MDA5, MAVS, STING, and NF-κB mRNA expression was upregulated after IFI6 overexpression and downregulated after IFI6 knockdown ($P < 0.05$, $P < 0.01$, or $P < 0.001$). The expression levels of IRF1 and IFN-β mRNA were significantly downregulated after IFI6 overexpression and significantly upregulated after IFI6 knockdown ($P<0.05$ or $P<0.001$). Therefore, we hypothesized that changes in IFI6 expression may affect the expression of MDA5, MAVS, STING, NF-κB, IRF1, and IFN-β in ARV infection but may not affect IRF7, TBK1, LGP2, IFN-α, and IFN-γ expression or that IFI6 may either enhance or attenuate the ARV-mediated regulation of IRF7, TBK1, LGP2, IFN-α, and IFN-γ expression.

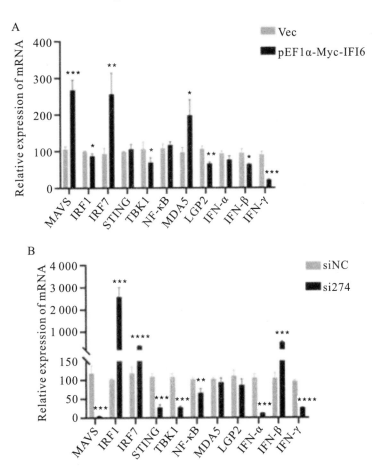

DF-1 cells were transfected with pEF1α-Myc-IFI6 or pEF1α-Myc (Vec) and then infected with ARV at a multiplicity of infection (MOI) of 1. Twenty-four hours after infection, the mRNA expression levels of MAVS, IRF1, IRF7, STING, TBK1, NF-κB, MDA5, LGP2, IFN-α, IFN-β, and IFN-γ were measured by real-time fluorescence quantitative PCR (A). DF-1 cells were transfected with short interfering RNA (si274) or untargeted control siRNA (siNC) and then infected with ARV at an MOI of 1. Twenty-four hours after infection, the MAVS, IRF1, IRF7, STING, TBK1, NF-κB, MDA5, LGP2, IFN-α, IFN-β, and IFN-γ mRNA levels were measured by real-time fluorescence quantitative PCR (B). The results are presented as the mean±SD deviations of three independent experiments. *$P<0.05$, **$P<0.01$, ***$P<0.001$, and ****$P<0.0001$.

Figure 4-2-7   Effects of IFI6 on the expression of innate immune signaling pathway-related genes during ARV infection

Discussion

Interferon-alpha inducible protein 6 is an ISG. In this study, we cloned the IFI6 gene and compared the full CDS of the cloned IFI6 gene with the corresponding nucleotide sequences of corresponding genes in other species. This analysis revealed that the cloned sequence was completely consistent with the corresponding sequence in *G. gallus*. The IFI6 nucleotide sequence showed 73.6% and 87.4% homology with the corresponding sequences in *M. nubicus* and *T. guttata*, respectively, and its homology with the corresponding amino acid sequences in *H. sapiens*, *B. taurus*, *O. aries*, *S. scrofa* and other mammals did not exceed 50%. Multiple alignment analysis of the deduced amino acid sequence indicated 67.6% and 83.1% similarity with the *M. nubicus* and *T. guttata* sequences, respectively, and its similarity with *H. sapiens, B. taurus, O. aries, S. scrofa* and other mammals did not exceed 30%. These findings were clearly reflected in the evolutionary tree. Mammals and birds were clearly divided into two branches. Chickens and other birds showed more similar evolutionary histories and greater homology. This finding is consistent with the recognized evolutionary history and reflects the high variability of the IFI6 gene, suggesting that the IFI6 gene may have undergone strong natural selection during evolution. In addition, the chicken IFI6 gene is located in a different branch than the mammalian IFI6 gene, suggesting that its function in birds during viral infection may be different from that in mammals. The protein structure of IFI6 was predicted to be similar to that of IFI27L1, with a coverage rate of 51%. IFI27L1 also known as FAM14B or ISG12c, it consists of 104 amino acids with a molecular weight of about 9.5 kDa and can encode a small hydrophobic protein. IFI27L1 belongs to the FAM14 family along with IFIT6[4, 5]. The sequence that is similar between IFI6 and IFI27L1 is the transmembrane region. Understanding the related functions of IFI27L1 is of guiding significance for the study of IFI6. A few literatures have shown that IFI27L1 is a protein that can promote apoptosis[24]. This provides clues for future research on the function of avian IFI6. The function of a protein is related not only to its structure but also to its distribution upon expression[25]. Recent studies have found that the human IFI6 protein is located in the endoplasmic reticulum in human hepatoma cells and blocks yellow fever virus replication by blocking the invagination of the endoplasmic reticulum membrane[9]. The porcine IFI6 protein is located in the membrane of muscle cells, and its mRNA is abundantly expressed in muscle, where it is expressed at a significantly higher level than it is in the brain. IFI6 is a candidate gene for use to increase the pork quality[26]. The shrimp IFI6 protein is localized to the cytoplasm of Drosophila embryonic cells, and its mRNA expression is highest in the intestine, followed by the hepatopancreas, stomach, heart and gills, whereas low expression is observed in the epithelium, pyloric cecum, blood cells, nerves, muscles and eye stalks, which indicates that this protein plays a key role in defense against white spot syndrome virus in shrimp[27]. We also studied the distribution of the expression of chicken IFI6. IFI6 was transfected into DF-1 cells and found to be a cytoplasmic protein. IFI6 mRNA expression was highest in the chicken lung, followed by the intestine, pancreas, liver, spleen, glandular stomach, thymus, bursa of Fabricius and trachea; weak signals were detected in the heart, kidney, brain, gizzard and joints; and the lowest expression was observed in the chicken muscle. These findings indicate that the IFI6 gene exhibits obvious tissue and species specificity, leading to the hypothesis that its functions also differ among different species.

Avian reovirus is mainly transmitted horizontally through the fecal-oral route but also vertically through eggs. After infection, ARV can replicate in the thymus, liver, spleen, bursa of Fabricius, intestine and other tissues and organs and can cause many diseases, such as hepatitis, enteritis, arthritis, myocarditis, and atrophy of the thymus and bursa of Fabricius[16, 19]. The distribution of innate immune-related genes in different parts of the host body is closely related to its antiviral effect[28, 29]. In this study, we examined the expression of IFI6 in various tissues and organs of specific-pathogen-free chickens. We found that IFI6 was highly expressed in immune-related organs (spleen, thymus, and bursa of Fabricius), respiratory tract-related organs (lung and trachea), and digestive tract-related organs (pancreas, liver, intestine, and glandular stomach). Therefore, we speculate that IFI6 may play an important role in the innate immune response against ARV infection in chickens.

In addition, the overexpression of IFI6 in human liver cancer cells promotes hepatitis C virus replication and weakens the antiviral activity of IFN-α[30], and the same findings have been observed in liver cancer cells. IFI6 inhibits hepatitis B virus replication by binding to the promoter of hepatitis B virus[31]. IFI6 plays different roles in different viral infections, and it has been speculated that the antiviral effect of IFI6 is specific.

To study the antiviral effect of avian IFI6, we overexpressed and inhibited IFI6 expression in DF-1 cells and then infected these cells with ARV. IFI6 overexpression inhibited viral replication, and IFI6 inhibition promoted viral replication (as determined by the σC expression level), indicating that avian IFI6 exhibits an antiviral function. To further understand the antiviral role of IFI6, we assessed the expression levels of innate immunity-related molecules and found that the overexpression and inhibition of IFI6 upregulated and downregulated the expression levels of various innate immunity-related molecules, respectively.

The host's pattern recognition receptors play an important role in sensing pathogen invasion. The current study found that the reovirus family is recognized by RIG-I or MDA5[32]. Only MDA5 and LGP2 are found in poultry, and RIG-I is missing[32, 33]; thus, in this study, we examined two nucleic acid receptors, MDA5 and LGP2. The results showed that the expression of MDA5 was upregulated after IFI6 overexpression and downregulated after IFI6 inhibition, and the expression of LGP2 was always downregulated. It is speculated that IFI6 may affect induction of the expression of MDA5 and MAVS.

In our study, overexpression or inhibition of IFI6 upregulated IRF7 expression and downregulated TBK1, LGP2, IFN-α, and IFN-γ expression after infection with ARV. The upregulated or downregulated expression of these factors was identified in comparison with the corresponding controls, which were different, and thus, the magnitude of the changes after overexpression or inhibition could not be compared to determine the involvement of IFI6 in the regulation of IRF7, TBK1, LGP2, IFN-α, and IFN-γ expression after infection with ARV. We hypothesize that changes in IFI6 expression may not affect the expression of IRF7, TBK1, LGP2, IFN-α, and IFN-γ in ARV infection and may also enhance or weaken the regulation of IRF7 and IFN-γ expression by ARV.

We were more interested in the changes in the expression of MAVS, IRF1and IFN-β. The differences in the MAVS, IRF1, and IFN-β levels were notable. After IFI6 was overexpressed in cells that were then infected with ARV, MAVS expression was significantly upregulated, and IRF1 and IFN-β expression was significantly downregulated. After IFI6 knockdown and ARV infection, MAVS expression was significantly downregulated, and the expression of IRF1 and IFN-β was significantly upregulated.

As an important linker molecule in the innate immune signal transduction pathway that mainly recognizes RNA viruses, MAVS plays a crucial role in the innate immune response[34]. For example, after RIG-I-like receptors recognize the virus, they interact with MAVS, and this interaction triggers a series of signal transduction cascades and ultimately leads to the expression of various proinflammatory factors and antiviral genes, such as IFNs and ISGs, which inhibit viral replication and spread[35-39]. MAVS exhibits an expression pattern similar to that of IFI6; presumably, ARV infection significantly upregulates the expression of IFI6, and after upregulation, IFI6 regulates the expression of MAVS and thus affects antiviral signaling cascades.

IRF1 is the first member of the IRF family found to activate type I IFN genes via their promoters[40]. Studies have shown that IRF1 drives reprogramming of bone marrow dendritic cells and macrophages and thus triggers an antiviral signaling pathway; that is, IRF1 interacts with myeloid differentiation factor 88 or IL-1 receptor-related kinase-1 to activate IFN-β[41]. In this study, IRF1 and IFN-β showed similar expression patterns, and in contrast to IFI6, ISGs negatively regulate the immune response of IFNs, such as SOCS1 and USP18, two well-characterized ISGs that negatively regulate IFN signaling by inhibiting the JAK-STAT signaling pathway[2, 42-45]. Therefore, we hypothesized that IFI6 negatively regulates the JAK-STAT signaling pathway by inhibiting the expression of IFN-β. These findings provide clues for improving our understanding of the function of IFI6, revealing the signaling pathways involving IFI6, and laying a theoretical foundation for understanding the pathogenesis of ARV infection and for blocking ARV replication. To acquire a greater understanding of IFI6 function, further investigation is needed.

References

[1] IVASHKIV L B, DONLIN L T. Regulation of type I interferon responses. Nat Rev Immunol, 2014,14(1):36-49.

[2] WILLIAM M S, MEIKE D C, CHARLES M R. Interferon-stimulated genes: a complex web of host defenses. Annual Review of Immunology, 2014,32(1).

[3] BORDEN E C, SEN G C, UZE G, et al. Interferons at age 50: past, current and future impact on biomedicine. Nat Rev Drug Discov, 2007,6(12):975-990.

[4] PARKER N, PORTER A C. Identification of a novel gene family that includes the interferon-inducible human genes 6-16 and ISG12. BMC Genomics, 2004,5(1):8.

[5] CHERIYATH V, LEAMAN D W, BORDEN E C. Emerging roles of FAM14 family members (G1P3/ISG 6-16 and ISG12/IFI27) in innate immunity and cancer. J Interferon Cytokine Res, 2011,31(1):173-181.

[6] TAHARA E J, TAHARA H, KANNO M, et al. G1P3, an interferon inducible gene 6-16, is expressed in gastric cancers and inhibits mitochondrial-mediated apoptosis in gastric cancer cell line TMK-1 cell. Cancer Immunol Immunother, 2005,54(8):729-740.

[7] CHERIYATH V, GLASER K B, WARING J F, et al. G1P3, an IFN-induced survival factor, antagonizes TRAIL-induced apoptosis in human myeloma cells. J Clin Invest, 2007,117(10):3107-3117.

[8] LIU Z, GU S, LU T, et al. IFI6 depletion inhibits esophageal squamous cell carcinoma progression through reactive oxygen species accumulation via mitochondrial dysfunction and endoplasmic reticulum stress. J Exp Clin Cancer Res, 2020,39(1):144.

[9] RICHARDSON R B, OHLSON M B, EITSON J L, et al. A CRISPR screen identifies IFI6 as an ER-resident interferon effector that blocks flavivirus replication. Nat Microbiol, 2018,3(11):1214-1223.

[10] VILLAMAYOR L, RIVERO V, LOPEZ-GARCIA D, et al. Interferon alpha inducible protein 6 is a negative regulator of innate immune responses by modulating RIG-I activation. Front Immunol, 2023,14:1105309.

[11] OLSON N O. Transmissible synovitis of poultry. Lab Invest, 1959,8:1384-1393.

[12] VAN DER HEIDE L. The history of avian reovirus. Avian Dis, 2000,44(3):638-641.

[13] ZHANG X, LEI X, MA L, et al. Genetic and pathogenic characteristics of newly emerging avian reovirus from infected chickens with clinical arthritis in China. Poult Sci, 2019,98(11):5321-5329.

[14] CZEKAJ H, KOZDRUN W, STYS-FIJOL N, et al. Occurrence of reovirus (ARV) infections in poultry flocks in Poland in 2010-2017. J Vet Res, 2018,62(4):421-426.

[15] SHARMA J M, KARACA K, PERTILE T. Virus-induced immunosuppression in chickens. Poult Sci, 1994,73(7):1082-1086.

[16] ROESSLER D E, ROSENBERGER J K. In vitro and in vivo characterization of avian reoviruses III host factors affecting virulence and persistence. Avian Dis, 1989,33(3):555-565.

[17] XIE L, XIE Z, WANG S, et al. Altered gene expression profiles of the MDA5 signaling pathway in peripheral blood lymphocytes of chickens infected with avian reovirus. Arch Virol, 2019,164(10):2451-2458.

[18] WANG S, XIE L, XIE Z, et al. Dynamic changes in the expression of interferon-stimulated genes in joints of SPF chickens infected with avian reovirus. Front Vet Sci, 2021,8:618124.

[19] WANG S, WAN L, REN H, et al. Screening of interferon-stimulated genes against avian reovirus infection and mechanistic exploration of the antiviral activity of IFIT5. Front Microbiol, 2022,13:998505.

[20] XIE L, WANG S, XIE Z, et al. Gallus NME/NM23 nucleoside diphosphate kinase 2 interacts with viral sigmaA and affects the replication of avian reovirus. Vet Microbiol, 2021,252:108926.

[21] ZHANG C, HU J, WANG X, et al. Avian reovirus infection activate the cellular unfold protein response and induced apoptosis via ATF6-dependent mechanism. Virus Res, 2021,297:198346.

[22] LOSTALE-SEIJO I, MARTINEZ-COSTAS J, BENAVENTE J. Interferon induction by avian reovirus. Virology, 2016,487:104-111.

[23] HE Y, XIE Z, DAI J, et al. Responses of the Toll-like receptor and melanoma differentiation-associated protein 5 signaling pathways to avian infectious bronchitis virus infection in chicks. Virol Sin, 2016,31(1):57-68.

[24] XIE Q, TANG Z, LIANG X, et al. An immune-related gene prognostic index for acute myeloid leukemia associated with regulatory T cells infiltration. Hematology, 2022,27(1):1088-1100.

[25] YANG H L, FENG Z Q, ZENG S Q, et al. Molecular cloning and expression analysis of TRAF3 in chicken. Genet Mol Res, 2015,14(2):4408-4419.

[26] KAYAN A, UDDIN M J, CINAR M U, et al. Investigation on interferon alpha-inducible protein 6 (IFI6) gene as a candidate for meat and carcass quality in pig. Meat Sci, 2011,88(4):755-760.

[27] LU K, LI H, WANG S, et al. Interferon-induced protein 6-16 (IFI6-16) from Litopenaeus vannamei regulate antiviral immunity via apoptosis-related genes. Viruses, 2022,14(5).

[28] LIN Z, WANG J, ZHU W, et al. Chicken DDX1 acts as an RNA sensor to mediate IFN-beta signaling pathway activation in antiviral innate immunity. Front Immunol, 2021,12:742074.

[29] CHENG Y, SUN Y, WANG H, et al. Chicken STING mediates activation of the IFN gene independently of the RIG-I Gene. J Immunol, 2015,195(8):3922-3936.

[30] CHEN S, LI S, CHEN L. Interferon-inducible Protein 6-16 (IFI-6-16, ISG16) promotes Hepatitis C virus replication in vitro. J Med Virol, 2016,88(1):109-114.

[31] SAJID M, ULLAH H, YAN K, et al. The functional and antiviral activity of interferon alpha-inducible IFI6 against hepatitis B virus replication and gene expression. Front Immunol, 2021,12:634937.

[32] KATO H, TAKEUCHI O, MIKAMO-SATOH E, et al. Length-dependent recognition of double-stranded ribonucleic acids by retinoic acid-inducible gene-I and melanoma differentiation-associated gene 5. J Exp Med, 2008,205(7):1601-1610.

[33] BARBER M R, ALDRIDGE J J, WEBSTER R G, et al. Association of RIG-I with innate immunity of ducks to influenza. Proc Natl Acad Sci USA, 2010,107(13):5913-5918.

[34] REN Z, DING T, ZUO Z, et al. Regulation of MAVS expression and signaling function in the antiviral innate immune response. Front Immunol, 2020,11:1030.

[35] MEYLAN E, CURRAN J, HOFMANN K, et al. Cardif is an adaptor protein in the RIG-I antiviral pathway and is targeted by hepatitis C virus. Nature, 2005,437(7062):1167-1172.

[36] SETH R B, SUN L, EA C K, et al. Identification and characterization of MAVS, a mitochondrial antiviral signaling protein that activates NF-kappaB and IRF 3. Cell, 2005,122(5):669-682.

[37] XU L G, WANG Y Y, HAN K J, et al. VISA is an adapter protein required for virus-triggered IFN-beta signaling. Mol Cell, 2005,19(6):727-740.

[38] KUMAR H, KAWAI T, KATO H, et al. Essential role of IPS-1 in innate immune responses against RNA viruses. J Exp Med, 2006,203(7):1795-1803.

[39] SONG N, QI Q, CAO R, et al. MAVS O-GlcNAcylation is essential for host antiviral immunity against lethal RNA viruses. Cell Rep, 2019,28(9):2386-2396.

[40] DOU L, LIANG H F, GELLER D A, et al. The regulation role of interferon regulatory factor-1 gene and clinical relevance. Hum Immunol, 2014,75(11):1110-1114.

[41] SCHMITZ F, HEIT A, GUGGEMOOS S, et al. Interferon-regulatory-factor 1 controls Toll-like receptor 9-mediated IFN-beta production in myeloid dendritic cells. Eur J Immunol, 2007,37(2):315-327.

[42] MALAKHOV M P, MALAKHOVA O A, KIM K I, et al. UBP43 (USP18) specifically removes ISG15 from conjugated proteins. J Biol Chem, 2002,277(12):9976-9981.

[43] MALAKHOVA O A, KIM K I, LUO J K, et al. UBP43 is a novel regulator of interferon signaling independent of its ISG15 isopeptidase activity. EMBO J, 2006,25(11):2358-2367.

[44] YOSHIMURA A, NAKA T, KUBO M. SOCS proteins, cytokine signalling and immune regulation. Nat Rev Immunol, 2007,7(6):454-465.

[45] YU C F, PENG W M, SCHLEE M, et al. SOCS1 and SOCS3 target IRF7 degradation to suppress TLR7-mediated type I IFN production of human plasmacytoid dendritic cells. J Immunol, 2018,200(12):4024-4035.

Screening of interferon-stimulated genes against avian reovirus infection and mechanistic exploration of the antiviral activity of IFIT5

Wang Sheng, Wan Lijun, Ren Hongyu, Xie Zhixun, Xie Liji, Huang Jiaoling, Deng Xianwen,

Xie Zhiqin, Luo Sisi, Li Meng, Zeng Tingting, Zhang Yanfang, and Zhang Minxiu

Abstract

Avian reovirus (ARV) infection can lead to severe immunosuppression, complications, and secondary diseases, causing immense economic losses to the poultry industry. In-depth study of the mechanism by which the innate immune system combats ARV infection, especially the antiviral effect mediated by interferon, is needed to prevent and contain ARV infection. In this study, ARV strain S1133 was used to artificially infect 7-day-old specific pathogen-free chickens. The results indicated that ARV rapidly proliferated in the immune organs, including the spleen, bursa of Fabricius, and thymus. The viral load peaked early in the infection and led to varying degrees of pathological damage to tissues and organs. Real-time quantitative PCR revealed that the mRNA levels of interferon and multiple interferon-stimulated genes (ISGs) in the spleen, bursa of Fabricius, and thymus were upregulated to varying degrees in the early stage of infection. Among the ISGs, IFIT5, and Mx were the most upregulated in various tissues and organs, suggesting that they are important ISGs for host resistance to ARV infection. Further investigation of the role of IFIT5 in ARV infection showed that overexpression of the IFIT5 gene inhibited ARV replication, whereas inhibition of the endogenously expressed IFIT5 gene by siRNA promoted ARV replication. IFIT5 may be a positive feedback regulator of the innate immune signaling pathways during ARV infection and may induce IFN-α production by promoting the expression of MAD5 and MAVS to exert its antiviral effect. The results of this study help explain the innate immune regulatory mechanism of ARV infection and reveal the important role of IFIT5 in inhibiting ARV replication, which has important theoretical significance and practical application value for the prevention and control of ARV infection.

Keywords

avian reovirus, interferon, interferon-stimulated genes, IFIT5, antiviral response

Introduction

Avian reovirus (ARV) is an important and prevalent avian pathogen that mainly infects chickens, turkeys, and a few other birds. ARV infection can cause viral arthritis, tenosynovitis, growth retardation, and malabsorption syndrome. It can also lead to severe immunosuppression, which in turn causes vaccine immunization failure and can lead to complications or secondary diseases, causing enormous economic losses to the poultry farming industry[1-3].

The innate immune response is the first line of defense against the virus. After the virus infects the host, the pathogen-related molecular pattern of the virus is specifically recognized by the host's pattern recognition receptor, and the host senses pathogen invasion, thereby initiating a series of downstream signaling pathways and inducing the production of various antiviral cytokines[4, 5]. The innate immune response is characterized by extensive function against non-specific antigens and a rapid response. The relationship between ARV and the host innate immune response is a current research area of intense interest. ARV infection can cause significant changes in host innate immune-related pattern recognition receptors and inflammatory cytokines at the transcriptional level, indicating that ARV infection is closely related to recognition by host innate immune-related pattern receptors and the production of inflammatory cytokines[6-9]. The interferon (IFN)-mediated antiviral effect is an extremely important link in the host's natural immune response. IFN itself does not inactivate the virus. After IFN is produced, it binds to the corresponding IFN receptor on cell surfaces and induces the transcriptional expression of many interferon-stimulated genes (ISGs) by activating the JAK-STAT signaling pathway to transmit extracellular signals to the cells[10, 11].

Interferon-stimulated genes are important antiviral molecules that play an important role in the host's defense against and elimination of foreign pathogens[12]. After reovirus infection in mammals, the expression of many ISGs is activated to resist viral infection[13, 14]. As a member of the Reoviridae family, infection by ARV can also induce the expression of IFN and ISGs. Lostale-Seijo et al.[7] found that ARV infection of chicken embryonic fibroblasts can efficiently induce the expression of IFN-α, IFN-β, Mx, and the double-stranded RNA (dsRNA) -dependent protein kinase (PKR) of the ISG family to exert antiviral effects. Research has also found that the ARV σA protein binds to viral dsRNA in an irreversible manner to inhibit activation of PKR, thereby enabling the virus to resist the antiviral effect of interferon and to defend against or escape the host immune system[15-17]. In our previous study, we found changes in the mRNA expression patterns of a variety of ISGs in the joints of specific pathogen-free (SPF) chickens infected with ARV S1133; these changes were essentially consistent with the trend of changes in the viral load in the joints, suggesting that IFN-α, IFN-β, and ISGs such as Mx, IFITM3, PKR, OAS, IFIT5, ISG12, Viperin, IFI6, and CD47 play important roles in the defense against ARV invasion, inhibition of ARV replication and proliferation, and viral clearance[18].

The immune organs of the body play an important role in maintaining the immune function of the body and preventing the invasion of pathogenic microorganisms. ARV infection can cause damage to tissues and organs such as the spleen, thymus, and bursa of Fabricius in poultry, which is an important cause of immunosuppression from ARV infection[19, 20]. Currently, no report is available on changes in ISG expression in immune organs after ARV infection.

In this study, 7-day-old SPF chickens were artificially infected with ARV strain S1133. By observing pathological changes in the immune organs after ARV infection, we detected the viral load of ARV in different immune tissues and organs, and real-time quantitative PCR was used to evaluate the effects of IFN-related genes and ISGs on expression levels during ARV infection. The immune responses induced by the immune organs after ARV infection were systematically analyzed. ISGs with an important role in ARV infection were screened out, and the antiviral effects of the selected ISGs were verified. The results of this study will further enrich our understanding of the regulatory mechanism of the innate immune response of ARV infection and lay the foundation for investigation of the antiviral molecular mechanism of ISGs.

Materials and methods

Virus strain

The ARV standard virulent strain S1133 was purchased from the China Institute of Veterinary Drug Control. Before use, ARV was inoculated into the yolk sacs of SPF chicken embryos for proliferation. After the virus was harvested, DF-1 cells were inoculated. Viral titers were determined by the Reed-Muench method.

Animal experiments

Specific pathogen-free White Leghorn chicken eggs were purchased from Beijing Boehringer Ingelheim Vital Biotechnology Co., Ltd. (Beijing, China) and were hatched using an automatic hatching machine. After hatching, the chicks were transferred to an SPF chicken isolator for rearing. Eighty 7-day-old SPF chickens with good growth were randomly divided into two groups of 40, which were housed in two different isolators. In the ARV infection group, each animal was challenged with 0.1 mL of ARV (10^4 of the 50% tissue culture infectious dose ($TCID_{50}$) /0.1 mL) by footpad injection. The control group was treated with an equal amount of phosphate-buffered saline (PBS) for footpad injection. After the challenge, the clinical symptoms of the chickens in the ARV infection group and the control group were observed every day. Samples were taken at 1, 2, 3, 4, 5, 6, 7, 10, 14, 21, 28, and 35 days after the challenge. Three chickens were randomly selected as three biological replicates, and the spleen, thymus, and bursa of Fabricius were collected for pathological observation and real-time fluorescence quantitative PCR detection.

Preparation of pathological sections

The tissue samples collected above were fixed in 10% neutral formalin, and tissue sections were prepared by Shuiyuntian Biotechnology Co., Ltd. (Guangzhou, China). The sections were stained with hematoxylin-eosin to observe pathological changes in the spleen, bursa of Fabricius, and thymus after ARV infection.

RNA extraction and cDNA synthesis

RNA was extracted from collected tissue samples using the RNA purification kit GeneJET RNA Purification Kit (Thermo Scientific, USA). cDNA was synthesized by reverse transcription using the Maxima™ H Minus cDNA Synthesis Master Mix with dsDNase (Thermo Scientific) and stored at −80 ℃ until real-time fluorescence quantitative PCR.

Real-time fluorescence quantitative PCR

Based on the gene sequence information in GenBank, primers for the ARV σC gene, IFN, and ISGs were synthesized by Invitrogen (Table 4-3-1). GAPDH was used as the internal reference gene for real-time fluorescence quantitative PCR. Real-time fluorescence quantitative PCR was performed using PowerUP SYBR Green Master Mix (Thermo Scientific) and a QuantStudio 5 real-time PCR system (Thermo Life Tech ABI, USA). Gene expression was compared by the $2^{-\Delta\Delta Ct}$ method.

Table 4-3-1 Primers used in this study

Gene	GenBank accession number	Primer sequences (5'-3')
ARVσC	L39002.1	F: CCACGGGAAATCTCACGGTCACT
		R: TACGCACGGTCAAGGAACGAATGT
IFN-α	AB021154.1	F: ATGCCACCTTCTCTCACGAC
		R: AGGCGCTGTAATCGTTGTCT
IFN-β	X92479.1	F: ACCAGGATGCCAACTTCT
		R: TCACTGGGTGTTGAGACG
OAS	NM_205041.1	F: GCGGTGAAGCAGACGGTGAA
		R: CGATGATGGCGAGGATGTG
IFITM3	KC876032.1	F: GGAGTCCCACCGTATGAAC
		R: GGCGTCTCCACCGTCACCA
MX	AY695797.1	F: AACGCTGCTCAGGTCAGAAT
		R: GTGAAGCACATCCAAAAGCA
PKR	AB125660.1	F: CCTCTGCTGGCCTTACTGTCA
		R: AAGAGAGGCAGAAGGAATAATTTGCO
Viperin	EU427332.1	F: CAGTGGTGCCGAGATTATGC
		R: CACAGGATTGAGTGCCTTGA
IFIT5	XM_421662	F: CTCCCAAATCCCTCTCAACA
		R: AAGCAAACGCACAATCATCA
ISG12	BN000222.1	F: TCCTCAGCCATGAATCCGAACA
		R: GGCAGCCGTGAAGCCCAT
ZFP313	AY604724.1	F: ATCGCTTTACCTTTCCTTG
		R: GTGCCATCGTATCATCTTCA
IFI6	NM 001001296.5	F: CACTCCTCAGGCTTTACC
		R: GACCGATGCTTCTTTCTATT
MDA5	NM 001193638	F: CAGCCAGTTGCCCTCGCCTCA
		R: AACAGCTCCCTTGCACCGTCT
LGP2	MF563595.1	F: CCAGAATGAGCAGCAGGAC
		R: AATGTTGCACTCAGGGATGT
MAVS	MF289560.1	F: CCTGACTCAAACAAGGGAAG
		R: AATCAGAGCGATGCCAACAG
TRAF3	XM_040672281.1	F: GGACGCACTTGTCGCTGTTT
		R: CGGACCCTGATCCATTAGCAT
TRAF6	XM_040673314.1	F: GATGGAGACGCAAAACACTCAC
		R: GCATCACAACAGGTCTCTCTTC
IKKε	XM_428036.4	F: TGGATGGGATGGTGTCTGAAC
		R: TGCGGAACTGCTTGTAGATG
TBK1	MF159109.1	F: AAGAAGGCACACATCCGAGA
		R: GGTAGCGTGCAAATACAGC
IRF7	NM_205372.1	F: CAGTGCTTCTCCAGCACAAA
		R: TGCATGTGGTATTGCTCGAT
NF-κB	NM_205129.1	F: CATTGCCAGCATGGCTACTAT
		R: TTCCAGTTCCCGTTTCTTCAC
GAPDH	NM_204305.1	F: GCACTGTCAAGGCTGAGAACG
		R: GATGATAACACGCTTAGCACCAC

Overexpression of the IFIT5 protein

With reference to the chicken IFIT5 gene sequence (accession number: NM_001320422.1), the IFIT5 recombinant plasmid pEF1α-Myc-IFIT5 was constructed. DF-1 cells were cultured in six-well plates overnight, and 2 μL Lipofectamine™ 3000 (Invitrogen, USA) was used to transfect 1 μg of pEF1α-Myc-IFIT5 plasmid into DF-1 cells to overexpress the IFIT5 protein. At 24 h after transfection, cells were infected with ARV strain S1133 at a multiplicity of infection (MOI) of 1. Cell samples and culture supernatants were collected 24 h after infection. RNA was extracted from the collected cell samples, and cDNA was synthesized. Real-time fluorescence quantitative PCR was run to measure ARV replication and changes in the expression of molecules related to the innate immune signaling pathways. The primer sequences of the ARV σC gene and molecules related to natural immune signaling are shown in Table 4-3-1 and were synthesized by Invitrogen. In addition, the culture supernatant was infected with DF-1 cells, and the viral titer was determined by the Reed-Muench method to measure ARV replication.

IFIT5 RNA interference assay

Three small interfering RNAs (siRNAs) targeting the IFIT5 gene were designed and synthesized by GenePharma (Table 4-3-2). DF-1 cells were cultured in six-well plates overnight, and 2 μL of Lipofectamine RNAiMAX (Invitrogen) was used to transfect 1 μg of 200 nM IFIT5 of each siRNA separately or the negative control (siNC) to inhibit IFIT5 protein expression. Cell samples and culture supernatants were collected 24 h after transfection and infected with ARV strain S1133 (MOI=1). We extracted RNA from the cells to detect ARV replication and changes in the expression of molecules related to innate immune signaling pathways. Viral replication was measured in the culture supernatant.

Table 4-3-2　siRNA sequences targeting the IFIT5 gene

siRNA	Sequences (5'-3')	Sequences (3'-5')
si225	GGAAGAAUCAAUAGAGGAUTT	AUCCUCUAUUGAUUCUUCCTT
si738	GCGUGCACUGAAACUGAAUTT	AUUCAGUUUCAGUGCACGCTT
si1453	GGUGAGAAGCUUGAAGCAATT	UUGCUUCAAGCUUCUCACCTT
siNC	UUCUCCGAACGUGUCACGUTT	ACGUGACACGUUCGGAGAATT

Statistical analysis

All data were plotted in GraphPad Prism 8 software. The results are expressed as the mean±standard deviation (SD). Statistical analysis was performed using IBM SPSS Statistics 2.0 software. Student's t-test was used to evaluate differences between groups. $P<0.05$ indicated statistically significant differences, which are marked with * in the figures; $P<0.01$ indicated very significant differences, which are marked with ** in the figures.

Results

Pathological changes in immune organs after avian reovirus infection

After ARV infection of SPF chickens, dissection results revealed edema in the spleen and atrophy of the bursa of Fabricius and thymus (Figure 4-3-1). Histopathology indicated that the three immune organs all had typical histopathological changes. The spleen mainly showed extensive lymphocyte degeneration and necrosis.

The bursa of Fabricius showed a loose structure of lymphoid follicles, lymphocyte degeneration and necrosis, heterophilic cell infiltration, interstitial porosity, and inflammatory cell infiltration, as well as local lymphoid follicular atrophy, interstitial widening, and fibrosis. The thymus showed atrophy, smaller cortex/medulla proportions, and few cortical lymphocytes (Figure 4-3-2).

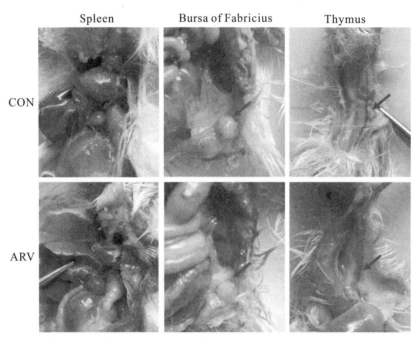

In the control group (CON), SPF chickens were injected with 0.1 mL of PBS into the footpad. In the experimental group (ARV), ARV strain S1133 (10^4 $TCID_{50}$/0.1 mL) was injected into SPF chickens through the footpad. After ARV infection of SPF chickens, dissection results revealed edema in the spleen and atrophy in the bursa of Fabricius and thymus.

Figure 4-3-1   Pathological changes in the spleen, bursa of Fabricius, and thymus in SPF chickens infected with ARV

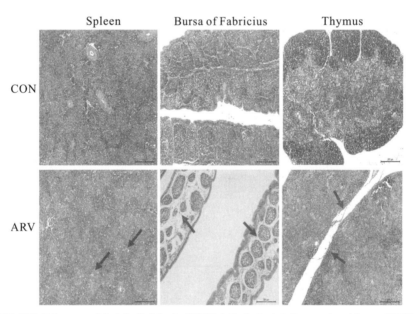

In the control group (CON), SPF chickens were injected with 0.1 mL of PBS into the footpad. In the experimental group (ARV), ARV strain S1133 (10^4 $TCID_{50}$/0.1 mL) was injected into SPF chickens through the footpad. The arrow in the spleen indicates extensive lymphocyte degeneration and necrosis. Arrows in the bursa of Fabricius indicate lymphatic follicular structural loosening, lymphocytic degeneration and necrosis, heterophilic cell infiltration, interstitial loosening, and inflammatory cell infiltration. Arrows in the thymus indicate atrophy, reduced cortex/medulla proportions, and decreased cortical lymphocytes (color figure in appendix).

Figure 4-3-2   Histopathological changes in the spleen, bursa of Fabricius, and thymus in SPF chickens infected with ARV

Viral load in immune organs after avian reovirus infection

To study ARV proliferation in the immune organs, the ARV viral load in the above tissues and organs was measured by real-time quantitative PCR. The results are shown in Figure 4-3-3. After ARV infection, ARV in the spleen began to proliferate rapidly at 1 day postinfection (dpi) until reaching the peak level, remained at a high level between 2 and 3 dpi, suddenly decreased at 4 dpi, and was not detectable after 6 dpi. ARV was detected only between 1 and 4 dpi in the bursa of Fabricius, but the viral copy number was low. ARV was detected in the thymus at 1 dpi, peaked at 3 dpi, and was undetectable after 10 dpi. Throughout the experiment, the ARV σC gene was detected by real-time quantitative PCR in all control samples, and the results were all negative.

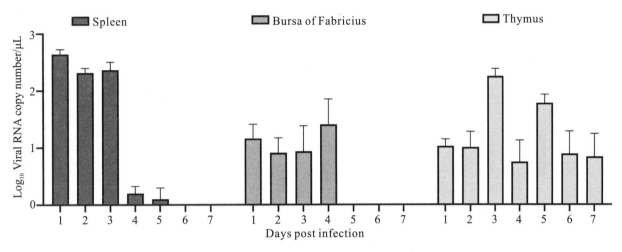

ARV strain S1133 (10^4 $TCID_{50}$/0.1 mL) was used to infect chickens through the footpad. At 1, 2, 3, 4, 5, 6, 7, 10, 14, 21, 28, and 35 days postinfection (dpi), three chickens were randomly collected from each group, and the ARV viral load was measured by real-time quantitative PCR. The data are the mean±SD of three independent experiments.

Figure 4-3-3   The ARV copy number in the spleen, bursa of Fabricius, and thymus in ARV-infected SPF chickens

Transcription level of IFN in immune organs after avian reovirus infection

As shown in Figure 4-3-4, IFN-α and IFN-β showed similar changes in the same immune organ after ARV infection. The expression levels of IFN-α and IFN-β in the spleen did not change significantly and were significantly downregulated only at 28 dpi ($P<0.05$). The mRNA levels of IFN-α and IFN-β in the bursa of Fabricius were both upregulated between 1 and 6 dpi, and the expression level of IFN-α was significantly higher than that of IFN-β. IFN-α peaked at 6 dpi at 14.39 times the expression level in the control group ($P<0.01$), and IFN-β peaked at 3 dpi at 5.89 times the control level ($P<0.05$). The mRNA levels of IFN-α and IFN-β in the thymus were significantly upregulated between 1 and 10 dpi, and both peaked at 3 dpi. The IFN-α level was 52.25 times that in the control group ($P<0.01$), and the IFN-β level was 36.67 times that in the control group ($P<0.01$).

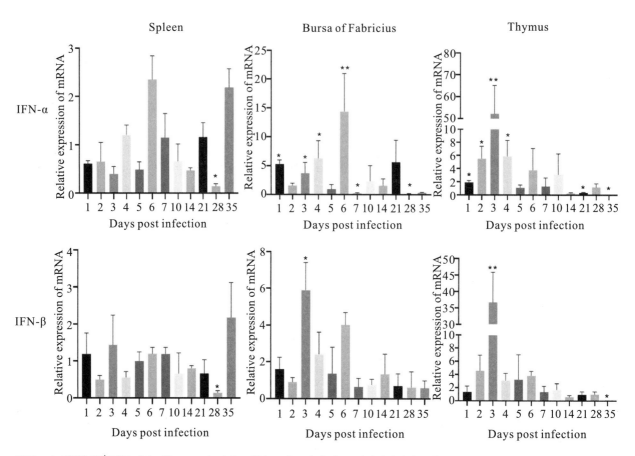

ARV strain S1133 (10^4 TCID$_{50}$/0.1 mL) was used to infect chickens through the footpad. At 1, 2, 3, 4, 5, 6, 7, 10, 14, 21, 28, and 35 days after infection (dpi), three chickens were randomly collected from each group, and changes in IFN gene expression were analyzed using real-time quantitative PCR. The data are the mean ± SD of three independent experiments. Asterisks indicate significant differences (*$P<0.05$, **$P<0.01$).

Figure 4-3-4　Changes in IFN transcription levels in the spleen, bursa of Fabricius, and thymus after ARV infection

Expression levels of interferon-stimulated genes mRNAs in immune organs after avian reovirus infection

The changes in the expression of ISGs in the spleen, bursa of Fabricius, and thymus are shown in Figures 4-3-5 to 4-3-7. ARV infection rapidly and significantly upregulated nine common avian ISGs in immune organs at the mRNA level to different degrees.

The changes in the mRNA levels of ISGs in the spleen after ARV infection are shown in Figure 4-3-5. The most significant increases were observed in Mx and IFIT5. Mx, IFIT5, IFI6, and Viperin were rapidly upregulated and reached their peak values at 1 dpi, which were 62.85, 32.11, 16.21, and 19.95 times the corresponding control levels, respectively (all $P<0.01$). OAS and ISG12 were also rapidly and significantly upregulated at 1 dpi. OAS peaked at 7 dpi at 23.51 times the control level, and ISG12 peaked at 5 dpi at 22.74 times the control level (both $P<0.01$). ZFP313 was downregulated between 1 and 4 dpi ($P<0.01$) and then increased to a level not significantly different from that in the control group. PKR and IFITM3 did not significantly change over time and were not significantly different from the control levels at most time points.

The changes in the mRNA levels of ISGs in the bursa of Fabricius are shown in Figure 4-3-6. The increase in IFIT5 was the most significant The changes in IFIT5, OAS, Viperin, and IFI6 were similar, which were rapidly upregulated at 1 dpi, decreased at 2 dpi, and peaked at 3 dpi at 34.61, 15.69, 13.64, and 12.67

times the corresponding control levels, respectively ($P<0.01$ or $P<0.05$). ISG12, IFITM3, Mx, and PKR were also slightly upregulated after infection. ZFP313 was slightly downregulated throughout the experiment.

Changes in the mRNA levels of ISGs in the thymus are shown in Figure 4-3-7. Mx and IFIT5 showed the most significant increases. Mx, IFIT5, Viperin, ISG12, IFI6, PKR, and IFITM3 exhibited similar changes in transcription levels, which were all rapidly upregulated at 1 dpi, peaked at 3 dpi, and remained high between 1 and 7 dpi. Mx and IFIT5 were the most upregulated at 62.57 and 51.57 times the control levels, respectively (both $P<0.01$). Viperin, ISG12, IFI6, and IFITM3 at 3 dpi were upregulated to 17.01, 19.96, 18.68, and 14.19 times the control levels, respectively (all $P<0.01$). OAS was rapidly upregulated at 1 dpi after infection and peaked at 4 dpi at 15.69 times the control level ($P<0.01$). PKR was slightly upregulated after infection. The expression level of ZFP313 did not change significantly during the infection process. Among the many upregulated ISGs, IFIT5 and Mx had the most significantly upregulated mRNA levels in these three tissues and organs, indicating that IFIT5 and Mx are important host ISGs against ARV infection.

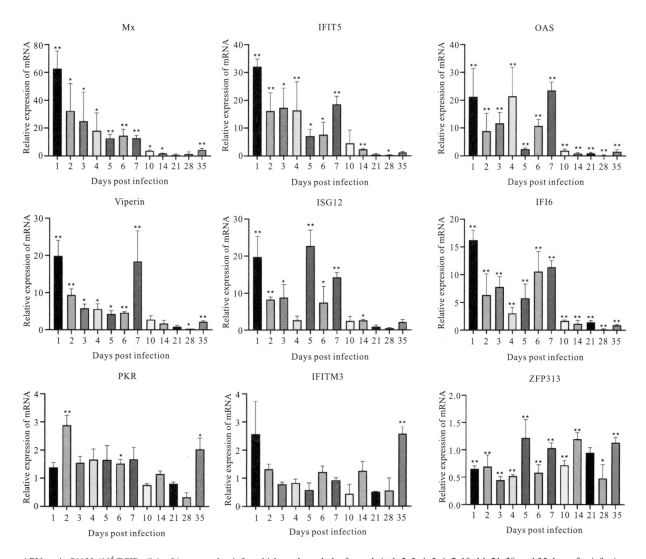

ARV strain S1133 (10^4 TCID$_{50}$/0.1 mL) was used to infect chickens through the footpad. At 1, 2, 3, 4, 5, 6, 7, 10, 14, 21, 28, and 35 days after infection (dpi), three chickens were randomly collected from each group, and changes in ISG mRNA levels in the spleen were measured by real-time quantitative PCR. The data are the mean±SD of three independent experiments. Asterisks indicate significant differences (*$P<0.05$, **$P<0.01$).

Figure 4-3-5   Changes in ISG transcription levels in the spleen after ARV infection

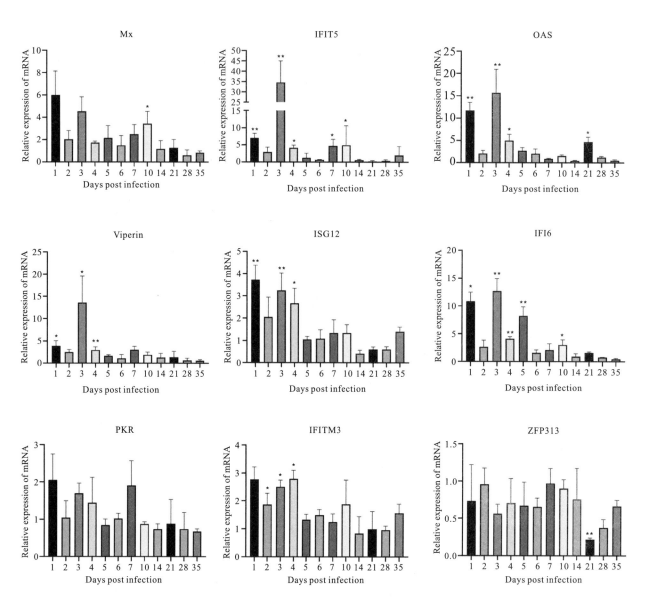

ARV strain S1133 (10^4 TCID$_{50}$/0.1 mL) was used to infect chickens through the footpad. At 1, 2, 3, 4, 5, 6, 7, 10, 14, 21, 28, and 35 days after infection (dpi), three chickens were randomly collected from each group, and changes in ISG mRNA levels in the bursa of Fabricius were measured by real-time quantitative PCR. The data are the mean ± SD of three independent experiments. Asterisks indicate significant differences (*$P < 0.05$, **$P < 0.01$).

Figure 4-3-6　Changes in ISG transcription levels in the bursa of Fabricius after ARV infection

High IFIT5 expression reduced avian reovirus replication

DF-1 cells were transfected with the pEF1α-Myc-IFIT5 recombinant plasmid. Real-time quantitative PCR results revealed that IFIT5 gene expression was significantly upregulated after transfection with the pEF1α-Mye-IFIT5 recombinant plasmid. Western blot detection using the Myc-tagged antibody indicated that the IFIT5 protein (approximately 56 kDa) was normally expressed in DF-1 cells transfected with the pEF1α-Myc-IFIT5 recombinant plasmid (Figure 4-3-8 A, B). After IFIT5 overexpression, cells were infected with ARV. Real-time quantitative PCR was used to measure the mRNA level of the ARV σC gene. The results showed that the mRNA level of the ARV σC gene was significantly reduced after IFIT5 overexpression (Figure 4-3-8 C). The measurement results for ARV titers in the cell supernatant also indicated that the viral titer of the

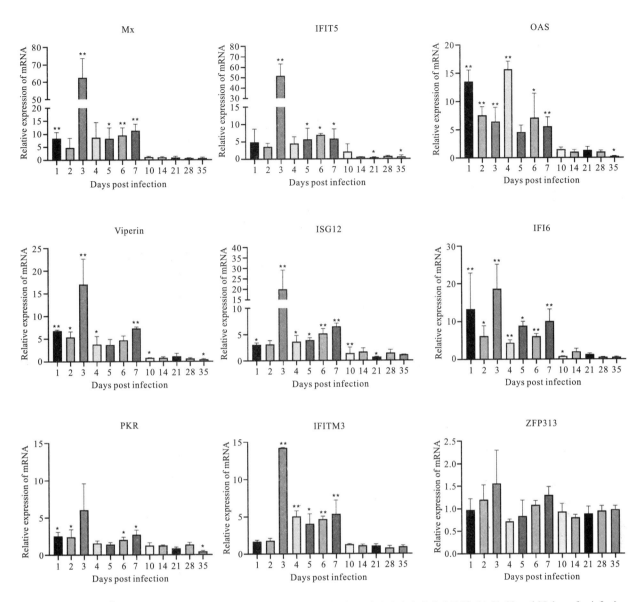

ARV strain S1133 (10^4 TCID$_{50}$/0.1 mL) was used to infect chickens through the footpad. At 1, 2, 3, 4, 5, 6, 7, 10, 14, 21, 28, and 35 days after infection (dpi), three chickens were randomly collected from each group, and changes in ISG mRNA levels in the thymus were measured by real-time quantitative PCR. The data are the mean ± SD of three independent experiments. Asterisks indicate significant differences (*$P<0.05$, **$P<0.01$).

Figure 4-3-7　Changes in ISG transcription levels in the thymus after ARV infection

cell supernatant was significantly reduced after IFIT5 overexpression (Figure 4-3-8 D).

Therefore, we speculated that IFIT5 could negatively regulate ARV replication and that downregulation of IFIT5 expression could promote ARV replication. We designed and synthesized three siRNAs targeting IFIT5, which were transfected into DF-1 cells to inhibit the expression of endogenous IFIT5. The results revealed that si225 had the best inhibitory effect (Figure 4-3-8 E). DF-1 cells were transfected with si225 and siNC and inoculated with ARV at 24 h after transfection. ARV replication was detected based on gene expression and viral titers. The results aligned with our expectations (Figure 4-3-8 F, G). Inhibition of IFIT5 expression promoted ARV replication.

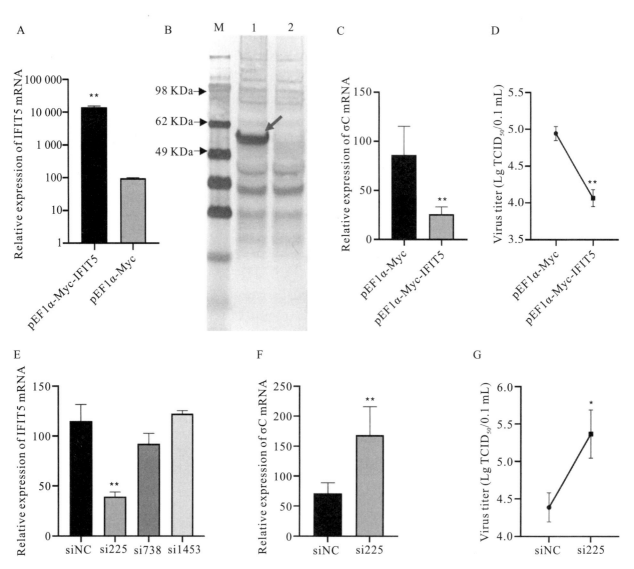

DF-1 cells were transfected with pEF1α-Myc-IFIT5 or pEF1α-Myc plasmids. Both real-time quantitative PCR (A) and Western blotting (B) confirmed that IFIT5 was highly expressed in DF-1 cells. DF-1 cells were transfected with pEF1α-Myc-IFIT5 or pEF1α-Myc plasmid and infected with ARV strain S1133 (MOI = 1). Viral replication was measured by real-time quantitative PCR (C) and viral titers (D) at 24 h after infection. The inhibitory efficiency of three IFIT5 siRNAs was compared by real-time quantitative PCR (E). DF-1 cells were transfected with si225 or siNC and infected with ARV strain S1133 (MOI = 1). Viral replication was measured by real-time quantitative PCR (F) and viral titers (G) at 24 h after infection. The data are the mean ± SD of three independent experiments. Asterisks indicate significant differences (*$P < 0.05$, **$P < 0.01$).

Figure 4-3-8　IFIT5 inhibits ARV replication in DF-1 cells

IFIT5 acts as a positive feedback regulator in natural immune signaling pathways

The above results indicate that IFIT5 can inhibit ARV proliferation, but the specific antiviral mechanism is still unclear. We overexpressed or knocked down IFIT5 in DF-1 cells and infected them with ARV to analyze the effect of IFIT5 on the expression of innate immune signaling pathway-related factors during ARV infection by real-time quantitative PCR. The results are shown in Figure 4-3-9. The mRNAs encoding MAD5 and MAVS were significantly upregulated after IFIT5 overexpression and significantly downregulated after IFIT5 knockdown. MAD5 and MAVS were upregulated 1.85- and 1.62-fold after IFIT5 overexpression, respectively, and MDA5 and MAVS were downregulated 0.52- and 0.33-fold after IFIT5 knockdown, respectively ($P < 0.01$). The mRNA levels of IRF7, TRAF3, and TRAF6 were significantly upregulated after IFIT5 overexpression or knockdown ($P<0.05$ or $P<0.01$). The mRNA levels of LGP2, TBK1, and NF-κB were significantly downregulated after IFIT5 overexpression ($P<0.05$), but no significant difference was observed after IFIT5 knockdown. The mRNA expression level of IKKε was significantly downregulated after IFIT5 overexpression and was significantly upregulated after IFIT5 knockdown ($P<0.05$ or $P<0.01$). The mRNA encoding IFN-α was significantly upregulated after IFIT5 overexpression at 1.48 times the control levels ($P<0.01$), but no significant difference was noted after IFIT5 inhibition. No significant difference in the mRNA expression level of IFN-β after overexpression or inhibition of IFIT5 was identifed. These results suggest that IFIT5 may be a positive feedback regulator of innate immune signaling pathways. IFIT5 may promote the expression of MAD5 and MAVS, thereby inducing IFN-α production to exert an antiviral effect.

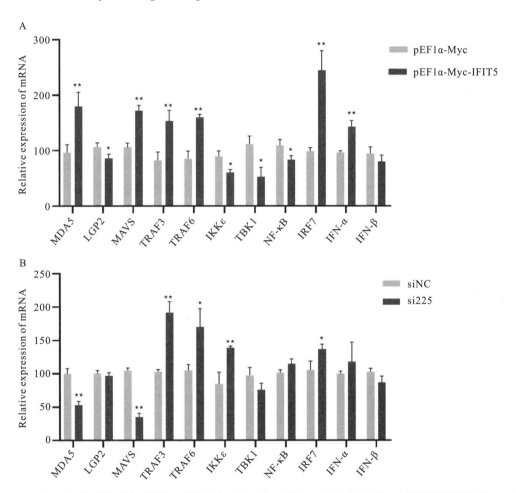

DF-1 cells were transfected with pEF1α-Myc-IFIT5 or pEF1α-Myc plasmids and infected with the ARV strain S1133 (MOI=1). Expression changes in genes related to the innate immune signaling pathway were measured by real-time quantitative PCR (A) at 24 h after infection. DF-1 cells were transfected with si225 or siNC and infected with ARV strain S1133 (MOI=1). Expression changes in genes related to the innate immune signaling pathway were measured by real-time quantitative PCR (B) at 24 h after infection. The data are the mean±SD of three independent experiments. Asterisks indicate significant differences (*$P<0.05$, **$P<0.01$).

Figure 4-3-9   IFIT5 acts as a positive feedback regulator in natural immune signaling pathways

Discussion

Avian reovirus is an important avian immunosuppressive pathogen, and innate immunity is the first line of defense against the virus[21]. Therefore, in-depth study of the innate immune response induced in immune organs after ARV infection will be valuable to reveal the pathogenic mechanism of ARV and the antiviral mechanism of the host. This study analyzed the effect of ARV infection on immune organs. Dissection results revealed that the spleen was swollen in the early stage of ARV infection, while the thymus and the bursa of Fabricius showed atrophy. Pathology indicated significant histopathological changes in the spleen, bursa of Fabricius, and thymus in the early stage of ARV infection, and these symptoms were gradually relieved in the late stage of infection, which is consistent with previous studies[22]. We further measured the ARV viral loads in the spleen, bursa of Fabricius, and thymus by real-time quantitative PCR. The results showed that ARV rapidly proliferated in several tissues and organs, the viral load peaked in the early stage of infection, and the pattern of change in the viral load was consistent with the pattern of histopathological changes in the spleen, bursa of Fabricius, and thymus, i.e., high proliferation of ARV leads to pathological damage, and this pathological damage gradually recovers when ARV is eradicated from the tissues in the later stages of infection. This finding indicates that the critical period of interaction of ARVs and hosts is mainly the early stage of infection, and ARV proliferation in the host is closely related to pathological damage. In addition, the viral load in the spleen was significantly higher than those in the bursa of Fabricius and the thymus, suggesting that the spleen is the main immune organ attacked by ARV.

Interferon is an important cytokine. IFN-α and IFN-β are type I interferons. After IFN is synthesized, it mainly exerts direct antiviral effects by inducing a series of ISGs. IFN and ISGs play important roles in the antiviral infection process of the body[23, 24]. ARV infection can induce upregulation of IFNs in chicken embryonic fibroblasts[7]. In this study, real-time quantitative PCR revealed differences in the expression levels of and changes in IFN-α and IFN-β in different tissues and organs after ARV infection. The mRNA levels of IFN-α and IFN-β were differentially upregulated in both the bursa of Fabricius and thymus at the early stage of ARV infection. IFN-α expression was significantly higher than IFN-β expression in the bursa of Fabricius. Both IFN-α and IFN-β were significantly upregulated in the thymus, while IFN-α and IFN-β were not significantly upregulated in the spleen. Therefore, ARV infection is closely related to IFN expression, but the antiviral mode of attack of different tissues and organs may differ after ARV infection. We also measured changes in the expression of ISGs after ARV infection by quantitative PCR. The expression levels of various ISGs in the spleen, bursa of Fabricius, and thymus showed changes at the transcriptionallevel; that is, all genes were rapidly upregulated at the early stage of infection and reached maximum expression at 1 to 5 days after infection. This study found similar results as previous studies. In the peripheral blood lymphocytes of ARV-infected SPF chickens, type I IFN (IFN-α, IFN-β) type II IFN (IFN-γ), and ISGs (Mx, IFITM1, and OAS) were all upregulated in the early stage of infection, with their expression peaking at 3 days after infection[8]. In our study, the changes in the mRNA levels of ISGs were essentially consistent with the changes in the viral load levels in tissues and organs. That is, when the ARV copy number increased, the mRNA levels of ISGs increased correspondingly, and after the proliferation of ARV was inhibited, the mRNA expression of ISGs also decreased correspondingly, which was associated with the dynamic pathological changes in tissues and organs. We speculate that ISGs play an important role in resisting ARV invasion, inhibiting ARV replication and proliferation, and promoting viral clearance. We found that the mRNA level of IFN in the spleen did not change significantly, but a variety of ISGs in the spleen were significantly upregulated, suggesting that the spleen might regulate the expression of ISGs in an IFN-independent manner. We also found some differences in the types and expression levels of ISGs regulated in different tissues and organs after ARV infection. Among the many upregulated ISGs, IFIT5 and Mx were the most highly upregulated in all three tissues: 62.85- (Mx) and 32.11-fold (IFIT5) in the spleen; 34.61-fold (IFIT5) in the bursa of Fabricius; and 62.57- (Mx) and 51.57-fold (IFIT5) in the thymus, respectively. In another study, we also found that in joints and target organs after ARV infection, Mx and IFIT5 increased the most significantly among the many ISGs, with the highest upregulation degrees of 748.05-fold for Mx and 660.88-fold for IFIT5, indicating that they are two important ISGs in host resistance to ARV infection[18]. Lostale-Seijo et al.[7] found that ARV can effectively induce Mx expression and exert antiviral effects after in vitro infection of chicken embryonic fibroblasts. No other study has reported on the role of IFIT5 in ARV infection and its mechanism.

The IFIT family contains important ISGs. To date, four typical IFIT family members have been identified in the human genome, namely, IFIT1, IFIT2, IFIT3, and IFIT5. IFIT family genes exist in different numbers in different species, and only IFIT5 exists in the avian genome. IFIT5 can be upregulated after stimulation by various viruses, poly (I : C), IFN-α, and IFN-β[25, 26]. IFIT5 is strongly induced by RNA viruses and much more weakly induced by DNA virus infection[27, 28]. A variety of avian viruses also upregulate IFIT5 expression. Avian-derived IFIT5 has a broad-spectrum antiviral effect[29-32]. Our study indicated that IFIT5 overexpression in DF-1 cells could inhibit ARV replication, whereas knockdown of endogenous IFIT5 expression in DF-1 cells by siRNA promoted ARV replication, indicating that IFIT5 genes are important ISGs for the host to resist ARV infection. IFIT5 has various antiviral mechanisms, as confirmed in mammals. IFIT5 can positively regulate type I IFNs to further exert its antiviral function[27, 32, 34]. IFIT5 can negatively regulate the type I IFN signaling pathway[28]. However, studies on poultry-derived IFIT5 are limited, and its mechanism of action still requires further study. In this study, real-time quantitative PCR was used to measure the effect of IFIT5 expression on the expression of factors related to the innate immune signaling pathway during ARV infection and found that IFIT5 overexpression signifcantly upregulated the mRNA levels of IFN-α to 1.48 times the control levels, but no significant difference was found after knockdown of IFIT5. In addition, IFIT5 overexpression significantly upregulated the mRNA levels of MAD5 and MAVS by 1.85- and 1.62-fold, respectively, while knockdown of IFIT5 significantly downregulated them by 0.52- and 0.33-fold, respectively. We speculate that IFIT5 is a positive feedback regulator of the innate immune signaling pathway during ARV infection. IFIT5 may exert its antiviral effects by promoting the expression of MAD5 and MAVS, which in turn induces IFN-α production. In this study, a dynamic analysis of ISGs in the spleen, bursa of Fabricius, and thymus after ARV infection was performed. We identified IFIT5 as a key antiviral candidate gene during ARV infection and conducted preliminary studies on the role of IFIT5 in the process of ARV infection. These research results provide new insights into the interaction between ARVs and hosts, which will facilitate the prevention and control of ARVs. The specific regulatory mechanism of IFIT5 during ARV infection and the mechanisms by which ARV antagonizes the antiviral effects of IFTT5 still require further research.

References

[1] VAN DER HEIDE L. The history of avian reovirus. Avian Dis, 2000,44(3):638-641.

[2] DANDAR E, BALINT A, KECSKEMETI S, et al. Detection and characterization of a divergent avian reovirus strain from a broiler chicken with central nervous system disease. Arch Virol, 2013,158(12):2583-2588.

[3] TENG L, XIE Z, XIE L, et al. Sequencing and phylogenetic analysis of an avian reovirus genome. Virus Genes, 2014,48(2):381-386.

[4] GREEN A M, BEATTY P R, HADJILAOU A, et al. Innate immunity to dengue virus infection and subversion of antiviral responses. J Mol Biol, 2014,426(6):1148-1160.

[5] IWASAKI A, PILLAI P S. Innate immunity to influenza virus infection. Nat Rev Immunol, 2014,14(5):315-328.

[6] LIN P Y, LIU H J, LIAO M H, et al. Activation of PI 3-kinase/Akt/NF-kappaB and Stat3 signaling by avian reovirus S1133 in the early stages of infection results in an inflammatory response and delayed apoptosis. Virology, 2010,400(1):104-114.

[7] LOSTALE-SEIJO I, MARTINEZ-COSTAS J, BENAVENTE J. Interferon induction by avian reovirus. Virology,

2016,487:104-111.

[8] XIE L, XIE Z, WANG S, et al. Altered gene expression profiles of the MDA5 signaling pathway in peripheral blood lymphocytes of chickens infected with avian reovirus. Arch Virol, 2019,164(10):2451-2458.

[9] XIE L, XIE Z, WANG S, et al. Study of the activation of the PI3K/Akt pathway by the motif of sigmaA and sigmaNS proteins of avian reovirus. Innate Immun, 2020,26(4):312-318.

[10] MORAGA I, HARARI D, SCHREIBER G, et al. Receptor density is key to the alpha2/beta interferon differential activities. Mol Cell Biol, 2009,29(17):4778-4787.

[11] SECOMBES C J, ZOU J. Evolution of interferons and interferon receptors. Front Immunol, 2017,8:209.

[12] SCHNEIDER W M, CHEVILLOTTE M D, RICE C M. Interferon-stimulated genes: a complex web of host defenses. Annu Rev Immunol, 2014,32:513-545.

[13] SHERRY B. Rotavirus and reovirus modulation of the interferon response. J Interferon Cytokine Res, 2009,29(9):559-567.

[14] ANAFU A A, BOWEN C H, CHIN C R, et al. Interferon-inducible transmembrane protein 3 (IFITM3) restricts reovirus cell entry. J Biol Chem, 2013,288(24):17261-17271.

[15] MARTINEZ-COSTAS J, GONZALEZ-LOPEZ C, VAKHARIA V N, et al. Possible involvement of the double-stranded RNA-binding core protein sigmaA in the resistance of avian reovirus to interferon. J Virol, 2000,74(3):1124-1131.

[16] VAZQUEZ-IGLESIAS L, LOSTALE-SEIJO I, MARTINEZ-COSTAS J, et al. Avian reovirus sigmaA localizes to the nucleolus and enters the nucleus by a nonclassical energy- and carrier-independent pathway. J Virol, 2009,83(19):10163-10175.

[17] GONZALEZ-LOPEZ C, MARTINEZ-COSTAS J, ESTEBAN M, et al. Evidence that avian reovirus sigmaA protein is an inhibitor of the double-stranded RNA-dependent protein kinase. J Gen Virol, 2003,84(Pt 6):1629-1639.

[18] WANG S, XIE L, XIE Z, et al. Dynamic changes in the expression of interferon-stimulated genes in joints of SPF chickens infected with avian reovirus. Front Vet Sci, 2021,8:618124.

[19] SOUZA S O, De CARLI S, LUNGE V R, et al. Pathological and molecular findings of avian reoviruses from clinical cases of tenosynovitis in poultry flocks from Brazil. Poult Sci, 2018,97(10):3550-3555.

[20] CRISPO M, STOUTE S T, HAUCK R, et al. Partial molecular characterization and pathogenicity study of an avian reovirus causing tenosynovitis in commercial broilers. Avian Dis, 2019,63(3):452-460.

[21] VAN DE ZANDE S, KUHN E M. Central nervous system signs in chickens caused by a new avian reovirus strain: a pathogenesis study. Vet Microbiol, 2007,120(1-2):42-49.

[22] LIN H Y, CHUANG S T, CHEN Y T, et al. Avian reovirus-induced apoptosis related to tissue injury. Avian Pathol, 2007,36(2):155-159.

[23] TARRADAS J, de la TORRE M E, ROSELL R, et al. The impact of CSFV on the immune response to control infection. Virus Res, 2014,185:82-91.

[24] SCHOGGINS J W, WILSON S J, PANIS M, et al. A diverse range of gene products are effectors of the type I interferon antiviral response. Nature, 2011,472(7344):481-485.

[25] SANTHAKUMAR D, ROHAIM M, HUSSEIN H A, et al. Chicken interferon-induced protein with tetratricopeptide repeats 5 antagonizes replication of RNA viruses. Sci Rep, 2018,8(1):6794.

[26] VLADIMER G I, GORNA M W, SUPERTI-FURGA G. IFITs: emerging roles as key anti-viral proteins. Front Immunol, 2014,5:94.

[27] ZHANG B, LIU X, CHEN W, et al. IFIT5 potentiates anti-viral response through enhancing innate immune signaling pathways. Acta Biochim Biophys Sin (Shanghai), 2013,45(10):867-874.

[28] ZHANG N, SHI H, YAN M, et al. IFIT5 negatively regulates the type I IFN pathway by disrupting TBK1-IKK epsilon-IRF3 signalosome and degrading IRF3 and IKKepsilon. J Immunol, 2021,206(9):2184-2197.

[29] MATULOVA M, VARMUZOVA K, SISAK F, et al. Chicken innate immune response to oral infection with Salmonella enterica serovar Enteritidis. Vet Res, 2013,44:37.

[30] LIU A L, LI Y F, QI W, et al. Comparative analysis of selected innate immune-related genes following infection of

immortal DF-1 cells with highly pathogenic (H5N1) and low pathogenic (H9N2) avian influenza viruses. Virus Genes, 2015,50(2):189-199.

[31] WANG B, CHEN Y, MU C, et al. Identification and expression analysis of the interferon-induced protein with tetratricopeptide repeats 5 (IFIT5) gene in duck (*Anas platyrhynchos domesticus*). PLOS ONE, 2015,10(3):e121065.

[32] LI J J, YIN Y, YANG H L, et al. mRNA expression and functional analysis of chicken IFIT5 after infected with Newcastle disease virus. Infect Genet Evol, 2020,86:104585.

[33] WU X, LIU K, JIA R, et al. Duck IFIT5 differentially regulates Tembusu virus replication and inhibits virus-triggered innate immune response. Cytokine, 2020,133:155161.

[34] ZHENG C, ZHENG Z, ZHANG Z, et al. IFIT5 positively regulates NF-kappaB signaling through synergizing the recruitment of IkappaB kinase (IKK) to TGF-beta-activated kinase 1 (TAK1). Cell Signal, 2015,27(12):2343-2354.

Dynamic changes in the expression of interferon-stimulated genes in joints of SPF chickens infected with avian reovirus

Wang Sheng, Xie Liji, Xie Zhixun, Wan Lijun, Huang Jiaoling, Deng Xianwen, Xie Zhiqin, Luo Sisi, Zeng Tingting, Zhang Yanfang, Zhang Minxiu, and Zhou Lei

Abstract

Avian reovirus (ARV) can induce many diseases as well as immunosuppression in chickens, severely endangering the poultry industry. Interferons (IFNs) play an antiviral role by inducing the expression of interferon-stimulated genes (ISGs). The effect of ARV infection on the expression of host ISGs is unclear. Specific-pathogen-free (SPF) chickens were infected with ARV strain S1133 in this study, and real time quantitative PCR was used to detect changes in the dynamic expression of IFNs and common ISGs in joints of SPF chickens. The results showed that the transcription levels of IFNA, IFNB, and several ISGs, including myxovirus resistance (MX), interferon-induced transmembrane protein 3 (IFITM3), protein kinase R (PKR), oligoadenylate synthase (OAS), interferon-induced protein with tetratricopeptide repeats 5 (IFIT5), interferon-stimulated gene 12 (ISG12), virus inhibitory protein (VIPERIN), interferon-alpha-inducible protein 6 (IFI6), and integrin-associated protein (CD47), were upregulated in joints on days $1\sim7$ of infection (the levels of increase of MX, IFIT5, OAS, VIPERIN, ISG12, and IFI6 were the most significant, at hundreds-fold). In addition, the expression levels of the ISGs encoding zinc finger protein 313 (ZFP313), and DNA damage-inducible transcript 4 (DDIT4) increased suddenly on the 1st or 2nd day, then decreased to control levels. The ARV viral load in chicken joints rapidly increased after 1 day of viral challenge, and the viral load remained high within 6 days of viral challenge. The ARV viral load sharply decreased starting on day 7. These results indicate that in SPF chicken joints, many ISGs have mRNA expression patterns that are basically consistent with the viral load in joints. IFNA, IFNB, and the ISGs MX, IFITM3, PKR, OAS, IFIT5, ISG12, VIPERIN, IFI6, and CD47 play important roles in defending against ARV invasion, inhibiting ARV replication and proliferation, and promoting virus clearance. These results enrich our understanding of the innate immune response mechanisms of hosts against ARV infection and provide a theoretical basis for prevention and control of ARV infection.

Keywords

avian reovirus, joints, interferon-stimulated genes, expression, chickens

Introduction

Avian reovirus (ARV) belongs to the genus *Orthoreovirus*, family Reoviridae. It mainly causes viral arthritis, chronic respiratory diseases, reduction in egg production, and runting-stunting syndrome and causes severe immunosuppression, all of which bring great economic losses to the poultry industry[1-3]. Chicken susceptibility to ARV infection is age-dependent, with older birds being more resistant to both infection and viral-induced lesions. Most birds appear to be infected via the fecaloral route, but also through the respiratory tract and through skin lesions[4]. Histological analyses have show that, major gross pathological lesions included marked swelling, edema, and hemorrhages. Serous exudate was present between the tendons and hock joint. Histological examination demonstrated necrosis and inflammation of muscle fibers, and mixed inflammatory infiltrate was observed in subcutaneous tissue and tendon sheaths[5, 6].

After infection with virus, the innate immune signaling pathways of host are first activated. Type I interferons (IFNs) are important cytokines that play important roles in inhibiting viral replication, regulating immune functions, and inhibiting malignant cell carcinogenesis[7, 8]. IFNs cannot directly exert antiviral functions, but they initiate the downstream Janus kinase (JAK)-signal transducer and activator of transcription (STAT) signaling pathway by binding to their receptors interferon alpha and beta receptor subunit 1 (IFNAR1) and IFNAR2 to induce the transcription of hundreds of interferon-stimulated genes (ISGs) to carry out antiviral functions[9, 10]. IFNs have been applied extensively in treatment of viral diseases. ISGs are the major antiviral executors of IFNs, so further investigation of the effect of IFN-induced ISGs on viral replication is important.

There are many types of ISG proteins with many different biological activities. Different ISGs can target different stages of viral replication to inhibit replication and achieve an antiviral effect[11]. Common avian ISGs include myxovirus resistance (MX), interferon-induced transmembrane protein 3 (IFITM3), interferon-induced protein with tetratricopeptide repeats 5 (IFIT5), oligoadenylate synthase (OAS), protein kinase R (PKR), virus inhibitory protein (VIPERIN), interferon-stimulated gene 12 (ISG12), zinc finger protein 313 (ZFP313), interferon-alpha-inducible protein 6 (IFI6), integrin-associated protein (CD47), and DNA damage-inducible transcript 4 (DDIT4). MX has inhibitory effects on various viruses, including influenza virus and vesicular stomatitis virus, and plays a role in the viral invasion stage. Its structure is highly conserved between species, and it has a broad specificity of antiviral activities through amino acids at specific sites[12-14]. The IFITM family proteins can affect the invasion and release of various viruses because they can exert antiviral effects by reducing membrane fluidity and stability[15, 16]. Different IFITM family proteins exhibit different inhibitory effects on different viruses. Chicken fibroblast cells with IFITM3 gene knockout are more susceptible to influenza A virus, indicating that IFITM3 can inhibit influenza A virus replication[16]. IFIT5 is the only member of the chicken IFIT family. It localizes to mitochondria, interacts with retinoic acid-inducible gene I (RIG-1) and mitochondrial antiviral signaling protein (MAVS), and can enhance the innate immune signaling pathway to enhance the antiviral effect[17]. OAS and PKR are the first-discovered enzymes that are activated after interaction with double-stranded RNA of viruses. These two enzymes rely on their own functions to inhibit viral protein synthesis and viral infection. Chicken OAS has excellent antiviral activities and localizes to the cytoplasm[18]. PKR can be activated by double-stranded RNA produced after viral invasion and replication in cells to exert antiviral functions through various routes[19]. The ARV σA protein can antagonize IFN-induced antiviral functions. σA protein exerts its function by downregulating the activity of PKR. σA protein can irreversibly interact with double-stranded RNA to block PKR activation and effectively prevent ARV from being recognized by the cellular immune system[20, 21]. VIPERIN has broad antiviral activities[22]. It mainly localizes to the endoplasmic reticulum (ER) and lipid droplets. *In vivo and in vitro* findings show that infectious bursa disease virus and influenza virus can induce significant upregulation of chicken VIPERIN. ISG12 protein is a newly discovered ISG protein. It mainly localizes to mitochondria and may exert antiviral functions by regulating cell apoptosis and reciprocally regulating type I IFN signaling pathways[23].

The typical symptom of ARV infection is arthritis. There is no report about the effect of ARV infection on ISG gene expression in host joints. The specific ISGs against various virus differ. Investigation of the changes in ISG transcription in host joints after ARV infection would be helpful for identifying the specific anti-ARV

ISG. Therefore, this study experimentally infected 7-day-old specific-pathogen-free (SPF) chickens with the ARV S1133 strain to observe the dynamic pathological changes in chicken joints after infection. Real time quantitative polymerase chain reaction (Real time quantitative PCR) was performed to detect the pattern of changes in the viral load in joints after infection and the changes in the expression of IFNs and ISGs at the transcription level. This study aimed to identify the ISG that may affect ARV replication and to provide a theoretical basis for the prevention and control of ARV.

Materials And methods

Virus

ARV strain S1133 was purchased from the China Institute of Veterinary Drug Control. Before use, ARV was propagated using SPF chicken embryos via yolk sac inoculation. Viruses were harvested and inoculated into LMH cells. The viral titer was measured by the Reed-Muench method.

Animal experiments

SPF White Leghorn chicken eggs were purchased from Beijing Boehringer Ingelheim Vital Biotechnology Co., Ltd. (Beijing, China). Eggs were kept in incubators for 21 days until hatching to obtain 72 1-day-old SPF chickens. Hatched 1-day-old SPF chickens were raised in SPF chicken isolators to 7 days of age. Chickens were randomly divided into the experimental group (group A) and the control group (group B), with 36 animals in each group. Each chicken in group A was challenged with 0.1 mL of ARV S1133 containing 10^4 tissue culture infection doses, 50% endpoint ($TCID_{50}$) through foot pad injection. Each chicken in group B was injected with the same volume of PBS in the foot pad. After virus challenge, the clinical symptoms of the two groups were observed and photographed. At 1, 2, 3, 4, 5, 6, 7, 10, 14, 21, 28, and 35 days postinfection (dpi), three chickens in each group were randomly collected for pathological observation and real time quantitative PCR detection.

Pathological observation

Pathological observation was performed on disease lesions of experimental chickens. Joint samples were collected, fixed in 10% formalin solution, and prepared into pathological sections by Guangxi University of Chinese Medicine. Pathological section preparation involved dehydrating the tissue samples, embedding them in paraffin, sectioning, and staining with hematoxylin and eosin to observe histopathological changes of chicken joints.

RNA extraction and cDNA synthesis

Total RNA was extracted from same-weight (30 mg) joint samples (tendon, synovium, and articular cartilage), using the GeneJET RNA Purification Kit (Thermo Scientific, USA) per the manufacturer's protocol. The concentration and purity of the total RNA were determined using a NanoDrop ND1000 spectrophotometer (Thermo Scientific, Boston, MA, USA). The extracted RNA of different samples were formulated to the same concentration (150 ng/μL), and was reverse-transcribed into cDNA with Maxima™ H Minus cDNA Synthesis Master Mix with dsDNase (Thermo Scientific) and stored at −80 ℃ for fluorescence quantitative detection.

Real time quantitative PCR

Based on gene sequence information in GenBank, specific primers for the IFN, ISG, and ARV σC genes were designed (Table 4-4-1). All primers were synthesized by Invitrogen. ARV σC genes were amplified and cloned into the pMD18-T vector (TaKaRa, Dalian, China), and the constructed recombinant plasmid was designated σC-pMD18-T. The σC-pMD18-T recombinant plasmid was used to generate the standard curve as described previously[30]. The transcription levels of IFN, ISG, and ARV σC genes were quantified by qPCR with PowerUp™ SYBR™ Green Master Mix (Thermo Scientific) in a QuantStudio 5 real-time PCR system (Thermo Lifetech ABI). GAPDH, a housekeeping gene, served as the internal control to normalize the relative expression levels of the detected genes. The real time quantitative PCR mix had 20 μL, including 10 μL of the SYBR mix, 1 μL of each of the upstream and downstream primers at the final concentration of 0.5 μM, and 2 μL of the cDNA template. The thermal cycling program was 50 ℃ activation for 2 min, 95 ℃ predenaturation for 5 min, and 40 cycles of 95 ℃ denaturation for 15 sec, annealing at 60 ℃ for 1 min, and extension at 72 ℃ for 10 sec. After the reaction was completed, melting curve analysis was performed at 95 ℃ for 15 sec, 60 ℃ for 1 min, and 95 ℃ for 1 sec.

Table 4-4-1 Pimers used in this study

Gene	Genbank accession number	Primer sequences (5'-3')	References
ARV σC	L39002.1	F: CCACGGGAAATCTCACGGTCACT, R: TACGCACGGTCAAGGAACGAATGT	
IFNA	AB021154.1	F: ATGCCACCTTCTCTCACGAC, R: AGGCGCTGTAATCGTTGTCT	[24]
IFNB	X92479.1	F: ACCAGGATGCCAACTTCT, R: TCACTGGGTGTTGAGACG	[25]
OAS	NM_205041.1	F: GCGGTGAAGCAGACGGTGAA, R: CGATGATGGCGAGGATGTG	[26]
GAPDH	NM_204305.1	F: GCACTGTCAAGGCTGAGAACG, R: GATGATAACACGCTTAGCACCAC	[26]
IFITM3	KC876032.1	F: GGAGTCCCACCGTATGAAC, R: GGCGTCTCCACCGTCACCA	[27]
MX	AY695797.1	F: AACGCTGCTCAGGTCAGAAT, R: GTGAAGCACATCCAAAAGCA	[28]
PKR	AB125660.1	F: CCTCTGCTGGCCTTACTGTCA, R: AAGAGAGGCAGAAGGAATAATTTGCC	[29]
VIPERIN	EU427332.1	F: CAGTGGTGCCGAGATTATGC, R: CACAGGATTGAGTGCCTTGA	
IFIT5	XM_421662	F: CTCCCAAATCCCTCTCAACA, R: AAGCAAACGCACAATCATCA	
ISG12	BN000222.1	F: TCCTCAGCCATGAATCCGAACA, R: GGCAGCCGTGAAGCCCAT	
ZFP313	AY604724.1	F: ATCGCTTTACCTTTCCTTG, R: GTGCCATCGTATCATCTTCA	
DDIT4	XM_015288240.1	F: CGACCTCTGCGTGGAGCA, R: CAGGGACTGGCCGAAAGC	
IFI6	NM_001001296.5	F: CACTCCTCAGGCTTTACC, R: GACCGATGCTTCTTTCTATT	
CD47	AY234188.1	F: GCTTTCAAGTTGTGGGTT, R: TGCAGTAGGTTCGGTCTC	

Statistical analysis

The cycle threshold (CT) value of the ARV σC gene was input into the standard curve to calculate the viral copy number as described previously[30], and analyze ARV replication in SPF chickens.

The CT values of the target genes (IFNs and ISGs) and the internal control gene (GAPDH gene) obtained from real time quantitative PCR were used to calculate relative expression levels of target genes at different time points using the $2^{-\Delta\Delta Ct}$ method[31]. First, sample differences of internal reference gene homogenization, ΔCt=Ct value (target gene)-Ct value (GAPDH gene). Second, comparison of infected samples and control samples, $\Delta\Delta Ct$=ΔCt values (infected samples)-ΔCt values (control samples). Third, calculate the target gene changes in transcription levels, times change=$2^{-\Delta\Delta Ct}$. The target gene times change were used for statistical analysis. Statistical analysis was performed within IBM SPSS Statistics 2.0. Student's t-test was run to assess differences. $P<0.05$ indicated that a difference between the experimental group and the control group was significant, in which case the corresponding figure was labeled with*. $P<0.01$ indicated that the difference between the experimental group and the control group was very significant (**in the figures). Figures were plotted by GraphPad Prism 8 software.

Results

Pathological changes in joints after ARV infection

The hock joints are the major target organ of ARV. The hock joints of SPF chickens in the infection group showed redness and swelling (Figure 4-4-1). Dissection results showed that exudates in the articular cavity of the hock joints of chickens in the infection group increased. Throughout the experimental process, the foot pads and hock joints of SPF chickens in the control group were all normal and did not change significantly.

The observation results of the pathological histology sections are shown in Figure 4-4-2. In the joint, tendon fibroblasts and synovial epithelial cells were degenerated and necrotic, and a large number of infiltrated inflammatory cells, mainly monocytes and macrophages, were noted.

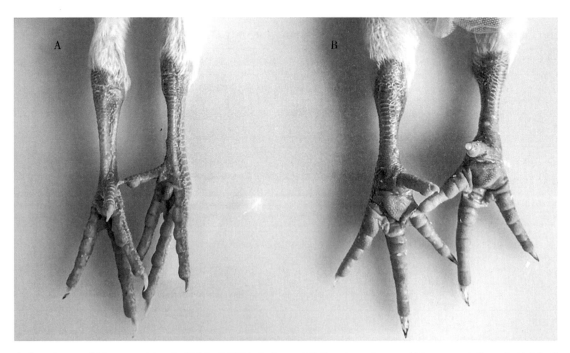

A: Control group, chickens were injected with 0.1 mL PBS in the foot pad; B: Experimental group, chickens were challenged with 0.1 mL of 10^4 median tissue culture infectious doses/0.1 mL ARV S1133 through foot pad injection.

Figure 4-4-1　Pathological changes in SPF chicken joints after ARV infection

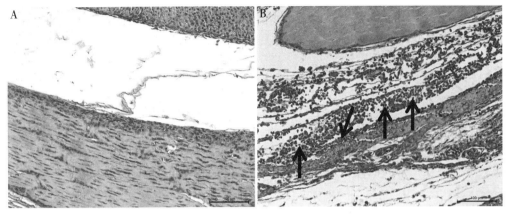

A: Control group, chickens were injected with 0.1 mL PBS in the foot pad; B: Experimental group, chickens were challenged with 0.1 mL of 10^4 median tissue culture infectious doses/0.1 mL ARV S1133 through foot pad injection. The arrow indicates tendon fibroblasts and synovial epithelial cells were degenerated and necrotic, and a large number of infiltrated inflammatory cells (color figure in appendix).

Figure 4-4-2   Histopathological changes of SPF chicken joints after ARV infection

Viral loads in joints after ARV infection

To analyze ARV proliferation in joints, the expression level of the ARV σC gene was detected by real time quantitative PCR to measure the viral load of ARV. The detection results (Figure 4-4-3) showed that ARV rapidly proliferated starting at 1 dpi. The viral load peaked by 3 dpi and was still high 6 dpi. The ARV viral load rapidly decreased starting on day 7. The control group was negative for the ARV σC gene throughout the experimental process.

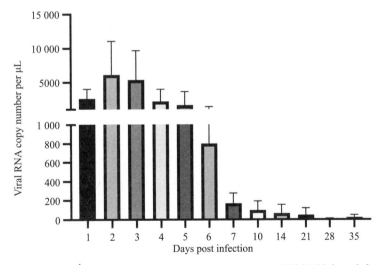

Chickens were challenged with 0.1 mL of 10^4 median tissue culture infectious doses/0.1 mL ARV S1133 through foot pad injection. At 1, 2, 3, 4, 5, 6, 7, 10, 14, 21, 28, and 35 days postinfection (dpi), three chickens in each group were randomly collected for ARV viral loads detection by real time quantitative PCR detection. Sampling was 30 mg joint (tendon, synovium, and 133 articular cartilage). The cycle threshold (CT) value of the ARV σC gene was input into the standard curve to calculate the viral copy number. Viral copy number expressed as ARV copies in 1 μL genome equivalent (150 ng RNA, about 1 mg joint tissues). Figures were plotted by GraphPad Prism 8 software.

Figure 4-4-3   Kinetics of ARV replication in SPF chicken joints after infection

Changes in transcription levels of IFNs in joints after ARV infection

IFNA expression in joints of the experimental group was upregulated between 1 and 7 dpi (except for 3 dpi, when it was downregulated) after artificial infection with ARV (Figure 4-4-4). The expression level peaked on 5 dpi at 20 times that in the control group ($P<0.01$). After 7 dpi, its expression went down.

The expression of IFNB was very similar to that of IFNA. It was upregulated between 1 and 7 dpi, peaking at 5 dpi at 118 times that in the control group ($P<0.01$). The expression levels of IFNB at 4 and 7 dpi were also very high, at 18 and 14 times those in the control group, respectively (both $P<0.01$). These genes were all downregulated after 7 dpi. These results indicate that, after ARV infection. The expression levels of IFNB and IFNB (type Ⅰ IFNs) were both upregulated, especially IFNB.

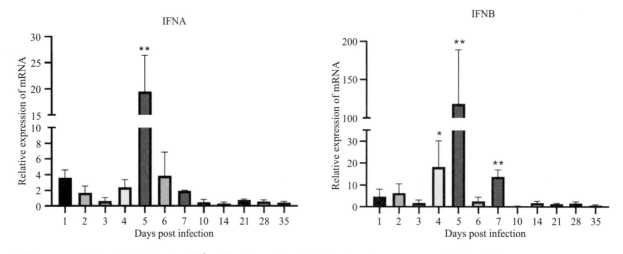

Chickens were challenged with 0.1 mL of 10^4 median tissue culture infectious doses/0.1 mL ARV S1133 through foot pad injection. At 1, 2, 3, 4, 5, 6, 7, 10, 14, 21, 28, and 35 days postinfection (dpi), three chickens in each group were randomly collected for IFNs detection by real time quantitative PCR detection. IFN gene changes in transcription levels$=2^{-\Delta\Delta Ct}$. Relative gene expression was expressed as fold of control group samples. Statistical analysis was performed within IBM SPSS Statistics 2.0. Student's t-test was run to assess differences. Figures were plotted by GraphPad Prism 8 software. Asterisks indicate significant differences ($*P<0.05$, $**P<0.01$).

Figure 4-4-4　Changes in the transcription levels of IFNs in joints after ARV infection

Changes in transcription levels of ISGs in joints after ARV infection

Changes in the transcription levels of ISGs induced by IFNs are shown in Figure 4-4-5. The results showed that ARV infection rapidly and significantly upregulated the mRNAs of 11 common avian ISGs in joints. The expression levels of ISGs MX, IFITM3, IFIT5, OAS, PKR, VIPERIN, ISG12, IFI6, and CD47 were upregulated on days 1~7 of infection. ZFP313 and DDIT4 expression suddenly increased on day 1 or 2 and then decreased. The expression levels of increase of MX, IFIT5, OAS, VIPERIN, ISG12, and IFI6 were the most significant (several hundred-fold), and their expression levels throughout the experimental process were all higher than those in the control group ($P<0.01$). MX expression stayed at a higher level between 1 and 28 dpi. Its expression on 1 dpi was 403 times that in the control group, and it was even higher afterward. Its expression level on 7 dpi was the highest, at 748 times that in the control group, and it began to decrease afterward. The expression levels of IFIT5 and ISG12 were both rapidly upregulated on 1 dpi. IFIT5 expression peaked at 5 dpi at 661 times the control level. ISG12 expression peaked at 2 dpi at 70 times the control level. The expression of IFIT5 and ISG12 were both upregulated between 2 and 7 dpi. OAS expression was also significantly upregulated at 1 dpi at 125 times higher than the control level. It peaked at 2 dpi at 350 times the control level. Except at 10 and 21 dpi, PKR expression was upregulated at all time points. Overall, its levels of upregulation were not very high, though interestingly, at 4 dpi it was at 331 times the control level. The expression levels of VIPERIN and IFI6 were upregulated at 1 dpi and stayed upregulated between 1 and 7 dpi. The expression levels of IFITM3, ZFP313, DDIT4, and CD47 were rapidly upregulated in the early stage of infection, though not very highly. They were downregulated after 1~3 days of infection to levels similar to those in the control group (except that the expression levels of ZFP313, DDIT4, and CD47 had significant downregulation at 10 dpi).

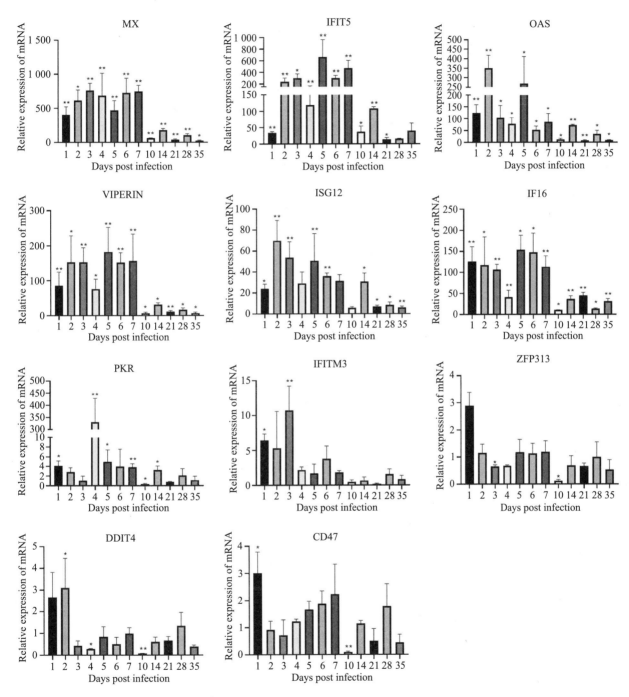

Chickens were challenged with 0.1 mL of 10^4 median tissue culture infectious doses/0.1 mL ARV S1133 through foot pad injection. At 1, 2, 3, 4, 5, 6, 7, 10, 14, 21, 28, and 35 days postinfection (dpi), three chickens in each group were randomly collected for ISGs detection by real time quantitative PCR detection. ISGs gene changes in transcription levels=$2^{-\Delta\Delta Ct}$. Relative gene expression was expressed as fold of control group samples. Statistical analysis was performed within IBM SPSS Statistics 2.0. Student's t-test was run to assess differences. Figures were plotted by GraphPad Prism 8 software. Asterisks indicate significant differences (*$P<0.05$, **$P<0.01$).

Figure 4-4-5   Changes in the transcription levels of ISGs in joints after ARV infection

Discussion

ARV was first isolated in 1954[32]. Current studies mainly focus on its etiology and epidemiology, and there have been relatively few studies about its pathogenesis and regulatory signaling pathways that induce innate immunity. After invasion of a host, a virus is recognized by pathogen-associated pattern molecules to induce the body to produce type I IFNs to further induce ISG production, thus eventually exerting the antiviral effect[33]. Therefore, analysis of changes in the transcriptional expression levels of ISGs after ARV infection is important to better understand the pathogenesis of ARV, and may be helpful for determining the ISGs that may be involved in the antiviral defense against ARV. IFN-α and IFN-β are type I IFNs that play important roles in the antiviral process. The main symptom of ARV infection is viral arthritis. Therefore, joints are the major target organ of this virus. Our real time quantitative PCR showed that both of these type I IFNs expression were upregulated in joints of SPF chickens after ARV S1133 infection, indicating that ARV infection can upregulate IFN expression. These results are consistent with previous results (changes in IFNs after ARV infection in CEF cells)[21]. This upregulated expression occurred in the early stage of viral infection. The expression level of IFNB (118.5-fold on 5 dpi) was significantly higher than that of IFNA (19.5-fold on 5 dpi). Therefore, we speculate that the body mainly exerts antiviral functions through IFN-β after ARV infection.

This study showed that mRNA expression levels of many ISGs in joints showed a pattern of changes after SPF chickens were infected with ARV S1133. The trend of changes was basically consistent with the viral load in joints (when the copy number of ARV increased, the mRNA transcription levels of the ISGs increased correspondingly; when ARV propagation was inhibited, the mRNA transcription levels of the ISGs decreased accordingly) and was associated with dynamic pathological changes in the joints. In SPF chicken joints, the transcription levels of IFNA, IFNB, and ISG genes, including MX, IFITM3, PKR, OAS, IFIT5, ISG12, VIPERIN, IFI6, and CD47, were upregulated between 1 and 7 dpi. The ARV viral load in chicken joints rapidly increased after 1 day of viral challenge. The viral load was high within the first 6 days of viral challenge but rapidly decreased starting on day 7. These results suggest that days 1～7 after ARV infection are the critical period of anti-ARV immune responses in the body. IFNA, IFNB, and ISGs, including MX, IFITM3, PKR, OAS, IFIT5, ISG12, VIPERIN, IFI6, and CD47, play important roles in defending against ARV invasion, inhibiting ARV replication and propagation, and viral clearance. Among the expression upregulated ISGs, MX, IFIT5, OAS, VIPERIN, ISG12, and IFI6 had the largest expression levels of upregulation (hundreds-fold), and these ISGs all showed a certain pattern of changes: persistent upregulation in the early stage of infection, peaking 3～5 days after infection. Accordingly, with expression increases in ISGs, the viral load of ARV began to decrease. These results indicate that these ISGs are important in the anti-ARV infection process in hosts, and these ISGs (MX, IFIT5, OAS, VIPERIN, ISG12, and IFI6) will be screened by overxpression and silencing experiments to determine the specific anti-ARV ISG in our further study.

The results of this study are similar to those of our previous study[26]considering the innate systemic immune responses. Our previous study[26] showed that expression levels of type I IFNs, including IFNA and IFNB, type II IFN IFN-γ, and the ISGs MX, IFITMI, and OSAL were all upregulated in peripheral-blood lymphocytes of SPF chickens in the early stage of ARV infection and peaked 3 days after infection. IFN-β induces significant upregulation of MX, OAS, and PKR expression[13, 18, 34] to defend against infection by various viruses. The expression levels of MX, OAS, and PKR were significantly upregulated in this study as well. These experimental results enrich our understanding of the innate immune response mechanisms of hosts in defending against ARV infection and provide a theoretical basis for prevention and control of ARV infection.

All ISG do not exhibit the same mechanism of antiviral activity. Indeed the ISGs that might be involved in antiviral response differ according to the virus. Shah et al. [35] confirmed that only the ISG protein VIPERIN was the anti-Newcastle disease virus ISG. In this study, the anti-ARV activity and the ARV replication inhibition activity of the ISGs that were significantly upregulated or downregulated and the mechanisms by which ARV antagonizes the antiviral effects of these ISGs still require further research.

An interesting phenomenon in this study was that the levels of changes in the expression of dsRNA-dependent protein kinase (PKR) did not change much throughout the experiment except that at 4 dpi it suddenly increased to 331 times the control level. This expression level of upregulation was far higher than those of other ISGs on 4 dpi. PKR is an interferon-induced and dsRNA-activated serine/threonine kinase. ARV σA, infectious bursa disease virus VP3, and influenza virus NS1 can all interact with dsRNA to inhibit PKR activation and hinder the antiviral function of IFNs[19, 20, 36]. The meaning of the sudden, high PKR expression upregulation in joints at 4 dpi after ARV infection requires further investigation.

References

[1] VAN DER HEIDE L. The history of avian reovirus. Avian Dis, 2000,44(3):638-641.

[2] DANDAR E, BALINT A, KECSKEMETI S, et al. Detection and characterization of a divergent avian reovirus strain from a broiler chicken with central nervous system disease. Arch Virol, 2013,158(12):2583-2588.

[3] TENG L, XIE Z, XIE L, et al. Sequencing and phylogenetic analysis of an avian reovirus genome. Virus Genes, 2014,48(2):381-386.

[4] JONES R C. Avian reovirus infections. Rev Sci Tech, 2000,19(2):614-625.

[5] SOUZA S O, De CARLI S, LUNGE V R, et al. Pathological and molecular findings of avian reoviruses from clinical cases of tenosynovitis in poultry flocks from Brazil. Poult Sci, 2018,97(10):3550-3555.

[6] CRISPO M, STOUTE S T, HAUCK R, et al. Partial Molecular Characterization and Pathogenicity Study of an avian reovirus causing tenosynovitis in commercial broilers. Avian Dis, 2019,63(3):452-460.

[7] GREEN A M, BEATTY P R, HADJILAOU A, et al. Innate immunity to dengue virus infection and subversion of antiviral responses. J Mol Biol, 2014,426(6):1148-1160.

[8] IWASAKI A, PILLAI P S. Innate immunity to influenza virus infection. Nat Rev Immunol, 2014,14(5):315-328.

[9] MORAGA I, HARARI D, SCHREIBER G, et al. Receptor density is key to the alpha2/beta interferon differential activities. Mol Cell Biol, 2009,29(17):4778-4787.

[10] SECOMBES C J, ZOU J. Evolution of Interferons and Interferon Receptors. Front Immunol, 2017,8:209.

[11] SCHNEIDER W M, CHEVILLOTTE M D, RICE C M. Interferon-stimulated genes: a complex web of host defenses. Annu Rev Immunol, 2014,32:513-545.

[12] BENFIELD C T, LYALL J W, KOCHS G, et al. Asparagine 631 variants of the chicken Mx protein do not inhibit influenza virus replication in primary chicken embryo fibroblasts or in vitro surrogate assays. J Virol, 2008,82(15):7533-7539.

[13] BENFIELD C T, LYALL J W, TILEY L S. The cytoplasmic location of chicken mx is not the determining factor for its lack of antiviral activity. PLOS ONE, 2010,5(8):e12151.

[14] NIGG P E, PAVLOVIC J. Oligomerization and GTP-binding Requirements of MxA for Viral Target Recognition and Antiviral Activity against Influenza A Virus. J Biol Chem, 2015,290(50):29893-29906.

[15] BRASS A L, HUANG I C, BENITA Y, et al. The IFITM proteins mediate cellular resistance to influenza A H1N1 virus, West Nile virus, and dengue virus. Cell, 2009,139(7):1243-1254.

[16] LI K, MARKOSYAN R M, ZHENG Y M, et al. IFITM proteins restrict viral membrane hemifusion. PLOS Pathog, 2013,9(1):e1003124.

[17] SANTHAKUMAR D, ROHAIM M, HUSSEIN H A, et al. Chicken Interferon-induced Protein with Tetratricopeptide Repeats 5 Antagonizes Replication of RNA Viruses. Sci Rep, 2018,8(1):6794.

[18] TAG-EL-DIN-HASSAN H T, SASAKI N, MORITOH K, et al. The chicken 2'-5' oligoadenylate synthetase A inhibits the replication of West Nile virus. Jpn J Vet Res, 2012,60(2-3):95-103.

[19] SCHIERHORN K L, JOLMES F, BESPALOWA J, et al. Influenza A Virus Virulence Depends on Two Amino Acids in the N-Terminal Domain of Its NS1 Protein To Facilitate Inhibition of the RNA-Dependent Protein Kinase PKR. J Virol, 2017,91(10).

[20] MARTINEZ-COSTAS J, GONZALEZ-LOPEZ C, VAKHARIA V N, et al. Possible involvement of the double-stranded RNA-binding core protein sigmaA in the resistance of avian reovirus to interferon. J Virol, 2000,74(3):1124-1131.

[21] LOSTALE-SEIJO I, MARTINEZ-COSTAS J, BENAVENTE J. Interferon induction by avian reovirus. Virology, 2016,487:104-111.

[22] ZHONG Z, JI Y, FU Y, et al. Molecular characterization and expression analysis of the duck viperin gene. Gene, 2015,570(1):100-107.

[23] LI X, JIA Y, LIU H, et al. High level expression of ISG12(1) promotes cell apoptosis via mitochondrial-dependent pathway and so as to hinder Newcastle disease virus replication. Vet Microbiol, 2019,228:147-156.

[24] KAPCZYNSKI D R, JIANG H J, KOGUT M H. Characterization of cytokine expression induced by avian influenza virus infection with real-time RT-PCR. Methods Mol Biol, 2014,1161:217-233.

[25] CHEN S, LUO G, YANG Z, et al. Avian Tembusu virus infection effectively triggers host innate immune response through MDA5 and TLR3-dependent signaling pathways. Vet Res, 2016,47(1):74.

[26] XIE L, XIE Z, WANG S, et al. Altered gene expression profiles of the MDA5 signaling pathway in peripheral blood lymphocytes of chickens infected with avian reovirus. Arch Virol, 2019,164(10):2451-2458.

[27] SMITH S E, GIBSON M S, WASH R S, et al. Chicken interferon-inducible transmembrane protein 3 restricts influenza viruses and lyssaviruses in vitro. J Virol, 2013,87(23):12957-12966.

[28] ABDALLAH F, HASSANIN O. Positive regulation of humoral and innate immune responses induced by inactivated avian influenza virus vaccine in broiler chickens. Vet Res Commun, 2015,39(4):211-216.

[29] DAVIET S, Van BORM S, HABYARIMANA A, et al. Induction of Mx and PKR failed to protect chickens from H5N1 infection. Viral Immunol, 2009,22(6):467-472.

[30] VAITOMAA J, RANTALA A, HALINEN K, et al. Quantitative real-time PCR for determination of microcystin synthetase e copy numbers for microcystis and anabaena in lakes. Appl Environ Microbiol, 2003,69(12):7289-7297.

[31] LIVAK K J, SCHMITTGEN T D. Analysis of relative gene expression data using real-time quantitative PCR and the 2(-Delta Delta C(T)) Method. Methods, 2001,25(4):402-408.

[32] FAHEY J E, CRAWLEY J F. Studies On Chronic Respiratory Disease Of Chickens Ⅳ. A Hemagglutination Inhibition Diagnostic Test. Can J Comp Med Vet Sci, 1954,18(7):264-272.

[33] SCHOGGINS J W, WILSON S J, PANIS M, et al. A diverse range of gene products are effectors of the type Ⅰ interferon antiviral response. Nature, 2011,472(7344):481-485.

[34] AMICI C, La FRAZIA S, BRUNELLI C, et al. Inhibition of viral protein translation by indomethacin in vesicular stomatitis virus infection: role of eIF2alpha kinase PKR. Cell Microbiol, 2015,17(9):1391-1404.

[35] SHAH M, BHARADWAJ M, GUPTA A, et al. Chicken viperin inhibits Newcastle disease virus infection *in vitro*: A possible interaction with the viral matrix protein. Cytokine, 2019,120:28-40.

[36] BUSNADIEGO I, MAESTRE A M, RODRIGUEZ D, et al. The infectious bursal disease virus RNA-binding VP3 polypeptide inhibits PKR-mediated apoptosis. PLOS ONE, 2012,7(10):e46768.

Gallus NME/NM23 nucleoside diphosphate kinase 2 interacts with viral σA and affects the replication of avian reovirus

Xie Liji, Wang Sheng, Xie Zhixun, Wang Xiaohu, Wan Lijun, Deng Xianwen, Xie Zhiqin, Luo Sisi, Zeng Tingting, Zhang Minxiu, Fan Qing, Huang Jiaoling, Zhang Yanfang, and Li Meng

Abstract

Our present study aimed to identify host cell proteins that may interact with avian reovirus (ARV) σA protein and their potential effect on ARV replication. The ARV structural protein σA has been demonstrated to suppress interferon production and confirmed to activate the PI3K/Akt pathway. However, host cell factors interacting with σA to affect ARV replication remain unknown. In current study, a cDNA library of chicken embryo fibroblasts (CEFs) was constructed, and host cell proteins interacting with σA were screened by a yeast two-hybrid system. We identified four candidate cellular proteins that interact with ARV σA protein. Among them, Gallus NME/NM23 nucleoside diphosphate kinase 2 (NME2) was further validated as a σA-binding protein through coimmunoprecipitation. The key interaction domain was identified at amino acids (aa) 121~416 in NME2 and at aa 71~139 in σA, respectively. We demonstrated that overexpression of NME2 substantially inhibited ARV replication. In addition silencing NME2 by small interfering RNAs (siRNAs) resulted in marked enhancement of ARV replication. Our work has demonstrated that NME2 is a σA-binding protein that may affect ARV replication in CEF cells.

Keywords

avian reovirus, σA protein, NME2 protein, protein-protein interactions

Introduction

Avian reovirus (ARV) is an important disease-causing pathogen in poultry. In particular, ARV can cause viral arthritis, retarded growth, chronic respiratory diseases, and malabsorption syndrome[1-3], resulting in heavy economic losses in poultry industry. Understanding the infection mechanism of ARV may allow for better design for vaccines and therapeutic medicines for prevention and control of ARV infection and spread.

ARV is an icosahedral, non-enveloped, double-stranded RNA (dsRNA) virus with a particle size of 70~80 nm and contains ten genome segments. ARV encodes 10 structural proteins (λA, λB, λC, μA, μB, μBC, μBN, σC, σA, and σB) and four non-structural proteins (μNS, P10, P17, and σNS)[4, 5]. The viral σA is a 46 kDa protein consisting of 416 aa. σA is encoded by the S2 gene, and is included in the inner viral capsid[6]. Moreover, it has been shown that ARV σA is a dsRNA-binding protein involved in development of interferon (IFN) resistance[6, 7]. In addition, σA protein could activate the PI3K/Akt signaling pathway[8]. Yin et al.[9] found that σA protein has non-specific nucleotidyl phosphatase activity, hydrolyzing four types of nucleoside triphosphates (NTPs) to their corresponding di- and monophosphates and free phosphate. More recently, Chi et al.[10] has demonstrated that σA is an energy activator and modulates suppression of lactate dehydrogenase and upregulation of glutaminolysis and the mTOC1/eIF4E/HIF-1α pathway to enhance glycolysis and the TCA cycle for virus replication.

However, it is unclear how the σA protein induces a series of downstream biological reactions in the host cells. Viral proteins could interact with cellular proteins and may influence viral infection, replication, or host cell functions[11-14]. It is likely that σA may interact with host cell proteins and trigger downstream intracellular biological reactions.

In our study, yeast two-hybrid (Y2H) system and coimmunoprecipitation (Co-IP) were employed to identify in chicken embryo fibroblasts (CEFs) proteins that interact with ARV σA. We found that four cellular proteins interacted with σA. We subsequently confirmed that gallus NME/NM23 nucleoside diphosphate kinase 2 (NME2) interacted and colocalized with σA in the tested cells. Functional analysis showed that over-expression of NME2 protein substantially inhibited ARV replication. In contrast, silencing NME2 by small interfering RNAs had significant enhancement effects.

Materials and methods

Virus and cells

The virus strain ARV S1133 used in this study was purchased from the China Institute of Veterinary Drugs Control, Beijing, China. CEF and DF-1 cells were cultured in Dulbecco's Modified Eagle Medium: Nutrient Mixture F-12 (DMEM/F-12; Gibco) supplemented with 10% fetal bovine serum (FBS; Gibco), 100 μg/mL streptomycin sulfate and 100 U/mL penicillin sodium (Beyotime Biotechnology).

Yeast two-hybrid (Y2H) screening

A Matchmaker gold yeast two-hybrid system from Clontech was utilized for screening host cell proteins that may interact with σA of ARV S1133. As the bait protein, the full-length σA sequence (forward primer σA-F: 5'-GCCATATGATGGCGCGTGCCATATACGAC-3', and reverse primer σA-R: 5'-GCGGATCCCTAGG CGGTAAAAGTGGCTAG-3') was ligated as an N-terminal fusion to the GAL4 DNA-binding domain (BD) in the pGBKT7 vector, and the plasmid was transformed into the yeast strain Y2HGold. The cDNA library of chicken embryo fibroblast (CEF), was constructed by the Make Your Own "Mate & Plate" Library System from Clontech, prey proteins from CEF cDNA libraries were expressed as fusions to the GAL4 activation domain (AD) in the pGADT7 vector, and the plasmid was transformed into the yeast strain Y187. The two transformed strains were co-cultured for 24 h. Selection was conducted on medium (DDO/X/A, SD lacked leucine and tryptophan, but containing X-alpha-galactosidase and aureobasidin A). The clones were then plated on quadruple-dropout medium (QDO/X/A, SD lacking adenine, histidine, leucine, and tryptophan, but containing X-alpha-galactosidase and aureobasidin A). Blue colonies were selected and cultured for plasmid DNA extraction. The isolated plasmid DNAs were sequenced to identify potential cellular interacting proteins.

Homologue searches were performed with BLAST in GenBank. To confirm the interaction between σA and host cells proteins, the Y2HGold yeast strain was co-transformed with the bait and prey plasmids. Cotransformation with BD-p53/AD-T (simian virus 40[SV40] large T antigen), BD-Lam (human lamin C protein)/AD-T, and BD/AD served as positive, negative, and blank controls, respectively.

Coimmunoprecipitation (Co-IP) assays

The σA gene was amplified from the genome of ARV via RT-PCR and ligated into pEF1α-HA (Clontech) to generate pEF1α-HA-σA. The NME2, Gallus nascent polypeptide associated complex subunit alpha (NACA), Gallus hypothetical protein (GHP) and Gallus mitochondrial ribosomal protein S9 (MRPS9) genes were amplified using cDNA from CEFs and cloned into pEF1α-Myc (Clontech), generating plasmids pEF1α-Myc-NME2, pEF1α-Myc-NACA, pEF1α-Myc-GHP, and pEF1α-Myc-MRPS9, respectively. The primers used in this study are listed in Table 4-5-1.

Table 4-5-1   Primers used in the study

Primer	Primer sequence (5'-3')	Usage
σA-1262F	GC*GTCGACC*ATGGCGCGTGCCATATACGAC (*Sal*I)	Amplification of σA
σA-1262R	GC*GCGGCCGC*CTAGGCGGTAAAAGTGGCTAGAAC (*Not*I)	
NACA648-F	GC*GAATTC*CTATGCCTGGCGAAGCTACAGAAAC (*Eco*RI)	Amplification of NACA
NACA648-R	GC*GCGGCCGC*CTACATCGTCAGCTCCATGAT (*Not*I)	
NME2-462F	GC*GAATTC*CTATGGCTGCCAACTGCGAGCGCAC (*Eco*RI)	Amplification of NME2
NME2-462R	GC*GCGGCCGC*TCACTCATAGACCCAGTCATG (*Not*I)	
GHP-501-F	GCG*AATTC*CGAAGCCGGGACCCACAGTAGG (*Eco*RI)	Amplification of GHP
GHP-501-R	GC*GCGGCCGC*TCTCGTTGCCCGCCGTTTATTGG (*Not*I)	
MRPS9-1178-F	GC*GAATTC*CGGTGGGCATGGCGGTGGTG (*Eco*RI)	Amplification of MRPS
MRPS9-1178-R	GC*GCGGCCGC*TCTTGGAGCATTCAGCCTTTCTTG (*Not*I)	
σA-F	TTACGCAGAGGCATTTCGCTTACG	qRT-PCR for detection of σA
σA-R	TTGCCCCTTCGCTGCTGACA	
NME2-F	CGGGGGCTGGTGGGGGAGAT	qRT-PCR for detection of NME2
NME2-R	GGATGGTGCCGGGCTTTGAGTC	
GAPDH-F	GCACTGTCAAGGCTGAGAACG	qRT-PCR for detection of GAPDH
GAPDH-R	GATGATAACACGCTTAGCACCAC	

Taking the Co-IP assay for σA and NME2 as an example, three groups were designed for the experiment. The experimental group was co-transfected with pEF1α-HA-σA and pEF1α-Myc-NME2. Two control groups were co-transfected with pEF1α-HA plus pEF1α-Myc-NME2 or pEF1α-HA-σA plus pEF1α-Myc. DF-1 cells were passaged under standard procedures in six-well plates. When the cells reached 80% confluence, transfection was performed using Lipofectamine™ 3000 (Invitrogen) with three replicate wells per group. The culture solution was discarded at 48 h post-transfection, and the plate was washed once with pre-chilled phosphate-buffered saline (PBS). The cells were harvested in lysis buffer, and Co-IP was performed according to instructions of the Capturem IP & Co-IP Kit (TaKaRa). A mouse monoclonal anti-HA antibody (Ab) (Abcam) was used for immobilization and adsorption in the forward Co-IP, and rabbit anti-Myc tag and rabbit anti-HA tag Abs (Abcam) were used as primary Abs in Western blot analyses. A mouse anti-Myc monoclonal antibody (Abcam) was used for immobilization and adsorption in the reverse Co-IP, and, rabbit anti-HA tag and rabbit anti-Myc tag Abs (Abcam) were used as primary Abs in Western blot analyses.

Confocal immunofluorescence microscopy

Four groups were designed for laser confocal microscopy of NME2 and σA: the experimental group was co-transfected with pEF1α-Myc-NME2 plus pEF1α-HA-σA. Three control groups were co-transfected with pEF1α-Myc-NME2 plus pEF1α-HA, pEF1α-Myc plus pEF1α-HA-σA, and pEF1α-Myc plus pEF1α-HA, respectively. DF-1 cells were passaged under normal conditions and sparsely spread in confocal Petri dishes. The cells were subjected to transfection 12 h later, and the media was removed at 48 h post-transfection. After washing once with pre-chilled PBS, the transfected cells were fixed with 4% paraformaldehyde for 30 min at room temperature. After washed three times with PBS, the cells were permeabilized with 0.1% Triton X-100 for 15 min at room temperature. The cells were subsequently washed three times with PBS and were blocked with 5% skim milk for 1 h at room temperature. After extra washed three times with PBS, the cell lysates were incubated with a rabbit anti-HA monoclonal Ab and a mouse anti-Myc monoclonal Ab (Abcam) at 37 ℃ for 2 h. Afterwards, the cells were washed five times with PBS containing 0.05% Tween-20 (PBST) and then incubated with an Alexa Fluor 488-labeled goat anti-mouse IgG (H + L) cross-absorbed secondary Ab and an Alexa Fluor 594-labeled goat anti-rabbit IgG (H + L) cross-absorbed secondary antibody (Invitrogen) at 37 ℃ for 1 h in the dark. Next, the cells were washed five times with PBST, and DAPI nuclear staining was performed at room temperature in the dark for 10 min. Following washed four times with PBST, 500 μL of 50% glycerol was added to the cells, and the confocal Petri dish was observed by laser confocal microscopy.

Identification of interaction domains between viral σA and cellular NME2 proteins using the Y2H assay

Secondary structure prediction for NME2 and σA proteins was performed using the online software SMART. Key domains of the NME2 and σA proteins were truncated from the amino- and carboxyl-terminal regions, respectively.

As shown in Figure 4-5-1 A, truncated NME2 sequences were expressed as fusions to GAL4 AD in yeast strain Y187 (truncated NME2-AD), full length σA sequences were expressed as fusions to GAL4 BD in yeast Y2HGold cells. The two transformed strains were co-cultured for 24 h, and subsequently selected in DDO/X/A medium. The colonies were then transferred to QDO/X/A medium to verify the interaction between the key domain of the NME2 protein and full-length σA.

As shown in Figure 4-5-1 B, truncated σA sequences were expressed as fusions to the GAL4 BD in yeast strain Y2HGold (truncated σA-BD), full length NME2 sequences were expressed as fusions to GAL4 AD in yeast Y187 competent cells. The two transformed strains were co-cultured for 24 h, and subsequently selected on DDO/X/A medium. The colonies were then transferred to QDO/X/A medium to verify the interaction between the key domain of the σA protein and full-length NME2.

Truncated NME2-AD and truncated σA-BD were co-cultured for 24 h, and subsequently selected on DDO/X/A medium. The colonies were then transferred to QDO/X/A medium to verify the key domain for the interaction between σA and NME2.

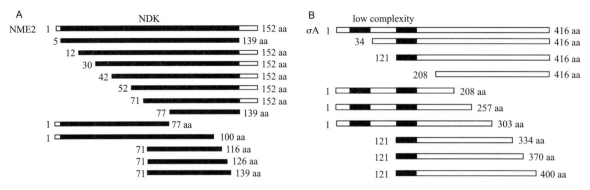

A: Schematic illustration of interaction between the truncated NME2 protein and the full-length σA; B: Schematic illustration of the interaction between the truncated σA and the full-length NME2 protein. NME2 was truncated as shown in the schematic and expressed as a fusion to the GAL4 activation domain (AD) in yeast strain Y187 for yeast two-hybrid assays with the full-length σA. The σA was truncated as shown in the schematic and expressed as a fusion to the GAL4 DNA-binding domain (BD) in yeast strain Y2HGold for yeast two-hybrid assays with full-length NME2.

Figure 4-5-1　Verification of the interaction domain between σA and NME2 proteins by a yeast two-hybrid assay

Overexpression of NME2 protein

DF-1 cells were transfected with the pEF1α-Myc-NME2 plasmid for overexpression of the NME2 protein. 24 h later, the transfected cells were infected with ARV at a multiplicity of infection (MOI) of 1. Cell and culture supernatants were collected, respectively, at 24 h post-infection. The cell samples were subjected to RNA detection and Western blot analysis. The culture supernatants were used for titration of viruses.

For RNA detection, total RNA was extracted from NME2-overexpressing cells using a GeneJET RNA Purification Kit (Thermo Scientific), and reverse transcribed into cDNA with Maxima™ H Minus cDNA Synthesis Master Mix with dsDNase (Thermo Scientific). RNA from empty vector-treated cells was used as a control. The transcriptional levels of NEM2 and σA were quantified via qPCR with PowerUp™ SYBR™ Green Master Mix (Thermo Scientific) in a QuantStudio 5 real-time PCR system (Thermo Lifetech ABI). The primers used in the qPCRs are listed in Table 4-5-1. GAPDH, a housekeeping gene, served as the internal control to normalize the relative expression level of NEM2 and σA gene. The relative transcript levels of NEM2 and σA gene expression were analyzed using the threshold cycle ($2^{-\Delta\Delta CT}$) method[15].

For Western blot analysis, a rabbit anti-Myc monoclonal Ab or a mouse anti-σA monoclonal Ab was used as the primary Ab (Abcam), with β-actin as the internal reference.

For viral replication analysis, $TCID_{50}$ was measured in triplicate by plaque assay on LMH cells.

RNA interference assay

Small interfering RNAs (siRNAs) against the NME2 were synthesized by GenePharma. The two synthesized siRNAs were siNME2-59F (5'-CCAACUGCGAGCGCACCUUTT-3') and siNME2-59R (5'-AAGGUGCGCUCGCAGUUGGTT-3'). The sequences of the negative control (siNC) were 5'-UUCUCCGAACGUGUCACGUTT-3' and 5'-ACGUGCCACGUUCGGAGAATT-3. DF-1 cells were transfected with 200 nM siRNAs or siNC using Lipofectamine RNAiMAX (Invitrogen) according to the manufacturer's instructions. At 48 h post-transfection, the supernatant was discarded, and the cells were washed three times with serum-free medium. Following infected with ARV at an MOI of 1, the cells were incubated for adsorption at 37 ℃ for 1 h. Afterwards, the culture media were discarded, and the cells were washed twice with serum-free medium and incubated in a fresh culture medium containing 1% FBS for another 24 h. Subsequently, cells were collected for RNA isolation and for Western blot analysis. Culture supernatants were used for virus titeration.

Cell viability assay

DF-1 cells seeded into opaque-walled 96-well microplates were transfected with various plasmids or siRNAs, and after each treatment, cell viability was measured with a WST-1 Cell Proliferation and Cyto-toxicity Assay Kit (Beyotime) according to the manufacturer's instructions.

Results

Identification of σA-binding proteins by Y2H assay

To better understand the impact of σA protein on ARV infection and to identify cellular proteins interacting with σA protein, we conducted a Y2H screening for a CEF cell cDNA library using a full-length σA as the bait. After removal of potential false-positive-interacting proteins, four cellular proteins were identified as candidate σA-binding partners with a high confidence (Figure 4-5-2). These four proteins are gallus nascent poly-peptide associated complex subunit alpha (NACA, GenBank accession number: NM_001276303.3), gallus NME/NM23 nucleoside diphosphate kinase 2 (NME2, GenBank accession: number NM_205047.1), gallus hypothetical protein (GHP, GenBank accession number AJ851804.1) and gallus mitochondrial ribosomal protein S9 (MRPS9, GenBank accession number: NM_001277757.1), respectively.

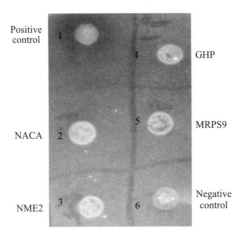

Yeast strain Y2HGold was co-transformed with one of the prey plasmids (AD-NACA, AD-NME2, AD-GHP, and AD-MRPS9) and with the bait plasmid BD-σA. Blue colonies indicates positive protein-protein interactions. Co-transformation with pGBKT7-53 (BD fused with murine p53, BD-P53) and with pGADT7-T (AD fused with the simian virus 40 large T antigen, AD-T) served as a positive control, because it is known that p53 protein and large T antigen interacts with each other in theY2H system. Co-transformation with pGBKT7-Lam (BD fused with human lamin C, BD-Lam), and with AD-T served as a negative control, because it is known that lamin C does not form protein-protein complexes with large T antigen, and also does not interact with most other proteins.

Figure 4-5-2   Interaction of σA with NME2 protein in the yeast two-hybrid system

NME2 protein co-precipitates with ARV σA protein

A Co-IP assay was carried out to identify protein interaction between σA and NME2 *in vitro*. When the anti-HA monoclonal antibody was immobilized by Co-IP, NME2-Myc was detected by Western blot analysis. When the anti-Myc monoclonal Ab was immobilized by Co-IP, σA-HA was detected by Western blot analysis. These results demonstrated that ARV σA protein was associated with NME2 protein (Figure 4-5-3). A Co-IP assay of σA protein with GHP, MRPS9, and NACA proteins, respectively, did not show interaction of σA with the above tested cellular proteins.

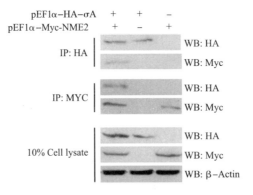

DF-1 cells were co-transfected with pEF1α-HA-σA and pEF1α-Myc-NME2 for 48 h. Cell lysates were immunoprecipitated with a mouse anti-HA Ab or a mouse anti-Myc Ab, and the immunoprecipitates were analyzed by Western blotting using a rabbit anti-HA Ab and rabbit anti-Myc Ab, respectively.

Figure 4-5-3   Co-immunoprecipitation of σA and NME2 proteins in DF-1 cells

Subcellular localization of σA and NME2

Immunofluorescence and confocal microscopy were employed to examine subcellular localization of σA and NME2 proteins in DF-1 cells. After co-transfection in DF-1 cells, σA and NME2 proteins were found to colocalize in the cytoplasm (Figure 4-5-4), with Pearson's coefficient of 0.937 and Overlap coefficient of 0.945. These results demonstrate that NME2 colocalized with σA protein in the same subcellular compartment, suggesting that NME2 is a σA-binding protein.

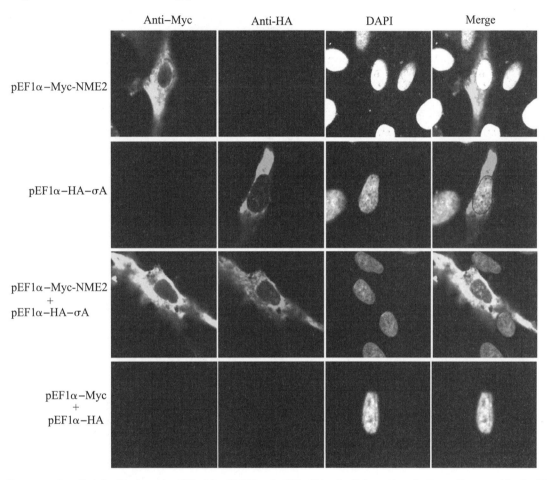

DF-1 cells were co-transfected with plasmids pEF1α-Myc-NME2 and pEF1α-HA-σA. 48 h post-transfection, cells were stained with rabbit anti-HA monoclonal antibody and mouse anti-Myc monoclonal antibody for 2 h, and then incubated with Alexa Flour 488 and Alexa Flour 594 labeled secondary antibody(HA-σA (red), Myc-NME2 (green)). Cell nuclei were stained blue with DAPI. Yellow indicates colocalization in the merged images. Data is representative of all cells that co-expressed both proteins (color figure in appendix).

Figure 4-5-4   Confocal analysis of the subcellular localization of σA and NME2 proteins in DF-1 cells

Idenfication of the interaction domain of NME2 with σA protein by the Y2H assay

The results of the interaction between truncated NME2 and full-length σA are shown in Figure 4-5-5. Truncated NME2 protein fragments (aa 5～139, aa 12～152, aa 30～152, aa 42～152, aa 52～152, aa 71～152, and aa 71～139) showed an interaction with the full-length σA protein, and aa 71～139 appeared to be the smallest truncated region that maintained an interaction in the experiment.

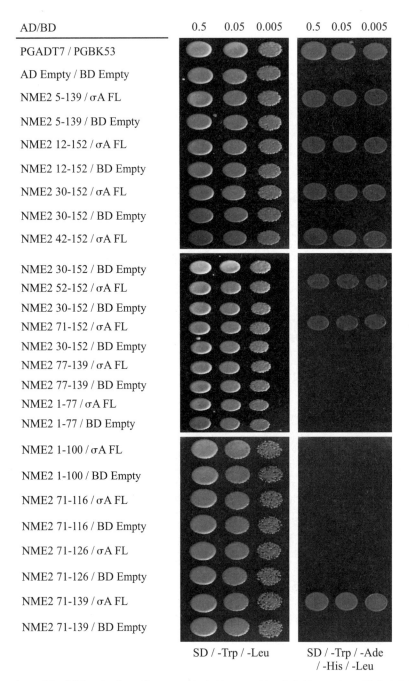

AD/BD	0.5	0.05	0.005	0.5	0.05	0.005
PGADT7 / PGBK53						
AD Empty / BD Empty						
NME2 5-139 / σA FL						
NME2 5-139 / BD Empty						
NME2 12-152 / σA FL						
NME2 12-152 / BD Empty						
NME2 30-152 / σA FL						
NME2 30-152 / BD Empty						
NME2 42-152 / σA FL						
NME2 30-152 / BD Empty						
NME2 52-152 / σA FL						
NME2 30-152 / BD Empty						
NME2 71-152 / σA FL						
NME2 30-152 / BD Empty						
NME2 77-139 / σA FL						
NME2 77-139 / BD Empty						
NME2 1-77 / σA FL						
NME2 1-77 / BD Empty						
NME2 1-100 / σA FL						
NME2 1-100 / BD Empty						
NME2 71-116 / σA FL						
NME2 71-116 / BD Empty						
NME2 71-126 / σA FL						
NME2 71-126 / BD Empty						
NME2 71-139 / σA FL						
NME2 71-139 / BD Empty						

SD / -Trp / -Leu   SD / -Trp / -Ade / -His / -Leu

The truncated NME2 proteins and the full-length σA protein were subjected to a yeast two-hybrid assay to verify their interaction. The truncated NME2 protein fragments (aa 5～139, aa 12～152, aa 30～152, aa 42～152, aa 52～152, aa 71～152, and aa 71～139) interacted with the full-length σA protein, and aa 71～139 is the smallest truncated region required for efficient interaction between the tested proteins.

Figure 4-5-5 Verification of interaction between the truncated NME2 and the full-length σA protein by a yeast two-hybrid assay

The verification results of the interaction between truncated σA and full-length NME2 are illustrated in Figure 4-5-6. Truncated σA protein fragments (aa 34～416 and aa 121～416) displayed interaction with the full-length NME2 protein, with amino acids 121～416 appearing to be the smallest truncated region maintaining the interaction.

In addition, interaction between truncated NME2 (aa 71～139) and truncated σA (aa 121～416) was detected.

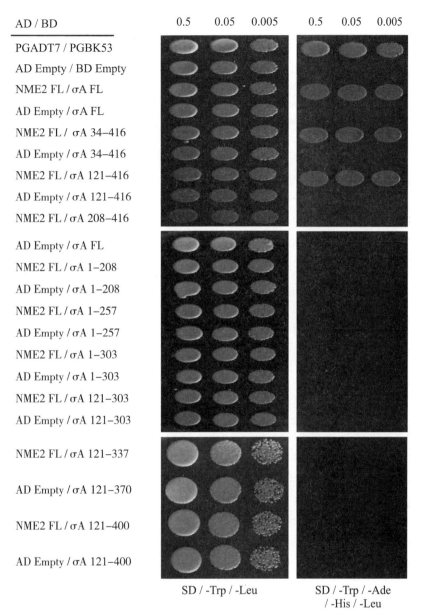

The truncated σA proteins and the full-length NME2 protein were subjected to a yeast two-hybrid assay to verify their interaction. The truncated σA protein fragments (aa 34~416 and aa 121~416) interacted with the full-length NME2 protein, and aa 121~416 is the smallest truncated region required for efficient protein-protein interactions.

Figure 4-5-6   Verification of interaction between the truncated σA and the full-length NME2 by a yeast two-hybrid assay

Overexpression of NME2 inhibits the proliferation of ARV

Although we have shown that σA protein associated with cellular protein NME2, the effect of NME2 on ARV replication remains unclear. Therefore, we evaluated the effect of NME2 overexpression on ARV replication in cultured cells.

After overexpression of the NME2 protein in DF-1 cells, significant reductions in the ARV gene copy number (Figure 4-5-7 A), viral protein expression (Figure 4-5-7 B), and virus titer (Figure 4-5-7 C) were observed compared with the control group. These results demonstrated that overexpression of NME2 protein inhibited ARV replication in susceptible cells.

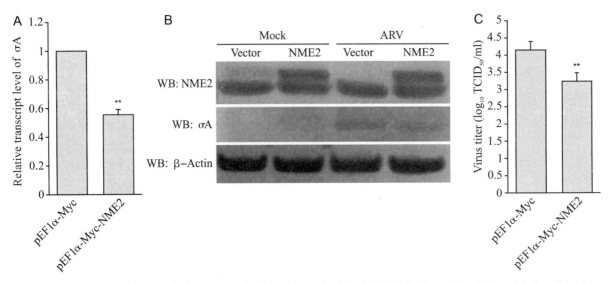

A: The transcription level of σA gene in the transfected cells was determined by qRT-PCR; B: Expression of σA protein in cell lysate was analyzed by Western blotting; C: 50% tissue culture infective dose (TCID₅₀) of ARV was measured using cell culture supernatants. DF-1 cells were transfected with pEF1α-Myc-NME2 or empty vector pEF1α-Myc for 24 h and then infected with ARV. Data are the means of triplicate determinations from three independent experiments (*$P < 0.05$, **$P < 0.01$).

Figure 4-5-7　Overexpression of NME2 inhibited ARV replication

Inhibition of NME2 protein expression with siRNAs enhances ARV proliferation

To further examine the effect of NME2 on ARV proliferation, specific 21-nt siRNAs was used to knockdown expression of NME2 protein.

NME2 protein expression was silenced by the siRNAs, resulting in a reduced expression of NME2 protein by 50% compared to the controls. After NME2 gene silencing, significant increases in ARV gene copy number (Figure 4-5-8 A), viral protein expression (Figure 4-5-8 B), and virus titer (Figure 4-5-8 C) were observed compared with the control group. These results demonstrated that silencing NME2 gene expression promoted ARV replication.

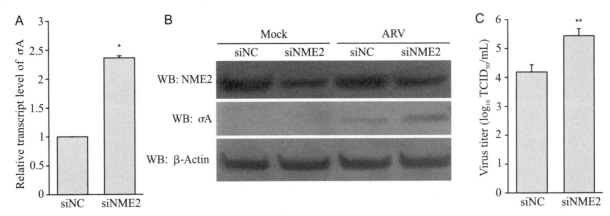

A: The transcription level of σA gene in the transfected cells was determined by qRT-PCR; B: Expression of σA protein in the cell lysate was analyzed by Western blotting; C: The TCID₅₀ of ARV was measured using cell culture supernatants. DF-1 cells were transfected with siNME2 and then infected with ARV; siNC was used as a negative control. Data are the means of triplicate determinations from three independent experiments (*$P < 0.05$, **$P < 0.01$).

Figure 4-5-8　Inhibition of NME2 protein expression enhances ARV replication

Cell viability assay

No notice substantial changes in cell proliferation or cytotoxicity were found.

Discussion

In this study, four candidate host cellular proteins interacting with the ARV structural protein σA were identified by a yeast two-hybrid system. Among the four proteins, we found that gallus NME/NM23 nucleoside diphosphate kinase 2 (NME2) is a crucial cellular protein with the ability to interact with σA protein, and to affect ARV replication.

Through overexpression and gene silencing assays, we verified that NME2 is a negative regulator of ARV replication. Additionally, qPCR, Western blotting, and TCID$_{50}$ analysis demonstrated that overexpression of NME2 protein in DF-1 cells substantially reduced the ARV replication. Inhibition of NME2 protein expression via siRNA interference markedly improved replication of the ARV. These results suggest that NME2 is a host cellular factor that may negatively affect ARV replication in the infected cells.

The mammalian NME2 gene encodes the B subunit of NDPK, which has nucleoside diphosphate kinase activity. It catalyzes the conversion of nucleoside diphosphate to the corresponding nucleoside triphosphate, participates in cellular processes (including energy metabolism and microtubule depolymerization), and may affect tumor infiltration and metastasis[16-18]. Currently, the protein function of avian-derived NME2 is unclear. Overexpression and silencing of NME2 in avian-derived DF-1 cells negatively affected ARV replication. Through microscopic observation, we found no apparent changes in cells when the NME2 protein is overexpressed or silenced compared to the control groups. In addition we do not notice substantial changes in cell proliferation or cytotoxicity measured by commercially available kits. Therefore, we suggest that the NME2 protein may not affect ARV replication by directly impairing cell function. However, the exact mechanism of action of NME2 protein on ARV replication should be further investigated.

In our study, overexpression of the NME2 protein inhibited ARV replication. These findings indicate that the interaction of NME2 protein with the ARV σA may affect ARV replication by impairing the function of σA protein in virus life cycle. The ARV σA protein could activate the PI3K/Akt intracellular signaling pathway and increase expression of phosphorylated Akt (P-Akt) in cells, thereby inhibiting apoptosis in infected cells, which is in favor for ARV infection and replication[8]. Following the interaction between the overexpressed NME2 protein and ARV σA protein, it is not known whether the σA protein loses its function of activating the PI3K/Akt signaling pathway, thereby inhibiting ARV replication.

By inducing the activation of dsRNA-dependent protein kinase, interferon (IFN) blocks viral protein synthesis using mRNA in host cells and produces an antiviral effect; this is an important mechanism by which cells can resist viral infection. The ARV σA protein binds to viral dsRNA in an irreversible manner, which inhibits the activation of dsRNA-dependent protein kinases, thereby allowing the virus to resist the antiviral effect of interferon and to defend against or escape the host's antiviral immune response[6, 7]. Nonetheless, it is unknown whether overexpressed NME2 protein could result in σA dysfunction unable to bind dsRNA, leading to activation of dsRNA-dependent protein kinases and inhibition of ARV replication. Further investigation is warranted to further elucidate the mechanism of action of NME2 protein on ARV replication.

The ARV S1 gene-encoded σC protein is located on the outer part of the capsid and functions in the identification and binding of the virus to the target cell[19, 20]. The σC protein is the most variable ARV protein in sequence and could induce production of neutralizing antibodies[21-23]. For this reason, in the over-expression and gene silencing assays, we analyzed viral mRNA expression of ARV genes by qPCR, detecting not only the mRNAs transcribed from σA gene, but also those from ARV σC gene (data not shown). The results were quite similar, indicating the high accuracy of our results.

In the overexpression assay, NME2 expression yielded two bands in Western immunoblotting. The smaller protein band represents an endogenous NME2 protein produced from DF-1 cells, whereas the larger protein band represents the NME2-Myc fusion protein. The NME2-Myc is a fusion protein and its molecular weight is slightly larger than that of the endogenous NME2. The presence of the two bands with different molecular weights did not affect determination of total NME2 protein expression.

Multiple pairs of siRNAs were designed and used in the gene silencing assays, and we found that siNME2-59F and siNME2-59R substantially inhibited NME2 protein expression, with the silencing effect reaching 50%. Therefore, siNME2-59F and siNME2-59R were used in our experiments to ensure the reliability.

Our results demonstrated that the negative effect on NME2 protein on ARV replication is specific and reproducible.

Reference

[1] JONES R C. Avian reovirus infections. Rev Sci Tech, 2000, 19(2): 614-625.

[2] VAN DER HEIDE L. The history of avian reovirus. Avian Dis, 2000, 44(3): 638-641.

[3] XIE L, XIE Z, WANG S, et al. Altered gene expression profiles of the MDA5 signaling pathway in peripheral blood lymphocytes of chickens infected with avian reovirus. Arch Virol, 2019, 164(10): 2451-2458.

[4] BODELON G, LABRADA L, MARTINEZ-COSTAS J, et al. The avian reovirus genome segment S1 is a functionally tricistronic gene that expresses one structural and two nonstructural proteins in infected cells. Virology, 2001, 290(2): 181-191.

[5] SCHNITZER T J. Protein coding assignment of the S genes of the avian reovirus S1133. Virology, 1985, 141(1): 167-170.

[6] YIN H S, SHIEN J H, LEE L H. Synthesis in Escherichia coli of avian reovirus core protein varsigmaA and its dsRNA-binding activity. Virology, 2000, 266(1): 33-41.

[7] MARTINEZ-COSTAS J, GONZALEZ-LOPEZ C, VAKHARIA V N, et al. Possible involvement of the double-stranded RNA-binding core protein sigmaA in the resistance of avian reovirus to interferon. J Virol, 2000, 74(3): 1124-1131.

[8] XIE L, XIE Z, HUANG L, et al. Avian reovirus sigmaA and sigmaNS proteins activate the phosphatidylinositol 3-kinase-dependent Akt signalling pathway. Arch Virol, 2016, 161(8): 2243-2248.

[9] YIN H S, SU Y P, LEE L H. Evidence of nucleotidyl phosphatase activity associated with core protein sigma A of avian reovirus S1133. Virology, 2002, 293(2): 379-385.

[10] CHI P I, HUANG W R, CHIU H C, et al. Avian reovirus sigmaA-modulated suppression of lactate dehydrogenase and upregulation of glutaminolysis and the mTOC1/eIF4E/HIF-1alpha pathway to enhance glycolysis and the TCA cycle for virus replication. Cell Microbiol, 2018, 20(12): e12946.

[11] CAO Y Q, YUAN L, ZHAO Q, et al. Hsp40 protein DNAJB6 interacts with viral NS3 and inhibits the replication of the Japanese encephalitis virus. Int J Mol Sci, 2019, 20(22).

[12] TIAN Y, HUANG W, YANG J, et al. Systematic identification of hepatitis E virus ORF2 interactome reveals that TMEM134 engages in ORF2-mediated NF-kappaB pathway. Virus Res, 2017, 228: 102-108.

[13] WANG R, Du Z, BAI Z, et al. The interaction between endogenous 30S ribosomal subunit protein S11 and Cucumber mosaic virus LS2b protein affects viral replication, infection and gene silencing suppressor activity. PLOS ONE, 2017, 12(8): e182459.

[14] ZHU P, LIANG L, SHAO X, et al. Host cellular protein TRAPPC6ADelta interacts with influenza A virus M2 protein and

regulates viral propagation by modulating M2 trafficking. J Virol, 2017, 91(1).

[15] LIVAK K J, SCHMITTGEN T D. Analysis of relative gene expression data using real-time quantitative PCR and the 2(-Delta Delta C(T)) Method. Methods, 2001, 25(4): 402-408.

[16] BIGGS J, HERSPERGER E, STEEG P S, et al. A Drosophila gene that is homologous to a mammalian gene associated with tumor metastasis codes for a nucleoside diphosphate kinase. Cell, 1990, 63(5): 933-940.

[17] DE LA ROSA A, WILLIAMS R L, STEEG P S. Nm23/nucleoside diphosphate kinase: toward a structural and biochemical understanding of its biological functions. Bioessays, 1995, 17(1): 53-62.

[18] TEE Y T, CHEN G D, LIN L Y, et al. Nm23-H1: a metastasis-associated gene. Taiwan J Obstet Gynecol, 2006, 45(2): 107-113.

[19] SCHNITZER T J, RAMOS T, GOUVEA V. Avian reovirus polypeptides: analysis of intracellular virus-specified products, virions, top component, and cores. J Virol, 1982, 43(3): 1006-1014.

[20] VARELA R, BENAVENTE J. Protein coding assignment of avian reovirus strain S1133. J Virol, 1994, 68(10): 6775-6777.

[21] LIU H J, GIAMBRONE J J. Amplification, cloning and sequencing of the sigmaC-encoded gene of avian reovirus. J Virol Methods, 1997, 63(1-2): 203-208.

[22] LIU H J, LEE L H, HSU H W, et al. Molecular evolution of avian reovirus: evidence for genetic diversity and reassortment of the S-class genome segments and multiple cocirculating lineages. Virology, 2003, 314(1): 336-349.

[23] WOOD G W, NICHOLAS R A, HEBERT C N, et al. Serological comparisons of avian reoviruses. J Comp Pathol, 1980, 90(1): 29-38.

Study of the activation of the PI3K/Akt pathway by the motif of σA and σNS proteins of avian reovirus

Xie Liji, Xie Zhixun, Wang Sheng, Deng Xianwen, and Xie Zhiqin

Abstract

The present study was conducted to determine whether avian reovirus (ARV) activates the phosphatidylinositol 3-kinase-dependent Akt (PI3K/Akt) pathway according to the PXXP or YXXXM motifs of σA and σNS proteins. Gene splicing by overlap extension PCR was used to change the PXXP or YXXXM motifs of the σA and σNS genes. Plasmid constructs that contain mutant σA and σNS genes were generated and transfected into Vero cells, and the expression levels of the corresponding genes were quantified according to immunofluorescence and Western blot analyses. The Akt phosphorylation (P-Akt) profile of the transfected Vero cells was examined by flow cytometry and Western blot. The results showed that the σA and σNS genes were expressed in the Vero cells, and P-Akt expression in the σA mutant groups (amino acids 110~114 and 114~117) was markedly decreased. The results indicated that the σA protein of ARV activates the PI3K/Akt pathway via the PXXP motif. The results of this study reveal the mechanisms by which ARV manipulates the cellular signal transduction pathways, which may provide new ideas for novel drug targets.

Keywords

avian reovirus, PI3K/Akt pathway, σA protein, σNS protein, PXXP motif

Introduction

Avian reovirus (ARV) is one of the most important avian viruses, causing clinical diseases in poultry worldwide and resulting in severe economic losses[1, 2]. ARV-affected flocks commonly suffer viral arthritis or tenosynovitis, runting-stunting syndrome (RSS), enteric disease, immunosuppression and malabsorption syndrome[2-5]. In the early stages of ARV infection, ARV activates the phosphatidylinositol 3-kinase-dependent Akt (PI3K/Akt) signalling axis in Vero cells[6]: a pathway that is associated with cell survival, proliferation, migration, differentiation and apoptosis[7-8].

PXXP or YXXXM/YXXM motifs are present in a number of viral and cellular proteins involved in PI3K signaling and form extended helices that bind to SH domains on the p85 subunit of PI3K[9-15]. Therefore, as our previous research review[16] showed that various amino acid sequence have PXXP/YXXXM/YXXM motifs, σA, σNS, μA, μB and μNS of ARV were speculated to be involved in PI3K signalling. The results showed that σA and σNS-expressing cells had higher P-Akt levels than the pcAGEN-expressing cells and that in the cells expressing other proteins (i. e. μA, μB and μNS), pretreatment with the PI3K inhibitor LY294002 inhibited Akt phosphorylation in σA- and σNS-expressing cells[16]. These results indicate that the σA and σNS proteins can activate the PI3K/Akt pathway.

Mutant the PXXP motif of NS1 protein from Influenza A virus or mutant the YXXM motif of envelope protein of avian leukosis virus results in loss of PI3K/Akt pathway activation[15, 17], According to amino acid sequence analysis of the σA and σNS genes, the PXXP and YXXXM motifs are conserved between different ARV strains. The aim of the current study was to determine whether the σA and σNS proteins affect the activation of the PI3K/Akt pathway *in vivo* via the PXXP or YXXXM motifs. To accomplish this objective, we mutated the PXXP or YXXXM motifs of σA and σNS genes. Plasmid constructs containing mutant σA and σNS genes were generated and transfected into Vero cells, and the expression levels of the σA and σNS genes were quantified according to immunofluorescence and Western blot analysis. The Akt phosphorylation (P-Akt) profiles of the transfected Vero cells were examined by flow cytometry and Western blot analysis.

Materials and methods

Plasmids and primers

The plasmids σA-pcAGEN and σNS-pcAGEN were generated by our lab[16]. Nine pairs of primers were designed to mutate the PXXP or YXXXM motifs of the σA and σNS genes (Table 4-6-1). The primer sequences are presented in Table 4-6-1. The red colours represent the mutant bases. The complete σA gene (1 248 bp) was amplified using the primers σA-F (5'-GATGAT*CTCGAG*GCCACCATGGCGCGTGCCATAT ACGAC-3') and σA-R (5'-ATC*GCGGCCGC*TTAGGCGGTAAAAGTGGCTAGAAC-3'), and the complete σNS gene (1 101 bp) was amplified using the primers σNS-F (5'-GATGAT*CTCGAG*GCCACCATGGACA ACACCGTGCGTGTT-3') and σNS-R (5'-ATC*GCGGCCGC*TTACGCCATCCTAGCTGGAGAGAC-3'). The underlined text indicates Kozak sequences, and the italicized text represents restriction sites (*Xho* I and *Not* I). All primers were synthesized by TaKaRa (Dalian, China).

Table 4-6-1   Primer sequences

Mutant gene	Primer name	Sequence (5'-3')	Amino acids	Amino acid sequence→mutated sequence
σA-MI	σA-M1-F	CAACGGCTGAAGCAGCCTATCCAGGT	55～58	PXXP→AXXA
	σA-M1-R	ACCTGGATAGGCTGCTTCAGCCGTTG		
σA-M2	σA-M2-F	ATGTTCCAAGAGTCTGCACTCCAC	65～69	YXXXM→FXXXA
	σA-M2-R	GTGGAGTGCAGACTCTTGGAACAT		
σA-M3	oA-M3-F	ATGCTGCCGCTGCCGCCGCGTATCAGCCAGC	110～114	PPXXP→AAXXA
	σA-M3-R	GCTGGCTGATACGCGGCGGCAGGGGGAGCAT		
σA-M4	σA-M4-F	ATCCTCCCCCTGCCGCCGCGTATCAGGCAGC	114～117	PXXP→AXXA
	σA-M4-R	GCTGCCTGATACGCGGCGGCAGGGGGAGGAT		
σA-M5	σA-M5-F	ACATGGCTGTTGAGGCTGATG	200～203	PXXP→AXXA
	σA-M5-R	CATCAGCCTCAACAGCCATGT		
σA-M6	σA-M6-F	ATTTCGATCAGCAGGCACGTGCT	207～211	YXXXM→FXXXA
	oA-M6-R	AGCACGTGCCTGCTGATCGAAAT		
σNS-M1	σNS-M1-F	GCTGCGAGTGGCGCTATTGACT	159～162	PXXP→AXXA
	σNS-M2-R	AGTCAATAGCGCCACTCGCAGC		
σNS-M2	σNS-M2-F	TCCCTTTCATGCTTGACGCAGTA	179～183	YXXXM→FXXXA
	σNS-M2-R	TACTGCGTCAAGCATGAAAGGGA		
σNS-M3	σNS-M3-F	CGGTCGTACTCGTCTTCTT	333～336	PXXP→AXXA
	σNS-M3-R	AAGAAGACGAGTACGACCG		

RNA extraction and RT-PCR amplification of the mutant σA and σNS genes

Genomic RNA was extracted from 200 μL of ARV using an EasyPure viral DNA/RNA kit (TransGen, Beijing, China) according to the protocol suggested by the manufacturer.

RT-PCR was performed using an RNA LA PCR kit (TaKaRa). Gene splicing by overlap extension PCR was used to mutate σA and σNS genes. To amplify the σA-M1 gene (σA gene mutant 1), the PCR protocol consisted of three rounds of PCR amplification. The first PCR amplification used σA-F and σA-M1-R primers, with σA-pcAGEN as the template; the second PCR amplification used σA-M1-F and σA-R primers, with σA-pcAGEN as the template. The PCR products of these two amplifications were obtained via gel extraction. The third PCR amplification used σA-F and σA-R primers with the gel extraction purification product of the first and second PCR used as the template. The PCR product was then obtained via gel extraction.

We used the same method to amplify the other mutant genes: σA-M2, σA-M3, σA-M4, σA-M5, σA-M6, σNS-M1, σNS-M2 and σNS-M3.

Recombinant plasmid construction

The mutant σA and σNS gene products were cloned into pMD18-T cloning vectors (TaKaRa) according to the manufacturer's instructions. The constructed recombinant plasmids were designated σA-M1-pMD18-T, σA-M2-pMD18-T, σA-M3-pMD18-T, σA-M4-pMD18-T, σA-M5-pMD18-T, σA-M6-pMD18-T, σNS-M1-pMD18-T, σNS-M2-pMD18-T and σNS-M3-pMD18-T. The plasmids were digested with *Xho* I and *Not* I enzymes (TaKaRa) and were then ligated into the corresponding sites of pcAGEN expression vectors before being transformed into competent *Escherichia coli* cells (DH5α). Positive colonies, which were designated σA-M1-pcAGEN, σA-M2-pcAGEN, σA-M3-pMD18-T, σA-M4-pcAGEN, σA-M5-pcAGEN, σA-M6-pcAGEN, σNS-M1-pcAGEN, σNS-M2-pcAGEN and σNS-M3-pcAGEN, were identified by PCR and double digestion and sequenced by Invitrogen (Guangzhou, China).

Expression of σA and σNS proteins

Plasmids (pcAGEN, σA-pcAGEN, σNS-pcAGEN, σA-M1-pcAGEN, σA-M2-pcAGEN, σA-M3-pMD18-T, σA-M4-pcAGEN, σA-M5-pcAGEN, σA-M6-pcAGEN, σNS-M1-pcAGEN, σNS-M2-pcAGEN and σNS-M3-pcAGEN) were extracted using a plasmid mini kit (Omega Bio-Tek, Norcross, GA). Vero cells were seeded onto 6- or 24-well cell culture plates and were then transfected with 2.5 or 0.5 μg of the appropriate expression vectors using Lipofectamine® 3 000 transfection reagent (Invitrogen) according to the manufacturer's instructions.

Immunofluorescence and Western blot analysis to quantify protein expression

After the cells were seeded onto 24-well plates and allowed to adhere, Vero cells were transfected with various plasmids for 6 h and were then washed three times with PBS. The cells were then fixed with cold methanol for 10 min at room temperature, washed three times with PBS and blocked for 1 h in 10% normal goat serum (Abcam, Cambridge, UK) in PBS containing 0.5% Triton X-100 (Sigma-Aldrich, St Louis, MO). The cells were then incubated with primary Abs against ARV[18] overnight at 4 ℃. Following three 5-min washes with PBST, the cells were incubated with fluorescently labelled secondary Abs (Alexa Fluor; Abcam) for 60 min at room temperature.

Vero cells seeded onto six-well plates were harvested, and their cell lysates were used for Western blot analysis to detect σA and σNS protein expression, as described by Xie et al.[18] Proteins were visualized using an enhanced chemiluminescence reagent (Bio-Rad, Hercules, CA) and were detected using a Bio-Rad ChemiDoc MP Imaging System.

Flow cytometry and Western blot analysis of P-Akt expression

For flow cytometry and Western blot analyses of P-Akt expression, Vero cells seeded onto six-well plates were harvested following transfection with the corresponding plasmids for 6 h. The cells were then subjected to flow cytometry and Western blot analysis to detect P-Akt expression.

For flow cytometry analysis, the transfected cells were detached from the culture plates via incubation with Accutase (Sigma-Aldrich) for 5 min. Then, the cells were washed with PBS and fixed and permeabilized by incubation with Cytofix/Cytoperm solution (BD Biosciences, Franklin Lakes, NJ) at 4 ℃ for 20 min. Next, the cells were washed twice with FACS buffer (0.5% BSA, 0.01% sodium azide in DPBS) and then incubated at 4 ℃ for 30 min with a rabbit Ab (Cell Signaling Technology, Danvers, MA) against P-Akt. After this incubation, the cells were washed twice with FACS buffer and then incubated at 4 ℃ for 30 min with goat anti-rabbit IgG conjugated to Alexa Fluor® 488 (Abcam). After staining, the cells were washed twice with FACS buffer and re-suspended in FACS buffer for analysis. The samples were then analysed via flow cytometry (Beckman Coulter, Brea, CA).

Vero cells seeded onto six-well plates were harvested, and their cell lysates were subjected to Western blot analysis to detect P-Akt expression, as described by Wang et al. [19].

All flow cytometry and Western blot analyses were repeated three times.

Results

Recombinant plasmid construction

The recombinant plasmids σA-M1-pcAGEN, σA-M2-pcAGEN, σA-M3-pMD18-T, σA-M4-pcAGEN, σA-M5-pcAGEN, σA-M6-pcAGEN, σNS-M1-pcAGEN, σNS-M2-pcAGEN and σNS-M3-pcAGEN were first subjected to PCR amplification (Figure 4-6-1) and double digestion (Figure 4-6-2). Then, DNA sequencing was performed to ensure that the recombinant plasmids contained intact mutant σA and σNS genes, validating that the recombinant plasmids were successfully constructed.

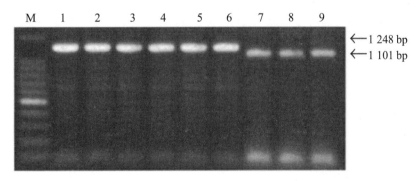

Lane M: 100-bp DNA ladder; lane 1: σA-M1-pcAGEN; lane 2: σA-M2-pcAGEN; lane 3: σA-M3-pcAGEN; lane 4: σA-M4-pcAGEN; lane 5: σA-M5-pcAGEN; lane 6: σA-M6-pcAGEN; lane 7: σNS-MI-pcAGEN; lane 8: σNS-M2-pcAGEN; lane 9: σNS-M3-pcAGEN.

Figure 4-6-1   Identification of the recombinant plasmids by PCR

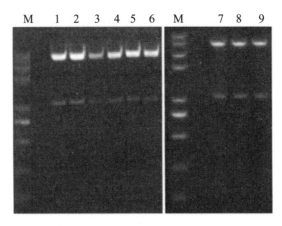

Lane M: Trans2K Plus II DNA Marker; lane 1: σA-M1-pcAGEN; lane 2: σA-M2-pcAGEN; lane 3: σA-M3-pcAGEN; lane 4: σA-M4-pcAGEN; lane 5: σA-M5-pcAGEN; lane 6: σA-M6-pcAGEN; lane 7: σNS-M1-pcAGEN; lane 8: σNS-M2-pcAGEN; lane 9: GNS-M3-pcAGEN.

Figure 4-6-2　Identification of the recombinant plasmids by double digests

σA and σNS protein expression

As shown by the immunofluorescence and Western blot analyses, the σA and σNS proteins were expressed in the Vero cells 6 h after transfection with the corresponding plasmids (σA-pcAGEN, σA-M1-pcAGEN, σA-M2-pcAGEN, σA-M3-pMD18-T, σA-M4-pcAGEN, σA-M5-pcAGEN, σA-M6-pcAGEN, σNS-pcAGEN, σNS-M1-pcAGEN, σNS-M2-pcAGEN and σNS-M3-pcAGEN; Figures 4-6-3 and 4-6-4). Cells that were not transfected cells and transfected with pcAGEN served as negative controls cells (Figure 4-6-3 A and B and Figure 4-6-4, lanes 1 and 2).

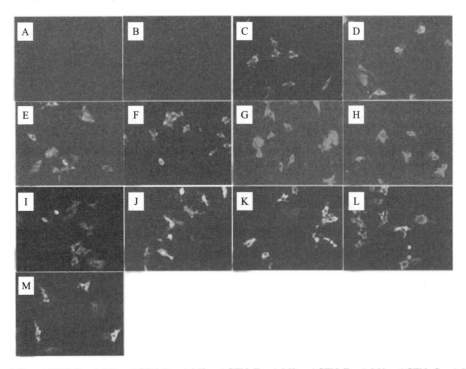

A: Negative control; B: pcAGEN; C: σA-M1-pcAGEN; D: σA-M2-pcAGEN; E: σA-M3-pcAGEN; F: σA-M4-pcAGEN; G: σA-M5-pCAGEN; H: σA-M6-pcAGEN; I : σA-pcAGEN; J: σNS-M1-pcAGEN; K: σNS-M2-pcAGEN; L: σNS-M3-pcAGEN; M: σNS-pcAGEN.

Figure 4-6-3　IFA analysis of the expression of recombinant plasmids in transfected Vero cells

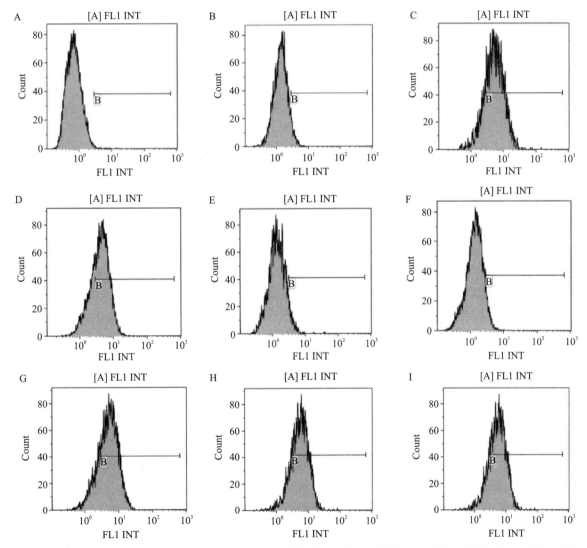

Lane 1: Negative control; lane 2: pcAGEN; lane 3: σNS-M1-pcAGEN; lane 4: σNS-M2-pcAGEN; lane 5: σNS-M3-pcAGEN; lane 6: σNS-pcAGEN; lane 7: σA-M1-pcAGEN; lane 8: σA-M2-pcAGEN; lane 9: σA-M3-pcAGEN; lane 10: σA-M4-pcAGEN; lane 11: σA-M5-pcAGEN; lane 12: σA-M6-pcAGEN; lane 13: σA-pcAGEN.

Figure 4-6-4   Western blot analysis of gene expression

σA and σNS proteins activate the PI3K/Akt signalling pathway

Figure 4-6-5 shows that σA- M3- and σA-M4- expressing cells had the lowest levels of P-Akt compared to σA-pcAGEN, σA-M1-pcAGEN, σA-M2-pcAGEN, σA-M5-pcAGEN and σA-M6-pcAGEN. However, these cells had the same levels of P-Akt as negative control and pcAGEN-expressing cells. Figure 4-6-6 shows the same results as Figure 4-6-4. After mutation at amino acids 110~114 (PPXXP → AAXXA) and 114~117 (PXXP → AXXA), the σA gene lost the capacity to increase P-Akt expression in Vero cells.

A: Negative control; B: pcAGEN; C: σA-pcAGEN; D: σA-M1-pcAGEN; E: σA-M2-pcAGEN; F: σA-M3-pcAGEN; G: σA-M4-pcAGEN; H: σA-M5-pcAGEN; I: σA-M6-pcAGEN.

Figure 4-6-5   Flow cytometry analysis of P-Akt expression levels

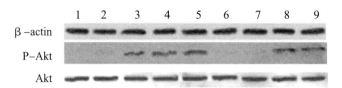

Lane 1: Negative control; lane 2: pcAGEN; lane 3: σA-pcAGEN; lane 4: σA-M1-pcAGEN; lane 5: σA-M2-pcAGEN; lane 6: σA-M3-pcAGEN; lane 7: σA-M4-pcAGEN; lane 8: σA-M5-pcAGEN; lane 9: σA-M6-pcAGEN.

Figure 4-6-6　Western blot analysis of P-Akt expression levels

The σNS-M1-, σNS-M2- and σNS-M3- expressing cells had similar P-Akt levels compared to σNS-pcAGEN, but these cells had increased levels of P-Akt expression compared to negative control cells and pcAGEN- expressing cells (data not shown). The mutations at amino acids 159~162 (PXXP→AXXA), 179~183 (YXXXM→FXXXA) and 333~336 (PXXP→AXXA) did not affect the capacity of the σNS gene to increase P-Akt expression in Vero cells.

Discussion

The PI3K/Akt pathway is involved in numerous cellular processes, such as cell proliferation, differentiation and survival[7, 8]. PI3K is activated by the binding of autophosphorylated tyrosine kinase receptors or non-receptor tyrosine kinases to the SH2 and SH3 domains of its regulatory/adaptor subunit p85[19]. This binding is mediated by PXXP and YXXXM/YXXM motifs and activates the PI3K/Akt pathway[9-15]. As reported by Shin[15], the Influenza A virus NS1 protein activates the PI3K/Akt pathway by direct interaction with the p85 subunit of PI3K (SH2-binding motif YXXXM and SH3-binding motif PXXP).

The S2-encoded protein σA, a component of the inner core shell, displays anti-IFN activity by preventing the activation of the dsRNA-dependent protein kinase PKR; this activity is likely linked to its capacity to bind and sequester dsRNA[20, 21]. The non-structural protein σNS is encoded by the ARV S4 genome segment. As a non-structural RNA-binding protein that accumulates in viral factories of ARV-infected cells, σNS is a likely candidate for playing key roles in RNA packaging and replication[22].

Our previous works showed that the σA and σNS proteins of ARV activate the PI3K/Akt pathway[16]. According to amino acid sequence analysis of the σA and σNS genes, σA contains four PXXP and two YXXXM motifs, and σNS contains one YXXXM and two PXXP motifs. Mutation of the PXXP motif of NS1 protein from influenza A virus or mutation of the YXXM motif of envelope protein from avian leukosis virus will result in loss of PI3K/Akt pathway activation[15, 17]. We speculate that the mutant PXXP and YXXXM motifs for the σA and σNS genes of ARV may also affect the activation of the PI3K/Akt pathway.

Gene splicing by overlap extension PCR was used to change the PXXP or YXXXM motifs of the σA and σNS genes and to construct the mutant σA and σNS recombinant plasmids. The mutant σA and σNS recombinant plasmids were transfected into Vero cells. The P-Akt profle of the transfected Vero cells was examined by flow cytometry and Western blot analysis.

In this study, after amino acids 110~114 (PPXXP→AAXXA) and 114~117 (PXXP→AXXA) were mutated, the σA gene lost its capacity to improve P-Akt expression in Vero cells, and the σA protein of ARV activated the PI3K/Akt pathway via the PXXP motif. PXXP motif of the σA protein of ARV is involved in PI3K signalling, and the mutant PXXP motif of σA affects PI3K/Akt pathway activation. These results are similar to those of a previous study[16]. However, whether this activation was a result of the σA protein binding to the p85 subunit via interactions with PXXP motifs, as occurs in other viruses[11, 15], requires further study.

We also demonstrated that the mutations at amino acids 159～162 (PXXP→AXXA), 179～183 (YXXXM→FXXXA) and 333～336 (PXXP→AXXA) did not affect the capacity of the σNS gene to increase P-Akt expression in Vero cells. The σNS protein may activate the PI3K/Akt pathway via more than one motif, other motifs or σNS interaction with the P85 subunit according to the middle protein. These hypotheses must be further studied.

Moreover, the relationship between the p110 catalytic subunit of PI3K and the PXXP or YXXXM/YXXM motifs of the σA/σNS genes will be further studied, and these research results will be published in the future.

Although the exact mechanisms by which PI3K/Akt regulates ARV replication and other biological functions of PI3K/Akt in virus infection remain uncharacterized, our study reveals the mechanism underlying PI3K/Akt activation and adds a novel aspect to the functions of the σA and σNS proteins of ARV.

References

[1] JONES R C. Avian reovirus infections. Rev Sci Tech, 2000,19(2):614-625.

[2] VAN DER HEIDE L. The history of avian reovirus. Avian Dis, 2000,44(3):638-641.

[3] HIERONYMUS D R, VILLEGAS P, KLEVEN S H. Identification and serological differentiation of several reovirus strains isolated from chickens with suspected malabsorption syndrome. Avian Dis, 1983,27(1):246-254.

[4] STERNER F J, ROSENBERGER J K, MARGOLIN A, et al. In vitro and in vivo characterization of avian reoviruses. II. Clinical evaluation of chickens infected with two avian reovirus pathotypes. Avian Dis, 1989,33(3):545-554.

[5] Van de ZANDE S, KUHN E M. Central nervous system signs in chickens caused by a new avian reovirus strain: a pathogenesis study. Vet Microbiol, 2007,120(1-2):42-49.

[6] LIN P Y, LIU H J, LIAO M H, et al. Activation of PI 3-kinase/Akt/NF-kappaB and Stat3 signaling by avian reovirus S1133 in the early stages of infection results in an inflammatory response and delayed apoptosis. Virology, 2010,400(1):104-114.

[7] DATTA S R, BRUNET A, GREENBERG M E. Cellular survival: a play in three Akts. Genes Dev, 1999,13(22):2905-2927.

[8] CANTLEY L C. The phosphoinositide 3-kinase pathway. Science, 2002,296(5573):1655-1657.

[9] THORPE L M, YUZUGULLU H, ZHAO J J. PI3K in cancer: divergent roles of isoforms, modes of activation and therapeutic targeting. Nat Rev Cancer, 2015,15(1):7-24.

[10] SONGYANG Z, SHOELSON S E, CHAUDHURI M, et al. SH2 domains recognize specific phosphopeptide sequences. Cell, 1993,72(5):767-778.

[11] PAWSON T. Protein modules and signalling networks. Nature, 1995,373(6515):573-580.

[12] KAY B K, WILLIAMSON M P, SUDOL M. The importance of being proline: the interaction of proline-rich motifs in signaling proteins with their cognate domains. FASEB J, 2000,14(2):231-241.

[13] ZHANG B, SPANDAU D F, ROMAN A. E5 protein of human papillomavirus type 16 protects human foreskin keratinocytes from UV B-irradiation-induced apoptosis. J Virol, 2002,76(1):220-231.

[14] STREET A, MACDONALD A, CROWDER K, et al. The Hepatitis C virus NS5A protein activates a phosphoinositide 3-kinase-dependent survival signaling cascade. J Biol Chem, 2004,279(13):12232-12241.

[15] SHIN Y K, LIU Q, TIKOO S K, et al. Influenza A virus NS1 protein activates the phosphatidylinositol 3-kinase (PI3K)/Akt pathway by direct interaction with the p85 subunit of PI3K. J Gen Virol, 2007,88(Pt 1):13-18.

[16] XIE L, XIE Z, HUANG L, et al. Avian reovirus sigmaA and sigmaNS proteins activate the phosphatidylinositol 3-kinase-dependent Akt signalling pathway. Arch Virol, 2016,161(8):2243-2248.

[17] FENG S, LI J, CAO W, et al. Effect of YXXM motif on viral replication of subgroup J avian leukosis virus. Acta Microbiological Sinica, 2011,51(12):1663-1668.

[18] XIE Z, QIN C, XIE L, et al. Recombinant protein-based ELISA for detection and differentiation of antibodies against avian reovirus in vaccinated and non-vaccinated chickens. J Virol Methods, 2010,165(1):108-111.

[19] WANG X, ZHANG H, ABEL A M, et al. Role of phosphatidylinositol 3-kinase (PI3K) and Akt1 kinase in porcine reproductive and respiratory syndrome virus (PRRSV) replication. Arch Virol, 2014,159(8):2091-2096.

[20] MARTINEZ-COSTAS J, GONZALEZ-LOPEZ C, VAKHARIA V N, et al. Possible involvement of the double-stranded RNA-binding core protein sigmaA in the resistance of avian reovirus to interferon. J Virol, 2000,74(3):1124-1131.

[21] GONZALEZ-LOPEZ C, MARTINEZ-COSTAS J, ESTEBAN M, et al. Evidence that avian reovirus sigmaA protein is an inhibitor of the double-stranded RNA-dependent protein kinase. J Gen Virol, 2003,84(Pt 6):1629-1639.

[22] TOURIS-OTERO F, MARTINEZ-COSTAS J, VAKHARIA V N, et al. Characterization of the nucleic acid-binding activity of the avian reovirus non-structural protein sigmaNS. J Gen Virol, 2005,86(Pt 4):1159-1169.

Altered gene expression profiles of the MDA5 signaling pathway in peripheral blood lymphocytes of chickens infected with avian reovirus

Xie Liji, Xie Zhixun, Wang Sheng, Huang Jiaoling, Deng Xianwen, Xie Zhiqin, Luo Sisi, Zeng Tingting, Zhang Yanfang, and Zhang Minxiu

Abstract

Avian reovirus (ARV) is a member of the genus *Orthoreovirus* in the family Reoviridae and causes a severe syndrome including viral arthritis that leads to considerable losses in the poultry industry. Innate immunity plays a significant role in host defense against ARV. Here, we explored the interaction between ARV and the host innate immune system by measuring mRNA expression levels of several genes associated with the MDA5 signaling pathway. The results showed that expression peaks for MDA5, MAVS, TRAF3, TRAF6, IRF7, IKKε, TBK1 and NF-κB occurred at 3 days postinfection (dpi). Moreover, type I IFN (IFN-α, IFN-β) and IL-12 expression levels peaked at 3 dpi, while type II IFN (IFN-γ), IL-6, IL-17 and IL-18 expression reached a maximum level at 1 dpi. IL-8 changed at 5 dpi, and IL-1β and TNF-α changed at 7 dpi. Interestingly, several key IFN-stimulated genes (ISGs), including IFITM1, IFITM2, IFITM5, Mx1 and OASL, were simultaneously upregulated and reached maximum values at 3 dpi. These data indicate that the MDA5 signaling pathway and innate immune cytokines were induced after ARV infection, which would contribute to the ARV-host interaction, especially at the early infection stage.

Keywords

ARV, MDA5 signaling pathway, mRNA expression level, peripheral blood lymphocytes

Introduction

Avian reovirus (ARV) belongs to the genus *Orthoreovirus*, one of 15 genera of the family Reoviridae. Clinically, disease in infected chickens is characterized by viral arthritis/tenosynovitis, chronic respiratory disease, malabsorption syndrome, and especially immunosuppression, resulting in secondary infection by other pathogens[1-3]. ARV is highly prevalent in domestic chicken flocks and results in substantial economic losses throughout the global poultry industry. Therefore, there is a critical need to better understand the pathogenic mechanism of this virus.

The innate immune system is the first line of defense against invading pathogens. It provides a more rapid immune response than adaptive immunity but lacks memory and specificity. Host cells recognize pathogens by sensing viral signature molecules such as ssRNA, dsRNA and envelope proteins via pattern recognition receptors (PRRs). Such receptors include Toll-like receptors (TLRs), RIG-I-like receptors (RLRs), and NOD-like receptors (NLRs)[4]. The RLRs include retinoic-acid-inducible gene I (RIG-I) and melanoma differentiation-associated gene 5 (MDA5), mediating signaling and transcriptional events initiated by various target cells, which are critical in responding to viral infection. Notably, RIG-I is absent in chickens[5]. Upon sensing viral infection, particular PRRs interact with interferon (IFN)-β promoter stimulator-1 (also known as MAVS) to activate members of the IKK protein kinase family. The canonical IKK family members IKKα and IKKβ mediate the phosphorylation and degradation of I-κB. The noncanonical IKK family members TBK1 and IKBKE activate IFN regulatory factors (IRF3/7) to form a functional homodimer or heterodimer[6, 7]. Then, the transcription factors IRF and NF-κB translocate to the nucleus to stimulate the expression of IFN and proinflammatory cytokines. Meanwhile, IFN induces the downstream synthesis of antiviral proteins encoded by IFN-stimulated genes (ISGs). ISG proteins such as IFITM, Mx1 and OASL play key roles in host immune defense against viral infections[8, 9]. Therefore, the IFN-activated signaling pathway is an important component of the innate immune system and has been targeted for clinical antiviral treatment.

Most studies of ARV have focused mainly on pathogen isolation and identification, genome and structural protein analysis, apoptosis, and diagnosis. However, the details of the molecular mechanism of ARV pathogenesis remain unknown. The outcome of viral infection is primarily determined by the interplay between the virus and the innate immune system. Therefore, insights into how ARV interacts with the host innate immune system are necessary to understand the nature of ARV pathogenesis.

Results from previous studies have suggested that RIG-I signaling plays a crucial role in restricting influenza A virus tropism and regulating the host immune responses[10]. Avian leukosis virus subgroup J (ALV-J) infection is primarily recognized by chicken TLR7 and MDA5 at the early and late infection stage, respectively[11]. In addition, avian Tembusu virus infection triggers a host innate immune response through MDA5- and TLR3-dependent signaling that controls IFN production and thereby induces effective antiviral immunity[12]. However, little information is available on the role of the innate immune system in the regulation of ARV infection. In this study, we investigated whether ARV induces MDA5 signaling activation and tried to obtain information about the molecular mechanism of the ARV-mediated MDA5 signaling pathway.

Materials and methods

Virus

The virus strain ARV S1133 (AV2311) used in this study was purchased from the China Institute of Veterinary Drug Control, Beijing, China. The virus was propagated in chicken embryos before infection. The virus was titrated in DF1 cells, and the viral titers were determined according to Reed-Muench method.

Animal experiments

The specific-pathogen-free (SPF) chicks used in this study were hatched from SPF white leghorn chicken eggs (Merial, Beijing, China) and housed in SPF chicken isolators (Suzhou Fengshi Laboratory Animal Equipment, Suzhou, China). The chicks were provided food and water *ad libitum* for 7 days after hatching. Thirty SPF chickens were randomly divided into two groups, with fifteen chickens in each group. The footpads of SPF chicks in the treatment group were injected with 10^4 $TCID_{50}$ of ARV S1133 in a volume of 100 μL. Chicks in the control group were injected with 100 μL of phosphate-buffered saline (PBS). After infection, chicks of both groups were observed for clinical signs throughout the course of the experiment. At 0, 1, 3, 5 and 7 days postinfection (dpi), three chickens from each group were sacrificed, and peripheral blood was collected for preparation of peripheral blood lymphocyte for further analysis.

Quantitative reverse transcription polymerase chain reaction (qRT-PCR)

Peripheral blood samples were obtained from each chicken at the desired time points post-ARV-infection for testing the course of RLR and cytokine expression and for analysis of ARV replication by assessment of ARV σC gene transcription. Peripheral blood lymphocytes (PBLCs) were prepared using Lymphocyte Separation Medium (Solarbio, Beijing, China). Total RNA was extracted from lymphocytes using TRIzol Reagent (Life Technologies, Carls-bad, CA, USA) according to the manufacturer's instructions. The concentration and purity of the total RNA were determined using a NanoDrop ND 1 000 spectrophotometer (Thermo Scientific, Boston, MA, USA). The integrity of the total RNA was assessed by visualization of the 28S and 18S rRNA bands. RNA samples were mixed with RNA loading buffer (+EB) (TaKaRa, Beijing, China), and electrophoresis was carried out at 80 V for 60 min using a 1% agarose gel. The RNA samples were stored at −80 ℃ until use.

Specific primer pairs for genes of the innate immune system were designed using sequence data from GenBank and synthesized by Invitrogen (Guangzhou, China). The specific primers for target genes used in this study are listed in Table 4-7-1. The 20-μL qRT-PCR mixture contained 10 μL of SYBR Premix Ex Taq Ⅱ system (TaKaRa, Beijing, China), 0.4 μM specific primers (final concentration) and 2 μL of cDNA template. The thermal cycling conditions were as follows: denaturation at 95 ℃ for 10 s, followed by 40 cycles of 95 ℃ for 10 s and annealing at 60 ℃ for 15 s, and extension at 72 ℃ for 15 s. A melting curve analysis was performed at 95 ℃ for 0 s, 65 ℃ for 20 s, and 95 ℃ for 0 s, and the temperature was increased at a rate of 0.1 ℃/s.

Table 4-7-1　Primers used in this study

Gene	Primer sequence (5'-3')	Reference
ARV σC	F: CGGAAACTCCACTGCCATCT, R: CGTGCCATCCACCGAAAT	
MDA5	F: CAGCCAGTTGCCCTCGCCTCA, R: AACAGCTCCCTTGCACCGTCT	He et al. [13]
MAVS	F: CCTGACTCAAACAAGGGAAG, R: AATCAGAGCGATGCCAACAG	He et al. [13]
TRAF3	F: GGACGCACTTGTCGCTGTTT, R: CGGACCCTGATCCATTAGCAT	He et al. [13]
TRAF6	F: GATGGAGACGCAAAACACTCAC, R: GCATCACAACAGGTCTCTCTTC	He et al. [13]
IKKε	F: TGGATGGGATGGTGTCTGAAC, R: TGCGGAACTGCTTGTAGATG	He et al. [13]
TBK1	F: AAGAAGGCACACATCCGAGA. R: GGTAGCGTGCAAATACAGC	He et al. [13]
NF-κB p65	F: CATTGCCAGCATGGCTACTAT, R: TTCCAGTTCCCGTTTCTTCAC	He et al. [13]
IRF3/7	F: ACACTCCCACAGACAGTACTGA, R: TGTGTGTGCCCACAGGGTTG	He et al. [13]
TRIF	F: TTCAGCCATTCTCCGTCCTC, R: GCCAATGATGCTTCCACAG	He et al. [13]
IFN-α	F: ATGCCACCTTCTCTCACGAC, R: AGGCGCTGTAATCGTTGTCT	Kapczynski et al. [14]
IFN-β	F: ACCAGGATGCCAACTTCT, R: TCACTGGGTGTTGAGACG	Chen et al. [12]
IFN-γ	F: ATCATACTGAGCCAGATTGTTTCG, R: TCTTTCACCTTCTTCACGCCAT	
IL-6	F: TGGTGTAAATCCCGATGAAG, R: GGCACTGAAACTCCTGGTCT	
IL-8	F: CACAGCTCCACAAAACCTCA, R: GTCCTACCTTGCGACAGAGC	He et al. [13]
IL-12	F: TCTGCTAAGACCCACGAGA, R: TTGACCGTATCATTTGCCCAT	He et al. [13]
IL-17	F: CTCCTCTGTTCAGACCACTGC, R: ATCCAGCATCTGCTTTCTTGA	
IL-18	F: CAGCGTCCAGGTAGAAGATAAG, R: TCCTCAAAGGCCAAGAACAT	
IL-1β	F: AGAAGAAGCCTCGCCTGGAT, R: CCTCCGCAGCAGTTTGGT	

continued

Gene	Primer sequence (5'-3')	Reference
TNF-α	F: GGACAGCCTATGCCAACAAG, R: TCTTTCACCTTCTTCACGCCAT	
IFITM1	F: AGCACACCAGCATCAACATGC, R: CTACGAAGTCCTTGGCGATGA	
IFITM2	F: AGGTGAGCATCCCGCTGCAC, R: ACCGCCGAGCACCTTCCAGG	
IFITM3	F: GGAGTCCCACCGTATGAAC, R: GGCGTCTCCACCGTCACCA	Smith et al. [15]
IFITM5	F: CCAAGCATTTCTGATAACC, R: TGAGCAGACCCTTTTATACT	Hang et al. [16]
Mxl	F: AACGCTGCTCAGGTCAGAAT, R: GTGAAGCACATCCAAAAGCA	Abdallah and Hassanin[17]
OASL	F: GCGGTGAAGCAGACGGTGAA, R: CGATGATGGCGAGGATGTG	
GAPDH	F: GCACTGTCAAGGCTGAGAACG, R: GATGATAACACGCTTAGCACCAC	

Statistical analysis

The Ct values of each target gene and the internal control were obtained from the qRT-PCR assay. The relative mRNA expression of each target gene at different time points was calculated based on the Ct values of the house-keeping gene glyceraldehyde 3-phosphate dehydrogenase (GAPDH) using the $2^{-\Delta\Delta Ct}$ method. The correlation between the measured gene expression level and ARV replication in cells was assessed by determining the correlation coefficient in Microsoft Excel software (Redmond, WA, USA). Statistical significance was analyzed using Student's t-test, and the values were considered statistically significant at $P < 0.05$ and very significant at $P < 0.01$. All of the experiments were performed in triplicate.

Results

Kinetics of viral loads in PBLCs in response to ARV S1133 infection

Initially, to investigate the replication kinetics of ARV S1133 in PBLCs, ARV σC gene expression was quantified by qRT-PCR. As shown in Figure 4-7-1, the viral mRNA expression level increased beginning at 1 dpi, peaked at 5 dpi (4.33-fold change compared to that at 0 dpi, $P < 0.01$), and subsequently decreased at 7 dpi (1.14-fold change, $P < 0.01$). The results demonstrate that ARV achieved infection and replication in SPF chickens at an infectious dose of 10^4 TCID$_{50}$.

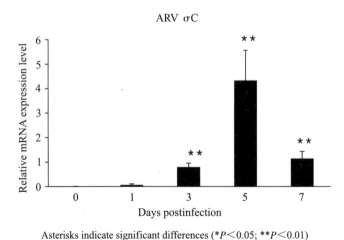

Asterisks indicate significant differences (*$P < 0.05$; **$P < 0.01$)

Figure 4-7-1　Kinetics of ARV replication in the PBLCs of SPF chickens

Changes in the mRNA expression of MDA5, signaling proteins and effector molecules

To assess whether ARV infection could induce host innate immune responses *in vivo*, each seven-day-old SPF chick was inoculated with 10^4 TCID$_{50}$ of ARV S1133 by footpad infection, and PBLCs were collected at the indicated times to examine the expression of MDA5 and associated signaling proteins. Notably, the expression of MDA5 was induced quickly at 1 dpi (4. 83-fold change, $P<0.01$), reached a maximum value at 3 dpi (13.64-fold change, $P<0.01$) and then declined (Figure 4-7-2). Similarly, the expression of some genes associated with the MDA5 signaling pathway, including MAVS, TRAF3, TRAF6, IKKε, TBK1, NF-κB, IRF3/7, and TRIF, peaked at 3 dpi and declined gradually (Figure 4-7-2).

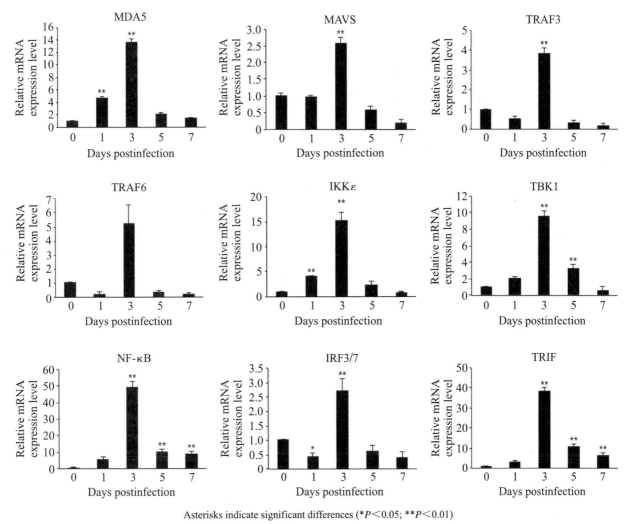

Asterisks indicate significant differences (*$P<0.05$; **$P<0.01$)

Figure 4-7-2   Relative mRNA expression levels of MDA5 and associated signaling molecules in ARV-infected PBLC specimens

Furthermore, we examined whether activation of MDA5 and signaling proteins regulated the production of type Ⅰ and type Ⅱ IFN and proinflammatory cytokines during ARV infection. As expected, the expression of IFN-α and IFN-β was significantly induced, peaked at 3 dpi (1.97- and 1.37-fold change, respectively) and gradually declined (Figure 4-7-3). Type Ⅱ IFN (IFN-γ) was also significantly induced but peaked at 1 dpi. The proinflammatory cytokine IL-12 also reached its maximum value at 3 dpi. Proinflammatory cytokines, including IL-6, IL-8, IL-17, IL-18, IL-1β and TNF-α, exhibited different expression patterns (Figure 4-7-3). IL-6, IL-17 and IL-18 reached their maximum values at 1 dpi. IL-8 reached its maximum value at 5 dpi. IL-1β and TNF-α reached their maximum values at 7 dpi.

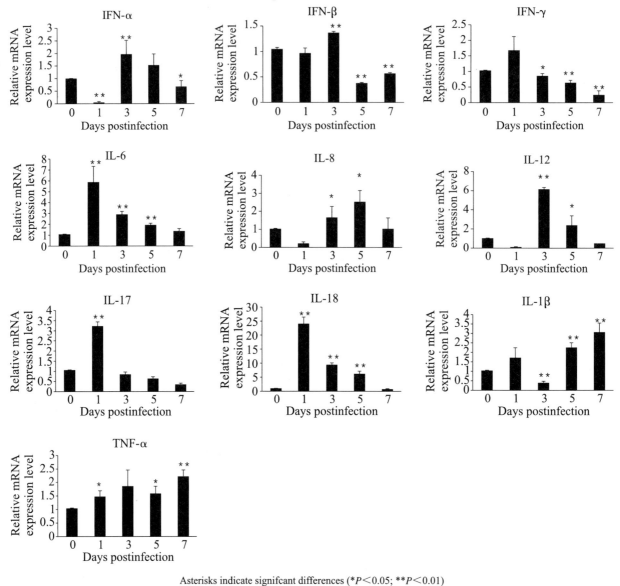

Asterisks indicate signifcant differences (*$P<0.05$; **$P<0.01$)

Figure 4-7-3　Changes in the gene expression levels of type Ⅰ IFN and cytokines in ARV-infected PBLCs

In addition, we also assessed whether the expression levels of multiple ISGs are regulated during ARV infection. Several key ISGs, including IFITM1, IFITM2, IFITM3, Mxl and OASL, were also examined by qRT-PCR (Figure 4-7-4). Similarly, ARV infection resulted in robust expression of these ISGs. The expression of all ISGs examined peaked at 3 dpi ($P<0.01$), except for IFITM2, which peaked at 5 dpi ($P<0.01$).

Taken together, these data suggest that ARV infection can trigger the innate immune response *in vivo*.

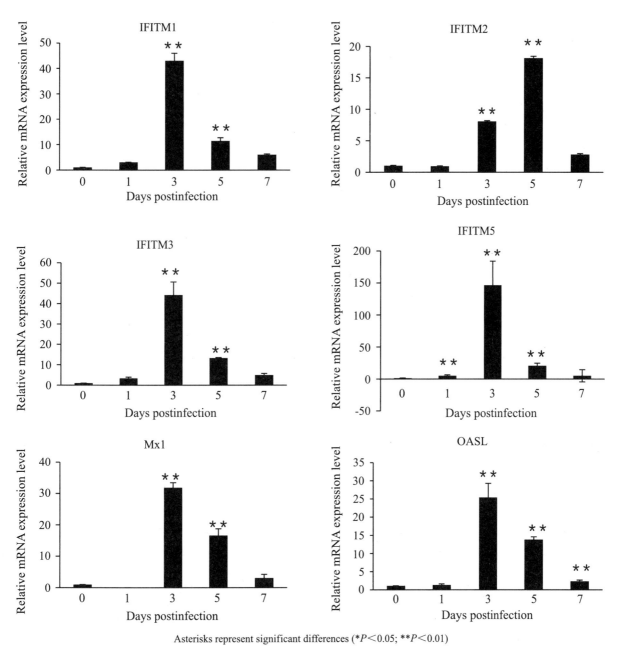

Asterisks represent significant differences ($*P < 0.05$; $**P < 0.01$)

Figure 4-7-4   Expression of ISGs in the PBLCs of SPF chickens infected with ARV

Discussion

Although ARV was identified long ago, its pathogenesis is poorly understood. In previous studies, ARV infection has been demonstrated to be associated with MAPK kinase (MKK) 3/6-MAPK p38 signaling, which is beneficial for ARV replication, and the p53 and mitochondria-mediated pathway, which is important for ARV-induced apoptosis, together with other signaling pathways involved in apoptosis[18]. However, information about the effects of the host innate immune system on ARV infection is scarce. During viral infection and replication, imnate host immune responses serve as the first line of defense and include the production of various IFNs and hundreds of ISGs. Thus, this study explored the host innate immune response following ARV infection. Our data show that ARV infection readily activates host innate immune signaling and causes robust expression of several critical IFNs and ISGs.

The signaling pathways mediated by RLRs, including two essential immune sensors, RIG-I and MDA5, that detect viral dsRNA, may be critical for host cells to detect and respond to virus infection. Whereas RIG-I is absent in chickens[5], MDA5 may be able to compensate and play a key role in host innate immunity. Here, the rapid increase in expression of MDA5 in ARV-infected PBLCs corresponding to the kinetics of ARV replication (as determined by σC expression) suggests that MDA5 can be activated by ARV infection and sense and recognize ARV. It has been established that MDA5 can be activated by diverse viruses, including influenza virus, hepatitis C virus and others[11, 12, 13]. After ubiquitination of MDA5 occurs, a pivotal cellular antiviral protein, MAVS, is activated and subsequently recruits MDA5 to the mitochondrial outer membrane as part of a macromolecular signaling complex. This event triggers the activation of IRF3/7 and NF-κB transcription factors, which in turn induce robust type I IFNs and proinflammatory cytokines[19]. We detected downstream adaptors of the MDA5 signaling pathway, including MAVS, TRAF3, TRAF6, IRF3/7, IKKε, TBK1, NF-κB and TRIF, and these adaptor proteins exhibited similar expression patterns, with peak values at 3 dpi.

Signaling induced by viruses can trigger type I IFN production and proinflammatory cytokine expression to orchestrate the immune response against the pathogen. The type I IFN system represents an important innate antiviral response to viral infection. It has been shown that type I IFN can be activated by various viruses, including avian Tembusu virus and dengue virus, and in turn restricts virus replication[9, 12]. The key host factors that are critical for antiviral immunity and for systemic inflammatory reactions include IL-6, IL-8, IL-12, IL-17, IL-18, IL-1β and TNF-α[14]. We detected increased expression of IFN-α, IFN-β and IL-12, with maximal values at 3 dpi, and also induction of IL-6, IL-8, IL-17, IL-18, IL-1β and TNF-α, with different expression patterns. Our data suggest that ARV infection might activate the host innate immune response through the production of IFN and proinflammatory cytokines. We also demonstrated that ARV infection triggers a robust protective host antiviral response associated with the expression of several ISGs, including IFITM1, IFITM2, IFITM3, Mxl and OASL, which have been characterized with respect to their inhibitory functions. All of the ISGs (except IFITM2) reached their peak values at 3 dpi, and this expression pattern is similar to that of the downstream adaptors of the MDA5 signaling pathway and that of type I IFN.

Here, we investigated changes in the expression of genes associated with natural immunity in PBLCs of SPF chickens infected with ARV only at the mRNA level. Despite this limited approach, the effects of ARV infection on kinetic changes in the expression of these genes could be observed to a degree. However, it will be necessary to assess the protein levels and functions of these tested genes to obtain more information on ARV pathogenesis in the future.

References

[1] VAN DER HEIDE L. The history of avian reovirus. Avian Dis, 2000,44(3):638-641.

[2] TENG L, XIE Z, XIE L, et al. Sequencing and phylogenetic analysis of an avian reovirus genome. Virus Genes, 2014,48(2):381-386.

[3] HUANG L, XIE Z, XIE L, et al. A duplex real-time PCR assay for the detection and quantification of avian reovirus and Mycoplasma synoviae. Virol J, 2015,12:22.

[4] HUSSER L, ALVES M P, RUGGLI N, et al. Identification of the role of RIG-I, MDA-5 and TLR3 in sensing RNA viruses in porcine epithelial cells using lentivirus-driven RNA interference. Virus Res, 2011,159(1):9-16.

[5] BARBER M R, ALDRIDGE J J, WEBSTER R G, et al. Association of RIG-I with innate immunity of ducks to influenza. Proc Natl Acad Sci U S A, 2010,107(13):5913-5918.

[6] KAWAI T, TAKAHASHI K, SATO S, et al. IPS-1, an adaptor triggering RIG-I- and Mda5-mediated type I interferon induction. Nat Immunol, 2005,6(10):981-988.

[7] BARBALAT R, EWALD S E, MOUCHESS M L, et al. Nucleic acid recognition by the innate immune system. Annu Rev Immunol, 2011,29:185-214.

[8] SCHOGGINS J W. Interferon-stimulated genes: roles in viral pathogenesis. Curr Opin Virol, 2014,6:40-46.

[9] BRASS A L, HUANG I C, BENITA Y, et al. The IFITM proteins mediate cellular resistance to influenza A H1N1 virus, West Nile virus, and dengue virus. Cell, 2009,139(7):1243-1254.

[10] KANDASAMY M, SURYAWANSHI A, TUNDUP S, et al. RIG-I signaling is critical for efficient polyfunctional T cell responses during influenza virus infection. PLOS Pathog, 2016,12(7):e1005754.

[11] FENG M, DAI M, XIE T, et al. Innate immune responses in ALV-J infected chicks and chickens with hemangioma in vivo. Front Microbiol, 2016,7:786.

[12] CHEN S, LUO G, YANG Z, et al. Avian Tembusu virus infection effectively triggers host innate immune response through MDA5 and TLR3-dependent signaling pathways. Vet Res, 2016,47(1):74.

[13] HE Y, XIE Z, DAI J, et al. Responses of the Toll-like receptor and melanoma differentiation-associated protein 5 signaling pathways to avian infectious bronchitis virus infection in chicks. Virol Sin, 2016,31(1):57-68.

[14] KAPCZYNSKI D R, JIANG H J, KOGUT M H. Characterization of cytokine expression induced by avian influenza virus infection with real-time RT-PCR. Methods Mol Biol, 2014,1161:217-233.

[15] SMITH S E, GIBSON M S, WASH R S, et al. Chicken interferon-inducible transmembrane protein 3 restricts influenza viruses and lyssaviruses in vitro. J Virol, 2013,87(23):12957-12966.

[16] HANG B, SANG J, QIN A, et al. Transcription analysis of the response of chicken bursa of Fabricius to avian leukosis virus subgroup J strain JS09GY3. Virus Res, 2014,188:8-14.

[17] ABDALLAH F, HASSANIN O. Positive regulation of humoral and innate immune responses induced by inactivated avian influenza virus vaccine in broiler chickens. Vet Res Commun, 2015,39(4):211-216.

[18] JI W T, LEE L H, LIN F L, et al. AMP-activated protein kinase facilitates avian reovirus to induce mitogen-activated protein kinase (MAPK) p38 and MAPK kinase 3/6 signalling that is beneficial for virus replication. J Gen Virol, 2009,90(Pt 12):3002-3009.

[19] HAYDEN M S, GHOSH S. Signaling to NF-kappaB. Genes Dev, 2004,18(18):2195-2224.

Avian reovirus σA and σNS proteins activate the phosphatidylinositol 3-kinase-dependent Akt signalling pathway

Xie Liji, Xie Zhixun, Huang Li, Fan Qing, Luo Sisi, Huang Jiaoling, Deng Xianwen, Xie Zhiqin, Zeng Tingting, Zhang Yanfang, and Wang Sheng

Abstract

The present study was conducted to identify avian reovirus (ARV) proteins that can activate the phosphatidylinositol 3-kinase (PI3K)-dependent Akt pathway. Based on ARV protein amino acid sequence analysis, σA, σNS, μA, μB and μNS were identified as putative proteins capable of mediating PI3K/Akt pathway activation. The recombinant plasmids σA-pcAGEN, σNS-pcAGEN, μA-pcAGEN, μB-pcAGEN and μNS-pcAGEN were constructed and used to transfect Vero cells, and the expression levels of the corresponding genes were quantified by immunofluorescence and Western blot analysis. Phosphorylated Akt (P-Akt) levels in the transfected cells were measured by flow cytometry and Western blot analysis. The results showed that the σA, σNS, μA, μB and μNS genes were expressed in Vero cells. σA-expressing and σNS-expressing cells had higher P-Akt levels than negative control cells, pcAGEN-expressing cells and cells designed to express other proteins (i. e., μA, μB and μNS). Pretreatment with the PI3K inhibitor LY294002 inhibited Akt phosphorylation in σA- and σNS-expressing cells. These results indicate that the σA and σNS proteins can activate the PI3K/Akt pathway.

Keywords

ARV, σA, σNS, PI3K/Akt pathway, activate

Introduction

Phosphatidylinositol 3-kinase (PI3K) is a heterodimeric enzyme that is composed of a p85 regulatory subunit and a p110 catalytic subunit[1]. PI3K is activated via the binding of the SH domain in its p85 subunit to autophosphorylated tyrosine kinase receptors, non-receptor tyrosine kinases, or select viral proteins in the cytoplasm[2, 3]. After activation, the p110 subunit of PI3K phosphorylates the lipid substrate phosphatidylinositol-4, 5-bisphosphate to produce phosphatidylinositol-3, 4, 5-trisphosphate[4].

The products of PI3K activate a number of downstream pathway components, including the serine/threonine protein kinase Akt. Akt is activated via the phosphorylation of Thr308 and Ser473[5], and phosphorylated Akt (P-Akt) plays a central role in modulating diverse downstream signalling pathways associated with cell survival, proliferation, migration, differentiation and apoptosis[6, 7].

Avian reovirus (ARV) induces apoptosis in host cells in the middle to late stages of infection[8, 9]. In the early stages of ARV infection, ARV activates the PI3K/Akt signalling axis. This activation contributes to virus-induced inflammation and an anti-apoptotic response[10].

The ARV genome expresses at least 15 different primary translation products, 10 of which are structural proteins (λA, λB, λC, μA, μB, μBc, μBN, σA, σB and σC). The remaining five proteins are non-structural (μNSC, μNS, σNS, P10 and P17)[11, 12]. The ARV proteins that are capable of activating the PI3K/Akt signalling pathway remain to be identified.

PXXP or YXXXM/YXXM motifs are present on a number of viral and cellular proteins involved in PI3K signalling and form extended helices that bind to SH domains on the p85 subunit of PI3K[3, 13-17]. According to amino acid sequence analysis of the 15 proteins encoded by ARV, five ARV proteins (σA, σNS, μA, μB and μNS) contain PXXP or YXXXM motifs. These PXXP and YXXXM motifs are conserved between different ARV strains. Therefore, σA, σNS, μA, μB and μNS may also bind to p85, thereby activating the PI3K/Akt pathway.

The aim of the current study was to determine whether the σA, σNS, μA, μB and μNS proteins could activate the PI3K/Akt pathway *in vivo*. To accomplish this, the recombinant plasmids σA-pcAGEN, σNS-pcAGEN, μA-pcAGEN, μB-pcAGEN and μNS-pcAGEN were constructed and used to transfect Vero cells, and the P-Akt profiles of the transfected cells were examined by flow cytometry and Western blot analysis.

Materials and methods

Primers

Five pairs of primers were designed according to the published sequences of the σA, σNS, μA, μB and μNS genes encoded by ARV (Table 4-8-1). In the primer sequences shown in Table 4-8-1, the italicized text represents restriction sites (*Xho*I and *Not*I), and the underlined text indicates Kozak sequences. All primers were synthesized by Invitrogen (Guangzhou, China).

Table 4-8-1   Primer sequences

Gene	Primer	Sequence (5'-3')	Target fragment size/bp
σA	σA-F	GATGATCTCGAGGCCACCATGGCGCGTGCCATATACGAC	1 248
	σA-R	ATCGCGGCCGCTTAGGCGGTAAAAGTGGCTAGAAC	
σNS	σNS-F	GATGATCTCGAGGCCACCATGGACAACACCGTGCGTGTT	1 101
	σNS-R	ATCGCGGCCGCTTACGCCATCCTAGCTGGAGAGAC	
μA	μA-F	GATGATCTEGAGGCCACCATGGCCTATCTAGCCACACCT	2 196
	μA-R	ATCGCGGCCGCTTAGTGCTCGCCTCCAACCGTCGA	
μB	μB-F	GATGATCTCGAGGCCACCATGGGAAACGCAACGTCTGTC	2 028
	μB-R	ATCGCGGCCGCTTACGATGGTTTGAACAACGTCTG	
μNS	μNS-F	GATGATCTCGAGGCCACCATGGCGTCAACCAAGTGGGGA	1 905
	μNS-R	ATCGCGGCCGCTTACAGATCATCCACCAACTCTTC	

RNA extraction and RT-PCR amplification

Genomic RNA was extracted from 200 μL of ARV using an EasyPure viral DNA/RNA kit (TransGen, Beijing, China) according to the protocol suggested by the manufacturer.

RT-PCR was performed using an RNA LA PCR kit (TaKaRa, Dalian, China). The cycling protocol used for RT-PCR consisted of an initial denaturation step at 94 ℃ for 5 min, followed by 35 cycles of denaturation at 94 ℃ for 1 min, annealing at 55 ℃ for 1 min and extension at 72 ℃ for 2 min. The sample was then heated to 72 ℃ for 10 min for a final extension.

Recombinant plasmid construction

Amplified σA, σNS, μA, μB and μNS gene products were cloned into pMD18-T cloning vectors (TaKaRa, Dalian, China) according to the manufacturer's instructions. The constructed recombinant plasmids were designated σA-pMD18-T, σNS-pMD18-T, μA-pMD18-T, μB-pMD18-T and μNS-pMD18-T. The plasmids were digested with *Xho*I and *Not*I enzymes (TaKaRa, Dalian, China) and then ligated into the corresponding sites of the pcAGEN expression vector before being introduced into competent *E. coli* (DH5α) cells by transformation. Positive colonies, designated σA-pcAGEN, σNS-pcAGEN, μA-pcAGEN, μB-pcAGEN and μNS-pcAGEN, were identified by PCR and then double-digested before being sequenced by Invitrogen (Guangzhou, China).

Expression of σA, σNS, μA, μB and μNS

Plasmids (pcAGEN, σA-pcAGEN, σNS-pcAGEN, μA-pcAGEN, μB-pcAGEN and μNS-pcAGEN) were extracted using a plasmid mini kit (Omega Bio-Tek, USA). Vero cells were seeded into 6-well or 24-well cell culture plates and then transfected with 2.5 μg or 0.5 μg of the appropriate expression vectors using Lipofectamine® 3 000 transfection reagent (Invitrogen, Guangzhou, China) according to the manufacturer's instructions.

Immunofluorescence and Western blot analysis to quantify protein expression

After seeding into 24-well plates and allowing them to become adherent, Vero cells were transfected with various plasmids for 2 h, 4 h, 6 h, 12 h and 24 h and then washed three times with phosphate-buffered saline (PBS). The cells were then fixed with cold methanol for 10 min at room temperature (RT), washed three times with PBS, and blocked for 1 h in 10% normal goat serum (Abcam) in PBS containing 0.5% Triton X-100 (Sigma). The cells were then incubated with primary antibodies against ARV[18] overnight at 4 ℃. After three 5-min washes with PBST, the cells were incubated with fluorescently labelled secondary antibodies (Alexa Fluor, Abcam) for 60 min at room temperature.

Vero cells that were seeded into 6-well plates were harvested, and their cell lysates were used for Western blot analysis to detect σA, σNS, μA, μB and μNS protein expression as described by Xie et al. [18]. Proteins were visualized using enhanced chemiluminescence (ECL) reagent (Bio-Rad) and detected using a Bio-Rad ChemiDoc MP Imaging System (Bio-Rad).

Flow cytometry and Western blot analysis of P-Akt expression

ARV activates the PI3K/Akt pathway during the early stages of infection[10]. According to the IFA and Western blot analysis, the σA, σNS, μA, μB and μNS proteins were all expressed in Vero cells at 6 h after transfection. For flow cytometry and Western blot analysis of P-Akt expression, Vero cells that were seeded into 6-well plates were harvested after transfection with the corresponding plasmids for 6 h. The cells were then subjected to flow cytometry and Western blot analysis to detect P-Akt expression.

For flow cytometry analysis, the transfected cells were detached from culture plates by incubation with Accutase (Sigma) for 5 min. Following this, the cells were washed with PBS and then fixed and permeabilised by incubation with Cytofix/Cytoperm solution (BD Biosciences) at 4 ℃ for 20 min. Next, the cells were washed twice with FACS buffer (0.5% BAS, 0.01% sodium azide in DPBS) and then incubated at 4 ℃ for 30 min with a rabbit antibody (CST) against P-Akt. After this incubation, the cells were washed twice with FACS buffer and then incubated at 4 ℃ for 30 min with goat anti-rabbit IgG conjugated to Alexa Fluor® 488 (Abcam). After staining, the cells were washed twice with FACS buffer and resuspended in FACS buffer for analysis. The samples were then analysed on a flow cytometer (Beckman).

Vero cells that had been seeded in 6-well plates were harvested, and their cell lysates were subjected to Western blot analysis to detect P-Akt expression as described by Wang et al. [19].

All flow cytometry and Western blot analyses were repeated three times.

The σA and σNS proteins induce PI3K-dependent survival signalling

To further establish the signal transmission pathway, Vero cells were transfected with σA-pcAGEN, σNS-pcAGEN, LY294002+σA-pcAGEN, LY294002+σNS-pcAGEN or pcAGEN plasmids. For the LY294002+σA-pcAGEN and LY294002+σNS-pcAGEN groups, the cells were pretreated with the PI3K-specific inhibitor LY294002 (CST) at a concentration of 20 μM for 1 h prior to plasmid transfection. P-Akt expression was analysed by flow cytometry.

Results

Recombinant plasmid construction

The recombinant plasmids σA-pcAGEN, σNS-pcAGEN, μA-pcAGEN, μB-pcAGEN and μNS-pcAGEN were first subjected to double enzyme digestion (Figure 4-8-1). Following this, DNA sequencing was performed to ensure that the recombinant plasmids contained intact σA, σNS, μA, μB and μNS genes, confirming that the recombinant plasmids were successfully constructed.

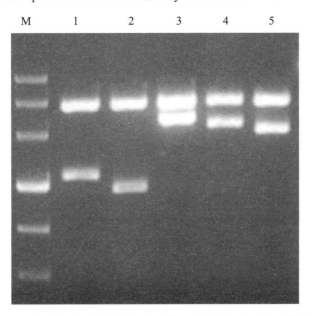

Lane M: DL4 500 bp ladder DNA marker; lane 1: σA-pcAGEN; lane 2: σNS-pcAGEN; lane 3: μA-pcAGEN; lane 4: μB-pcAGEN; lane 5: μNS-pcAGEN.

Figure 4-8-1   Validation of recombinant plasmids by double digestion

σA, σNS, μA, μB and μNS protein expression

According to immunofluorescence and Western blot analyses, the σA, σNS, μA, μB and μNS proteins were expressed in Vero cells 6 h after transfection with the corresponding plasmids (oA-pcAGEN, σNS-pcAGEN, μA-pcAGEN, μB-pcAGEN and μNS-pcAGEN) (Figure 4-8-2 and Figure 4-8-3). Cells transfected with pcAGEN and cells that were not transfected served as negative controls cells (Figure 4-8-2 A and B; Figure 4-8-3, lanes 1 and 2).

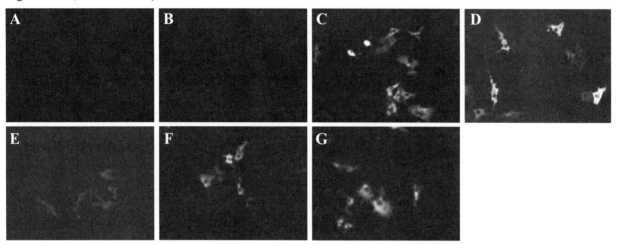

A: Negative control; B: pcAGEN; C: σA-pcAGEN; D: σNS-pcAGEN; E: μA-pcAGEN; F: μB-pcAGEN; G: μNS-pcAGEN.

Figure 4-8-2　Expression of recombinant plasmids in transfected Vero cells

Lane 1: Negative control; lane 2: pcAGEN; lane 3: σA-pcAGEN; lane 4: σNS-pcAGEN; lane 5: μA-pcAGEN; lane 6: μB-pcAGEN; lane 7: μNS-PcAGEN.

Figure 4-8-3　Western blot analysis of protein expression

The σA and σNS proteins activate the PI3K/Akt signalling pathway

Figure 4-8-4 shows that σA- and σNS-expressing cells had higher levels of P-Akt than negative control cells, pcAGEN-expressing cells and cells constructed to express other proteins (i.e., μA, μB and μNS). Figure 4-8-5 shows that the negative control cells, pcAGEN-expressing cells and cells constructed to express other proteins (i.e., σA, σNS, μA, μB and μNS) had the same levels of Akt and β-actin. The σA- and σNS-expressing cells had higher levels of P-Akt expression.

Flow cytometry analysis showed that σA- and σNS-expressing cells that were pre-treated with LY294002 had lower levels of P-Akt expression than σA- and σNS-expressing cells that were not pre-treated (data not shown). In the presence of the PI3K inhibitor LY294002, Akt phosphorylation was inhibited in σA- and σNS-expressing Vero cells. This result confirmed that Akt activation occurs in a PI3K-dependent manner and that the σA and σNS proteins could activate the PI3K/Akt signalling pathway.

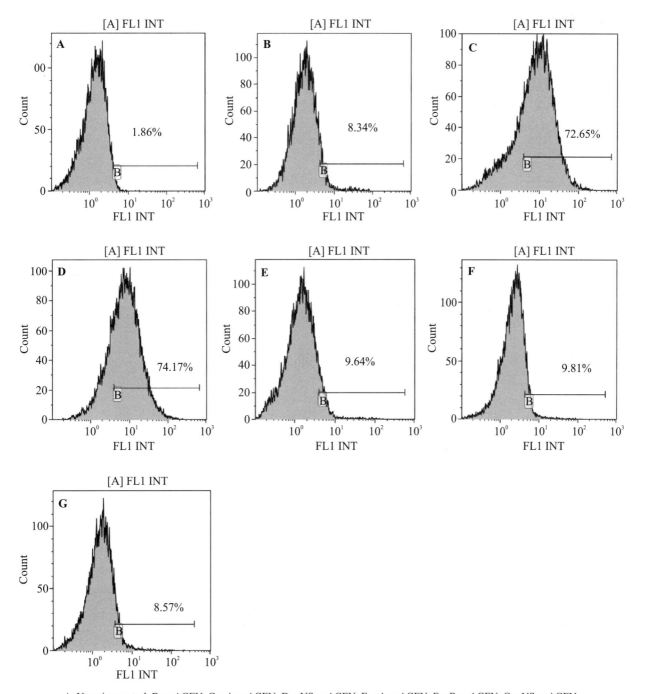

A: Negative control; B: pcAGEN; C: σA-pcAGEN; D: σNS-pcAGEN; E: μA-pcAGEN; F: μB-pcAGEN; G: μNS-pcAGEN.

Figure 4-8-4   Flow cytometry analysis of P-Akt expression

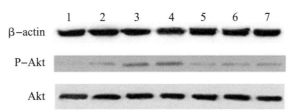

Lane 1: Negative control; lane 2: pcAGEN; lane 3: σA-pcAGEN; lane 4: σNS-pcAGEN; lane 5: μA-pcAGEN; lane 6: μB-pcAGEN; lane 7: μNS-pcAGEN.

Figure 4-8-5   Western blot analysis of P-Akt expression

Discussion

The data presented here demonstrate that Vero cells transfected with vectors expressing σA or σNS have much higher P-Akt levels than Vero cells transfected with the pcAGEN vector or negative control cells. dsRNA stimulation has been shown to trigger PI3K/Akt activation[20, 21]. One function of σA is to bind to dsRNA: during ARV infection, dsRNA-bound σA serves to activate PI3K. σA- and σNS-expressing cells pre-treated with LY294002 showed lower levels of P-Akt expression than σA- and σNS-expressing cells that were not pre-treated and had levels of P-Akt expression that were similar to those in pcAGEN-expressing cells. These results indicated that the σA and σNS proteins induce PI3K-dependent survival signalling and could activate the PI3K/Akt pathway.

A number of viral and cellular proteins have been reported to bind to the SH3 and SH2 domains of the p85 subunit. This binding is mediated by PXXP and YXXXM/YXXM motifs. After binding, these proteins activate the PI3K/Akt pathway[3, 13-17]. Although we demonstrated here that the σA and σNS proteins can activate the PI3K/Akt pathway, whether this activation is a result of these proteins binding to the p85 subunit via interactions with PXXP or YXXXM/YXXM motifs requires further study.

We also demonstrated that cells transfected with μA, μB or μNS expression vectors did not show changes in P-Akt expression relative to cells transfected with the pcAGEN vector or negative control cells. This indicates that the μA, μB and μNS proteins do not directly activate the PI3K/Akt pathway. It remains unknown whether μA, μB, μNS or other ARV proteins enhance or inhibit σA- and σNS-mediated PI3K/Akt pathway activation. It has been reported that the ARV σC protein induces apoptosis in cultured cells[9]; however, whether this protein inhibits σA- and σNS-mediated PI3K/Akt pathway activation requires further study.

Subsequent to transfection with the pcDNA3.1 (-) vector, strong P-Akt expression has been detected in vector-control samples as well as in samples transfected with the influenza A virus NS1 protein[15]. This enhanced P-Akt expression may be because the cationic lipid reagents increased signal transduction, leading to marked activation of insulin receptor kinase activity due to the formation of hexagonal phases in the cell membrane[22, 23]. Similar to the above results, we found that transfection with the pcAGEN vector slightly increased P-Akt expression compared to that found in negative control cells. Additionally, replacing the cationic lipid transfection reagent Lipofectamine 2 000 with Lipofectamine 3 000, using culture medium without serum and antibiotics after transfection, and harvesting cells no more than 6 h after transfection reduced this increase in P-Akt expression to a certain extent in pcAGEN-transfected cells (these data will be published in a future study). Additionally, although transfection with the pcAGEN vector slightly increased P-Akt expression compared to that measured in our negative control cells, cells transfected with σA or σNS expression vectors showed over 9-fold higher levels of P-Akt expression than cells transfected with pcAGEN. Therefore, our observations that the σA and σNS proteins activate the PI3K/Akt pathway are valid.

References

[1] VIVANCO I, SAWYERS C L. The phosphatidylinositol 3-Kinase AKT pathway in human cancer. Nat Rev Cancer, 2002, 2(7): 489-501.

[2] CARPENTER C L, AUGER K R, CHANUDHURI M, et al. Phosphoinositide 3-kinase is activated by phosphopeptides that bind to the SH2 domains of the 85-kDa subunit. The Journal of Biological Chemistry, 1993, 268(13): 9478-9483.

[3] STREET A, MACDONALD A, CROWDER K, et al. The hepatitis C virus NS5A protein activates a phosphoinositide 3-Kinase-dependent survival signaling cascade. The Journal of Biological Chemistry, 2004, 279(13): 12232-12241.

[4] TOKER A, CANTLEY L C. Signalling through the lipid products of phosphoinositide-3-OH kinase. Nature (London), 1997, 387(6634): 673-676.

[5] ALESSI D R, ANDJELKOVIC M, CAUDWELL B, et al. Mechanism of activation of protein kinase B by insulin and IGF-1. The EMBO journal, 1996, 15(23): 6541-6551.

[6] CANTLEY L C. The phosphoinositide 3-Kinase pathway. Science (American Association for the Advancement of Science), 2002, 296(5573): 1655-1657.

[7] DATTA S R, BRUNET A, GREENBERG M E. Cellular survival: a play in three Akts. Genes & development, 1999, 13(22): 2905-2927.

[8] LABRADA L, BODELÓN G, VIÑUELA J, et al. Avian reoviruses cause apoptosis in cultured cells: viral uncoating, but not viral gene expression, is required for apoptosis induction. Journal of Virology, 2002, 76(16): 7932-7941.

[9] SHIH W L, HSU H W, LIAO M H, et al. Avian reovirus sigmaC protein induces apoptosis in cultured cells. Virology (New York, N Y), 2004, 321(1): 65-74.

[10] LIN P Y, LIU H J, LIAO M H, et al. Activation of PI 3-kinase/Akt/NF-kappaB and Stat3 signaling by avian reovirus S1133 in the early stages of infection results in an inflammatory response and delayed apoptosis. Virology, 2010, 400(1): 104-114.

[11] LIU H J, LEE L H, HSU H W, et al. Molecular evolution of avian reovirus: evidence for genetic diversity and reassortment of the S-class genome segments and multiple cocirculating lineages. Virology (New York, N Y), 2003, 314(1): 336-349.

[12] TENG L, XIE Z, XIE L, et al. Sequencing and phylogenetic analysis of an avian reovirus genome. Virus Genes, 2014, 48(2): 381-386.

[13] KAY B K, WILLIAMSON M P, SUDOL M. The importance of being proline: the interaction of proline-rich motifs in signaling proteins with their cognate domains. FASEB J, 2000, 14(2): 231-241.

[14] PAWSON T. Protein modules and signalling networks. Nature (London), 1995, 373(6515): 573-580.

[15] SHIN Y K, LIU Q, TIKOO S K, et al. Influenza A virus NS1 protein activates the phosphatidylinositol 3-kinase (PI3K)/Akt pathway by direct interaction with the p85 subunit of PI3K. J Gen Virol, 2007, 88(Pt 1): 13-18.

[16] SONGYANG Z, SHOELSON S E, CHAUDHURI M, et al. SH2 domains recognize specific phosphopeptide sequences. Cell, 1993, 72(5): 767-778.

[17] ZHANG B, SPANDAU D F, ROMAN A. E5 protein of human papillomavirus type 16 protects human foreskin keratinocytes from UV B-irradiation-induced apoptosis. J Virol, 2002, 76(1): 220-231.

[18] XIE Z, QIN C, XIE L, et al. Recombinant protein-based ELISA for detection and differentiation of antibodies against avian reovirus in vaccinated and non-vaccinated chickens. J Virol Methods, 2010, 165(1): 108-111.

[19] WANG X, ZHANG H, ABEL A M, et al. Role of phosphatidylinositol 3-kinase (PI3K) and Akt1 kinase in porcine reproductive and respiratory syndrome virus (PRRSV) replication. Arch Virol, 2014, 159(8): 2091-2096.

[20] EHRHARDT C, MARJUKI H, WOLFF T, et al. Bivalent role of the phosphatidylinositol-3-kinase (PI3K) during influenza virus infection and host cell defence. Cell Microbiol, 2006, 8(8): 1336-1348.

[21] SARKAR S N, PETERS K L, ELCO C P, et al. Novel roles of TLR3 tyrosine phosphorylation and PI3 kinase in double-stranded RNA signaling. Nat Struct Mol Biol, 2004, 11(11): 1060-1067.

[22] GIORGIONE J R, HUANG Z, EPAND R M. Increased activation of protein kinase C with cubic phase lipid compared with liposomes. Biochemistry, 1998, 37(8): 2384-2392.

[23] PRAMFALK C, LANNER J, ANDERSSON M, et al. Insulin receptor activation and down-regulation by cationic lipid transfection reagents. BMC Cell Biol, 2004, 5: 7.

Analysis of chicken IFITM3 gene expression and its effect on avian reovirus replication

Ren Hongyu, Wang Sheng, Xie Zhixun, Wan Lijun, Xie Liji, Luo Sisi, Li Meng, Xie Zhiqin, Fan Qing, Zeng Tingting, Zhang Yanfang, Zhang Minxiu, Huang Jiaoling, and Wei You

Abstract

Interferon-inducible transmembrane protein 3 (IFITM3) is an antiviral factor that plays an important role in the host innate immune response against viruses. Previous studies have shown that IFITM3 is upregulated in various tissues and organs after avian reovirus (ARV) infection, which suggests that IFITM3 may be involved in the antiviral response after ARV infection. In this study, the chicken IFITM3 gene was cloned and analyzed bioinformatically. Then, the role of chicken IFITM3 in ARV infection was further explored. The results showed that the molecular weight of the chicken IFITM3 protein was approximately 13 kDa. This protein was found to be localized mainly in the cytoplasm, and its protein structure contained the CD225 domain. The homology analysis and phylogenetic tree analysis showed that the IFITM3 genes of different species exhibited great variation during genetic evolution, and chicken IFITM3 shared the highest homology with that of *Anas platyrnynchos* and displayed relatively low homology with those of birds such as *Anser cygnoides* and *Serinus canaria*. An analysis of the distribution of chicken IFITM3 in tissues and organs revealed that the IFITM3 gene was expressed at its highest level in the intestine and in large quantities in immune organs, such as the bursa of Fabricius, thymus and spleen. Further studies showed that the overexpression of IFITM3 in chicken embryo fibroblasts (DF-1) could inhibit the replication of ARV, whereas the inhibition of IFITM3 expression in DF-1 cells promoted ARV replication. In addition, chicken IFITM3 may exert negative feedback regulatory effects on the expression of TBK1, IFN-γ and IRF1 during ARV infection, and it is speculated that IFITM3 may participate in the innate immune response after ARV infection by negatively regulating the expression of TBK1, IFN-γ and IRF1. The results of this study further enrich the understanding of the role and function of chicken IFITM3 in ARV infection and provide a theoretical basis for an in-depth understanding of the antiviral mechanism of host resistance to ARV infection.

Keywords

IFITM3, avian reovirus, bioinformatics analysis, antiviral, innate immunity

Introduction

Avian reovirus (ARV) is a pathogen that circulates widely in poultry and can cause viral arthritis, tenosynovitis and malabsorption syndrome. This virus can also induce severe immunosuppression, which can easily lead to complications or secondary infection with other diseases. These effects lead to reduced production performance and increased mortality in chickens[1-3], resulting in major economic losses in the poultry industry. At present, the prevention and control of ARV mainly involve vaccination, but due to continuous mutation of the strain, the expected immune protection effect is not achieved[4-6]. Therefore, in-depth study of the innate immune regulatory mechanism of ARV infection is highly important for its prevention and control.

Innate immunity is the body's first line of defense against viral infections. After viral invasion, the pattern recognition receptor of the host cell specifically recognizes the molecular pattern associated with the pathogen and thereby activates specific signaling pathways and induces the production of antiviral cytokines such as interferon (IFN) and interleukin, which causes the body to enter an antiviral state[7, 8]. Among these, the interferon-mediated antiviral effect is an important part of the host antiviral response[9]. IFN induces the production of many interferon-stimulated genes (ISGs) by activating the JAK/STAT signaling pathway[10], and ISGs are the main executors of IFN antiviral functions[11]. Studies have shown that ARV infection can induce the transcriptional expression of IFN-α, IFN-β and ISGs such as IFITM1, IFITM3, IFIT5, Mx, ISG12 and other cytokines in various tissues and organs; at the early stage of ARV infection, IFITM3 is significantly upregulated in peripheral blood lymphocytes, joints, the thymus and the bursa of Fabricius and shows a consistent trend with the viral load of ARV in these tissues and organs, which suggests that IFITM3 plays a crucial role in ARV infection[12-14].

The IFITM is an important ISG. Different species have different varieties of IFITM proteins. The human IFITMs include IFITM1-3, IFITM5 and IFITM10. Similar to humans, chickens also have five IFITM genes. IFITMs play a significant role in biological activities such as tumorigenesis, cell adhesion and immune signal transduction[15, 16]. According to previous research findings, IFITM1, IFITM2 and IFITM3 are related to immune regulatory processes in the body, and the expression of IFITM3 exerts a certain limiting effect on the replication of a variety of highly pathogenic viruses[17-19]. IFITM3 is expressed in fish, amphibians, poultry and mammals, and its antiviral activity is relatively conserved from prokaryotes to vertebrates[20, 21]. Early studies have found that IFITM3 can effectively inhibit the replication of a variety of enveloped viruses, such as influenza A viruses (IAVs), dengue virus (DENV) and Ebola virus (EBOV)[21-23], mainly by affecting the fusion of the virus to the endosomal membrane to prevent the virus from entering cells[24, 25]. Additional studies in this research area revealed that IFITM3 effectively inhibits nonenveloped viruses, such as foot-and-mouth disease virus (FMDV), norovirus (NoV) and mammalian orthoreoviruses[18, 26, 27]. Although nonenveloped viruses cannot mediate membrane fusion via proteins on the envelope as do enveloped viruses, they still need to enter cells through endosomes. IFITM3 inhibits viral replication by inhibiting the process of virus entry from endosomes into the cytoplasm. The antiviral mechanism of IFITM3 in mammalian orthoreovirus infection has been well described, and IFITM3 restricts the entry of the virus into host cells by altering the acidic environment of endosomes and reducing protease activity[18]. However, the mechanism of action of chicken IFITM3 has been relatively poorly studied, and the role of IFITM3 in ARV infection has not been reported.

Therefore, in this study, we conducted further investigations on the biological role of chicken IFITM3 in preventing ARV infection. First, we cloned the chicken IFITM3 gene and performed bioinformatic analysis and analyses of its subcellular localization and tissue-organ distribution. Subsequently, the effect of IFITM3 on ARV replication and the regulatory effect of IFITM3 on the expression of correlated molecules in the innate immune signaling pathway were analyzed via overexpression or RNA inhibition assays. The results of this study will provide new ideas for further exploration of the mechanism of the innate immune response to host resistance during ARV infection.

Materials and methods

Animals, virus and cells

The "white leghorn" specific-pathogen-free (SPF) chicken embryos used in this study were purchased from Beijing Boehringer Ingelheim Vital Biotechnology Co., Ltd. (Beijing, China). The ARV S1133 strain was purchased from the China Institute of Veterinary Drug Control. DF-1 cells were preserved in our laboratory and cultured in DMEM (Gibco, Grand Island, NY, USA) supplemented with 10% fetal bovine serum (Gibco).

Cloning and bioinformatics analysis of the IFITM3 gene

The nucleotide sequences of IFITM3 genes were downloaded from the National Center for Biotechnology Information (NCBI) database. Sequence alignment analysis was performed with DNAStar software (DNAstar 7.1) to design primers (Table 4-9-1). Total RNA was extracted from DF-1 cells and reverse-transcribed into cDNA, and the resulting cDNA was subsequently used as a template for the amplification of the IFITM3 gene.

Table 4-9-1　PCR primers used in this study

Primers	Primer sequences (5'-3')	Usage
IFITM3-1	F: GCGTCGACCATGCAGAGCTACCCTCAGCAC R: GCGCGGCCGCTCAGGGCCTCACAGTGTACAA	RT-PCR
IFITM3-2	F: GGAGTCCCACCGTATGAAC R: GGCGTCTCCACCGTCACCA	RT-qPCR
ARV σC	F: CCACGGGAAATCTCACGGTCACT R: TACGCACGGTCAAGGAACGAATGT	RT-qPCR
MAVS	F: CCTGACTCAAACAAGGGAAG R: AATCAGAGCGATGCCAACAG	RT-qPCR
IRF1	F: GCTACACCGCTCACGA R: TCAGCCATGGCGATTT	RT-qPCR
IRF7	F: CAGTGCTTCTCCAGCACAAA R: TGCATGTGGTATTGCTCGAT	RT-qPCR
STING	F: TGACCGAGAGCTCCAAGAAG R: CGTGGCAGAACTACTTTCAG	RT-qPCR
TBK1	F: AAGAAGGCACACATCCGAGA R: GGTAGCGTGCAAATACAGC	RT-qPCR
NF-κB	F: CATTGCCAGCATGGCTACTAT R: TTCCAGTTCCCGTTTCTTCAC	RT-qPCR
MDA5	F: CAGCCAGTTGCCCTCGCCTCA R: AACAGCTCCCTTGCACCGTCT	RT-qPCR
LGP2	F: CCAGAATGAGCAGCAGGAC R: AATGTTGCACTCAGGGATGT	RT-qPCR
IFN-α	F: ATGCCACCTTCTCTCACGAC R: AGGCGCTGTAATCGTTGTCT	RT-qPCR
IFN-β	F: ACCAGGATGCCAACTTCT R: TCACTGGGTGTTGAGACG	RT-qPCR
IFN-γ	F: ATCATACTGAGCCAGATTGTTTCG R: TCTTTCACCTTCTTCACGCCAT	RT-qPCR
GAPDH	F: GCACTGTCAAGGCTGAGAACG R: GATGATAACACGCTTAGCACCAC	RT-qPCR

The conserved domain was predicted based on the NCBI CD-Search database. SOPMA software was used for secondary structure prediction analysis of the chicken IFITM3 protein. SWISS-MODEL was used to predict the tertiary structure of the protein. DNAStar 7.1 and MEGA 11 were used for homology analysis and phylogenetic tree construction. The GenBank accession numbers of the IFITM3 genes from different species are shown in Table 4-9-2.

Table 4-9-2   GenBank accession numbers of the IFITM3 genes used in this study

Name of species	GenBank accession number
Homo sapiens	BC070243.1
Gorilla gorilla gorilla	KU570011.1
Capra hircus	KM236557.1
Gallus gallus	KC876032.1
Serinus canaria	XM_009102512.1
Anas platyrhynchos	KJ739866.1
Mus musculus	BC010291.1
Anser cygnoides	KX594327.1
Sus scrofa	JQ315416.1

Overexpression of the IFITM3 protein

The recombinant plasmid pEF1α-Myc-IFITM3 was constructed and transfected into DF-1 cells using Lipofectamine™ 3 000 (Invitrogen, Carlsbad, CA, USA) to overexpress the IFITM3 protein. 24 h after transfection, the cells were infected with ARV S1133 (MOI = 1), and cell samples and culture medium supernatant were collected 24 h later. RNA from cell samples was extracted and reverse-transcribed into cDNA. The changes in the expression of ARV σC gene and innate immune signaling pathway-correlated molecules were detected by real-time fluorescence quantitative PCR (RT-qPCR). The utilized primers[12, 14, 28] were described previously (Table 4-9-1). In addition, the above-mentioned culture medium was diluted for the infection of DF-1 cells. The lesions of the cells were observed and recorded, and the $TCID_{50}$ of the virus was calculated by the Reed-Muench method.

IFITM3 RNA interference assay

Three small interfering RNAs (siRNAs) for the chicken IFITM3 gene were designed (Table 4-9-3), and the utilized primers were synthesized by GenePharma (Suzhou, China). siR-NAs or siNCs (30 pmol) were transfected separately into DF-1 cells using Lipofectamine™ RNAiMAX (Invitrogen) to inhibit the expression of IFITM3 protein. 24 h after transfection, the cells were infected with the ARV S1133 strain, and cell samples and culture medium supernatant were collected 24 h later. The cell samples were used to detect the changes in the expression of ARV σC gene and innate immune signaling pathway-correlated molecules, and the culture supernatant was used for the detection of viral replication.

Table 4-9-3   siRNA sequences targeting the IFITM3 gene

siRNA	Sequences	Sequences
siIFITM3-35	GCAUCAACAUGCCUUCUUATT	UAAGAAGGCAUGUUGAUGCTT
siIFITM3-200	GGAUCAUCGCCAAGGACUUTT	AAGUCCUUGGCGAUGAUCCTT
siIFITM3-242	GGACAGCGAAGAUCUUUAATT	UUAAAGAUCUUCGCUGUCCTT
siNC	UUCUCCGAACGUGUCACGUTT	ACGUGACACGUUCGGAGAATT

RNA extraction and RT-qPCR

Total RNA was extracted from the samples using a TRIzol kit (Invitrogen). The RNA was reverse-transcribed to cDNA using Maxima™ H Minus cDNA Synthesis Master Mix (Thermo Fisher Scientific, Boston, MA, USA) and stored at −80 ℃ for subsequent assays.

Based on the gene sequence information in GenBank, primers for ARV σC, IFITM3 and innate immune signaling pathway-related molecules were designed and synthesized (Table 4-9-1). RT-qPCR was performed using PowerUp SYBR Green Master Mix (Thermo Fisher Scientific), and the GAPDH gene served as an internal control. The reaction program was as follows: 94 ℃ for 2 min and 40 cycles of 94 ℃ for 15 s and 60 ℃ for 30 s. The detection results were analyzed by the $2^{-\Delta\Delta Ct}$ method.

Confocal microscopy analysis of the subcellular localization of the IFITM3 protein

The cells were transfected with the recombinant plasmids pEF1α-Myc-IFITM3 and pEF1α-Myc, respectively. After 24 h of incubation, the culture medium was discarded. The cells were subsequently washed three times with phosphate-buffered saline (PBS) (Solarbio, Beijing, China) and fixed with 4% paraformaldehyde (Solarbio) for 30 min at room temperature. After three washes with PBS, the cells were infiltrated with 0.1% Triton X-100 (Solarbio) for 15 min and blocked with 5% BSA (Solarbio) for 1 h at room temperature. The cells were incubated with mouse anti-Myc monoclonal antibody (Invitrogen) as the primary antibody at 37 ℃ for 2 h and then with Alexa Fluor 488-labeled goat antimouse IgG (Invitrogen) as the secondary antibody at 37 ℃ while protected from light for 1 h. The nuclei were then stained with DAPI (Solarbio) for 10 min at room temperature while protected from light. After washing with PBS, 50% glycerol was added to the cell plates, and the results were observed by laser confocal microscopy.

Western Blotting

The cells transfected with the recombinant plasmids were washed with PBS and lysed on ice for 30 min using lysis buffer supplemented with protease inhibitors (Sangon Biotech, Shanghai, China). The lysate was then boiled at 100 ℃ for 10 min and centrifuged to obtain protein samples. The proteins were separated by SDS-PAGE and then transferred to polyvinylidene difluoride membranes (Millipore, Billerica, MA, USA). The membranes were blocked overnight with 5% skim milk at 4 ℃ and incubated with primary antibody at 37 ℃ for 2 h and then with the secondary antibody for 1 h. Mouse anti-Myc monoclonal antibody (Invitrogen) and mouse anti-β-actin antibody (Invitrogen) were used as primary antibodies. AP-labeled goat anti-mouse IgG (H+L) (Beyotime Biotechnology, Beijing, China) was used as the secondary antibody. The proteins were then visualized using a BCIP/NBT alkaline phosphatase color development kit (Beyotime Biotechnology).

Statistical analysis

All the data were statistically analyzed using Student's t-test and graphed using Graph-Pad Prism 8. The data were obtained from biological replicates and technical replicates. The results are expressed as the mean ± standard deviation (SD) of three independent experiments. Each sample was measured three times during RT-qPCR. * indicates $P<0.05$, ** indicates $P<0.01$, *** indicates $P<0.001$, and **** indicates $P<0.0001$.

Results

Cloning, bioinformatics analysis and subcellular localization of IFITM3

The full-length sequence of chicken IFITM3 (approximately 342 bp) was successfully cloned using the IFITM3-1 primers (Figure 4-9-1). The sequence was uploaded to the NCBI-BLAST online website for comparison, and the results confirmed that the cloned sequence was the full-length sequence encoded by the IFITM3 gene of *Gallus gallus*, which consists of 342 bases and encodes a total of 113 amino acids. Based on the NCBI CD-Search, the CD225 conserved domain in the chicken IFITM3 protein was predicted (Figure 4-9-2 A). The secondary structure analysis of the IFITM3 protein showed that alpha helices accounted for 42.48%, beta turns accounted for 1.77%, random coils accounted for 40.71%, and extended strands accounted for 15.04% (Figure 4-9-2 B). Tertiary structure prediction showed that the global model quality estimation (GMQE) of the chicken IFITM3 protein and IFITM3 derived from Northern Bobwhite equaled 0.61, and the coverage rate was 80.91% (Figure 4-9-2 C). Homology analysis revealed that chicken IFITM3 exhibited the highest homology (99.4%) with that of *Anas platyrhynchos*. The homologies between chicken IFITM3 and those of *Anser cygnoides* and *Serinus canaria* were 46% and 45.7%, respectively, and the homologies between chicken IFITM3 and those of *Homo sapiens*, *Gorilla gorilla gorilla*, *Capra hircus*, *Sus scrofa* and *Mus musculus* were 50.4%, 53.5%, 51%, 49% and 46.5%, respectively. We constructed a phylogenetic tree to explore the genetic relationships between chicken IFITM3 and IFITM3s from other species (Figure 4-9-3). The results showed that chicken IFITM3 is most closely related to IFITM3 in *A. platyrhynchos*. *A. cygnoides* and *S. canaria* are found in the same group of birds as chickens, but their IFITM3s are distantly related to chicken IFITM3. Chicken IFITM3 is most distantly related to IFITM3s in mammals, such as *H. sapiens* and *G. gorilla* gorilla. These results are consistent with the results of the homology analysis described above. The subcellular localization of the IFITM3 protein in DF-1 cells was analyzed by immunofluorescence and laser confocal microscopy. The nuclei were labeled with blue fluorescence, and the IFITM3 protein was labeled with green fluorescence. As shown in Figure 4-9-4, DF-1 cells transfected with the pEF1α-Myc-IFITM3 plasmid exhibit green fluorescence in the cytoplasm, whereas control cells transfected with the pEF1α-Myc vector do not show green fluorescence, indicating that the IFITM3 protein is localized in the cytoplasm of DF-1 cells.

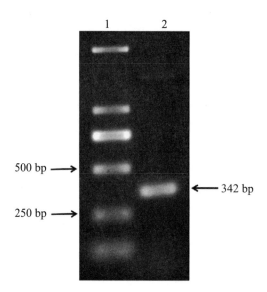

1: DNA marker; 2: Amplification product of the IFITM3 gene. The size of the amplified IFITM3 gene fragment is 342 bp.

Figure 4-9-1   Analysis of the PCR product of the IFITM3 gene via agarose gel electrophoresis

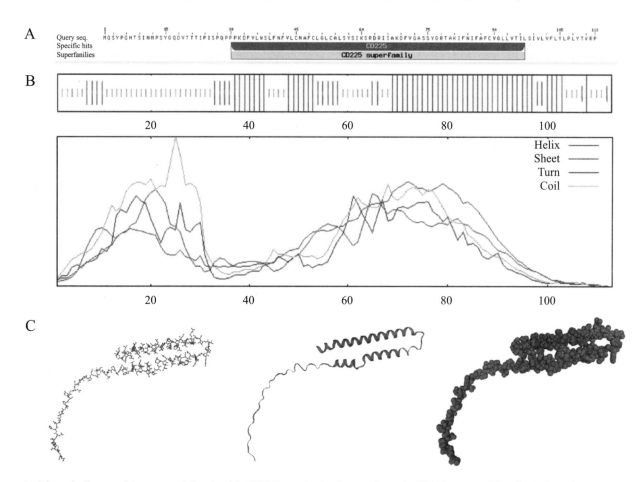

A: Schematic diagram of the conserved domain of the IFITM3 protein, the diagram shows the CD225 conserved domain; B: Secondary structure of the IFITM3 protein, the longest lines represent alpha helices (Hh), the second longest lines represent extended strands (Ee), the third longest lines represent beta turn (Tt), and the shortest lines represent random coils (Cc); C: Tertiary structure of the IFITM3 protein, the global model quality estimation (GMQE) of the chicken IFITM3 protein and IFITM3 derived from Northern Bobwhite equaled 0.61, and the coverage rate was 80.91% (color figure in appendix).

Figure 4-9-2   Structural analysis of the IFITM3 protein

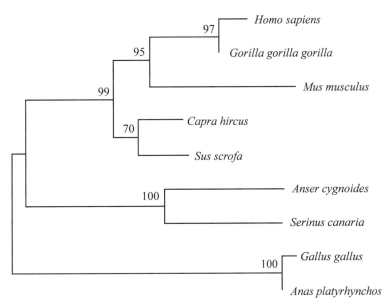

A phylogenetic tree was constructed using MEGA 11 via the neighbor-joining method. The scale bar indicates the length of the branches, and the bootstrap confidence values are shown on the nodes of the tree.

Figure 4-9-3　Phylogenetic tree analysis of IFITM3 genes in different species

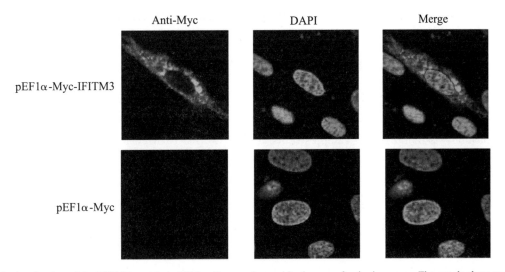

The subcellular localization of the IFITM3 protein in DF-1 cells was observed by laser confocal microscopy. The panels show nuclei stained with DAPI (blue), the IFITM3 protein labeled with Alexa Fluor 488-labeled goat anti-mouse IgG (green) and a merged image (color figure in appendix).

Figure 4-9-4　Subcellular localization of the IFITM3 protein (63×magnification)

Distribution characteristics of IFITM3 in chicken tissues and organs

The distribution of the IFITM3 gene in different tissues and organs of 14-day-old SPF chickens was determined by RT-qPCR. The results showed that IFITM3 was widely expressed in a variety of tissues and organs of chickens, and its highest expression was found in the intestine, followed by the bursa of Fabricius, blood, lung, pancreas, trachea, thymus and spleen. IFITM3 was expressed at low levels in the liver, skin, heart, glandular stomach, gizzard, joint and kidney, and its relative expression in muscle and brain tissues was extremely low (Figure 4-9-5).

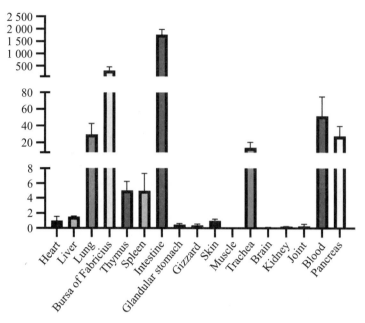

RT-qPCR was used to measure the IFITM3 mRNA levels in the heart, liver, lung, bursa of Fabricius, thymus, spleen, intestine, glandular stomach, gizzard, skin, muscle, trachea, brain, kidney joint, blood and pancreas of 14-day-old SPF chickens. The data are presented as the mean ± SD of three independent experiments.

Figure 4-9-5　Analysis of the IFITM3 gene expression in different tissues and organs

High-level expression of IFITM3 reduces ARV replication

The effect of IFITM3 on ARV replication was analyzed via overexpression and interference assays. First, the overexpression of chicken IFITM3 in DF-1 cells was verified by Western blotting and RT-qPCR. Western blot analysis of DF-1 cell samples transfected with pEF1α-Myc-IFITM3 revealed that a specific band of approximately 13 kDa could be detected by Myc-tagged antibody, whereas no specific bands were detected in cell samples transfected with the empty vector (Figure 4-9-6 A). The RT-qPCR results showed that, compared with that in the control group, the expression of IFITM3 in D-F1 cells transfected with pEF1α-Myc-IFITM3 was significantly upregulated, and its expression increased by approximately 155-fold (Figure 4-9-6 B). Subsequently, IFITM3 was overexpressed in DF-1 cells, the resulting cells were subsequently infected with ARV, and the mRNA level of the ARV σC gene was then detected by RT-qPCR to determine the changes in the viral load. The results showed that the viral load of ARV was significantly reduced after IFITM3 overexpression (Figure 4-9-6 C). Moreover, the detection of viral titers in cell culture supernatants also showed that the viral titer of ARV after the overexpression of IFITM3 was significantly lower than that in the control group (Figure 4-9-6 D). Therefore, we inferred that the overexpression of chicken IFITM3 could effectively inhibit the replication of ARV, and based on this finding, we speculated that inhibition of the expression of IFITM3 may be beneficial for ARV replication. In the following experiments, three siRNAs were designed and synthesized to inhibit the expression of IFITM3. As shown in Figure 4-9-6 E, si242 exerted the greatest inhibitory effect. The expression of IFITM3 in DF-1 cells was inhibited by transfection with si242, and ARV infection was performed 24 h after transfection. The mRNA level of the ARV σC gene and the viral titer in the cell supernatant were subsequently measured. The results showed that the level of ARV replication increased significantly after inhibition of the expression of IFITM3 (Figure 4-9-6 F and G). The results were consistent with the expectations.

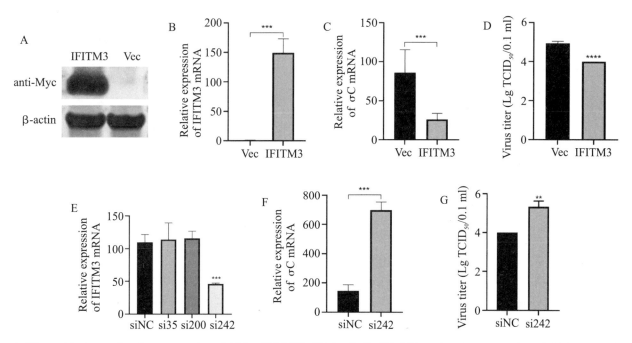

A: Western blotting results of overexpression of chicken IFITM3 in DF-1 cells; B: RT-qPCR results of overexpression of chicken IFITM3 in DF-1 cells; C: RT-qPCR results of the effect of IFITM3 overexpression on ARV replication; D: Virus titer results of the effect of IFITM3 overexpression on ARV replication; E: RT-qPCR results of the inhibition efficiency of the three siRNAs on IFITM3; F: RT-qPCR results of the effect of inhibition of IFITM3 expression on ARV replication; G: Virus titer results of the effect of inhibition of IFITM3 expression on ARV replication (**$P<0.01$, ***$P<0.001$, ****$P<0.000\,1$).

Figure 4-9-6　IFITM3 inhibits the replication of ARV in DF-1 cells

Effect of IFITM3 on innate immune signaling pathway-correlated molecules during ARV infection

The above-described test results showed that chicken IFITM3 exerts an inhibitory effect on the replication of ARV. To further explore the antiviral mechanism of IFITM3 in the process of ARV infection, we studied the regulatory effect of IFITM3 on the innate immune response after ARV infection. IFITM3 was overexpressed or inhibited in DF-1 cells, and 24 h later, the cells were infected with ARV. The changes in the expression of molecules related to the innate immune signaling pathway were then detected by RT-qPCR, and the results are shown in Figure 4-9-7. After infection, the expression of MAVS, IRF7, STING, NF-κB, MAD5, LGP2, IFN-α and IFN-β was upregulated compared with that in the control group, regardless of whether IFITM3 was overexpressed or inhibited. Interestingly, the expression levels of IRF1, TBK1 and IFN-γ were significantly downregulated after IFITM3 overexpression ($P<0.05$ or $P<0.01$). However, the expression levels of IRF1, TBK1 and IFN-γ were significantly upregulated after IFITM3 inhibition ($P<0.05$ or $P<0.001$). It is speculated that changes in the expression of IFITM3 during ARV infection may affect the expression of IRF1, TBK1 and IFN-γ.

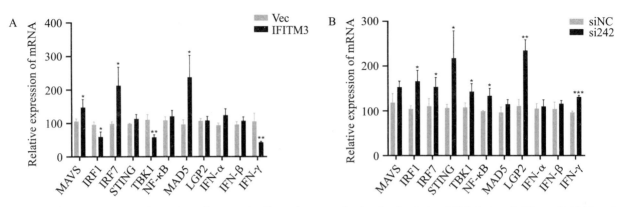

A: Effect of IFITM3 overexpression on innate immune signaling pathway-correlated molecules during ARV Infection; B: Effect of inhibition of IFITM3 expression on innate immune signaling pathway-correlated molecules during ARV Infection.

Figure 4-9-7   Effect of IFITM3 on the expression of molecules associated with innate immune signaling pathways during ARV infection

Discussion

IFITM3 is an important effector of the innate immune system and plays an important role in host resistance to viral infections. Gene structure is often closely related to biological function. In this study, we cloned the chicken IFITM3 gene and analyzed this gene via bioinformatic approaches. Multiple comparison analyses revealed that the homology between chicken IFITM3 and *A. platyrhynchos* was as high as 99.4%. Furthermore, chicken IFITM3 did not exhibit more than 50% homology with that of *A. cygnoides* or *S. canaria*. The homologies of chicken IFITM3 with those of other mammals did not exceed 55%. The same results were obtained via phylogenetic tree analysis, which revealed substantial genetic variation in the IFITM3 genes of different species. The homology between chicken IFITM3 and *A. platyrhynchos* IFITM3 was high, whereas chicken IFITM3 exhibited low homology with bird IFITM3s, such as those of *A. cygnoides* and *S. canaria*. Studies have shown that IFITM proteins belong to the CD225 superfamily, and their members share a highly conserved region of amino acids, the CD225 domain[29]. The CD225 domain was also identified in the protein structure analysis of chicken IFITM3. The CD225 domain consists of an intramembrane domain (IMD), cytoplasmic intracellular loop (CIL) and transmembrane domain (TMD)[30]. Previous studies have shown that the CD225 domain contains multiple key regions associated with antiviral effects, which are also closely correlated with the antiviral effects of IFITM3[31]. The first intramembrane domain (IM1) contains two critical residues, F75 and F78, which are decisive factors affecting the interaction of the IFITM3 protein with the host[29]. GxxxG is an oligomeric motif in the CD225 domain. The glycine-95 in GxxxG is closely related to the oligomerization of IFITM3 and its antiviral activity[32]. Therefore, the conserved structure of the CD225 superfamily may cause IFITM3 proteins derived from different species to exhibit certain similarities in their biological functions. Additionally, the question of whether IFITM3 proteins from different species have antiviral specificity deserves in-depth study, as do their mechanisms of action.

The antiviral function of proteins is closely related to their subcellular localization in cells and their distribution in tissues and organs. The subcellular localization of the chicken IFITM3 protein in DF-1 cells was analyzed by laser confocal microscopy, and the results showed that IFITM3 was localized in the cytoplasm of DF-1 cells. Some previous analyses have investigated the subcellular localization of the IFITM3 protein. S. E. Smith[33] reported that the IFITM3 protein in chickens localizes to the perinuclear area of DF-1 cells, and the human IFITM3 protein also localizes to the perinuclear area of human-derived A549 cells. The distribution of IFITM3 in chicken tissues and organs was then analyzed, and the highest expression of the IFITM3 gene was found in the chicken intestine. This expression pattern is similar to that of human IFITM3, which is most highly expressed in the ileum and cecum in the human digestive system[34]. Moreover, IFITM3 is significantly upregulated in the intestines of pigs infected with porcine circovirus type 2 (PCV2) and porcine parvovirus virus (PPV)[35]. Zoya Alteber et al. [15] experimented with IFITM3-deficient mice and revealed that the IFITM3 gene is involved in regulating the stability of the intestinal environment. IFITM3 also significantly improves the incidence of colitis and prevents inflammation-associated tumorigenesis. After ARV infection, the virus replicates primarily in the host's gut and subsequently spreads through the fecal-oral route and respiratory tract[1, 36, 37]. However, whether ARV infection further induces the expression of the IFITM3 gene in the gut is unknown. Furthermore, IFITM3 was found to be abundantly expressed in immune organs such as the bursa of Fabricius, thymus and spleen. Studies have shown that high expression of IFITM3 can be induced in a variety of immune organs after ARV infection, which is generally consistent with the trend found for the expression of ARV[12, 14].

Further experimental results showed that overexpression of IFITM3 in DF-1 cells could inhibit the replication of ARV, whereas the inhibition of IFITM3 expression in DF-1 cells could promote the replication of ARV, indicating that IFITM3 is an important antiviral factor against ARV infection. Recent studies have shown that IFITM3 can also inhibit the replication of a variety of avian-derived viruses. Stable expression of the duck IFITM3 protein in DF-1 cells can significantly limit the replication of H6N2 and H11N9 IAV strains[38]. Both chicken and duck IFITM3 can effectively inhibit the replication of avian Tembusu virus (ATMUV)[39].

To date, studies on the antiviral mechanism of IFITM proteins have focused mainly on their ability to block contact between viruses and cells and less on their ability to regulate innate immune signaling pathways. Therefore, in this study, the effect of IFITM3 on the expression of innate immune-related molecules after ARV infection was further investigated. The results showed that the expression of MAVS, IRF7, STING, NF-κB, MAD5, LGP2, IFN-α and IFN-β was upregulated compared with that in the control group, regardless of whether IFITM3 was overexpressed or inhibited after infection. Previous studies have shown that ARV infection induces the upregulation of these cytokines[12]. It is hypothesized that changes in IFITM3 expression during ARV infection may not affect the expression of these molecules. In addition, previous studies revealed that the expression of TBK1 and IFN-γ is significantly upregulated after ARV infection[12]. However, in the present study, the expression of TBK1, IFN-γ and IRF1 was significantly downregulated after the overexpression of IFITM3 and significantly upregulated after the inhibition of IFITM3. This interesting phenomenon deserves more in-depth discussion. Researchers have shown that IFITM3 is induced by type I IFNs and can also negatively regulate the production of type I IFNs[40], indicating that IFITM3 may play a role in innate immunity as a negative feedback regulator. TBK1 is an important linker molecule that connects upstream receptor signaling and downstream gene activation in apoptosis, inflammation and immune responses[41, 42]. TBK1 has been found to be involved in lipopolysaccharide (LPS)-induced IFITM3 expression[43]. This finding suggests a potential link between IFITM3 and TBK1 in the body's inflammatory response. Nevertheless, in ARV infection, IFITM3 may exert a negative regulatory effect on TBK1. IFN-γ is a cytokine with antiviral activity and immunomodulatory functions that can act on different types of immune cells to regulate innate and adaptive immunity[44]. IRF1 plays a very important role in the innate immune response induced by IFN-γ[45, 46]. IFN-γ mainly regulates transcription factors such as IRF1 through the JAK/STAT signaling pathway and thus drives subsequent transcriptional regulation[47]. In this study, the overexpression of IFITM3 during ARV infection significantly downregulated the expression of

IFN-γ and IRF1, whereas the inhibition of IFITM3 significantly upregulated the expression of IFN-γ and IRF1. It is hypothesized that IFITM3 may be involved in the body's innate immune response by negatively regulating IFN-γ and IRF1 during ARV infection.

In this study, the chicken IFITM3 gene was cloned and bioinformatically analyzed, and its role in ARV infection was then further analyzed. The results of this study lay a theoretical foundation for obtaining an in-depth understanding of the antiviral mechanism of host resistance to ARV and provide new ideas for the development of new ARV prevention measures. However, the specific regulatory mechanism of IFITM3 on innate immunity during ARV infection needs to be further studied.

References

[1] JONES R C. Avian reovirus infections. Rev Sci Tech, 2000, 19(2): 614-625.

[2] VAN DER HEIDE L. The history of avian reovirus. Avian Dis, 2000, 44(3): 638-641.

[3] TENG L, XIE Z, XIE L, et al. Sequencing and phylogenetic analysis of an avian reovirus genome. Virus Genes, 2014, 48(2): 381-386.

[4] SOUZA S O, De CARLI S, LUNGE V R, et al. Pathological and molecular findings of avian reoviruses from clinical cases of tenosynovitis in poultry flocks from Brazil. Poult Sci, 2018, 97(10): 3550-3555.

[5] TANG Y, LU H. Whole genome alignment-based one-step real-time RT-PCR for universal detection of avian orthoreoviruses of chicken, pheasant and turkey origins. Infect Genet Evol, 2016, 39: 120-126.

[6] LU H, TANG Y, DUNN P A, et al. Isolation and molecular characterization of newly emerging avian reovirus variants and novel strains in Pennsylvania, USA, 2011-2014. Sci Rep, 2015, 5: 14727.

[7] IWASAKI A, PILLAI P S. Innate immunity to influenza virus infection. Nat Rev Immunol, 2014, 14(5): 315-328.

[8] GREEN A M, BEATTY P R, HADJILAOU A, et al. Innate immunity to dengue virus infection and subversion of antiviral responses. J Mol Biol, 2014, 426(6): 1148-1160.

[9] HALLER O, WEBER F. The interferon response circuit in antiviral host defense. Verh K Acad Geneeskd Belg, 2009, 71(1-2): 73-86.

[10] TAKEUCHI O, AKIRA S. Pattern recognition receptors and inflammation. Cell, 2010, 140(6): 805-820.

[11] SECOMBES C J, ZOU J. Evolution of interferons and interferon receptors. Front Immunol, 2017, 8: 209.

[12] XIE L, XIE Z, WANG S, et al. Altered gene expression profiles of the MDA5 signaling pathway in peripheral blood lymphocytes of chickens infected with avian reovirus. Arch Virol, 2019, 164(10): 2451-2458.

[13] WANG S, XIE L, XIE Z, et al. Dynamic changes in the expression of interferon-stimulated genes in joints of SPF chickens infected with avian reovirus. Front Vet Sci, 2021, 8: 618124.

[14] WANG S, WAN L, REN H, et al. Screening of interferon-stimulated genes against avian reovirus infection and mechanistic exploration of the antiviral activity of IFIT5. Front Microbiol, 2022, 13: 998505.

[15] ALTEBER Z, SHARBI-YUNGER A, PEVSNER-FISCHER M, et al. The anti-inflammatory IFITM genes ameliorate colitis and partially protect from tumorigenesis by changing immunity and microbiota. Immunol Cell Biol, 2018, 96(3): 284-297.

[16] RANJBAR S, HARIDAS V, JASENOSKY L D, et al. A Role for IFITM Proteins in Restriction of Mycobacterium tuberculosis Infection. Cell Rep, 2015, 13(5): 874-883.

[17] HUANG I C, BAILEY C C, WEYER J L, et al. Distinct patterns of IFITM-mediated restriction of filoviruses, SARS coronavirus, and influenza A virus. PLOS Pathog, 2011, 7(1): e1001258.

[18] ANAFU A A, BOWEN C H, CHIN C R, et al. Interferon-inducible transmembrane protein 3 (IFITM3) restricts reovirus cell entry. J Biol Chem, 2013, 288(24): 17261-17271.

[19] JIANG D, GUO H, XU C, et al. Identification of three interferon-inducible cellular enzymes that inhibit the replication of hepatitis C virus. J Virol, 2008, 82(4): 1665-1678.

[20] WANG A, SUN L, WANG M, et al. Identification of IFITM1 and IFITM3 in goose: gene structure, expression patterns, and

immune responses against Tembusu virus infection. Biomed Res Int, 2017, 2017: 5149062.

[21] FEELEY E M, SIMS J S, JOHN S P, et al. IFITM3 inhibits influenza A virus infection by preventing cytosolic entry. PLOS Pathog, 2011, 7(10): e1002337.

[22] BRASS A L, HUANG I, BENITA Y, et al. IFITM proteins mediate the innate immune response to influenza A H1N1 virus, west nile virus and dengue virus. Cell, 2009, 139(7): 1243.

[23] PANG Z, HAO P, QU Q, et al. Interferon-inducible transmembrane protein 3 (IFITM3) restricts rotavirus infection. Viruses, 2022, 14(11): 2407.

[24] PERREIRA J M, CHIN C R, FEELEY E M, et al. IFITMs restrict the replication of multiple pathogenic viruses. J Mol Biol, 2013, 425(24): 4937-4955.

[25] DIAMOND M S, FARZAN M. The broad-spectrum antiviral functions of IFIT and IFITM proteins. Nat Rev Immunol, 2013, 13(1): 46-57.

[26] XU J, QIAN P, WU Q, et al. Swine interferon-induced transmembrane protein, sIFITM3, inhibits foot-and-mouth disease virus infection in vitro and in vivo. Antiviral Res, 2014, 109: 22-29.

[27] XING H, YE L, FAN J, et al. IFITMs of African Green Monkey can inhibit replication of SFTSV but not MNV in vitro. Viral Immunol, 2020, 33(10): 634-641.

[28] He Y, Xie Z, Dai J, et al. Responses of the Toll-like receptor and melanoma differentiation-associated protein 5 signaling pathways to avian infectious bronchitis virus infection in chicks. Virol Sin, 2016, 31: 57-68.

[29] JOHN S P, CHIN C R, PERREIRA J M, et al. The CD225 domain of IFITM3 is required for both IFITM protein association and inhibition of influenza A virus and dengue virus replication. J Virol, 2013, 87(14): 7837-7852.

[30] COOMER C A, RAHMAN K, COMPTON A A. CD225 proteins: a family portrait of fusion regulators. Trends Genet, 2021, 37(5): 406-410.

[31] KIM Y C, JEONG M J, JEONG B H. Genetic characteristics and polymorphisms in the chicken interferon-induced transmembrane protein (IFITM3) gene. Vet Res Commun, 2019, 43(4): 203-214.

[32] RAHMAN K, COOMER C A, MAJDOUL S, et al. Homology-guided identification of a conserved motif linking the antiviral functions of IFITM3 to its oligomeric state. Elife, 2020, 9: e58537.

[33] SMITH S E, GIBSON M S, WASH R S, et al. Chicken interferon-inducible transmembrane protein 3 restricts influenza viruses and lyssaviruses in vitro. J Virol, 2013, 87(23): 12957-12966.

[34] SEO G S, LEE J K, YU J I, et al. Identification of the polymorphisms in IFITM3 gene and their association in a Korean population with ulcerative colitis. Exp Mol Med, 2010, 42(2): 99-104.

[35] ANDERSSON M, AHLBERG V, JENSEN-WAERN M, et al. Intestinal gene expression in pigs experimentally co-infected with PCV2 and PPV. Vet Immunol Immunopathol, 2011, 142(1-2): 72-80.

[36] KIBENGE F S, GWAZE G E, JONES R C, et al. Experimental reovirus infection in chickens: observations on early viraemia and virus distribution in bone marrow, liver and enteric tissues. Avian Pathol, 1985, 14(1): 87-98.

[37] WAN L, WANG S, XIE Z, et al. Chicken IFI6 inhibits avian reovirus replication and affects related innate immune signaling pathways. Front Microbiol, 2023, 14: 1237438.

[38] BLYTH G A, CHAN W F, WEBSTER R G, et al. Duck interferon-inducible transmembrane protein 3 mediates restriction of influenza viruses. J Virol, 2016, 90(1): 103-116.

[39] CHEN S, WANG L, CHEN J, et al. Avian interferon-inducible transmembrane protein family effectively restricts avian Tembusu virus infection. Front Microbiol, 2017, 8: 672.

[40] JIANG L Q, XIA T, HU Y H, et al. IFITM3 inhibits virus-triggered induction of type I interferon by mediating autophagosome-dependent degradation of IRF3. Cell Mol Immunol, 2018, 15(9): 858-867.

[41] CHAU T L, GIOIA R, GATOT J S, et al. Are the IKKs and IKK-related kinases TBK1 and IKK-epsilon similarly activated? Trends Biochem Sci, 2008, 33(4): 171-180.

[42] KOOP A, LEPENIES I, BRAUM O, et al. Novel splice variants of human IKKepsilon negatively regulate IKKepsilon-

induced IRF3 and NF-kB activation. Eur J Immunol, 2011, 41(1): 224-234.

[43] NAKAJIMA A, IBI D, NAGAI T, et al. Induction of interferon-induced transmembrane protein 3 gene expression by lipopolysaccharide in astrocytes. Eur J Pharmacol, 2014, 745: 166-175.

[44] MASUDA Y, MATSUDA A, USUI T, et al. Biological effects of chicken type III interferon on expression of interferon-stimulated genes in chickens: comparison with type I and type II interferons. J Vet Med Sci, 2012, 74(11): 1381-1386.

[45] KIM E J, LEE J M, NAMKOONG S E, et al. Interferon regulatory factor-1 mediates interferon-gamma-induced apoptosis in ovarian carcinoma cells. J Cell Biochem, 2002, 85(2): 369-380.

[46] KANO A, HARUYAMA T, AKAIKE T, et al. IRF-1 is an essential mediator in IFN-gamma-induced cell cycle arrest and apoptosis of primary cultured hepatocytes. Biochem Biophys Res Commun, 1999, 257(3): 672-677.

[47] SCHRODER K, HERTZOG P J, RAVASI T, et al. Interferon-gamma: an overview of signals, mechanisms and functions. J Leukoc Biol, 2004, 75, 163-189.

The knob domain of the fiber-1 protein affects the replication of fowl adenovirus serotype 4

Li Xiaofeng, Xie Zhixun, Wei You, Xie Zhiqin, Wu Aiqiong, Luo Sisi, Xie Liji, Li Meng, and Zhang Yanfang

Abstract

Fowl adenovirus serotype 4 (FAdV-4) outbreaks have caused significant economic losses in the Chinese poultry industry since 2015. The relationships among viral structural proteins in infected hosts are relatively unknown. To explore the role of different parts of the fiber-1 protein in FAdV-4-infected hosts, we truncated fiber-1 into fiber-1-Δ1 (73~205 aa) and fiber-1-Δ2 (211~412 aa), constructed pEF1α-HA-fiber-1-Δ1 and pEF1α-HA-fiber-1-Δ2 and then transfected them into leghorn male hepatocyte (LMH) cells. After FAdV-4 infection, the roles of fiber-1-Δ1 and fiber-1-Δ2 in the replication of FAdV-4 were investigated, and transcriptome sequencing was performed. The results showed that the fiber-1-Δ1 and fiber-1-Δ2 proteins were the shaft and knob domains, respectively, of fiber-1, with molecular weights of 21.4 kDa and 29.6 kDa, respectively. The fiber-1-Δ1 and fiber-1-Δ2 proteins were mainly localized in the cytoplasm of LMH cells. Fiber-1-Δ2 has a greater ability to inhibit FAdV-4 replication than fiber-1-Δ1, and 933 differentially expressed genes (DEGs) were detected between the fiber-1-Δ1 and fiber-1-Δ2 groups. Functional analysis revealed these DEGs in a variety of biological functions and pathways, such as the phosphoinositide 3-kinase-protein kinase b (PI3K-Akt) signaling pathway, the mitogen-activated protein kinase (MAPK) signaling pathway, cytokine–cytokine receptor interactions, Toll-like receptors (TLRs), the Janus tyrosine kinase-signal transducer and activator of transcription (Jak-STAT) signaling pathway, the nucleotide-binding oligomerization domain (NOD)-like receptors (NLRs) signaling pathway, and other innate immune pathways. The mRNA expression levels of type I interferons (IFN-α and INF-β) and proinflammatory cytokines (IL-1β, IL-6 and IL-8) were significantly increased in cells overexpressing the fiber-1-Δ2 protein. These results demonstrate the role of the knob domain of the fiber-1 (fiber-1-Δ2) protein in FAdV-4 infection and provide a theoretical basis for analyzing the function of the fiber-1 protein of FAdV-4.

Keywords

FAdV-4; fiber-1; knob; innate immune signaling pathway; type I interferon; cytokines

Introduction

Fowl adenovirus (FAdV) belongs to the adenovirus genus of the Adenoviridae family and can be divided into five genotypes (FAdV-A~FAdV-E) and 12 serotypes (1~7, 8a, 8b, 9~11)[1-3]. After the first FAdV infection occurred in Pakistan in 1987, subsequent outbreaks occurred worldwide, resulting in high mortality rates in chickens[4]. Once a flock is infected with FAdV, the virus spreads throughout the flock for a long time, posing a serious threat to the poultry industry worldwide[4,5]. In 2015, an outbreak of fowl adenovirus serotype 4 (FAdV-4) disease occurred in Shandong, China[6]. FAdV-4 is highly pathogenic in chickens, especially in 3~6-week-old broilers, and causes severe hydropericardial hepatitis syndrome (HHS). FAdV-4 primarily affects the heart and liver of chickens and can be isolated from the liver[6,7]. The typical symptoms include liver swelling, local necrosis and hemorrhage, cardiac cyst enlargement, and the presence of transparent yellow fluid[4]. Although FAdV-4 has been active for 30 years, few studies have examined the effects of its viral structural proteins on infected hosts.

FAdVs are nonenveloped double-stranded DNA viruses with virion diameters of 70~90 nm and icosahedral structures, and each virion is composed of 252 capsids[8]. Each capsid contains the major structural proteins hexon, penton base, and fiber and the minor proteins X, VI, VII, and VIII. The hexon protein is often targeted in molecular epidemiology studies because of its epitope and serum neutralizing properties[9]. During infection, the penton base binds to host cells and relies on host cell surface integrins to mediate endocytosis[10], and studies have shown that the use of the penton base as a subunit vaccine in 2-week-old SPF chickens provides a 90% protection rate against infection, indicating that the penton base is a candidate protein for subunit vaccines[10].

The fiber protein consists of three domains, the knob, shaft and tail, and varies among serotypes. FAdV-1, FAdV-4, and FAdV-10 have two fibrous proteins, fiber-1 and fiber-2, which play different roles in the infection process of FAdV-4. Recent studies have shown that fiber-1, but not fiber-2, directly causes FAdV-4 infection[11,12]. However, in highly pathogenic FAdV-4 strains, fiber-2 and hexon, rather than fiber-1, are virulence determinants[13-15]. Studies have shown that during FAdV-4 infection, the fiber-1 protein attaches the viral capsid to the host cell surface through its knob domain interaction with the D2 domain of the CAR cell receptor to mediate viral infection[11]. Fiber-2 promotes viral replication. The tail of the fiber-1 protein is linked to the penton base. The shaft and knob domains of fiber-1 are considered key factors in viral infection[16]. The use of serum antibodies against ffiber-1 is effective enough to neutralize the virus and prevent FAdV-4 infection[12].

After the virus infects the host, the innate immune system is activated to recognize the virus through numerous pattern recognition receptors (PRRs). The PRRs include Toll-like receptors, melanoma differentiation-associated gene 5 (MDA5), and cyclic guanosine monophosphate-adenosine monophosphate synthase (cGAS) receptors[17,18]. PRRs recognize viral nucleic acids and activate downstream signaling pathways to induce the production of type I interferons (IFN-α/β) and inflammatory cytokines (IL-1β, IL-6, IL-8, and IL-15) to inhibit viral replication. However, FAdV-4 can evade the innate immune response of the host and even develop various defense mechanisms to ensure its own survival and replication. Proteins often perform their physiological functions by forming protein complexes or interacting with nucleic acids. During viral replication, the interaction between viral proteins and host proteins plays a key role in disease development and prognosis.

The results of our previous study revealed that fiber-1 protein overexpression in LMH cells after infection with FAdV-4 significantly upregulated the mRNA expression of TLR receptors (TLR1b, TLR3, TLR7, and TLR21) and related signaling pathways (myeloid differentiation factor 88 (MyD88), IRF7, and IFN-β)[19]. In this study, we cloned fiber-1-Δ1 (73~205 aa) and fiber-1-Δ2 (211~412 aa) to construct eukaryotic expression vectors, investigated their effects on FAdV-4 replication, and analyzed their molecular mechanisms via transcriptome sequencing and real-time quantitative fluorescence PCR (qPCR). The results of this study provide a basis for understanding the structure and function of the fiber-1 protein and a reference for further elucidating the pathogenic mechanism of FAdV-4 and host immune stress during infection.

Materials and methods

Virus and cells

The FAdV-4 virus strain used in this study was isolated from the liver of a chicken infected with HHS in Nanning, Guangxi, and maintained at the Guangxi Veterinary Research Institute. LMH cells were cultured in Dulbecco's modified Eagle's medium (DMEM)/F-12 (Gibco, Grand Island, NY, USA) supplemented with 10% fetal bovine serum (FBS, Gibco) and 1 × penicillin-streptomycin mixture (Solarbio, Beijing, China).

Identification of fiber-1 protein domains and their functionality via prediction software

The domain of the fiber-1 protein was predicted via the online software SMART (version 9, accessed on 24 August 2023). The prediction results are given below. According to the functional annotation of the truncated protein, 73~205 aa (fiber-1-Δ1) and 211~412 aa (fiber-1-Δ2) were selected for the experiment. The antigenic peptide fragments of fiber-1-Δ1 and fiber-1-Δ2 were predicted by DNAMan version 6 (accessed on 5 September 2024), and the hydrophilicity, antigenicity, flexibility and surface accessibility of the potentially dominant antigenic epitopes in B cells were predicted by DNAStar Protean version 7.1.0 (accessed on 5 September 2024).

Cloning genes and construction plasmids

The primers were synthesized with reference to two gene sequences. The fiber-1-Δ1 and fiber-1-Δ2 genes were cloned from DNA extracted from the FAdV-4 virus, separately inserted into the pEF1α-HA vector, and verified by sequencing by RayBiotech (Guangzhou, China). The specific primers used for cloning the fiber-1-Δ1 and fiber-1-Δ2 genes are listed in Table 4-10-1.

Table 4-10-1   Fiber-1-Δ1 and fiber-1-Δ2 gene cloning primers

Primer Name	Sequence (5'-3')	Amplified Sequence Length
Fiber-1-Δ1-U	CCGGAATTCGGATGGGTGGCGGAGGAGGAGGT	399 bp
Fiber-1-Δ1-D	CCCGGTACCTCAGGGTCCCACGGAGCTG	
Fiber-1-Δ2-U	CCGGAATTCGGTCAGGGTCCCACGGAGCTG	606 bp
Fiber-1-Δ2-D	CCCGGTACCTCAGATTGGGCCCGTGGTCA	

Note: Fiber-1-Δ1-F (217~615 bp); fiber-1-Δ2-R (631~1236 bp).

Transfection and verification by Western blotting

LMH cells were cultured in 6-well plates, followed by transfection with pEF1α-HA (mock), pEF1α-HA-fiber-1-Δ1 and pEF1α-HA-fiber-1-Δ2 plasmids (1, 1.25, or 2.0 μg per well) via a Lipofectamine™ 3 000 Transfection Kit (Invitrogen, Carlsbad, CA, USA). Twenty-four hours after transfection, in our previous experiments, the $TCID_{50}$ of this strain was known to be 10-7.4/mL[19]. Cells were infected with FAdV-4 (MOI = 0.01), and cell samples and culture medium supernatants were collected and lysed with lysis buffer 24 h later. The cell samples were mixed with 1 × sodium dodecyl sulfate (SDS) loading buffer (Solarbio), boiled and then incubated on ice. Protein samples were separated by sodium dodecyl sulfate polyacrylamide gel electropheresis (SDS-PAGE, Solarbio) and transferred to polyvinylidene fluoride (PVDF) membranes. The membranes were incubated with Western blot blocking solution (Solarbio) at 4 ℃ overnight due to a lack of antibodies against the fiber-1-Δ1 and fiber-1-Δ2 proteins and then incubated with an anti-HA monoclonal antibody (Invitrogen).

The primary antibody was discarded, and the membrane was washed three times (10 min each time) with 1 × phosphate-buffered saline with Tween-20 (PBST, Solarbio), incubated with goat anti-mouse IgG antibody (Invitrogen, Code No. F-2761) for 1 h, and washed four times with 1 × PBST buffer. Before imaging, the membranes were incubated with a 3,3'-diaminobenzidine (DAB) horseradish peroxidase color development kit (Beyotime Biotechnology, Beijing, China) working solution for 3~10 min at room temperature.

Immunofluorescence assay

After the pEF1α-HA-fiber-1 (plasmids kept in laboratory), pEF1α-HA-fiber-1-Δ1 and pEF1α-HA-fiber-1-Δ2 plasmid transfection and FAdV-4 infection, the culture medium supernatant was removed, and the cells were incubated sequentially with 4% paraformaldehyde (Solarbio) for 30 min at room temperature, Triton X-100 (Solarbio) for 30 min, and 5% bovine serum albumin (BSA) blocking buffer (Solarbio, Code No. SW3015) for 1 h. The cells were then incubated with an anti-HA monoclonal antibody (Invitrogen) at 4 ℃ overnight. After removal of the primary antibody, the cells were washed with PBS, incubated with a fluoresceine isothiocyanate (FiTC)-conjugated rabbit anti-mouse antibody (Invitrogen), and stained with 4',6-diamidino-2-phenylindol (DAPI, Solarbio). The DAPI was discarded, the cells were washed three times with PBS, and the results were observed via inverted fluorescence microscopy and laser confocal microscopy.

Measurement of FAdV-4 growth in LMH cells

After the plasmid transfection and FAdV-4 infection, the cells were harvested at different time points (6, 12, 18, 24, 36, and 48 hpi). Viral DNA was extracted via a Universal Genomic DNA Kit (Cowin Biotechnology, Beijing, China). The DNA concentration was normalized to 100 ng/μL. qPCR was performed with 2 × SYBR Green Mix (10 μL, Thermo Fisher Scientific, Boston, MA, USA), 10 μmol/L hexon-U and hexon-D primers (1 μL, Table 4-10-2), a DNA template (2 μL), and RNase-free water (6 μL, Solarbio, Code No. R1600). The absolute quantitative PCR program was as follows: 94 ℃ for 2 min and 40 cycles of 94 ℃ for 15 s and 60 ℃ for 30 s. This procedure was repeated 3 times for each sample. The results were calculated via a standard curve ($y=-3.362x+36.081$; $R^2=0.999$) previously established by our laboratory for the detection of the FAdV-4 viral load. The data are expressed as the mean ± SD.

Table 4-10-2　Primers used for the qPCR

Primer Name	Sequence (5'-3')	Login Number
TLR3-U	ACAATGGCAGATTGTAGTCACCT	NM_001011691.3
TLR3-D	GCACAATCCTGGTTTCAGTTTAG	
TLR7-U	TCTGGACTTCTCTAACAACA	NM_001011688
TLR7-D	AATCTCATTCTCATTCATCATCA	
IFN-α-U	ATGCCACCTTCTCTCACGAC	AB021154.1
IFN-α-D	AGGCGCTGTAATCGTTGTCT	
INF-β-U	CCTCAACCAGATCCAGCATT	KF741874.1
INF-β-D	GGATGAGGCTGTGAGAGGAG	
IL-1β-U	GTTAATGATGAAGATGTTGATAGC	NM_204305.1
IL-1β-D	GTTCCAGACACAGCAATC	
IL-6-U	TGGTGATAAATCCCGATGAAG	NM_204628.2
IL-6-D	GGCACTGAAACTCCTGGTCT	
IL-8-U	CCATCTTCCACCTTTCACA	HM179639.1
IL-8-D	ATCCCACAGCACTGACCAT	

continued

Primer Name	Sequence (5'-3')	Login Number
TRIF-U	AGCCTGATGGAGAGAGACAGAG	NM_001081506.1
TRIF-D	GATAGACGAGAGGAACTGACCTG	
IRF7-U	ACACTCCCACAGACAGTACTGA	NM_205372.1
IRF7-D	TGTGTGTGCCCACAGGGTTG	
Hexon-U	ACGATCAGACCTTCGTGGAC	KY379035.1
Hexon-D	GGTGTGCGAGAGGTAGAAGC	
GADPH-U	GCACTGTCAAGGCTGAGAACG	KC294567
GADPH-D	GATGATAACACGCTTAGCACCAC	

Transcriptome sequencing analysis

The cells were harvested 48 h post-transfection and sent to Sangon Biotech (Shanghai, China) for transcriptome sequencing. DESeq2 (version 1.12.4) was used to determine the DEGs between two samples. The gene expression differences were visualized via volcano plots.

Functional analysis of DEGs

To functionally classify the DEGs, the DEGs were subjected to Gene Ontology (GO) and Kyoto Encyclopedia of Genes and Genomes (KEGG) analyses. The GO and KEGG analyses were performed with the assistance of Shanghai Sangon Biotech.

RNA extraction and qPCR

After transfection, the cells were infected with FAdV4 at an MOI of 0.01, and the cell cultures were collected at 48 h. The RNA was extracted via the Gene JET RNA Purification Kit (Invitrogen). After the concentration was determined, the RNA was reverse transcribed into cDNA, and the reaction mixture was 20 μL: 4 μL of 5 × PrimeScript RT Master Mix (Takara Bio Inc., Beijing, China), 1 μg of RNA, and nuclease-free water. The reverse transcription procedure was performed as follows: 37 ℃ for 15 min and 85 ℃ for 5 s. The resulting cDNA was diluted 10-fold and used as a template for the qPCR. The qPCR setup and reaction procedure are described above in section *Measurement of FAdV−4 growth in LMH cells*. Three replicates were performed for each sample. GAPDH was used as the reference gene. The expression levels of TLR3, TRIF, IRF7, IFN-α, INF-β, IL-1β, IL-6 and IL-8 were measured, and the primer sequences are shown in Table 4-10-2. The expression levels of the genes were calculated via the $2^{-\Delta\Delta Ct}$ method.

Data processing and statistical analysis

The data were plotted with GraphPad Prism 8.0 software. Differences between groups were analyzed by the t test with IBM SPSS Statistics version 29 (accessed on 5 September 2024). Here, * indicates $P<0.05$ and ** indicates $P<0.01$.

Results

Identification of fiber-1 protein domains and cloning of the genes

The predicted results are shown below (Figure 4-10-1 A). According to the functional annotation of the truncated protein, 73～205 aa (fiber-1-Δ1) and 211～412 aa (fiber-1-Δ2) were selected for the experiment (Figure 4-10-1 B). According to the three parts of fiber-1, knob, shaft and tail, fiber-1-Δ1 was the shaft domain of fiber-1 according to GenBank accession no. MH392486.1, and fiber-1-Δ2 was the knob domain of fiber-1 according to GenBank accession no. MH392487.1. The fiber-1-Δ1 and fiber-1-Δ2 gene segments were amplified via DNA extracted from viruses as a PCR template. The PCR amplification products were separated on an agarose gel, with bands at approximately 399 bp and 606 bp (Figure 4-10-1 C), which were consistent with the expected results.

We used DNAMan to predict the antigenic peptide fragments of fiber-1-Δ1 and fiber-1-Δ2, and as shown in Table 4-10-3, there were 5 antigenic peptide fragments in fiber-1-Δ1 and 10 antigenic peptide fragments in fiber-1-Δ2. The hydrophilicity, antigenicity, flexibility and surface accessibility of the potentially dominant antigenic epitopes in the B-cells are shown in Figure 4-10-1 D.

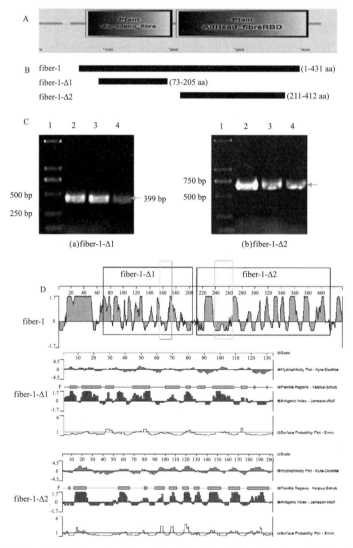

A: Prediction of the fiber-1 protein domain. B: Schematic of the amino acids of the full-length or truncated fiber-1 molecule, fiber-1 (1～431 aa), fiber-1-Δ1 (73～205 aa), and fiber-1-Δ2 (211～412 aa). C: Analysis of the PCR products of the fiber-1-Δ1 and fiber-1-Δ2 genes via agarose gel electrophoresis; (a) The results for fiber-1-Δ1, lane 1: DL2 000 DNA marker, lanes 2-4: Fiber-1-Δ1 gene PCR amplification products, the size of the fiber-1-Δ1 gene fragment is 399 bp; (b) The results for fiber-1-Δ2, lane 1: DL2 000 DNA marker, lanes 2-4: Fiber-1-Δ2 gene PCR amplification products; the size of the fiber-1-Δ2 gene fragment was 606 bp. D: Complete fibre-1 protein map and prediction of the hydrophilicity, antigenicity, flexibility and surface accessibility of potential dominant B-cell epitopes of fiber-1-Δ1 and fiber-1-Δ2. The light grey boxes represent the highest scoring antigenic epitopes.

Figure 4-10-1　Prediction of fiber-1 major structural domains, PCR amplification and antigenic peptide prediction analysis

Table 4-10-3　Prediction of the B-cell epitopes

Antigen	Position	Antigenic Peptide	Score
Fiber-1-Δ1	85~104	LDSVTGVLKVLVDSQGPLQA	1.208
	116~130	QDFVVNNGVLALASSPSSCLQD	1.148
	32~46	PIYVSDRAVSLLIDD	1.141
	57~66	ALMVKTAAPL	1.095
	9~15	QIAVDPD	1.080
	17~27	PLELTGDLLTLE	1.044
Fiber-1-Δ2	82~95	EVNLSLIVPPTVSP	1.172
	4~12	YEVTPVLGI	1.145
	29~55	IGYYIYMVSSAGLVNGLITLELAHDLT	1.140
	171~182	DAIAFTVSLPQT	1.123
	110~119	DVGYLGLPPH	1.122
	98~106	QNHVFVPNSF	1.117
	68~78	NFTFVLSPMYP	1.144
	153~159	LGYCAAT	1.111
	124~139	WYVPIDSPGLRLVSFM	1.105
	193~199	PDTVVTT	1.110

Expression of fiber-1-Δ1 and fiber-1-Δ2 in LMH cells

In the FAdV-4-infected cells at an MOI of 0.01, the virus proliferated rapidly at 12~48 h, then proliferated slowly, and reached peak growth at 96 h (Figure 4-10-2 A). To investigate the optimal transfection efficiency, we used different transfection plasmid concentrations (1, 1.25, and 2.0 μg per well). The results revealed that the transfection efficiency increased with an increasing concentration of the transfected plasmid (Figure 4-10-2 B). On the basis of the test results and the plasmid concentration recommended by the kit, the cells were transfected with 2.0 μg/well.

After the recombinant plasmids pEF1α-HA-fiber-1-Δ1 and pEF1α-HA-fiber-1-Δ2 were transfected into LMH cells, the fiber-1-Δ1 and fiber-1-Δ2 proteins were verified by Western blotting. Owing to the lack of fiber-1-Δ1 and fiber-1-Δ2 monoclonal antibodies, anti-HA monoclonal antibodies were used to verify their expression in LMH cells. The results revealed specific bands of approximately 21.4 kDa and 29.6 kDa (Figure 4-10-2 C) in the cells transfected with pEF1α-HA-fiber-1-Δ1 and pEF1α-HA-fiber-1-Δ2, whereas specific bands were not detected in the cells transfected with pEF1α-HA (mock). To further verify that the fiber-1-Δ1 and fiber-1-Δ2 proteins were correctly expressed in LMH cells, the subcellular localization of the fiber-1-Δ1 and fiber-1-Δ2 proteins in the cells was analyzed by immunofluorescence and laser confocal microscopy. The cell nuclei were labeled with blue fluorescence, and the recombinant proteins were labeled with green fluorescence. As shown in Figure 4-10-2 D, green fluorescence was mainly observed in the cytoplasm of LMH cells, whereas control cells transfected with the pEF1α-HA vector did not show green fluorescence, indicating that the fiber-1-Δ1 and fiber-1-Δ2 proteins were localized in the cytoplasm of LMH cells.

To confirm that fiber-1-Δ1 and fiber-1-2 proteins were localized in the cytoplasm of LMH cells, we simultaneously transfected the full-length fiber-1 protein into LMH cells, and the results showed that the fiber-

1 protein was expressed in the cytoplasm (Figure 4-10-2 D). Meanwhile, the amino acid sequences of fiber-1, fiber-1-Δ1 and fiber-1-Δ2 were analyzed by GenScript online, the results showed that there were no nuclear localization signals in the fiber-1-Δ1 and fiber-1-Δ2 protein sequences. fiber-1, fiber-1-Δ1 and fiber-1-Δ2 were predicted to be expressed in the cytoplasm by Reinhardt's cytoplasmic/nuclear discrimination method with a reliability of 70.6%, 94.1% and 76.7%, respectively (Figure 4-10-2 E).

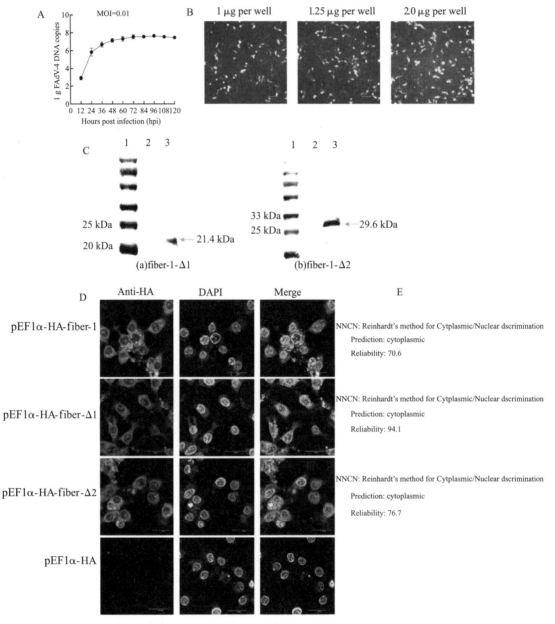

A: Replication curves of FAdV-4 on LMH, MOI = 0.01. B: LMH cells were transfected with different plasmid concentrations (10 ×). The plasmid concentrations were 1, 1.25, and 2.0 μg per well, and the transfection efficiency increased with increasing concentration of the transfected plasmid. C: The results for fiber-1-Δ1 and fiber-1-Δ2 were verified by Western blotting; (a) The results of fiber-1-Δ1, lane 1: protein marker (11∼245 kDa), lane 2: LMH cells transfected with the empty pEF1α-HA plasmid as a control, lane 3: Fiber-1-Δ1 protein was overexpressed in LMH cells. The size of the fiber-1-Δ1 protein containing the HA label was approximately 21.4 kDa; (b) The results for fiber-1-Δ2, lane 1: Protein marker (11∼245 kDa), lane 2: LMH cells transfected with the empty pEF1α-HA plasmid as a control, lane 3: Fiber-1-Δ2 protein was overexpressed in LMH cells, the size of the fiber-1-Δ2 protein containing the HA label was approximately 29.6 kDa. D: The fiber-1, fiber-1-Δ1 and fiber-1-Δ2 proteins were localized mainly in the cytoplasm of LMH cells (60 ×), the scale bar was 20 μm. E: Prediction of cellular sublocalization of fiber-1, fiber-1-Δ1 and fiber-1-Δ2 proteins.

Figure 4-10-2   Fiber-1-Δ1 and fiber-1-Δ2 were expressed in LMH cells

High-level expression of fiber-1-Δ1 and fiber-1-Δ2 inhibits FAdV-4 replication

To investigate the effects of fiber-1-Δ1 and fiber-1-Δ2 on FAdV-4 replication in LMH cells, LMH cells were transfected with pEF1α-HA-fiber-1-Δ1 and pEF1α-HA-fiber-1-Δ2 plasmids and then infected with FAdV-4. At different time points (6, 12, 18, 24, 36, and 48 h) after FAdV-4 infection, viral replication was determined via absolute quantitative PCR. The results are calculated by log10/μL, as shown in Figure 4-10-3 below.

When the cells were transfected with pEF1α-HA, the virus grew rapidly. Compared with those in the mock group, the number of LMH cells infected with pEF1α-HA-fiber-1-Δ1 was lower, but the effect was not significant ($p > 0.05$). The LMH cells transfected with pEF1α-HA-fiber-1-Δ2 had substantially reduced viral loads from 24 hpi ($P < 0.05$) to 36 hpi ($P < 0.05$) to 48 hpi ($P < 0.01$), suggesting that compared with the fiber-1-Δ1 protein, the fiber-1-Δ2 protein has a strong ability to inhibit FAdV-4 replication (Figure 4-10-3).

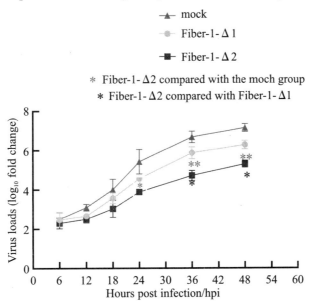

LMH cells were transfected with pEF1α-HA, pEF1α-HA-fiber-1-Δ1 or pEF1α-HA-fiber-1-Δ2 and then infected with FAdV-4 (MOI = 0.01). Viral replication was determined at 6, 12, 18, 24, 36, and 48 hpi via absolute quantitative PCR. The results are expressed as the means ± SDs of three independent experiments. pEF1α-HA-fiber-1-Δ2 compared with the mock group, significant differences are denoted by blue* pEF1α-HA-fiber-1-Δ2 compared with pEF1α-HA-fiber-1-Δ1, significant differences are denoted by black*. One * indicates $P < 0.05$, and two * indicate $P < 0.01$.

Figure 4-10-3　Fiber-1-Δ1 and fiber-1-Δ2 inhibit FAdV-4 replication in LMH cells

Identification and clustering of DEGs

The aforementioned results showed that fiber-1-Δ2 has a more obvious inhibitory effect on viruses than fiber-1-Δ1. Next, the underlying molecular mechanisms and functional differences between these two proteins were explored.

After LMH cells were transfected with pEF1α-HA-fiber-1-Δ1, pEF1α-HA-fiber-1-Δ2, or FAdV-4 for 48 h, a total of 933 DEGs were identified between the pEF1α-HA-fiber-1-Δ1 group (A) and the pEF1α-HA-fiber-1-Δ2 group (B). In the pEF1α-HA-fiber-1-Δ1 group (A), compared with the pEF1α-HA-fiber-1-Δ2 (B) group, 194 of these genes were upregulated and 739 genes were downregulated (Figure 4-10-4).

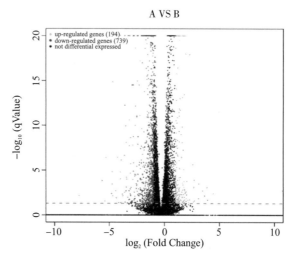

A: Fiber-1-Δ1; B: Fiber-1-Δ2.

Figure 4-10-4　DEGs between fiber-1-Δ1 and fiber-1-Δ2 overexpression in LMH cells

GO enrichment and KEGG enrichment analyses

To further analyze the DEGs, we classified the genes into functional categories. GO analysis of the DEGs was conducted to elucidate the functions of the DEGs. The DEGs were annotated on the basis of the GO biological process, cellular component, and molecular function categories. The 30 most enriched GO terms ($P < 0.05$) are shown in Figure 4-10-5 A. The GO analysis revealed that the DEGs were enriched in cell communication, response to chemicals, cellular response to stimulus, negative regulation of biological processes, regulation of multicellular organisms, and response to organic substances.

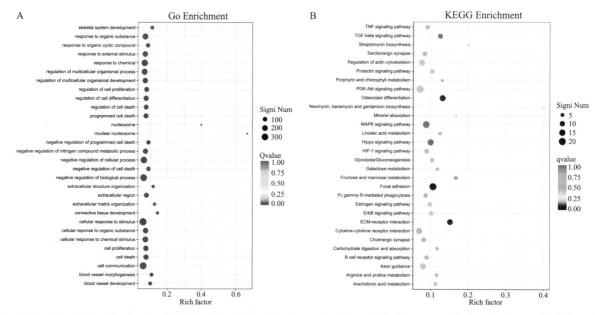

A: The top 30 enriched GO categories; B: The top 30 enriched KEGG categories. The vertical axis represents the functional annotation information, and the horizontal axis represents the enrichment factor corresponding to the function (the number of DEGs annotated to the function divided by all the genes). In terms of the number of genes annotated to a function, the size of the q value is indicated by the color of the dot, so the smaller the q value, the redder it is. The number of DEGs associated with each function is indicated by the size of the dots (only the 30 GO terms with the highest enrichment were selected).

Figure 4-10-5　GO and KEGG enrichment analyses of DEGs

The DEGs were subjected to Kyoto Encyclopedia of Genes and Genomes (KEGG) pathway enrichment analysis to explore the biological processes and molecular functions involved in FAdV-4 infection. As shown in Figure 4-10-5 B, 30 KEGG pathways ($P<0.05$) were enriched among the DEGs between the two groups. Among the 30 KEGG pathways, 6 were associated with immune function, such as the tumor necrosis factor (TNF) signaling pathway, the transforming growth factor-beta (TGF-beta) signaling pathway, the PI3K-Akt signaling pathway, the MAPK signaling pathway, cytokine-cytokine receptor interactions, and the B-cell receptor signaling pathway.

Activation of several immune-related pathways after FAdV-4 infection

The genes enriched in canonical pathways after fiber-1-Δ1 and fiber-1-Δ2 overexpression in LMH cells are listed in Table 4-10-4. These canonical pathways are mainly related to the host immune response. The RIG-I-like helicase receptors (RIG-I-like receptor) signaling pathway, NOD-like receptor signaling pathway, Toll-like receptor (TLR) signaling pathway, Jak-STAT signaling pathway, and cytokine-cytokine receptor interaction pathway are associated with the innate immune response and inflammatory response. Compared with those in the fiber-1-Δ1 overexpression group, the results revealed that, in the RIG-I-like receptor signaling pathway, only IL8L2 was downregulated, whereas in the NOD-like receptor signaling pathway, receptor-interacting protein kinase 3 (RIPK3) and IL8L2 were downregulated. FOS, JUN, PIK3R3, IL8L2 and CCL4 were downregulated in the Toll-like receptor signaling pathway. CISH, CSF3R, PIK3R3, OSMR, CDKN1A, SOCS3, SOCS2, PIM1, and STAT3 were downregulated in the Jak-STAT signaling pathway. BMP7, TNFRSF21, CSF3R, PDGFB, OSMR, IL8L2, TNFRSF8, TGFB1, TGFB2, TNFSF15, CCL4, CXCL14 and LOC10705392 were downregulated in the cytokine-cytokine interaction pathway. Some genes were enriched in one or more pathways. Only the PI3K-Akt signaling pathway, MAPK signaling pathway, and cAMP signaling pathway had upregulated genes, and upregulated genes were not detected in the other pathways. These results suggest that fiber-1-Δ2 inhibits FAdV-4 replication possibly by activating innate immunity during FAdV-4 infection.

Table 4-10-4   Canonical pathways related to the immune response in the fiber-1-Δ1 and fiber-1-Δ2 groups

Pathway ID	KRGG pathway	Downregulated genes	Upregulated genes
ko04151	PI3K-Akt signaling pathway	ITGA8, ITGA7, EPHA2, COL9A2, ITGB3, ITGB4, ITGB6, TNR, GNG2,CSF3R, PIK3R3, PDGFB, OSMR, FN1, CDKN1A, GNG10, TNC,LAMC2, PIK3AP1, THBS1, PPP2R2B	GYS2, PPP2R2C
ko04010	MAPK signaling pathway	NFKB2, FOS, MRAS, JUND, FINC, MAP3K12, JUN, PDGFB, PTPN5, FINB, LOC107055388, HSPA2, TGFB1, TGFB2, DUSP10, PRKCB, HSPA8, DUSP8, DUSP4	CD36, GYS2, PPP2R2C
ko04060	Cytokine-cytokine receptor interaction	BMP7, TNFRSF21, CSF3R, PDGFB, OSMR, IL8L2, TNFRSF8, TGFB1, TGFB2, TNFSF15, CCL4, CXCL14, LOC10705392	
ko04024	cAMP signaling pathway	FOS, JUN, PIK3R3, PTGER2, SOX9, HCN2, ATP2B4	NPY, GABBRR2, CAMK4
ko04350	TGF-beta signaling pathway	NBL1, BMP7, SMAD6, TGFB1, TGFB2, CDKN2B, ID4, ID1, GDF6, THBS	
ko04630	Jak-STAT signaling pathway	CISH, CSF3R, PIK3R3, OSMR, CDKN1A, SOCS3, SOCS2, PIM1, STAT3	
ko04668	TNF signaling pathway	FOS, CEBPB, EDN1, JUN, PIK3R3, MLKL, PTGS2, SOCS3	
ko04620	Toll-like receptor signaling pathway	FOS, JUN, PIK3R3, IL8L2, CCL4	
ko04310	Wnt signaling pathway	CTBP2, JUN, WNT5A, APC2, PRKCB	
ko04621	NOD-like receptor signaling pathway	RIPK3, IL8L2	
ko046222	RIG-I-like receptor signaling pathway	IL8L2	

Effects on Type I interferons and cytokines

According to the above results, the fiber-1-Δ1 and fiber-1-Δ2 proteins exert different inhibitory effects on the replication of FAdV-4 through different innate immune signaling pathways. To further explore and validate the role of fiber-1-Δ1 and fiber-1-Δ2 in the FAdV-4 infection process, the relative mRNA transcript levels of TLR3, TLR7, TRIF, IRF7, IFN-α, IFN-β, IL-1β, IL-6 and IL-8 were measured via qPCR.

The levels of TLR3, TLR7, TRIF, IFN-α, IFN-β, IL-1β and IL-6 were greater in fiber-1-Δ1-overexpressing cells than in control cells, but the difference was not significant ($P>0.05$). Compared with that in the control group, the level of IRF7 in the fiber-1-Δ1-overexpressing group was lower, but the difference was not significant ($P>0.05$).

Compared with those in the control group, in the fiber-1-Δ2 group, the levels of TLR3, IFN-β and IL-8 were significantly increased ($P<0.05$), and the levels of TIRF, IFN-α and IL-1β were significantly increased ($P<0.01$). The levels of IRF7 and IL-6 were significantly increased, but the differences were not significant ($P>0.05$).

Compared with those in fiber-1-Δ1-overexpressing cells, IFN-β, IL-6, IL-8 (all $P<0.05$), IFN-α and IL-1β (both $P<0.01$) were significantly upregulated in fiber-1-Δ2-overexpressing cells. The levels of other mRNAs were not significantly different between the groups ($P>0.05$) (Figure 4-10-6).

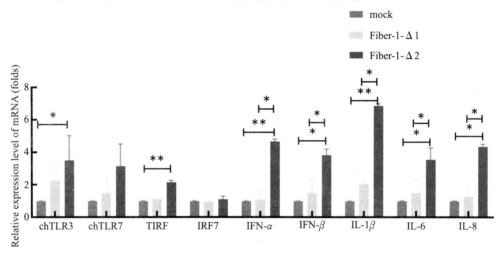

The results are expressed as the means ± SDs of three independent experiments. * indicates $P<0.05$ and ** indicates $P<0.01$. In the mock group, LMH cells were transfected with the pEF1α-HA plasmid; in the fiber-1-Δ1 group, LMH cells were transfected with the pEF1α-HA-fiber-1-Δ1 plasmid; and in the fiber-1-Δ2 group, LMH cells were transfected with the pEF1α-HA-fiber-1-Δ2 plasmid.

Figure 4-10-6　Effects of fiber-1-Δ1 and fiber-1-Δ2 on the expression of innate immune signaling pathway-related genes during FAdV-4 infection

Discussion

FAdV-4 infection can lead to cases of severe HHS in chickens. The high number of cases of FAdV-4 infection on farms has caused considerable economic losses and hindered the development of the poultry industry[6, 20]. To develop highly effective novel HHS vaccines, a complete understanding of FAdV-4 infection is needed. However, the mechanisms of FAdV infection have not yet been determined.

Fiber-1, an adenoviral spike protein of FAdV-4, plays a crucial role in mediating viral infection and inducing neutralizing antibodies[21, 22]. Previous studies have shown that the fiber-2 protein acts as a subunit vaccine, and a single immunization of 10-day-old chickens induces humoral and cellular immunity that is protective against the virus but does not induce neutralizing antibody production[23]. The fiber proteins of FAdV-4 and FAdV-11 were transformed into a chimeric subunit vaccine, which was subsequently used to immunize SPF chickens, which were then challenged with strains that cause HHS or inclusion body hepatitis (IBH), and the development of neutralizing antibodies limited against FAdV-11 and absent against FAdV-4 indicated that the protection conferred by such an antigen may be linked to different immunization pathways[24]. Moreover, studies have reported that when fiber-1 was used as a subunit vaccine, the first immunization of 2-week-old chickens did not induce neutralizing antibodies; rather, neutralizing antibodies appeared after the second immunization of 4-week-old chickens, and antibody production was maintained for 10 weeks. Therefore, the age of the chickens immunized with FAdV-4 fibrin may affect the production of neutralizing antibodies[25]. The fiber-1 protein is divided into tail, shaft, and knob segments. In this study, fiber-1-Δ1 was the shaft domain of fiber-1, and fiber-1-Δ2 was the knob domain of fiber-1. The shaft domain of fiber-1 represents the N-terminal domain of avian adenovirus fiber proteins, which have been linked to variations in virulence[26]. Avian adenoviruses possess penton base capsomers that consist of a pentameric base associated with two fibers[27]. The shaft of the fiber plays an important role in the entry of adenovirus into host cells, primarily by facilitating the binding of fibers to cell receptors, as well as the interaction of the penton base with cellular integrins[28]. The knob domain of fiber-1 represents the C-terminal part of the head domain of the dsDNA viruses, not the RNA-stage adenovirus. This is a globular head domain with an antiparallel beta-sandwich fold formed by two four-stranded beta-sheets with the same overall topology as human adenovirus fiber heads. This C-terminal domain is the receptor-binding domain of the avian adenovirus long fiber[29]. Moreover, many adenoviruses (including human adenoviruses and avian adenoviruses) contain the spike protein. The knob domain contains many antigenic sites and has an epitope against specific neutralizing antibodies. Studies have shown that serum targeting the knob domain can block FAdV-4 infection and that a fusion protein containing the knob domain has an effective protective effect on a lethal attack of FAdV-4 in chickens[12]. In our study, we utilized in silico methods to predict the epitopes/peptides of fiber-1-Δ1 and fiber-1-Δ2. These antigenic sites, or epitopes, are specific chemical groups on the surface of antigen molecules that activate the immune system and facilitate antibody interactions. We evaluated key characteristics of the potentially dominant B-cell epitopes for both fiber-1-Δ1 and fiber-1-Δ2, including hydrophilicity, antigenicity, flexibility, and surface accessibility. Our findings may provide ideas to explain the inhibition of viral replication by fiber-1-Δ2 in terms of natural immunity, as well as valuable insights for the development of novel diagnostic methods and related vaccines. However, it is important to note that this in silico approach is less definitive than in vitro or in vivo testing, which represents a limitation of this study.

To test the role of two fragments of the fiber-1 protein in FAdv-4 replication, we transfected LMH cells with pEF1α-HA-fiber-1-Δ1 and pEF1α-HA-fiber-1-Δ2 and then infected them with FAdV-4. The LMH cell line is an epithelial cell line of chicken hepatocellular carcinoma[30]. It once served as an analytical model for transcriptomic studies on hosts infected with FAdV-4[31,32]. These findings suggest that LMH cells can serve as a suitable model for investigating FAdV-4 infection and host–FAdV-4 interactions.

In addition, the dynamics of virus replication at 6, 12, 18, 24, 36, and 48 hpi after FAdV-4 infection were determined via an absolute relative quantification technique, and the results revealed that fiber-1-Δ1 and fiber-1-Δ2 inhibited the replication of FAdV-4 in LMH cells. Fiber-1-Δ2 had a more obvious inhibitory effect on viruses than fiber-1-Δ1, and the viral load of the cells in each group did not differ at 6～18 h. We hypothesized that it took a period for the virus to enter the host cells and that the host had not yet responded. After 24 h, the viral load of the cells in the fiber-1-Δ2 group was significantly lower than that in the control group, possibly because fiber-1-Δ2 stimulated the innate immune immunity of LMH cells. High-throughput RNA-seq analysis was performed on leghorn male hepatocytes (LMHs) at 12, 24, and 48 h after LMF infection with FAdV-4, and the top 10 GO categories were significantly enriched in biological processes without a response to stimulus and in the cellular response to stimulus not detected in the analysis[31]. The roles of fiber-1-Δ2 and fiber-1-Δ1 in the replication of FAdV-4 differ and are worthy of further exploration.

To explore the reasons for the above results, LMH cells were transfected with pEF1α-HA-fiber-1-Δ1 or pEF1α-HA-fiber-1-Δ2, and the cells were collected after FAdV-4 infection for transcriptome sequencing. A total of 933 DEGs were identified; compared with those in fiber-1-Δ2-overexpressing cells, 194 of these genes were upregulated and 739 of these genes were downregulated in fiber-1-Δ1-overexpressing cells. Further analysis revealed that these DEGs were involved in a range of biological processes, including cell communication, response to chemicals, negative regulation of biological processes, cellular response to stimuli, regulation of multiple organelles, and response to organic substances. The majority of the genes were involved in cell communication, response to chemicals, and the cellular response to stimuli. These results indicate that fiber-1-Δ1 and fiber-1-Δ2 have different responses to FAdV-4 infection in LMH cells. Further analysis revealed that the following signaling pathways were enriched in the DEGs: focal adhesion, the TNF signaling pathway, the TGF-beta signaling pathway, the PI3K–Akt signaling pathway, the MAPK signaling pathway, cytokine–cytokine receptor interaction, the B-cell receptor signaling pathway, and axon guidance.

The activated immune signaling pathways were further analyzed. The RIG-I-like receptor signaling pathway, NOD-like receptor signaling pathway, TLR signaling pathway, Jak-STAT signaling pathway, and cytokine-cytokine receptor interaction pathway were associated with the innate immune response and inflammatory response.

Innate immunity is a general protective mechanism that has evolved in organisms as the first line of defense against pathogens, and innate immunity is the basis of acquired immunity. Pathogen-associated molecular patterns (PAMPs), such as the nucleic acids of viruses, are first recognized by the PRRs of host cells and induce a series of signaling cascades that limit the proliferation and spread of the virus. During the process of virus infection in host cells, nucleic acids can be recognized by RIG-I-like receptors, NOD-like receptors and TLRs. The transcriptome sequencing results revealed differences in gene expression across different signaling pathways: in the fiber-1-Δ1 group compared with the fiber-1-Δ2 group, in the RIG-I-like receptor signaling pathway, only IL8L2 was downregulated, and in the NOD-like receptor signaling pathway, RIPK3 and IL8L2 were downregulated.

Compared with those in the fiber-1-Δ2 group, in the fiber-1-Δ1 group, the TLR signaling pathway, FOS, JUN, PIK3R3, IL8L2 and CCL4 were downregulated. In the Jak-STAT signaling pathway, CISH, CSF3R, PIK3R3, OSMR, CDKN1A, SOCS3, SOCS2, PIM1 and STAT3 were downregulated. With respect to the cytokine–cytokine receptor interactions, BMP7, TNFRSF21 CSF3R, PDGFB, OSMR, IL8L2, TNFRSF8, TGFB1, TGFB2, TNFSF15, CCL4, CXCL14, and LOC10705392 were downregulated. Some genes were enriched in one or more pathways, suggesting that these genes play a role in multiple pathways. These results suggest that FAdV-4 infection activates innate immune and inflammatory responses and that the virus uses the material of the host for replication while also being inhibited by the host.

TLRs play crucial roles in the innate immune response by recognizing highly conserved microbial structural motifs, including PAMPs, and by activating a series of downstream signals that induce the secretion of inflammatory cytokines and type I interferons[33]. Ten TLRs have been identified in chickens[34-37]. TLR3 is an intracellular TLR that is expressed at varying levels in the liver, kidney, cecal tonsils, and intestines, as well as in CD8+ and TCR1 cells [38]. Mammalian TLR3 can be activated by viral dsRNA or poly(I:C) analogs, which bind to each other to activate the adaptor protein TRIF[33,39] and subsequently promote the production of interferon through signal transduction, promoting viral infection and replication. In our study, compared with those of the control cells, the mRNA transcript levels of TLR3 and TIRF were significantly increased in fiber-1-Δ2-overexpressing cells. The level of TLR3 significantly increased in these cells, which is consistent with the finding of Zhang et al.[22] that FAdV-4 infects LMH cells and causes TLR3 upregulation. Like TLR3, TLR7 is an intracellular TLR that is expressed in a variety of tissues and cells. Stimulation with TLR7 receptor agonists (R848, poly(U)) upregulates the expression of the inflammatory cytokines IL-1β and IL-8 and type I interferon[35,40]. In our study, the TLR7 levels were increased in the fiber-1-Δ1 and fiber-1-Δ2 groups, but the difference was not significant. It is hypothesized that fiber-1-Δ2 activates downstream type I interferon and cytokine expression through the TLR3 and TIRF signaling pathways.

IFNs and other innate immune-related factors constitute the first line of host defense against pathogenic infection. They activate factors that aid in inhibiting viral replication and transmission. Compared with those in the fiber-1-Δ1 group, the levels of IFN-β, IL-6 and IL-8 were significantly upregulated ($P < 0.05$) in the fiber-1-Δ2 group. Similarly, the levels of IL-6 and IL-8 were consistent with those reported in a study of FAdV-4 infection in SPF chickens by Li Rong[41], indicating that our results are reliable. These results also suggest that fiber-1-Δ2 inhibits FAdV-4 replication by upregulating these cytokines. Compared with those in the fiber-1-Δ2 group, the levels of IFN-α and IL-1β were significantly greater in the fiber-1-Δ2 group ($P < 0.01$). Infection with the highly pathogenic FAdV-4 has been shown to induce inflammatory damage in many tissues, accompanied by high secretion levels of the proinflammatory cytokine IL-1β[41-43]. In this study, overexpression of the fiber-1-Δ2 protein upregulated the expression of type I interferon (IFN-α, IFN-β), IL-1β, IL-6 and IL-8 during FADV-4 infection of LMH cells and inhibited viral replication.

From the above results, it can be seen that fiber-1-Δ2 activated the natural immune signaling pathway to inhibit the replication of FAdV-4 in LMH cells. In this experiment, fiber-1-Δ2 was obtained by truncating fiber-1, which is rich in B-cell antigenic epitopes, the key to successful vaccine development. At the same time, the truncated protein can effectively avoid unnecessary antigenic components and reduce potential toxic reactions. Meanwhile, fiber-1-Δ2 may have better antigenicity than full-length fiber-1 and can be used as a detection antigen for the development of the FAdV-4 ELISA antibody assay with higher sensitivity to react with specific antibodies, the development of more valuable assays and more to understand all the aspects of viral pathogenesis in order to propose novel antiviral strategies. Finally, fiber-1-Δ2 is much smaller than full-length fiber-1, which facilitates the in vitro expression of FAdV-4 used to study the FAdV-4–host relationship.

In conclusion, our results showed that fiber-1-Δ2 (211～412 aa), which is the knob domain of fiber-1, significantly inhibits virus replication by activating innate immune signaling pathways. These findings lay a theoretical foundation for the function of the fiber-1 protein of FAdV-4.

References

[1] HESS M. Detection and differentiation of avian adenoviruses: a review. Avian Pathol, 2000, 29(3): 195-206.

[2] STEER P A, KIRKPATRICK N C, O'ROURKE D, et al. Classification of fowl adenovirus serotypes by use of high-resolution melting-curve analysis of the hexon gene region. Journal of Clinical Microbiology, 2009, 47(2): 311-321.

[3] BENKO M, AOKI K, ARNBERG N, et al. ICTV virus taxonomy profile: adenoviridae 2022. J Gen Virol, 2022, 103(3).

[4] SCHACHNER A, MATOS M, GRAFl B, et al. Fowl adenovirus-induced diseases and strategies for their control—a review

on the current global situation. Avian Pathol, 2018, 47(2): 111-126.

[5] HAIYILATI A, ZHOU L, LI J, et al. Gga-miR-30c-5p enhances apoptosis in fowl adenovirus serotype 4-infected leghorn male hepatocellular cells and facilitates viral replication through Myeloid cell Leukemia-1. Viruses, 2022, 14(5): 990.

[6] YE J, LIANG G, ZHANG J, et al. Outbreaks of serotype 4 fowl adenovirus with novel genotype, China. Emerging Microbes & Infections, 2016, 5(1): 1-12.

[7] LIU Y, WAN W, GAO D, et al. Genetic characterization of novel fowl aviadenovirus 4 isolates from outbreaks of hepatitis-hydropericardium syndrome in broiler chickens in China. Emerging Microbes & Infections, 2016, 5(1): 1-8.

[8] RUSSELL W C. Adenoviruses: update on structure and function. Journal of General Virology, 2009, 90(1): 1-20.

[9] LIU J, MEI N, WANG Y, et al. Identification of a novel immunological epitope on Hexon of fowl adenovirus serotype 4. AMB Express, 2021, 11(1).

[10] WANG X, TANG Q, CHU Z, et al. Immune protection efficacy of FAdV-4 surface proteins fiber-1, fiber-2, hexon and penton base. Virus Res, 2018, 245: 1-6.

[11] PAN Q, WANG J, GAO Y, et al. Identification of chicken CAR homology as a cellular receptor for the emerging highly pathogenic fowl adenovirus 4 via unique binding mechanism. Emerging Microbes & Infections, 2020, 9(1): 586-596.

[12] WANG W, LIU Q, LI T, et al. Fiber-1, not fiber-2, directly mediates the infection of the pathogenic serotype 4 fowl adenovirus via its shaft and knob domains. J Virol, 2020, 94(17).

[13] ZHANG Y, LIU R, TIAN K, et al. Fiber2 and hexon genes are closely associated with the virulence of the emerging and highly pathogenic fowl adenovirus 4. Emerging Microbes & Infections, 2018, 7(1): 1-10.

[14] LIU R, ZHANG Y, GUO H, et al. The increased virulence of hypervirulent fowl adenovirus 4 is independent of fiber-1 and penton. Res Vet Sci, 2020, 131: 31-37.

[15] XIE Q, WANG W, LI L, et al. Domain in fiber-2 interacted with KPNA3/4 significantly affects the replication and pathogenicity of the highly pathogenic FAdV-4. Virulence, 2021, 12(1): 754-765.

[16] ZOU X, RONG Y, GUO X, et al. Fiber-1, but not fiber-2, is the essential fiber gene for fowl adenovirus 4 (FAdV-4). Journal of general virology, 2021, 102(3).

[17] YAMAMOTO M, TAKEDA K. Current views of toll-like receptor signaling pathways. Gastroenterol Res Pract, 2010, 2010: 240365.

[18] ZHANG T, REN M, LIU C, et al. Comparative analysis of early immune responses induced by two strains of Newcastle disease virus in chickens. Microbiologyopen, 2019, 8(4): e701.

[19] YONGLI S. Study on the effect of fowl adenovirus type 4 and its main encoding proteins on the mRNA transcription levels of Toll-like receptors. Nanning: Guangxi University, 2022.

[20] WANG P, ZHANG J, WANG W, et al. A novel monoclonal antibody efficiently blocks the infection of serotype 4 fowl adenovirus by targeting fiber-2. Vet Res, 2018, 49(1): 29.

[21] HENRY L J, XIA D, WILKE M E, et al. Characterization of the knob domain of the adenovirus type 5 fiber protein expressed in Escherichia coli. J Virol, 1994, 68(8): 5239-5246.

[22] RUAN S, ZHAO J, YIN X, et al. A subunit vaccine based on fiber-2 protein provides full protection against fowl adenovirus serotype 4 and induces quicker and stronger immune responses than an inactivated oil-emulsion vaccine. Infect Genet Evol, 2018, 61: 145-150.

[23] SCHACHNER A, MAREK A, JASKULSKA B, et al. Recombinant FAdV-4 fiber-2 protein protects chickens against hepatitis–hydropericardium syndrome (HHS). Vaccine, 2014, 32(9): 1086-1092.

[24] De LUCA C, SCHACHNER A, HEIDL S, et al. Vaccination with a fowl adenovirus chimeric fiber protein (crecFib-4/11) simultaneously protects chickens against hepatitis-hydropericardium syndrome (HHS) and inclusion body hepatitis (IBH). Vaccine, 2022, 40(12): 1837-1845.

[25] WATANABE S, YAMAMOTO Y, KUROKAWA A, et al. Recombinant fiber-1 protein of fowl adenovirus serotype 4 induces high levels of neutralizing antibodies in immunized chickens. Archives of Virology, 2023, 168(3).

[26] PALLISTER J T U C, WRIGHT P J, SHEPPARD M. A single gene encoding the fiber is responsible for variations in virulence in the fowl adenoviruses. Journal of Virology, 1996, 70(8): 5115-5122.

[27] HESS M, CUZANGE A, RUIGROK R W, et al. The avian adenovirus penton: two fibres and one base. J Mol Biol, 1995, 252(4): 379-385.

[28] WU E, PACHE L, Von SEGGERN D J, et al. Flexibility of the adenovirus fiber is required for efficient receptor interaction. J Virol, 2003, 77(13): 7225-7235.

[29] GUARDADO-CALVO P, LLAMAS-SAIZ A L, FOX G C, et al. Structure of the C-terminal head domain of the fowl adenovirus type 1 long fiber. J Gen Virol, 2007, 88(Pt 9): 2407-2416.

[30] KAWAGUCHI T, NOMURA K, HIRAYAMA Y, et al. Establishment and characterization of a chicken hepatocellular carcinoma cell line, LMH. Cancer Research (Chicago, Ill), 1987, 47(16): 4460-4464.

[31] ZHANG J, ZOU Z, HUANG K, et al. Insights into leghorn male hepatocellular cells response to fowl adenovirus serotype 4 infection by transcriptome analysis. Vet Microbiol, 2018, 214: 65-74.

[32] REN G, WANG H, HUANG M, et al. Transcriptome analysis of fowl adenovirus serotype 4 infection in chickens. Virus Genes, 2019, 55(5): 619-629.

[33] ACHEK A, YESUDHAS D, CHOI S. Toll-like receptors: promising therapeutic targets for inflammatory diseases. Arch Pharm Res, 2016, 39(8): 1032-1049.

[34] TEMPERLEY N D, BERLIN S, PATON I R, et al. Evolution of the chicken Toll-like receptor gene family: a story of gene gain and gene loss. BMC genomics, 2008, 9(1): 62.

[35] BROWNLIE R, ALLAN B. Avian toll-like receptors. Cell and Tissue Research, 2011, 343(1): 121-130.

[36] BOYD A, PHILBIN V J, SMITH A L. Conserved and distinct aspects of the avian Toll-like receptor (TLR) system: implications for transmission and control of bird-borne zoonoses. Biochem Soc Trans, 2007, 35(Pt 6): 1504-1507.

[37] SMITH J, SPEED D, LAW A S, et al. In-silico identification of chicken immune-related genes. Immunogenetics (New York), 2004, 56(2): 122-133.

[38] IQBAL M, PHILBIN V J, SMITH A L. Expression patterns of chicken Toll-like receptor mRNA in tissues, immune cell subsets and cell lines. Vet Immunol Immunopathol, 2005, 104(1-2): 117-127.

[39] KEESTRA A M, de ZOETE M R, BOUWMAN L I, et al. Unique features of chicken Toll-like receptors. Dev Comp Immunol, 2013, 41(3): 316-323.

[40] NEERUKONDA S N, KATNENI U. Avian pattern recognition receptor sensing and signaling. Vet Sci, 2020, 7(1): 14.

[41] LI R, LI G, LIN J, et al. Fowl adenovirus serotype 4 SD0828 infections causes high mortality rate and cytokine levels in specific pathogen-free chickens compared to ducks. Front Immunol, 2018, 9: 49.

[42] ZHAO W, LI X, LI H, et al. Fowl adenoviruse-4 infection induces strong innate immune responses in chicken. Comp Immunol Microbiol Infect Dis, 2020, 68: 101404.

[43] NIU Y, SUN Q, ZHANG G, et al. Fowl adenovirus serotype 4-induced apoptosis, autophagy, and a severe inflammatory response in liver. Veterinary Microbiology, 2018, 223: 34-41.

Appendix

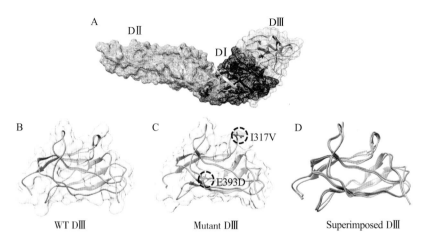

A: Domain mapping of ZIKV E protein; B: Wild-type DⅢ of the E protein; C: Mutant (I317V, E393D) DⅢ; D: Superimposition of wild-type DⅢ on mutant DⅢ to calculate the RMSD.

Figure 1-5-3　Domains architecture of ZIKV E protein and mutant modeling

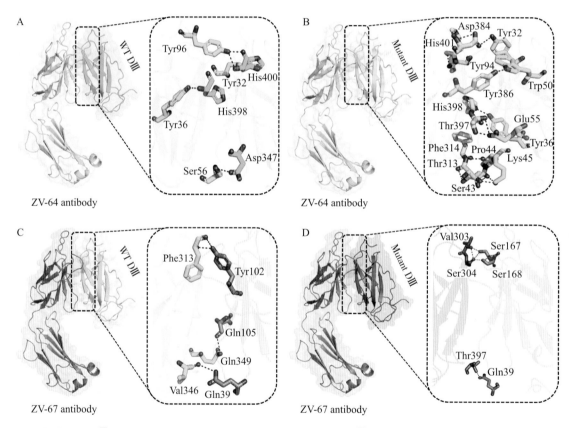

A: Docking of wild-type DⅢ with ZV-64 neutralizing antibody; B: Docking of mutant DⅢ with the ZV-64 neutralizing antibody; C: Docking of wildtype DⅢ with the ZV-67 neutralizing antibody (DⅢ); D: Docking of mutant DⅢ with ZV-67 neutralizing antibody.

Figure 1-5-4　Docking of wild-type and mutant DⅢ with ZV-64 and ZV67 neutralizing antibodies

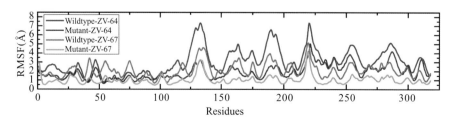

Figure 1-5-7　RMFS calculated for the residual flexibility of all four complexes

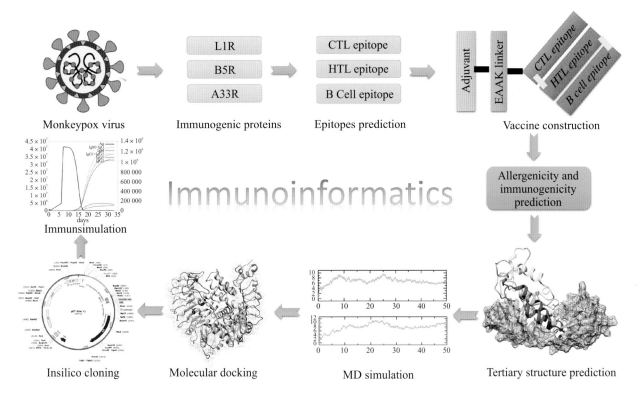

Figure 1-7-1   Overall workflow of the development of a multiepitope subunit vaccine targeting monkeypox virus using the immunoinformatic approach

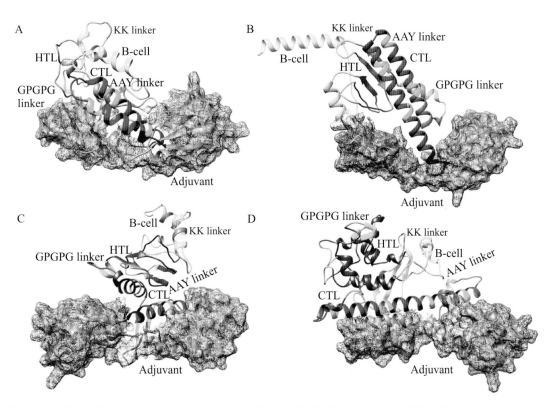

A: The 3D structure of the L1R inal vaccine construct generated by Robetta; B: The 3D structure of the B5R inal vaccine construct generated by Robetta; C: The 3D structure of the A33R inal vaccine construct generated by Robetta; D: The 3D structure of the proteome-wide inal vaccine construct generated by Robetta.

Figure 1-7-3   Robetta-generated 3D structures for the inal vaccine constructs of the monkeypox virus

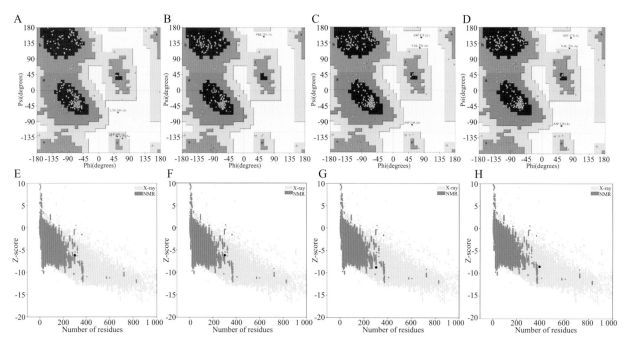

A-D: Structural validation of the L1R, B5R, A33R, and proteome-wide vaccines by Ramachandran plots; E-H: Structural validation of the L1R, B5R, A33R, and proteome-wide vaccines by ProSA -web. Uppercase and lowercase: This is the international farmate for this sever. ProSA-web (protein structure analysis-web).

Figure 1-7-4　Quality analysis of Robetta-generated models by Ramachandran plots and ProSA-web

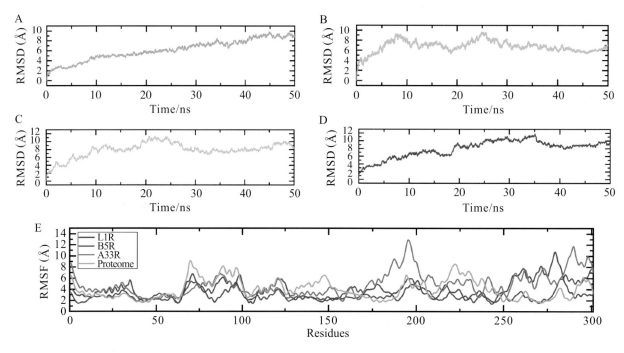

A: RMSD value for the L1R vaccine construct; B: RMSD value for the B5R vaccine construct; C: RMSD value for the A33R vaccine construct; D: RMSD value for the proteome-wide vaccine construct; E: RMSF values for the L1R, B5R, A33R and proteome-wide vaccine constructs.

Figure 1-7-5　Molecular dynamics simulations of the constructed vaccines

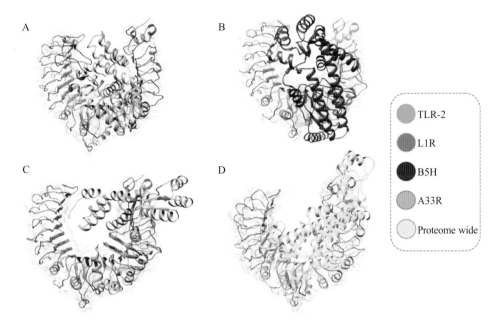

A: L1R-TLR-2 complex; B: B5R-TLR-2 complex; C: A33R-TLR-2 complex; D: proteome-wide construct-TLR-2 complex.The designed vaccines and TLR-2 are shown in different colors.

Figure 1-7-6　Complexes of the L1R, B5R, A33R, and proteome-wide vaccine constructs with human TLR-2
(The designed vaccines and TLR-2 are shown in different colors)

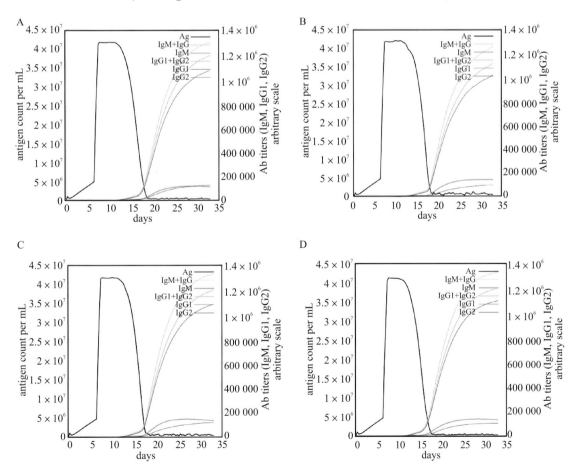

A: Immune responses to the L1R vaccine; B: Immune responses to the B5R vaccine; C: Immune responses to the A33R vaccine; D: Immune responses to the proteome-wide vaccine.

Figure 1-7-8　Immune response simulation for a constructed vaccine against monkeypox virus

A Circular tree was created by using the maximum likelihood tree based on the Tamura-Nei substitution model and 1 000 bootstrap replicates. Different genotypes are highlighted with different shapes and colors. Black circles: G Ⅰ lineages (1-31), Meroon square: G Ⅱ, Blue square: G Ⅲ, Green rhombus: GⅣN, Yellow square: G Ⅴ, Grey inverted-triangle: GⅥ, Purple square: GⅦ, Red circle: GⅧ, Green square: GⅨX.

Figure 1-11-2　Molecular phylogenetic analysis of the S1 gene of the reference IBV genotypes

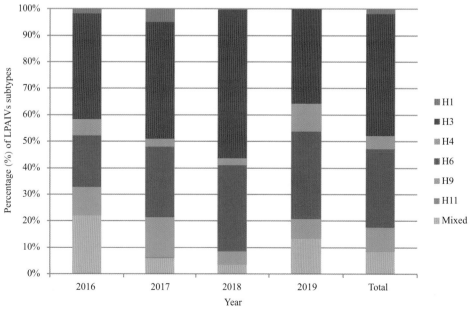

Figure 2-3-3　Percentage of the isolated HA subtypes from 2016 to 2019

C

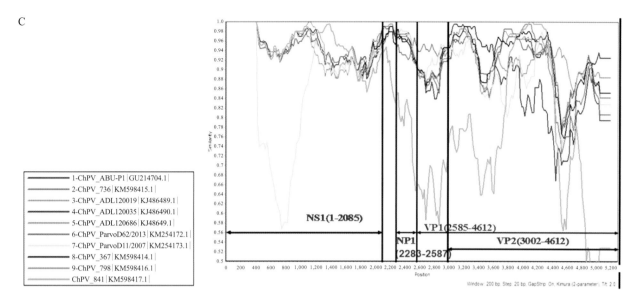

SimPlot analysis based on full-length nucleotide sequences using GX-CH-PV-7 as the query sequence. Shaded letters indicate the structural motif composed of fivefold cylinder regions. The conserved leucine (shown in a box) represents the structural motif for the constriction of the pore in cylindrical projections at each fivefold axis of symmetry. Both the nucleotide and amino acid positions are from the ABU-P1 strain[2].

Figure 2-9-1(C)   Structural motifs in ORF2 and SimPlot analysis of ChPVs

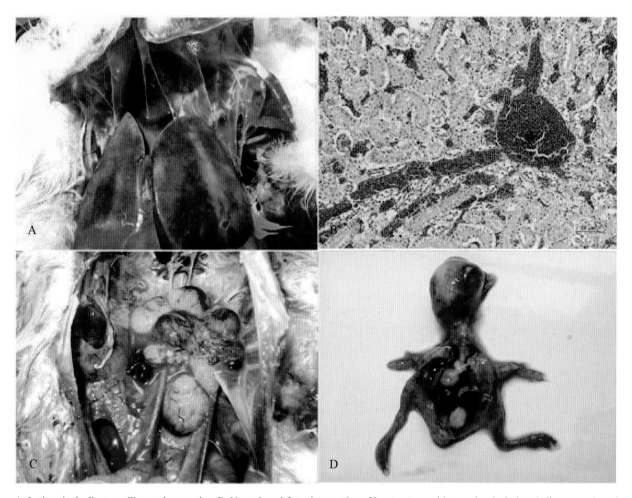

A: Lesions in the liver, swelling, and congestion; B: Necrosis and fatty degeneration of hepatocytes, and intranuclear inclusions in liver parenchymal cells; C: Lesions in the ovary, follicular dysplasia, necrosis, and rupture; D: An infected embryo with subcutaneous hemorrhage.

Figure 2-10-1   Gross lesions and histopathologic changes in ducks and embryos

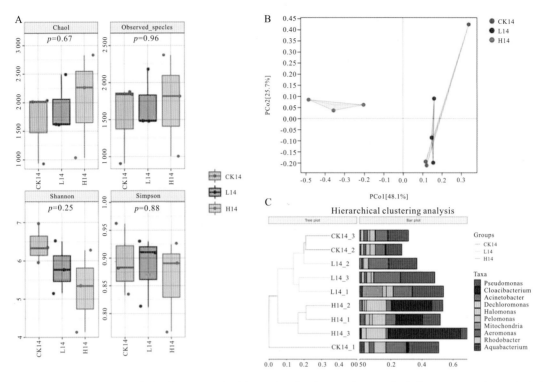

A: α-diversity comparisons in the intestinal microflora among the CK14, L14, and H14 groups; B: Bray-Curtis distances were calculated and visualized through Principal Coordinate Analysis (PCoA) (Ellipses were drawn with 95% confidence intervals); C: Hierarchical cluster analysis of the Bray-Curtis distances generated from taxa tables showed Amplicon Sequence Variant (ASV) similarity across microbial communities among different groups.

Figure 2-13-2　Intestinal microbiome diversity in the control (CK14), 0.5 mg/L (L14) and H14 (1.5 mg/L) groups

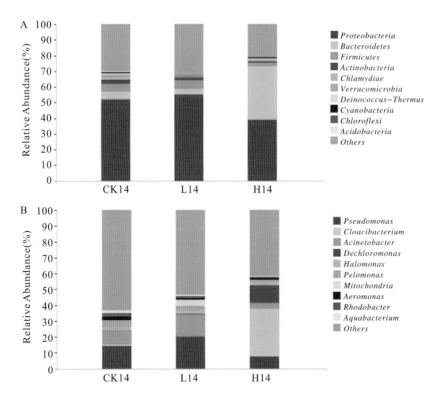

A: Compositions of the intestinal microflora at the phylum level. B: Compositions of the intestinal microflora at the genus level. The top ten abundant genera (higher than 1% in at least one sample) are shown in the figure and the rest are indicated as "Others".

Figure 2-13-3　Compositions of the intestinal microflora among the control (CK14), 0.5 mg/L (L14), and H14 (1.5 mg/L) treatments

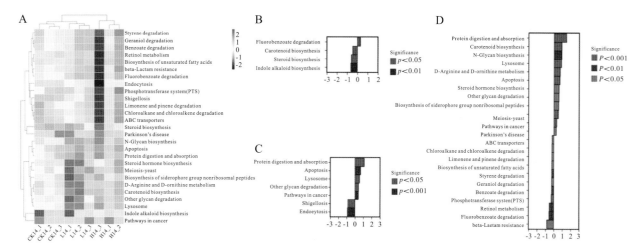

A: Heatmap of the significant differential pathways among the control (CK14), 0.5 mg/L (L14), and H14 (1.5 mg/L) treatments. B: Significant different pathways between the control and L14 treatments. C: Significantly different pathways between the control and H14 treatments. D: Significantly different pathways between the L14 and H14 treatments.

Figure 2-13-5　Intestinal microbiota predictive metabolic functions from the Kyoto Encyclopedia of Genes and Genomes (KEGG) database in all samples

A: Depressed mental state, fluffed feathers, anorexia, and somnolence in the diseased chickens; B: Yellow-green watery stools; C: Presence of effusion in the pericardium, enlarged liver, liver yellowing, and bleeding spots; D: Congested and enlarged kidneys; E: Congested and enlarged spleen.

Figure 2-14-1　Clinical symptoms and gross lesions in chickens infected with GX2020-019

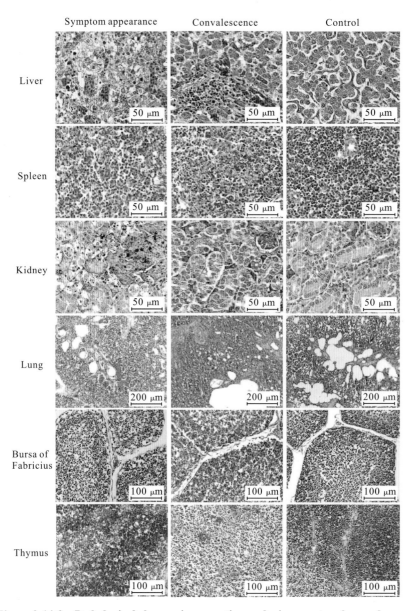

Figure 2-14-3   Pathological changes in organ tissues during onset and convalescence

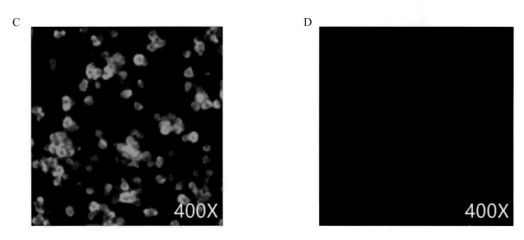

C: GX2020-019 was identified using an anti-FAdV-4 f iber-2 monoclonal antibody by IFA; D: Negative control.

Figure 2-14-5(C and D)   Detection and purity testing of FAdV-4 virus

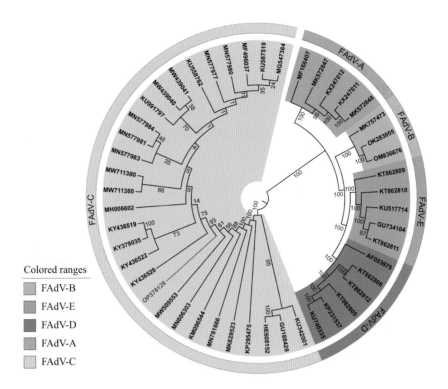

The phylogenetic tree was created by neighbor-joining analysis with 500 replicates for bootstrapping, and evolutionary distances were calculated using the maximum composite likelihood method through MEGA 11 software. The GX2020-019 strain is indicated in red.

Figure 2-14-7   Phylogenetic analysis of the whole gene nucleotide sequence of strain GX2020-019 of Guangxi

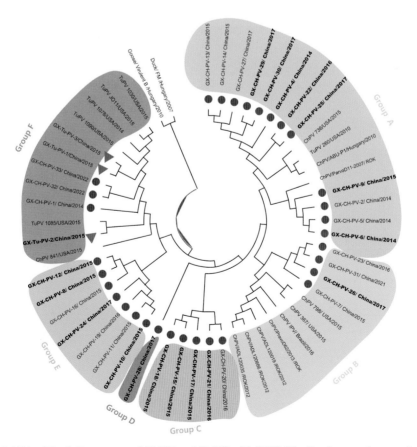

Figure 2-15-1(D)   The full genomes of ChPV and TuPV, with DPV (GenBank accession number U22967) and GPV (GenBank accession number NC_001701) as outgroups

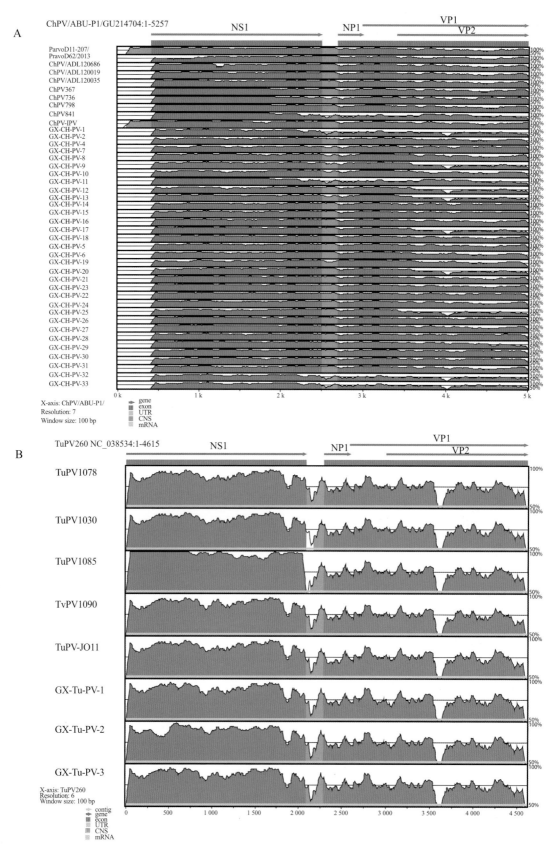

A: Guangxi ChPV field strains; B: Guangxi TuPV field strains.Colour coding: the coloured regions represent a similarity＞90%, and the white regions represent a similarity＜90%. The height of the shaded area at any sampling point is proportional to the genetic relatedness.

Figure 2-15-2　mVISTA whole-genome alignments comparing the nucleotide sequences of the Guangxi ChPV and TuPV field strains with representative ABU-P1 and 260, 1078, 1030, 1085, 1090 and JO11 TuPV strains

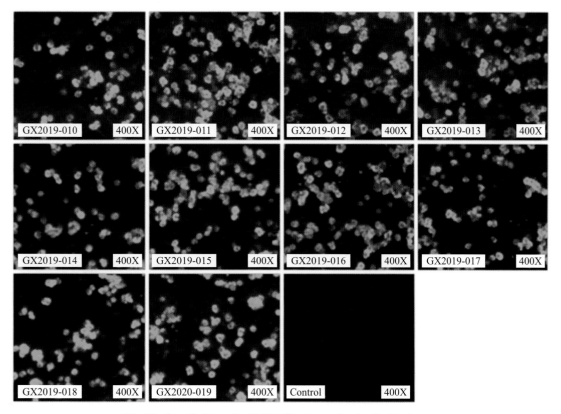

Identification of isolates using FAdV-4 fiber-2 monoclonal antibodies via IFA.

Figure 2-16-1(B)　Isolation and identification of FAdV-4

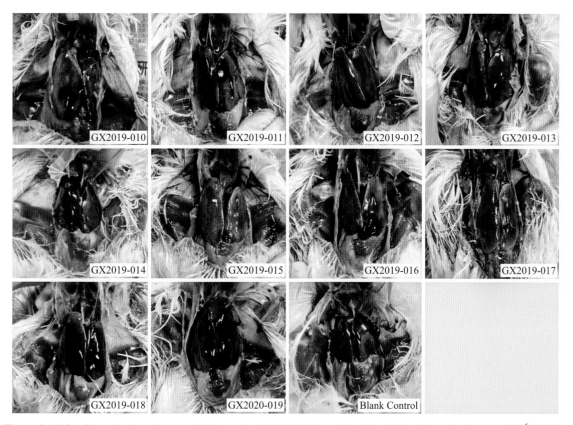

Figure 2-16-3　Gross lesions observed in 4-week-old SPF chickens infected with isolates at a dosage of 10^6 TCID$_{50}$

A

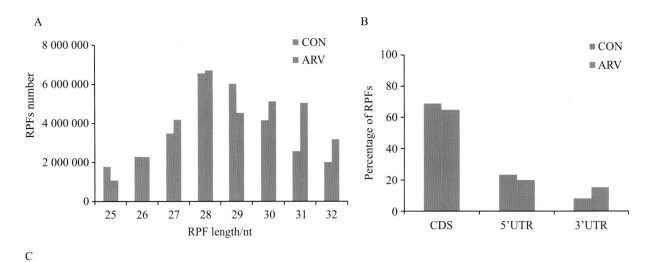

C

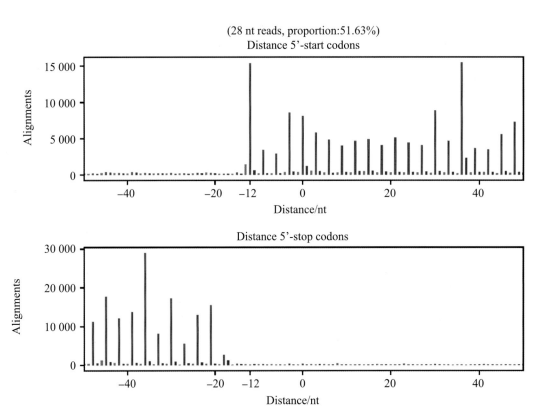

A: Length distribution of RPFs; B: The percentage of RPFs located in CDS, 5' UTR and 3' UTR; C: The total number of RPFs along CDS start and stop codon regions in ARV group. A codon contains three bases, which are represented by red, blue and green.

Figure 4-1-1 Characteristics of ribosome profiling data in the CON and ARV groups

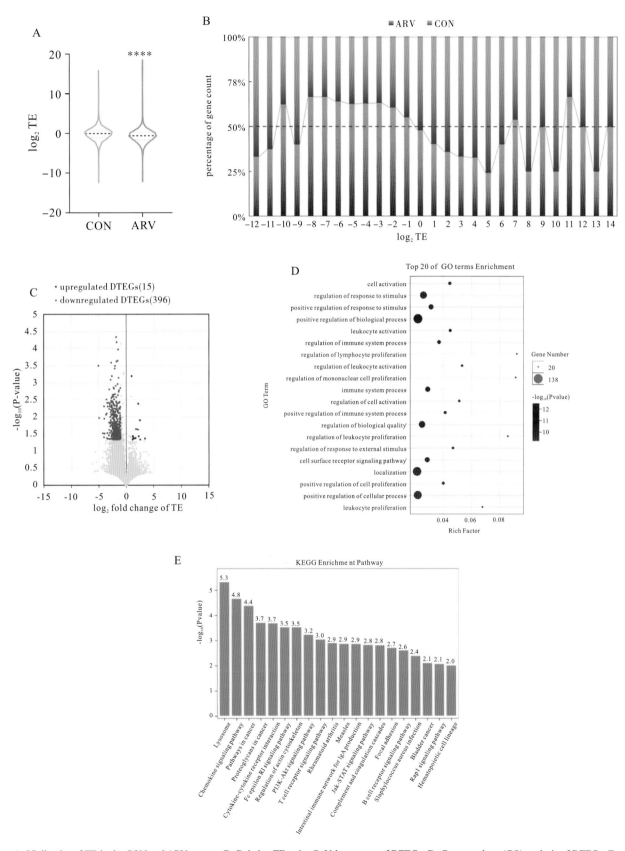

A: Violin plot of TE in the CON and ARV groups; B: Relative TE ratio; C: Volcano map of DTEGs; D: Gene ontology (GO) analysis of DTEGs; E: Kyoto Encyclopedia of Genes and Genomes (KEGG) analysis of DTEGs. Asterisk indicates significant difference (****$P < 0.000\ 1$). Genes were classified based on their rounded log2 TE values. The gray dot represents the nondifferentially expressed translation efficiency genes.

Figure 4-1-2　Translation efficiency analysis

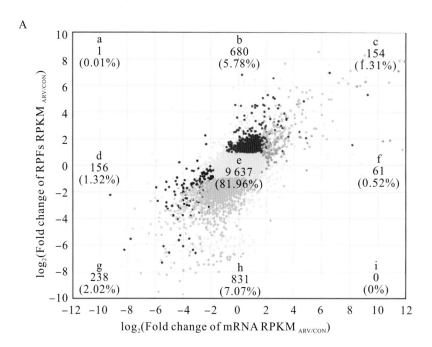

- a: Downregulation for transcription and upregulation for translation;
- b: No change for transcription and upregulation for translation;
- c: Up-regulation for both transcription and translation;
- d: Down-regulation for transcription and no change for translation;
- e: No change for both transcription and translation;
- f: Up-regulation for both transcription and no change for translation;
- g: Down-regulation for both transcription and translation;
- h: No change for transcription and downregulation for translation;
- i: Up-regulation for transcription and downregulation for translation.

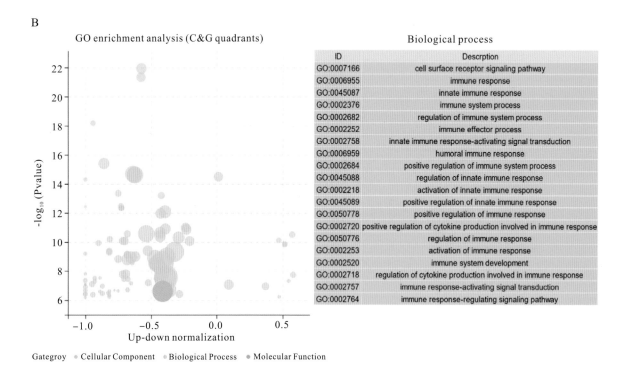

A: scatter plot of the fold change in the ARV/CON group at the transcriptional and translational levels; B: GO enrichment analysis of genes in quadrants C and G.

Figure 4-1-3   Avian reovirus altered gene expression at both the transcriptional and translational levels

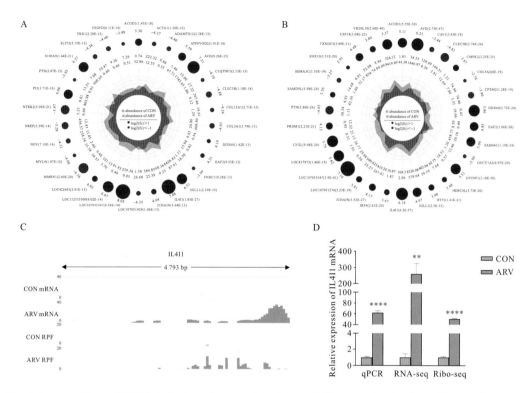

A and B: The circle maps of the top 30 DEGs in the transcriptome and translatome; C: The expression abundance of IL4I1 at the transcriptional and translational levels; D: the relative level of IL4I1 mRNA expression. Asterisks indicate significant differences (**$P<0.01$, ****$P<0.000\ 1$).

Figure 4-1-4   Functional gene screening

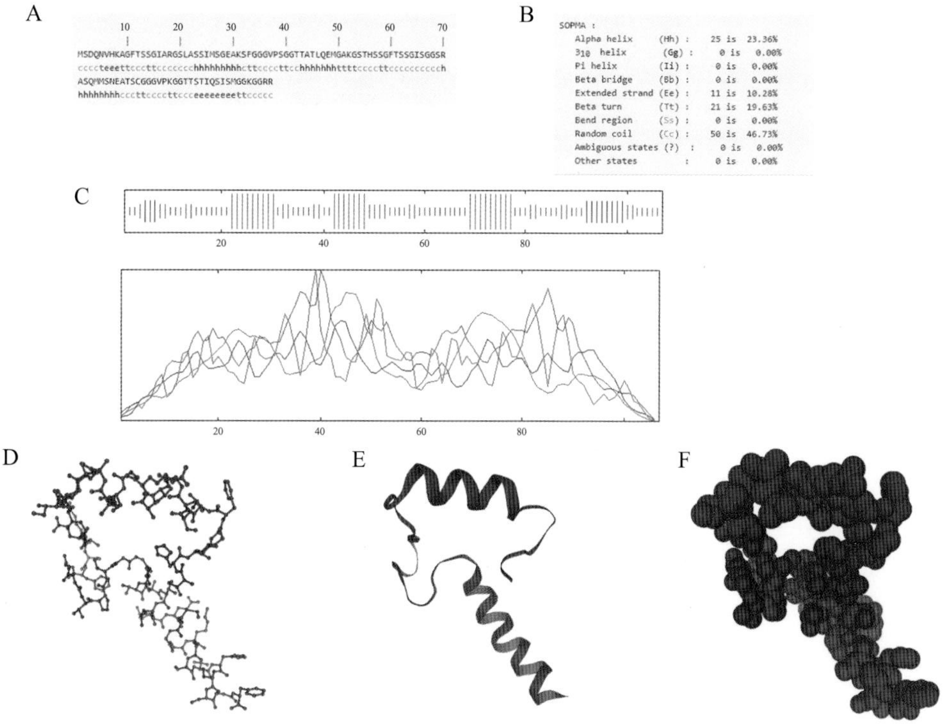

A: Primary structure showing the polypeptide chain encoded by the gene consisting of 107 amino acids; B: Secondary structure; C: Amino acid sequence analysis (blue represents alpha helices, purple represents random coils, red represents extended strands, and green represents beta turns); D-F: Prediction of the three-dimensional structural model of the IF I6 protein: backbone structure (D), ribbon structure (E), and spherical structure (F). The global model quality estimation (GMQE) score for the predicted IF I6 protein structure and interferon alpha-inducible protein 27-like protein 1 (IF I27L1) structure was 0.30 and the coverage was 51%.

Figure 4-2-2   Structural prediction of the IF I6 protein

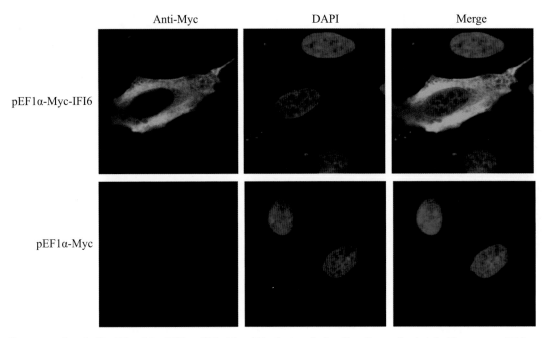

DF-1 cells were transfected with pEF1 α -Myc-IF I6 or pEF1α-Myc. 48 h after transfection, the cells were incubated with a mouse anti-Myc monoclonal antibody for 2 h and then incubated with an Alexa F luor 488 (green)-labeled goat anti-mouse IgG (H+L) cross-adsorbed secondary antibody for 1 h. Nuclei were stained with DAPI (blue), and the fluorescence intensity was detected by confocal microscopy and used to assess the location of expressed IF I6.

Figure 4-2-4　Subcellular localization of the IF I6 protein (63×magnification)

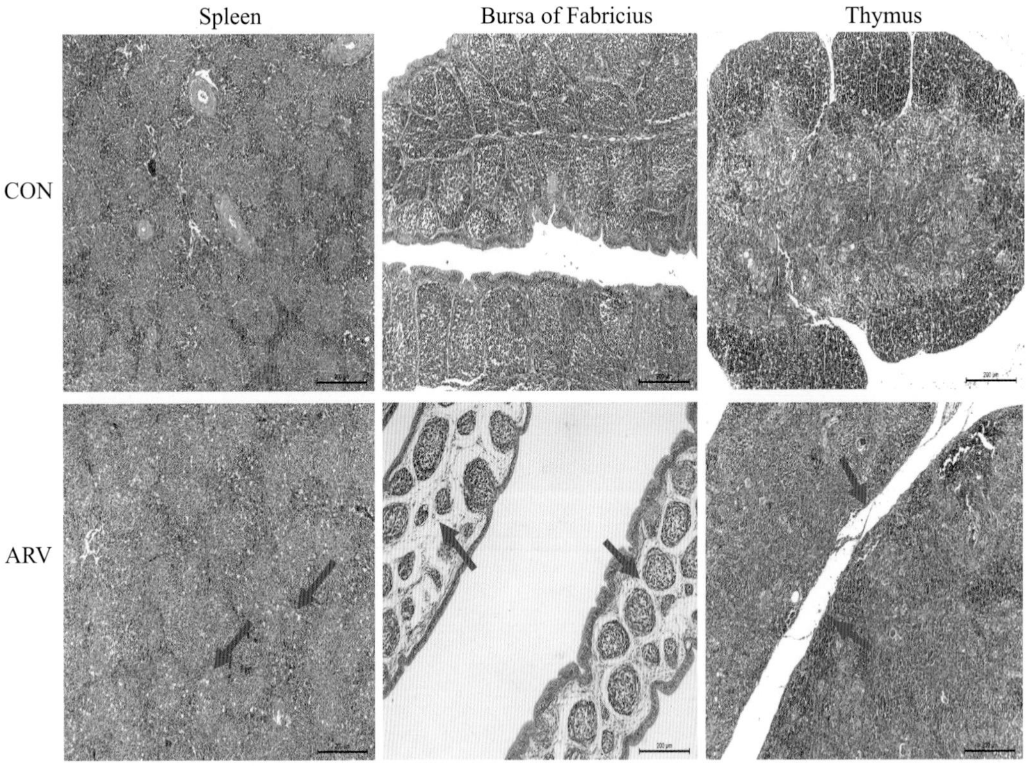

In the control group (CON), SPF chickens were injected with 0.1 mL of PBS into the footpad. In the experimental group (ARV), ARV strain S1133 (10^4 $TCID_{50}/0.1$ mL) was injected into SPF chickens through the footpad. The arrow in the spleen indicates extensive lymphocyte degeneration and necrosis. Arrows in the bursa of Fabricius indicate lymphatic follicular structural loosening, lymphocytic degeneration and necrosis, heterophilic cell infiltration, interstitial loosening, and inflammatory cell infiltration. Arrows in the thymus indicate atrophy, reduced cortex/medulla proportions, and decreased cortical lymphocytes.

Figure 4-3-2　Histopathological changes in the spleen, bursa of Fabricius, and thymus in SPF chickens infected with ARV

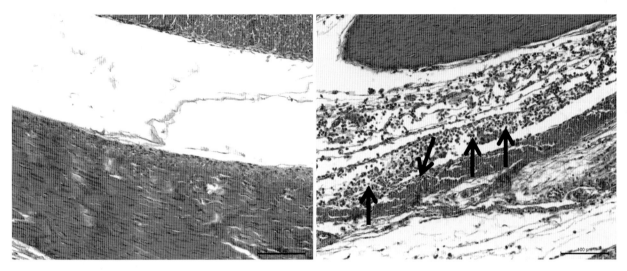

A: Control group, chickens were injected with 0.1 mL PBS in the foot pad; B: Experimental group, chickens were challenged with 0.1 mL of 10⁴ median tissue culture infectious doses/0.1 mL ARV S1133 through foot pad injection. The arrow indicates tendon fibroblasts and synovial epithelial cells were degenerated and necrotic, and a large number of infiltrated inflammatory cells.

Figure 4-4-2　Histopathological changes of SPF chicken joints after ARV infection

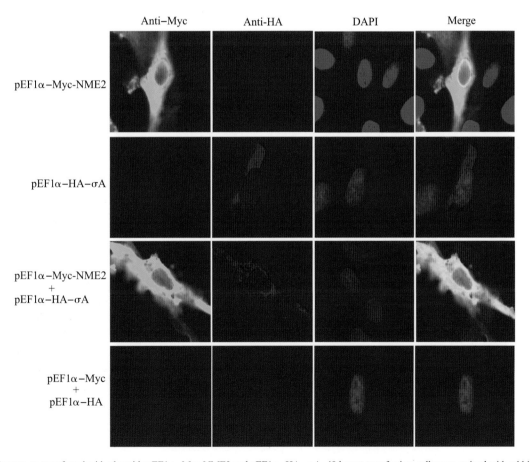

DF-1 cells were co-transfected with plasmids pEF1α-Myc-NME2 and pEF1α-HA-σA. 48 h post-transfection, cells were stained with rabbit anti-HA monoclonal antibody and mouse anti-Myc monoclonal antibody for 2 h, and then incubated with Alexa Flour 488 and Alexa Flour 594 labeled secondary antibody(HA-σA (red), Myc-NME2 (green)). Cell nuclei were stained blue with DAPI. Yellow indicates colocalization in the merged images. Data is representative of all cells that co-expressed both proteins.

Figure 4-5-4　Confocal analysis of the subcellular localization of σA and NME2 proteins in DF-1 cells

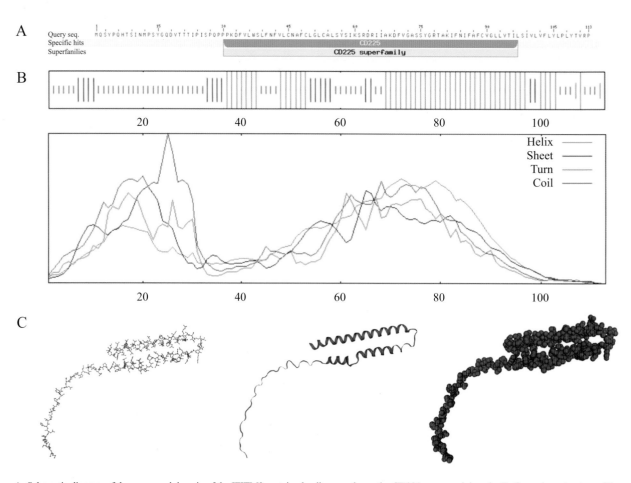

A: Schematic diagram of the conserved domain of the IFITM3 protein, the diagram shows the CD225 conserved domain; B: Secondary structure of the IFITM3 protein, the longest lines represent alpha helices (Hh), the second longest lines represent extended strands (Ee), the third longest lines represent beta turns (Tt), and the shortest lines represent random coils (Cc); C: Tertiary structure of the IFITM3 protein, the global model quality estimation (GMQE) of the chicken IFITM3 protein and IFITM3 derived from Northern Bobwhite equaled 0.61, and the coverage rate was 80.91%.

Figure 4-9-2　Structural analysis of the IFITM3 protein

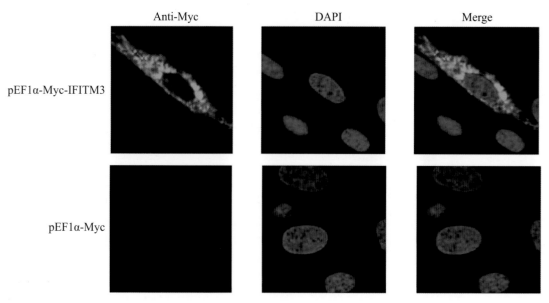

The subcellular localization of the IFITM3 protein in DF-1 cells was observed by laser confocal microscopy. The panels show nuclei stained with DAPI (blue), the IFITM3 protein labeled with Alexa Fluor 488-labeled goat anti-mouse IgG (green) and a merged image.

Figure 4-9-4　Subcellular localization of the IFITM3 protein (63×magnification)

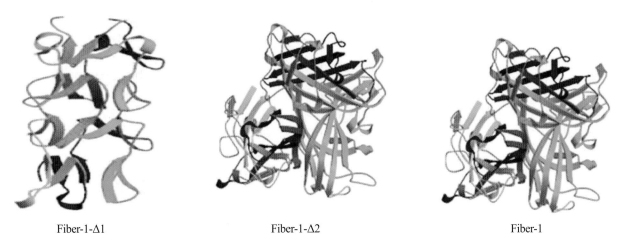

Fiber-1-Δ1   Fiber-1-Δ2   Fiber-1

Schematic diagrams of the tertiary structures of the fiber-1-Δ1 and fiber-1-Δ2 proteins.

Figure 4-10-1(D)   Schematic diagrams of the domains of fiber-1 predicted using SMART software

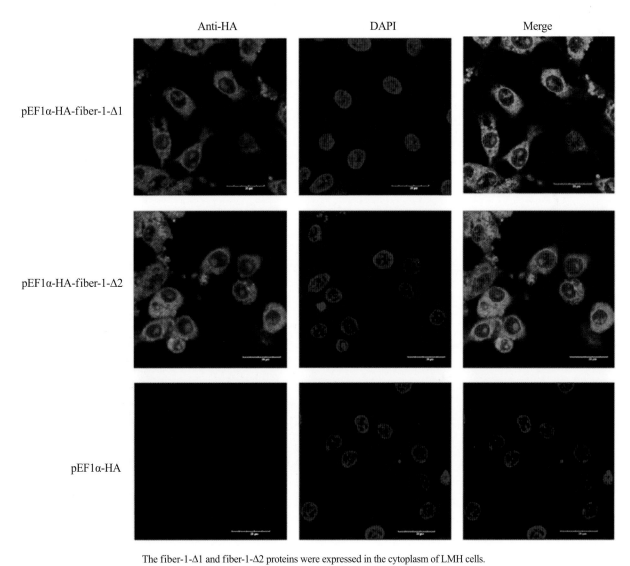

The fiber-1-Δ1 and fiber-1-Δ2 proteins were expressed in the cytoplasm of LMH cells.

Figure 4-10-3(C)   Fiber-1-Δ1 and fiber-1-Δ2 overexpression in LMH cells